Pathology and Biology of Human Germ Cell Tumors

Francisco F. Nogales • Rafael E. Jimenez
Editors

Pathology and Biology of Human Germ Cell Tumors

Springer

Editors
Francisco F. Nogales
Departamento de Anatomía Patológica
Universidad de Granada Facultad de Medicina
Granada
Spain

Rafael E. Jimenez
Department of Laboratory Medicine and Pathology
Mayo Clinic
Rochester, Minnesota
USA

ISBN 978-3-662-53773-2 ISBN 978-3-662-53775-6 (eBook)
DOI 10.1007/978-3-662-53775-6

Library of Congress Control Number: 2017931581

Printed on acid-free paper

This Springer imprint is published by Springer Nature
The registered company is Springer-Verlag GmbH Germany
The registered company address is: Heidelberger Platz 3, 14197 Berlin, Germany

To Gordon Barry Pierce (1925–2015) who found the key to unlocking the magic of germ cell tumors.

Preface

August 2016

Since their first documented descriptions, germ cell tumors have always been a leading exhibit in an imaginary cabinet of scientific curiosities. The reason is their extraordinariness, resulting from the finding of distorted bodily structures in unexpected organs and at different ages. This distortion of the body image has prompted the evolution of thought, ideas, and theories about their origin and meaning ranging from the purely magical to the current ideas of their pathogenesis and differentiation potentials.

However, the key to the understanding of their multiple and complex differentiations lies in the concepts of pluripotentiality and tumor stem cells, an original idea developed by G. Barry Pierce (1925–2015), mentor and kind friend to some of the older contributors (FFN and ID) of this monograph. His farsighted work with L. Kleinsmith on murine teratocarcinoma [1] demonstrated, for the first time, the existence of pluripotent tumor stem cells. Thus, the study of these curious tumors turned out to be a biological Rosetta stone, linking embryonic and neoplastic development [2] in what Rupert A. Willis called a borderland of embryology and pathology, [3] eventually becoming the key to understanding tumor biology and especially pluripotentiality as the paradigm in the origin and histology of germ cell tumors.

Since Willis' [4] book in the early 1950s, no other reference book has provided a complete and integrated picture of germ cell tumors in the various organs and ages of life including their pathogenesis and surgical pathology, having been usually partially analyzed from pediatric, gynecologic, uropathologic, and other similarly specialized viewpoints. In this monograph, we have brought together the current knowledge on gonadal and extragonadal germ cell tumors and analyzed them from a broader perspective that includes basic clinical features and management, epidemiology, molecular biology, and an extensive clinicopathologic analysis, with emphasis on their most frequent locations: testicular, ovarian, and mediastinal. Since germ cell tumors exhibit a stereotyped histology in the various organs, a certain degree of overlap and repetition is unavoidable.

The characteristic histology of germ cell tumors in various organs has led to the general belief that tumors with a similar morphology share the same origin. This assumption obscures the understanding of their biology, since germ cell tumors behave differently depending on the age of the patient and the organ they arise from. The explanation for their diverse behavior lies not in a *generic* germ cell origin but in the developmental state of the precursor stem cells, which are biologically different in the various anatomical sites and ages of life.

In this monograph, we attempt to analyze germ cell tumors, not only histopathologically but under the developmental perspective outlined by Oosterhuis and Looijenga in Chap. 3, thus lending support to G. B. Pierce's notion of germ cell tumors as caricatures of progressive stages of embryonal development. Their approach, focusing on the developmental potential of embryonic stem and germ cells, provides a unifying model for all germ cell tumors. This concept crystallizes in a pathogenetic classification of germ cell tumors into seven types, each of them reflecting a defined stem cell potency state. This classification also includes, for the first time, germ cell tumor patterns derived from somatic tumors that are the result of induced pluripotency of tumor stem cells. Their proposal provides a good explanation for the clinicopathologic diversity of germ cell tumors, answering many extant questions about their epidemiology, morphology, and behavior. Consequently, Oosterhuis and Looijenga's proposed classification will be followed in most chapters, especially in gonadal germ cell tumors.

Histopathologic terminology is updated to the recently proposed changes in the World Health Organization blue books in the testis, ovary, and mediastinum.

A brief summary of the contents follows:

Dr. Ivan Damjanov is a leading figure in the experimental pathology of germ cell tumors. In his introductory chapter, he reviews the flow of clinicopathologic and experimental knowledge, to which he has been an important contributor, leading to the present concepts and terminology.

A European-wide study on the epidemiology of germ cell tumors is presented by Drs. Trama and Berrino in Chap. 2. This study complements recent studies from the UK and USA.

As previously mentioned, Chap. 3 integrates a wealth of clinicopathologic with cytogenetic and basic stem cell research data to provide a rationale for a comprehensive biological classification of germ cell tumors.

Chapter 4 complements Chap. 3 with a practical approach, the analysis of antibody expression, reviewing current data on diagnostic immunohistochemistry and analyzing both stage-specific, pluripotency markers and organ-specific ones. The genes and developmental role of each antibody are discussed and a hands-on approach to the use of commercially available antibodies is provided.

In Chap. 5, the Mayo Clinic Medical Oncology team summarizes the management of germ cell tumors using testicular germ cell tumors as the prototype example.

Chapters 6, 7, 8, and 9 provide an extensive coverage of both histopathologic and clinicopathologic findings of germ cell tumors in their more frequent locations: gonads and mediastinum.

Chapter 6 is an update of the current histopathology of ovarian germ cell tumors, emphasizing the expression of characteristic pluripotency markers as a mandatory diagnostic tool for differential diagnosis. Yolk sac tumors are reconsidered as primitive endodermal tumors applying a diagnostic immunohistochemical panel able to distinguish between extraembryonal and somatic variants. Prognostically relevant histologic grading of immature teratomas is reanalyzed, taking into account the presence of immature endodermal structures and the expression of pluripotency markers. Finally, an emerging category of highly malignant germ cell tumors originating not from germ cells but from somatic müllerian tumors in older patients (endometrioid carcinomas and clear cell tumors) is analyzed in depth.

Chapter 7 focuses on postpubertal testicular tumors. It incorporates recent terminology and classification of testicular neoplasms recently introduced by the World Health Organization. These include the new terminology of germ cell neoplasia in situ and spermatocytic tumor. The concepts of prepuberal- and postpubertal-type teratomas are defined and contrasted, concepts that are highly analogous to the premises of Oosterhuis and Looijenga's classification. The pathology of these tumors is analyzed in the context of their clinical implications. Sadly, the senior author of this chapter, Dr. Thomas J. Sebo, passed away during the preparation of this manuscript. The chapter pays tribute to his outstanding skills as diagnostic urologic pathologist.

Chapter 8 focuses on features of mediastinal GCT that might differ from their gonadal counterparts including imaging, immunophenotype, cytogenetic and molecular characteristics, and prognosis. Important differential diagnoses that should be considered before establishing a diagnosis of primary mediastinal GCT are also discussed.

Chapter 9 summarizes current knowledge about the clinicopathologic, phenotypic, and molecular characteristics of intracranial germ cell tumors, highlighting specific properties of intracranial sites.

Chapter 10 emphasizes differential findings relevant in the extensive and fascinating morphologic spectrum of pediatric germ cell tumors, particularly those associated with disorders of sex development.

Germ cell tumors found in miscellaneous sites are reviewed in Chap. 11 with special emphasis on their organ-related particularities and their differential diagnoses.

Finally, Chap. 12 covers, for the first time, another emerging category of tumors: somatic-type malignancies that develop in pre-existing germ cell tumors. The topic is presented in the context of the different types of germ cell tumors according to Oosterhuis and Looijenga's classification.

With its wide multiorganic and biopathologic approach, we hope that the present monograph will prove useful to the understanding of the pathology and biology of germ cell tumors.

The editors would like to thank the authors for their generosity with their time and knowledge and patience to bear innumerable and persistent requests from the editors. Ms. M. Himberger, project coordinator from Springer, stoically bore with us the delays due to the late appearance of the new WHO classifications of tumors. Dr. Heather Fulwood was a daily inspiration and great help throughout the edition of this book.

We are sad indeed that Dr. G. Barry Pierce did not live quite long enough to see the publication of this monograph which is our homage to his brilliant commitment and contribution to the understanding of germ cell tumors.

Finally, we would like to thank the partners and family of the authors and editors for their understanding of the time we have often robbed from family life.

Zahara de los Atunes, Spain — Francisco F. Nogales
Rochester, MN, USA — Rafael E. Jimenez

References

1. Kleinsmith LJ, Pierce GB Jr., Multipotentiality of single embryonal carcinoma cells. Cancer Res. 1964;24:1544–51.
2. Damjanov I. In Memoriam - Prof. G. Barry Pierce (1925–2015). Int J Dev Biol. 2016;60(1–2–3):1–3. doi:10.1387/ijdb.160014id.
3. Willis RA. The Borderland of embryology and pathology. The Middleton-Goldsmith Lecture. Bull NY Acad Med. 1950;26(7):440–60.
4. Willis RA. Teratomas. Atlas of tumor pathology. Section 3, pt. 9. Washington, DC: Armed Forces Institute of Pathology; 1951.

Contents

Contributors

Franco Berrino, MD Department of Preventive and Predictive Medicine, Fondazione Istituto Nazionale dei Tumori, Milan, Italy

Brian A. Costello, MD Mayo Clinic, Rochester, MN, USA

Ivan Damjanov, MD, PhD Department of Pathology & Laboratory Medicine, Kansas City, KS, USA

Nooshin K. Dashti, MD, MPH Anatomic and Clinical Pathology, Department of Laboratory Medicine and Pathology, Mayo Clinic, Rochester, MN, USA

Danielle De Stefano, MD Department of Pathology, Stanford University Medical Center, Stanford, CA, USA

Sounak Gupta, MBBS, PhD Anatomic and Clinical Pathology, Department of Laboratory Medicine and Pathology, Mayo Clinic, Rochester, MN, USA

Loren P. Herrera-Hernandez, MD Department of Laboratory Medicine and Pathology, Mayo Clinic, Rochester, MN, USA

Rafael E. Jimenez, MD., MHA Laboratory Medicine and Pathology, Mayo Clinic College of Medicine, Rochester, MN, USA

Manish Kohli, MD Mayo Clinic, Rochester, MN, USA

Leendert H.J. Looijenga, MD Department of Pathology, Erasmus MC – University Medical Center, CE Rotterdam, Netherlands

Francisco F. Nogales, MD, PhD University of Granada Medical School, Granada, Spain

J. Wolter Oosterhuis, MD Department of Pathology, Erasmus MC – University Medical Center, CE Rotterdam, Netherlands

Ovidiu Preda, MD Master Diagnóstica S.L., Granada, Spain

Juan A. Retamero, MD, MRCPsych (UK) Department of Pathology, Hospital Universitario del PTS, Granada, Spain

Miguel Reyes-Múgica, MD Children's Hospital of Pittsburgh of UPMC, Department of Pathology B256, Pittsburgh, PA, USA

Fausto J. Rodriguez, MD Department of Pathology, Division of Neuropathology, Johns Hopkins Hospital, Baltimore, MD, USA

Anja Roden, MD Department of Laboratory Medicine and Pathology, Mayo Clinic, Rochester, MN, USA

Alejandro Rubio-Fernández, MD Department of Pathologist, Hospital San Pedro de Alcántara, Cáceres, Spain

Maolly Schuldt Nogales-Ortiz Fellow in Pathology, Facultad de Medicina, Granada, Spain

Thomas J. Sebo, MD, PhD Department of Laboratory Medicine and Pathology, Mayo Clinic, Rochester, MN, USA

Annalisa Trama, MD Department of Preventive and Predictive Medicine, FondazioneIstitutoNazionaledeiTumori, Milan, Italy

M. Adelita Vizcaino, MD Department of Cellular and Tissue Biology, Faculty of Medicine, UNAM, Mexico City, Mexico

Ben Y. Zhang, MD Division of Medical Oncology, Mayo Clinic, Rochester, MN, USA

Eduardo Zambrano, MD, MSc Pathology and of Pediatrics, Lucile Packard Children's Hospital Stanford Stanford University Medical Center, Stanford, CA, USA

Germ Cell Tumors: Classifications, Definitions, and Terminology

1

Ivan Damjanov

1.1 Introduction

The history of human ovarian, testicular, and extragonadal teratomas and, by extension, human germ cell tumors in general has been extensively reviewed on several occasions [1–10]. From these reviews it seems that the first fully documented case of teratoma of the ovary was reported in 1658 by Johann Scholz [also known under his Latin surname *Scultetus*] [4], while the first teratoid tumor of the testis was reported in 1696 by Saint Donat [10]. The first description of sacrococcygeal and retroperitoneal teratomas and other germ cell tumors cannot be determined exactly, although it seems that the Chaldean Babylonian cuneiform papyri 4000 years ago contained a description of child with a sacrococcygeal teratoma [2]. According to the literature review by Lamphier [8], the first case of a mediastinal dermoid cyst was reported by J. A. Gordon in an address to the Medico-Chirurgical Society of London in November 1823 and published later in the society proceedings in 1827. According to Wheeler [4], Maier reported a cerebral dermoid in a two-week-old boy, and Weigert in 1875 provided histologic evidence for a teratoma of the pineal gland. These two cases would qualify as the first intracranial germ cell tumors on record. According to the review of Lynch and Blewett [9], it was Askenazy in 1906 who first described an intracranial choriocarcinoma.

These early reports of human germ cell tumors were based on a vague understanding of their biology and histogenesis and usually included a lot of speculation and verbosity. Actually, in many cases the germ cell nature of tumors described was not even mentioned. The confusing terminology used by these pioneers of pathology was reviewed by James Ewing [10], who himself did not lack strong opinions and a tendency for speculation. Ewing also contributed to the confusing view that seminoma and embryonal carcinoma (EC) are more or less the same.

In summary, although I have tried to decipher some of the early thinking that went into these efforts to make the terminology of germ cell tumors consistent with the clinical requirements, I must admit that in retrospect that it was almost impossible for me to reconstruct the thinking of our predecessors. To this end I tend to agree with Rupert Willis [5] who in his very influential book wrote the following: "It would be of little use to recount the confused views which were held of the nature of testicular tumors during the later part of last century, when "sarcomas" and "endotheliomas"

I. Damjanov, MD, PhD
Department of Pathology & Laboratory Medicine, The University of Kansas School of Medicine, 3901 Rainbow Blvd., Mail Stop 3045 2017 Wahl Hall West, Kansas City, KS 66160-7410, USA
e-mail: idamjano@kumc.edu;
http://www2.kumc.edu/pathology/damjanov.html

F.F. Nogales, R.E. Jimenez (eds.), *Pathology and Biology of Human Germ Cell Tumors*, DOI 10.1007/978-3-662-53775-6_1

abounded and when teratomas were called "chondromas", "chondro-carcinomas", "rhabdo-myomas","myo-chondromas", etcetera, according to the tissues which had been detected in them." Accordingly, I will continue my discussion with the contributions of Friedman and Moore, who in essence initiated the modern era of germ cell research with their classical 1946 paper [11].

1.2 Classification of Testicular Germ Cell Tumors

All the modern classifications of testicular tumors can be traced to the groundbreaking work of Friedman and Moore who examined more than 900 testicular tumors removed from male members of the US military during the World War II [WWII] period [11]. These authors recognized four fundamental structural patterns corresponding to seminoma, embryonal carcinoma, chorioepithelioma, and teratomas and proposed that the germ cell tumors be classified in four groups: (1) seminoma [germinoma]; (2) embryonal carcinoma, including a subset of chorioepithelioma; (3) teratoma; and (4) teratocarcinoma (Fig. 1.1). As pointed out by Young [6], the most important contribution of these authors was their recognition of embryonal carcinoma [EC] as a neoplasm distinct from seminoma. I would also add that they recognized the capacity of embryonal carcinoma cells to differentiate into the chorioepitheliomatous elements and the somatic tissues forming the teratomatous part of mixed germ cell tumor (Fig. 1.2). Their insight was followed up by G. Barry Pierce [12] who proposed and experimentally proved that EC cells represent the developmentally pluripotent malignant stem cells of teratocarcinomas or mixed germ cell tumors as we usually call them today. Like other stem cells, EC cells can self-renew on one hand side and differentiate on the other. If unchecked, they propagate until they kill the host. They can metastasize but also differentiate into benign nonproliferating somatic tissues or extraembryonic structures such as yolk sac or trophoblast. The reversibility of malignancy of EC cells, a concept pioneered by Pierce in 1950 and later [13], has been by now generally accepted and confirmed in the murine models of teratocarcinoma and EC, pioneered by Ralph Brinster, Gail Martin, and Martin Evans, as reviewed by Davor Solter [14], and also at a recent international conference attended by most scientists who have experimentally contributed to this field [15]. In clinical practice these principles are confirmed repeatedly by the finding of mature teratomas in lymph nodes of testicular cancer patients treated by modern cis-platinum-based chemotherapy.

The original classification of Friedman and Moore [11] was first modified by Dixon and Moore and printed in the first edition of atlases of the Armed Forces Institute of Pathology [AFIP] in 1952 [16]. In that treatise, Dixon and Moore have in de facto divided testicular germ cell tumors in two groups: seminomas and all others, stating that embryonal carcinoma, choriocarcinoma, and teratomas are closely related tumors, as indicated in their drawing (Fig. 1.3). They also stated that there are numerous mixed forms, calculating at least 15 possible varieties and combinations which could be encountered in various tumors. To simplify the matters and to make the classification as clinically relevant as possible, Dixon and Moore [16] divided the germ cell tumors in five groups: (I) seminoma, pure; (II) embryonal carcinoma, pure or with seminoma; (III) teratoma, pure or with seminoma; (IV) teratoma with either embryonal carcinoma or choriocarcinoma or both and with or without seminoma; and (V) choriocarcinoma, pure or with either seminoma or embryonal carcinoma or both.

The classification proposed by Dixon and Moore [16] was modified several times over the last 60 some years [17–19], resulting in a clinically useful approach combined with an easily reproducible microscopic subdivision of testicular tumors. The 2004 classification sponsored by the World Health Organization [WHO] and recommended by the panel of WHO experts for general use has been further disseminated in the widely circulated in the AFIP series of atlases of pathology [19]. Even though this classification is primarily based on microscopic morphology of tumors [Table 1.1], it correlates well with the clinical requirements for subdividing testicular germ cell tumors into two major

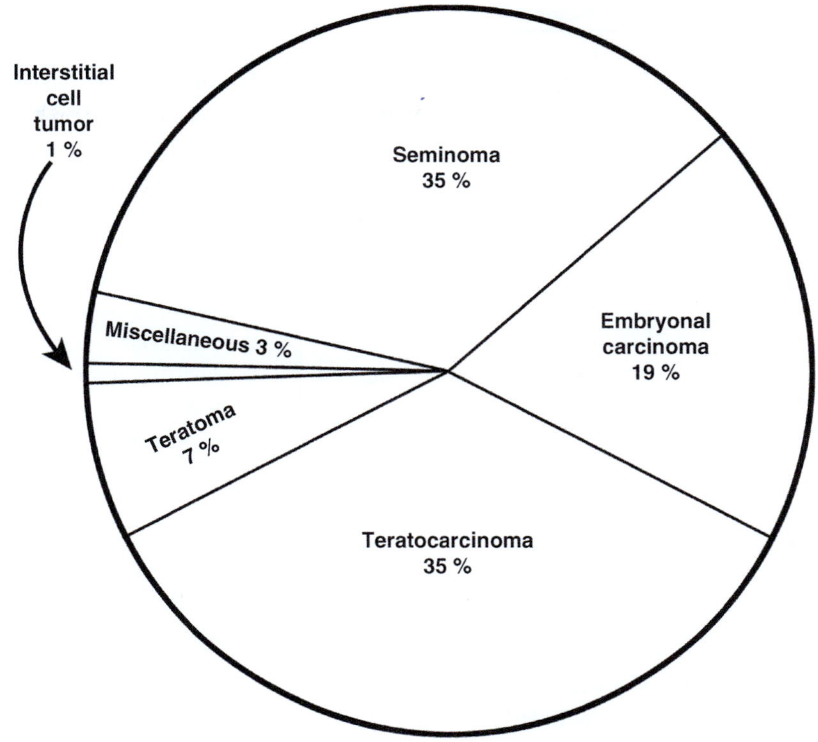

Fig. 1.1 The relative incidence of testicular tumors in the study of Friedman and Moore [11] (Reproduced with permission of the publisher)

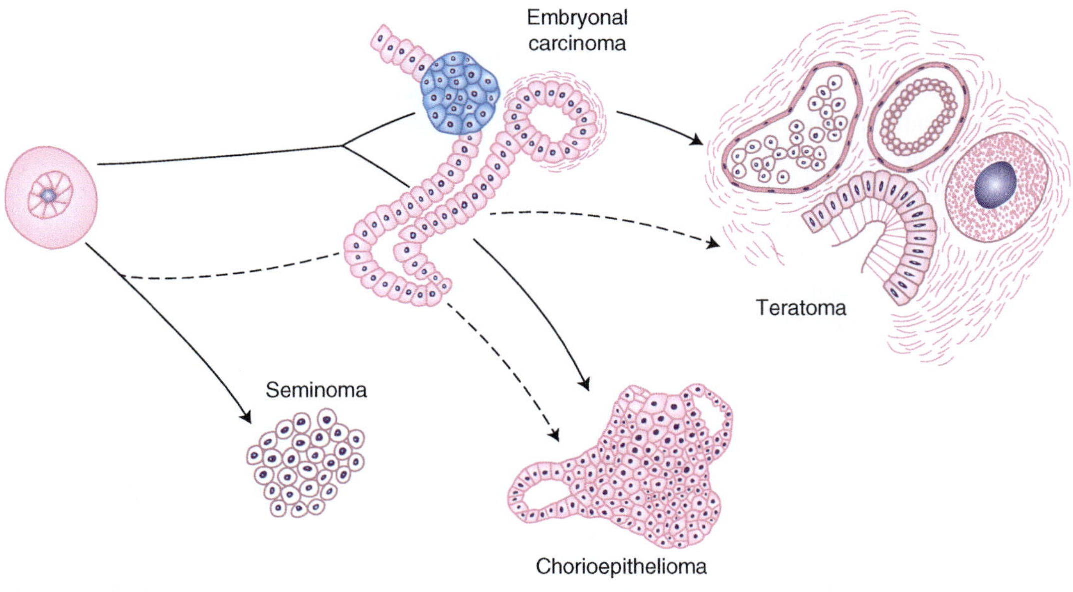

Fig. 1.2 Schematic drawing illustrating the developmental potential of embryonal carcinoma according to Friedman and Moore [11] (Reproduced with the permission of the publisher)

groups: seminomas and nonseminomas, also known as nonseminomatous germ cell tumors [NSGCT]. It also incorporates some of the basic histogenetic tenets and discoveries that were made over the years since the WWII. For epidemiologic research, for international studies, and for billing purposes, the International Classification of Diseases has produced its

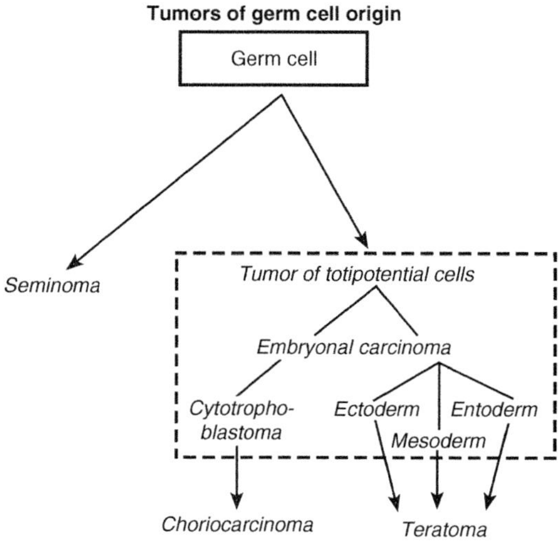

Fig. 1.3 Histogenesis of testicular tumors according to Dixon and Moore [16] (Reproduce with the permission of the publisher)

ICD-10 listing of testicular tumors, which is for completeness' sake included as Table 1.2.[1]

The limitations of space do not allow me to enumerate all the major figures who have played a critical role in advancing our knowledge about germ cell tumors during the last few decades. Thus, by necessity I will mention only five physician-scientists, realizing that this is a rather subjective choice, with a strong personal bias. Nevertheless, I feel that I would be remiss for not mentioning the comparative studies of Gunnar Teilum, which led to the better understanding of the similarities between ovarian and testicular germ cell tumors and the recognition of yolk sac carcinoma as a distinct entity [20]. Robert E. Scully was a true giant of urogenital and gynecologic pathology, whose contributions were recently lovingly reviewed by Oliva and Young [21–23]. His contributions are too many to list, but in essence his work over half a century helped us all to conceptualize and delineate some of the basic aspects of testicular and ovarian pathology. F. K. Mostofi was instrumental in defining the basic approaches to classifying testicular and other urogenital tumors promoting worldwide discussions under the aegis of the WHO [17, 24, 25]. The astute observations and persistence of Niels E. Skakkebæk led to the recognition of intratubular germ cell neoplasia [26]. G. Barry Pierce, my mentor and longtime friend, performed some of the fundamental experiments on mouse germ cell tumors and provided a new

[1]The latest 2016 WHO classification is commented in Chap. 3 [Eds].

Table 1.1 WHO 2004-based classification of testicular germ cell tumors

Testicular germ cell tumors
Precursor lesions
Intratubular germ cell neoplasia, unclassified
Intratubular germ cell neoplasia, specific types
Germ cell tumors of one histologic type
Seminoma
Spermatocytic seminoma
Embryonal carcinoma
Yolk sac tumor
Trophoblastic tumors
Teratoma
Germ cell tumors of more than one histologic type
Mixed germ cell tumors
Polyembryoma
Diffuse embryoma
Regressed ("burnt-out") germ cell tumors
Germ cell sex cord-stromal tumors
Gonadoblastoma
Unclassified

Abbreviated and slightly modified from Ulbright and Young [19]

Table 1.2 Classification of testicular germ cell tumors

2014 ICD-10-CM diagnosis code C62.90
Malignant neoplasm of testis NOS
Cancer of the testis
Cancer of the testis, choriocarcinoma
Cancer of the testis, nonseminomatous germ cell
Cancer of the testis, seminoma
Choriocarcinoma of testis
Mixed germ cell tumor of testis
Nonseminomatous germ cell neoplasm of testis
Primary malignant neoplasm of testis
Seminoma of testis
Testicular cancer
Testis, mixed germ cell tumor

According to the International Coding of Diseases (ICD) http://www.icd10data.com/

insight into the basic biology of human germ cell tumor based on sound scientific principles [13].

1.3 Classification of Ovarian Germ Cell Tumors

Ovarian germ cell tumors are basically equivalent to those originating from male germ cells. Yet there are some important biological and clinical differences between these two groups of tumors [27–29]. For example, in contrast to the malignant nature of the vast majority of testicular tumors, most ovarian tumors are benign, presenting clinically as mature teratomas.

Experimental data obtained in mice indicate that ovarian teratomas are formed from parthenogenetically activated ovarian germ cells [30]. Human parthenotes isolated from the ovaries can rise to embryonic stem cells [31], and thus by extrapolation, one can assume that these cells could give rise to teratomas and other germ cell tumors as well.

The histogenesis of teratoma can be readily explained by parthenogenetic activation of ovarian germ cells. The histogenesis of malignant ovarian germ cell tumors is a bit more complicated, and several histogenetic schemes have been proposed, as reviewed by J. Prat in the monograph which he has edited with G. Mutter [31]. Despite many attempts to modify our understanding of malignant ovarian germ cell, histogenesis is still incomplete. The panel of experts of WHO has thus decided to base the latest WHO classification of malignant germ cell tumors on the most popular model of histogenesis of these tumors dating back to the work of Teilum [20]. According to this scheme, the malignant germ cell can form either dysgerminoma or embryonal carcinomas, which in turn could give rise to choriocarcinoma, yolk sac carcinoma, or teratoid tumors. In the expanded histogenetic algorithm presented here (Fig. 1.4), we propose that the tumor formation depends in all cases on parthenogenetic activation of the ovarian germ cells [ova], which may give rise as such to benign tumors, i.e., teratomas. Alternatively, if the germ cells undergo malignant transformation, they may give rise to dysgerminoma or form embryonal carcinoma cells. EC cells may form either a monotypic tumor-embryonal carcinoma or by differentiating into somatic and extrasomatic cells and tissues form the malignant stem cells of a malignant mixed germ cell tumor. Even monotypic EC tumors contain syncytiotrophoblastic cells, which may be present in most of such tumors. EC cells can sometimes differentiate into yolk sac or choriocarcinoma cells, which as such may form tumors of the same name, overgrowing the EC component, which may be hard to find. Alternatively, EC cells may remain part of the mixed germ cell tumors which in such cases will contain several other distinct components: EC cells, teratomatous tissues, yolk sac tumor, and choriocarcinoma, or various combinations of these elements. There is also some evidence that dysgerminomas may give rise to embryonal carcinoma, but there is no doubt that they can be part of mixed germ cell tumors. An abbreviated most recent classification of all these ovarian germ cell tumors is presented in Table 1.3.

1.4 Definitions and Terminology of Testicular and Ovarian Germ Cell Tumors

For the sake of consistency and uniformity, most if not all definitions listed here were taken with slight modifications from the publications of the World Health Organization [18, 28], the fascicles of the Armed Forces Institute of Pathology [19, 27], and the recent textbook of pathology of the female reproductive tract [29]. These terms are listed following the order of their appearance in Tables 1.1, 1.2, and 1.3. Equivalent tumors arising in the testis as well as in the ovary are usually discussed under the same heading unless indicated otherwise.

Intratubular germ cell neoplasia, unclassified, is a preinvasive form of most testicular germ cell tumors, composed of malignant germ cells, which appear enlarged, which have a centrally located "boxy" nucleus surrounded by clear cytoplasm. Originally, it was named carci-

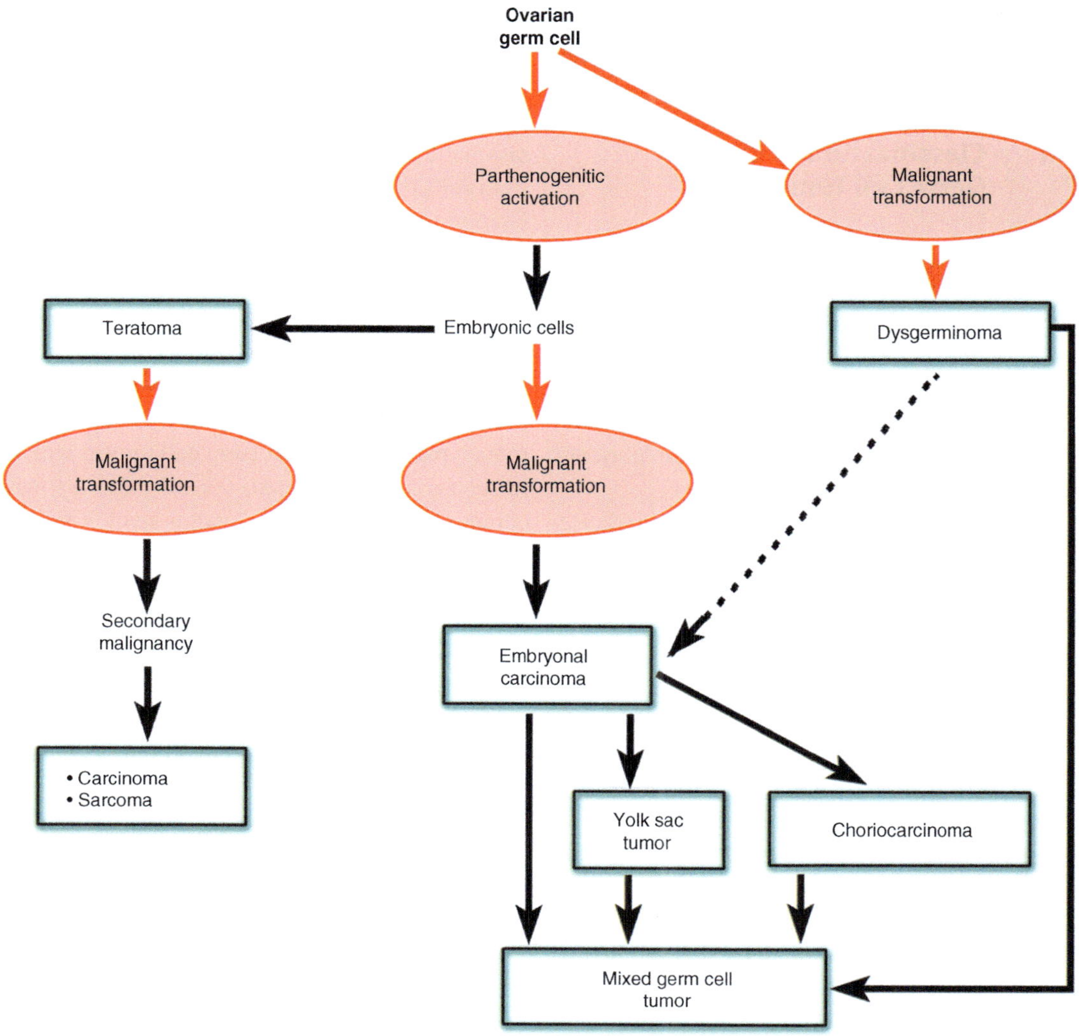

Fig. 1.4 Hypothetical histogenesis of ovarian germ cell tumors, based on the original concepts of Gunnar Teilum [20]. The key processes in the formation of various germ cell tumors, such as parthenogenetic activation of the oocytes, and the malignant transformation of their descendants (marked in *red*) remain still incompletely understood

noma in situ [CIS] by Skakkebaek who observed the atypical cells in the testicular biopsies of two infertile Danish men [32, 33]. Skakkebaek was the first to suggest that these cells are precursors of invasive germ cell tumors. Subsequently, CIS was replaced in 1980 by a more acceptable name *intratubular germ cell neoplasia, unclassified* [ICGNU], as it currently appears in the 2004 WHO classification [18, 19]. In the latest 2016 WHO classification it has been renamed and it has been listed as *germ cell neoplasia in situ*. An equivalent preinvasive malignancy has not been identified in the ovary.

Seminoma is a malignant germ cell tumor composed of uniform, round to polygonal cells measuring on average 15–25 μm in diameter, containing an enlarged vesicular nucleus with one or two nucleoli. Each cell has a well-developed, mostly clear, glycogen-filled cytoplasm with well-defined borders. Tumor cells are arranged into sheets, usually subdivided into smaller nests by fibrous septa, infil-

Table 1.3 WHO 2014 histologic classification of ovarian germ cell tumors

Dysgerminoma
Yolk sac tumor (primitive endodermal tumor)
Embryonal carcinoma
Polyembryoma
Non-gestational choriocarcinoma
Teratomas
Immature
Mature
Solid
Cystic
With secondary tumor
Monodermal
Struma ovarii
Carcinoid tumor
Neuroectodermal tumors
Sebaceous tumors
Mixed germ cell tumors
Gonadoblastoma
With mixed germ cell tumor
Mixed germ cell sex cord-stromal tumor

Abbreviated and slightly modified from Mutter and Prat [29]

trated by lymphocytes. The tumor was named *séminome* by the French surgeon-urologist Maurice-Auguste Chevassu in his book *Tumeurs du testicule*, Paris: G. Steinheil, 1906 [34]. Bell [35] introduced later on the term spermatocytoma, but it did not gain much popularity. Other synonyms used historically for this tumor are embryoma, embryonal carcinoma with lymphoid stroma, large cell carcinoma testis, and germinoma. Several histologic variants of seminoma have been described but such morphologic subtyping is of limited clinical significance. Seminoma can be admixed to other components of mixed germ cell tumors. Seminomas can occur in extragonadal sites. Morphologically, testicular seminomas are equivalent to ovarian dysgerminomas or germinomas of the mediastinum, pineal region, and other midline locations.

Spermatocytic seminoma is a germ cell tumor composed of three cell types [small, intermediate, and large to giant] ranging in size from 6 to 100 μm. Tumor cells grow in a diffuse or edematously nodular manner forming sheets that lack fibrous septa and lymphocytic infiltrates of classical seminoma. The tumor was first recognized by Pierre Masson [36], who separated "*le séminome spermatocytaire*" from the classical seminoma. In the latest edition of the WHO classification is was renamed and it is listed as *spermatocytic tumor*. Most spermatocytic seminomas have an indolent clinical course [18], but some can undergo sarcomatous transformation [37]. Spermatocytic seminomas do not occur outside of the testis.

Embryonal carcinoma is a malignant germ cell tumor composed of undifferentiated anaplastic epithelial cells with scant to well-developed cytoplasm and indistinct cell borders. Tumor cells grow in several patterns such as solid, papillary/tubular, or gland-like. EC cells may form the entire tumor, which is then classified as embryonal carcinoma composed of a one single cell type. In the British classification of testicular germ cell tumors, the term *malignant teratoma, undifferentiated* [MTU], is used as a synonym for embryonal carcinoma [38]. EC cells can also differentiate into somatic embryonic and also extraembryonic tissues (yolk sac and trophoblastic elements) and thus form mixed germ cell tumors. In these mixed germ cell tumors, embryonal carcinoma cells act as the rapidly proliferating malignant stem cells. EC cells account for the malignant nature of the tumor and its metastatic potential. Pure embryonal carcinoma is a rare tumor of the ovary, but it may be admixed to other germ cell tumoral elements and form the mixed germ cell tumors of the ovary. EC cells are the malignant equivalent of human embryonic stem cells (ESC) isolated from early human embryos [39]. Equivalent malignant stem cells have been isolated from murine teratocarcinomas [14].

Yolk sac tumor is a malignant germ cell neoplasm composed of cells and structures reminiscent of embryonic/fetal yolk sac, allantois, and extraembryonic mesenchyme. It is also known as primitive endodermal tumor and yolk sac carcinoma [41]. Older names such as endodermal sinus tumor or *mesoblastoma vitellinum* introduced by Gunnar Teilum [20, 40] or previous terms such as orchioblastoma and adenocarcinoma of the infant testis, polyvesicular vitelline tumor, extraembryonic mesoblastoma, malignant endothelioma of perithelioma type, and several others have been more or less abandoned [41]. *Yolk sac tumors of the prepubertal type* occur in the testis of infants. In adults yolk sac elements are rarely forming a pure yolk sac tumor and are more often part of mixed germ cells tumors of the testis. Pure malignant yolk sac tumors occur as such in the ovary or admixed to the mixed germ cell tumors of the ovary [28, 29, 41].

Choriocarcinoma is a highly malignant tumor composed of syncytiotrophoblastic and mononuclear cytotrophoblastic cells. Other synonyms are chorionepithelioma, chorioma, chorioteratoblastoma, carcinoma syncytial, and trophoblastic carcinoma. In the British classification, choriocarcinoma is called *malignant teratoma trophoblastic* [MTT] [38]. Pure choriocarcinomas are extremely rare tumors. Choriocarcinoma elements may be found in testicular mixed germ cell tumors and their metastases. Equivalent tumors occur in the ovary and the extragonadal sites as well. Ovarian tumors are labeled as non-gestational choriocarcinomas to be distinguished from malignant tumors originating from the placenta.

Teratoma is a germ cell tumor composed of somatic tissues derived from all three embryonic germ layers, i.e., ectoderm, endoderm, and mesoderm. *Monodermal teratomas* are composed of derivatives of only one germ layer. In the testis teratomas may occur in a pure form, typically in infancy and childhood, or as part of mixed germ cell tumors, which are typically found in postpubertal persons. In general, the prepubertal testicular teratomas are benign, in contrast to those in postpubertal tumors which are malignant. Pure teratomas of the ovary are the most common germ cell tumor in that organ. They are also called dermoid cysts, which is not entirely correct because most of them contain not only skin and skin appendages but other tissues as well. Teratomas of the ovary may be further divided into cystic and solid tumors and mature and immature teratomas. Benign mature teratomas may undergo malignant transformation and give rise to various somatic type malignant tumors, such as carcinoma or sarcomas. Monodermal teratomas of the ovary may present as struma ovarii, carcinoid tumor, neuroectodermal tumor, or sebaceous tumor.

Mixed germ cell tumors are malignant germ cell tumors which contain more than one germ cell tumor component. Most of these tumors contain embryonal carcinoma cells which serve as their malignant stem cells. The four most common combinations within testicular germ cell tumors are as follows: (1) embryonal carcinoma and teratoma; (2) embryonal carcinoma and seminoma; (3) embryonal carcinoma, teratoma, and yolk sac tumor; and (4) embryonal carcinoma, teratoma, and choriocarcinoma. In the original classification of Friedman and Moore [11], such mixed germ cell tumors were mostly classified as teratocarcinomas, but that term has not been used recently in human clinical pathology.

Mouse teratocarcinomas are composed of embryonal carcinoma cells which serve as their stem cells and various somatic tissues. These tumors may also contain yolk sac components but only exceptionally trophoblastic elements. Experimental murine teratocarcinomas have been used extensively as laboratory equivalents of human mixed germ cell tumors of the testis [14].

Polyembryoma is a rare variant of mixed germ cell tumors composed of embryonal carcinoma and yolk sac tumor cells and usually teratoma. Embryonal carcinoma cells and yolk sac components form embryoid bodies resembling presomitic human embryos.

Diffuse embryoma is a rare malignant tumor composed of embryonal carcinoma and yolk sac tumor components arranged in form of numerous embryoid bodies.

Gonadoblastoma is a mixed germ cell tumor composed of germ cells resembling seminoma or dysgerminoma and sex cord-stromal cells resembling immature granulosa or Sertoli cells [27]. These tumors typically develop in dysgenetic gonads in individuals who have a Y chromosome. Invasive germ cell tumors developing from gonadoblastomas are most often diagnosed as germinoma [seminoma or dysgerminoma], but in a few instances, the tumors which developed from gonadoblastoma were histologically classified as embryonal carcinoma or yolk sac tumor or teratoma [28].

References

1. Pantoja E, Noy MA, Axtmayer RW, Colon FE, Pelegrina I. Ovarian dermoids and their complications. Comprehensive historical review. Obstet Gynecol Surv. 1975;30:1–20.
2. Pantoja E, Rodriguez-Ibañez I. Sacrococcygeal dermoids and teratomas: historical review. Am J Surg. 1976;132:377–83.
3. Pantoja E, Llobet R, Gonzalez-Flores B. Retroperitoneal teratoma: historical review. J Urol. 1976;115:520–3.
4. Wheeler JE. History of teratomas. In: Damjanov I, Knowles BB, Solter D, editors. The human teratomas. Clifton: Humana Press; 1983. p. 1–22.
5. Willis RA. Pathology of tumours. 3rd ed. London: Butterworths; 1960.
6. Young RH. A brief history of the pathology of the gonads. Mod Pathol. 2005;18:S3–S17.
7. Damjanov I, Wewer-Albrechtsen N. Testicular germ cell tumors and related research from a historical point of view. Int J Dev Biol. 2013;57:197–200.
8. Lamphier TA. Mediastinal teratoma simulating adenomatous goiter. Am J Surg. 1957;94:455–61.
9. Lynch MJG, Blewett GL. Choriocarcinoma arising in the male mediastinum. Thorax. 1953;8:157–61.
10. Ewing J. Teratoma testis and its derivatives. Surg Gynecol. 1911;12:230–61.
11. Friedman NB, Moore RA. Tumors of the testis: a report on 922 cases. Milit Surgeon. 1946;99:573–93.
12. Pierce GB. Teratocarcinoma: model for a developmental concept of cancer. Curr Top Dev Biol. 1967;2:223–46.
13. Pierce GB. On the boundary between development and neoplasia. An interview with Professor G. Barry Pierce. Interview by Juan Arechaga. Int J Dev Biol. 1993;37:5–16.
14. Solter D. From teratocarcinomas to embryonic stem cells and beyond: a history of embryonic stem cell research. Nat Rev Genet. 2006;7:319–27.
15. Barbaric I, Harrison NJ. Rediscovering pluripotency: from teratocarcinomas to embryonic stem cells. Cardiff, 10–12 October 2011. Int J Dev Biol. 2012;56:197–206.
16. Dixon FJ, Moore RA. Tumors of the male sex organs, Atlas of tumor pathology, 1st series, fascicle 31b and 32. Washington, DC: Armed Forces Institute of Pathology; 1952.
17. Mostofi FK, Price Jr EB. Tumors of the male genital system, Atlas of tumor pathology, 2nd Series, Fascicle 8. Washington DC: Armed Forces Institute of Pathology; 1973.
18. Eble JN, Sauter G, Epstein JI, Sesterhenn IA, editors. World Health Organization classification of tumors. Pathology of tumours of the urinary system and male genital organs. Lyon: IARC press; 2004.
19. Ulbright TM, Young MD. Tumors of the testis and adjacent structures, AFIP atlas of tumor pathology, 4th series, fascicle 18. Silver Spring: American Registry of Pathology; 2013.
20. Teilum G. Special tumors of the ovary and testis. Comparative pathology and histological identification. Philadelphia: JB Lippincott; 1976.
21. Oliva E, Prat J, Young RH. Dr. Robert E. Scully: a tribute. Virchows Arch. 2013;462:255–6.
22. Young RH, Oliva E. Selected topics in gynecologic pathology honoring Robert E. Scully, M.D. His life and legacy. Int J Gynecol Pathol. 2014;33:323–4.
23. Oliva E, Young RH. Endocrine pathology of the ovary: in tribute to Robert E Scully, MD. Endocr Pathol. 2014;25:102–19.
24. Mostofi FK, Sobin LH. Histological typing of testis tumours. Geneva: World Health Organization; 1977.
25. Mostofi FK. Pathology of germ cell tumors of testis. Cancer. 1980;45:1735–54.
26. Skakkebaek NE. Carcinoma in situ of the testis: frequency and relationship to invasive germ cell tumours in infertile men. Histopathology. 1978;2:157–70.
27. Scully RE, Young RH, Clement PB. Tumors of the ovary, maldeveloped gonads, fallopian tube, and broad ligament, Atlas of tumor pathology, 3rd series, fascicle 23. Washintgon, DC: Armed Forces Institute of Pathology; 1998.
28. Kurman RJ, Carcangiu ML, Herrington CS, Young RH, editors. WHO classification of tumours of female reproductive organs. Lyon: International Agency for Research on Cancer; 2014.
29. Mutter GL, Prat J. Pathology of the female reproductive tract. Edinburg: Churchill Livingstone Elsevier; 2014.
30. Stevens LC, Varnum DS. The development of teratomas from parthenogenetically activated ovarian mouse eggs. Dev Biol. 1974;37:369–80.
31. Daughtry B, Mitalipov S. Concise review: parthenote stem cells for regenerative medicine: genetic, epigenetic, and developmental features. Stem Cells Transl Med. 2014;3:290–8.
32. Skakkebaek NE. Possible carcinoma-in-situ of the testis. Lancet. 1972;2(7776):516–7.
33. Skakkebaek NE. Abnormal morphology of germ cells in two infertile men. Acta Pathol Microbiol Scand A. 1972;80:374–8.
34. Androutsos G. Maurice Chevassu [1877–1957]: a pioneer of the study and therapy of testicular neoplasms. J Buon. 2003;8:185–90.
35. Bell FG. Tumours of the testicle: the spermatocytoma group. Br J Surg. 1925;13:282–301.
36. Masson P. Étude sur le séminome. Rev Can Biol. 1946;5:361–87.
37. True LD, Otis CN, Delprado W, Scully RE, Rosai J. Spermatocytic seminoma of testis with sarcomatous transformation. A report of five cases. Am J Surg Pathol. 1988;12:75–82.
38. Pugh RCB, editor. Pathology of the testis. Oxford: Blackwell; 1976.
39. Thomson JA, Itskovitz-Eldor J, Shapiro SS, Waknitz MA, Swiergiel JJ, Marshall VS, Jones JM. Embryonic stem cell lines derived from human blastocysts. Science. 1998;6(282):1145–7.
40. Talerman A. Gunnar Teilum. Am J Surg Pathol. 1981;5:719–23.
41. Nogales FF, Preda O, Nicolae A. Yolk sac tumours revisited. A review of their many faces and names. Histopathology. 2012;60:1023–33.

The Epidemiology of Malignant Germ Cell Tumors: The EUROCARE Study

2

Annalisa Trama and Franco Berrino

2.1 Introduction

Germ cell tumors (GCTs) comprise a heterogeneous group of tumors in terms of histology, age at diagnosis, anatomical site and prognosis. Here we describe the epidemiology of GCT on the basis of data from European population-based cancer registries (CRs) analysed in the framework of the EUROCARE (European Cancer Registry-based study on survival and care of cancer patients) project, which has monitored cancer patients' survival since 1978 (www.eurocare.it).

2.2 Materials

We have chosen to use the EUROCARE data because, even if CRs collect data on the basis of the International Classification of Diseases for Oncology (ICD-O3) [1] which includes morphology and topography, cancer statistics are usually provided for broad cancer categories, based on the anatomic site of the malignancies. Thus, the current statistics do not provide specific information of germ cell tumors. We have used the morphology and topography data collected by different CRs and analysed by EUROCARE to describe the epidemiology of GCT in Europe. EUROCARE includes only overt malignant tumors, while in situ tumors are requested only for screening-target cancers (breast, cervix, colon-rectum and skin melanoma) and benign tumors are collected only for central nervous system and urinary bladder. The International Agency for Research on Cancer (IARC) provides, for selected cancers, the age-standardised incidence rates of microscopically verified cases by histological type and by gender in Cancer Incidence in 5 Continents (CI5X) [2]. Since information on GCT was only available for testis and ovary, in this chapter, we have used CI5X [2] data to analyse testicular and ovarian GCT outside Europe.

EUROCARE is the widest collaborative research project on cancer survival attempted in Europe. The project started in 1989, and the fifth edition, EUROCARE-5, includes data on more than 21 million cancer diagnoses provided by 116 CRs in 30 European countries (www.eurocare.it). The data analysed in this chapter are from EUROCARE-5 and consequently included only malignant tumors. CRs cover the whole national population in European countries such as Austria, Bulgaria, Croatia, Czech Republic, Estonia, Ireland, Latvia, Lithuania, Malta, Finland, Iceland, Norway, Sweden, Slovakia, Slovenia, the Netherlands and the UK, while in

A. Trama (✉) • F. Berrino
Department of Preventive and Predictive Medicine, Fondazione IRCCS Istituto Nazionale dei Tumori, ViaVenezian 1, 20133 Milan, Italy
e-mail: annalisa.trama@istitutotumori.mi.it; franco.berrino@istitutotumori.mi.it

F.F. Nogales, R.E. Jimenez (eds.), *Pathology and Biology of Human Germ Cell Tumors*,
DOI 10.1007/978-3-662-53775-6_2

others CRs cover only one or several regions [3]. In 2000–2007 (the study period of EUROCARE-5), 54,000 GCTs were registered in the countries included in the incident analyses (Table 2.1). Table 2.1 shows the number of GCT cases contributed by the different countries. Please note that the differences by country might be due to the different CR coverage (national vs regional) and to the different incidence years that CR have contributed. Not all included the full period 2000–2007.

Table 2.1 Cancer registration coverage in EUROCARE-5 and the number of germ cell tumor cases registered in the countries included to the incident analyses

	Proportion of population covered by cancer registries included in EUROCARE-5 (%)[a]	Number of germ cell tumor cases registered in 2000–2007
Austria	100	2.610
Belgium (Flanders)	58	1.215
Bulgaria	100	1.457
Croatia	100	900
Czech Republic	100	3.491
Estonia	100	190
Finland	100	944
France	23	1.213
Germany	23	6.950
Iceland	100	79
Ireland	100	1.210
Italy	35	3.450
Latvia	100	267
Lithuania	100	268
Malta	100	76
Norway	100	2.185
Poland	13	904
Portugal	76	698
Slovakia	100	1.683
Slovenia	100	812
Spain	17	778
Switzerland	30	923
The Netherlands	100	5.078
UK	100	16.626
Total		54.007

[a]Cancer registration is continuously improving since EUROCARE-5

GCTs include different histological subtypes, internationally grouped as seminomas and non-seminomas. Throughout this chapter, the generic term seminoma and non-seminoma will be used. Seminoma, dysgerminoma and germinoma are histopathologic equivalent terms for a neoplasm of identical morphology in testis, ovary and extragonadal locations. Seminoma includes all seminoma histological types (ICD-O3 codes 9060–9064); non-seminomas in their turn include embryonal carcinoma (ICD-O3 codes 9070, 9072), yolk sac tumor (ICD-O3 code 9071), choriocarcinoma (ICD-O3 codes 9100, 9102), teratoma (ICD-O3 codes 9080,9082,9083), mixed germ cell tumors (ICD-O3 codes 9081,9085,9101), malignant struma ovarii (ICD-O3 code 9090), cystic teratoma with somatic malignant transformation (ICD-O3 code 9084) and other non-seminomatous germ cell tumors (ICD-O3 codes 9065). Spermatocytic tumor, an exclusively testicular neoplasm, is clinically and pathologically distinct from classic seminoma; thus data are provided separately for this specific type.

2.3 Incidence

The crude and age-adjusted (European standard population) incidence rates of GCT in Europe were both equal to 34/1.000.000 with marked differences between male (64/1.000.000) and female (4/1.000.000). In the USA, the incidence rate was 56/1.000.000 in white males contrasting with 3.2/1.000.000 in white females, over a period of more than 30 years from 1973 to 2007 [4]. In the same country, the incidence in black males was much lower (10/1.000.000), due to a lower incidence of seminomas, while no difference was reported between white and black females [4]. More than 90 % of testicular tumors were indeed GCT. Figure 2.1 shows incidence of testicular cancer across different continents in 2012. White males living in Western industrialised countries, particularly in Northern and Western Europe, showed the highest incidence rates of testicular tumors (12/100.000 in Denmark,

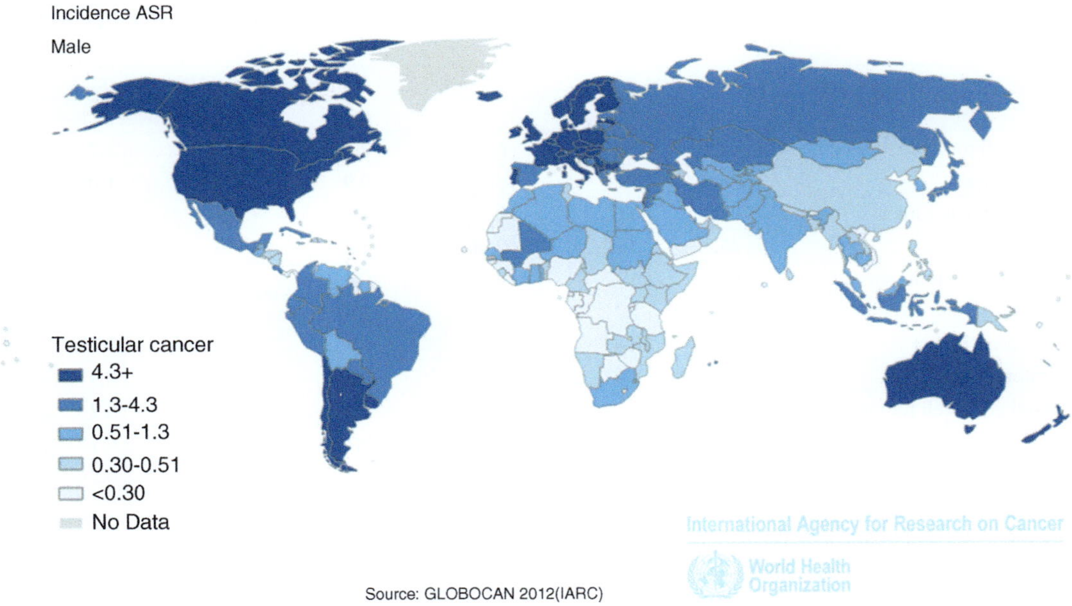

Fig. 2.1 Testicular cancer age-standardised (world) incidence rate per 100.000 (Source http://globocan.iarc.fr/Pages/fact_sheets_cancer.aspx)

Norway and Switzerland), whereas black males in Africa showed the lowest (<0.5/100.000 in the majority of African countries). In Australia and New Zealand, the incidence was 7/100.000. In North America (USA and Canada), it was 5/100,000; in South and Central America, it was 2/100.000 with differences among countries (Chile 7/100.000, Uruguay 6/100.000, Argentina and Costa Rica 5/100.000, Mexico and Colombia 3/100.000, Brazil 2/100.000 and remaining countries <1/100.000). In Japan, the incidence was 2/100.000; in South, Eastern and Central Asia, it was <1/100.000, being higher in Western Asia, 1.7/100.000, with regional countries (5/100.000 in Israel, 3/100.000 in Georgia and <1 in Oman, Qatar, Iraq and Azerbaijan). In China, the incidence was 0.5/100.000 [5].

Incidence of malignant ovarian GCT was low in all continents: ≤0.9/100.000 in Japan, ≤0.7/100.000 in Central and South America and in China, ≤0.5/100.000 in Australia and Asia, 0.4/100.000 in Canada and <0.4/100.000 in Africa except Malawi where the incidence was 1.3/100.000 [2].

Gonadal GCT (GGCT) Most GCTs arise in the gonads. The incidence in Europe is 33/1.000.000 being substantially higher in males than in females (62/1.000.000 vs 2.5/1.000.000, respectively). Histologic differences are observed between both genders: in males, seminomas are more common than non-seminomas, contrary to women who have more non-seminomas than seminomas (Table 2.2). In males, non-seminomas are mixed germ cell tumors, embryonal carcinoma and teratoma, while in females they are (immature) teratomas and yolk sac tumors (Table 2.2). In males, spermatocytic tumor is very unusual (Table 2.2).

Testicular GCTs have an early incidence peak in the age group 0–4 years followed by a second peak in adolescents and young adults (15–19 and 25–29 and 30–34 years) (Fig. 2.2). The first peak is due to non-seminomas (incidence 1.5/1.000.000 vs 0.07/1.000.000 of seminomas) and mainly due to yolk sac tumour and teratoma, which have an incidence of 1/1.000.000 and 0.3/1.000.000, respectively.

Non-seminomas are more common than seminomas until the age of 30 years; however, the histologic types of those between 15 and 30 years

Table 2.2 (EUROCARE) Gonadal germ cell tumors incidence rate per million, in Europe by histological type, and sex (age adjusted) with 95 % confidence interval (CI)

	Male			Female		
	Rate	95 % CI		Rate	95 % CI	
Germ cell tumors	62.1	61.6	62.7	2.5	2.4	2.7
Seminomas	**36.9**	36.4	37.3	**0.9**	0.8	1.0
Spermatocytic tumor	**0.6**	0.6	0.7	–	–	–
Non-seminomas	**25.3**	24.9	25.6	**1.7**	1.6	1.8
Embryonal carcinoma	**6.8**	6.7	7.0	0.1	0.0	0.1
Yolk sac tumor	1.3	1.2	1.3	**0.4**	0.4	0.5
Choriocarcinoma	0.4	0.4	0.5	>0.1	0.0	0.1
Teratoma	**5.6**	5.4	5.8	**0.8**	0.7	0.9
Mixed germ cell tumors	**10.6**	10.4	10.9	0.2	0.2	0.2
Struma ovarii, malignant	0.0	0.0	0.0	0.1	0.0	0.1
Cystic teratoma with somatic malignant transformation	0.0	0.0	0.0	0.1	0.1	0.2

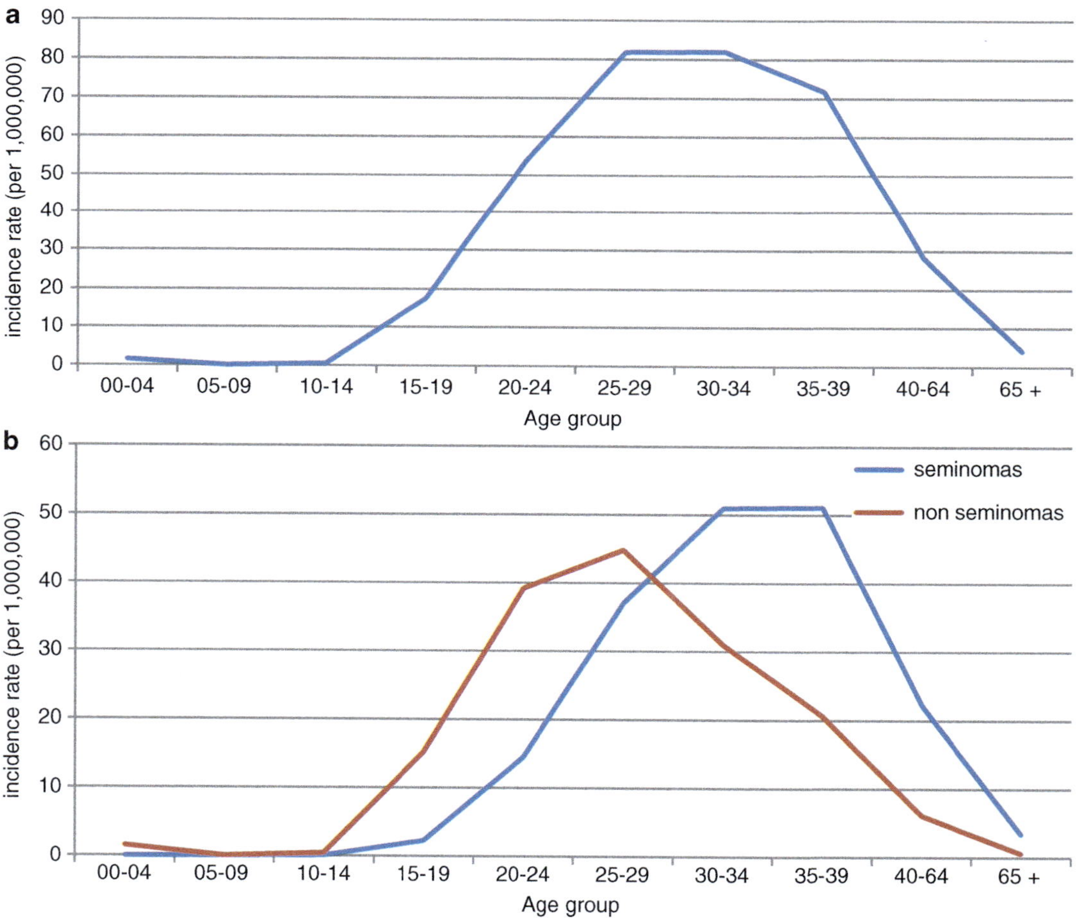

Fig. 2.2 (EUROCARE) Testicular germ cell tumor incidence per 1.000.000 by age group overall (**a**) and by age group and histology (**b**)

are mainly embryonal carcinoma and mixed germ cell tumors and only to a lower extent teratoma and yolk sac tumor.

After the age of 30, seminomas are predominant. Overall, seminomas presented an incidence peak 10 years later than non-seminomas (Fig. 2.2).

Ovarian GCTs have a small peak in children under 5 years and a clear peak in the 15–19-year age group (Fig. 2.3). These results are coherent with those reported in the USA [4]. Both seminomas and non-seminomas have a peak at 15 to 19 years; non-seminomas are more common than seminomas in almost all age groups and in particular in young females (>25 years) (Fig. 2.3). Yolk sac tumour and teratoma represent the most common histologic types in the 0–4-year age group; teratoma incidence increases with age, and it is the most common type in all age groups as from 5 years old.

Extragonadal Germ Cell Tumors (EGGCTs) Only 4 % of all GCTs were extragonadal and arose mainly in midline locations such as the central nervous system (CNS), the mediastinum and the pelvis. Gonadal GCTs were mainly seminomas (59 %

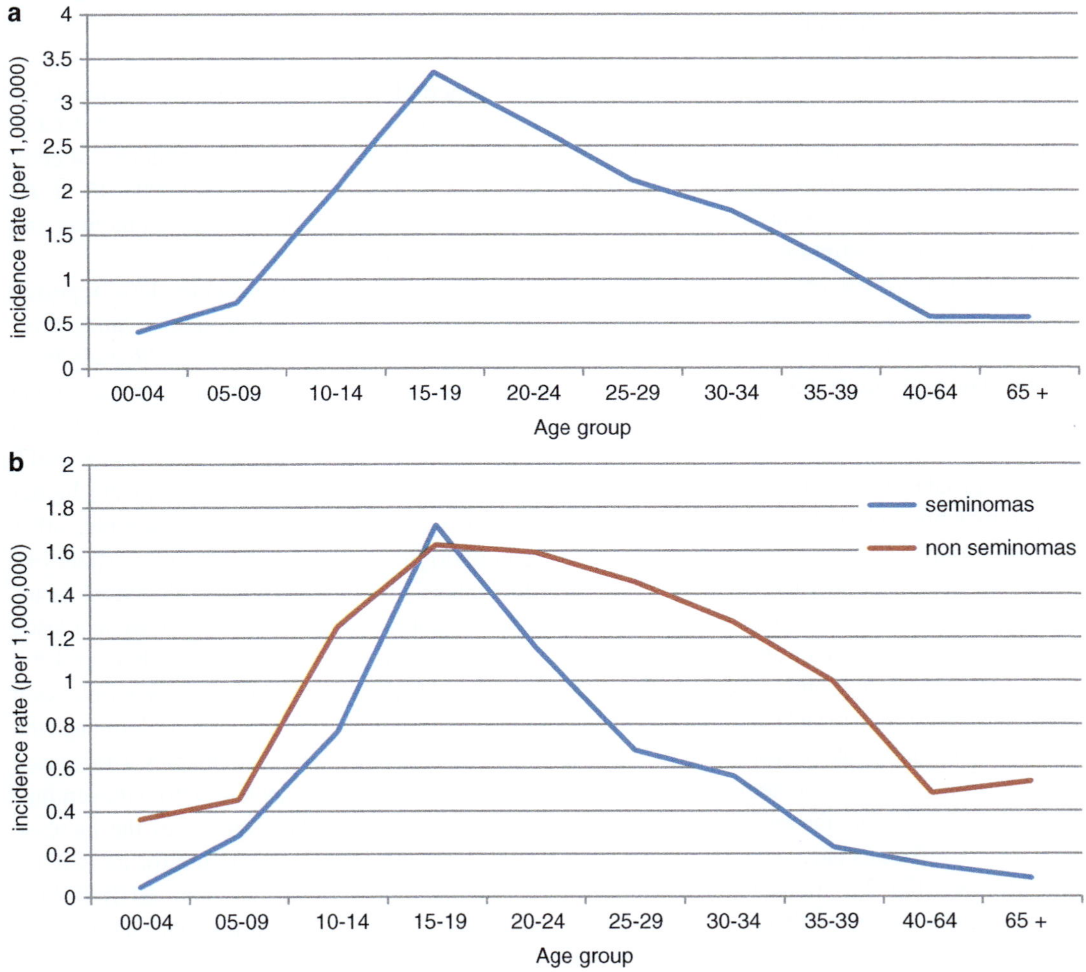

Fig. 2.3 (EUROCARE) Ovarian germ cell tumors incidence per 1.000.000 by age group overall (**a**) and by age group and histologies (**b**)

Table 2.3 (EUROCARE) Incidence rate of germ cell tumors per 1.000.000 by sex (age adjusted), extragonadal sites and histological types with standard error (SE)

	Male						Female					
	All		Seminomas		Non-seminomas		All		Seminomas		Non-seminomas	
	Rate	SE	Rate	SE	Rate	SE	Rate	SE	Rate	SE	Rate	SE
Extragonadal GCTs	1.81	0.05	0.90	0.03	0.91	0.04	1.19	0.04	0.22	0.02	0.97	0.04
Central nervous system	**0.62**	0.03	**0.51**	0.03	0.12	0.01	**0.18**	0.02	**0.14**	0.01	0.03	0.01
Mediastinum and thorax	**0.51**	0.03	**0.21**	0.02	**0.30**	**0.02**	**0.07**	0.01	0.01	> 0.01	**0.05**	0.01
Abdomen and pelvis	**0.38**	0.02	**0.10**	0.01	**0.27**	**0.02**	**0.79**	0.03	0.03	0.01	**0.76**	0.03

seminomas vs 41 % non-seminomas), while 62 % of EGGCT were non-seminomas vs 38 % seminomas. No significant differences in the EGGCT incidence were observed between males and females.

Among males, the most frequent extragonadal sites, in decreasing order, were the CNS, mediastinum and thorax and abdomen and pelvis, while in females locations were the abdomen and pelvis, CNS and mediastinum and thorax. In both sexes, EGGCTs of the brain were mainly seminomas, while in females, non-seminomas were predominant in other sites, especially in the pelvis and abdomen (Table 2.3).

In males, the proportion of EGGCT was lower (3 %) than in females (32 %); in males, the proportion of EGGCT was highest in the 5–14-year age group, while in females, it peaked in the 0–4-year age group (Fig. 2.4).

The EGGCT had an early incidence peak in children of less than 5 years; the incidence increased from 5 to 20 years and started to decline again in the older age groups (age older than 30). The first peak was due to GCT of the abdomen and pelvis. The second peak corresponded to GCT of the mediastinum and of the abdomen and pelvis. GCTs of the CNS were the most common in children older than 5 years old and in adolescents from 15 to 19 years (Fig. 2.5). In the extragonadal sites, the age distribution was similar for seminomas and non-seminomas. The age incidence pattern was similar for males and females. Teratoma was the most common non-seminomatous histological type among the extragonadal sites.

2.4 Incidence Trends of Gonadal and Extragonadal Germ Cell Tumors

In the period 1996–2007, there was a statistically significant increase in the incidence of GGCT from 28.6/1.000.000 (95 % confidence interval (CI) 27.2–28.9) to 35.5/ 1.000.000 (95 % CI 34.5–36.5) overall and for both seminomas and non-seminomas (Table 2.4).

Several groups of investigators have reported an increasing incidence of testicular cancer over the last 20 years [4, 6–8]. A recent CR study from the USA revealed a significant increase in the incidence of testicular GCT in both white and black males in the period 1973–2007. Nonetheless, the incidence was much higher among whites. Because of small sample, other ethnic origins were excluded from the study [4]. In Finland, the incidence rates of several histological subtypes of testicular GCT have increased over the last four decades, particularly from 1990 onwards; yet the increasing trend was only seen in men aged 15–44 years [7]. In Germany, the annual incidence rate of testicular GCT increased over the entire period (1998–2008) with seminomas accounting for the majority of the increase [5]. Similar results were reported for the UK for the period 1979–2003 [8]. Our EUROCARE analyses (Table 2.4) are in line with these previous studies.

In comparison with testicular GCT, the overall incidence trends of ovarian GCT differed. In Finland, even if a significant increase in the inci-

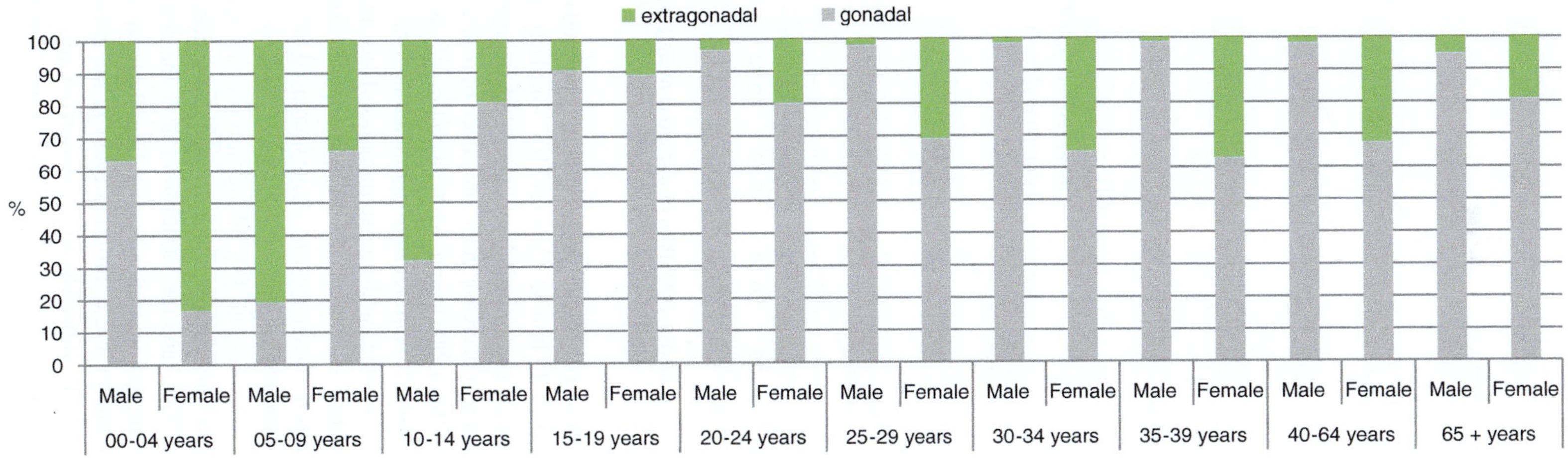

Number of cases in male (M) and female (F) by age group

	00-04		05-09		10-14		15-19		20-24		25-29		30-34		35-39		40-64		65 +	
	M	F	M	F	M	F	M	F	M	F	M	F	M	F	M	F	M	F	M	F
EGCTs	79	170	42	33	121	45	184	41	202	70	182	102	145	110	114	86	266	142	57	34
GGCTs	135	34	10	64	57	189	1,736	331	5,530	282	8,764	227	9,355	203	8,640	145	14,544	294	1,042	142

EGCTs=extragonadal germ cell tumours

GGCT=gonadal germ cell tumours

Fig. 2.4 (EUROCARE) Age-related variation in the proportion of gonadal and extragonadal germ cell tumors in male and female. *EGCTs* extragonadal germ cell tumors, *GGCTs* gonadal germ cell tumors

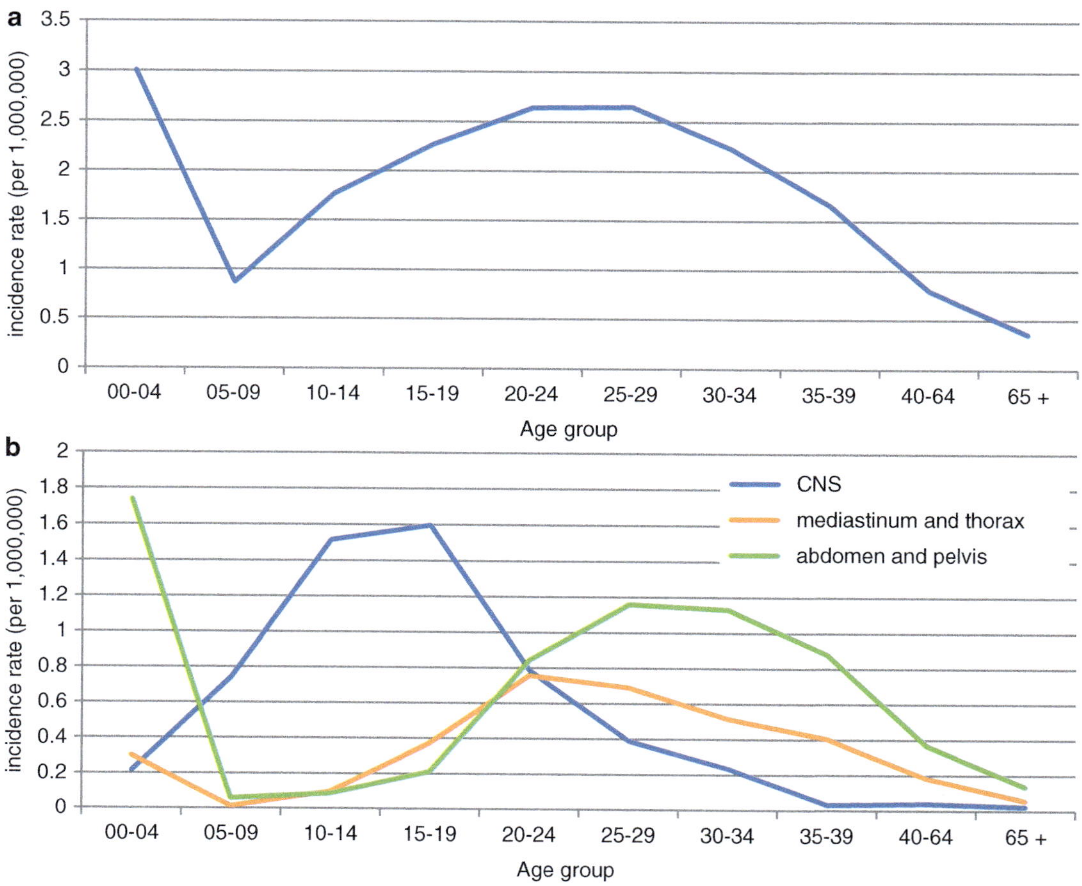

Fig. 2.5 (EUROCARE) Extragonadal germ cell tumors incidence rate (per 1,000,000) by age group overall (**a**) and by age group and extragonadal sites (**b**)

dence of ovarian non-seminomas occurred in the age group of 15–44 years between the first and the last study decade (1969–2008), it remained stable during the last three decades. Similarly, a study on ovarian cancer incidence by histological type carried out in Osaka, Japan, in 1975–1998 [9] reported that the incidence of ovarian GCT also remained stable. In the UK, the incidence of ovarian GCT increased from 1979 to 2003, due to a higher frequency of non-seminomas between the period of time 1980–2000, and in the age group of 10–49 years. However, in the last decade, the incidence has remained stable [8]. In Germany, the incidence of GGCT stayed constant over the period 1998–2008 [6]. In contrast, in the USA, a study covering more than 1200 cases of malignant ovarian GCT concluded that incidence rates have declined over the last 30 years, with the decrease being confined (nearly 30 %) to non-seminomas [10]. Similarly, a more recent study carried out in the USA demonstrated a slight decrease in the incidence of ovarian GCT in both black and white women [4]. We have observed a minor non-significant decline of both seminomas and non-seminomas. Thus, the incidence of ovarian GCT in industrialised countries has not been shown to have increased along with testicular GCT.

Arora et al. [8], taking into account similarities between the shapes of age-incidence curves of GCT and the variation in peak incidence and longitudinal trends by site, hypothesised about a common GCT initiation event in the embryonal period followed by a progression of tumorigenesis conditioned by site-specific events during the foetal and/or postnatal period.

Table 2.4 Gonadal germ cell tumors (GGCTs) per 1.000.000 and age-adjusted incidence (95 % confidence interval), EUROCARE 1996–2007 with average percent change (APC)

	GCTs of testis				GCTs of ovary		
Years	All	Seminomas	Non-seminomas		All	Seminomas	Non-seminomas
1996	26.74 (25.9–27.6)	15.64 (15.0–16.3)	11.09 (10.6–11.6)		1.32 (1.2–1.5)	0.44 (0.3–0.6)	0.88 (0.7–1.0)
1997	27.49 (26.7–28.3)	15.97 (15.4–16.6)	11.52 (11.0–12.1)		1.44 (1.3–1.6)	0.46 (0.4–0.6)	0.98 (0.8–1.2)
1998	28.75 (27.9–29.6)	16.90 (16.3–17.5)	11.85 (11.3–12.4)		1.39 (1.2–1.6)	0.44 (0.3–0.6)	0.96 (0.8–1.1)
1999	29.73 (28.9–30.6)	17.52 (16.9–18.2)	12.21 (11.7–12.8)		1.35 (1.2–1.6)	0.42 (0.3–0.5)	0.93 (0.8–1.1)
2000	29.86 (29.0–30.7)	17.87 (17.2–18.5)	11.99 (11.5–12.6)		1.40 (1.2–1.6)	0.48 (0.4–0.6)	0.92 (0.8–1.1)
2001	30.88 (30.0–31.8)	18.50 (17.8–19.2)	12.39 (11.8–13.0)		1.24 (1.1–1.4)	0.39 (0.3–0.5)	0.84 (0.7–1.0)
2002	30.59 (29.7–31.5)	17.99 (17.3–18.7)	12.60 (12.1–13.2)		1.47 (1.3–1.7)	0.41 (0.3–0.5)	1.05 (0.9–1.2)
2003	31.24 (30.4–32.1)	18.19 (17.5–18.9)	13.05 (12.5–13.6)		1.28 (1.1–1.5)	0.49 (0.4–0.6)	0.80 (0.7–1.0)
2004	32.04 (31.2–32.9)	18.78 (18.1–19.5)	13.25 (12.7–13.8)		1.30 (1.1–1.5)	0.45 (0.4–0.6)	0.85 (0.7–1.0)
2005	33.79 (32.9–34.7)	19.62 (18.9–20.3)	14.17 (13.6–14.8)		1.30 (1.1–1.5)	0.37 (0.3–0.5)	0.92 (0.8–1.1)
2006	34.00 (33.1–34.9)	20.23 (19.5–20.9)	13.77 (13.2–14.4)		1.28 (1.1–1.5)	0.48 (0.4–0.6)	0.81 (0.7–1.0)
2007	34.29 (33.4–35.3)	20.43 (19.7–21.2)	13.86 (13.3–14.5)		1.22 (1.1–1.4)	0.37 (0.3–0.5)	0.85 (0.7–1.0)
APC	2.21[a]	2.28[a]	2.12[a]		–0.9	–0.7	–1.0

[a]The APC is significantly different from zero ($p < 0.05$)

Regarding EGGCT in the period 1996–2007, we found a statistically significant increase in the incidence of GCT of the thymus (APC 2.68) and uterus (an extremely rare location) (APC 1.04). However, the latter result should be considered with caution since uterine mixed mesodermal Mullerian tumors, which are relatively common, are frequently mistaken for teratomas. The incidence was fairly constant for GCT of the brain (APC 0.46), while no changes were observed for GCT of the mediastinum and of retroperitoneum.

Similar results were observed in the UK with an increasing incidence of CNS GCT but not of those of the mediastinum, abdomen and pelvis [8]. In Germany, the incidence of EGGCT in males remained virtually constant, while among females, the incidence of EGGCT declined [6]. Similar results were observed in the USA where the incidence of EGGCT among white males remained fairly constant over the entire time period. The incidence of EGGCT among black males increased from 1973 to 1992 and then reached a plateau before declining in the latest time period; however, rates were based on small numbers, making the interpretation difficult. Among white and black females, incidence rates for EGGCT registered small decreases over time. However, the fluctuation of rates among black females was large, due to small numbers [4].

2.5 Survival

The 5-year relative survival for GCT was very good (93 %), with seminomas having better survival than non-seminomas: 95 % (95%CI 93.5–95.6) and 90 % (95%CI 88–91), respectively.

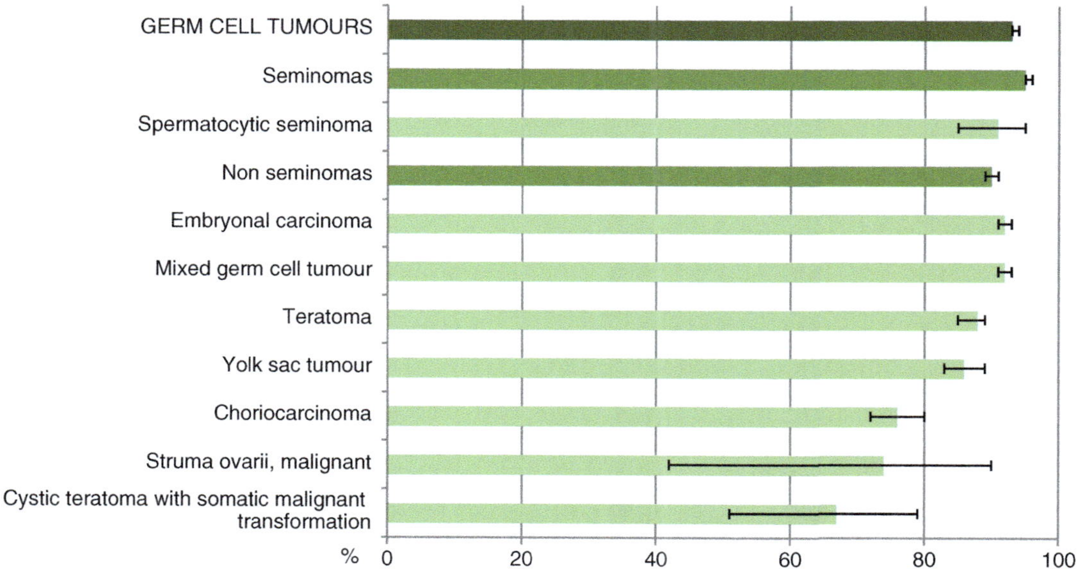

Fig. 2.6 (EUROCARE) Germ cell tumors 5-year relative survival (period 2000–2007) by histotype. Error bars are 95 % confidence interval

Table 2.5 (EUROCARE) Germ cell tumors 5-year relative survival (%) period of diagnosis by site

	Relative survival (%)	95 % CI	
Gonadal germ cell tumors	94.0	93.6	94.3
Testis	94.4	94.0	94.7
Ovary	83.9	81.1	86.3
Extragonadal germ cell tumors	74.2	71.2	76.9
Mediastinum	50.7	43.0	57.9
Retroperitoneum	76.2	64.6	84.4
Brain	81.1	72.0	87.4
Pineal gland	79.9	69.7	86.9
Pituitary gland	96.3	76.2	99.5

Among non-seminomas, the histological type with worst prognosis was mature cystic teratoma with somatic malignant transformation, although the survival estimate must be considered with caution as it was based on a limited number of cases (Fig. 2.6).

Gonadal GCT had better survival than EGGCT: 94 % (95 %CI 93.6–94.3) vs 74 % (95 %CI 71 %–77 %), respectively. Among EGGCTs, survival was good for the brain 81 % (95%CI 72–87), pituitary gland 96 % (95 %CI 76–99), pineal gland 80 % (95 %CI 70–87) and retroperitoneum 76 % (95 %CI 65–84) and lower, 51 % (95 %CI 43–58), for the mediastinum.

Males had better survival than females: 94 % (95 %CI 93.3–94) vs 84 % (95 % 82–86).

Survival decreased with increasing age with the worst survival observed in those older than 40 years for both seminomas and non-seminomas. However, differences were observed between GCT of testis and ovary. For ovarian GCT, survival was 90 % before 40 years of age decreasing to 63 % in the 40–64-year age group and to 29 % in women older than 65. For testicular GCT, survival was 94 % before 40 years and 78 % after 65+ years. These results were in line with those previously published [4, 11].

Interestingly, Verhoeven et al. [12] noted that in 2003–2007, despite the improvements in the relative survival of non-seminoma patients aged ≥50 years, survival remained markedly lower than the survival of seminoma patients of the same age. There is little room for survival improvement among testicular seminoma patients, especially for those aged <50 years. Older testicular cancer patients remain at increased risk of death, which seems mainly attributable to the lower survival of non-seminoma patients [12].

According to the RARECARE [13] definition of rare cancers (incidence <6/100,000), GCTs with an incidence of 34/1.000.000 are rare cancers in Europe. However, germ cell cancer incidence varies considerably in different geographical areas. GCTs include seminomas and non-seminomas and arise mainly in the gonads. However, GCT can be diagnosed also in the CNS, mediastinum, thorax, abdomen and pelvis. The distribution of seminomas and non-seminomas differs by sex and by site of origin. Overall, GCTs are typical tumors of adolescents and young adults and are characterised by a good prognosis, especially gonadal GCT. The incidence is stable except for testicular GCT (Table 2.5).

References

1. Percy C, Fritz A, Jack A, Shanmugarathan S, Sobin L, Parkin DM, Whelan S. International Classification of Diseases for the Oncology (ICD-O). 3rd ed. Geneva: World Health Organisation; 2000.
2. Forman D, Bray F, Brewster DH, Gombe Mbalawa C, Kohler B, Piñeros M, Steliarova-Foucher E, Swaminathan R and Ferlay J eds 2013. Cancer incidence in five continents, Vol. X (electronic version). Lyon: IARC. http://ci5.iarc.fr. Last Accessed on 29 Dec 2014.
3. De Angelis R, Sant M, Coleman MP, Francisci S, Baili P, Pierannunzio D, et al. Cancer survival in Europe 1999–2007 by country and age: results of EUROCARE-5—a population-based study. Lancet Oncol. 2014;15:23–34.
4. Stang A, Trabert B, Wentzensen N, Cook MB, Rusner C, Oosterhuis JW, et al. Gonadal and extragonadal germ cell tumors in the United States, 1973–2007. Int J Androl. 2012;35(4):616–25.
5. Ferlay J, Soerjomataram I, Ervik M, Dikshit R, Eser S, Mathers C, Rebelo M, Parkin DM, Forman D, Bray F. GLOBOCAN 2012 v1.0, cancer incidence and mortality worldwide: IARC cancer base no. 11 [Internet]. Lyon: International Agency for Research on Cancer; 2013. Available from: http://globocan.iarc.fr. Accessed on 29 Dec 2014.
6. Rusner C, Trabert B, Katalinic A, Kieschke J, Emrich K, Stang A, et al. Incidence patterns and trends of gonadal and extragonadal germ cell tumors in Germany, 1998-2008. Cancer Epidemiol. 2013;37(4):370–3.
7. Pauniaho SL, Salonen J, Helminen M, Vettenranta K, Heikinheimo M, Heikinheimo O. The incidences of malignant gonadal and extragonadal germ cell tumors in males and females: a population-based study covering over 40 years in Finland. Cancer Causes Control. 2012;23:1921–7.
8. Arora RS, Alston RD, Eden TO, Geraci M, Birch JM. Comparative incidence patterns and trends of gonadal and extragonadal germ cell tumors in England, 1979 to 2003. Cancer. 2012;118(17):4290–7.
9. Ioka A, Tsukuma H, Ajiki W, Oshima A. Ovarian cancer incidence and survival by histologic type in Osaka. Japan Cancer Sci. 2003;94(3):292–6.
10. Smith HO, Berwick M, Verschraegen CF, Wiggins C, Lansing L, Muller CY, et al. Incidence and survival rates for female malignant germ cell tumors. Obstet Gynecol. 2006;107(5):1075–85.
11. Trama A, Mallone S, Nicolai N, Necchi A, Schaapveld M, Gietema J, et al. Burden of testicular and extragonadal germ cell tumors in Europe. Eur J Cancer. 2012;48(2):159–69.
12. Verhoeven RH, Gondos A, Janssen-Heijnen ML, Saum KU, Brewster DH, Holleczek B, et al. Testicular cancer in Europe and the USA: survival still rising among older patients. Ann Oncol. 2013;24:508–13.
13. Gatta G, van der Zwan JM, Casali PG, et al. Rare cancers are not so rare: The rare cancer burden in Europe. Eur J Cancer. 2011;47(17):2493–511.

Germ Cell Tumors from a Developmental Perspective: Cells of Origin, Pathogenesis, and Molecular Biology (Emerging Patterns)

3

J. Wolter Oosterhuis and Leendert H.J. Looijenga

3.1 Introduction

Germ cell tumors (GCT) are a seemingly heterogeneous family of neoplasms, whose histological composition likely reflects the developmental potential of the cells from which they are derived.

Recent discoveries on the regulation of developmental states of cells in the early embryo and the germline allow a deeper understanding of the origin and developmental potential of GCT and provide a biologically plausible and clinically relevant basis for their classification.

3.2 Developmental States of Early Embryonic Cells

3.2.1 Restriction Versus Maintaining Developmental Potential

Multicellular organisms develop from a single omnipotent cell, the zygote, through a tightly regulated program of restriction of pluripotency [1], yet for maintaining their kind, they have to preserve totipotency in the germ cell lineage [2]. For full developmental potential, the zygote of placental mammals needs a biparental genomic imprinting (GI) [3–5] and a specifically methylated intact genome with X-inactivation in female cells [6, 7].

Here a brief explanation of GI is appropriate; changes of (global) methylation in the early embryo and the germline will be discussed later in this section. GI is the phenomenon whereby in mammals the expression of some genes depends on maternal or paternal origin, due to parental-specific DNA methylation and histone modification [8]. The GI cycle starts with erasure of the original biparental imprinting pattern of the zygote early in the germ lineage through replacement of methylcytosine by unmethylated cytosine via the base excision repair pathway [9]. Later, during oogenesis and spermatogenesis, respectively, fresh maternal and paternal imprinting patterns are established by de novo methylation of the relevant targets, an estimated 100–200 genes including noncoding RNAs, about 1 % of the genome [10–15] (Fig. 3.1). A variety of human diseases is caused by aberrations of specific imprinted genes through genetic and epigenetic mechanisms (for review [16]).

Only blastomeres after the first few cleavage divisions, up to the eight-cell stage, may have the full developmental potential (omnipotency) of the zygote [17]. In fact, in the mouse, it is lost beyond the two-cell stage [18]. Later,

J.W. Oosterhuis, MD, PhD (✉)
L.H.J. Looijenga, PhD
Department of Pathology, Erasmus University Medical Center, Rotterdam, The Netherlands
e-mail: j.w.oosterhuis@erasmusmc.nl; l.looijenga@erasmusmc.nl

F.F. Nogales, R.E. Jimenez (eds.), *Pathology and Biology of Human Germ Cell Tumors*, DOI 10.1007/978-3-662-53775-6_3

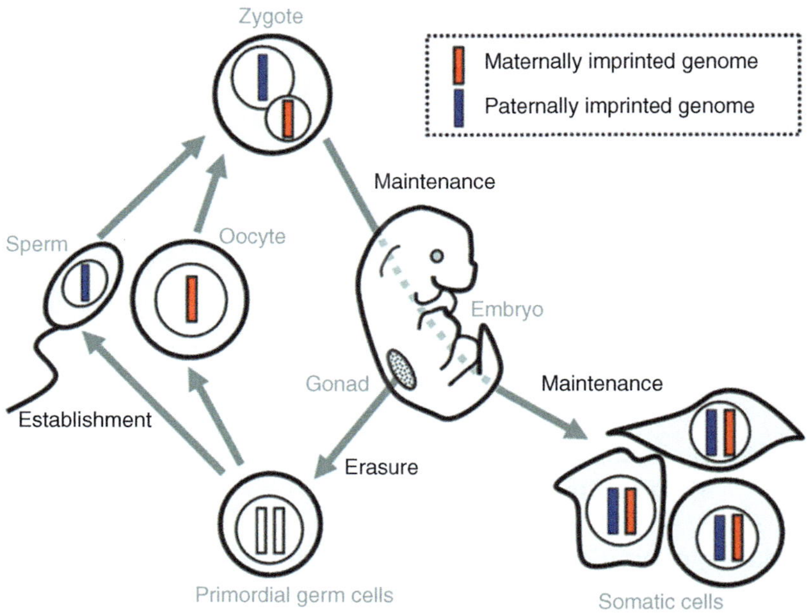

Fig. 3.1 Cycle of genomic imprinting (GI). Upon fertilization, the zygote acquires a haploid set of paternally imprinted chromosomes from the father and a haploid set of maternally imprinted chromosomes from the mother; the cells of the embryo therefore have a biparental GI pattern. In the germline, GI is erased; during spermatogenesis and oogenesis, respectively, paternal and maternal imprinting is reestablished [746]

blastomeres and the embryonal stem cells (ESC) of the inner cell mass (ICM) and epiblast undergo further restriction step by step of their developmental potential, as stem cells are generated with commitment to developing specialized organs and tissues [1], including the germ lineage, which is specified to transfer omnipotency to the next generation [2, 12]. Probably all cells of the ICM and the epiblast are in principle germline competent and thus potentially totipotent/omnipotent [2, 19], although with different efficiency [20].

3.2.2 OCT4: Key Protein in Pluripotency

From studies mainly in the mouse, Oct4 (also known as Pouf1, Oct3, Oct4, and Otf3), a member of the POU-domain family of octamer-binding transcription factors, emerges as the key component in the regulatory network that maintains pluripotency [21–24]. Although not indispensable for the establishment of omnipotency in the zygote, it is required for maintaining pluripotent states in the developing embryo [25]. From the two-cell stage onward [19, 23], after the genome of the zygote is activated, Oct4 is expressed in all cells through the morula stage. Later, in the preimplantation embryo, Oct4 is confined to the ICM and the epiblast, while after implantation, its expression is limited to the primitive ectoderm (Fig. 3.2). Simultaneously with downregulation of Oct4 in the primitive ectoderm during gastrulation, primordial germ cells (PGC), the stem cells of gametogenesis in later life, are formed, with continued expression of Oct4 [26] that is maintained in the developing germ lineage until entry in meiosis. Meiotic oocytes are negative for Oct4; in the mouse, Oct4 is reexpressed in oocytes of the postnatal ovary. In the mouse testis, it is only expressed in type A spermatogonia [27]. In the human embryo, OCT4 expression starts somewhat later than in the mouse, in the eight-cell embryo [28]. OCT4 is normally not expressed in the testis beyond the age of 6 months and thus negative in spermatogonia, i.e., in male germ cells from mitotic arrest onward [29]. In contrast to the mouse, in humans, OCT4 is not expressed in meiotic germ cells both in males and females, and it thus remains also negative after birth in the pre- and postpubertal ovary [29–31]. OCT4 is specific for normal and neoplastic pluripotent cells and not expressed in normal adult human tissues and the large majority of cancers derived from adult tissues [32].

Notably, in the cells of the ICM and the epiblast of the preimplantation mouse embryo and the germline (collectively, the totipotent ESC) which efficiently contribute to the germline in chimeric embryos [33], Oct4 expression is driven by its distal enhancer. In contrast, in ESC from primitive ectoderm of the postimplantation embryo, which are pluripotent and contribute to the germline with low efficiency, it is driven from its proximal enhancer [26]. In the ESC, both of the pre- and postimplantation embryos and in the germline Oct4 physically partners with Sox2, regardless of the driving enhancer [20, 34]. In humans, OCT4 is coupled with SOX2 in ESC, however, with SOX17 in the germline [35].

OCT4 (6p21–22) [36] is involved in a network of pluripotency factors including among others SOX2 (3q26–27) [37] and NANOG, STELLAR, and GDF3 (12p13) [38] that induce and maintain pluripotency of ESC, repress development of somatic lineages, and regulate cell fate decisions in the early embryo [20, 39].

In various animal models, factors involved specification of early lineages, orchestrated by the pluripotency network, have been identified, such as Ezh2, Sox21, and Cdx2 for trophectoderm [40–42], Sox17 [43] (SOX17 in humans) [44] for primitive endoderm, and TBX3 for mesoderm (in *Xenopus*) [45]. Oct4 switches partner from Sox2 to Sox17 in the primitive endoderm [34] and in humans also in the germline [35].

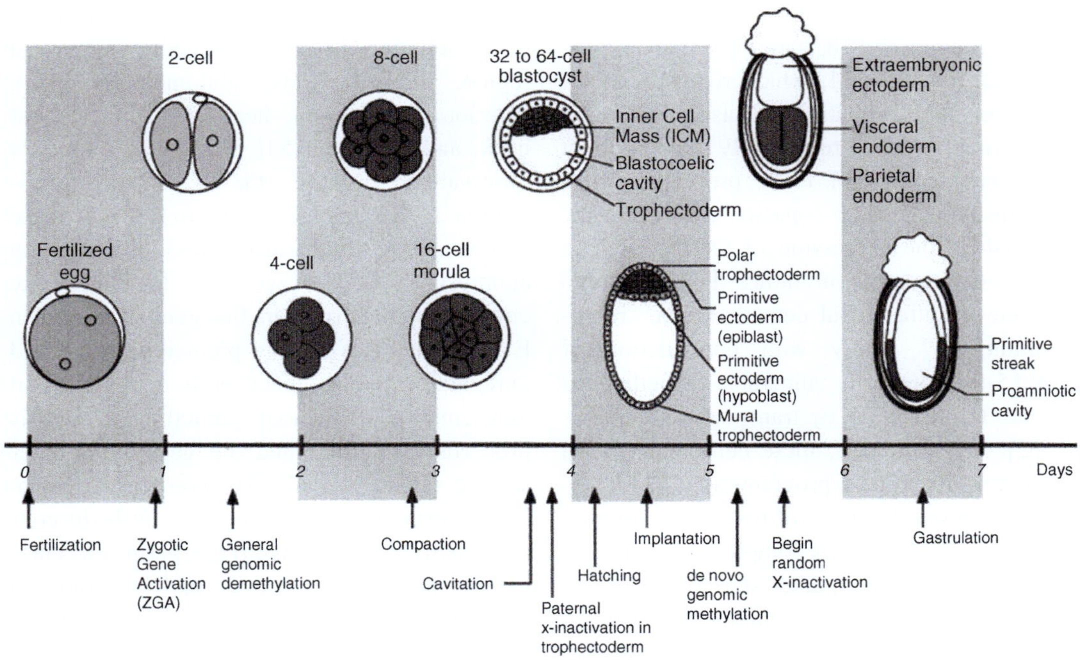

Fig. 3.2 Oct4 expression in the early mouse embryo. The progressive stages of murine preimplantation development, through implantation and gastrulation (embryonic days 0.5–6.5), are schematically represented. Critical genetic and epigenetic events initiated during this period are indicated at the appropriate time points. The expression pattern of Oct4 mRNA and protein in the developing embryos is represented by shading, with the intensity of color reflecting the level of expression. Oct4 is present in the nuclei of all cells through the morula stage. At day 3.5, Oct4 becomes restricted to the inner cell mass (ICM) and, later, at day 4.5, to migrating cells of differentiating primitive endoderm. Following implantation, Oct4 expression is limited to primitive ectodermal cells. Expression in primordial cells is detectable at day 8.5 (not shown) [23]

3.2.3 Specification and Maintenance of the Germline

In mice timeline at embryonic day 6 (E6), Bmp4 initiates the specification of the germline by inducing Blimp1, Prdm14, and Ap2ɣ in proximal epiblast cells. These three proteins act to repress somatic genes and induce expression of PGC proteins, such as nanos3, re-induce pluripotency genes, and start the epigenetic reprogramming ([35] for review). At E7.25, they form a cluster of 40–50 cells at the base of the allantois due to the homotypic adhesion molecule fragilis. The cells in the center of the cluster with the highest expression of fragilis start to express stella(r) (also known as Dpp3a) and Tnap (the mouse homolog of PLAP) and become recognizable as the first PGC, which on E 8.5, after downregulation of fragilis, start to migrate to the genital ridges, the future gonads [46, 47].

Different from mice, in humans SOX17 is a critical specifier of PGC fate [35], inducing the expression of BLIMP1, which represses endodermal and other somatic genes as in the mouse. SOX17 and BLIMP1 are probably also important in the maintenance of PGC, preventing them from displaying their capacity to totipotency endowed by the expression of OCT4. Both in mice and humans, migrating PGC undergo germline-specific global demethylation, "reprogramming 1," jointly with upregulation of PRMT5 to protect the vulnerable demethylated genome from damage by transposable elements [7, 35]. In the gonads, these cells, now called gonocytes, undergo "reprogramming 2," including completion of erasure of parental imprinting [48, 49], a process that is completed within 24 h in the mouse, however, takes several weeks in a locus-specific manner in the human embryo [6]. In the process of global demethylation and erasure of GI of PGC, 5mC is replaced by 5hmC in mouse [50] and man [6].

In the mouse, PGC start migration at E8.5 from the base of the allantois, as mentioned, and reach the genital ridges at E10.5 by passive movement due the folding of the embryo and active migration partly guided by chemotactic factors from the genital ridge [2, 12, 51]. In humans, PGC expressing OCT4 (see above), as well as cKIT (membrane receptor for KIT ligand (KITLG), also known as stem cell factor, crucial for survival and proliferation of PGC), can be recognized in the yolk sac wall from 3 to 4 week postconception (wpc) [52]. PGC are present in the hindgut epithelium, in the mesenchyme of the dorsal mesentery, and in the developing gonadal ridge in wpc 4–6. In wpc 4–5, they leave the gut epithelium by a process resembling epithelial mesenchymal transition (EMT). KITLG activates KIT signaling in the PGC and facilitates their further migration [2], but after establishment of connections between the enteric and sympathetic nervous systems, PGC follow sympathetic nerve fibers toward the gonads. Numerous PGC are still present in the nervous system by wpc14. PGC failing to exit the nerve branches at the gonadal site may continue along the sympathetic trunk along the midline of the body and may end up in other distant localizations including the retroperitoneum (suprarenal region, adrenal glands), abdomen (stomach), anterior mediastinum, heart, lungs, head and neck, and CNS [52, 53]. This is an important observation because these so-called mis-migrated PGC may give rise to GCT in these various extragonadal sites, unless eliminated by apoptosis [54–57] (Fig. 3.3). In the mouse embryo, upon arrival in the genital ridges on E12.5, gonocytes enter a premeiotic stage and upregulate meiotic genes both in female and male embryos. In the male genital ridge, meiosis proceeds no further and the germ cells enter mitotic arrest as G0/G1 prespermatogonia, which resume mitosis only after birth. In contrast, in the female genital ridge, germ cells enter meiotic prophase as oocytes and pass through leptotene, zygotene, and pachytene stages before arresting in diplotene at the time of birth. Germ cells enter meiotic prophase at about the same time not only in the female genital ridge, but also in extragonadal localizations, such as the adrenal gland, in female and male embryos [58, 59].

Mouse and human PGC are mortal, nullipotent cells; in vitro exposure to KITLG, LIF, and bFGF reprograms them into totipotent stem cells

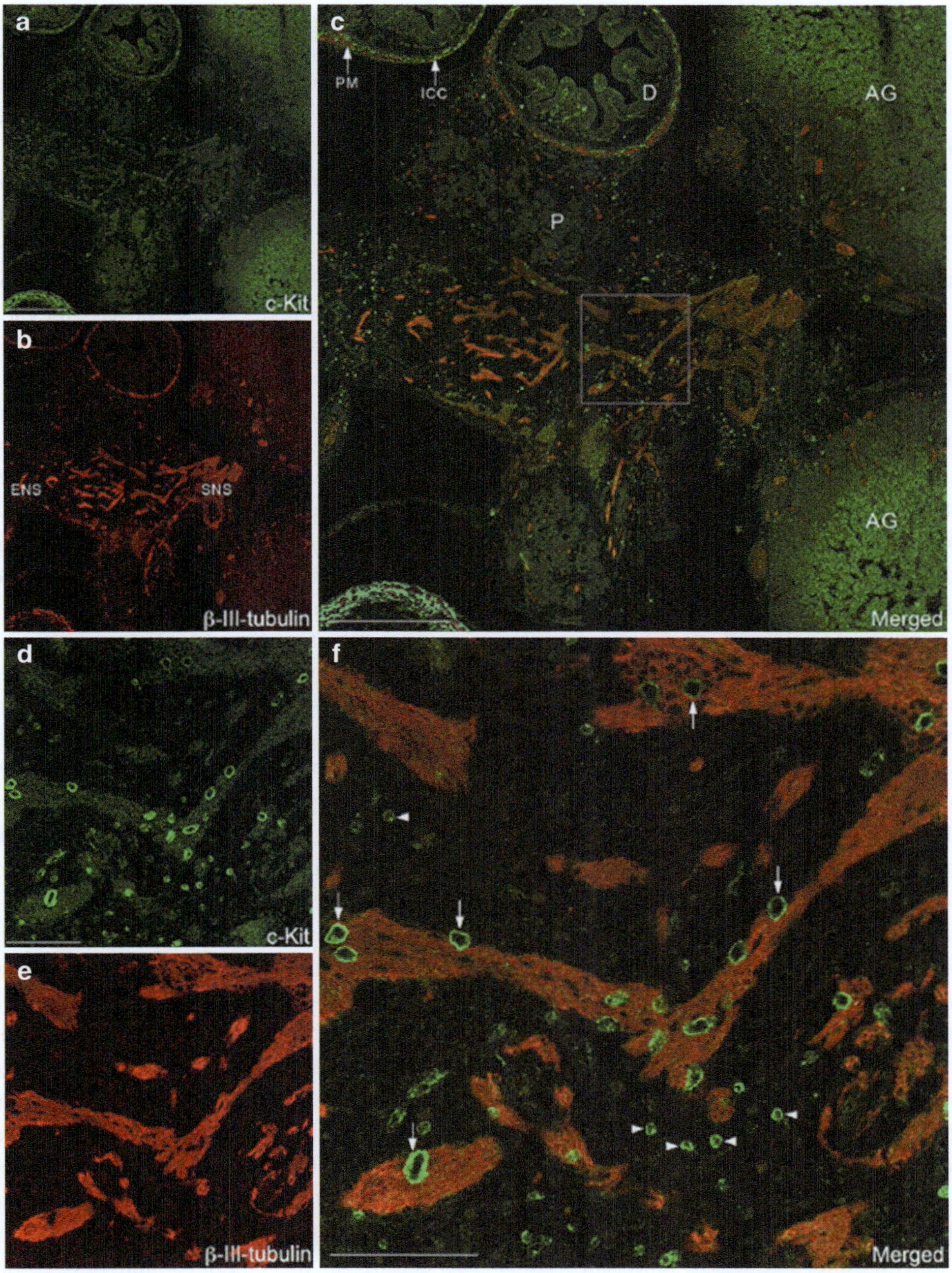

Fig. 3.3 Horizontal section through abdomen of a human embryo, 8 wpc. Horizontal section of human embryo, CRL= 30 mm, 7 weeks and 6 days pc immunofluorescent labeled against cKit and b-III-tubulin antibody. Survey depicted (**a**–**c**), with commencing connectivity of the enteric (*ENS*) and sympathetic (*SNS*) nervous system (**b**). (**c**) Some sympathetic nerve fibers are found in the adrenal glands (*AG*), the pancreas (P), and especially in the dorsal mesentery located in the middle of the section. Positive b-III-tubulin reactivity is seen in nerve fibers of ENS, in general in the plexus myentericus (*PM*), and similar reactivity is observed in the duodenum (**d**). Furthermore, cKit is also observed in the interstitial cells of Cajal (*ICC*) of PM. (**d**) The PGCs in the nerve fibers demonstrate cKit reactivity. (**e**) Higher magnification of (**b**) demonstrating b-III-tubulin reactivity of SNS. (**f**) Higher magnification of boxed area in (**c**). The larger PGCs, with strong membranous cKit reactivity, are located in close correspondence to the periphery of the individual nerve fibers of the SNS (**f**, *arrows*). The small, densely labeled cKit-positive cells outside of the nerve fibers are mast cells (**f**, *arrowheads*). Scale bars: (**a**, **b**) 500 mm, (**c**) 200 mm, (**d**, **e**) 100 mm, (**f**) 50 mm [52]

(EGCs) that can grow indefinitely [60] and can enter the germline efficiently [33, 61].

3.2.4 Plasticity of Pluripotent States

From recent research papers and reviews [17, 19, 20, 62–66], a model emerges of the spectrum of developmental states of the different types of stem cells in the early embryo (mouse and human), how they are regulated at the molecular level in vivo, and how these developmental states can be modeled in vitro depending on culture conditions.

The term "pluripotency," often used in a more general sense in the quoted papers, as in the legend of Fig. 3.4, is replaced by "developmental potential" throughout this chapter, to avoid confusion with the more specific application of the term "pluripotency" to indicate the developmental potential of cells in the primed state.

The 2C state represents the full developmental potential (omnipotency) of the zygote, still present in the blastomeres of the two-cell stage of the embryo. These cells have not yet undergone global demethylation, erasure of parental imprinting, and X-inactivation (the latter in female cells) and do not (yet) express Oct4 and Sox2 [19]. In fact, this corresponds to the omnipotent state.

ESC derived from the preimplantation embryo (ICM and epiblast) have the broadest developmental potential, compared to other ESC, with a permissive epigenetic signature, including two active X chromosomes (in female cells), capable of forming embryonal and extraembryonal tissues and efficiently contributing to the germline. They can continuously self-renew; *Oct4* expression is driven from the distal enhancer, and Oct4 partners with Sox2. These ESC represent what is called the ground state, naïve state, or totipotent state.

ESC derived from the primitive ectoderm of the postimplantation embryo exhibit reduced/absent expression of many ancillary pluripotency factors, including Klf4, Klf5, Prdm14, Rex1, and Esrrb, due to the attenuated Nanog expression [67]. These cells accumulate epigenetic barriers incompatible with the naïve state, such as female X-inactivation and promoter methylation at pluripotency genes, and thereby resemble the anterior primitive streak [68]. They give rise to somatic lineages and do not readily contribute to extraembryonic tissues and the germline. Their self-renewal capacity is limited, as they progressively differentiate toward stem cells committed to organs and tissues of the embryo proper; *Oct4* is driven from the proximal enhancer, and Oct4 partners with Sox2. These ESC represent the so-called primed state or pluripotent state.

The developmental potential of PGC upon reprogramming depends on their epigenetic status. Early PGC, prior to completion of erasure of GI, give rise to EGC with the developmental potential of pluripotent ESC in the primed state. Late PGC with completed erasure of GI give rise to EGC with characteristics of naïve state, totipotent ESC, including a permissive epigenetic signature, the absence of X-inactivation, the activation of *Oct4* expression from the distal enhancer, and the combination of Oct4 with Sox2. In human PGC, OCT4 partners with SOX17; upon reprogramming to EGC, OCT4 switches partner with SOX2. In parallel with their changing epigenetic status, PGC will have developmental potentials ranging from the primed to the naïve state.

In vivo, these developmental states are tightly controlled partly by cell autonomous factors (such as retroviral regulatory sequences) [19] but probably more by external cues, like position of ESC in the embryo. Plasticity of the developmental states in vivo is demonstrated by transplantation experiments, for example, cells from the tip of the epiblast become committed to the germline when transplanted in the proximal epiblast [12]. In vitro, naïve state and primed state can alternate depending on the culture conditions [20] (Fig. 3.4). A startling example of plasticity is the phenomenon that probably all ESC from the ICM transiently acquire the omnipotency of two-cell stage embryonic cells, the 2C state [19].

Apart from these physiological pluripotent cells, there are now somatic cells induced to pluripotency (iPSC) by the very factors involved in regulation of pluripotency in the embryo and the germline. This feat was first reported by

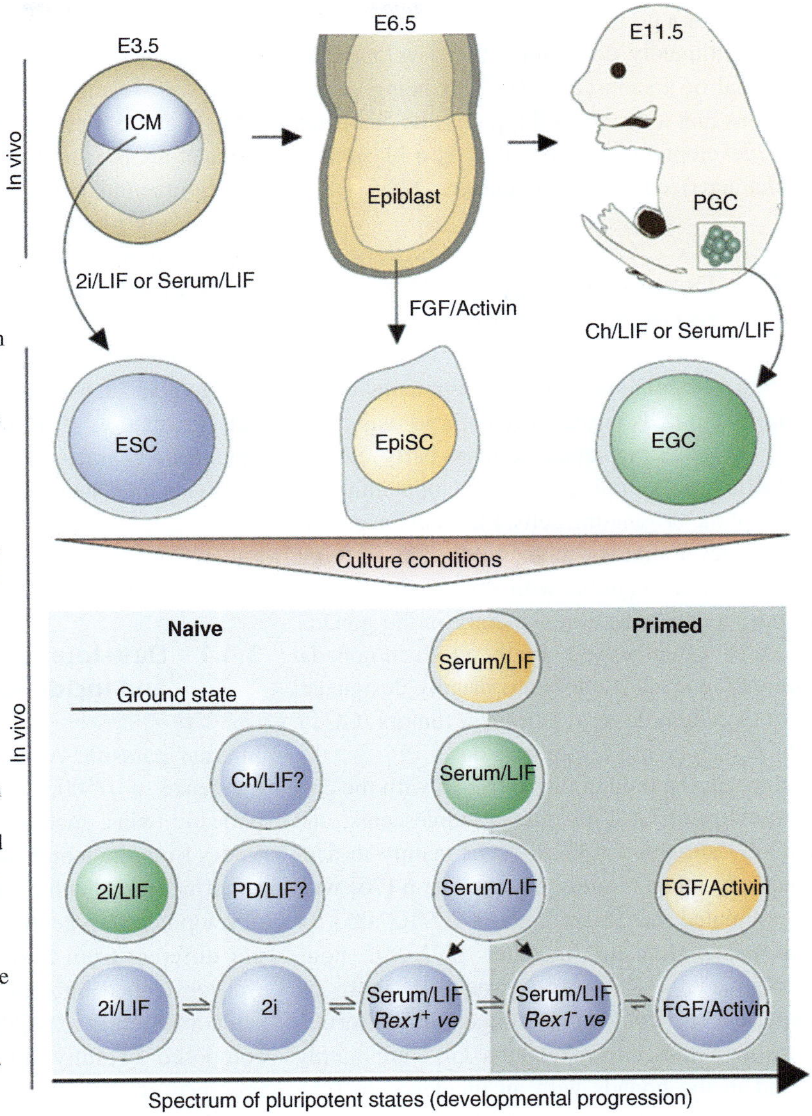

Fig. 3.4 Embryonic origin and spectrum of pluripotent stem cell states. The pluripotent cells of a blastocyst between E3.5 and E4.5 can give rise to functionally naive ESC (*blue*). Between E5.5 and E8.0 postimplantation epiblast can establish EpiSC (*orange*), which occupy a primed pluripotent state. Additionally, primordial germ cells (PGC), which are the founders of the germline lineage, can give rise to naive EGC (*green*), which are highly comparable to ESC. Depending on the culture/derivation conditions, these pluripotent stem cells occupy discrete molecular states that can be broadly classed as naive or primed. The most optimized state of naive pluripotency, which closely recapitulates the naive epiblast cells of the blastocyst, is termed ground state. An interchangeable spectrum of pluripotent states may arise that ranges from ground state to primed pluripotency. The state of pluripotency adopted in vitro is primarily dictated by the combination extrinsic signals in the culture environment rather than the developmental source of the pluripotent cells. *CH* Chiron, *PD* PD03 [20]

Takahashi et al., using the same cocktail of pluripotency transcription factors, consisting of Oct4, Sox2, Klf4, and c-Myc, for mouse [69] and human somatic cells [70]. Shortly thereafter, Kim et al. demonstrated that mouse [71] and human [72] neural stem cells (NSC) can be induced to pluripotency by OCT4 alone, probably because these cells endogenously express SOX2, c-MYC, and KLF4. ESC and NSC appear to have many similarities at the transcriptional level [73]. Pluripotent stem cells can also be generated with embryonic stem cell-specific cell cycle regulating miRNAs [74].

iPSC, including those derived from NSC, resemble human ESC as to developmental potential, which means that they produce teratomas in vivo. By proper in vitro conditions, iPSC can be made germline competent [20].

In iPSC, the genomic imprint of the somatic cells from which they are derived is stably retained; however, a low frequency of loss of imprinting can be found, probably acquired in the process of reprogramming [75].

The high degree of plasticity of the developmental potential of stem cells, including iPSC, implies that the actual state of developmental potential of a given cell, rather than the cell type itself, ultimately determines the developmental potential of a stem cell [20]. This being said, it remains that a certain cell type has its characteristic developmental potential, e.g., a blastomere is characterized by omnipotency.

3.3 Developmental Potential of Germ Cell Tumors

Failure of regulation of the developmental potential of stem cells in the early embryo may result in mainly extragonadal tumors early in life reflecting the overall somatic developmental program of the originating cells. Flaws in the control of the developmental potential in the germline may give rise to tumors with a broad spectrum of developmental capacities, mainly in the gonads, and most often beyond childhood. Such gonadal and extragonadal tumors are usually designated with the umbrella-term germ cell tumors (GCT), which shall be used from here on.

Indeed, the predictions above fit with the epidemiology of GCT in infants, adolescents, and adults. Extragonadal GCT occur mainly in neonates and infants, rarely beyond age 6 [76] with an estimated incidence of about 1.5/100,000 for males and females together [77]. Of note, extragonadal GCT are associated with an increased risk for various congenital malformations. In adolescents and adults, GCT are mainly found in the gonads with an incidence of 0.5–12/100,000 for the testis, virtually always malignant, and an incidence of up to 15/100,000 for the ovary, most often benign [78]. Overall GCT are rare, and even in high-incidence countries like Denmark, the lifetime risk for a testicular GCT is only 1 % [79]. It is noteworthy that GCT of the gonads are associated with a risk for impaired fertility.

The rarity of these tumors in any anatomical site in humans, the mouse [80, 81], and other animal species, perhaps with exception of the horse [80, 82, 83], demonstrates how successfully the hazards of dealing with embryonic stem cells and germ cells are coped with, probably because these cells are highly apoptosis prone. To illustrate this point, targeted loss of OCT4 as well as Nanog in PGC in the developing mouse results in apoptosis of these cells [84, 85]. It could well be that these cells can only escape apoptosis if their normally repressed developmental potential unfolds. This mechanism likely plays a role in the origin of many GCT in humans.

GCT can be classified according to their developmental potential [86]. Tumors of a certain developmental type appear to have more features in common, such as age of presentation, anatomical distribution, (cyto)genetic aberrations, and epigenetic characteristics including global methylation and GI status [87] (Table 3.1, Fig. 3.5).

3.4 Type 0 GCT

3.4.1 Developmental Potential and Incidence

Internal parasitic twins (fetus in fetu) with an incidence of 1/500,000 births [88] and external parasitic twins, such as the epignathus that protrudes from the mouth, are extremely rare. These abnormal growths have the highest, omnipotent, developmental potential of all GCT, essentially not different from a zygote. They may contain well-developed internal organs, limbs [55], and often a vertebral axis [89] and are histologically composed of fully differentiated tissues. The presence of immature tissue or yolk sac tumor (YST) is exceptional [90–93], as is recurrence as YST [92, 94].

3.4.2 Anatomical Distribution

Fetus in fetu is in 80 % of the cases localized in the retroperitoneum and often enclosed in an amniotic sac, sometimes with rudiments of an umbilical cord [95] and extremely rarely placental tissue [92]. Other sites are the skull, hard palate, liver, sacrum, scrotum, and attached to ovary [94] and undescended testis [95]. External parasitic twins are localized at the same sites where conjoined twins are united [96].

Table 3.1 Characteristics of seven defined types of germ cell tumors (GCT)

Type GCT	Age (years)	Sex	Anatomical site	Phenotype/ developmental potential	Developmental state	Precursor cell	Genomic imprinting; methylation	Karyotype	Animal model
0	Neonates	F/M	Retroperitoneum/ sacrum/skull/hard palate	Included and parasitic twins	2C state (omnipotent)	Blastomere	Biparental	Normal diploid	Not available
I	Neonates and children <6; rarely beyond childhood	F/M	Testis/ovary/sacral region/ retroperitoneum/ anterior mediastinum/neck/ midline brain/other rare sites	(Immature) teratoma (TE)/ yolk sac tumor (YST)	Primed state (pluripotent)	Methylated PGC/ gonocyte	Biparental to partially erased	Diploid (TE)/aneuploid (YST): Gain: 1q,12(p13),20q Loss: 1p,4,6q	Mouse teratoma
II	After start of puberty; in DSD, Klinefelter's and Down's syndrome rarely before puberty	≫M	Dysgenetic gonad/ testis/ovary/anterior mediastinum (thymus)/midline brain (pineal gland)	Seminoma/ dysgerminoma/ germinoma Reprogrammed to non-seminoma/ non-dysgerminoma/ non-germinoma	Naïve state (totipotent)	Hypomethylated PGC/gonocyte	Erased	Aneuploid (+/- triploid) Gain: X,7,8,12p,21 Loss: Y,1p,11,13,18 In mediastinum and midline brain also (near)diploid and (near) tetraploid with gain of 12p	Not available
III	Older men, usually >55	M	Testis	Spermatocytic tumor	Spermatogonium to premeiotic spermatocyte	Spermatogonium/ spermatocyte	Partially to completely paternal	Gain: 9	Canine seminoma
IV	After puberty	F	Ovary	Dermoid cyst	Maternally imprinted 2C state	Oogonia/oocyte	Partially to completely maternal	(Near)diploid Diploid/tetraploid Peritriploid gain: X,7,12,15	Mouse gynogenote
V	After puberty	F	Placenta/uterus	Hydatidiform mole	Paternally imprinted 2C state	Empty ovum/ spermatozoa	Completely paternal	Diploid (XX and XY)	Mouse androgenote
VI	Older age, usually >60	F/M	Ovary and atypical sites for GCT	Resembling type I or non-seminoma components of type II	Primed state or non-seminoma lineages of naïve state	Somatic cell induced to pluripotency	Imprinting pattern of originating cell	Depending on precursor cell	Xenografts derived from iPSC

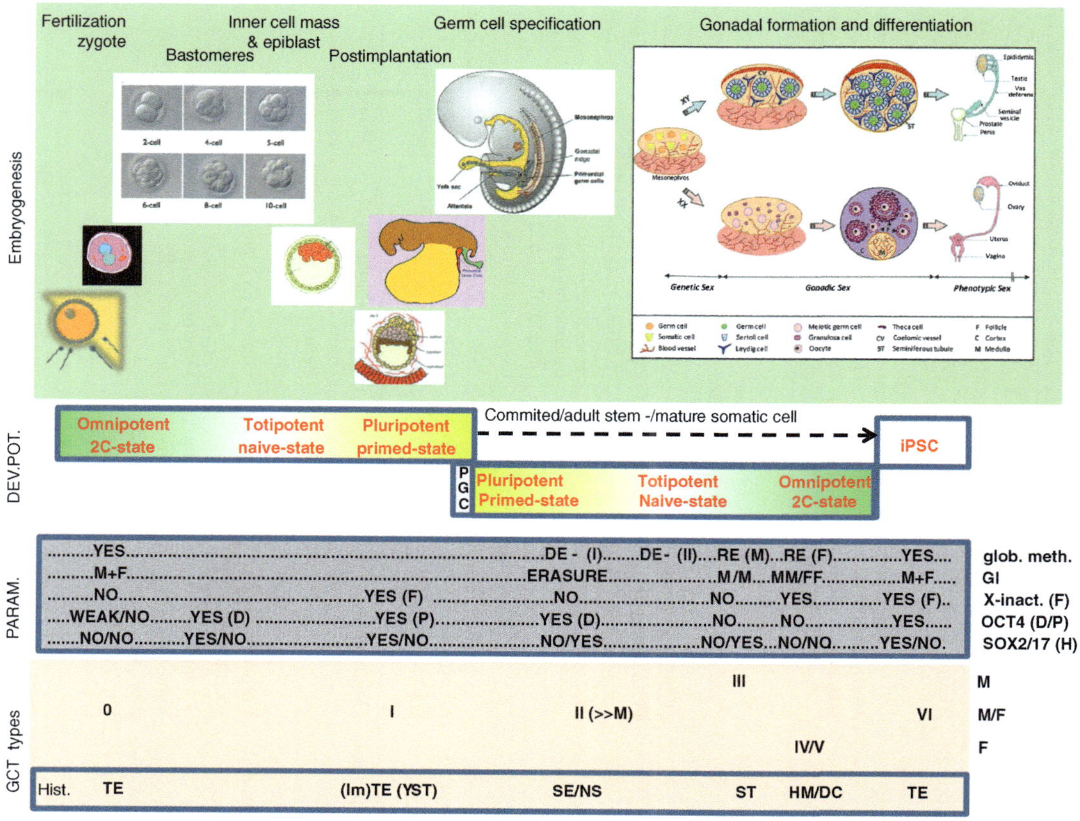

Fig. 3.5 Unifying model of the pathogenesis of GCT based on the hypothesis that the developmental potential of GCT is determined by the developmental state (2C, naïve, primed) of the originating cell. Juxtaposed in the figure are stages of embryogenesis (*upper panel*), developmental potential of stem cells in subsequent stages of embryonic development and the germline (*second panel*), critical features of the involved stem cells (*third panel*), and corresponding GCT types with gender distribution and their histology, reflecting developmental potential (*bottom panel*, linking Fig. 3.5 to Table 3.1) (abbreviations in order of appearance: *DEV.POT.* developmental potential, *PGC* primordial germ cell, *iPSC* induced pluripotent stem cell, *PARAM.* parameters, *M* male, *F* female, *DE-(I)* first wave of demethylation, *DE-(II)* second wave of demethylation, *RE* remethylation, *glob. Meth.* global methylation, *GI* genomic imprinting, *X-inact.* X-inactivation, *D* distal enhancer, *P* proximal enhancer, *H* human, *GCT* germ cell tumor, *TE* teratoma, *Im* immature, *YST* yolk sac tumor, *SE* seminoma, *NS* non-seminoma, *ST* spermatocytic tumor, *HM* hydatidiform mole, *DC* dermoid cyst)

3.4.3 Genetics and Pathogenesis

Genetic analyses in some of the more recent cases have with rare exceptions failed to demonstrate differences with the host [97]. These observations are consistent with fetus in fetu and external parasitic twins being monozygotic diamniotic twins [95] lacking a heart and deriving their circulation from the host. Apart from the heart, the brain is also usually missing; in fact, most of the rostral part of the embryo is poorly developed [96].

There are features, such as common anatomical localization and female preponderance, suggesting a continuum and common pathogenesis of conjoined twins, parasitic twins, fetus in fetu, acardiacs, which are considered parasitic twins attached via the placenta, and teratomas [96, 98, 99]. Multiple pregnancies could be the far end of this continuum, as each of the mentioned conditions as well as type I GCT of various anatomical sites [94] and perhaps also dermoid cysts (type IV GCT) [100, 101] is associated with a family history of multiple pregnancies

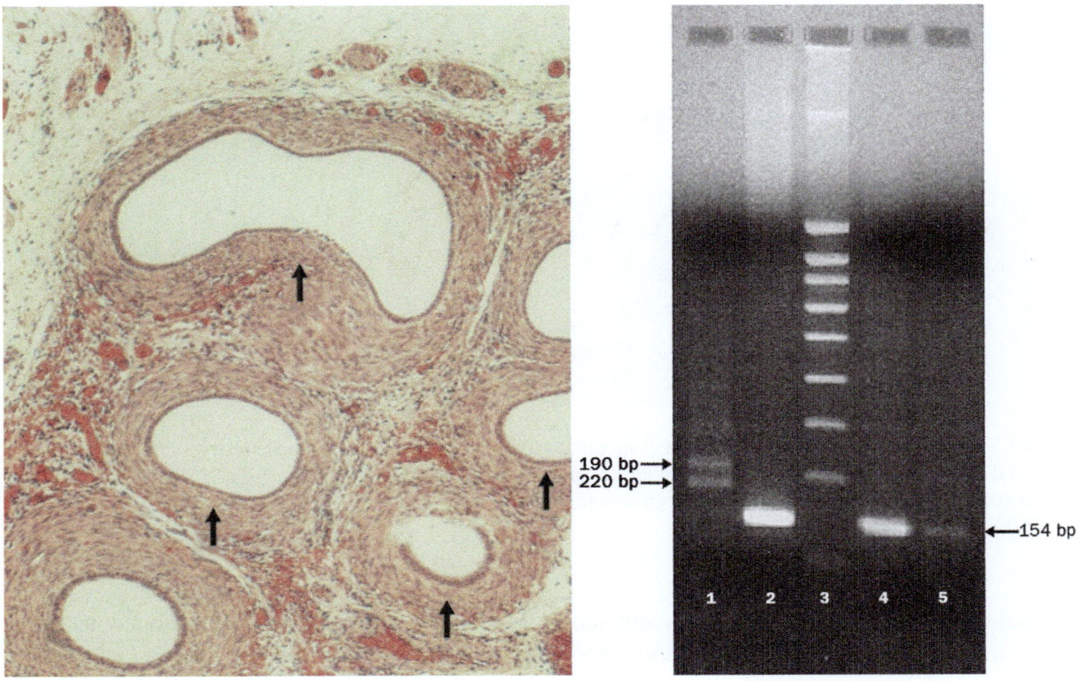

Fig. 3.6 Histology and PCR of male epignathus, disguised as teratoma, in a female neonate. *Left panel*, histology of teratoma with epididymal ducts (*arrows*); *right panel*, PCR-amplification of Y-chromosomal DNA. *Lane 1*: control DNA, female. *Lane 2*: control DNA, male. *Lane 3*: 100 bp ladder. *Lane 4*: DNA extracted from 10 μm thick slides cut from paraffin-embedded teratoma tissue showing ductus epididymis on light microscopy. *Lane 5*: DNA extracted from paraffin-embedded teratoma tissue showing no ductus epididymis on light microscopy [102]

[96] (see below). In fact, over 15 % of cases of fetus in fetu have a family history of twins or double fetus in fetu [94]. The basic defect then would be an increased risk of multiple pregnancies or, more mechanistically phrased, proneness of blastomeres in the 2C state to escape the organizing influence of the developing embryo or rather escape from control of their developmental potential. If the twin fails to develop a functional heart, it will either die or, if it succeeds in getting its circulation from the host, develop as a parasitic twin or a teratoma [96]. The latter may seem far-fetched; however, there is a case report on an oral mature teratoma in a female neonate that contained epididymal tissue. In the tumor, Y-chromosomal DNA was demonstrated by PCR, which was lacking in the peripheral blood of the girl who had a normal female karyotype in peripheral blood lymphocytes (Fig. 3.6). This extraordinary teratoma is probably best regarded as a poorly organized epignathus originating from dizygotic twinning [102], illustrating an exceptional mechanism of origin of teratoma.

3.5 Type I GCT

3.5.1 Type I GCT General

3.5.1.1 Developmental Potential

The natural history of type I GCT, emerging from numerous clinical and pathological observations [55, 94, 103–109], is that regardless of anatomical site, they begin during embryonic life as immature teratoma, probably arising from one pluripotent progenitor cell, which may evolve to mature teratoma with trilaminar derivatives. However, an immature teratoma component may contain foci of YST easily overlooked on microscopic examination [106, 110–113], which may eventually overgrow the

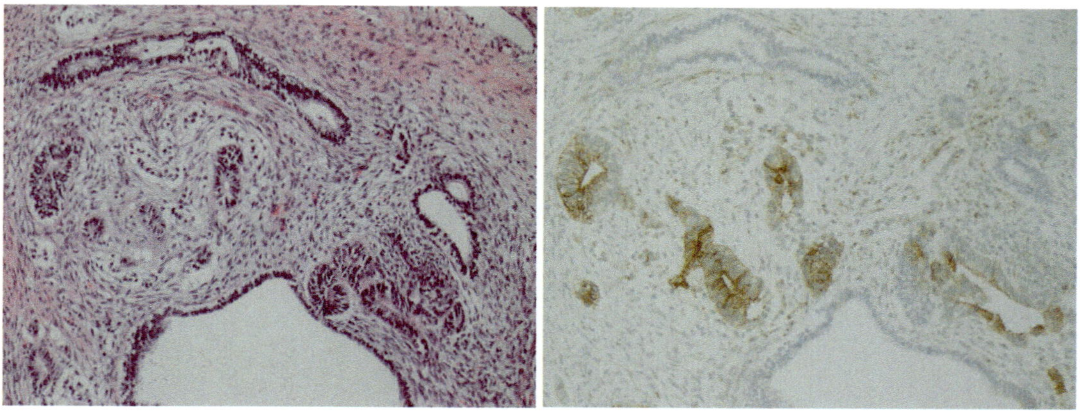

Fig. 3.7 Testicular type I GCT, in infant of 5 months, composed of mature and immature teratoma with deceptive microscopic foci of YST, difficult to recognize without the aid of immunohistochemistry. (*left*, H and E ×200; *right* glypican 3, ×200)

original teratoma (Fig. 3.7). In fetuses and neonates, and in prenatally resected tumors, YST is virtually always associated with immature teratoma, while pure YST is rare [94, 113]. Thus, these tumors come in three histological variants: pure (immature) teratoma; pure YST, whereby the original teratoma component is probably overgrown by the more aggressive YST component; and combinations of (immature) teratoma and YST. The younger the infant, the more often an immature component is present and the lesser the chance that an overt YST component has developed, irrespective of gender (Table 3.2). YST can take the form of both intraembryonic endodermal derivatives, such as the primitive gut and liver, and extraembryonic structures such as allantois and yolk sac [114], reason for Nogales to advocate the name primitive endodermal tumor instead of YST (Chap. 6).

The frequency of the different histological variants differs per anatomical site; however, in population-based registries, teratomas are the most frequent at all sites. In fact, the large majority of type I GCT have a favorable course regardless of degree of immaturity, with the exception of high-grade immature teratomas of the ovary. A YST component, overall present in about 5–10 % of the cases at birth [94, 109], is the only predictor of recurrence at any site [106, 109]. Type I GCT typically lack a seminomatous component, EC, and choriocarcinoma, which are indicative for a type II GCT (see below). Choriocarcinoma may rarely occur in infants, in association with an intracranial type I teratoma [115] or metastatic from placental/gestational choriocarcinoma [116, 117].

Type I GCT may contain OCT4-positive cells, usually in higher-grade immature teratoma components [118, 119], which however are negative for SOX2 and CD30. This is in contrast to EC cells, the stem cells of type II GCT, which typically express these two proteins in addition to OCT4 and other pluripotency proteins, such as NANOG and STELLAR [57, 120]. These OCT4-positive cells may be the stem cells of type I GCT. The lack of expression of CD30 may be explained by the cells being diploid. It was shown that in vitro ESC cells, the normal counterparts of EC cells, only start to express CD30 when they become aneuploid [121]. The rarity of these stem cells in type I GCT suggests that they do not readily self-renew but are rather poised to differentiation, particularly into the various somatic lineages, explaining the usually benign character of these tumors. In fact, the precursor cells seem to be in the primed state.

Type I GCT may contain SOX2-positive cells; however, these are not the OCT4-positive putative stem cells [122], shown in Fig. 3.8.

Table 3.2 Histology of 96 sacral type I GCT related to sex and age

Histology		Male	Female	Total
Mature teratoma	Number	17	47	64
	Age (average)	230 day	97 day	
	Age (median, range)	12 day (1 day – >3 year)	8 day (1 day–2.5 year)	
Immature teratoma	Number	3	13	16
	Age (average, SD)	6 day	8 day	
	Age (median, range)	4 day (1–14 day)	4 day (1–45 day)	
(Immature) teratoma Plus yolk sac tumor	Number	0	5	5
	Age (average, SD)		60 day	
	Age (median, range)		12 day (1 day to > 6 month)	
Yolk sac tumor	Number	3	8	11
	Age (average, SD)	20 month	21 month	
	Age (median, range)	18 month (15–27 month)	21 month (12 to >34 month)	

Based on original data from De Backer [77]

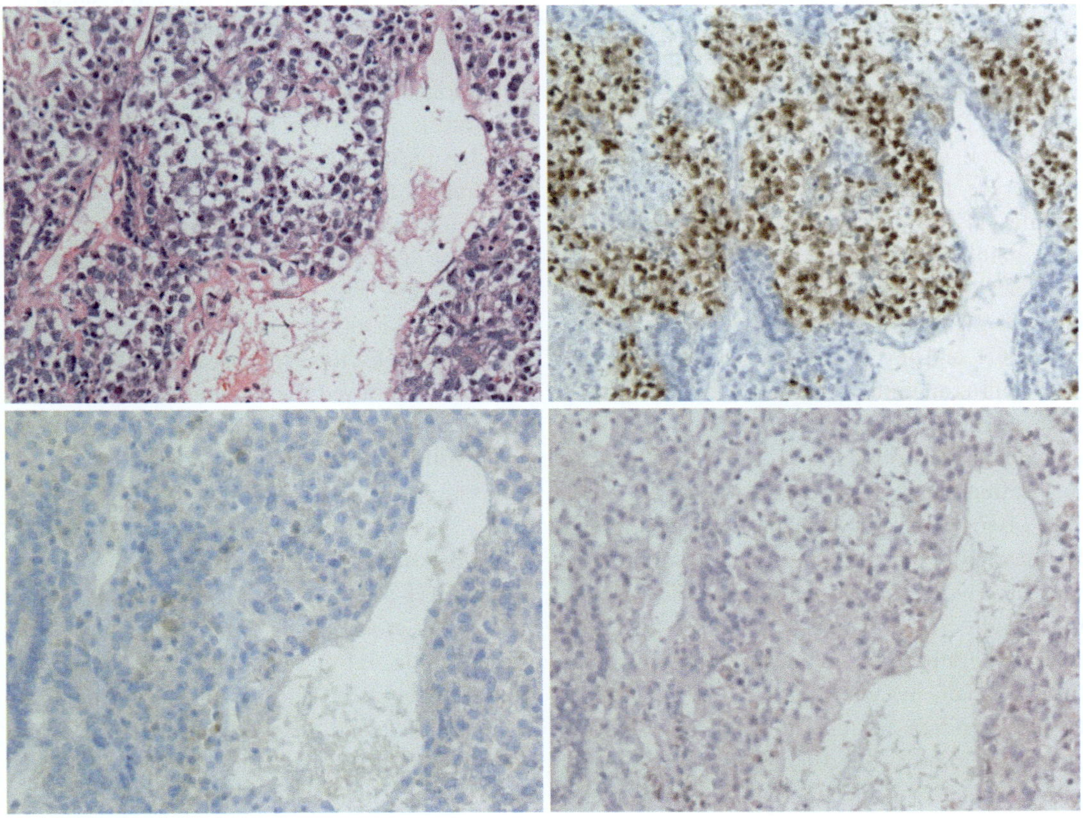

Fig. 3.8 Type I immature teratoma with OCT4-positive, SOX2- and CD30-negative stem cells. (clockwise, HE, OCT4, SOX2, and CD30; original magnification, ×200)

3.5.1.2 Epidemiology

The age distribution of GCT shows a neonatal peak in the sacrococcygeal area, retroperitoneum, mediastinum, head and neck, brain (apart from the pineal gland), and testis. In the ovary, the early peak is missing; however, GCT do occur from birth through adulthood without interruption. The tumors represented by the early peak are type I GCT, rare tumors, most often occurring in the fetus, neonates, and children under the age of two and seldom beyond age six, with an overall predilection for girls, mainly due to the about 3.5:1 female to male ratio of sacrococcygeal tumors [76]. This skewed sex distribution may be due to global demethylation of PGC taking place earlier in males than in females. As a result, in males, the PGC are more fragile and prone to apoptosis upon mis-migration, whereas in females, the PGC are more robust and therefore have a greater chance to undergo reprogramming, giving them survival advantage outside a proper niche [57]. Such a mechanism would explain the overall slightly increased risk of extragonadal type I and perhaps also type 0 GCT in females [123], not answering the question of why mainly in the sacrococcygeal region.

Exact incidence figures are hard to get because in most cancer registries, only malignant GCT and not teratomas are included. The best approximation is achieved by combining cancer registry data with population-based figures on all GCT obtained in centers specializing in the treatment of these tumors [55, 94, 103, 105, 107, 109, 114]. The Netherlands and Belgian National Cancer Registries report, respectively, 5.2 and 5.4 malignant GCT per million children less than 15 years of age [77]; this figure is 4/million in Germany [107]. Assuming that over half, in fact, up to 70 %, of the type I GCT in children are teratomas, the overall incidence would be 1–1.5/100,000 [77].

Multiplicity of type I GCT is exceedingly rare: one child with a bilateral pure YST of the testis [124]; four cases of bilateral teratoma of the testis [125–127], two of which were brothers with Klinefelter's syndrome [127]; no bilateral stage I ovarian YST [128, 129]; 1–2 % of ovarian immature teratomas is bilateral [130, 131]; and no published multiple extragonadal cases to the best of our knowledge.

Combinations in one individual of type I GCT with other GCT types do occur: type I GCT may rarely be combined with fetus in fetu (type 0 GCT) among others in the cranial region [132]; ovarian type I GCT may be combined with a type IV GCT in the same ovary and in 11 % in the contralateral ovary [130].

Familial cases of type I GCT have not been described for the testis, apart from the two brothers, both infants, with Klinefelter's syndrome, mentioned above, with bilateral testicular teratomas [127]. In view of the rarity of bilateral testicular type I teratoma [125, 126] as well as Klinefelter's syndrome, this is probably not a chance occurrence, suggesting that this syndrome is a risk factor also for type I GCT, in addition to being an established risk factor for type II GCT of the mediastinum and brain.

Ovarian type I GCT may cluster with dermoid cysts of the ovary (type IV GCT). Since the latter may have a familial component, this is probably also true for the type I GCT [130, 133–135].

Finally, there is the family of a mother with an immature teratoma of the ovary coexisting with a newborn baby with an intracranial immature teratoma [132, 136]; Poremba et al. excluded that the tumors were clonally related. Giambartolomei et al. [137] retrieved four families from the literature in which an ovarian type I GCT was combined with one or more type II GCT of the testis (five cases) or ovary (one case).

3.5.1.3 Anatomical Distribution

Type I GCT are most often localized in extragonadal sites along the midline of the body: the sacral region, retroperitoneum (cranially of the kidneys), stomach, anterior mediastinum, heart, head and neck, and brain [94, 138]. This peculiar distribution along the midline, including the brain, is attributed to the migration route of PGC during embryonic development [52, 54, 56, 57]. Others explain it by the relative abundance of ESC (for review [55]) or NSC [139] along the midline of the developing embryo. Type I GCT occur also in the testis, the second most frequent site after the sacral region, and in the ovary. They have never been described in dysgenetic gonads in keeping with the different pathogenesis of type I and type II GCT of the gonads, the latter being derived from transformed, virtually always aneuploid gonocytes in the naïve state, while the former probably originate through direct reprogramming of essentially normal, still methylated, and pre-erased diploid gonocytes to ESC in the primed state.

3.5.1.4 (Cyto)Genetics

Type I (immature) teratoma, either pure or combined with YST, is virtually always diploid, lacking chromosomal rearrangements. However, YST of type I, pure or combined with teratoma, is most often aneuploid, usually (near)diploid, with multiple gains and losses of (parts) of chromosomes. Involved in gains are 1q, 3, 3p, 8q24, 12p13, 20q, and 22; involved in losses are 1p (1p36), 4, 4q, 6q (6q24-qter), 16q, and 20p. Overrepresentation of the whole of 12p or the region 12p11.2–p12.1, typical for type II GCT, is not a feature of type I GCT. As mentioned above, the distal part of 12p and in particular 12p13 may be overrepresented ([111, 140–150], for review [151]). Some of these changes, such as gain of 1q and loss of 1p and 6q, are shared by type II YST and may be related to the phenomenon of progression/differentiation toward YST rather than being specific for type I GCT [152].

Although highly speculative, for some of the chromosomal gains and losses, possibly involved genes have been suggested; gain of 8q24 has been associated with amplification of *MYC* [142]; gain of 12p13 might involve the pluripotency genes *STELLAR*, *NANOG*, and *GDF3* [38]; and loss of 1p36 [140] might involve *CHD5*, a tumor suppressor gene deleted from 1p36.31 in neuroblastoma [153].

Sporadic case reports describe specific balanced chromosomal translocations in type I GCT. Two infantile sacral teratoma cases had constitutional balanced translocations involving 12q13 probably affecting different genes [146, 154]. In one of the two patients, the genes involved in the translocation t(12;15)(q13;q25) were identified as SUMO-/Sentrin-specific protease 1 gene (*SENP1*) and the embryonic polarity-related mesoderm development gene (*MESDC2*) [155]. The resulting fusion protein SEME interferes with the function of MESDC2 as a chaperone for the WNT co-receptors LRP5 and/or LRP6. It is suggested that in both patients, the constitutional translocation was predisposing to the development of the sacral teratoma. In an intrathoracic mature teratoma in a 15-year-old girl, a balanced chromosomal translocation, t(8;22)(p21;q12), was the sole cytogenetic aberration. It resulted in fusion of the genes *PPP2R2A* and *CHEK2*, supposedly the initiating event in this teratoma [156].

A genome-wide association study involving type I and II GCT suggests that a variant in *BAK1* involved in suppression of apoptosis [157] is associated with gonadal GCT of both types. Type I GCT were not associated with variants in *KITLG*, *SPRY4*, and *DMRT1* (doublesex and mab-3 related transcription factor 1), which confer an increased risk for type II GCT of the testis [158].

The Wnt/beta-catenin [159] and the TGFbeta/BMP signaling [160] pathways are strongly expressed in type I and II YST, compared to seminoma/dysgerminoma, EC, and choriocarcinoma ([161] for review). Methylation of APC and LOH at 5q21-22 suggests that APC might be involved in the activation of the Wnt pathway [162]. In general, the transcriptome of pediatric YST is enriched for genes associated with a differentiation and proliferation phenotype as compared to seminomatous (type II) GCT [163]. It is likely that the expression patterns of the various GCT types are mostly determined by the cell type(s) present and much less by pathways activated in the process of tumorigenesis. Seminomatous GCT and EC express pluripotency genes, and the various extraembryonic and somatic tissues express genes characteristic for the involved cell lineages [161, 163].

3.5.1.5 Epigenetics, including GI

Type I GCT usually have a biparental GI pattern as in somatic cells. In a small proportion, GI is partially erased, in particular in type I GCT of the testis and ovary [144, 164–167]. These findings on GI support the hypothesis that extragonadal type I GCT may originate from PGC, which are pre-erased or partially erased, corresponding to the methylation status of the genome (see Sect. 3.2.3). Reprogramming of these cells will produce an ESC with the developmental potential of the primed state. It does not exclude their derivation directly from ESC in the primed state, which are also characterized by a biparental GI pattern. Also, a somatic cell with induced pluripotency (iPSC), for example, by reactivation of pluripotency genes, in particular OCT4, as demonstrated for human NSC [72], could theoretically be the precursor of type I GCT. This has been suggested by Scotting and co-workers for GCT of the brain [139, 168, 169], without making the essential distinction between type I and type II GCT [86]. Such iPSC would endow the derived tumors with their own GI pattern [56], as it is stably transmitted to daughter cells [75, 170]. Any degree of loss of imprinting in a type I GCT tumor would suggest that the tumor is derived from a germ cell precursor [166], unless, as reported [169], NSC may also have partial loss of imprinting.

3.5.1.6 Animal Models and Pathogenesis

Apart from humans, pluripotent tumors have been most extensively studied in the mouse. Relevant for human testicular type I GCT are the spontaneous [80, 81] and experimental [171] testicular teratomas in the 129 mouse strain. The spontaneous ovarian teratomas in LT mice [172] are probably a model for human ovarian type I GCT. Teratomas derived from pre- and postimplantation embryos transplanted to various organs, in particular the testis [173] or the kidney [174], might be a model for extragonadal type I GCT in man. Spontaneous extragonadal teratomas in mice [83, 175, 176] are too rare to be practically useful for animal experiments.

The difference between the spontaneous and experimental testicular teratomas as compared to the embryo-derived teratomas is that the gonocytes from which the testicular tumors originate are committed to the germ lineage and not themselves pluripotent [61]. They have to be reprogrammed before being able to form pluripotent tumors [60], a process similar to what happens in the human type I GCT, and to reprogramming of a somatic cell to an iPSC, by converting the nucleus from nullipotent to pluripotent [61].

These different mouse models have a similar developmental potential; when fully developed, they are mainly composed of mature somatic tissues derived from the three germ layers. Immature teratoma and EC cells are less frequent and often minor components; rarer still are extraembryonic lineages. Late takes of embryo transplantation under the kidney capsule consist of parietal YST and occasionally trophoblastic giant cells [177]; these tumors are most often aneuploid [178], just as human type I YST (Fig. 3.9). The observation that in chimeric blastocysts polyploid murine ESC only give rise to extraembryonal lineages (yolk sac and placenta), while the embryo proper is derived from the diploid ESC [179, 180], might explain the restricted developmental potential of aneuploid tumor cells in type I GCT: probably, they are no longer capable to form somatic tissues.

The testicular teratomas in the mouse models originate when a luminal gonocyte or a prespermatogonium in its niche is reprogrammed to an ESC in the primed state, either directly or via an EGC that apparently loses its naïve-state developmental potential [60, 61, 80, 181, 182]. Initially, when proliferating within the seminiferous tubules, the tumor cells stay undifferentiated, as EC cells. When the EC distends and disrupts the tubular wall and invades the testicular interstitium, it starts to differentiate into immature somatic tissues, which gradually develop into mature teratomas [81, 183]. A minority of the tumors will maintain immature teratoma and EC and will be retransplantable in syngeneic hosts. Rare tumors are pure EC from the start, which can readily be transplanted. This same evolution is seen in tumors derived from pre- or postimplantation embryos up to E8. The percentage of tumors with an EC component depends on the strain of the transplanted embryos and is usually higher than in the testicular teratomas [184]. These mouse models, as well as the ovarian teratomas in LT strain mice, resemble human type I GCT. They have the same cells of origin (PGC/

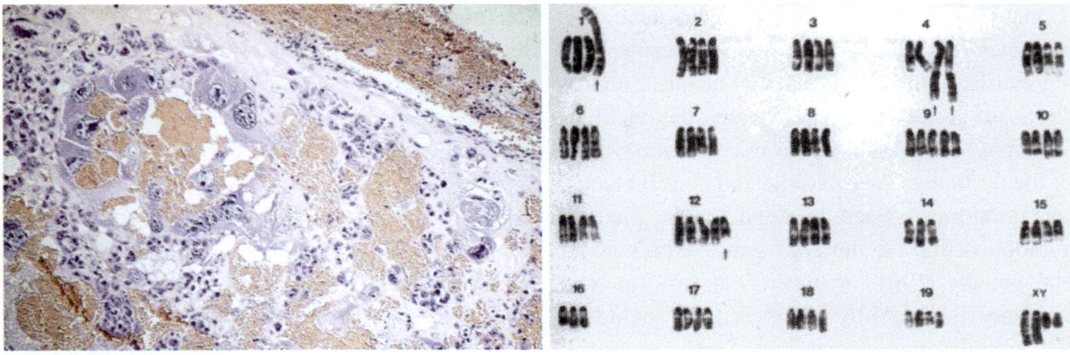

Fig. 3.9 Late take of transplantation of snowman stage mouse embryo under kidney capsule of syngeneic mouse. *Left panel*: parietal YST with trophoblastic giant cells in hemorrhagic context (H and E, ×200) [177]; *Right panel:* aneuploid karyotype of same tumor [178]

gonocytes and ESC), histological evolution, and developmental potential. Whatever the cells of origin, they seem to have or acquire the primed state in view of the developmental potential of the derived tumors.

Probably the most important lesson to be learnt from these models is that disruption of the microenvironment of the pluripotent cell itself suffices to initiate a pluripotent tumor. For gonocytes/prespermatogonia in the developing testis of 129 strain mice, this principle is demonstrated by genital ridge transplantation, as will be discussed in the following.

In 129/Sv mice carrying the loss of function *steel* mutation (*steel* or *Kitlg* is the mouse homolog of *KITLG*), spontaneous testicular teratomas occur in about 4 % of the animals, twofold the spontaneous rate (2 %) in 129 strain mice lacking this mutation. When from the same animals the genital ridges are transplanted, teratomas develop in over 80 % of the grafted genital ridges, often at multiple sites [171]. This is counterintuitive: loss of PGC with the steel mutation, and even more so by the procedure of the genital ridge transplantation, increases the yield of teratomas. The steel mutation, in the membrane-bound *Kitlg* [185], and the transplantation procedure are not carcinogenic events acting on the PGC but rather factors that disturb the niche of the PGC, promoting reprogramming of the surviving PGC. Apparently, cell-intrinsic mechanisms for repression of the developmental capacity of gonocytes/prespermatogonia, such as those via Blimp1, Prdm14, and AP2γ [186–188] and Dmrt1 [189], are not sufficient when the restraints of the normal tubular environment are disturbed, like Nanos2-/Dmrt-dependent GDNF signaling by Sertoli cells [189]. Remarkably, only male genital ridges produce teratomas; female genital ridges never do [173], probably because in the female genital ridge, the germ cells are blocked in meiosis I and few in numbers. In the male genital ridges, the germ cells are more numerous, premeiotic, and arrested in G0/G1 of mitosis [12]. Contrasting patterns of Dnd1 expression in female and male gonads may also contribute to the different susceptibility to teratoma formation [190].

As for embryo-derived tumors, perfectly normal embryos may turn into teratomas when transplanted into a testis or a kidney [173, 184]. EC cells derived from such tumors, when introduced into the ICM of a blastocyst, can contribute to the normal tissues of the resulting chimeric mouse, demonstrating that the tumor cells when restored to their proper environment may normalize, as well as cause malignant tumors [191].

The importance of genetic factors conditioning the micro-milieu of the niche was demonstrated by crossing susceptibility genes for testicular teratomas into 129 strain mice. As already mentioned, in the original strain, about 2 % of the mice had spontaneous testicular teratomas, introducing the loss of function *steel* mutation, which reduces the number of PGC and spermatogenesis [192], doubled this percentage [171], and by adding the *ter* mutation, one third of the mice developed spontaneous testicular teratomas [81]. The gene *ter* is a recessive gene that causes germ cell deficiency in mice, and in 129/Sv-ter mice, it also enhances the yield of teratomas. Male 129/Sv-ter mice, homozygous for the *ter* mutation, are sterile and almost always have teratomas, often bilateral [193]. The *ter* mutation occurs in the *Dnd* gene, expressed in fetal gonads [190]; in mice, Dnd isoform α is necessary for viability of germ cells including PGC from E8 and for viability of embryos [194]. Specifically, in 129 strain mice homozygous for this mutation, PGC die apoptotically or when they escape apoptosis may be reprogrammed to ESC, which form teratomas. It is even more likely the other way round that some PGC escape apoptosis because they have been reprogrammed [190]. As a corollary, the proneness of 129 mice – and not of other strains [194] – to form teratomas is due to the ease with which PGC of 129 mice PGC are reprogrammed to an ESC in the primed state. This may be due to incompetence of 129 strain mice to adequately suppress reprogramming to pluripotency in germ cells, a process in which among others Dmrt1 expressed in PGC/gonocytes is involved [189].

The variants in *KITLG* that increase the susceptibility for testicular type II GCT in humans [195, 196] do not seem to affect the incidence of type I GCT [158].

Although derived from PGC, *KIT* mutations are probably exceptional in type I GCT. In support of this contention, none of the pure immature and mature teratomas of the brain studied by Wang [197], almost certainly type I GCT, had *KIT* mutations and also very rarely other mutations. This is indeed remarkable since KIT is the crucial survival and proliferation factor for PGC.

In the mouse, germ cells that do not reach the genital ridges die through apoptosis caused by the proapoptotic protein Bax. In *Bax*-null embryos, large numbers of ectopic (extragonadal) germ cells fail to die [57]. A similar mechanism of impairment of apoptosis of mis-migrated PGC might enhance the development of extragonadal human type I GCT; however, this has not been demonstrated.

The available evidence points to the pathogenesis of type I GCT being foremost "developmental" and not driven by somatic mutations. This implies that the p53-dependent DNA damage response is intact in these tumors, explaining their favorable response to cisplatin-based chemotherapy, just like type II GCT.

3.5.1.7 Summary of the Pathogenesis of Type I GCT

The most likely cells of origin of extragonadal type I GCT are mis-migrated PGC, as these cells have been demonstrated along the midline of the body, indeed in large numbers at the typical sites of these tumors [52]. Most of these PGC die apoptotically; probably only those that are reprogrammed to an ESC manage to survive outside the niches in the gonads, the thymus, and the midline of the brain suitable for PGC. Reprogramming occurs when the mechanisms, with a key role for SOX17, BLIMP1, and OCT4 [35], maintaining the phenotype and suppressing the developmental potential of PGC, break down, probably because of lack of a suitable niche. Since the PGC are pre-erased, reprogramming will result in an ESC in the primed state capable of forming immature somatic tissues that will usually differentiate to fully mature teratoma. YST and very rarely choriocarcinoma are the only other components, which develop from tumor cells that have become aneuploid.

In the testis and ovary, type I GCT originate when diploid, pre-erased gonocytes, and oogonia are reprogrammed to ESC in the primed state due to failure of control of developmental potential by germ cell-intrinsic (DMRT1 in addition to SOX 17, BLIMP1, and OCT4) and niche factors (such as GDNF) [189]. Reprogramming to an ESC can occur directly or via an EGC in which the naïve state is rapidly dismantled [182].

Pathogenesis is mainly developmental; somatic mutations probably play a minor role.

3.5.2 Site-Specific Aspects of Type I GCT

3.5.2.1 Sacral Region

Sacrococcygeal type I GCT, with a frequency of 1/35,000 live births, constitute about 40–50 % of extragonadal type I GCT and are the most frequent neonatal tumor. They are rarely diagnosed beyond the age of 2 years and virtually do not occur after age six [76, 109, 198, 199]. The fact that there are practically no GCT at all in the sacral region past the age of six is in accordance with the absence of type II GCT at this anatomical site. The rare sacrococcygeal type I teratomas in adults probably had their inception before birth and went undetected [55]. There is a strong predilection for girls with a male to female ratio of 1:3.5.

Other congenital disorders occur in up to 25 % of patients with sacrococcygeal type I GCT, including trisomy 21/Down's syndrome (implying a higher risk for type I GCT in Down's syndrome), genitourinary malformations, congenital hip dislocation, esophageal atresia and congenital heart disease [114, 198], and duplication of pelvic organs attributable to hindgut twinning [200]. There is a well-documented association with multiple pregnancies, either within the same pregnancy or as a family history of multiple pregnancies [94, 96].

The evolution of these tumors is typical for type I GCT. Starting as immature teratomas prior to birth, they become more mature with time. When completely removed at this stage, which entails removal of the coccyx bone in continuity with the tumor ([114] for review), the child is cured. Incomplete or delayed surgery may allow the tumor to recur as mature or immature teratoma, or by means of tumor progression, to develop a YST component in the primary tumor or in a recurrence. A YST component is found in 5–10 % of the tumors removed before the age of 2 months; thereafter, this figure increases rapidly, and by the age of three, most sacrococcygeal type I GCT are malignant, in principle due to progression to YST [94, 201, 202]. The tendency for malignancy seems somewhat greater in males than in females [201]. Rarely, a somatic-type malignancy may develop such as Wilms' tumor [203, 204]. Also the type I teratomas of adults may in some 10 % develop a malignant component [205]. Metastases can be local or visceral and are usually composed of YST or less frequently immature teratoma [198, 199]. In contrast to teratoma, YST is aneuploid with the chromosomal aberrations characteristic for YST progression in type I GCT, as discussed.

Sacral teratomas so highly developed that they have a vertebral axis should according to the definition of Willis [206] be classified as parasitic twins. A somewhat less strict definition [96, 98] considers a sacral teratoma with clearly developed limbs as a parasitic twin; in view of the site of attachment, they should be classified as a parasitic pygopagus [96]. Indeed, a personal case, published as sacral teratoma with a classical clinical history, including recurrence as YST (with the characteristic chromosomal aberrations) upon incomplete surgery [207], should probably be reclassified as a parasitic pygopagus, a conjoined twin parasite attached to sacrococcygeal area (Fig. 3.10). This case illustrates the continuum between twinning and the development of a type I GCT and the difficulty pinpointing the cells of origin of these growths. Indeed, some deem it possible that all extragonadal teratomas have originated as twins [96], and at the other end of the spectrum, others consider them as derived from mis-migrated PGC, which have a preference for the rostral and caudal part of the sympathetic nervous system [52]. In between are those who favor the idea that they are derived from an ESC.

Cases like ours [207], and an almost identical one reported by Chen et al. [208], blur the distinction between parasitic twin and teratoma or rather between type 0 and type I GCT. The two types of GCT may be derived from the same or different precursor cells in the 2C, respectively, primed state.

3.5.2.2 Retroperitoneum

In the retroperitoneal region, all GCT under the age of six are probably type I GCT [76]. Perhaps some may be poorly organized included twins (type 0 GCT), as the retroperitoneum is the most common site of fetus in fetu [95].

Five to ten percent of extragonadal type I GCT occur in the retroperitoneum, most of them in the left or right suprarenal region consistent with lateral migration of PGC toward the gonadal ridges [52]. The sex distribution is about equal when several smaller series are combined [104, 209, 210]. Over 10–20 % of tumors are partly or

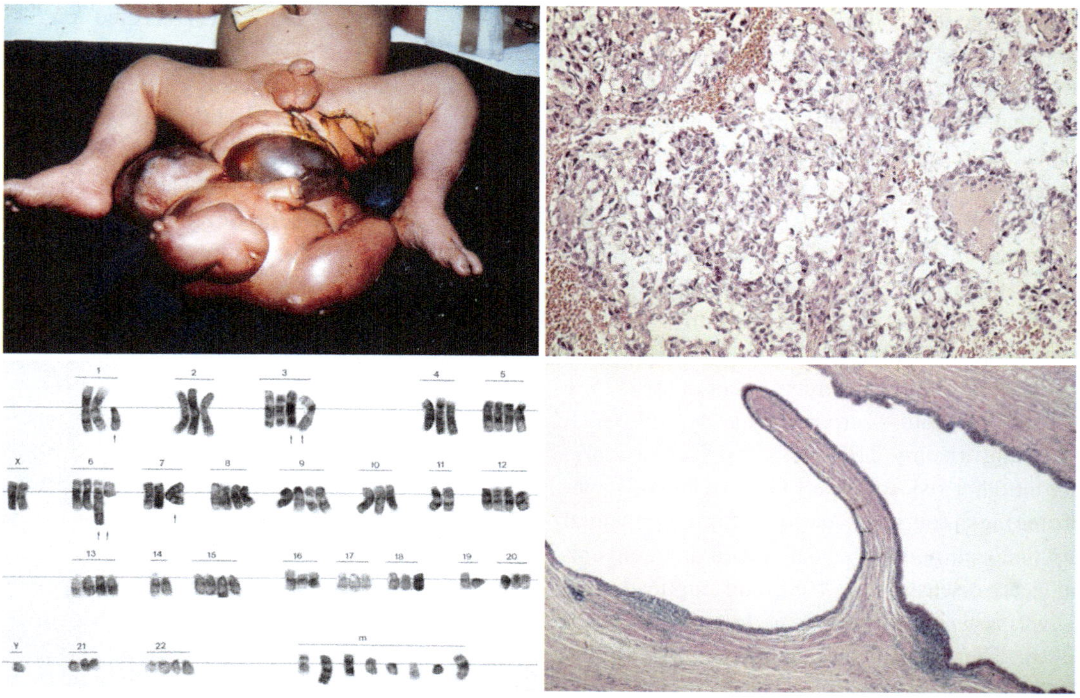

Fig. 3.10 Neonate with diploid sacral teratoma/parasitic twin (pygopagus) with a clearly recognizable foot; upon irradical removal recurring as aneuploid YST (*right top*; H and E, ×100); karyotype with typical loss of 1p and gain of 6q (*left bottom*); after chemotherapy a small residual mature teratoma was resected (*right bottom*; H and E, ×40) [207]

wholly composed of YST, of the remainder, about half have an immature teratoma component and half are completely mature teratomas. The relatively high figure for YST is probably due to the fact that most of the tumors are diagnosed a couple of months after birth.

In postpubertal males, retroperitoneal GCT are virtually always metastatic from unrecognized testicular type II GCT [108, 211–213]. In postpubertal females, retroperitoneal GCT are very rare, usually benign, and probably type I GCT that have remained undetected until after puberty.

3.5.2.3 Stomach

Of the type I GCT, 2–3 % are located in the stomach with a male-to-female ratio of 1:3.7; progression to YST is rare [94]; however, like the type I GCT of the neck, they may metastasize in the form of immature teratoma [94].

3.5.2.4 Mediastinum

The mediastinal type I GCT constitute 2–3 % of the total, most are located in the anterior mediastinum, originating in the thymus [114], and only rarely in the posterior mediastinum. There is a slight preponderance of females [94, 138]. Progression to YST occurs in up to 30 % probably due to surgery several months after birth [214].

3.5.2.5 Heart

Type I GCT of the heart are relatively frequent, 4–7 % of the total, most often located in the pericardial cavity, attached to the great vessels at the base of the heart, and only rarely within the heart itself, very much in accordance with the sites where mis-migrated PGC are found [52]. Males and females are equally affected. Progression to YST occurs in about 5 % [94].

3.5.2.6 Head and Neck

Type I GCT of the head and neck occur in less than 40,000 live births and constitute 10–20 % of all extragonadal type I GCT; the sex distribution is roughly equal. Anatomical localizations are the neck including the thyroid gland (35 %); face

(8 %); oro- and nasopharynx and surrounding structures, in particular hard palate and nasopharynx (45 %); and orbit (12 %) [94, 138, 215–217].

They develop during embryonic life and are often diagnosed before birth. The histology is most often mature teratoma, about one third of the cases contain immature teratoma. Immature neural tissue may rarely metastasize to regional lymph nodes and the lungs and on very rare occasions spontaneously mature [218]. About 3 % of the tumors present as pure YST or as teratomas with microscopic foci of YST [94, 216]. In the series of 16 cases described by Lack [215], there were three YST, respectively, in the oropharynx, the nasopharynx, and the floor of the mouth. In two cases, surgery was not carried out immediately after birth but after 6 and 10 months, respectively. Progression to a somatic-type malignancy may occur, in particular neuroblastoma [218]; squamous cell carcinoma has been reported as well [219].

Progression to YST and metastasis did not occur in 51 cases occurring in the neck [217], probably because surgery is done shortly after delivery, preventing the tumors to progress. The low progression rate might raise the suspicion that many of the teratomas are in fact parasitic cephalopagus [96]. Indeed the oral mature teratoma, mentioned before, diagnosed prenatally in a female baby, most likely was a disorganized dizygotic twin [102], which in retrospect should have been classified as epignathus or more formally as parasitic cephalopagus. This is yet another example of a case that blurs the distinction between type 0 and type I GCT.

Oro-nasopharyngeal and cervical teratomas are associated with other congenital disorders in 12 and 6 %, respectively [94].

The highly aggressive sinonasal pluripotent tumors in adults [220–222] are often characterized by chromosomal translocations and will be discussed as type VI GCT.

3.5.2.7 Brain

Intracranial type I GCT constitute about 10–15 % of all type I GCT with an equal sex ratio and 3 % associated with YST [94, 109, 132]. In one third of the cases, the size of tumor obscures the original anatomical localization. When the site can be determined, it is most often cerebral hemisphere (25.5 %), followed by the suprasellar region (23 %), third ventricle (5.6 %), and pineal region (4.4 %) [132]. The tumors may extend into the orbit, neck, face, mouth, or pharynx [94, 132, 138].

3.5.2.8 Testis

Under the age of six, GCT of the testis are practically always of type I [76], amounting to 5–10 % of all type I GCT [94, 109, 223]. Eleven out of the 19 tumors described by De Backer et al. [223] were teratomas, confirming that teratomas are more frequent than YST in unbiased institutional registries [224]; indeed, under the age of 1.5 years, no YST was diagnosed. Four of the 11 teratomas had immature areas; however, none of the tumors was combined with YST or a raised serum alpha-fetoprotein (AFP). Mixed type I GCT, combining teratoma with YST, are rare but do occur also in the testis [224].

In view of the supposed pathogenesis of type I GCT, it is remarkable that mixed tumors are so rare in the testis, at least ten times less frequent than pure YST [224–226]. The presence of immature teratoma may increase the risk of progression toward YST [106, 198]. Probably, when progression occurs early, in a microscopic immature teratoma, the tumor appears as pure YST at clinical presentation; progression in an established teratoma results in a mixed type I GCT combining teratoma with YST. Teratomas may very rarely, also by way of progression, develop PNET as a somatic-type malignancy [227].

Type I GCT of the testis are not associated with germ cell neoplasia in situ (GCNIS) [228] and testicular dysgenesis syndrome (TDS) [229] and do not share the risk factors of testicular type II GCT nor their increasing incidence. Familial susceptibility for prepubertal YST has not been demonstrated [230, 231]. Unlike the type II GCT of the testis, there is no association with single nucleotide polymorphism (SNP) variants of *KITLG*, *SPRY4*, and *DMRT1*, among others. There seems to be an association with a SNP variant of *BAK1* [158], suggesting that resistance to apoptosis of primitive germ cells might play a role in the pathogenesis of prepubertal GCT. This is in line with the hypothesis that testicular type I GCT originate through reprogramming of a diploid, methylated, pre-erased, premeiotic PGC to an ESC in the primed state.

3.5.2.9 Ovary

In the ovary, the early neonatal peak in the age distribution of GCT, representing type I GCT, is not apparent [76]; however, it is unlikely that they do not exist. Rather their age distribution is probably broader and overlaps with types II and IV, as shown below.

Among 158 reviewed cases of pure and mixed dysgerminomas of the ovary, by definition type II GCT, the youngest was 4 years old; 6 % were in the age group 0–9 years and 41 % between 10 and 19 [232]. A review of 517 dermoid cysts, type IV GCT of the ovary, showed an almost Gaussian age distribution with no cases under age 10 and 1.5 % under age 15 [233]. From these figures, it can be deduced that the large majority of the 66 pediatric patients through age 15, reported by De Backer et al. [234], had a type I GCT. Six tumors were purely cystic, thus probably type IV GCT, and 12 were type II GCT on histological grounds, leaving 48 type I GCT. Of these, three were pure YST, consistent with the rate of about 5 % YST in other anatomical sites. This makes the ovary the second most frequent site of type I GCT after the sacrococcygeal region, accounting for 15–25 % of all type I GCT.

Apparently, teratomas of the ovary can be of three types: I, II, and IV and taking the type VI teratomas associated with clear cell carcinoma of the ovary (Chap. 6) also into account, four types. The overlapping age distributions and morphological resemblance may pose problems separating them. A morphologically typical dermoid cyst in a patient over 10 years of age is almost certainly a type IV GCT. A solid teratoma or a pure YST, or the combination of the two under age five, is most probably a type I GCT. Any GCT with a dysgerminoma, EC, or choriocarcinoma component, with or without other components, is a type II GCT regardless of age. Teratomas associated with epithelial cancers of the ovary are of type VI. Cases composed of teratoma and/or YST over age 5 could be type I or type II GCT. Separating these malignant GCT is probably not so important clinically. However, for a (partly) solid pure teratoma, in a patient over 5 years, it is crucial to make the distinction, since a type I teratoma is benign, whereas a type II teratoma is malignant. In such cases, the diagnosis needs (cyto)genetic confirmation.

Like for the testicular ones, the assumption is that ovarian type I GCT originate through reprogramming of a diploid, methylated, pre-erased, premeiotic PGC to an ESC in the primed state. Such mitotic germ cells persist in the periphery of the ovary through week 20 gestational age [235, 236]. Oogonia can be present in the cortex of the ovary in the two first years of life before they are finally cleared [237].

3.5.3 Type I GCT Beyond Infancy

In general, type I GCT occur neonatally and in early infancy, in prepubertal individuals. However, GCT with essentially the same developmental potential may become clinically manifest at older ages, also in postpubertal patients. This is obvious for the ovary where the neonatal incidence peak is lacking, and the type I GCT have a broad age range, overlapping with the age distribution of the type II and type IV GCT of the ovary. The existence of prepubertal type I teratomas in the postpubertal testis was recently established [228, 238, 239]. Typically, these teratomas are highly differentiated, lack (cyto)genetic abnormalities in particular gain of the complete short arm of chromosome 12 (12p), and are not associated with GCNIS. Remarkably, they may grossly present as dermoid cysts, sometimes containing hair [238], as the type IV dermoid cysts of the postpubertal ovary almost invariably do. Like type I teratomas, they may, albeit rarely, progress to YST [120]. Zhang et al. [238] have proposed that they have the same pathogenesis as type I, prepubertal teratomas. However, the possibility that they arise later in life from "dormant" germ cells arrested in meiotic prophase, like extragonadal mis-migrated PGC, which can be reprogrammed to the primed state, cannot be excluded [120]. It is likely that

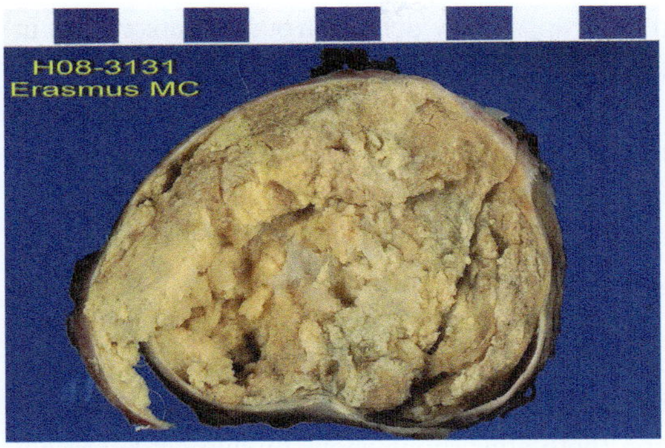

Fig. 3.11 Mediastinal teratoma, late type I GCT, with intermediate phenotype between types I and IV: cyst filled with sebaceous material and hairs

type I teratomas may also occur beyond infancy at extragonadal sites, like the mediastinum [55, 240, 241] and brain [242]. Particularly in the mediastinum, postpubertal mature teratomas may have the gross appearance of a dermoid cyst, grossly containing hair and even tooth structures [55, 240, 241] (Fig. 3.11). Microscopically, the cysts are lined by squamous epithelium with pilosebaceous structures and may have glial tissue in the solid parts of the cyst wall. It seems that in these anatomical localizations, ovary, mediastinum, other extragonadal sites, and perhaps also testis, the developmental potential of the teratomas may have intermediate phenotypes between typical type I and type IV GCT. In each site, teratomas occur that are partly dermoid cysts and partly solid teratoma sometimes with immature components. In the ovary, typical type I teratomas may occur side by side with type IV teratomas, both uni- and bilaterally [130]. Remarkably, the incidence of type I GCT beyond infancy in the mediastinum [241] and brain [243] is rather similar in males and females as opposed to type II GCT, which are much more frequent in males than females.

These clinical observations on early and late type I GCT may be explained by the phenomenon that PGC in females and males regardless of anatomical site enter meiotic prophase by default [58, 59]. The only exception are gonocytes in the testis, which within the seminiferous tubules, under the influence of Sertoli cells, undergo mitotic arrest until puberty. The various phenotypes of these tumors, ranging from typical, solid type I GCT to mainly cystic teratomas closely resembling type IV GCT, may be due to epigenetic differences between the originating PGC. It is hypothesized that pre-erased PGC of early infancy, reprogrammed to ESC in the primed state, will give rise to the typical type I GCT phenotype, while PGC that later in life, beyond infancy, have entered meiotic prophase and concomitantly have undergone partial erasure of GI and possibly some degree of maternal imprinting, may form tumors resembling type IV GCT. In fact, these GCT have intermediate phenotypes between type I and type IV GCT. This assumption is supported by the observation that in mice, EGC are totipotent when derived from PGC but that this phenotype is gradually lost in EGC derived from more mature germ cells [61]. Fully fledged type IV GCT seem to occur only in the postpubertal ovary [233].

3.6 Type II GCT

3.6.1 Type II GCT General

3.6.1.1 Developmental Potential

Type II GCT are malignant tumors that come in two variants: first, seminomas (named dysgerminoma in the ovary; germinoma in the brain; seminoma or germinoma in the mediastinum), which are homogeneous neoplasms composed of neoplastic PGC/gonocytes, the default development of type II GCT; second, non-seminomas, which are caricatures of embryonic development, including both somatic and extraembryonic lineages [244]. Non-seminomas arise when a neoplastic PGC/gonocyte is reprogrammed to become an EGC in the naïve state, or, in pathological terms, when a seminomatous cell is reprogrammed to a totipotent EC cell, the stem cell of non-seminomas [245], as originally demonstrated for mouse EC cells by Kleinsmith and Pierce [246]. EC cells may give rise to all lineages of embryogenesis: YST (secreting AFP) and choriocarcinoma (secreting beta-human chorionic gonadotropin (β-HCG)) represent the extra-embryonic tissues; immature and mature teratomas represent somatic tissues of

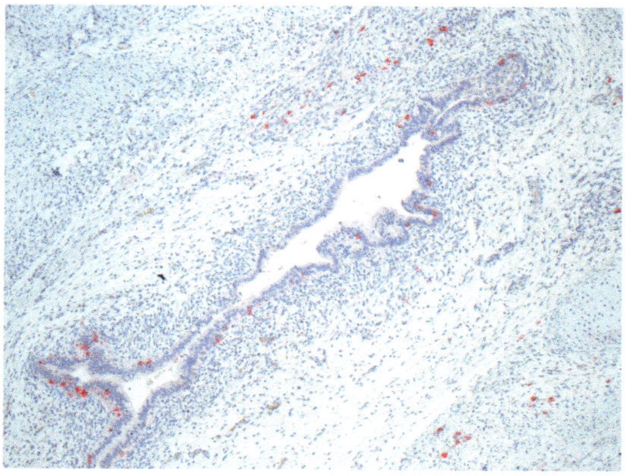

Fig. 3.12 Germ cell differentiation in non-seminoma: TSPY-positive cells within epithelium of primitive gut-like structure and dispersed in surrounding mesenchymal tissue (TSPY, original magnification 100×) [56]

the three germ layers of the embryo in varying degrees of maturation; occasional primitive germ cells represent the germ lineage [247] (Fig. 3.12). These elements are characterized by lineage-specific mRNA [248, 249] and protein expression profiles [161]. In non-seminomas, so-called embryoid bodies can be encountered, which strongly resemble 10-day-old, early presomite human embryos. They show the same expression patterns of both mRNA and proteins as during normal development, like OCT4. Beyond that particular stage, corresponding to the time that in a pregnancy implantation is completed [250], development becomes disorganized, with embryoid bodies turning into patches of EC, YST, trophoblastic giant cells/choriocarcinoma, or teratoma, and disorderly combinations thereof. A possible explanation is that the neoplastic embryo lacks the biparental imprinting pattern of the zygote that is required for proper development of extraembryonic structures and concomitant vascular supply. Mature teratoma may be highly differentiated at the tissue level and even contain organoid structures closely resembling the gut, bronchi, etc., but never fully developed organs as in type 0 GCT or hair and teeth as in type IV GCT. The complete gamut of differentiation lineages, in particular the capacity to develop both embryonic and extraembryonic lineages, the germline competence, and the high capacity of self-renewal of its stem cells (EC cells) characterize type II GCT indeed as totipotent, apparently derived from precursor cells in the naïve state [251].

The mechanism of reprogramming of a seminomatous tumor cell (including the cells of GCNIS) is unknown. It is likely that microenvironmental factors play an important role, suggested by the observation that in the cryptorchidism, the percentage of seminoma depends on the location of the testis: about 90 % in abdominal, about 80 % in inguinal, and about 50 % in scrotal position (both after spontaneous or surgical/hormonal correction of cryptorchidism) [252–254]. The age of clinical presentation of the tumor was the same as in patients with scrotal tumors without a history of cryptorchidism [253]. Recently, it was suggested that interstitial stromal factors like NOGGIN might inhibit bone morphogenetic protein (BMP) in the tumor cell, whereupon reprogramming is initiated via NODAL signaling in two stages [255]. During a maturation phase, a fast-acting NODAL loop stimulates its own activity and temporarily inhibits BMP signaling. During the stabilization phase, a slow-acting NODAL loop, involving WNT signaling [159], reestablishes BMP signaling and the pluripotency circuitry [255]. This is in line with the observations on Cripto, the co-receptor for Nodal [256, 257], which is highly expressed in GCNIS, seminoma, EC, and YST, associated with hypomethylation of the promoter and absent in teratoma where the promoter is hypermethylated [257].

Interestingly, inhibition of BMP is the opposite mechanism from initiation of germline specification in the mouse embryo via expression of Bmp4 ([35] for review). This NODAL-mediated mechanism of reprogramming implies that the tumor cells are exposed to interstitial stromal cells, which is not the case in the intratubular environment, suggesting that within the seminiferous tubule, other factors are involved in reprogramming of GCNIS or intratubular seminoma cells. Moreover, stromal factors are usually not sufficient as primary seminoma is reprogrammed in only about 15 %, giving rise to a mixed non-seminoma with a seminoma component. It seems there is more to be learned about reprogramming of a seminomatous precursor cell to a totipotent EGC.

In all anatomical sites, over half of all primary type II GCT are pure seminomatous tumors. In fact, the younger the patient population, the higher the proportion of seminoma: in dysgenetic gonads and the brain about 80 %, in the ovary 60 %, in the mediastinum 55 %, and in the testis about 50 %. Reprogramming continues even in metastatic seminoma of the testis: 44 % of seminoma metastases eventually develop non-seminoma components [258]. Thus, in the natural history of testicular type II GCT, only 30 % maintain their seminoma phenotype until demise of the patient. These observations suggest that reprogramming is a chance event accumulating over time, whereby in a non-scrotal testis, the chance of reprogramming is deminished, as discussed above.

Seminomatous tumors are by definition pure; the only cells other than neoplastic gonocytes are scattered trophoblastic cells occurring in less than 10 % of the cases [244]. Dysgerminomas in the ovary, mediastinal seminomas, and germinomas of the brain may also contain trophoblastic giant cells in a small percentage [243, 259, 260]. Non-seminomas are often composed of more than one differentiation lineage, in all possible combinations including seminoma, so-called mixed non-seminomas. EC is almost always present and may be the only component, like its derived lineages, thus accounting for pure EC, YST, choriocarcinoma, and teratoma. The frequency of EC attests to the high capacity for self-renewal of these totipotent stem cells of non-seminomas and likely explains the more rapid evolution and earlier clinical manifestation of non-seminomas than seminomas. This is well documented for the testicular type II GCT, where the age distribution for seminomas peaks at 35 years and for the non-seminomas at 25 years. Mixed non-seminomas with a seminoma component, in which reprogramming is delayed because it occurs in already invasive seminoma, peak at the median age of 30 in between non-seminoma and seminoma [261, 262] (Fig. 3.13). Primary type II GCT of the brain, mediastinum, and ovary show the same order in age distribution: for brain, the mean age for germinomas, mixed tumors with a germinoma component, and EC is, respectively, 18, 15, and 12 years [263]; for mediastinum, the mean age for seminomas is about 30 [264] and for non-seminomas 25 years [265]; and for the ovary, the median age of dysgerminoma is 22 years [266], for EC 14 years [267], and for mixed non-dysgerminomas with a dysgerminoma component in between.

Somatic tissues of non-seminomas may progress to form somatic-type malignancies that closely resemble their somatic counterparts, as will be discussed per primary site (see also Chap. 12) (for review [244, 268]).

Seminomas and non-seminomas may metastasize to regional lymph nodes and from then on to distant organs, so-called visceral metastasis, in order of frequency: lungs, liver, brain, and bone [258]. Choriocarcinoma has a propensity for blood-borne metastases, which may cause the first clinical manifestation of the tumor [244]. Seminoma cells may at metastatic sites be reprogrammed to non-seminoma in up to 44 %, as mentioned [258]. In non-seminoma, EC cells are the principal metastatic cells; likewise, in somatic cancers, cancer stem cells are the ones that frequently metastasize [269]. This is microscopically apparent as tumor emboli are virtually always composed of EC cells [244]. At the site where these tumor stem cells will eventually lodge, they may resume differentiation, mimicking the histology of the primary tumor. This phenomenon is demonstrated in mouse models of type I GCT, which in this respect are also valid for type II GCT [270] (Fig. 3.14). The level of differentiation at the metastatic site is in general less than in the primary tumor [271], which may be related to the different microenvironment or due to selection of more metastatic EC cells. In support of the latter, the more distant the metastases, the lesser the level of differentiation; in particular, visceral metastases rarely contain teratoma components but often consist only of EC cells that apparently in the metastatic process have progressively been selected toward stemness and loss of differentiation capacity [272].

3.6.1.2 Epidemiology

Over 90 % of type II GCT occur in the testis [76], being the most frequent cancer of males aged 25–45 in Western white Caucasian populations

Fig. 3.13 Distribution of age of presentation of testicular non-seminoma (*dashed line*), seminoma (*solid line*), and non-seminoma with a seminoma component (*dotted line*) [261]

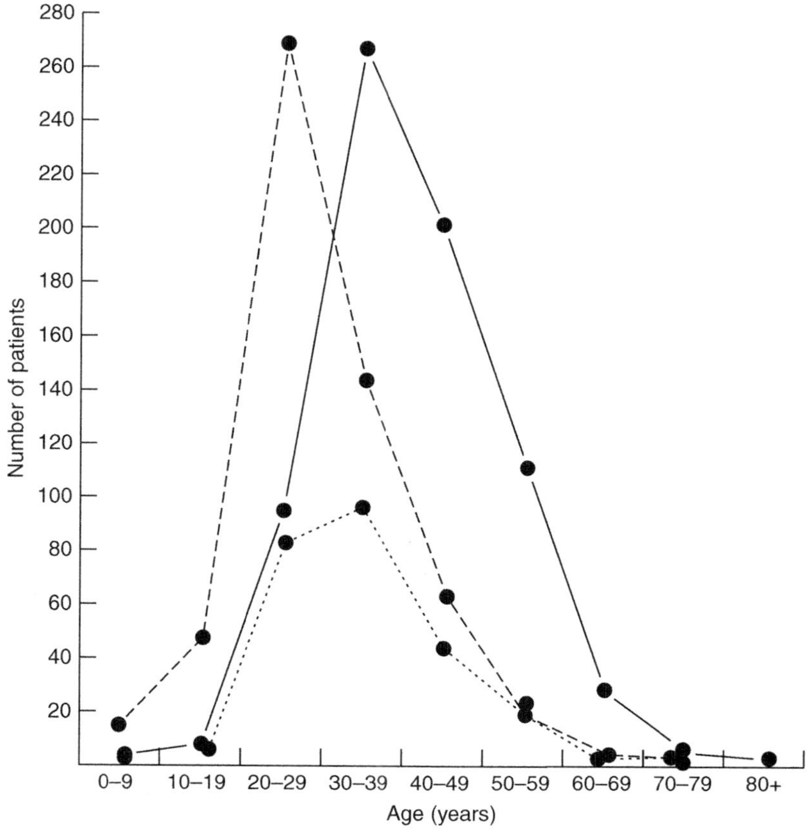

Fig. 3.14 Inguinal lymph node metastasis of retransplantable embryo-derived teratocarcinoma in the thigh of mouse; *left panel*, early metastasis, composed of EC cells only (H and E, original magnification ×200); *right panel*, later metastasis shows somatic differentiation in addition to EC cells (H and E, original magnification ×100) [270]

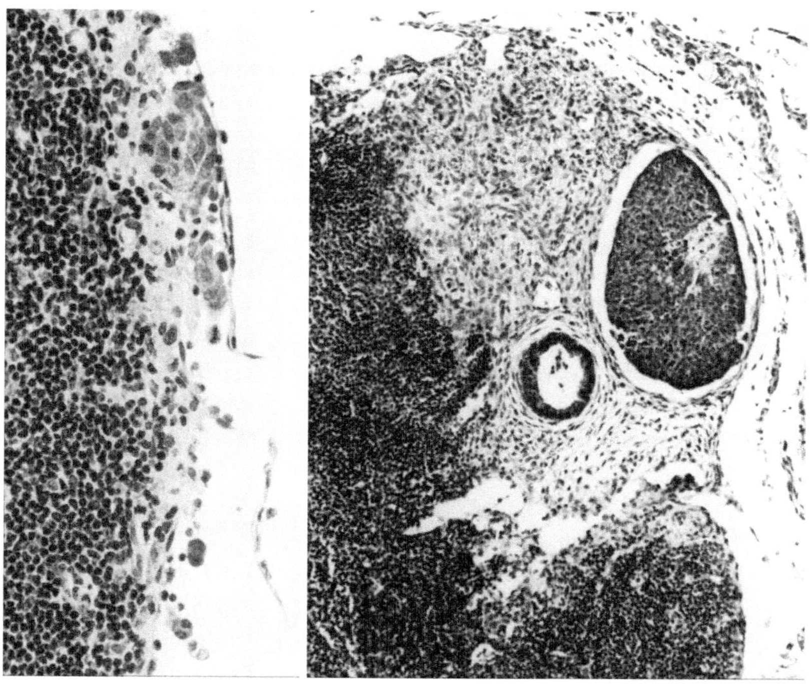

[273]; the remaining develop in dysgenetic gonads/ovary (about 4 %) and in the extragonadal sites: anterior mediastinum/thymus and brain midline/pineal gland (about 3 %) [274]. The youngest age of presentation is in dysgenetic gonads/ovary (from age four with a broad age distribution) [232, 266], followed by brain (children from age 10 to adolescence) [263], mediastinum, and testis (adolescents and adults) [264, 265]. Remarkably, overall the peak age is about 30 for testicular type II GCT: 15 years later than the peak age for those of the brain and 5 years later than for those in the ovary and mediastinum. Type II GCT occur in patients in whom puberty has started or is completed, except for rare cases associated with disorders of sex development (DSD) [275], Down's syndrome [276], and Klinefelter's syndrome [277].

In each of the extragonadal sites, males greatly outnumber females with regard to type II GCT [76]. Apparently, type II GCT is very much a disease of adolescent and adult males, probably related to the presence of the TSPY gene on the Y chromosome, as will be explained later on.

The overall global incidence is 1.5/100,000 with a 20-fold difference between areas with the lowest and highest incidence [273]. The global incidence differences and the rising incidence are attributable to the testicular type II GCT [273, 278]. Geographic incidence differences for the other anatomical sites are dwarfed by those of the testis [76].

Testicular and ovarian type II GCT are bilateral, respectively, in 3–5 % [279, 280] and 10–15 % [281, 282]. Rarely, gonadal tumors may be combined with extragonadal type II GCT in the same individual, like in the patient who had a testicular seminoma and a germinoma of the pineal gland [283] and a patient with GCNIS of the testis simultaneous with a mediastinal nonseminoma [212].

Type II GCT have a strong familial component: over 5 % of patients with a testicular type II GCT have a relative with a similar tumor [284]; an estimated 25 % of testicular cases is due to familial susceptibility [285, 286]. Familial clustering is also documented for ovarian tumors, and testicular may cluster with ovarian type II GCT [137]. In one family, a woman with an ovarian dysgerminoma had a brother with a mediastinal EC [287], suggesting a common etiology in these cases. The only difference between sporadic and familial cases is a younger age of clinical manifestation: 2–3 years for testicular tumors [288] and about 7 years for ovarian cases [137].

Gonadal type II and I GCT may cluster in families as discussed under the type I GCT. An intriguing combination was reported by Heimdal et al. [284] of two brothers, one with a nonseminoma (type II) and the other with a spermatocytic tumor (type III). Spermatocytic tumor being so rare, this combination is probably not by chance but due to a pathogenetic commonality.

3.6.1.3 Anatomical Distribution

From the epidemiological data, it appears that type II GCT occur only in the testis, ovary, dysgenetic gonads, anterior mediastinum, most likely arising in the thymus, and midline of the brain with a preference for the pineal gland [76]. Type II GCT localized in the retroperitoneum are not primary tumors as suggested [289] but metastases from unrecognized primary testicular tumors [108, 211–213].

The occurrence of type II GCT in the mediastinum and the midline of the brain is explained by the migration route of PGC, also the explanation proposed for the anatomical distribution of type I GCT [54]. Clearly, the anatomical distribution of type II GCT is much more limited than for type I GCT. Probably initiation and development of the former require specific conditions of the microenvironment, only offered by certain cell types in these sites [56]. The PGC/gonocytes giving rise to type II GCT are hypomethylated, erased, and apoptosis prone and therefore in need of specific supportive cells for their survival: Sertoli cells in the testis, granulosa cells in the dysgenetic gonad/ovary, and perhaps equivalent cells in the thymus and pineal gland. What these supportive cells probably have in common is expression of soluble and membrane-bound KITLG that may activate the KIT receptor expressed on PGC/gonocytes, thereby enhancing their survival and proliferation [290–292]. Moreover, assuming a similar role for AKT in PGC/gonocytes as in EC,

KIT signaling may through phosphorylation of OCT4 by AKT be involved in maintenance of the undifferentiated PGC phenotype [293].

Type I GCT occur at all sites of type II GCT and in addition in many more anatomical localizations along the midline of the body and occasionally in organs outside the midline. It seems that the requirements for the development for a type I GCT are less demanding than for a type II GCT, as mentioned. This may be explained by the PGC/gonocytes giving rise to type I GCT still being in an earlier stage and therefore still methylated, pre-erased, or rarely partially erased, and thus less fragile than the more mature, hypomethylated, erased PGC from which the type II GCT originate. Finally, and perhaps most importantly in view of the animal models of type I GCT discussed earlier, the PGC giving rise to type I GCT, because they lack a proper niche, probably do not survive as such but only if reprogrammed to pluripotent, primed state-ESC.

3.6.1.4 (Cyto)Genetics

Testicular type II GCT are virtually always peritriploid [294, 295], whereas in the ovary [296], mediastinum [297], and brain [197, 298], type II GCT may be (near)diploid or (near)tetraploid, reportedly in up to 50 % in the brain [197]. The consistent peritriploidy of the testicular tumors is probably due to the older age of clinical manifestation than at the other sites and the long preceding period of intratubular development with concomitant karyotype evolution.

Regardless of anatomical site, type II GCT are characterized by gain of (parts of) the short arm of chromosome 12, usually in the form of an isochromosome of 12p (i(12p)) [299, 300] (Fig. 3.15). In the testis, it occurs in virtually 100 % [295, 301–304], in the ovary/dysgenetic gonad in about 75 % [296, 305, 306], in the mediastinum in 87 % [307, 308], and in the midline of the brain in 60 % [309, 310]. Also just the more proximal parts of 12p may be involved, as an amplicon, in particular 12p11.2-p12.1, specifically in the invasive components [311–314]. It looks as if the proportion of tumors with 12p gain is inversely related to the age of clinical presentation. The very high proportion of 12p gain in testicular tumors is probably, like their peritriploidy, due to the long period of intratubular karyotype evolution.

Isochromosome 12p arises from an erroneous centromeric division during mitotic anaphase preceded by tetraploidization [315] and is of uniparental origin [316]. Among the genes involved in the 12p aberrations are *NANOG*, *STELLAR*, *GDF3*, and *EDR1*, necessary for maintaining pluripotency; *cyclin D2* and *KRAS* providing proliferative advantage; genes involved in glucose or glycolytic metabolism, including *GLUT3*, *GAPDH*, and *TPI1* for energy metabolism in a low-oxygen environment [304, 313, 317, 318]; and genes involved in suppression of apoptosis such as *EKI1*, *SOX5*, and *DAD-R* [312, 313]. Expression of these genes maintains the PGC/gonocyte-phenotype of the tumor cells and allows them to survive and proliferate in the proper niches.

With rare exceptions [306], the tumors have over- and underrepresentation of (parts of) chromosomes other than 12p. Gains involve chromosomes

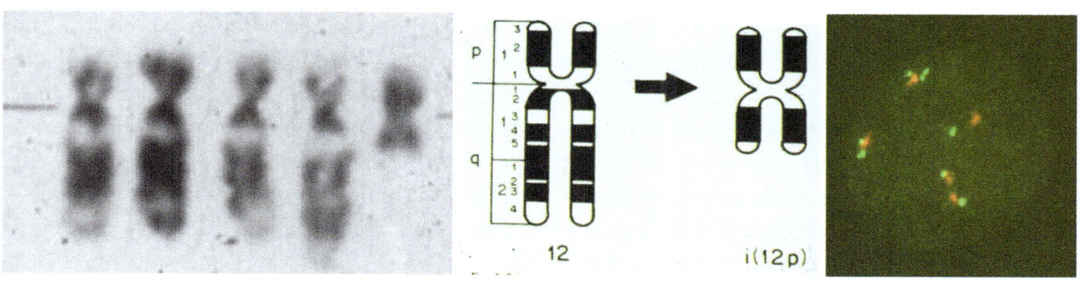

Fig. 3.15 Partial karyotype showing four copies of chromosome 12 and one copy of i(12p) (*left panel*); schematic drawing of chromosome 12 and i(12p) (*middle panel*); in situ hybridization on cytopreparation of interphase nucleus showing three copies of chromosome 12 and two copies of i(12p) [244]

X, 7, 8, 12, and 21 and losses the chromosomes Y, 1p, 11, 13, and 18. The overall pattern is consistent with early tetraploidization of the tumor cells, possibly as a result of malfunction of the mitotic-meiotic switch [319], followed by net loss of chromosomes due to nonrandom losses and gains of (parts of) chromosomes [294, 320]. The large stretches involved, often entire chromosome arms, suggest that aberrant meiotic division may have a role in the evolution of the chromosomal aberrations [197].

Type II GCT are chromosomally instable, probably due to their hypomethylated [321] and polyploid genome, and therefore subject to a continuous reallocation of chromosomal material between chromosomes (Fig. 3.16). Upon cell division, this may cause unequal distribution of chromosomal material over the two daughter cells, resulting in different gene dosage [304]. Chromosomal instability likely drives tumor progression of type II GCT, exemplified by the increasing gain of entire chromosome 12, 12p (among others *KRAS*), 12q (*KITLG*, located on 12q) [296], and 4q12 (*KIT*) [322], which renders the neoplastic gonocytes ultimately feeder cell independent and endows them with invasive capacity. Interestingly, in non-seminoma where because of reprogramming KITLG is no longer advantageous to the tumor, 12q13-q22 including *KITLG* is often deleted [323]. Chromosomal instability may also drive cell fate decisions in the tumor cells, e.g., by influencing the balance of signals promoting germline (BMP) versus embryonal phenotype (NODAL) [255], and thereby the reprogramming of a seminomatous cell to a totipotent EC cell. Finally, in a non-seminoma, it could be involved in lineage determination, e.g., by tipping the balance of normally maternally and paternally expressed genes favoring somatic and extraembryonic/trophoblastic differentiation, respectively.

Mutations and amplifications of oncogenes are rare in type II GCT, with 0.5 mutations per Mb [295] lower than in any other solid cancer of adults [324]. *KIT* is most frequently mutated and mainly involved in seminomatous GCT: seminoma of the testis, about 30 % [295, 324–329]; dysgerminoma of the ovary, up to 50 % [326, 330–332]; seminoma of the mediastinum, 38 % [333]; and germinomas of the brain, over 50 % [197, 334]. In non-seminomatous tumors, *KIT* mutations are rare, less than 1.5 %; the same low figure was reported in gonadoblastoma and derived germinomas [332]. Functional studies have shown the *KIT* mutations to be activating [326, 327], occurring predominantly in the activation loop (exon 17, usually in

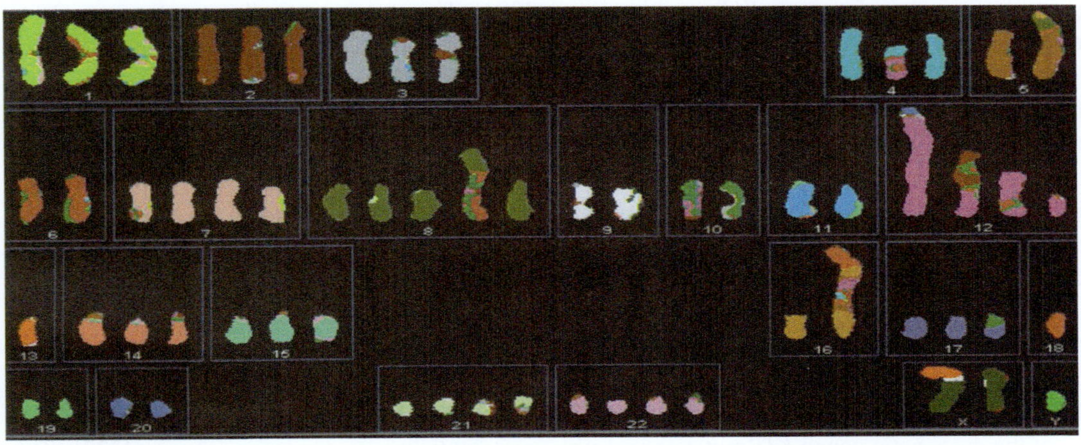

Fig. 3.16 Spectral karyotype of type II GCT. Normally, each chromosome should have one specific color, e.g., chromosome 12 stains pink. In this case of a type II GCT, chromosome 12 contains fragments of other chromosomes, and fragments of chromosome 12 are present in other chromosomes, such as chromosomes 4 and 10. Asymmetric distribution of chromosomes over daughter cells upon cell division may result in different gene dosage

codons 816, 820, 822, 823, and 825) of the second TK domain [335].

In seminomatous tumors, mutations of *KIT* appear to be only one of the mechanisms of activation of KIT signaling and its downstream pathways: KRAS/RAF/MEK/ERK and AKT/mTOR. In fact, in these tumors, KIT signaling is always activated either by upregulation of expression or genetically by mutation or amplification [322]. In the TCam-2 seminoma cell line, siRNA-induced reduction of KIT expression reduced the viability of the cells, although only marginally [329].

The quoted studies report that activating *KRAS* mutations occur in a few percent of type II GCT, more or less at the same rate in seminomatous and non-seminomatous GCT. Earlier studies had found higher figures in testicular seminomas, with subclonal activating *KRAS* and *NRAS* mutations in 40 % [325] and activating *KRAS* mutations in 2/15 cases (15 %) [336] (Fig. 3.17).

KIT and *KRAS* mutations are mutually exclusive in type II GCT of the brain [334] and perhaps also in those of the testis [324]. Consistent with this observation, "large-scale" gain of 12p (harboring *KRAS*) seems also to be mutually exclusive with mutations of *KIT* both in seminomas of the testis [295] and germinomas of the brain [197]. This phenomenon may explain why in non-testicular seminomatous tumors, where gain of 12p is less frequent, *KIT* mutations are more common.

The fact that *KIT* mutations are predominantly found in seminomatous tumors and only rarely in non-seminomas suggests that they, like upregulation and amplification of the gene, are in general involved in the progression of seminomatous tumors, rather than in initiation of type II GCT. Further evidences that *KIT* mutations are most often related to progression of seminomatous tumors include:

In gonadoblastoma, the precursor of type II GCT of the dysgenetic gonad *KIT* mutations occurs in 0.6 %, the same rate as the derived dysgerminomas [332].

The copy number of *KIT* (4q12) is greater in seminoma than in non-seminoma, and high-level amplification of the gene found in a couple of seminomas was not present in adjacent GCNIS [322].

In a case of bilateral seminoma, a *KIT* mutation was found in only one of the two tumors [337].

The report that virtually all bilateral type II GCT, including non-seminomas, had *KIT* mutations, usually in codon 816 and often with the same mutation on both sides, has strongly incriminated *KIT* mutations as an initiating event in migrating PGC prior to their reaching the genital ridges [338]. Later studies could not reproduce these findings neither in bilateral tumors of the testis [283, 322] nor of the ovary [330]. Other studies did show more (14/22, 64 %) [339] or less (2/7, 28 %) [340] preference of KIT mutations for bilateral tumors. Whole exome sequencing of 42 testicular type II GCT [295] demonstrated *KIT* mutations in 3/9 bilateral cases (33 %) and in 3/33 unilateral cases (9 %). It seems, after all, that bilateral cases do have *KIT* mutations at a somewhat higher rate than unilateral cases, in support of a possible initiating role of this genetic event. Notably, the most frequent mutation in bilateral cases and unilateral cases is probably different, respectively, Y823D and D816V [339]. Also in favor of initiation, the same KIT mutation (A816V) was found in the testicular seminoma and the pineal germinoma of the same patient [283]. This pathogenetically revealing case demonstrates two significant points: initiating *KIT* mutations may occur in migrating PGC and extragonadal type II GCT of the brain may be derived from mis-migrated PGC.

Indeed, upregulation of KIT signaling [341] and mutant *KIT* [342] can transform cells; thus, it is very likely that *KIT* mutations may occasionally be the initiating event in PGC, which during migration depend on KIT signaling for survival and proliferation.

It has been proposed that when a *KIT* mutation is the initiating event, the transformed PGC will preferentially develop as seminoma [327]. The rarity of *KIT* mutations in non-seminomas then could be due to the unlikeliness that a seminomatous tumor cell with a *KIT* mutation is reprogrammed to an EC cell. If this were true, one

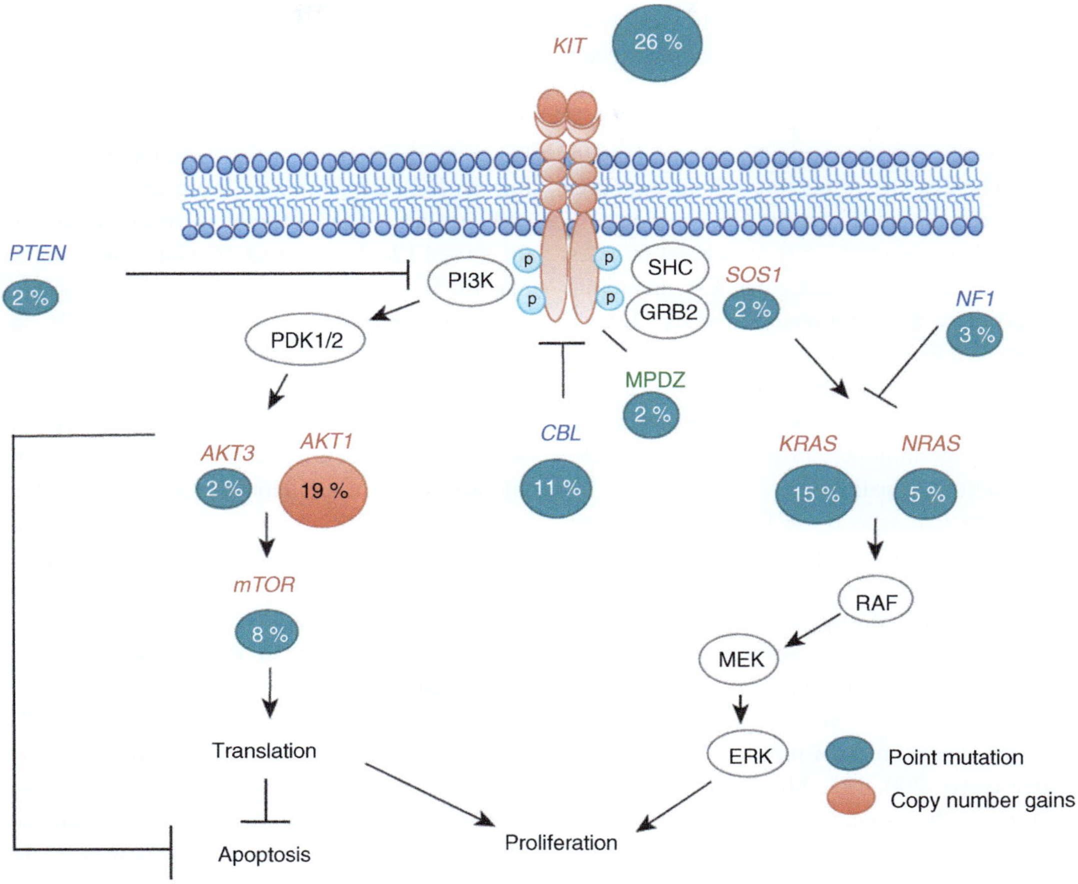

Fig. 3.17 KIT/RAS and AKT/mTOR pathway interactions showing frequencies of somatic alterations in key genes. Alteration frequencies are expressed as a percentage of all intracranial GCT patients. *Red text*, protein positively regulates signaling; *blue text*, protein negatively regulates signaling; *green text*, physically interacting protein [197]

would expect to find *KIT* mutations preferentially in GCNIS adjacent to seminomas with a *KIT* mutation and vice versa; this has not yet been investigated. This hypothesis would also predict the absence of *KIT* mutations in the seminoma-component of mixed non-seminomas where reprogramming of an invasive seminoma cell has given rise to the non-seminoma component. In the brain 4/8 (50 %) mixed GCT including a germinoma component showed a *KIT* mutation [197], suggesting that a *KIT* mutation is compatible with reprogramming in a seminomatous GCT.

Tumor suppressor genes seem to play a modest role, perhaps not surprising in view of the early tetraploidization in the pathogenesis of type II GCT, making loss of heterozygosity less likely to develop. In addition, inactivation of tumor suppressor genes by promoter hypermethylation is rare in type II GCT [343, 344].

The overwhelming male predominance of type II GCT may be explained by the role of co-expression of OCT4 and TSPY [345–347] in their pathogenesis, as it appears from the study of these GCT in DSD patients [290, 291], cryptorchid testis [348], and complete androgen insensitivity (CAIS) [349] (Sect. 3.6.2.1). Notably, *TSPY* is present in about 35 copies [350] in the GBY region of the male-specific region on Y [351].

Similarly, in the proper niches in the thymus and midline of the brain where sporadic mis-migrated, but not yet demonstrated PGC [52], may survive, co-expression of these two proteins may drive transformation of PGC. In view of the predominance of males, this pathogenetic mechanism is probably the most important but not the only one to explain the origin of type II GCT, as they do occur in phenotypically normal females [352], whereby the possibility of constitutional or chimeric mosaicism for the GBY region of Y has to be kept in mind [353]. In normal females, somatic mutations, in particular *KIT* mutations, may indeed be the initiating event of type II GCT as illustrated by the study of Hersmus et al. [332], where *KIT* mutations were found in 53 % of ovarian dysgerminomas of normal women and in only 6 % of dysgerminomas originating in gonadoblastoma in DSD patients.

It seems that type II GCT have at least two fundamentally different pathways of origin: first and foremost, if GBY/*TSPY* is present, the "developmental pathway," due to disturbed maturation of PGC leading to co-expression of embryonal proteins in particular OCT4 and early differentiation genes in particular TSPY (followed by overexpression of KITLG in the supportive cells) [290, 291, 348, 349]; second, and much rarer, the "somatic mutation pathway," in which mutations in oncogenes, in particular in *KIT* and *RAS*, are initiating events. In view of the mutation frequency of these two genes in non-seminomatous tumors, the "somatic mutation pathway" seems to occur in less than 2 % of the cases. The frequent mutations in *KIT* in seminomatous GCT are in general progression related, as discussed. In normal females, obviously lacking GBY/TSPY, initiation of type II GCT by somatic mutations is probably a more common mechanism.

A brief summary of the progression of type II GCT, with emphasis on the various roles of KIT activation, is at this point appropriate. Early progression is sustained by upregulation of KITLG, followed by tetraploidization, whereby probably gain of chromosome 12, with *KITLG* on 12q and among many others *KRAS* on 12p, is most significant. In the in situ precursor lesions prior to the development of an invasive tumor, best studied in GCNIS, progression is driven by nonrandom gains and losses of parts of chromosomes harboring genes that promote feeder cell independence, most conspicuously gain of 12p [315]. Activation of KIT (gene on 4q12) signaling via upregulation, mutation, or amplification of the gene is essential in the progression to seminoma. Upon reprogramming of a seminomatous cell into an EC cell, factors favoring seminomatous cells are no longer relevant, as exemplified by the loss of 12q, harboring *KITLG* in non-seminomas [323].

Mutations common in adult cancers, like in *p53*, appear in type II GCT when they acquire resistance to cisplatin-based chemotherapy [354] or progress to somatic-type malignancies, most often in (late) recurrences [244, 268, 355]. Of interest is the observation that expression of a specific set of miRNAs, i.e., 372 and 373, might function as an alternative for inactivation of *p53* in the pathogenesis of type II GCT [356].

3.6.1.5 Epigenetics: Including GI and miRNAs

Except for teratoma components, type II GCT, including GCNIS and gonadoblastoma, are characterized by global demethylation, erasure of parental imprinting, and the presence of permissive histone modifications [87, 357–364]. Only Alu repeats have been reported to be methylated in non-seminomas [344].

Specifically the developmentally important miRNAs, miR-371-373 as well as miR-302 and miR-367 [356, 365], which have a crucial role in development of embryonic stem cells, are highly expressed in type II GCT, including GCNIS [366].

Global demethylation and this typical expression pattern of miRNAs are part of the phenotype of PGC, underscoring the origin of type II GCT from these germ cells committed to totipotency and confirming the phenotypic similarity of PGC and the cells of the precursors GCNIS [367–370] and gonadoblastoma [290–292], respectively, in the testis and the dysgenetic gonad/ovary.

3.6.1.6 Sensitivity/Resistance/Residual Teratoma/Further Progression

Type II GCT are the solid tumors in adults with the highest sensitivity to DNA-damaging agents, where, e.g., for testicular primary tumors, cure rates >80 % are achieved in disseminated disease [371]. Both seminoma cells and EC cells, the stem cells non-seminoma, are probably highly accessible for DNA-targeting drugs because of their open chromatin structure [344, 360, 372]. EC cells are a factor two to four more sensitive to cisplatin than the clonogenic cells of cell lines derived from common adult cancers [372]. Seminomas are exquisitely sensitive to radiotherapy and cisplatin-based chemotherapy. The typical failure of seminoma cells to repair radiation-induced double-stand breaks, requiring homologous recombination, may be related to premeiotic and embryonic characteristics of the neoplastic gonocyte. A significant change brought about by the reprogramming from seminoma to non-seminoma is that this high radiosensitivity is lost.

Progression from a gonocyte to a type II GCT, particularly in the "developmental pathway," requires fewer genetic changes than in the development of virtually any other type of cancer. Significantly, the most stringent barrier to immortalization and thereby carcinogenesis is progressive shortening of telomeres due to repression of telomerase activity [373]. This barrier is absent in PGC and ESC, as telomerase is indeed active, and thus immortality is part of their normal [374] and neoplastic phenotype, except for mature teratoma [375–377]. Accordingly, mutations are rare in testicular type II GCT [295, 324]; there is no selective pressure for loss of function of p53 [356], and the DNA-damage response remains intact (for review [167, 354, 378]). In fact, it is hypersensitive, reflecting the physiological situation in germ cells and embryonal cells, characterized by inducibility of wild-type p53 jointly with the absence of p21-induced cell cycle arrest [379], whereby cells with damaged DNA are not repaired but rather eliminated due to a low threshold for apoptosis. This preference of apoptosis over DNA repair is a physiologic mechanism protecting against propagation of repair errors via germ cells into the next generation or via ES cells into the developing embryo.

The open chromatin structure and the hypersensitive germ cell/embryonal phenotype get lost upon somatic differentiation into teratoma [380, 381, 382] of which the adult tissue stem cells are well equipped to survive DNA damage by prolonged G1 and G2 arrest and proficient repair (for review [167, 354, 378]). The phenomenon that residual teratoma after chemotherapy is usually associated with primary tumors with a teratoma component demonstrates that it results from selective survival of somatically differentiated cells rather than induction of somatic differentiation due to chemotherapy [271, 383].

Primary and acquired resistance is relatively rare in type II GCT, probably because of the low mutation rate in these tumors [324, 378]. Mutations involved in resistance have been found in *BRAF*, correlated with MSI [384], and in the DNA repair gene *XRXX2*, promoting cisplatin resistance in animal studies [295]. Also mismatch repair deficiency was found correlated with treatment failure [384]. Further molecular mechanisms of (cisplatin)resistance, partly the same as occurring in spontaneous somatic differentiation, are: somatic differentiation accompanied by downregulation of OCT4 (e.g., as a result of hypoxia or treatment with retinoic acid); failure to induce the apoptotic factors Puma and Noxa; changes in the expression levels of miRNAs such as miR-17, miR-106b, and miR-302a or miR-371-373; elevated levels of MDM2 and cytoplasmic translocation of p21 by phosphorylation; activation of the PDGFRβ/PI3K/pAKT pathway [354]. Jointly, these molecular mechanisms explain only the minority of therapy-resistant cases. Development of resistance usually occurs without obvious deviation from typical GCT morphology, mainly as YST but also as EC, choriocarcinoma, and seminoma [385].

A final important mechanism causing resistance of type II GCT is further progression of teratoma and YST due to accumulation of mutations commonly found in adult cancers. Twenty percent of late relapses of testicular type II GCT, defined as recurrences more than 2 years after initial complete response, contain histological elements resulting from further tumor progression with morphologies not typical of type II GCT [386], so-called somatic-type malignancies, which may be derived from teratoma or YST (see also Sect. 3.6.2.2).

In addition to 12p aberrations [387–389], the somatic-type malignancies may have the genetic characteristics of their somatic counterparts, like 2q37 rearrangements in rhabdomyosarcoma, *p53* mutations in sarcomas, and t(11;22) translocations in PNET [355]. A somatic-type malignancy confined to a stage I non-seminoma, occurring in about 5 %, does not adversely affect prognosis [390].

Cisplatin-based chemotherapy of non-seminoma is, in fact, a model for stem cell therapy of a solid tumor showing that eradication of the stem cell population does not necessarily cure the patient [268, 391]. The surviving committed stem cells and differentiated cells may possess or acquire clonogenic potential and appear as therapy-resistant recurrence.

3.6.1.7 Animal Models and Cell Lines

Type II GCT are probably unique for humans, as no convincing examples of spontaneous or induced type II GCT have been reported in animals. Neoplasms of primitive germ cells in mice [392] and fish, e.g. experimentally induced in zebrafish [393–395], have been reported; however, reprogramming of a neoplastic primitive germ cell into a totipotent GCT, a key feature of type II GCT, has to the best of our knowledge not been described.

The only experimental model for seminoma is the TCam-2 cell line, derived from a human primary testicular type II GCT with a seminoma component, which can be propagated in vitro and as xenograft in immune-compromised mice [396] due to a *BRAF* mutation making it more apoptosis resistant than seminoma cells normally are [397, 398]. TCam-2 cells were at the molecular and epigenetic level characterized as seminoma cells [329, 398], including the expression of OCT4 in combination with SOX17 [399]. As expected, TCam-2 cells resemble human PGC, cells committed to totipotency, whose fate is determined by SOX17 and by BLIMP1 that represses differentiation into endodermal and other somatic lineages in PGC by repressing SOX2 [35]. Indeed, by inhibition of BMP, TCam-2 cells could be reprogrammed to an EC phenotype (among others expression of SOX2 instead of SOX17; genome-wide DNA methylation) via NODAL signaling [255, 400, 401], demonstrating in vitro what was first hypothesized from pathological observation [245].

The absence of type II GCT in mice may be explained by the molecular mechanisms of specification and epigenetic modification of the early germline being different in mice and humans. Particularly relevant could be that Oct4 in murine PGC is co-expressed with Sox2, like in mouse ESC, whereas OCT4 in human PGC is co-expressed with SOX17 [35]. Moreover, as mentioned before, epigenetic reprogramming of mouse PGC during embryonic development takes place in 24 h [8], whereas in humans, this process takes several weeks [6, 236]. This situation in humans creates a longer time frame for neoplastic transformation of early germ cells with a totipotent developmental potential, which in addition have the obstacle that they have to switch from SOX17 to SOX2 before being able to revert to an EGC phenotype. The very frequent step of tetraploidization early in the pathogenesis of type II GCT suggests that it is important for maintenance and survival of totipotent tumor cells in the naïve state.

In mice, and possibly also in other animals, the time frame for generating neoplastic totipotent cells is short, probably not long enough for the necessary steps, including polyploidization, to give rise to a tumor of transformed PGC/gonocytes. In mice, PGC/gonocytes can only give rise to neoplasms if they are directly reprogrammed to a diploid, pluripotent ESC in the primed state, probably because it expresses Oct4 in tandem with Sox2, as in ESC. As mentioned before, PGC that are not reprogrammed die apoptotically. The various teratoma models in mice are not representative for type II GCT but rather for type I GCT (Sect. 3.5.1.6). Several features of the models that do help to understand type II GCT are addressed throughout this chapter, where appropriate.

In humans, the PGC/gonocyte phenotype is indeed fairly stable in type II GCT, as at each anatomical site, over half are seminomas, the default state of these tumors. At the same time, it is obvious from the biology and histology of these tumors that reprogramming of a seminomatous cell (GCNIS or seminoma, primary or metastatic) to a totipotent EGC-like cell giving rise to a non-seminoma is a regular, though poorly understood event. To study this phenomenon, more typical seminoma cell lines would be helpful.

Non-seminoma cell lines, derived from in vivo reprogrammed seminomatous precursors, have been less difficult to establish than seminoma cell lines and are readily available. The first clonal cell line, and probably best characterized and intensely studied in vitro and in xenografts, is NT2D1, derived from Tera-2 [402]. It has been widely used, among others non-seminoma cell lines, to study differentiation of EC cells as a model for human embryonic development before human ESC became available [403, 404].

Because of the paucity of models for type II GCT, their pathogenesis has been studied primarily in human tumors in a multidisciplinary approach, combining epidemiology, pathology, (cyto)genetics, cell biology, and molecular approaches, as will be discussed per anatomical site in the following paragraphs.

3.6.2 Specific Aspects of Pathogenesis per Anatomical Site

3.6.2.1 Dysgenetic Gonad

Much of the pathogenesis of type II GCT has been learnt from the study of gonadal dysgenesis in DSD and its typical type II precursor lesion gonadoblastoma, in which the inception of these GCT can be closely followed. It has provided crucial insight into the interactive role of supportive cells and tumor cell-intrinsic factors in the pathogenesis of type II GCT.

Developmental Potential

Gonadoblastoma, originally described by Scully ([405], for review [406]) is composed of two cell types: nonneoplastic immature granulosa cells and gonocytes. The granulosa cells serve as feeder cells, offering a niche for the gonocytes, of which some, actual gonadoblastoma cells have the same atypical morphology as the cells of GCNIS. Gonadoblastoma cells will eventually outgrow the nonneoplastic gonocytes and by way of further progression become feeder-independent invasive dysgerminoma, the counterpart of seminoma of the testis. Most type II GCT of dysgenetic gonads are dysgerminomas (80 %); apparently, reprogramming to non-seminoma occurs in only 20 % of the cases [406].

Epidemiology/Risk Factors

Gonadoblastoma is a rare lesion that develops in patients with certain forms of DSD, bilaterally in 40 % of the cases [405–408]. At high risk are 46,XY patients with mutations in *WT1* (including Denys-Drash, Fraser, and WAGR syndromes) [409–413], *SRY*, *SOX9*, *DHH*, *ARX*, *RR5A1*, or *TSPYL1*, resulting in a dysgenetic testis in the presence of Y-chromosomal sequences, although the male gonadal initiation/differentiation pathway is disrupted [275, 408, 414, 415]. Patients with 46,XY/45,XO mosaicism with a high risk for streak gonads [416, 417] may develop gonadoblastoma in up to 50 %. Cases reported in 45,XO patients with a Turner phenotype probably have undetected mosaicism for Y-chromosomal material containing the GBY region with the candidate gene TSPY within the dysgenetic gonad [353, 418, 419].

Anatomical Distribution

Gonadoblastoma and its derived invasive GCT occur at the sites of dysgenetic gonads: intra-abdominal, inguinal, and sometimes scrotal.

(Cyto)Genetics/Epigenetics

The cytogenetic changes in gonadoblastoma are similar to those of GCNIS: early polyploidization followed by the same nonrandom losses and gains of chromosomes. Overrepresentation of 12p material, usually as i(12p), occurs when gonadoblastoma progresses to invasiveness. As opposed to GCNIS, tetraploidization does not always take place [290, 306]. KIT mutations have been identified in dysgerminoma [296, 330], however only once in a dysgerminoma originated from gonadoblastoma [332]. Gonadoblastoma cells have the hypomethylated genome with (partially) erased GI of PGC [360].

Pathogenesis

Typically, gonadoblastoma develops in those forms of DSD, where the presence of the GBY region [351] including the candidate TSPY gene [345–347] is combined with a disturbed gonadal development due to mutations or deletions in genes necessary for the physiological male (46,XY) pathway (initiated by SRY, followed by SOX9/WT1, SF1, and downstream targets) [275, 408, 411, 414], which may be hereditary [420–422]. Disturbed expression of members of this pathway may result in immature supportive stromal cells, resembling granulosa cells [423, 424]. Multiple genes can be affected within a single patient [425]. In this hypovirilized condition, maturation of the gonocytes is delayed, thereby creating a window for co-expression of embryonal genes and early differentiation genes, in particular OCT4 and TSPY [290, 291, 426], which, jointly with enhanced KIT/KITLG signaling [292], promote neoplastic transformation of gonocytes into gonadoblastoma cells in the dysgenetic gonad. A combination of GCNIS and GB has also been reported within a single gonad [427, 428] (Fig. 3.18).

Again, the important lesson to be learned from gonadoblastoma is that there are probably two pathogenetic pathways for the origin of dysgerminoma: the first, as in DSD, merely by disturbance of the normal development of the gonad, the "developmental pathway," related to GBY enabling co-expression of TSPY and OCT4, and the second, much rarer pathway, the somatic mutation pathway involving mutations, particularly in *KIT*. The somatic mutation pathway is probably more frequent in normal females, as discussed earlier (Sect. 3.6.1.4) [332].

3.6.2.2 Testis

Developmental Potential

Slightly over 50 % of testicular type II GCT are pure seminomas, about 15 % combine seminoma with a non-seminoma component, and the remainder lack seminoma and are composed of one or more embryonal (immature and mature somatic tissues) or extraembryonal (YST and choriocarcinoma) lineages; primitive germ cells are occasionally encountered [247]. Most testicular non-seminomas combine two or more lineages; the most frequent pure non-seminoma is pure EC, followed by pure teratoma; pure choriocarcinoma and pure YST are rare [244]. Somatic-type malignancies occur in primary testicular non-seminomas in 3–6 % [390], in postchemotherapy RPLND in 8 % [429]. In late recurrences, the percentage is over 20 % [386]. Over half of somatic-type malignancies are sarcomas [388], the most frequent types being rhabdomyosarcoma, followed by angiosarcoma, and leiomyosarcoma [430]. Next in frequency are carcinomas of various types and small blue round cell tumors [388, 390, 430–433]. Rarely two somatic-type malignancies develop simultaneously in late recurrences [268]. About 75 % are derived from teratoma ([268] for review); in 25 %, progression has occurred in YST, giving rise to glandular and sarcomatoid YST, mimicking somatic-type malignancy [434, 435].

Epidemiology

The lowest incidence figures of <0.5 for testicular type II GCT are found in Africa and parts of Asia. The incidence is a factor 10–20 higher among most white Caucasian populations in Western societies, like in Europe, North America, Australia, New Zealand, and parts of South America. The highest figures of >12 are recorded for Denmark, Norway, and Switzerland [273, 436–438]. Worldwide, the incidence has increased in the last four decades, in fact more than doubled in most Western and Northern European countries. In earlier low-incidence countries like Spain, Slovakia, and Slovenia, the rates are increasing rapidly and approaching those of Western Europe [273, 278]. In the US, the incidence among Caucasians is 6.6 versus 1.2 in blacks. In both groups, the rates have increased in the past 30 years [76, 439].

The large incidence differences among ethnic groups within the same society, e.g., Caucasians and blacks in the USA [76], demonstrate the

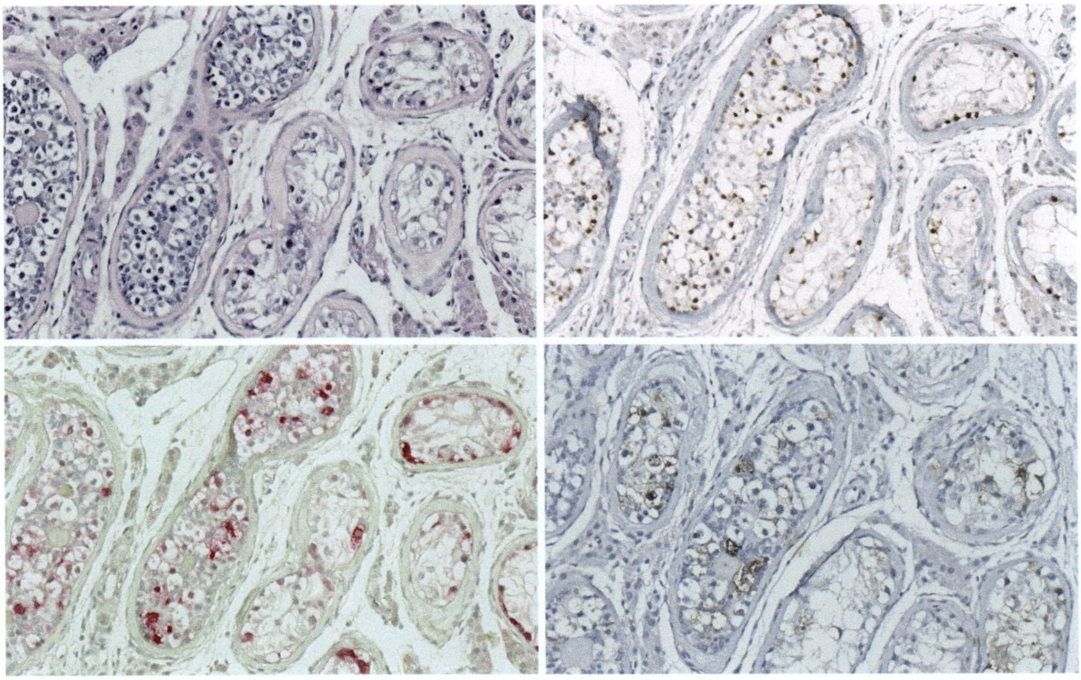

Fig. 3.18 Co-expression of OCT4, TSPY, and KITLG in gonad with gonadoblastoma (*left half of the photo's* with granulosa cells and Call-Exner bodies between the neoplastic gonocytes) and GCNIS (*right half of the photo's* with neoplastic gonocytes in the spermatogonial niches); clockwise: H and E, OCT4 (*brown*), KITLG (*brown*), and TSPY (*red*), ×200 [427]

importance of genetic factors. On the other hand, the geographic pattern of increasing incidence of testicular type II GCT, the changing incidence among certain immigrant populations [440–442], points to an important causative role of environmental factors, associated with modern, Western lifestyle that probably favor hypovirilization of the developing male embryo [167, 370, 443]. Indeed, the young age of the patients, the bell-shaped age distribution, the decreased risk for testicular type II GCT in Danish men born during World War II [444] and similar birth cohort effects in other populations, and the evidence that type II GCT are derived from gonocytes; all these observations point to their initiation during embryonal development.

Risk Factors

Risk factors for type II GCT of the testis have recently been comprehensively reviewed [445, 244]. Briefly, features of TDS [229] confer a higher risk: cryptorchidism (OR 4.3) [446]), previous inguinal hernia (OR 1.63), as well as hypospadias [447], previous testicular cancer (testicular type II GCT are bilateral in 3–5 %) [448], impaired spermatogenesis [449], and a family history of testicular type II GCT [286] which encompasses urogenital anomalies as in TDS [450]. TDS is considered as a relatively mild disturbance of sex differentiation, due to hypovirilizing factors in utero. This is consistent with the observation that gene mutations that cause DSD in 46,XY individuals, such as *AR*-mutations, confer a (somewhat) higher risk for type II GCT. Also in individuals with normal sexual development variants in AR, ESR2, HSD17B4, and CYP19A1, involved in steroid signaling or metabolism, confer an increased risk of testicular type II GCT [451, 452].

Factors increasing risk with a lower OR, in the order of 1.3 or less, which also one way or another may relate to disturbed hormonal conditions in utero are maternal bleeding, low birthweight, short gestational age, twin, tall stature, and being first born child. Sibship size is protective, the more siblings the lower risk, OR 0.80, as is late puberty, OR 0.81 [453]. There is sufficient evidence to support a relation between testicular type II GCT and three widely used hormone disruptive organochlorine insecticides (dichlorodiphenyldichloroethylene, cis-nonachlor, and trans-nonachlor), which may have a hypovirilizing influence on a developing male embryo in a specific window of development (masculinization window) [454].

The two identified occupational risks (fire fighting and aircraft maintenance) [455] and cannabis smoking [456] obviously affect males postnatally, in adolescence and adulthood. It might be assumed that these risk factors modify the development of already existing precursor lesions. Cannabis smoking is particularly interesting because it selectively increases the risk of non-seminoma, suggesting that it might promote reprogramming of a seminomatous precursor cell.

Genetic Susceptibility

Familial risk is among the highest in cancers [457, 458]: having a brother with a testicular type II GCT confers a three to eight times and a father a two to four times higher risk. For comparison, having a brother with colon cancer increases ones risk by a factor two. Yet, despite substantial efforts in international collaborations, only few low-penetrance gene mutations were identified by comparing familial and sporadic cases. The first identified risk locus was the gr/gr deletion in azoospermia factor c region of Y [459], and recently a deleterious probably causative germline mutation in *PDE11A* was discovered in familial and sporadic cases [460]; both mutations explain only a few percent of the familial cases.

Indeed, the small size of affected families, usually a father and a son or two brothers, and the high risk in monozygotic compared to dizygotic twins [457] are consistent with multiple autosomal recessive low-penetrance susceptibility genes [286, 288, 461–463].

The polymorphic gene variants increasing susceptibility to type II GCT shown in recent genome-wide association studies (GWAS) are also consistent with genetic susceptibility being the result of multiple common, relatively low-penetrance gene variants. The first variants (KITLG, BAK1, SPRY4) were demonstrated in 2009 by two independent studies [195, 196]. As of 2015, over 30 variants of genes for proteins which are plausibly involved in the biology of testicular type II GCT have been published, e.g., KIT/KITLG signaling (*KITLG*, *SPRY4*, *BAK1*, *PDE11A*) in relation to PGC survival and proliferation; *DMRT1* variants, involved in sex determination and regulation of meiotic division; genes involved in telomere maintenance, testis differentiation, and sex determination (such as *HPGDS*), among others [288, 376, 464–467]. Together, these account for about 15 % of the excess familial risk to brothers of testicular type II GCT patients and 22 % of the excess to sons of testicular type II GCT patients [468]. Remarkably, these variants have no association with the established phenotypic risk factors: family history, cryptorchidism, inguinal hernia, age at diagnosis, and bilateral testicular type II GCT [469]. TDS, including cryptorchidism, hypospadias, male infertility, impaired testicular development, and testicular type II GCT [229] may be associated with variants of *TGFBR3* and *BMP7* [470] and variants in genes involved in steroid signaling but not with the established risk variants for testicular type II GCT.

Importantly, different distribution of variants in *KITLG* and *AR* in Caucasian and black populations may partly explain the 20-fold ethnic difference in the incidence of testicular type II GCT [195, 196].

It seems that some of the variants primarily target the PGC/gonocyte, in particular the KIT/KITLG pathway and others primarily the supportive cells (Sertoli and Leydig cells in the testis), and that homozygosity for risk alleles in the two pathways confers the highest risk. For example, men with testicular type II GCT have a 14 times higher chance than controls to be homozygous for the two risk alleles, *KITLG* (PGC/gonocyte survival and proliferation) and *DMRT1* (testicular development) [464].

Pathogenesis: Initiation

There are striking similarities between the pathogenesis of type II GCT in dysgenetic gonads and the testis. Testicular type II GCT are considered part of TDS [229, 471], which as mentioned above also includes cryptorchidism, hypospadias, impaired testicular development, and male infertility, broadly, features of hypovirilization. In fact, the changes in the testis can be considered as a mild form of gonadal dysgenesis [411, 445, 472], supposedly resulting from an interplay between genetic factors, such as mutations in the *AR* [473], and endogenous and environmental hypovirilizing factors exerting their influence in utero, foremost on the stromal cells of the developing male gonad, so-called genvironmental interactions [474]. Of note, hypovirilization may not only occur during embryonic development, as DMRT1 is required to prevent female reprogramming of the postnatal mammalian testis [475, 476], conceivably with impact on the postnatal development of testicular type II GCT.

In keeping with the above observations, the association between increased risk for testicular type II GCT and impaired fertility might be due to variants within *DMRT1* [464], comparable to the role of *Dmrt1* in fertility and testicular teratoma formation in 129Sv mice [189]. Similarly, there might be a parallel between the loss of function mutation of *steel/Kitlg* in this mouse model, impairing fertility and enhancing teratoma formation [185] and variants in *KITLG* in men. Variants in *KITLG* resulting in loss of function may disturb the function of Sertoli cells and the germ cell niche, thereby impairing fertility and promoting initiation of type II GCT.

Specifically, the disturbed Sertoli cells/niche might interfere with downregulation of OCT4 in gonocytes relocated from the center of the tubules to the prespermatogonial niche, thereby creating a window for co-expression of OCT4 and TSPY, assumed to respectively protect the germ cells from apoptosis and to stimulate their proliferation [84, 477]. Co-expression of these proteins is in due time accompanied by increased expression of KITLG by the stromal component and/or via an autocrine loop, which supposedly further stimulates the neoplastic transformation. The association of a higher risk for testicular type II GCT with SNP variants in *KITLG* [195, 196], for example, might be mechanistically explained by interference with KIT signaling in Leydig cells and, as mentioned above, a disturbed interaction between Sertoli cells and gonocytes [348, 377]. Interestingly, one of the likely related variants within *KITLG* concerns a binding site for *p53* [478], whereby the expression level of this allele of *KITLG* might increase under conditions of stress during early development, such as TDS/DSD [415], via upregulation of p53.

Pathogenesis: Early Development

The early morphological changes in gonocytes undergoing neoplastic transformation have been studied in patients with various degrees of androgen insensitivity, including complete insensitivity [349], and in infants with cryptorchid testes [348]. In both conditions, a continuum from delayed maturation of gonocytes, via pre-GCNIS to GCNIS is observed. In delayed maturation, gonocytes located centrally in the tubules still express OCT4 beyond the normal age limit of 6 months [29]; in pre-GCNIS, gonocytes that have migrated into the spermatogonial niche at the basement membrane fail to switch off OCT4 expression and start to co-express OCT4 and TSPY in a heterogeneous pattern, accompanied by focal expression of KITLG (Fig. 3.19); in GCNIS, gonocytes located in the spermatogonial niche consistently express OCT4, usually with co-expression of TSPY and combined with diffuse expression of KITLG. The GCNIS cells meet the established morphological criteria, including enlarged, angulated nuclei. The change of nuclear morphology in the progression step from pre-GCNIS to GCNIS suggests that polyploidization takes place at this stage. These observations are consistent with the hypothesis that gonocytes moving from the center of the tubules into the prespermatogonial niche do not turn off *OCT4*, because the malfunctioning Sertoli cells do not give the proper licensing signal toward male gametogenesis [479].

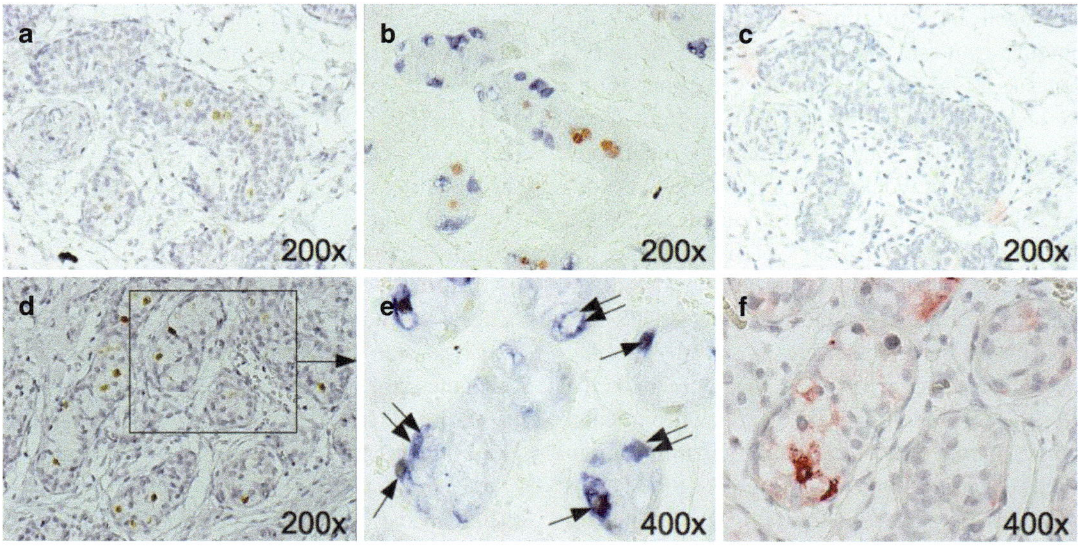

Fig. 3.19 Morphology and expression of OCT3/4, TSPY, and KITLG in delayed-matured gonocytes and pre-GCNIS. Delayed maturation and pre-GCNIS puts gonocytes at risk for malignant transformation without the need of mutations by creating a window for co-expression of OCT3/4, TSPY, and KITLG. (**a**) Delayed maturation with OCT3-/4-positive germ cells (*brown*) all in the center of the tubules, gonad of 10-month-old individual. (**b**) Same area as in A double stained for OCT3/4 (*orange*) and TSPY (*blue*) which are not co-expressed within the same cells. (**c**) Same area as in A and B negative for KITLG. (**d**) Pre-GCNIS with most of the OCT3-/4-positive germ cells are attached to the basal lamina, gonad of 9-year-old individual. (**e**) Detail of the same area as in D double stained for OCT3/4 and TSPY, heterogeneity of the germ cells within particular tubules – cells are either positive for both OCT3/4 and TSPY (*arrow*) or only for TSPY (*double arrow*). (**f**) Same area as in E strongly positive for KITLG (*red*) [349]

Pathogenesis: GCNIS, Progression to Seminoma and Non-seminoma

GCNIS is the common precursor of seminoma and non-seminoma of the testis [480, 481], which is bilateral in 3–5 % of the patients [448, 482, 483]. This high propensity for bilaterality is probably because the germ cell niche is disturbed in both testes as a consequence of TDS. Consistent with this assumption is the fact that in DSD, where the disturbance of the niche is more severe, bilaterality of gonadoblastoma may occur in up to 40 % [405]. GCNIS has a heterogeneous phenotype due to the plasticity of GCNIS cells reflecting different stages of maturation of primitive germ cells [484–487]; a subset of GCNIS cells expresses spermatogonial markers [488]. GCNIS will probably always progress to an invasive type II GCT if left untreated [482, 489], although not proven so far. As yet, no features have been found that predict whether it will progress to (intratubular) seminoma or non-seminoma.

As in dysgenetic gonads, the neoplastic transformation of gonocytes does not require somatic mutations but rather results from a disturbed timing of expression of critical embryonal and differentiation proteins during development, as discussed. Therefore, type II GCT of the testis can be considered developmental tumors [370]. Indeed, somatic mutations are rare and limited to *KIT* and *KRAS* [295, 322–326, 490], whereby activating *KIT* mutations are found in about 25 % of seminomas [327] and rarely in non-seminomas [340]. In seminomas, the KIT pathway is always activated via mutation or amplification of *KIT* or via overexpression of the protein [322, 491]. *KRAS* mutations are in a few percent found both in seminomas and non-seminomas. Mutation of *KRAS* does not seem to occur in combination with high-level amplification of the gene [324] or overexpression of the protein [329]. In testicular type II GCT, *KIT* and *RAS* mutations may be mutually exclusive [324].

The much higher frequency of activating *KIT* mutations in seminoma than non-seminoma suggests that this genetic event does not take place in an early stage of tumor evolution shared by seminoma and non-seminoma. Rather, as discussed before, mutation of *KIT* is part of progression of seminoma, as is amplification of the gene, and upregulation of its expression [322] resulting in KIT activation, characteristic for seminoma. Since it is related to seminoma progression, it is understandable that non-seminomas rarely harbor *KIT* mutations, for they offer no advantage after reprogramming of a seminomatous tumor cell into an EC cell. In the rare non-seminomas with a *KIT* mutation, reprogramming probably took place in a seminomatous tumor cell that already had acquired a *KIT* mutation. Indeed, one of the first two type II GCT in which an activating *KIT* mutation was demonstrated was an ovarian mixed dysgerminoma/YST, with the mutation in both components [326].

The observation that *KIT* mutations are usually found in seminomas that lack large-scale 12p amplifications [295] could mean that for seminomas, KIT mutations and 12p gain are alternative pathways of tumor progression.

The occurrence of the same *KIT* mutations in bilateral testicular tumors demonstrates that a *KIT* mutation can be the initiating event taking place in PGC prior to their arrival in the gonadal ridges [283, 339, 340]. Even more convincing, as already referred to, the same *KIT* mutation was found in the testicular seminoma and the pineal germinoma of the same patient [283].

GCNIS stays more or less dormant until resuming further progression upon hormonal stimulation at puberty [492]. In prepubertal GNCIS, about a quarter of the tumor cells express DMRT1, a key regulator of the mitosis-meiosis switch, whereas in adult cases, this figure drops to a few percent. GCNIS cells that express DMRT1 are not mitotically active and considered dormant [486].

In so-called isolated GCNIS that has not yet given rise to an invasive type II GCT [493] and in GCNIS confined to the spermatogonial niche [304], there is no overrepresentation of 12p (Fig. 3.20). Gain of 12p coincides with the GCNIS cells becoming feeder independent, apparent from their leaving the spermatogonial niche and their capacity to float in the lumen of the seminiferous tubules [304], usually adjacent to an invasive GCT [493] (Fig. 3.21). The next step in the "default development" of a testicular type II GCT

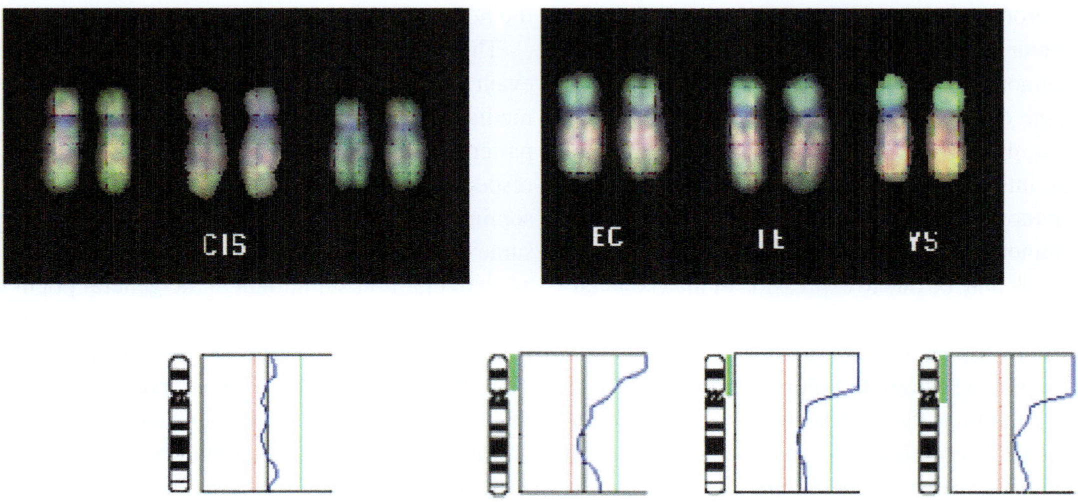

Fig. 3.20 Comparative genomic hybridization on GCNIS cells in the niche (*left panel*) and on three invasive type II GCT, from left to right EC, teratoma, and YST (*right panel*). Gain of 12p is absent in GCNIS, while it is present in invasive type II GCT [244]

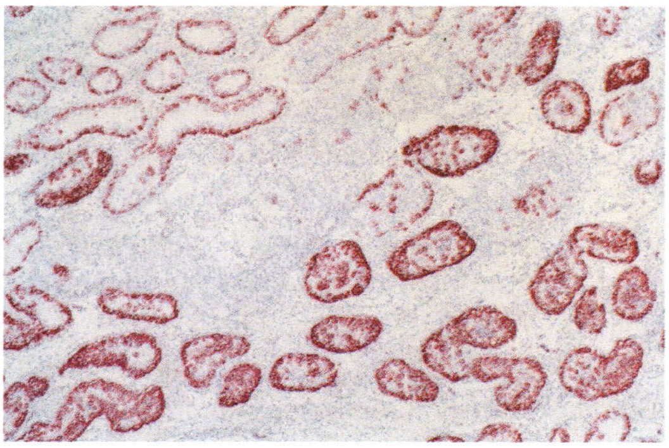

Fig. 3.21 GCNIS cells in the spermatogonial niche and floating in the seminiferous tubules (direct alkaline phosphatase, ×100)

is the formation of intratubular seminoma, whereby the GCNIS cells completely fill and distend the seminiferous tubules and oust the Sertoli cells. Intratubular seminoma may contain lymphocytes like invasive seminoma. Though highly proliferative it virtually never shows necrosis, which is practically always present in intratubular non-seminoma. Apparently, intratubular seminoma cells, like normal gonocytes, are well adapted to the intratubular low-oxygen environment [244].

The trigger for invasive growth of intratubular seminoma is not known. Morphologically, invasion of seminoma has different appearances: one whereby intratubular seminoma extends the seminiferous tubule to the point of breaching the tubular wall [262] and another, microinvasive seminoma in which the tumor cells appear as single cells in the interstitial stroma of the testis [244, 494–496], and combinations of both. Just like intratubular seminoma [262], microinvasive seminoma can be found adjacent to a non-seminoma in about 20 % [494]. Microinvasive growth may be due to expression of matrix metalloproteinase 9 and plasminogen activator, urokinase, by the tumor cells [495].

Upon invasion, seminoma invariably elicits an inflammatory host response, usually composed of lymphocytes, macrophages, plasma cells, and often a granulomatous reaction. Its significance remains controversial. Recently, it was suggested that it is not involved in active immune surveillance [497]. An earlier study demonstrated clonally expanded cytotoxic T cells and evidence of specific and functional T-cell responses operating in seminoma, indicating that the inflammatory infiltrate is indeed involved in the immunological control of the tumor; however, class I MHC molecules could not be demonstrated on the seminoma cells [498], making them invisible to cytotoxic T cells. From histology, the infiltrating lymphocytes seem capable to cause complete regression of seminoma, leaving the scar of a so-called burnt-out tumor. GCNIS and intratubular seminoma may undergo regression as well, whereby the tubules become atrophic and in the end completely fibrosed. This host reaction, and probably the older age of the patients, explains why GCNIS is usually much less extensive in association with seminoma than non-seminoma and even absent in up to 15 % of the cases [262]. The few lymphocytes accompanying GCNIS adjacent to non-seminoma, as opposed to the many adjacent to seminoma, suggest that the host response is indeed elicited by the seminoma, which upon invasion disturbs the mechanisms of immune privilege in the testis [499]. GCNIS, composed of tumor cells that are phenotypically similar to seminoma cells, is probably secondarily involved. It seems less likely that the GCNIS cells themselves, within the intact immunologically privileged testis, trigger a reaction of the host.

The host response is probably clinically relevant in view of the 10 years difference of the median age of presentation of seminoma in patients with AIDS and the general population: respectively, 25 and 35 years. In AIDS patients, seminoma and non-seminoma present at the same age, 25 years, also the age of presentation of testicular non-seminoma in the general population. In addition, a higher risk for disseminated seminoma has been reported in patients with AIDS, suggesting a protective role of an intact immune system [500]. Invasive non-seminoma also contains inflammatory cells; however, the surrounding parenchyma is much less involved than with seminoma, probably explaining why GCNIS is often very extensive and rarely absent [262].

Pathogenesis: Non-seminoma Due to Reprogramming Seminomatous Progenitor Cell

Deviation from the default development of seminoma occurs when a seminomatous cell, either a GCNIS cell or an invasive or metastatic seminoma cell, is reprogrammed to an EC cell, the stem cell of non-seminoma. How often a primary testicular non-seminoma is due to reprogramming in an invasive seminoma (Fig. 3.22) can be estimated from the percentage of mixed non-seminomas with a seminoma component, which is about 15 %. This figure could be an underestimation, since microinvasive seminoma may be overlooked [495, 496]; on the other hand, because a mixed non-seminoma could be a collision tumor of separately developed seminoma and non-seminoma, 15 % may be too high an estimate. The remaining non-seminomas are probably due to reprogramming of an intratubular seminomatous cell. Indeed, the intratubular non-seminoma stage can be demonstrated in the parenchyma adjacent to a non-seminoma in about 15 % of the cases, most often adjacent to small tumors, suggesting that large tumors have overgrown their intratubular precursor [262]. Intratubular reprogramming cannot be explained by downregulation of BMP in GCNIS cells due to interstitial stromal factors [255]. Here, the possibility of unequal distribution of chromosomal material over the two daughter cells [501] might be considered, which would result in a low gene dosage of BMP in one of them, starting off NODAL signaling and thereby reprogramming. In invasive seminoma, the mechanism could be the same or BMP could be downregulated by exposure to interstitial stromal factors like NOGGIN [255].

With sporadic exceptions, intratubular non-seminoma is composed of pure EC that is partly necrotic and often calcified. The invariable necrosis indicates that EC cells are less adapted to the intratubular low-oxygen environment than GCNIS/seminoma cells. It is conceivable that hypoxia-induced factors like MET trigger invasion of intratubular non-seminoma. Remarkably, only upon invasion, the EC cells start to differentiate and display the totipotent nature of naïve-state EGC due to differentiation-inducing factors in the stroma of the testis and/or loss of differentiation-inhibiting factors in the intratubular microenvironment (Fig. 3.23). Possible stromal factors are TGF-β, FGF, and BMP, which

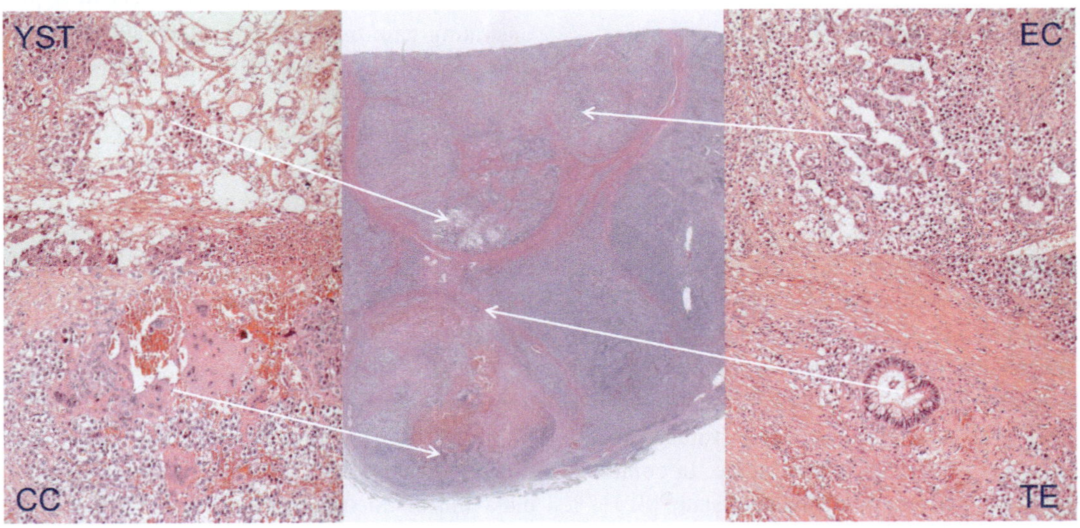

Fig. 3.22 Foci of *EC*, teratoma (*TE*), *YST*, and choriocarcinoma (*CC*) in an otherwise typical testicular seminoma demonstrating the phenomenon of reprogramming to non-seminoma in an invasive seminoma. The reprogramming at multiple sites and into different lineages may be due to aneuploidy of the tumor cells, which upon cell division give rise to daughter cells with different chromosomal constitutions (*middle panel*, H and E ×1; YST, EC, TE, and CC H and E ×200)

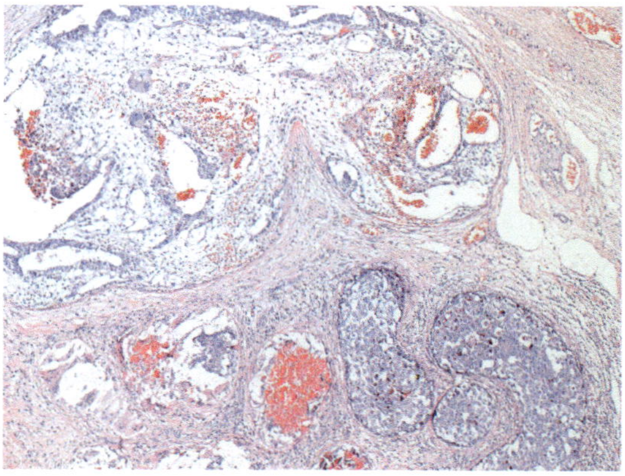

Fig. 3.23 Differentiation of intratubular non-seminoma upon invasion. Intratubular component consists exclusively of EC (*right lower corner*); upon invasion development of teratoma elements, YST, and trophoblastic giant cells, in addition to EC. (H and E, ×100)

may derepress differentiation-promoting genes by removing polycomb repressive complexes recruited to the promotor sites of these genes by the pluripotency proteins OCT4, SOX2, and NANOG [502]. Noteworthy, spontaneous and experimental teratomas in 129 mice also start as intratubular EC and begin to differentiate when the seminiferous tubules are extended and disrupted [81, 183] (Fig. 3.24).

Pathogenesis: Summary

Most of the testicular type II GCT have a developmental origin in the context of TDS, a condition of mild undervirilization, whereby inadequate signaling by Sertoli cells interferes with normal maturation of gonocytes, creating a window for co-expression of embryonal and differentiation proteins, in particular OCT4 and TSPY, combined with upregulation of KITLG, resulting in transformation of gonocytes. Polyploidization is an early event, possibly due to dysfunction of the mitotic to meiotic switch [319, 486], providing survival advantage to the gonocytes in the suboptimal niche. In combination with their hypomethylated state [321], it endows the transformed gonocytes with chromosomal instability, which drives tumor progression through nonrandom gains and losses of (parts of) chromosomes, most conspicuously gain of 12p. This chromosomal region harbors a cluster of genes whose products via various mechanisms convey further proliferative and survival advantage to the neoplastic gonocytes, resulting in GCNIS and intratubular seminoma that by default develops into seminoma. Seminomas may harbor mutations predominantly in *KIT* in up to 25 %, probably most often as a genetic mechanism of tumor progression and only rarely as initiating event. Non-seminoma originates when a neoplastic gonocyte, usually within a seminiferous tubule, is reprogrammed to an EC cell, the totipotent stem cell of non-seminoma, giving rise to intratubular EC that upon invasion of the testicular interstitial tissue may give rise to all extraembryonal and somatic lineages and occasionally the germ lineage.

3.6.2.3 Ovary

Developmental Potential

Ovarian GCT containing dysgerminoma, EC, or choriocarcinoma, either alone or in various combinations with or without YST and/or teratoma, are type II GCT with totipotent developmental potential and can be classified as such on histological grounds. Solid tumors solely composed of (immature) teratoma and/or YST can only be classified with certainty as a type II GCT by demonstrating gain of 12p [503]. However, age and histology do give clues, since in general, pure immature and/or mature solid teratomas are of type I, as they typically lack gain of 12p ([296, 503] for review). Of the pure YST, about 40 % have 12p gains and are therefore type II, while the remaining 60 % are best classified as type I GCT [296]. Tumors combining teratoma and YST can also be of either type; those in infants are likely progressed type I teratomas lacking 12p gain, while the postpubertal ones are likely non-germinomatous type II GCT with 12p gain [296, 305, 503]. As mentioned earlier, in the case of a pure mature solid teratoma, it is important to make this distinction because a type I teratoma is benign and a type II teratoma is malignant.

The ratio between dysgerminoma and non-dysgerminoma, and thus the rate of reprogramming of dysgerminoma, is difficult to establish because usually epidemiologic studies do not

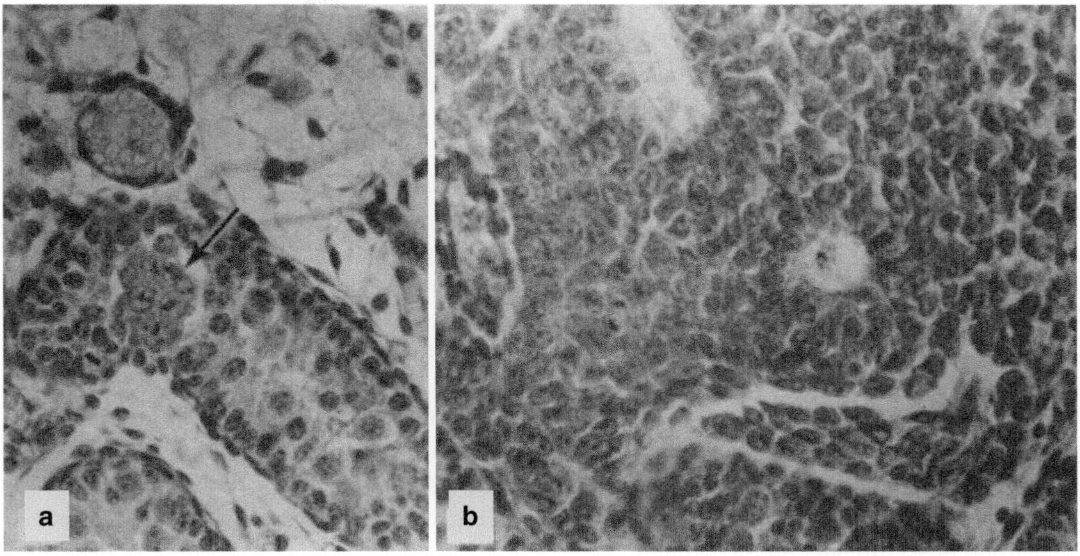

Fig. 3.24 (a) Intratubular GCT, composed of EC (*arrow*) in testicular tubule of 19-day fetus of 129 strain mouse. (b) GCT (in 19-day fetus) which has enlarged and ruptured seminiferous tubule in which it arose and is composed of EC and also more differentiated cells forming tubules [81]

specify the histology to the degree that the distinction between type I, II, and IV GCT is possible. The report by Smith et al. [504] is a notable exception: among 1262 malignant ovarian GCT, the 449 pure immature teratomas and at least 110 pure YST (60 % of 183, based on Kraggerud [296]) should be considered as type I GCT, while the 37 teratomas with malignant degeneration should be type IV GCT with a somatic-type malignancy. That leaves 666 type II GCT: 414 dysgerminomas (62 %) and 252 non-dysgerminomas (73 pure YST, 67 mixed GCT, 52 EC, 27 choriocarcinomas, and 33 EC plus teratoma, so-called teratocarcinomas). The percentage of pure dysgerminoma is higher than the slightly over 50 % seminomas within testicular type II GCT. It suggests that reprogramming in ovarian dysgerminoma, which is convincingly documented [505–508] and also apparent from the mixed GCT with a dysgerminoma component [504], occurs less frequently than in GCNIS and seminoma of the testis. The spectrum of histological types in ovarian and testicular non-seminomatous GCT is similar; however, the distribution is different: mixed non-seminoma is the most frequent histology in the testis and pure YST in the ovary; pure choriocarcinoma seems to be more frequent in the ovary than in the testis. In mixed tumors, the percentage of tumors combining dysgerminoma with just immature teratoma or YST seems to be higher than the combination of seminoma with only immature teratoma or YST in the testis (Chap. 6).

Epidemiology

In the ovary, the second most frequent site of type II GCT after the testis, the incidence is about 20-fold lower than in the testis [76, 137, 509, 510], with reportedly a slight decrease over the past 30 years [76, 504]. Data from South East England for about the same period suggest a rate of increase comparable to that of testicular type II GCT [509]. However, the quoted data on incidence trends are not representative for type II GCT, because in neither of the studies, the trends were specified for the different types of GCT of the ovary. The incidence of dysgerminoma in England was reported stable between 1971 and 1984 [511] and in the same period increasing in Los Angeles County [512]. Therefore, the pathogenetically important question whether or not ovarian type II GCT parallel the increasing incidence of those of the testis in recent decades, due to comparable gene-environment interactions, remains open. In the US, dysgerminoma seems to be more frequent in whites and other nonwhites than in blacks [76], suggesting an ethnic influence, just as in testicular type II GCT.

Type II GCT of the ovary may arise before puberty, likely in DSD [275, 513] (see also Sect. 3.6.3). DSD may remain unrecognized, as in the case of two phenotypically normal females having a 46,XX karyotype, who well after puberty were diagnosed, respectively, with gonadoblastoma with a germinoma [514] and gonadoblastoma with a mixed type II GCT [515]. Ninety-five percent of ovarian type II GCT become manifest after puberty in women with a normal 46,XX karyotype [516], a couple of years earlier than in males, in accordance with the earlier onset of puberty in females [232, 266, 517]. Age of presentation is in the typical order: first, the pure non-dysgerminomas, followed by the mixed non-dysgerminomas/dysgerminomas, and then the pure dysgerminomas with a median age of the latter close to 20 years [511]. Dysgerminomas and non-dysgerminomas are bilateral in over 6 % [129, 131, 518–520], a slightly higher figure than for bilateral type II GCT of the testis, probably due to the contribution of gonadoblastoma-related tumors, as 40 % of gonadoblastomas are bilateral [405]. In fact, in a retrospective study, one out of three bilateral dysgerminomas was associated with bilateral gonadoblastoma [521].

Familial cases are rare: among 18 families retrieved from the literature with at least one female with a GCT [137, 522, 523], eight involved only females and ten both females and males (in about 0.2 % of the pedigrees of familial testicular cancer, a female member has a GCT [524]). The information on the histology of the tumors provided in the reviews and the quoted original case reports allows the distinction between GCT of types I and II. In 12 of the families, at least one case concerned a dysgerminoma, combined with various type II GCT of the gonads in close relatives of both sexes; remarkably, one relative had a mediastinal EC. Also noteworthy, in three families, type II GCT clustered with at least one ovarian type I GCT: three pure (immature) teratomas and in two families with pure YST, which may have been of type I. (The remaining family consisted of a woman with an ovarian type I immature teratoma and her baby with a type I immature teratoma of the brain [136].) These families demonstrate two significant points: type II GCT of the ovary may cluster with similar testicular and extragonadal GCT and also with type I GCT of the ovary.

Risk Factors

An established risk for (bilateral) ovarian type II GCT are the various forms of DSD in phenotypic females who have the GBY region, containing *TSPY*, in their genome, as discussed earlier.

Less certain is that a disturbed hormonal milieu in the mother increases the risk of malignant ovarian GCT in daughters, comparable to the role of hormone disruption in the etiology of TDS and type II GCT of the testis [229, 525]. Reported risks are maternal use of exogenous hormones during pregnancy (OR 3.6), maternal obesity (OR 2.7), early regular menstruation after menarche (OR 1.8), and age at index pregnancy under 20 years (OR 2.8) [526]. Other studies have also reported association of malignant ovarian GCT with reproductive risk factors such as parity, use of contraceptives, ages at first and last births, and time since last birth [527–529]. A more recent study could not link levels of circulating sex hormones with risk of ovarian germ cell cancers [530]. Most studies do not distinguish between type I and II GCT, making the results difficult to interpret.

As discussed above, type II GCT of the ovary lack the strong familial risk of those of the testis. Sporadically, type II and I GCT cluster in families. Both tumor types are so rare that the clustering is probably not by chance, rather these families have susceptibility for both type I and II GCT, pointing to a common cell of origin in different states of developmental potential. The obvious target cell is the PGC/gonocyte, and the familial susceptibility factor might have bearing, for example, on resistance to apoptosis of PGC, such as variants in *BAK1*, which are associated with a higher risk for type II GCT [195, 196] and perhaps also for type I GCT [158].

In one of the families reported by Huddart [524], a male with a testicular type II GCT clustered with a female with bilateral dermoid cysts (type IV GCT) of the ovary. More recently, a similar family was identified, with a father having a seminoma and his daughter metachronous bilateral dermoid cysts [531]. In both families, it may be a chance occurrence in view of the high frequency of dermoid cysts of the ovary, with bilaterality in 10–15 % of the patients [233, 532] (Chap. 6). It is intriguing though that in both females, the dermoid cysts were bilateral, because in familial cases of dermoid cysts where laterality was stated, 11/28 (39 %) were bilateral, and among the patients who were twins or triplets, 9/12 (75 %) were bilateral [134, 135]. One of identical twins [533] had over the years seven dermoid cysts removed from her left and one from her ovary. These case histories are suggestive of a genetic risk factor for bilaterality of type IV GCT of the ovary, which perhaps could also increase the risk for type II GCT.

(Cyto)Genetics/Epigenetics

Over ninety percent of dysgerminomas are aneuploid, often close to tetraploid [534], and in 77 % have gain of 12p and similar gains and losses of (parts) of other chromosomes as testicular type II GCT (for review [86, 296]). Mixed non-dysgerminomas are also most often aneuploid and have gain of 12p in 68 %. This somewhat lower figure as compared to dysgerminoma is probably due to the fact that part of the mixed GCT combining teratoma and YST is type I GCT. Of the pure YST, 41 % have gain of 12p, indicating that less than half are type II GCT. Pure immature teratomas and mature teratomas have gain of 12p in 5 and 9 %, respectively, and are therefore most often type I GCT (for review [296]). Conversely, one out of ten solid mature teratomas of the ovary is of type II and therefore malignant.

As for the other chromosomes in dysgerminoma, the most common changes are gains from chromosome arms 1p (33 %), 6p (33 %), 12q (75 %), 15q (42 %), 20q (50 %), 21q (67 %), and 22q (58 %); gains of the whole of chromosomes 7 (42 %), 8 (42 %), 17 (42 %), and 19 (50 %); and losses from 13q (58 %) [535]. The strong predominance of gains over losses might be explained by the (near)tetraploidy of most dygerminomas.

Somatic mutations of *KIT* in exon 17 codon 816 have been found in 27–33 % of dysgerminomas [330–332] and in codon 822 in an additional 20 % [332], adding up to activating mutations in 53 %, always in unilateral and not in bilateral cases. Among 16 DSD patients with GBY in their genome who developed a dysgerminoma, only one case had a *KIT* mutation, in codon 820; the same mutation was found in the gonadoblastoma from which the dysgerminoma was derived [332]. This may be part of the explanation for the absence of *KIT* mutations in bilateral dysgerminoma, as bilateral tumors are probably in a substantial proportion derived from gonadoblastoma. *KIT* mutations are absent in other malignant ovarian GCT, i.e., immature teratoma, YST, and tumors combining these two components. Gain of 12q in dysgerminoma may be related to the localization of *KITLG* on 12q22, with possible involvement of KITLG in an autocrine loop with KIT. *KRAS* has hardly been investigated in ovarian type II GCT; in two studied dysgerminomas, it was not mutated [536].

Pathogenesis

Type II GCT of the ovary are derived from hypomethylated, erased, premeiotic PGC/gonocytes in the naïve state (totipotent), present in the early developing gonad, as has been convincingly demonstrated for GCT arising in dysgenetic gonads [275]. It is very likely that the phenotypically identical GCT outside the context of DSD have the same cell of origin. Morphology, immunohistochemistry, and expression studies of dysgerminoma have shown the same profiles as in seminomatous GCT of other anatomical sites and in PGC, with high expression of pluripotency factors, in particular expression of OCT4 in combination with SOX17. The non-germinomas have the same profiles as the type II non-seminomas at other sites, with EC cells expressing OCT4 in combination with SOX2 in addition to other pluripotency factors and the derived lineages showing tissue-specific expression patterns (for review [296]).

Five percent of ovarian type II GCT develop in the context of DSD by the "developmental" pathway, related to the presence of GBY and co-expression of OCT4 and TSPY, consistent with the low rate (0.6 %) of *KIT* mutation in these tumors [332]. In a molecular analysis of 45 malignant ovarian GCT in patients without signs of DSD, 32 can be classified as type II GCT. In four of these (13 %, two dysgerminomas and two immature teratomas), *TSPY* was demonstrated in tumor tissue [353]. This would mean that the origin is "developmental," in association with *TSPY*, in about 20 % of ovarian type II GCT (5 % DSD and 13 % clinically silent mosaicism for *TSPY*).

As discussed in the general section on (cyto) genetics of type II GCT (Sect. 3.6.1.4), the *KIT* mutations found in up to 50 % of dysgerminomas [330, 332] may be initiating, but they are probably most often progression related. In view of the absence of KIT mutations in type II nongerminomatous tumors of the ovary [330], it is likely that the situation is not much different from the testis and that also in the ovary, very few type II GCT are caused by somatic mutations. Thus the initiation pathway in over 80 % of ovarian type II GCT in 46,XX phenotypically normal females remains to be explained.

It is tempting to speculate that in the ovary, the majority of type II GCT develop in the context of mild dysgenesis, comparable to TDS of the male, mainly caused by imbalances of factors regulating gonadal development. It has been shown in a mouse model that downregulation of FOXL2, the key protein for maintenance of the female identity of the gonad, results in reprogramming of the ovary in the male direction, even in adults. Granulosa cells are transformed into Sertoli cells forming tubular structures. Remarkably, if in this situation the expression of FOXL2 is restored, the ovarian identity is repaired [537]. If this would happen during gonadal development, even only transiently, in the stage before oocytes become arrested in the prophase of meiosis I, it would create a hypovirilized testis-like environment favoring disturbance of maturation oogonia/gonocytes and a window for the development of a type II GCT. However, due to the absence of GBY/TSPY, at a much lower rate than in males or in DSD patients carrying GBY in their genome, who have an up to 70-fold risk of developing a type II GCT as compared to normal females [275]. Indeed, insight into the factors involved in the development and maintenance of sexual identity of the gonads (DMRT1, FOX9, and SOXL2) lends credence to the hypothesis that as yet pathogenetically unexplained type II GCT of the ovary have their origin in mild forms of ovarian dysgenesis, possibly even transient, which leave no obvious further phenotypical traces (Fig. 3.25). If this hypothesis were true, it would mean that the large majority ovarian type II GCT have a "developmental" origin.

The much lower incidence of type II GCT in the ovary as compared to the testis has been explained by the lower number of susceptible germ cells in the ovary, and the fact that they are blocked in meiosis I, whereas the gonocytes are more numerous in the testis, are arrested in mitosis [509]. This is probably only part of the explanation in view of the epidemiology of the type II GCT of the mediastinum and brain. One might assume that in these sites, the number of target cells is similar in men and women; moreover, the mis-migrated PGC are blocked in meiosis I in both genders [58, 59]. Yet type II GCT of the mediastinum [241] and brain [539] are much more frequent in males than females. As alluded to earlier, this may be due to men having the GBY region on Y and thus being able to express TSPY in combination with OCT4 in the critical stage of neoplastic transformation of erased PGC. This suggestion is consistent with the high risk of developing a type II GCT in dysgenetic gonads containing GBY/TSPY [275].

Also relevant in this context is the observation that type I GCT have a roughly similar frequency in the ovary and the testis and that the sex distribution, apart from the sacral region, is equal for most extragonadal sites, in particular for the mediastinum and brain [94]. It suggests that indeed the number of susceptible cells, probably mis-migrated, pre-erased PGC for type I GCT, is similar in males and females and that GBY has no role in the pathogenesis of type I GCT.

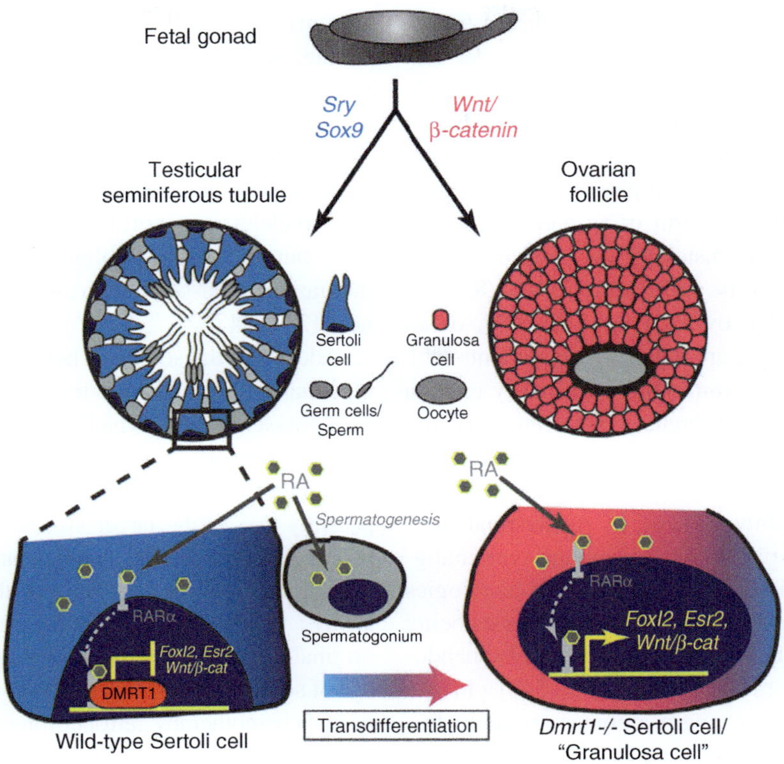

Fig. 3.25 Role of DMRT1, FOX9, and SOXL2 in development and maintenance of ovarian identity. DMRT1 silences RA-dependent feminization genes to ensure postnatal sex maintenance. During fetal sex determination (*1*), the bipotential gonad makes a choice between male (*blue*) and female (*pink*), largely guided by the presence or absence of Sry. The sexual differentiation machinery downstream of sex determination transforms the undifferentiated gonad into a mature testis or ovary (*2*), manifested in the formation of Sertoli-cell-containing seminiferous tubules in the male and granulosa-cell-containing ovarian follicles in the female. Postnatal sex maintenance within Sertoli cells (*3*) is achieved via the silencing of RA signaling-dependent feminization genes (such as Foxl2) by the transcriptional regulator DMRT1. RA is thereby allowed to act in adjacent spermatogonia to promote spermatogenesis within the seminiferous tubule. In Dmrt1 mutant Sertoli cells, however, RA acting through RARa activates feminizing genes and reprograms the Sertoli cell into a granulosa-like cell through the process of transdifferentiation [538]

3.6.2.4 Mediastinum

Developmental Potential

Of the mediastinal GCT in postpubertal male patients, seminomas constitute 32 %, mixed GCT 16 %, EC 4 %, choriocarcinoma 3 %, YST 10 %, and teratoma 35 % [241]. In CGH analysis [165], most mixed GCT in patients older than six had gain of 12p and are therefore type II GCT; in one out of three YST and none of the teratomas in that age group 12p was involved. Thus it can be estimated that in early- and postpubertal patients, seminomas constitute about 55 % and non-seminomas 45 % of mediastinal type II GCT (mixed 28 %, EC 7 %, YST 5 %, choriocarcinoma 5 %). Apart from a higher proportion of pure YST and pure choriocarcinoma, the distribution of histological subtypes is similar to that of the type II GCT of the testis.

Of note, in patients older than six [76, 165], pure (immature) teratoma, pure YST, and mixed GCT combining the two can only with certainty be classified as type I or type II GCT by analysis of 12p status. This is clinically relevant as a type I teratoma is benign and type II teratoma is malignant.

Most intriguingly, 41 patients with Klinefelter's syndrome who had a type II GCT based on histology and secretion β-HCG showed a very different distribution of histological types from the general adult male population, with seminoma 0 %, mixed non-seminoma 44 %, EC 10 %, choriocarcinoma 15 %, YST 2 %, and (immature) teratoma (29 %) [277]. Not only there were no pure seminomas but also only two of the 12 mixed GCT had a seminoma component. Choriocarcinoma/trophoblastic differentiation, either in pure form or as part of a mixed tumor based on histology or elevated serum β-HCG, was very common: 35/41 (85 %). Considering that over 20 % [540, 541], of all mediastinal GCT, are diagnosed in patients with Klinefelter's syndrome, this may partly explain the histological differences between testicular and mediastinal type II GCT in adolescent and adult males.

Ten to twenty percent of mediastinal non-seminomas develop a solid somatic-type malignancy [542, 543]. The distribution of histologies is largely similar as in the testis, sarcomas being the most frequent type and among them rhabdomyosarcoma ranking first [430, 543]. In the mediastinal cases, angiosarcomas are more frequent than in the testis [241, 544]. The associated GCT are most often mixed non-seminomas with a (immature) teratoma component and rarely pure YST or seminoma [430]. Patients with Klinefelter's syndrome may develop somatic-type malignancies other than hematopoietic malignancies.

Hematologic malignancies develop in 2–6 % [543, 545–547], sometimes combined with a sarcoma [388]. These hematological somatic-type malignancies are uniquely associated with mediastinal YST, either pure or as part of a mixed non-seminoma [265, 543, 547–549]; the most frequent types being megakaryoblastic leukemia, followed by malignant and benign histiocytosis and myelomonoblastic leukemia among many other types encompassing virtually all hematopoietic lineages [543]. Except associated with mediastinal non-seminomas, hematologic malignancy has been reported only once, in association with a suprasellar dysgerminoma [550]. Hematologic malignancies are usually diagnosed at a median time of 6 month after primary treatment [549]; about 40 % are synchronous with the primary mediastinal non-seminoma [543]. The rate is comparable in patients with and without Klinefelter's syndrome, as about 20 % of the hematologic malignancies are diagnosed in patients with this syndrome, indicating that it is indeed the mediastinal localization that predisposes to this somatic-type malignancy. These hematologic malignancies are not treatment related [546, 549] but a peculiar biologic characteristic of mediastinal non-seminomas with a YST component [241, 547].

Added up, the solid and hematologic somatic-type malignancies develop in a quarter of mediastinal non-seminomas, which is about sixfold of the 3–6 % in primary testicular non-seminomas [390]. The higher rate of somatic-type malignancies is possibly due to the larger size at surgery of the non-seminomas in the mediastinum than in the testis [551] and due to the fact that surgery is always preceded by chemotherapy. Indeed, in originally testicular non-seminomas, the percentage of somatic-type malignancies increases to 8 % in postchemotherapy retroperitoneal lymph node dissection specimens [429] and to >20 % in late relapses [386], approaching the rate of somatic-type malignancies in mediastinal non-seminomas.

The poorer prognosis of mediastinal than testicular non-seminomas may be because of the larger size at clinical manifestation and the overall higher rate of somatic-type malignancies, which are largely resistant to the chemotherapy given for non-seminomas [354].

Epidemiology/Risk Factors

With an incidence of about 0.12 in white and 0.05 in black males, which has not increased in the past decades [76], the mediastinal type II GCT constitute 50–70 % of extragonadal type II GCT [552]. Over 95 % occur in men [241, 308], which is true for whites and blacks [76]. The mean age for seminomas is about 30 [264] and for non-seminomas 25 years [265]. The mean age of patients with Klinefelter's syndrome, always with mediastinal non-seminomas, is 17 years (range 4–31), substantially younger than in non-seminoma patients without Klinefelter's syndrome. In all Klinefelter cases younger than 12, there was precocious puberty [553], due to β-HCG produced by the tumor [277].

In a large Danish cohort, mediastinal non-seminoma was the only cancer for which Klinefelter's syndrome conferred a higher risk compared to males without this syndrome, with a relative risk of 67 [554]. Over 20 % of mediastinal type II GCT are associated with Klinefelter's syndrome [540, 541], meaning that roughly half of the mediastinal type II non-seminomas occur in this context, as pure seminomas do not seem to occur in Klinefelter's syndrome. The age distribution of Klinefelter patients with mediastinal GCT is bimodal. However, the early peak (between age 4 and 10) is later than in typical type I GCT; moreover, the tumors in the early peak are not the usual type I GCT but more like type II GCT, based on the histological composition and the secretion of β-HCG causing precocious puberty [277]. Remarkably, the two mixed GCT with a seminoma component occurred in two 8-year-old boys. Klinefelter's syndrome also increases the risk of type II GCT of the brain [555, 556].

Sporadically, mediastinal seminomas have been diagnosed in individuals with Down's syndrome [557, 558]. Neurofibromatosis 1 [559, 560] and Li-Fraumeni syndrome [561, 562] are not associated with mediastinal type II GCT.

There is one report of a patient with mediastinal type II non-seminoma associated with only GCNIS in one testis, suggesting that the tumors were two independent primary type II GCT [212]. Indeed, contrary to retroperitoneal type II GCT, which are metastatic from unrecognized testicular tumors, mediastinal type II GCT are normally not associated with GCNIS of the testis [211, 563]. König et al. [564] report a mediastinal non-seminoma ("teratocarcinoma") and a metachronous pituitary stalk germinoma in a patient with Klinefelter's syndrome, probably both related to the underlying syndrome.

There is one patient, mentioned earlier, with a mediastinal EC who had a sister with an ovarian dysgerminoma [287], demonstrating that in some families, mediastinal type II GCT may cluster with type II GCT at other anatomical sites. To the best of our knowledge, clustering with other GCT types has not been reported [137].

Anatomical Distribution

Primary mediastinal type II GCT are only localized in the anterior or anterosuperior mediastinum in association with the thymus [241]. Occasionally, small tumors are completely localized within the thymus, showing that they had their origin in the thymus itself [241]. Concurring with this conclusion is the observation that in about a quarter of all mediastinal seminomas, remnants of the thymus can be identified in the periphery of the tumor [565]. Thymic cysts have been found in 10 % of seminomas [566] and occasionally in YST of the mediastinum [567], again, indicating that these tumors originate in the thymus. Apparently, in most cases, the thymic origin is obscured by tumor overgrowth.

Mediastinal type II GCT are indeed primary tumors, as testicular type II GCT only rarely metastasize to the anterior mediastinum and only simultaneous with metastases in the visceral mediastinum [568].

(Cyto)Genetics

In a study of 19 malignant mediastinal GCT, 14 were definite type II GCT, based on histology and expression of β-HCG, with the expected distribution of histological types [297]. Four tumors were (near)diploid, six (near)tetraploid, three had a (near)diploid plus a (near)tetraploid stem line, and one tumor had two hypertriploid stem lines. This pattern is completely different from type II GCT of the testis, where on average, the seminomas (and GCNIS) are hypertriploid (DNA index 1.61) and the non-seminomas hypotriploid (DNA index 1.40) [294, 569]. This suggests that the precursor cells do not necessarily undergo tetraploidization and that following that event they undergo less extensive of karyotype evolution, whereby the original (near)diploid tumor cells still may coexist with the derived (near)tetraploid clone.

Results of karyotyping are in agreement with these ploidy data, with chromosome numbers being (near)diploid [307, 570–572], (near)tetraploid [573], (near)diploid plus (near)tetraploid [572], and hypertriploid [574]. Karyotyping, CGH analysis [165], and FISH [308] show i(12p) as the most common structural aberration in type II GCT of the mediastinum; other recurrent changes are gains of chromosomes 21 and X and loss of chromosome 13, similar to type II GCT of other sites. In the study by Schneider [165], two patients had Klinefelter's syndrome, as did the case karyotyped by Mann [571], partly explaining the extra copies of X. Different from testicular cases is that gain of 12p may be lacking [571, 574] and that fewer chromosomes are involved in gains and losses, again consistent with less extensive karyotype evolution. Mediastinal non-seminomas in patients with Klinefelter's syndrome may [165] or may not have i(12p) [571].

The demonstration of gain of 12p (in particular i(12p)) in solid and hematologic somatic-type malignancies, often in combination with genetic hallmarks of the somatic cancer, proves their origin from the GCT [575, 576].

Three out of eight mediastinal seminomas (38 %) had activating *KIT* mutations (exon 17 and codons 818, 820, and 822) [333]; 1/13 seminomas (8 %) had a *KRAS* mutations (exon 1, codon 13) [336]. Essentially the same pattern of somatic mutations as demonstrated for testicular seminomas [295]). Non-seminomas of the thymus have not been studied for the presence of mutations.

There are no published studies addressing the epigenetics of mediastinal type II GCT.

Pathogenesis

The occasional finding of mediastinal type II GCT entirely within the thymus and the frequent thymus rests in these neoplasms indicates that they have their origin in the thymus. Apparently, this organ offers a niche in which hypomethylated, erased, premeiotic [571] PGC, the precursor cells of type II GCT, may survive [56] as already proposed by Teilum [54]. The fact that mis-migrated PGC have been demonstrated in large numbers in the anterior mediastinum but not in the thymus itself [52] does not rule out this possibility. The tumors probably originate from very few PGC that manage to escape their normal fate of apoptosis by ending up in the thymus. There is circumstantial evidence to suggest that thymic epithelium may have the capacity to support erased, premeiotic PGC. Seminoma cells, the neoplastic counterparts of PGC, tend to home in thymic epithelium [56]. More convincing still is the finding of "seminoma-like" cells enclosed by thymic epithelium in the absence of an accompanying invasive type II GCT [241], resulting in lesions resembling gonadoblastoma of the dysgenetic gonad. Moreover, like Sertoli and granulosa cells, thymic epithelium produces KITLG, the survival and growth factor of PGC [56] (Fig. 3.26).

The assumption that mediastinal type II GCT arise from PGC is more credible than the proposed alternatives. Origin from a primordial cell of the thymus (for review [577]) is unlikely in view of the close phenotypic and genetic resemblance of the mediastinal type II GCT with their counterparts in the gonads, of which the origin from PGC/gonocytes is undisputed.

The hypothesis of origin through dissemination of early gonadal lesions, which recapitulate embryonal memory and reverse migrate to thymus [578], was immediately refuted [579]. This idea is conflicting with the absence of testicular type II GCT in patients with Klinefelter's syndrome combined with their 67-fold risk of mediastinal type II GCT [554]. In addition, the presumed testicular precursor lesions are typically absent in patients with mediastinal type II GCT [211, 563]. Finally, Chaganti et al. [578] stressed the identity of karyotypic changes of testicular and mediastinal type II GCT, which they are not. As discussed, mediastinal type II GCT are characterized by a shorter karyotype evolution than their testicular counterparts. For a mediastinal GCT to be derived from a testicular precursor lesion, one would expect the reverse pattern, with the testicular lesions being in an earlier stage of karyotype evolution than the mediastinal tumors.

Mediastinal type II GCT, like those in the gonads, usually originate via the "developmental" pathway. The almost exclusive male patient population is consistent with a crucial role of TSPY as in the testis and gonadal dysgenesis in 46,XY DSD, in combination with KITLG stimulation. It cannot be excluded that some tumors are initiated by somatic mutations in *KIT* or in genes

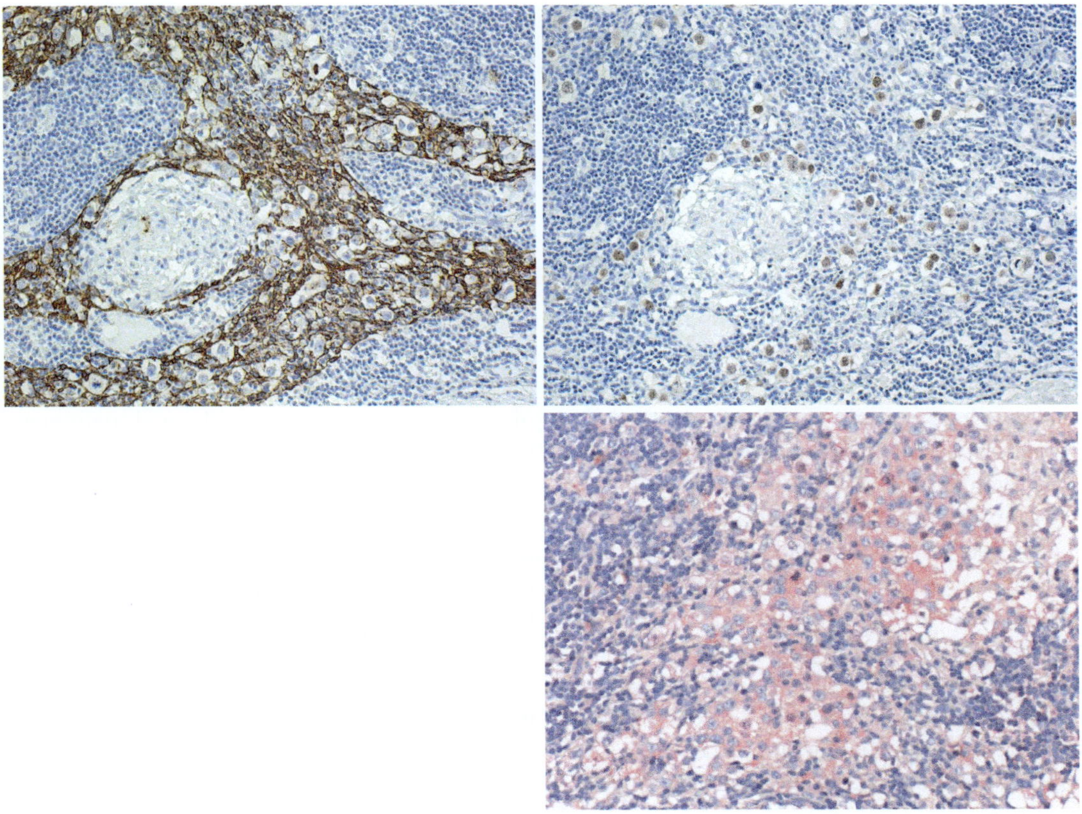

Fig. 3.26 Seminoma cells homing in thymus epithelium; clockwise: cytokeratin-positive thymus epithelium (*brown*) enclosing seminoma cells; same area with OCT4-positive nuclei (*brown*) of seminoma cells; same case expressing KITLG (*red*) in thymus epithelium (original magnification ×200) [56]

of its downstream signaling proteins; however, most *KIT* mutations are probably engaged in the progression of seminoma. Non-seminomas, also in the mediastinum, result from reprogramming of a seminomatous precursor cell, i.e., a transformed PGC, into an EC cell.

The 67-fold risk of mediastinal type II GCT in Klinefelter's syndrome [554] supports the notion that these tumors usually have a developmental origin. The syndrome, affecting 1 in 500 males and diagnosed in only a quarter of the cases [277], is characterized by hypergonadotropic hypogonadism, small testes, infertility, gynecomastia, abnormal body habitus, and mild developmental abnormalities due to an abnormal sex chromosomal complement, usually 47,XXY. Relative androgen deficiency at least at the testicular level accelerates degeneration of the testis at the onset of puberty [580]. In adolescents and adults, most tubules become atrophic with very few or no germ cells left. This may explain why testicular type II GCT are rare in these patients with only sporadic cases reported: one seminoma and two non-seminomas [581–583]. It seems that like the normal spermatogenic cells, precursor cells of GCT cannot survive in the defective spermatogonial niche.

At the same time, it is hypothesized that in Klinefelter's syndrome, the increased levels of gonadotropins, which physiologically stimulate germ cell proliferation, promote malignant transformation of PGC in the thymus [165, 277]. Indeed, the tumors develop at a much younger age than in the testis, before or at the onset of puberty and thus before degeneration of the precursor cells. Hypothetically, the absence of pure seminomas [553] may be due to a similar process of degeneration as in the testis that eventually destroys all precursor cells except those that have undergone reprogramming to EC cells, the more apoptosis-resistant totipotent stem cells of non-seminoma. Supporting this hypothesis, the only mixed GCT that contained a seminoma component were in two 8-year-old boys.

The reportedly higher incidence of type II GCT of the brain in Klinefelter's syndrome [556] is possibly also caused by the hypergonadotropic stimulation. It remains elusive why as opposed to the mediastinal ones, the brain GCT in Klinefelter's syndrome have a similar age distribution and histology as in the general population, with more than 80 % seminomas [584].

From a different angle, it has been suggested that the extra copy/copies of chromosome X in Klinefelter's syndrome play a direct role in the pathogenesis of the type II GCT because the region Xq27 harbors a susceptibility locus for testicular germ cell neoplasms [585]. However, linkage to this locus was not confirmed in a larger set of pedigrees, which included the original 66 families [586], and a candidate gene has not been identified.

Solid somatic-type malignancies are most often due to further tumor progression of somatic components of non-seminomas but may also arise from progression of YST, angiosarcoma, for example [544]. The unique association of mediastinal mixed non-seminomas containing YST, or pure YST, with hematologic malignancies is as yet unexplained.

It has been suggested that the angiosarcomas may arise in myxoid/mesenchymal foci of YST, called magma reticulare by Teilum (quoted in [241]). This tissue has vasoformative capacity, whereby dysplastic spindled and epithelioid cells condense into vessels. In the mouse, the first adult hematopoietic stem cells arise from the endothelium of the major vasculature, in particular the aorta [587, 588]. It is tempting to speculate that the higher incidence of angiosarcoma and hematological malignancies in mediastinal YST (component) are related and that endothelial cells developing in magma reticulare are the source not only of angiosarcoma but also of hematopoietic stem cells [297, 576], which may progress into hematopoietic malignancies (Fig. 3.27). The question remains: why only in mediastinal non-seminomas? Could it be that local factors in the anterior mediastinum, which regulate the specification of adult hematopoietic stem cells in normal embryogenesis, also, in the early stages of tumor development, promote the development of neoplastic hematopoietic stem cells from suitably primed endothelial precursors? Human thymic epithelial cells could be a source of such factors as they reportedly produce granulocyte and macrophage colony-stimulating factors [589].

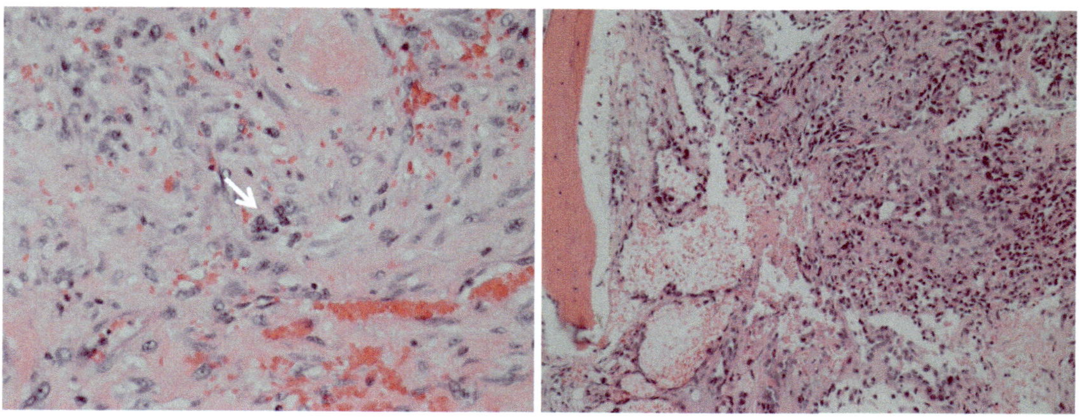

Fig. 3.27 Mediastinal non-seminoma. *Left*: primary tumor with florid vascular proliferation (magma reticulare) with groups of blast-like cells suggestive of hematopoietic cells (*arrow*) (H and E, ×400); *right*: bone metastasis of same patient with the histology of angiosarcoma 7 months after initial treatment (H and E, ×200)

3.6.2.5 Central Nervous System

Developmental Potential

Type II GCT of the brain have the same totipotent developmental potential as those of other anatomical sites. The proportion of seminomas (called germinomas in the CNS), 80 %, is higher than in other sites, apart perhaps from the dysgenetic (intra-abdominal) gonad. This figure may be a slight overestimation because the histological diagnosis of GCT of the brain is often made on small biopsies not always representing all components of the tumors. It explains the occasional event of recurrences having a different histology from the original tumor, such as a germinoma recurring as YST [590], a YST as growing teratoma [591], and in particular teratoma as germinoma, sometimes after a long interval [592–596].

Most of the non-germinomatous tumors are mixed (54 %), followed by mature teratoma (21 %), immature teratoma (8 %), EC (6 %), YST (3 %), and choriocarcinoma (3 %). Seventy-five percent of the mixed GCT have a germinoma component, most often combined with, usually immature, teratoma, YST and/or EC; a choriocarcinoma component is rare [263]. This observation is consistent with the pathogenetic mechanism, whereby non-germinomas result from reprogramming of a germinomatous precursor cell.

Somatic-type malignancies, sarcoma among others, develop rarely; the reported squamous cell carcinomas may have developed in late type I GCT [263]. There is an exceptional case report, mentioned before, of a black female who presented with a mixed lineage acute myeloid leukemia 4 months after treatment of a suprasellar dysgerminoma, obviously not related to Klinefelter's syndrome [550].

GCT of the brain (except those of the spinal cord) in patients with Klinefelter's syndrome have about the same distribution of histological subtypes as in non-Klinefelter cases, with 6/7 being dysgerminoma (86 %), at the median age of 15 years (range 12–35) [584]. This is remarkably different from the mediastinal type II GCT in Klinefelter's syndrome, which are exclusively non-seminomas, at variance with the general population [553], and may appear before puberty [277]. Their predominant germinoma histology by itself makes it less likely that leukemias develop in association with GCT of the brain in Klinefelter patients. Indeed, this phenomenon has, to our knowledge, not been reported.

In Down's syndrome, type II GCT of the brain have a lower percentage of germinomas than in the general population (6/11, 55 %), in agreement with the atypical anatomical localization of the tumors, with only one tumor in the pineal gland; of the five non-germinomas, there were four YST and one teratoma [597].

Epidemiology

Historically, for GCT of the brain, a five- to eightfold higher frequency in the Far East than in Western countries has been reported [598–600]. The last WHO classification [243, 601] quotes a twofold difference (0.17 and 0.09 in Japan and the USA, respectively). A recent comprehensive epidemiological study based on four large databases, two from Japan and two from the USA, demonstrates that the incidence is not significantly different in both countries: 0.143 for males and 0.046 for females in Japan and 0.118 for males and 0.030 for females in the USA [539]. In these figures, type I and II GCT are combined; however, >90 % are probably type II, as less than 3 % of the registered cases, often perinatal (immature) teratomas, are below the age of 5 years [602]. Virtually all pineal tumors are type II; indeed, they hardly occur under age six [76]. In Japan, the incidence of CNS GCT has reportedly increased in the 1980s, mostly in males, and has plateaued since [603]. The median age is about 15 years both for males and females [243, 539], whereby patients with germinomas, mixed GCT, and non-germinomas are on average 18, 15, and 12 years, respectively [263].

The overall male-to-female ratio of CNS GCT is about 4:1 [243, 539, 601]. In Japan, all histological types occur predominantly in males [602]. The incidence of non-malignant GCT is similar in males and females, 0.029 and 0.020, respectively. The male-to-female ratio for malignant GCT is 16:1 in the pineal region and 2.1:1 in the rest of the CNS. More than half of all malignant GCT are located in the pineal region [539].

Multifocal tumors in the brain usually involve the pineal gland and the suprasellar region (so-called bifocal tumors) or rarely both basal ganglionic regions, simultaneously or sequentially [243, 604, 605]. Without molecular analysis, it is virtually impossible to prove that the tumors are independent primaries; locoregional metastasis or recurrence is usually the more likely explanation [594].

A series of case reports demonstrates that type II GCT of the brain may be combined with primary, usually type II, GCT at other anatomical sites. The brain tumors were always germinomas located in the pineal gland or immediate vicinity and followed by a seminoma of the testis (three cases) [283, 606, 607], a non-seminoma of the testis [608], a mediastinal non-seminoma [609], or preceded by a mediastinal seminoma [610]. In one case, the brain germinoma was followed by a mediastinal type I YST, in view of the demonstrated 1p36 deletion and the absence of gain of 12p in the latter [611]. In one patient with Klinefelter's syndrome, the germinoma of the brain was simultaneous with a non-seminoma of the mediastinum, and in another, it was followed by a mediastinal pure choriocarcinoma [612]. In a patient with Down's syndrome, the germinoma of the brain was simultaneous with an EC of the testis [613].

In view of the relative frequency of testicular and mediastinal type II GCT, the latter are clearly overrepresented, which is plausible: if the conditions are favorable for the development of extragonadal type II GCT, e.g., by hypergonadotropism, they could stimulate their development both in the mediastinum and brain.

In the patient described by Coffey [283], the pineal germinoma had the same KIT mutation as the testicular seminoma diagnosed 6 month later, suggestive for a common cell of origin, as mentioned earlier.

There are occasional reports of familial clustering of type II GCT of the brain: two brothers with a teratoma with germinoma elements in the pineal region [614]; three brothers, one with an EC and two with a dysgerminoma in the pineal region [615]; and a boy with a germinoma of the basal ganglia and his sister with a germinoma of the suprasellar region [616].

Risk Factors

In Klinefelter's syndrome, type II GCT, which constitute only few percent of brain tumors in the general population, are the most frequent tumors of the brain (median age 16), implicating an increased risk in this syndrome [555]. This is perhaps also true for spinal cord germinomas, although the high age of the patients (median 29 year) suggests that the spinal tumors might be cerebrospinal fluid-borne metastases of subclinical germinomas of the brain [584].

Patients with Down's syndrome have an increased risk of leukemias and a lower risk of solid cancers [557, 558]; the latter is attributed to overexpression of the *DSCR1* and *DYRK1a* genes on chromosome 21, which suppress the production of VEGF and thereby angiogenesis, which sustains solid tumor growth [617]. Exceptions among the solid tumors are lymphomas and GCT of the brain [557] and possibly the testis [276, 558].

Isolated case reports have associated type II GCT of the brain with neurofibromatosis 1 [618] and multiple congenital melanocytic nevi [619].

Anatomical Distribution

Over 80 % of type II GCT of the brain are located in the pineal gland, suprasellar region (neurohypophysial axis; occasionally within the neurohypophysis), hypothalamus, and the wall of the third ventricle. In these midline structures, the large majority are in male patients, malignant, and germinomas. Germinomas occur also in the basal ganglia, cerebral hemispheres, and in the posterior fossa; however, in these and other atypical, non-midline anatomical sites, the proportion of females, non-germinomas and benign GCT (the latter probably of type I), is somewhat higher [243, 539]. In fact, type I GCT of the brain have an anatomical distribution resembling that of the type II non-germinomas [132]. It seems that germinoma involving the basal ganglia is more frequent in Asian than Western children [604, 605].

The anatomical distribution of type II GCT in Klinefelter's syndrome, 6/7 in the pineal, suprasellar, and hypothalamic region, is similar as in the general population [584].

In Down's syndrome, where there is a high percentage of non-germinomas, the tumors lie most often outside the typical midline sites [597].

(Cyto)Genetics

Half of the type II GCT of the brain are (near) diploid and half (near)tetraploid, which is true for germinomas, mixed GCT, and non-germinomas [197]. This finding is consistent with the young age of clinical manifestation of these tumors (median 15 years) and therefore shorter period of karyotype evolution than in their testicular and mediastinal counterparts. The pattern of gains and losses of (parts of) chromosomes is largely similar to that in type II GCT of other anatomical sites: in order of frequency, gain of 12p, 1q, 8, 21, and X, with more defined regions of gain being 12p12, 1q11-q24, and 8q11-q21, and loss of 11q, 18q, and 13 [310]. In addition, Wang et al. [197] found gain of 14q and loss of 10q and 17p, thereby making the picture even more similar to type II GCT of other sites. A comprehensive analysis by SNP microarray further confirms and details this data. The most frequently observed copy number gains were regions on chromosomes 1q (44 %), 2p (37 %), 7q (37 %), 8q (41 %), 12p (59 %), 14 (33 %), 20q (30 %), 21 (63 %), 22 (41 %), and Xq (44 %). Frequently observed copy number losses were regions on chromosomes 1p (26 %), 4q (26 %), 5q (33 %), 9q (30 %), 10q (37 %), 11q (41 %), and 13 (48 %) [298].

Gain of 12p is present in 9/17 (53 %) type II GCT studied by Schneider et al. [310], in 5/15 (33 %) unequivocal type II GCT (germinomas and mixed GCT) in the material of Wang et al. [197], and in 59 % in the study by Terashima et al. [298]. These figures are significantly lower than in other sites, again consistent with less extensive karyotype evolution.

Activating *KIT* or *RAS* mutations occur frequently in germinomas (60 %) and less so in non-germinomas (9 %); they are mutually exclusive in both [334]. By next-generation sequencing, Wang et al. [197] confirmed and extended these data. It appears that activating mutations in *KIT* (47 %) (in order of frequency in exons 17, 11, 18, and 13), *KRAS* (18 %), and *NRAS* (6 %) and inactivating mutations in *CBL* (6 %), a negative regulator of KIT expression, are all mutually exclusive and occur most often in germinomas and mixed GCT that lack gain of 12p. The complementary character of these genetic events and their preferred occurrence in germinomas and mixed GCT indicate that they are probably not initiating but engaged in the progression of germinoma. The AKT/mTOR signaling pathway is activated in about 20 % of cases, mostly by focal amplification of 14q32.33, containing the *AKT1* locus, often in tumors lacking gain of 12p. Less frequently, loss of function mutations were identified in *BCORL1*, *MTOR*, *TP53*, *SPTA1*, *KDM2A*, and *LAMA4* [197]. BCORL1 is a tumor suppressor and a transcriptional corepressor, of which the mutation might interfere with the function of nuclear receptors such as the AR [620].

The study by Terashima et al. [298] highlights frequent gain of *PRDM14* on 8q13, which was earlier identified as a susceptibility locus for testicular type II GCT and also frequent aberrations of *CCND2* (12p13) and *RB1* (13q14) suggesting that the cyclin/CDK-RB-E2F pathway might be involved in the pathogenesis of type II GCT of the brain. Finally, Wang et al. [197] found in their cohort a rare germline variant of *JMJD1C* in Japanese patients that functions as a chromatin modifier gene interacting with the AR in humans [621] and that in mice is involved in long-term maintenance of male germ cells [622]. This gene variant is reportedly enriched in the Japanese population and fivefold higher in the Japanese patients with GCT of the brain compared to the general population. The authors propose that this variant of *JMJD1C* might explain the higher incidence of GCT of the brain in Japan [197]. It seems, however, in view of recent incidence figures [539], that there is no significant difference to be explained.

Pathogenesis

The strong phenotypic [352] and genotypic [197, 298, 310, 334, 623] resemblance of type II GCT of the brain and the gonads makes it plausible that they share the same cell of origin: a hypomethylated, erased, premeiotic, totipotent PGC [56], as already proposed by Teilum [54]. Indeed, in 7–14 wpc, human embryos and fetuses mis-migrated PGC that have escaped elimination by apoptosis can be seen in the midline of the CNS. Probably, these PGC have arrived here because they failed to exit the sympathetic trunk at the gonadal site and continued migration in cranial direction along other nerve branches from the sympathetic trunk [52]. Apparently, the midline of the brain, in particular the pineal gland and the suprasellar region, offers a niche with conditions where some of these PGC can survive long enough to give rise to germinoma, which upon reprogramming may go on to develop non-germinoma. The high proportion of pure germinomas (80 %) may be due to the young age of the patients (median age 15 years). Consequently, the time for progression to non-germinoma via reprogramming is short, as in patients with gonadal dysgenesis. This model may also explain why type II GCT in sites away from the midline are rarer and more often non-germinomas. In these non-midline sites, the conditions are supposedly less suitable for neoplastic PGC, favoring precursors that have undergone reprogramming to an ESC-like precursor, with a developmental potential in between the totipotent and the pluripotent state. Indeed, the spectrum of histologies of type II non-germinomas has resemblance to that of type I GCT: relatively high proportions of pure (immature) teratoma, pure YST, and pure choriocarcinoma and rarely pure EC. Yet these tumors occur most often beyond the age of six and have the (cyto)genetic characteristics of type II GCT. In fact, in the brain, particularly away from the midline, there seems to be a gray area with a gradual transition between GCT of type I and II, featuring tumors that are type II by genotype and age but resembling type I by phenotype.

An alternative to the hypothesis that all type II GCT develop from PGC proposes that only germinoma stems from PGC and that the other tumor types develop from corresponding embryonic rests that get incorporated in the developing neural tube through folding errors. In this manner, choriocarcinoma would arise from misplaced trophoblast, YST from patches of yolk sac, and EC and teratoma from fragments of the embryo proper [624]. However, such misplaced elements, contrary to PGC, have never been detected in the developing CNS. Reprogramming of germinoma cells, the neoplastic recapitulation of physiological process in embryonic development, is a more plausible explanation for the origin of the different types of non-germinoma, if only because this mechanism applies also to gonadal type II GCT where misplacement of embryonic rests has no bearing. Nevertheless, the embryonic misplacement hypothesis cannot be totally dismissed. Sporadically, growths have been reported in the head and neck region, including the brain [96], with a morphology in principle compatible with type I or II (immature) teratoma that turned out to be mono- or even dizygotic twins [102]. In such cases, misplacement of a zygote or a blastomere during embryonic development is a plausible pathogenetic mechanism.

More recently, it was proposed by Scotting and colleagues that all GCT of the brain have their origin in NSC that during embryonic development have been induced to pluripotency by activation of OCT4 through demethylation of its promoter region [139, 168]. Indeed, Kim et al. have shown that mouse [71] and human [72] NSC can be induced to pluripotency by activation of OCT4 alone, as mentioned earlier, and when grafted into mice, these stem cells give rise to teratomas. According to Scotting et al., teratomas may develop in the brain via the same mechanism, which subsequently can give rise to all other types: germinoma and the various mixed and pure non-germinomas. To support their hypothesis, they quote a series of case reports to show that each type of GCT can recur as any of the other types of GCT, except germinoma recurring after resection of a YST. None of the cases, however, prove the crucial point of teratoma or rather the stem cells of teratoma, giving rise to germinoma. In two of the quoted cases, the authors of the case reports themselves conclude that the pineal tumor, originally diagnosed as teratoma, was probably mixed and that the germinoma component was missed in the initial intervention. This was histologically likely in the case described by Janzarik et al. [595] and clinically suspected in the case reported by Mao et al. [596]. In the three cases where the original teratoma and the later diagnosed germinoma were anatomically separate, the authors interpret the germinoma as a second primary tumor [592–594], admitting that a late relapse cannot be excluded, as seminoma/germinoma usually has a long protracted clinical course.

The biology of metastasis of type II GCT, of the testis in particular, shows that the seminoma component in a mixed primary non-seminoma disappears over time in the evolution of the tumor [271]. Residual mature teratoma after chemotherapy for non-seminomatous germ cell tumors of the testis occurs significantly less often in the lung than in retroperitoneal lymph node metastases [272]. On the other hand, a pure seminoma may undergo reprogramming to all non-seminomatous components [244], even in metastatic sites, in up to 44 % of the cases [258], as mentioned earlier.

Consistent with the lack of clinical evidence of germinoma originating from teratoma, this phenomenon has never been reported in the decades of research in mouse models of teratoma, starting with Stevens' seminal work [80]. Also, the many iPSC that have been tested for their developmental potential by grafting into mice have never produced seminomatous tumors [321, 625]. iPSC appear to be in the primed state by default, with a pluripotent development potential, typical for type I GCT. It is therefore very unlikely that induced pluripotent NSC would give rise to type II GCT. They may perhaps be the origin of perinatal/infantile, type I teratoma GCT of the brain, if it is true that NSC, particularly in the midline of the brain (not the typical site of the type I GCT), have partial loss of imprinting [169]. The quoted observations are consistent with specification of the germ lineage being a tightly controlled process that is unlikely to spontaneously occur in a teratoma developed from an iPSC (for review [35]). Reprogramming of PGC to pluri- or totipotency on the other hand is a kind of default pathway for which molecular mechanisms are in place to prevent it from happening ([39], for review [20]). Finally, if all germinomas of the brain were to originate via a teratoma stem cell, one would expect a smaller percentage of the GCT of the brain being pure germinoma/seminoma than in other sites, while in fact in the brain, this figure is the highest. It is suggested that *KIT* mutations could bias induced pluripotent NSC toward developing germinoma [139]. However, *KIT* mutations are effect rather than cause of germinoma development, as they are mostly engaged in progression of seminomatous GCT and rarely in initiation of type II GCT, as appears from molecular analysis of GCT of the testis [295, 324] and brain [197].

Like in other sites, probably the majority of type II GCT of the brain have a "developmental" origin with the same arguments in support. The overwhelming preponderance of male patients, like in the thymus, suggests a similar role for co-expression of OCT4, TSPY, and KITLG in maturation-disturbed PGC, as in the hypovirilized conditions in the testis in TDS and in 46,XY gonadal dysgenesis.

The increased risk of type II GCT of the brain in patients with Klinefelter's and Down's syndrome is consistent with a developmental origin of these tumors. In both conditions, disturbed development of the gonads causes increased secretion of gonadotropins in the diencephalic centers at the inception of puberty. These hormones are supposed to stimulate the neoplastic transformation of mis-migrated PGC in the midline of the brain [165, 243, 277].

In addition, it has been speculated that overdose of certain, as yet unidentified, genes on chromosomes X and 21 might favor the development of type II GCT. Indeed, both are among the most frequently overrepresented chromosomes in type II GCT in the general population, regardless of anatomical site. Patients with Klinefelter's and Down's syndrome seem to have a constitutional "chromosomal advantage" for developing type II GCT. Indeed, the only malignancies for which Klinefelter patients have an increased risk are mediastinal [554] and brain type II GCT [555, 556]. In Down's syndrome, apart from leukemias and lymphomas, there is only an increased risk for type II GCT of the brain and probably the testis [276] [557], [558].

Probably some tumors are initiated by somatic mutations, as in the earlier quoted patient with the same activating *KIT* mutation in his testicular seminoma and pineal germinoma [283]. In this case, the mutation has likely occurred in migrating PGC thereby enabling them to not only reach the gonads but also the pineal gland.

Somatic mutations are rare in type II GCT of the brain with 0.50 non-silent mutations per Mb [197], the same figure as in testicular type II GCT [295]. The majority of mutations occur in germinoma and can be explained as involved in the progression of germinoma rather than as initiating events. Initiating mutations should be as frequent in germinoma as in non-germinoma, as the PGC is the precursor for both. The mutation rate in non-germinomas, less than 10 % [197], is therefore a fair indication for the maximum rate of type II GCT of the brain initiated by a somatic mutation. Probably the rare type II GCT in females without mosaicism for Y, where TSPY obviously is not involved in the pathogenesis [352], are more often caused by somatic mutations, particularly in *KIT*, like in the ovary [332].

Incidental reports on association of NF1 with type II GCT of the brain underscore the pivotal role in the development of these tumors of the KIT/RAS signaling pathway, of which NF1 is a negative regulator (Fig. 3.17).

3.6.3 Type II GCT Before Puberty

Typically, type II GCT develop after puberty. There are three situations in which type II GCT occasionally may occur before puberty: in patients with DSD in the gonads (for review [275], for age distribution of dysgerminomas [232]); in Down's syndrome, e.g., a seminoma of the testis in a boy of 2 years ([626], for review [276]); and in Klinefelter's syndrome in the mediastinum, e.g., two 8-year-old boys with a mixed GCT having a seminoma component ([564, 627], for review [277]). What these conditions broadly have in common is the severity of the disturbance of the niches where PGC/gonocytes may home. This is apparent from the gonads in DSD and Klinefelter's and Down's syndrome where gonocytes can barely survive and only rarely (in DSD and Down's syndrome) may differentiate into functional gametes. When gonocytes do survive, in DSD and Down's syndrome, they have an increased risk for neoplastic transformation; in Klinefelter's syndrome, transformed PGC probably degenerate before they can produce manifest tumors. What applies to the gonads is probably also true for extragonadal niches: PGC surviving there have a higher risk of neoplastic transformation, particularly when they are reprogrammed to ESC. These precursors may have a developmental potential ranging from primed-state-like, as in early mediastinal GCT in Klinefelter's syndrome, to fully fledged naïve state, depending on their methylation status at the time of neoplastic transformation.

This generalizing hypothesis does not explain the observation that in Klinefelter's syndrome mediastinal and not brain type II GCT may occur before puberty and in Down's syndrome, those of the testis but not those of mediastinum and brain.

3.7 Type III GCT

3.7.1 Developmental Potential

The cells of a spermatocytic tumor, the name proposed in the fourth edition of the WHO classification instead of spermatocytic seminoma [244], resemble postpubertal germ cells with nuclei in three distinct size classes. Those with the smallest nuclei with dense chromatin look like A-dark spermatogonia (considered reserve spermatogonial stem cells); the cells with intermediate and large paler nuclei with finely granular filamentous chromatin resemble A-pale spermatogonia (self-renewing stem cells), B spermatogonia, and leptotene spermatocytes [244, 628–630]. Transcript and protein analyses of spermatocytic tumor cells, reviewed by Waheeb and Hofmann [631] and summarized in Table 3.3, show that they lack markers of embryonic gonocytes and postmeiotic germ cells and express markers of prespermatogonia, spermatocytic stem cells/undifferentiated spermatogonia,

Table 3.3 Germ cell markers in spermatocytic tumor [631]

Marker	Spermatocytic tumor	Gonocytes	Spermatogonia	Spermatocytes
MAGEA4	+	+	+	+/-
SSX	+	+	+	+
DAZ family	+	+	+	+
CHK2	+	+	+	−
KIT	−	+[a]	+/-[a]	−
PLAP	−	+[a]	−	−
OCT4	−	+[a]	−	−
NSE	+	+/-	+	−
p19[INK4d]	−	−	−	+
UTF1	+/-	+/-	+	−
DMRT1	+	−	+	+
NY-ESO-1	+	+	+	+
FGFR3	+	+	+	−
RAS	+	+	+	+
pERK1/2	+	+	+	ND
REX-1	+/-	+/-	+/-	+
SYCP1	+	−	−	+
LDHc	+	−	−	+
CLGN	+	−	−	+
TCFL5	+	−	−	+

Table 3.1 in Waheeb and Hofmann 2011 [631]
+/- weak or variable staining
[a]Modified according to Oosterhuis et al. 2011 [348]

and spermatocytes in various combinations, suggesting the phenotype of a postnatal germ cell arrested at any stage of maturation between prespermatogonium and primary spermatocyte. OCT2 expression seems confined to tumor cells resembling A-dark spermatogonia [632]. Rare spermatocytic tumors are composed only of OCT2-expressing tumor cells and thus of neoplastic A-dark spermatogonia with blocked differentiation [632]. In fact, the stem cells of spermatocytic tumors, type III GCT, are committed to spermatogenesis with differentiation capacity limited to premeiotic cells.

So-called anaplastic spermatocytic tumor, a rare variant, has morphological features in common with seminoma [633]. One report describes a metastasizing anaplastic tumor [634]; however, overall this variant behaves as benign as the usual spermatocytic tumor, which only sporadically metastasizes [633]. Exceptionally, with less than 20 published cases, the tumor is associated with a sarcomatous component, usually undifferentiated sarcoma, and rarely rhabdomyo- or chondrosarcoma [635–637]. It is highly malignant and readily metastasizes to regional lymph nodes or, bloodborne, to visceral organs. The sarcoma component is probably the result of progression of the spermatocytic tumor, similar to progression in low-grade leukemias, lymphomas, and sarcomas [636]. The possibility that it has its origin in a germ cell that is reprogrammed to rudimentary somatic differentiation cannot be dismissed.

3.7.2 Epidemiology/Risk Factors

The only population-based epidemiological study finds an incidence of 0.4 per million for spermatocytic tumors, constituting 0.6 % of all testicular cancers in Australia. An increasing incidence over the past 20 years is suggested but has not been found statistically significant; risk factors have not been identified [638]. In about 9 %, the tumor is bilateral, more often metachronous than synchronous [244, 630]. The median age of clinical manifestation is 54 (range 19–92) [638].

3.7.3 Anatomical Distribution

Spermatocytic tumor occurs only in the postpubertal testis [630]. There is one report on a tumor originated in a maldescended testis [639]. Apparently, the tumor develops only if the conditions in the testis are compatible with survival of postnatal germ cells and induction of spermatogenesis.

3.7.4 (Cyto)Genetics

Most spermatocytic tumors are (near)diploid, the second largest group is (near)tetraploid, and a small number is peritriploid [640–642]. The most consistent cytogenetic aberration, present in all studied tumors, is an extra copy of chromosome 9 [643, 644], in which subsequent CGH analysis demonstrated a small amplified region on 9p, containing *DMRT1* as the most likely candidate gene involved in tumorigenesis [645]. In passing, DMRT1 has been shown to be an immunohistochemically detectable, useful marker for diagnosing spermatocytic tumor, apparent from Fig. 3.28 [645]. A small number of tumors, usually in the oldest half of the patients, have mutually exclusive, paternal age-related mutations in *FGFR3* or *HRAS* [646]. p53, not expressed in normal postpubertal germ cells, is demonstrated in 80 % of spermatocytic tumors, supposedly related to genomic instability [647].

The metastasizing anaplastic spermatocytic tumor published by Mikuz [634] resembled seminoma morphologically but lacked expression of PLAP and OCT4; cytogenetically, it had gain of both chromosome 9 and 12p. This tumor seems a hybrid between seminoma and spermatocytic tumor, whereby the phenotype is partly determined by overexpression of DMRT1, partly by the overdose of the pluripotency genes on 12p, perhaps not adequately repressed by DMRT1 [189]. This is yet another example of the plasticity of developmental states of precursors of GCT, blurring, in this case, the line between type II and type III GCT.

3.7.5 Epigenetics Including GI

An immunohistochemical study found a heterogeneous pattern of DNA methylation and histone modification in spermatocytic tumors, quite different from the regular patterns in normal spermatogenesis, probably because the regulatory signals conveyed by the niche toward the spermatogenetic cells are lacking in the tumors [648].

3.7.6 Pathogenesis and Animal Models

Spermatocytic tumor is not associated with GCNIS [481, 640]; however, it has its own intratubular precursor, at the luminal side of the tight junctions connecting the Sertoli cells, with essentially the same morphology as the adjacent invasive tumor [244, 628–630]. The occasional finding of exclusively intratubular spermatocytic tumor without an invasive component proves that the intratubular part is not due to intratubular extension of the invasive component [244, 630] and indeed the precursor lesion. Unlike GCNIS, it shows no obvious accumulation of precursor cells in the spermatogonial niche, neither as stacking of multiple layers of precursor cells nor as pagetoid extension within the seminiferous tubules to the detriment of spermatogenesis. It is possible that the precursor cells are phenotypically so similar to normal spermatogonia that expansion in the niche is not recognized with light microscopy, including immunohistochemistry. Alternatively, only the spermatogonial tumor stem cells remain in the niche, and upon the earliest differentiation, the tumor cells, like their normal counterparts, move to the lumen of the tubule. Finally, it is possible that the initiated cell lies at the luminal side of the tight junctions.

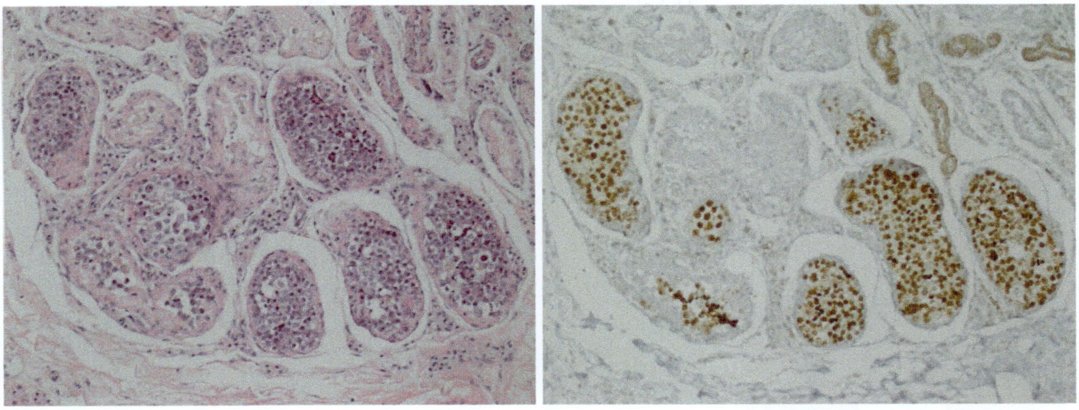

Fig. 3.28 Exclusively intratubular spermatocytic tumor, the precursor of spermatocytic tumor; no pagetoid involvement of adjacent tubules as in GCNIS (*left* H and E, ×200; *right* DMRT1, ×200)

Whatever the initial development, eventually the tumor cells become invasive and apparently independent from the tubular micro-milieu. It is indeed remarkable that fragile, apoptosis-prone cells like spermatogenic cells manage to survive in conditions so alien to them.

Studies in mouse models and human tumors begin to untangle the molecular mechanisms, both in the germ and niche cells, involved in maturation of postnatal male germ cells and controlling the mitosis versus meiosis switch, and how these might bear on the development of spermatocytic tumors.

Glial cell line-derived neurotrophic factor (GDNF), a distant member of the transforming growth factor superfamily, is secreted by Sertoli cells as paracrine factor involved in the regulation of spermatogonial self-renewal and differentiation in mouse and men ([649], for review [631]). Spermatogonial stem cells express the GFRA1/RET receptor complex at the cell surface. Binding of GDNF to this complex upregulates MYCN transcription factor via the PI3K/AKT pathway and FOS transcription factor via the RAS/ERK1/2 pathway (Fig. 3.29), as well as FGFR2 in spermatogonial stem cells. Other niche factors are FGF2, the ligand for FGFR2, produced by Sertoli cells, and CSF-1 secreted by Leydig cells (Fig. 3.30). Downregulation of GDNF in mice causes a Sertoli cell-only phenotype with complete absence of spermatogenic cells. Overexpression causes accumulations of undifferentiated spermatogonia in seminiferous tubules, resembling intratubular spermatocytic tumor, abrogation of spermatogenesis, and tumors in older animals, which are bilateral in over 50 %. By geno- and phenotype, the tumors have intermediate phenotypes between type II seminoma and spermatocytic tumors [650, 651], like the tumor described by Mikuz [634].

DMRT1 is the transcriptional gatekeeper controlling the mitosis versus meiosis decision in male germ cells [652]. It prevents differentiation and meiosis of spermatogonial cells by blocking the transcription of STRA8 and rendering these cells less sensitive to RA-induced meiosis. At the same time, it upregulates SOHLH1, a factor stimulating proliferation of spermatogonial stem cells, and suppresses the pluripotency genes *NANOG*, *SOX2*, and *OCT4*. Decreasing the level of DMRT1 disrupts GDNF signaling, cell cycle control, and pluripotency regulation. In 129Sv mice, but not in other mouse strains, it causes teratoma formation in a dose-dependent manner, probably due to failure to repress pluripotency regulators and reduced GDNF signaling. Postnatally elevated DMRT1 and GDNF signaling blocks differentiation of spermatogonial stem cells, resulting in tumors resembling spermatocytic tumors [189].

From these studies, it appears that both niche factors and factors intrinsic to spermatogonial stem cells may contribute to the formation of spermatocytic tumors. Overexpression of GDNF

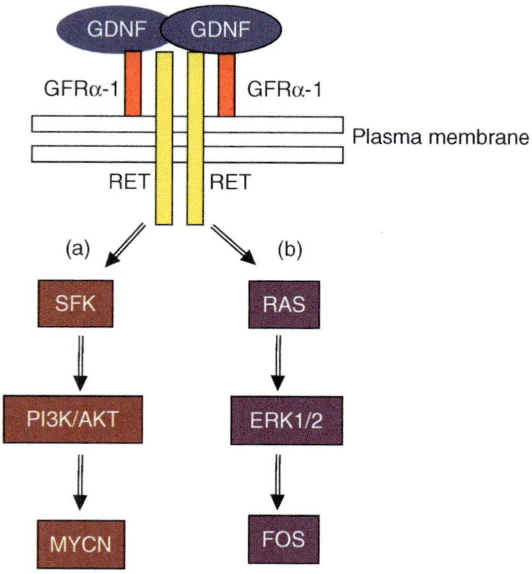

Fig. 3.29 Signaling pathways triggered by GDNF in spermatogonial stem cells. GDNF dimerizes and binds to the GFRA1/RET receptor complex. (**a**) Binding of GDNF activates RET, which triggers SRC kinase phosphorylation and the downstream activation of PI3K/AKT. Ultimately, the transcription factor MYCN is upregulated. (**b**) Binding of GDNF also can activate the RAS-mediated signaling pathway, which triggers ERK1/2 phosphorylation and upregulation of the transcription factor FOS [631]

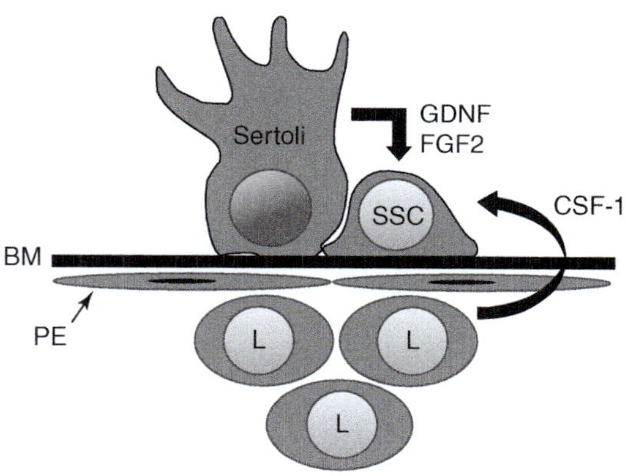

Fig. 3.30 A simplified view of the spermatogonial stem cell niche showing the main extrinsic factors driving SSC maintenance and self-renewal. Sertoli cells and spermatogonial stem cells (SSCs) are both attached to the basement membrane (BM). Sertoli cells provide for structural support and produce glial cell line-derived neurotrophic factor (GDNF) and basic fibroblast growth factor (bFGF) which are crucial for SSC self-renewal in vitro and in vivo. Leydig cells (L) and peritubular cells (PE) produce colony-stimulating factor-1 (CSF-1), also essential for self-renewal [631]

by Sertoli cells and CSF-1 by Leydig cells, perhaps in response to reduced spermatogenesis in aging men, might explain the old age of clinical manifestation and the relatively high risk of bilateral tumors, like in the mouse model [650, 651]. The niche factors may synergize with cell-intrinsic factors, also acquired with increasing age, such as elevated levels of DMRT1 through gain of chromosome 9 [645], and accumulation of paternal age-related mutations in HRAS and FGFR3 [646].

It is an emerging pattern: like in the other types of GCT, in spermatocytic tumors, initiation is probably primarily due to a developmental deregulation rather than somatic mutations, the latter being mainly progression related.

So-called seminomas have been described as spontaneous tumors in a variety of animals, like the dog [653] and rhinoceros [654], and have been experimentally produced, e.g., in *C. elegans* [655], zebrafish [393–395], and mice [651]. None of these can be reprogrammed to totipotency like type II seminomas in men and are therefore best regarded as models for spermatocytic tumors, sharing some features with seminomas [629, 653]. In dogs, the often bilateral, DMRT1-positive [656], spermatocytic tumors are frequently combined with nodular hyperplasia or even benign tumors of Leydig and Sertoli cells [653], supporting the idea that disturbance of the hormonal regulation of the spermatogonial niche is a crucial factor in the origin of spermatocytic tumors.

3.8 Type IV GCT

3.8.1 Developmental Potential

Dermoid cysts, type IV GCT, are unique for the ovary as they are derived from meiotic oocytes [657]. Typically, completely mature teratomas, they present as a thin-walled cyst lined with epidermis with attached appendages and filled with sebaceous material and hairs. Usually one solid nodule (Rokitansky's protuberance) protrudes from the wall into the lumen of the cyst. It is often composed of fat tissue, bone, teeth (with intermediate shapes between deciduous and permanent teeth) [658], and glial tissue and covered with skin with well-developed appendages including hair follicles forming hairs. The nodules mainly contain cranial tissues, suggesting that they represent the rostral part of an attempted embryo; however, virtually any adult tissue can be present. Exceptional tumors are predominantly solid with highly organized structures resembling a fetus, lacking extraembryonic tissues, as is the case in typical dermoid cysts. Benign tumors, such as struma ovarii and carcinoids, may arise within a dermoid cyst. Probably so-called monodermal teratomas similarly have their origin in dermoid cysts, eventually obscured by overgrowth of one tissue type. Epidermoid cysts may be a variant of monodermal teratoma [659], as in the testis. Monodermal teratomas could also originate in type I and II GCT of the ovary, as discussed in the relevant sections. Somatic-type malignancies develop reportedly in 0.2–3 % of dermoid cysts, including, in order of frequency, squamous cell carcinoma (80–90 %), adenocarcinomas (7 %), and sarcomas (7 %). Among the many rare types, PNET is also described (Chap. 6). A peculiar association with dermoid cysts is gliomatosis peritonei [660, 661], which will be discussed in the section on type VI GCT.

Sporadically, dermoid cysts contain immature foci, even more exceptionally combined with YST [133], prompting Yanai-Inbar and Scully to study the relationship between immature teratoma and the dermoid cyst [130]. Among 350 immature teratomas of the ovary submitted to the authors for second opinion, 92 (26 %) contained one or more grossly visible cysts lined by squamous epithelium with pilosebaceous structures. The figure of 26 % is most probably an overestimation because unusual cases are sent for consultation. In 10 % of all 350 cases, there was a dermoid cyst in the contralateral ovary, which is not much different from the percentage of bilaterality for dermoid cysts. In nine cases (aged 17–28, mean 23 years) of immature teratoma, there was a history of prior removal of a dermoid cyst in the same ovary. Four of these nine cases had a dermoid cyst in the other ovary, and three cases had multiple dermoid cysts in the same ovary. In addition, the authors had ten referral cases (aged 15–30, mean 23 years), collected over a period of 23 years, of otherwise typical dermoid cysts with minor immature areas, which did not recur after surgery, as usual for a dermoid cyst.

These cases illustrate a continuum between the typical dermoid cyst, type IV GCT, and type I immature teratomas of infancy. The ten dermoid cysts with small foci of immature teratoma are probably type IV GCT of which not all tissues had fully matured, consistent with the young age of these patients. The nine dermoid cysts recurring as immature teratoma belong probably to the same group as the 92 immature teratomas with macroscopically visible dermoid cysts. These tumors have an intermediate behavior and phenotype between type I and type IV, with an age of presentation in between that of pure immature teratomas and dermoid cysts and bilaterality like in dermoid cysts, whereas bilaterality is rare in pure immature teratomas [130, 131, 662]. These features justify the classification of these GCT as type I tumors beyond infancy, which occur also in the testis and extragonadal sites, in particular in the anteroposterior mediastinum, as discussed earlier.

3.8.2 Epidemiology/Risk Factors

Cancer registries do not provide data on the incidence of dermoid cysts as they are benign tumors. The best approximation is achieved by multiplying the frequency of dermoid cysts relative to cancers of the ovary (from hospital registries) with the incidence of the latter (from cancer registries) in the same region/country, as follows. Among 861 self-referred patients with an ovarian tumor, all of whom were treated in a single hospital (Women's and Children's Hospital, Los Angeles, CA), and therefore without obvious referral bias [663], dermoid cysts were the most common tumors with 379 cases (44 %); 211 tumors were malignant (25 %). The incidence of ovarian cancer in Northern America being 8.1 in

2012 [664], the incidence of dermoid cysts of the ovary can be estimated at about 13, making it overall the most common GCT, more frequent even than testicular type II GCT in high-incidence countries.

As for age of presentation, in a review of 517 dermoid cysts from a single institution, the two youngest patients were 10 and 13 years of age, the median age was 30, and the oldest patient 90. There were no prepubertal cases, and the large majority of dermoid cysts occur in the reproductive age between the onset of puberty and menopause [233]. The tumors diagnosed after the menopause probably had their origin during reproductive age [130]. Bilaterality occurred in 10.8 %; in seven cases, the dermoid cyst was associated with a malignant or benign epithelial surface tumor. Upon long-term follow-up, Anteby et al. [133] found 18 bilateral cases among 99 patients; multiple dermoid cysts in a single ovary were found in 9/18 bilateral cases as opposed to 1/81 unilateral cases. The mean age was similar for uni- and bilateral cases: 32.4 and 34.6 years, respectively. However, the age distribution was significantly different: 18/19 cases (95 %) were between 20 and 40 years old in unilateral compared to 61/80 (76 %) in bilateral cases, implying that overall bilateral cases are diagnosed at a younger age. Bilateral cases had a significantly higher risk of developing a recurring dermoid cyst. In familial cases where laterality was stated, 11/28 (39 %) were bilateral, and among the patients who were twins or triplets, 9/12 (75 %) were bilateral [134, 135]. One of the identical twins [533] had over the years seven dermoid cysts removed from her left and one from her right ovary. The age of diagnosis of the first tumor in the familial bilateral cases was known in ten patients; all presented between age 7 and 26, with a median age of 22.5, substantially younger than in nonfamilial bilateral cases [133]. These case histories strongly suggest a genetic predisposition for bilaterality and familial occurrence of type IV GCT of the ovary, which may also confer an increased risk for type II GCT, in view of two families with clustering of bilateral dermoid cysts with testicular seminoma [524, 531].

3.8.3 Anatomical Distribution

Dermoid cysts occur exclusively in the ovaries; on rare occasions, they may become detached from the ovary and reimplanted in either omentum [665], fallopian tube [666], or Douglas' pouch [667]. In four of the 31 omental and one of the 29 tubal cases, there was also a dermoid cyst in one of the ovaries.

Tumors in the testis and extragonadal sites resembling dermoid cysts have been discussed earlier as type I GCT beyond infancy, with a developmental potential in between that of type I and type IV GCT, probably explained by the arrest of PGC in the prophase of meiosis I at all anatomical sites, except within the seminiferous tubule [58, 59]. As discussed above, GCT with an intermediate phenotype between type I and IV occur also in the ovary.

3.8.4 (Cyto)Genetics

In the study by Surti et al. [668], 93 % of dermoid cysts were diploid, and the remaining 7 % had chromosomal abnormalities including trisomy for chromosomes 7, 8, 12, 15, and X; one case was tetraploid in mosaic form; there were no recurring structural aberrations. Trisomy for chromosomes 8, 12, and X is shared with immature teratomas [668]. Somatic-type malignancies developing in dermoid cysts have the same genetic profiles as the adjacent dermoid cysts, confirming their origin in the teratoma [669]. In general, the somatic-type malignancies have the same genetic changes as their somatic counterparts. For example, malignant struma ovarii has the same *BRAF* point mutations as papillary carcinomas of the thyroid [670, 671].

Most significantly, the genomic profile and pattern of imprinting of dermoid cysts reflects the stage of meiosis of the oocytes from which they are derived, as will be discussed in the following paragraphs.

3.8.5 Epigenetics Including GI

The process of erasure of biallelic imprinting in the germ lineage and the establishment of the maternal imprint during oogenesis has bearing on the pathogenesis of type IV GCT, the dermoid cysts of the ovary [16, 86]. Pronuclear transplantation experiments [3–5] have demonstrated that gynogenotes, with two haploid sets of maternally imprinted chromosomes, have a relatively good development of the embryo proper but very poor development of trophoblast, particularly the placenta. In contrast, androgenotes with two haploid sets of paternally imprinted chromosomes have a relatively normal development of the placenta but a very poor development of embryonic tissues (Fig. 3.31). The strong preference of the mouse gynogenote for developing somatic tissues is mirrored by the dermoid cyst, which is typically only composed of highly organized mature somatic tissues. Like in the gynogenote, the chromosomes of dermoid cysts lack paternal imprinting but share the maternal imprinting pattern of the oocytes from which they are derived [672]. The further the stage of meiosis of the oocyte, the better the maternal imprinting pattern is established in the derived dermoid cyst, suggesting that GI is a progressive process throughout oogenesis [672].

3.8.6 Pathogenesis and Animal Models

Ever since Linder [673] discovered the parthenogenetic origin of dermoid cysts by looking at allelic loss of isozymes, this mechanism has been reinvestigated with state-of-the-art technology, including study of chromosomal polymorphic markers and various DNA polymorphisms [668, 672, 674–680]. For various reasons, such as inadequate sample preparation, selective growth of host fibroblasts in tissue culture, and a poor noise to signal ratio in the applied assays, contaminating host cells have been interpreted as tumor cells along with genuine neoplastic cells. Therefore, the genomic profile of premeiotic oogonia and host cells being the same, a proportion of dermoid cysts have been misclassified as derived from premeiotic oogonia [657], e.g., in 12 % by Ohama [680] and 25 % by Deka et al. [681]. Kaku et al. [657] have addressed this problem by careful sample preparation and quantitative measurement of the signals from tumor cells and contaminating host cells, enabling them to demonstrate allelic conversions (from hetero- to homozygous and vice versa) even in the presence of somatic cell contamination. They found allelic conversion in all samples taken from Rokitansky's protuberance; therefore, none of the 64 analyzed dermoid cysts could have been derived from a premeiotic oogonium, 33 stemmed from primary oocytes, 16 from secondary oocytes, and 15 from ova (with endoreduplication) (Fig. 3.32). Apart from the absence of oogonium-derived dermoid cysts, the distribution over the stages of oogenesis was similar as in previous studies [668, 678–680]. The equivalence of numbers of dermoid cysts derived from primary oocytes, secondary oocytes, and ova is unexpected, considering that nearly 300,000 primary oocytes exist through adolescence [682] and only about 20 primary follicles begin to form secondary follicles during the menstrual cycle, in which subsequently secondary oocytes and ova develop. In view of these numbers, one would expect that almost all teratomas would be derived from primary oocytes. This obviously not being the case, Kaku et al. [657] propose that dermoid cysts are not derived from functional oocytes but from primary oocytes that have escaped meiotic arrest and start uncontrolled meiosis. This could explain the lack of premeiotic teratomas and the almost even distribution over the later stages of oogenesis (Fig. 3.33). As meiotic arrest is hormonally regulated [683], this mechanism also clarifies why dermoid cysts occur during reproductive age [233]. Moreover, it is conceivable that the assumed genetic factor underlying proneness for bilaterality and familial occurrence of type IV GCT is also involved in the regulation of meiotic arrest. Thus, it seems probable that, as in other GCT, the origin of type IV GCT is mainly determined by developmental factors, with a minor role for genetic events.

The rare but well-established phenomenon of ovarian teratomas developing as a fetiform

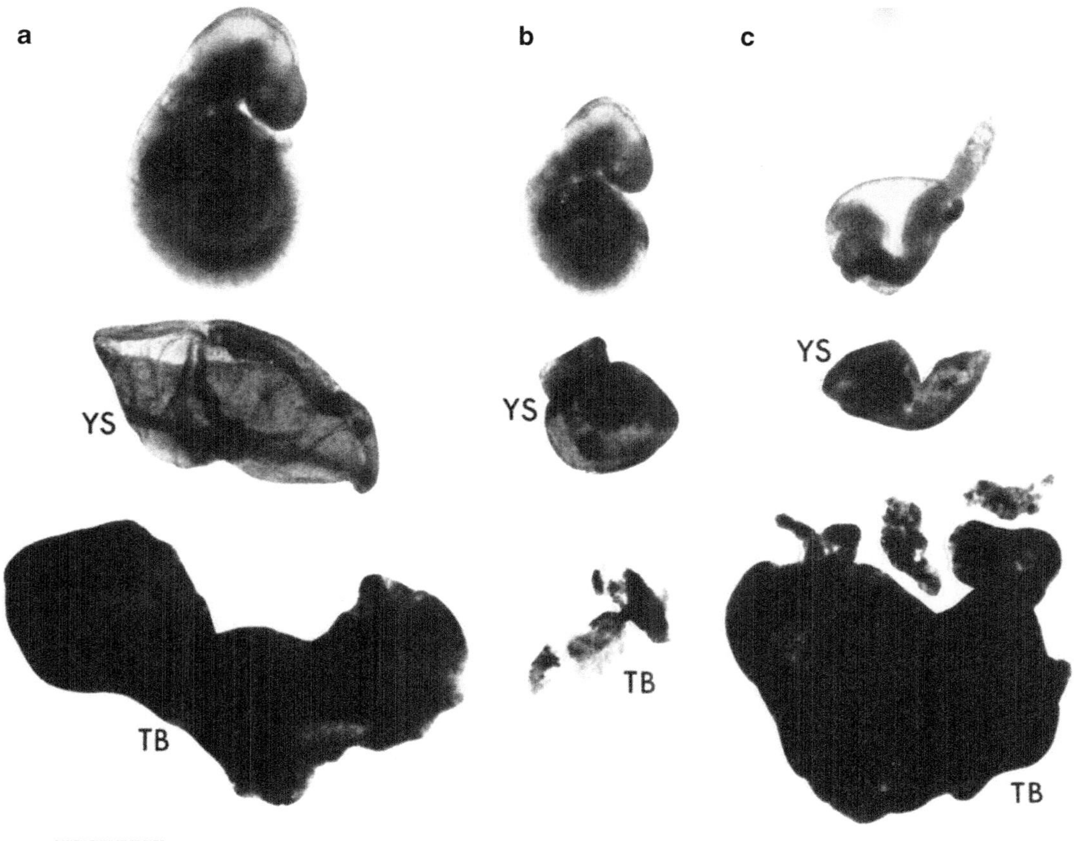

Fig. 3.31 Compare development of control embryo (**a**) with that obtained from eggs with two maternal genomes (**b**) in which a small but well-advanced 25-somite embryo was the maximum development but with poor extraembryonic tissues. The eggs with two paternal nuclei developed maximally to about the 6- to 8-somite stage but with extensive trophoblast development (**c**). *YS* yolk sac, *TB* trophoblast. Scale bar, 1 mm [5]

structure (homunculus) [684–686] attests to the close to omnipotent developmental potential (2C state) of some of the precursor cells of ovarian mature teratomas, except for the ability to form trophoblastic tissue. In fact, the principal difference with a type 0 GCT, a parasitic or included twin, is the absence of extraembryonic structures, in agreement with the absence of a paternal imprint. It is conceivable that homunculi develop from precursor cells with the most complete maternal imprinting, closely resembling a zygote. The usual dermoid cysts, mainly composed of tissues from the rostral part of the embryo, with the skin turned inside, might develop from precursor cells with incomplete maternal imprinting. These observations on type IV GCT lead to the speculation that the spatial-temporal organization of embryonic development is somehow related to the progression of maternal imprinting and that a complete maternal imprint is required for developing the entire embryo proper.

At the other end of the spectrum are the poorly developed dermoid cysts, prone to rupture and combined with immature teratoma [130], in the gray zone between type I and type IV GCT, which are probably derived from the least mature primary oocytes next to premeiotic oogonia. The latter cells are the precursors of type I immature teratomas of the ovary as discussed.

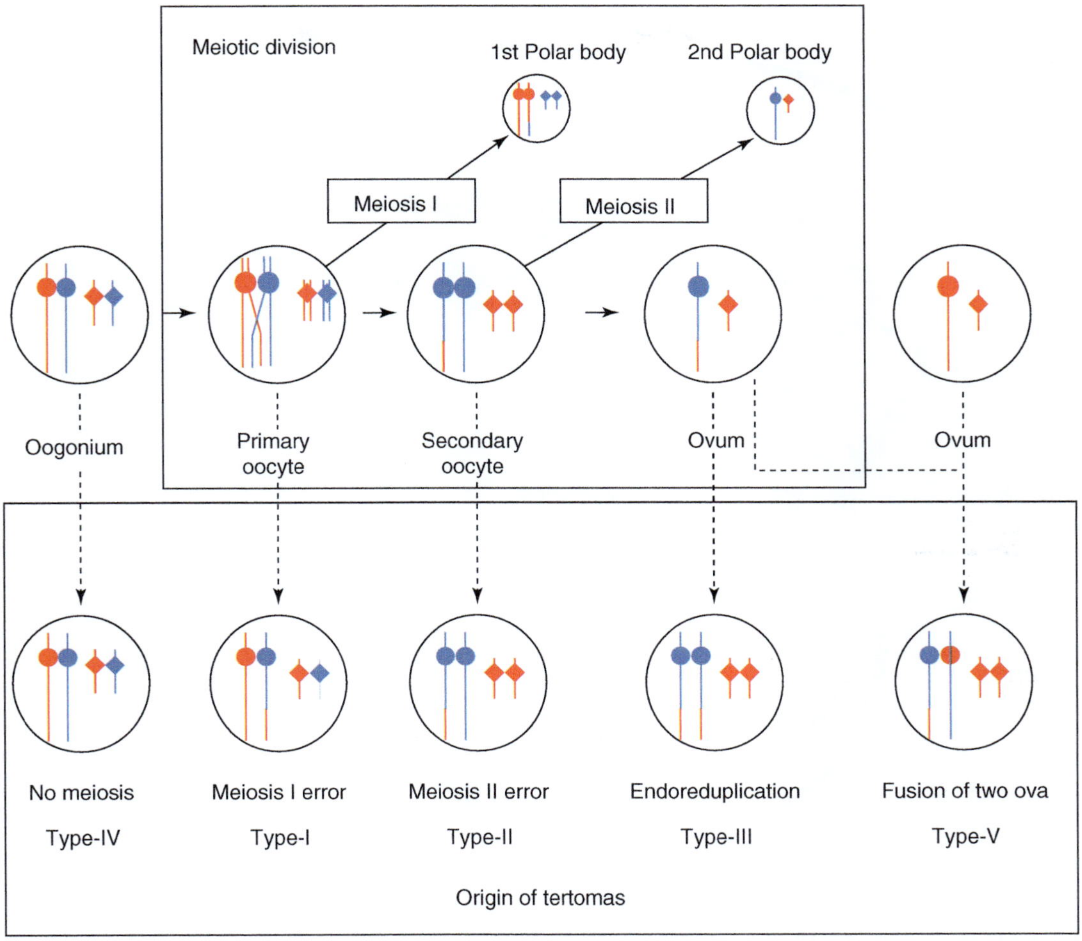

Fig. 3.32 Graphical representation of the origins of mature ovarian cystic teratomas and the relevance of meiotic division. The origins of mature cystic teratomas are conceptually classified into five types. Type I teratomas result from a meiosis I error, in which the segregation of sister chromatids occurs without a preceding mono-oriented separation of bivalent chromosomes, generating biparental diploid cells with homologous DNA recombination. Type II teratomas result from meiosis II errors, in which the nondisjunction of all sister chromatids gives rise to diploid cells with homologous DNA recombination. Type III teratomas occur via endoreduplication of a haploid ovum, which is entirely mono-allelic. Type IV teratomas arise from oogonia. The constitution of chromosomes from type IV teratomas is identical to those from somatic cells. Type V teratomas involve a fusion of two normal haploid ova [657]

3.9 Type V GCT

3.9.1 Developmental Potential

Complete hydatidiform moles, type V GCT, consist of placental tissue only, lacking somatic tissues of the embryo proper. A comprehensive discussion of these abnormal growths of the placenta is beyond the scope of this chapter (for review [687–689]). They are briefly mentioned here to show that complete hydatidiform moles are in the opposite side of the spectrum of developmental potential from dermoid cysts, which are composed of somatic tissues and lack trophoblastic tissue.

Grossly, a complete hydatidiform mole resembles a bunch of grapes, whereby the individual grapes represent enlarged placental villi covered by hyperplastic trophoblast with cyto-nuclear atypia and thus to be considered as dysplastic. The

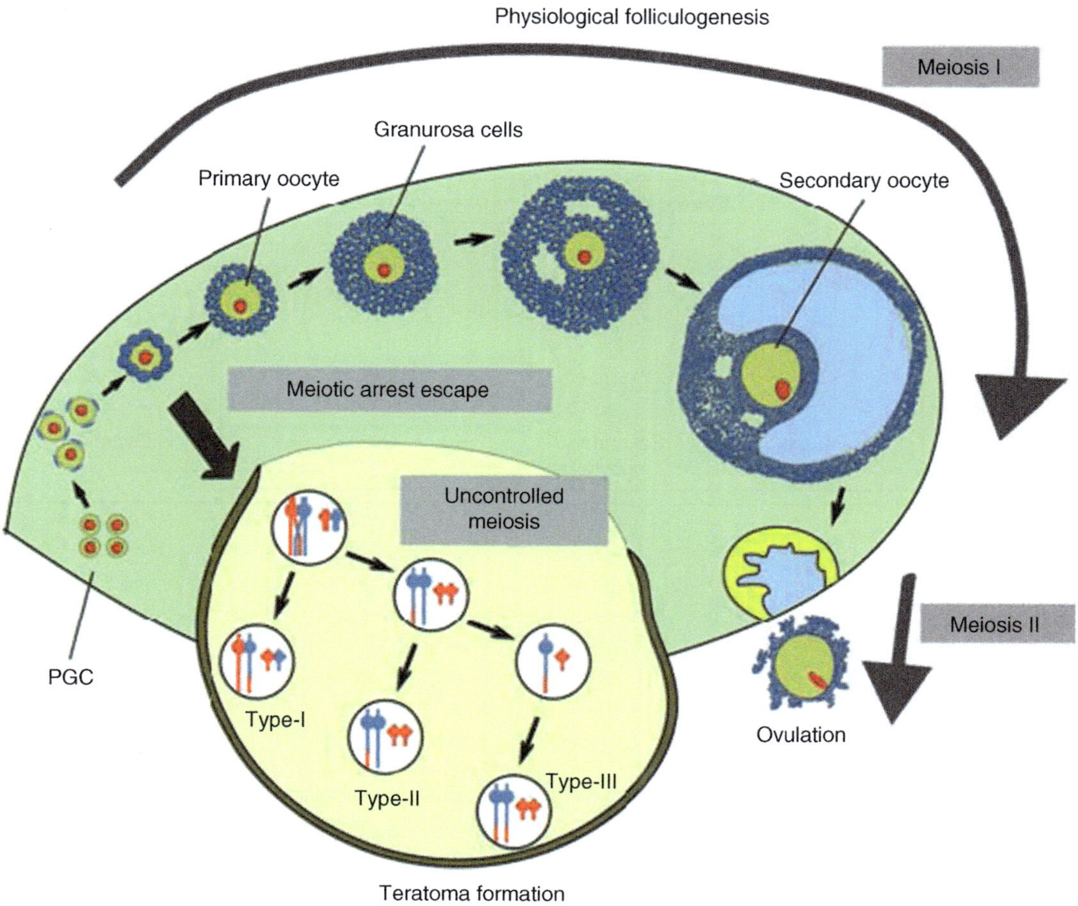

Fig. 3.33 The postulated mechanism of human ovarian teratoma formation. The source cells are proposed to be primary oocytes that escaped from meiotic arrest. Subsequent uncontrolled meiotic division could produce ovarian teratomas. *PGC* primordial germ cell [657]

increased size is due to accumulation of fluid, probably caused by defective vasculogenesis and apoptotic degeneration of villous stromal components [690, 691]. Rarely, molar tissue metastasizes to the vagina or lungs. In agreement with its dysplastic nature, it may progress to choriocarcinoma in 2–3 % of cases [692]. Malignant transformation is possibly driven by hypomethylation-associated genomic instability [693], as trophoblast and placenta are hypomethylated compared to somatic tissues (for review) [694]. More specifically, promoter hypermethylation of p16 alone or combined with E-cadherin is associated with progression of hydatidiform mole to choriocarcinoma [695]. In fact, 50 % of all gestational choriocarcinomas originate from complete hydatidiform moles [696].

3.9.2 Epidemiology/Risk Factors

The incidence of gestational trophoblastic disease, mostly complete hydatidiform moles, is 1 in 120 pregnancies in some parts of Asia and South America, more than tenfold higher than in Western societies. Risk factors are pregnancy at young or old age, prior gestational trophoblastic disease, Asian ethnicity, and possibly dietary deficiencies and low socioeconomic status (for review [689]).

A pathogenetically informative genetic risk factor is the presence of maternal mutations of *NALP7/NLRP7* on 19q13.4; the protein NALP is a member of the CATERPILLER protein family involved in inflammation and apoptosis. The mutation causes abnormal imprinting with overexpression of the paternal genome, resulting in recurrent familial biparental complete hydatidiform moles and reproductive wastage [697].

3.9.3 Anatomical Distribution

Complete hydatidiform moles develop where pregnancies occur, virtually always in the uterus and occasionally in the fallopian tube as an ectopic pregnancy [698].

3.9.4 (Cyto)Genetics/Epigenetics Including GI

Complete hydatidiform moles are generally diploid with a 46,XX (90 %) or 46,XY (10 %) karyotype [699, 700]; rare cases are tetraploid with four haploid sets of paternal chromosomes [701]. As a consequence of this chromosomal constitution, the genome of these lesions has an exclusively paternal GI, as will be further discussed in the following section.

3.9.5 Pathogenesis and Animal Models

Complete hydatidiform moles are caused by so-called androgenesis with two haploid sets of chromosomes from the father and none from the mother [702]. 46,XX complete hydatidiform moles arise from fertilization of an anuclear empty ovum by one 23,X sperm that replicates its chromosomes; when a 23,Y sperm is involved, this event results in a nonviable zygote. Complete hydatidiform moles with a 46,XY karyotype are the result of fertilization of an empty ovum by two sperm, respectively, with a Y and an X chromosome. Both mechanisms create a zygote with an exclusively paternal imprint, which gives rise to placental tissue only and no somatic tissues of the embryo proper, similar to experimentally produced mouse androgenotes [3, 4].

A partial mole constitutes an intermediate phenotype between a complete mole and a normal pregnancy; it arises when a normal ovum is fertilized by two sperm (69,XXY in 70 %; 69,XXX in 27 %; 69,XYY in 3 %) [703, 704]. Partial moles have enlarged hydropic placental villi in addition to normal villi combined with some development of tissues of the embryo proper [705].

The partial mole and maternal mutations of *NALP7/NLRP7* demonstrate the critical importance of the dose of paternally and maternally imprinted genes. Apparently, an overdose of paternally imprinted genes favors placental and severely impairs embryonic development. The apoptosis of stromal cells and the defective vessels (both derived from the embryo proper) observed in very early complete hydatidiform moles [691] and the poorly developed fetal tissues in a partial mole suggest that somatic tissues may be formed but degenerate in an embryo with an exclusively or predominantly paternally imprinted genome.

The precursor cell of the hydatidiform mole has the 2C-state developmental potential, except for the ability to form/maintain somatic tissues of the embryo proper.

3.10 Type VI GCT

3.10.1 Definition

Type VI GCT are defined as neoplasms derived from mature somatic cells or committed stem cells, which resemble GCT as to their developmental potential.

The observation that genetically engineered iPSC may form tumors with the developmental potential of GCT is experimental support for this concept.

3.10.2 Developmental Potential of Genetically Engineered iPSC

The first human iPSC derived from somatic cells [70] and NSC [72] when grafted in mice reportedly produced teratomas with mature somatic derivatives from the three germ layers. In the meantime, it has become clear that iPSC also may give rise to other tumor types such as EC, immature teratoma (often primitive neuroectodermal tissues), YST, and somatic-type malignancies. The composition of the tumors depends on the induced cell type (with varying numbers of somatic mutations) and the genes combined in the transducing vectors (vectors including *MYC* carry a significant risk of developing malignant GCT) (for review [706, 707, 321, 625]). Each of the consecutive steps in the procedure of induction of pluripotency may contribute to carcinogenesis: integration of gene delivery vectors and transgenes into genomes of the host cells; chromosomal damage during reprogramming; clonal selection for transformed colonies during iPSC expansion; incomplete reprogramming; failure to silence pluripotent networks in differentiated progeny; DNA damage accumulated during cell culture, so-called culture adaptation; and aberrant regulation of the imprinting process [321]. Culture adaptation involves gain of (parts of) chromosomes in particular the chromosomes 12 (12p) [708, 709], 17 (17q) [708–710], 20 (20q11.21) [710–713], and X [708, 714]. In fact, the gains are remarkably similar to those seen in type II GCT [303].

Notably, premature termination of reprogramming in vivo was shown to cause the development of a pediatric cancer (Wilms' tumor) through altered epigenetic regulation [715]. Also the recipient tissue for the graft is an important factor: human ESC transplanted into mice gave teratomas; however, when transplanted into human fetal tissue grafts in mice, they gave rise to pure EC [716]. The developmental potential of human iPSC with an intact genome is similar to that of human ESC, matching with the primed state of mouse ESC derived from the primitive ectoderm [706] and thus with type I GCT. Genetic aberrations and epigenetic modifications acquired in the derivation of iPSC may result in a higher capacity of self-renewal of the stem cells and thus the development of tumors containing EC or consisting of pure EC [321, 706, 717]. The developmental potential of these iPSC acquires features of the naïve state with totipotent developmental potential corresponding with human erased PGC, mouse ESC derived from the ICM, and the non-seminomatous variants of type II GCT. Seminoma, pure or as part of a mixed tumor, has not been reported. In fact, human iPSC often have a developmental potential somewhere in between that of type I GCT and type II non-seminomas and may be accompanied with somatic-type malignancies, such as can occur in both types of GCT. MYC with its central role in core pluripotency networks, involving NANOG, OCT4, and SOX2, and at the same time being an oncogene [321] is probably crucial for both the change in developmental potential of the iPSC and the causation of somatic malignancies.

3.10.3 Developmental Potential of Spontaneous Type VI GCT

It is well established that in humans there are somatic malignant tumors, apparently not derived from germ cell precursors, in which GCT components develop. Examples are sinonasal teratocarcinosarcomas (for review [220, 718]), cancer of the stomach [719], urothelial cancer [720], and endometrioid adenocarcinomas [721]. GCT arising in association with endometriosis and epithelial cancers of the ovary are comprehensively discussed in Chap. 6. So is gliomatosis peritonei, a rare condition often associated with immature teratoma of the ovary, characterized by the presence of mature glial tissue in the peritoneum, considered implants from the teratoma [661, 722]. It has been suggested that gliomatosis might in rare cases develop directly from peritoneal cells, in particular when associated with endometriosis [661, 723]. Possibly in support of a broad developmental capacity of mesothelial cells, ovarian surface epithelium, scraped from the ovary of postmenopausal women, reportedly expressed early embryonic developmental markers such as stage-specific embryonic antigen-4 (SSEA-4), OCT4, NANOG, and SOX2. When grown in culture, they were claimed to form oocyte-like cells, expressing markers of oocytes, as well as blastocyst-like structures expressing OCT4, SOX2, and NANOG; however when grafted in SCID mice, the ultimate test for pluripotency, no teratomas were formed [724, 725].

There are scattered reports in the literature of highly malignant GCT [726] sometimes in combination with somatic malignancies [727] that for various reasons do not readily fit into the types 0 to V of GCT. They lack the treatment sensitivity of GCT, occur usually at a much higher age than GCT, or at sites that are not compatible with parasitic twinning or mismigration of PGC, such as the foot [728] and the upper arm [729].

The developmental potential of these tumors is not unambiguously that of a type I or II GCT and resembles the potential of tumors produced upon grafting of iPSC. GCT arising in ovarian cancer, for example, may contain EC cells (expressing OCT4 and SOX2 and inconsistently CD30) in addition to polyembryoma, somatic lineages, YST, and choriocarcinoma [730] (Chap. 6). Indeed, strongly suggesting that these GCT components result from induction of pluripotency in somatic (cancer) cells, as will be discussed hereafter.

3.10.4 Epidemiology/Risk Factors

There are no epidemiological studies, as type VI GCT is emerging; the numbers of cases are small, and the patient material is heterogeneous. What the patients have in common is their high age (median 50–60), much higher than usual for GCT, apart from type III, spermatocytic tumor. This is true for the GCT associated with ovarian cancer (Chap. 6), the sinonasal teratocarcinosarcomas [718], and the GCT described by Van Echten et al. [726] and Noguera et al. [727].

3.10.5 Anatomical Distribution

The anatomical distribution is in accordance with the various types of cancer in which development of GCT components occurs, like cancer of the ovary and stomach as mentioned. The sinonasal teratocarcinosarcomas are virtually always located in the nasal cavity and/or ethmoid sinus with occasionally extension into the maxillary sinus or orbit [220]. The location of the tumors described by Van Echten et al. and Noguera et al. was atypical for primary GCT, certainly considering the old age of the patients and included retroperitoneum, posterior mediastinum, and inside the sacrum.

Scotting and colleagues [139, 168] have proposed that GCT of the brain are derived from NSC, induced to pluripotency by activation of OCT4, like the iPSC produced by Kim et al. [72]. However, in terms of developmental potential, epidemiology, anatomical localization, and (cyto) genetics, they fit into the overall pattern of type I and II GCT. Moreover mis-migrated PGC, the most likely precursor cells of these tumors, have been demonstrated in the brain of human embryos [52]. In fact, there are no convincing arguments to assume another cell of origin for GCT of the brain, than the current hypothesis that they are derived from PGC.

3.10.6 (Cyto)Genetics

Little is known on the genetics of GCT originated in somatic cancers. There was no gain of 12p in three sinonasal teratocarcinosarcomas [731]. Thomas et al. [732] demonstrated by ISH an extra copy of 12p13 (a feature of type I GCT) in a subpopulation of cells in a nasal teratocarcinosarcoma.

The three atypical GCT described by Van Echten et al. [726] and Noguera et al. [727] were karyotyped and showed complex balanced translocations, with 6p21 being a common breakpoint in each of them; chromosome 12 was not involved in these cases. The two cases described by Van Echten shared two chromosomal fusions: 6p21::11q13 and 6p22::6q23. Despite considerable efforts, the breakpoints were never fully characterized at the molecular level [733]. The nasal immature teratoma described by Houri et al. [221] was diploid with a balanced translocation t(1;11)(q12;p15).

3.10.7 Epigenetics Including GI

There are no specific data on the epigenetics of type VI GCT; however, in view of the often high age of the patients, it may be assumed that gradual loss of DNA methylation may have resulted in aberrant gene activation [734]. This applies even more to advanced cancers, where epigenetic changes may disrupt the stem cell program [735]. Moreover, mis-regulation of imprinted genes, so-called loss of imprinting, is a frequent and early phenomenon in a large variety of human tumors [736]. In particular, the imprinting of *H19* and *IGF-II* is often lost, e.g., in colorectal cancer [737] and in the normal mucosa of the affected individuals [738] due to hypomethylation. The same genes are also frequently hypomethylated in epithelial cancers of the ovary [739], suggesting that development of a GCT component in epithelial cancers of the ovary might be due to derepression of pluripotency genes, such as OCT4 and SOX2, as has been demonstrated immunohistochemically in these tumors (see Chap. 6) (Fig. 3.34).

3.10.8 Pathogenesis and Animal Models

For GCT developing in somatic cancers, various combinations of genetic and epigenetic changes could result in activation of repressed pluripotency genes. This could be a random event; however, it could also specifically target cancer cells with particular mutations or stem cell characteristics. The latter mechanism might apply to nasal teratocarcinosarcoma suggested to originate from transformed stem cells of the sinonasal mucosa [732]. Similarly, the low efficiency of induction of pluripotency in normal somatic cells has been explained as either due to slow stochastic accumulation of events in random cells or due to the targeting of rare ("elite") cells, probably stem cells [707]. MYC could play a crucial role, since it is a central player in oncogenesis and pluripotency; indeed, more aggressive cancers express both the core pluripotency genes (*OCT4*, *NANOG*, *SOX2*, and *KLF4*) and *MYC*-centered networks [740, 741].

The few cytogenetically characterized atypical GCT [221, 726, 727] suggest that breakpoints in certain chromosomal regions might activate the pluripotency program. Most conspicuously, the breakpoint in 6p21–22 in these tumors could involve *OCT4* [36]. A breakpoint in 11p15, possibly involving *IGF2*, was demonstrated in a nasal immature teratoma [221]. Of note, 6p22 was also involved in an atypical teratoid/rhabdoid tumor, a pediatric cancer of the brain. In this tumor, a breakpoint was found in 11p15 as well, likely involving *IGF2* implicated in various childhood cancers (Wilms' tumor, hepatoblastoma, and rhabdomyosarcoma) [742–744]. The overrepresentation of 12p13 in a sinonasal teratocarcinosarcoma [732] might lead to overexpression of the pluripotency cluster *NANOG*, *STELLAR*, and *GDF3* [38, 745].

The localization of these tumors and the overlap of their chromosomal rearrangements with those of pediatric cancers suggest that committed stem cells in which pluripotency genes are activated due to chromosomal aberrations could be the originating cells.

In terms of morphology and cytogenetics, there is also an overlap with bona fide type I GCT raising the question whether the atypical GCT described here as type VI GCT could be exceptional manifestations of type I GCT. In two sacral type I teratomas where balanced translocations, t(12;15)(q13;q25), were demonstrated, these were constitutional [154, 155]. These are cases relevant for this discussion because they demonstrate that balanced translocations per se can very probably give origin to neoplasms with GCT morphology. The translocation t(8;22)(p21;q12) in an intrathoracic mature teratoma, described by Jin et al. [156], concerned a girl aged 15 and could thus be best considered a type I GCT beyond infancy. It cannot be excluded that balanced translocations are more frequent in these GCT than thus far documented, because they are only detected with dedicated approaches. Further research into this category of tumors will clarify what is now a gray zone between type I GCT beyond infancy and the atypical GCT in the category of type VI GCT, in whose pathogenesis balanced chromosomal translocations seem to play an important role.

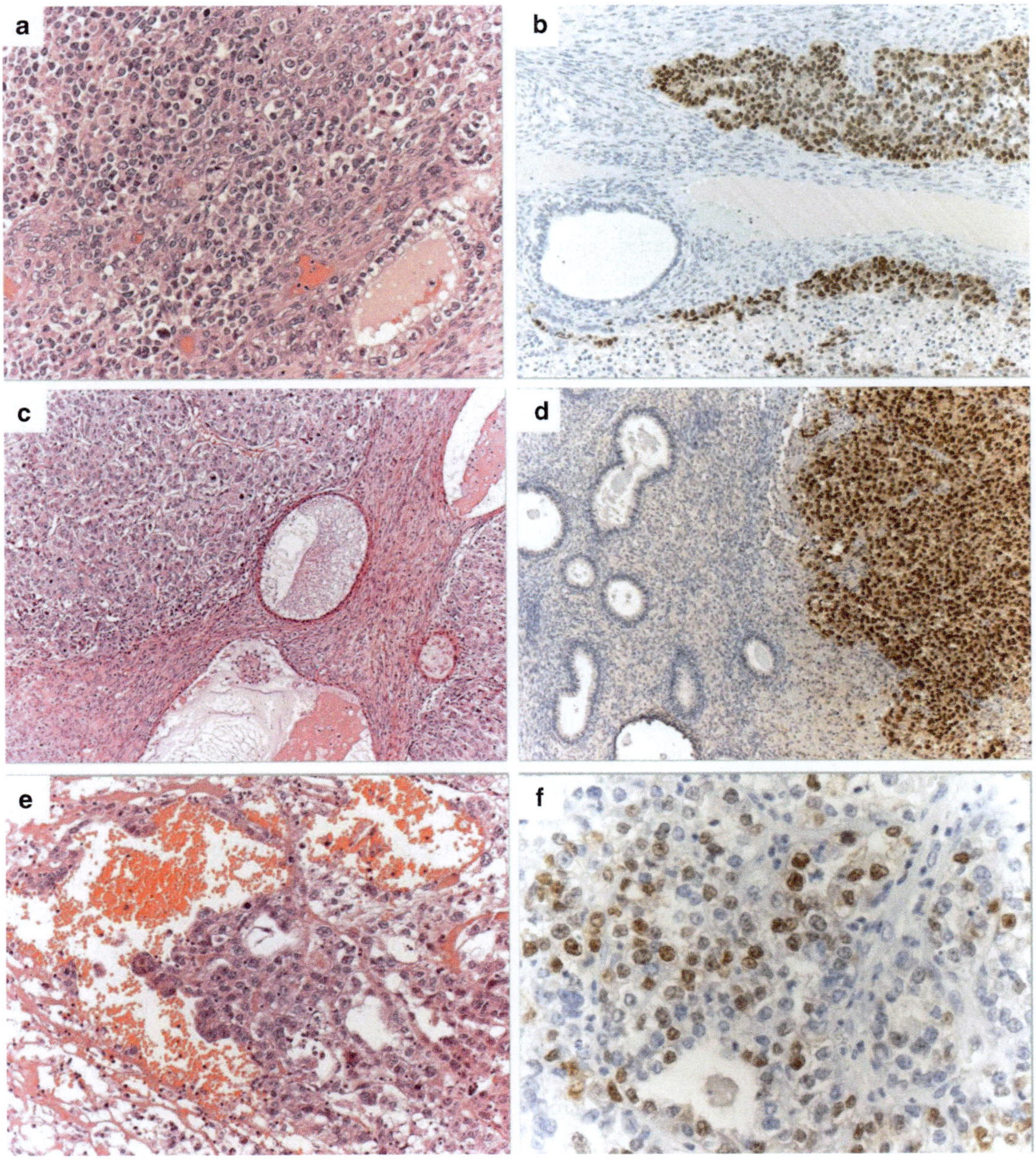

Fig. 3.34 Histology of GCT originated in clear cell carcinoma of ovary showing EC and teratoma (**a**, **c**) and extraembryonic tissue surrounding an embryoid structure (**e**); EC cells expressing OCT4 (**b**, **d**, and **f**) (From Nogales and Schuldt Chap. 6)

Summarizing, type VI GCT are neoplasms that share morphological features with GCT but do not originate from germ cell precursors. They may develop from somatic cells, most often in aggressive cancers, in which by various epigenetic and genetic changes, among others translocations, pluripotency is induced. They come in three variants: GCT as part of somatic neoplasia, de novo by induction of pluripotency in nonneoplastic somatic cells, and most likely in the future as a complication of therapeutic application of human iPSC.

3.11 Integrated View and Summary

3.11.1 GCT, States of Developmental Potential, and Precursor Cells Matched

Type 0 GCT, parasitic and included twins, approach the developmental potential of the zygote and, therefore, must be derived from omnipotent precursor cells in the 2C state, probably blastomeres or ESC similar to mouse ESC from the ICM that happened to be in the 2C state. Familial clustering with twinning supports the hypothesis that these growths are derived from blastomeres that have escaped the organizing influence of the developing embryo or rather the molecular mechanisms that check the omnipotency of these cells.

Type I GCT consist of teratomas with somatic tissues at various levels of maturation representing the three germ layers. YST only develops by way of tumor progression in aneuploid cells that have lost their ability to contribute to somatic lineages of the embryo proper. Tumor stem cells are rarely encountered in these tumors: only in immature components occasional OCT4-positive cells are found, which do not express SOX2. OCT4 is probably driven from the distal enhancer. Germ cell differentiation has not been demonstrated. This developmental potential, the poor self-renewing capacity of the stem cells, which show reduced expression of pluripotency proteins and the germline incompetence of these tumors, is in accordance with a pluripotent precursor cell in the primed state, corresponding to mouse ESC derived from the primitive ectoderm.

In view of the anatomical distribution of the extragonadal type I GCT along the midline of the body, their most likely precursor is a migrating diploid PGC in an early, methylated, pre-erased stage, which has escaped apoptosis because it was reprogrammed to an EGC that acquired the primed state in accordance with its epigenetic status. Neoplastic growth starts during fetal development in keeping with clinical presentation of these tumors at birth or in early infancy, usually before age six. Type I GCT of the gonads, likewise, are derived from methylated, pre- or partially erased, diploid, premeiotic PGC via EGC reprogrammed to ESC in the primed state.

Type II GCT have the broadest developmental potential of human GCT comprising both seminomas, composed of neoplastic, hypomethylated (including both X chromosomes in females), partially to completely erased, premeiotic PGC/gonocytes, as well as non-seminomas, which are caricatures of early embryonic development. The latter develop when the developmental potential of a neoplastic gonocyte is unleashed by reprogramming to an EC cell, the totipotent stem cell of non-seminoma, which may give rise to YST and choriocarcinoma representing the extraembryonic tissues, and also somatic tissues from the three germ layers, from immature to fully mature, and occasionally early germ cell differentiation. EC cells have a high capacity of self-renewal and express many pluripotency markers, such as OCT4, SOX2, NANOG, and LIN28. OCT4 is likely expressed from the proximal enhancer. These characteristics are compatible with the developmental potential of the totipotent or naïve state corresponding to mouse ESC derived from the ICM and preimplantation epiblast.

The precursor cells are more mature PGC (hypomethylated, partially to completely erased, and premeiotic), which can only survive in suitable niches in the gonads, thymus, and midline of the brain. Outside these niches, such cells die apoptotically, explaining the absence of type II GCT at other anatomical sites. When in the gonads the niche functions properly, the gonocytes will differentiate into germ cells. This will not happen in the extragonadal niches, which are incapable of sustaining germ cell development beyond the prophase of meiosis I. A disturbed niche, whether gonadal or extragonadal, may result in delayed maturation of the gonocytes creating, when GBY is present, a window for co-expression of OCT4 and TSPY and accumulation of chromosomal rearrangements, particularly gain of 12p, which maintain the PGC/gonocyte phenotype and totipotent developmental potential of the precursor cells of type II GCT. The crucial role of GBY/TSPY explains the overwhelming male preponderance of type II GCT.

Type III GCT, so-called spermatocytic tumors, which occur only in the testis, have a developmental potential that is limited to postpubertal, premeiotic, spermatogenic cells: A-dark and A-pale spermatogonia, B spermatogonia, and leptotene spermatocytes. The most likely precursor cell is a postpubertal, paternally imprinted spermatogonial cell.

Type IV GCT, dermoid cysts, which occur only in the ovary, are composed of mature somatic tissues mainly from the rostral part of the embryo, often containing teeth; occasional solid variants may resemble a complete fetus; extraembryonic tissues are typically absent. This developmental potential is consistent with a parthenogenetically activated oocyte or ovum with an exclusively maternal genomic imprint as precursor cell that is incapable to support the development of extraembryonic tissues: the "maternal half" of the 2C state.

Type V GCT, hydatidiform moles, are hyperplastic, dysplastic growths composed of placental tissue only. The precursor cell is an empty zygote, fertilized by one sperm, followed by endoreduplication or by two sperm, resulting in a genome that has an exclusively paternal imprint, incapable of sustaining the development of somatic tissues of the embryo proper: the "paternal half" of the 2C state.

Type VI GCT are derived from spontaneous or genetically engineered iPSC that may form somatic tissues with varying degrees of maturation, EC, YST, and occasionally choriocarcinoma. The stem cells of tumors derived from spontaneously induced somatic cells in humans resemble those of type I GCT, with reduced expression of pluripotency genes and limited self-renewal capacity, as in the primed state. Human somatic cells induced to pluripotency in vitro, when assayed in the proper context, may contain large amounts of EC, up to 100 %, in particular if *MYC* was included in the inducing cocktail. The stem cells of these tumors share characteristics with the naïve-state stem cells of type II GCT; however, PGC or seminoma-like components have never been reported.

3.11.2 Intermediate Phenotypes

Consistent with the plasticity of the developmental states of embryonic stem cells, there are, between the different defined types of GCT, intermediate phenotypes, which will be briefly summarized here.

There is continuum between multiple pregnancies, conjoined twins, parasitic twins, and type I GCT, with intermediate types between type 0 and type I GCT, which could arbitrarily be classified in either type.

Type I GCT and type II GCT have gradual transitions, in particular among prepubertal GCT of the mediastinum in Klinefelter's and among GCT of the brain in patients with Down's syndrome, with tumors that are genotypically type II but phenotypically resemble type I, suggesting that the genomic changes typical for type II, particularly gain of 12p, have occurred in a PGC that is still too heavily methylated to allow the full spectrum of the naïve-state developmental potential of a type II GCT.

Type I and type IV GCT may cluster in the same families with an increased risk of bilaterality and multiplicity. The occurrence of tumors composed of immature teratoma with dermoid cysts embedded in the immature teratoma component likely represent a transition form between type I and type IV GCT. It may be hypothesized that such tumors are derived from a precursor cell somewhere in between an oogonium and a type I oocyte, in which maternal GI is not yet completed.

There is at least one published case of a spermatocytic tumor [634] that is intermediate between a type II and a type III GCT, in terms of morphology, chromosomal composition, and behavior: a seminomatous morphology, lacking lymphocytes, gain of 12p and chromosome 9, and metastasis.

The partial mole has an intermediate phenotype between a complete mole (Type V GCT) and a normal pregnancy due to an overdose of paternally imprinted genes in the presence of maternal imprinting.

Type VI tumors have developmental characteristics with features of type I and type II GCT.

Conclusion

GCT should be related to a developmental state rather than a specific cell of origin, as a particular originating cell may assume different developmental states, and different cell types may have the same developmental potential in agreement with the plasticity of developmental states. PGC are a good example: they give rise to type I and type II GCT, respectively from the primed/pluripotent and the naïve/totipotent developmental state, as apparent from the increased risk in Klinefelter's and Down's syndrome for both type I and II GCT. As yet, it cannot be excluded that some type I GCT originate from ESC in the primed state. It would not make any difference as to the developmental potential of the tumor. This principle justifies the inclusion of neoplasms, such as those derived from iPSC, in the classification of GCT, even when they are not derived from germ cells, because they share the developmental state of genuine GCT.

The intermediate phenotypes between some of the defined types of GCT attest to the plasticity of the different developmental states from which they had their origin.

References

1. Slack JM. Origin of stem cells in organogenesis. Science. 2008;322(5907):1498–1501. doi: 322/5907/1498 [pii] 10.1126/science.1162782.
2. Wylie C. Germ cells. Cell. 1999;96(2):165–174. doi:S0092-8674(00)80557-7 [pii].
3. Surani MA, Barton SC, Norris ML. Development of reconstituted mouse eggs suggests imprinting of the genome during gametogenesis. Nature. 1984; 308(5959):548–50.
4. McGrath J, Solter D. Completion of mouse embryogenesis requires both the maternal and paternal genomes. Cell. 1984;37(1):179–183. doi:0092-8674(84)90313-1 [pii].
5. Surani MA, Barton SC, Norris ML. Nuclear transplantation in the mouse: heritable differences between parental genomes after activation of the embryonic genome. Cell. 1986;45(1):127–136. doi:0092-8674(86)90544-1 [pii].
6. Gkountela S, Li Z, Vincent JJ, Zhang KX, Chen A, Pellegrini M, et al. The ontogeny of cKIT+ human primordial germ cells proves to be a resource for human germ line reprogramming, imprint erasure and in vitro differentiation. Nat Cell Biol. 2013;15(1):113–122. doi:ncb2638 [pii] 10.1038/ncb2638.
7. Kim S, Gunesdogan U, Zylicz JJ, Hackett JA, Cougot D, Bao S, et al. PRMT5 protects genomic integrity during global DNA demethylation in primordial germ cells and preimplantation embryos. Mol Cell. 2014;56(4):564–579. doi:S1097-2765(14)00787-4 [pii] 10.1016/j.molcel.2014.10.003.
8. Hajkova P, Erhardt S, Lane N, Haaf T, El-Maarri O, Reik W, et al. Epigenetic reprogramming in mouse primordial germ cells. Mech Dev. 2002;117(1–2):15–23. doi:S0925477302001818 [pii].
9. Hajkova P, Jeffries SJ, Lee C, Miller N, Jackson SP, Surani MA. Genome-wide reprogramming in the mouse germ line entails the base excision repair pathway. Science. 2010;329(5987):78–82. doi:329/5987/78 [pii] 10.1126/science.1187945.
10. Morison IM, Reeve AE. A catalogue of imprinted genes and parent-of-origin effects in humans and animals. Hum Mol Genet. 1998;7(10):1599–1609. doi:ddb178 [pii].
11. Surani MA. Reprogramming of genome function through epigenetic inheritance. Nature. 2001; 414(6859):122–128. doi:10.1038/35102186 35102186 [pii].
12. McLaren A. Primordial germ cells in the mouse. Dev Biol. 2003;262(1):1–15. doi:S0012160603002148 [pii].
13. Durcova-Hills G, Burgoyne P, McLaren A. Analysis of sex differences in EGC imprinting. Dev Biol. 2004;268(1):105–110. doi:10.1016/j.ydbio.2003.12.018 S0012160604000144 [pii].
14. Morison IM, Ramsay JP, Spencer HG. A census of mammalian imprinting. Trends Genet. 2005;21(8):457–465. doi:S0168-9525(05)00166-6 [pii] 10.1016/j.tig.2005.06.008.
15. Diplas AI, Lambertini L, Lee MJ, Sperling R, Lee YL, Wetmur J, et al. Differential expression of imprinted genes in normal and IUGR human placentas. Epigenetics. 2009;4(4):235–240. doi:9019 [pii].
16. Hall JG. Genomic imprinting: review and relevance to human diseases. Am J Hum Genet. 1990; 46(5):857–73.
17. Surani A, Tischler J. Stem cells: a sporadic super state. Nature. 2012;487(7405):43–45. doi:487043a [pii] 10.1038/487043a.
18. McGrath J, Solter D. Inability of mouse blastomere nuclei transferred to enucleated zygotes to support development in vitro. Science. 1984;226(4680):1317–9.
19. Macfarlan TS, Gifford WD, Driscoll S, Lettieri K, Rowe HM, Bonanomi D, et al. Embryonic stem cell potency fluctuates with endogenous retrovirus activity. Nature. 2012;487(7405):57–63. doi:nature11244 [pii] 10.1038/nature11244.
20. Hackett JA, Surani MA. Regulatory principles of pluripotency: from the ground state up. Cell Stem Cell. 2014;15(4):416–430. doi:S1934-5909(14)00406-8 [pii] 10.1016/j.stem.2014.09.015.
21. Scholer HR, Balling R, Hatzopoulos AK, Suzuki N, Gruss P. Octamer binding proteins confer transcriptional activity in early mouse embryogenesis. EMBO J. 1989;8(9):2551–7.
22. Scholer HR, Hatzopoulos AK, Balling R, Suzuki N, Gruss P. A family of octamer-specific proteins present during mouse embryogenesis: evidence for germline-specific expression of an Oct factor. EMBO J. 1989;8(9):2543–50.

23. Ovitt CE, Scholer HR. The molecular biology of Oct-4 in the early mouse embryo. Mol Hum Reprod. 1998;4(11):1021–31.
24. Nichols J, Zevnik B, Anastassiadis K, Niwa H, Klewe-Nebenius D, Chambers I, et al. Formation of pluripotent stem cells in the mammalian embryo depends on the POU transcription factor Oct4. Cell. 1998;95(3):379–391. doi:S0092-8674(00)81769-9 [pii].
25. Wu G, Han D, Gong Y, Sebastiano V, Gentile L, Singhal N, et al. Establishment of totipotency does not depend on Oct4A. Nat Cell Biol. 2013;15(9):1089–1097. doi:ncb2816 [pii] 10.1038/ncb2816.
26. Yeom YI, Fuhrmann G, Ovitt CE, Brehm A, Ohbo K, Gross M, et al. Germline regulatory element of Oct-4 specific for the totipotent cycle of embryonal cells. Development. 1996;122(3):881–94.
27. Pesce M, Wang X, Wolgemuth DJ, Scholer H. Differential expression of the Oct-4 transcription factor during mouse germ cell differentiation. Mech Dev. 1998;71(1–2):89–98.
28. Niakan KK, Eggan K. Analysis of human embryos from zygote to blastocyst reveals distinct gene expression patterns relative to the mouse. Dev Biol. 2013;375(1):54–64. doi:S0012-1606(12)00674-4 [pii] 10.1016/j.ydbio.2012.12.008.
29. Honecker F, Stoop H, de Krijger RR, Chris Lau YF, Bokemeyer C, Looijenga LH. Pathobiological implications of the expression of markers of testicular carcinoma in situ by fetal germ cells. J Pathol. 2004;203(3):849–57. doi:10.1002/path.1587.
30. Rajpert-De Meyts E, Hanstein R, Jorgensen N, Graem N, Vogt PH, Skakkebaek NE. Developmental expression of POU5F1 (OCT-3/4) in normal and dysgenetic human gonads. Hum Reprod. 2004;19(6):1338–1344. doi:10.1093/humrep/deh265 deh265 [pii].
31. Stoop H, Honecker F, Cools M, de Krijger R, Bokemeyer C, Looijenga LH. Differentiation and development of human female germ cells during prenatal gonadogenesis: an immunohistochemical study. Hum Reprod. 2005;20(6):1466–1476. doi:deh800 [pii] 10.1093/humrep/deh800.
32. Looijenga LH, Stoop H, de Leeuw HP, de Gouveia Brazao CA, Gillis AJ, van Roozendaal KE, et al. POU5F1 (OCT3/4) identifies cells with pluripotent potential in human germ cell tumors. Cancer Res. 2003;63(9):2244–50.
33. Stewart CL, Gadi I, Bhatt H. Stem cells from primordial germ cells can reenter the germ line. Dev Biol. 1994;161(2):626–628. doi:10.1006/dbio.1994.1058 S0012-1606(84)71058-X [pii].
34. Aksoy I, Jauch R, Chen J, Dyla M, Divakar U, Bogu GK, et al. Oct4 switches partnering from Sox2 to Sox17 to reinterpret the enhancer code and specify endoderm. EMBO J. 2013;32(7):938–953. doi:emboj201331 [pii] 10.1038/emboj.2013.31.
35. Irie N, Weinberger L, Tang WW, Kobayashi T, Viukov S, Manor YS, et al. SOX17 is a critical specifier of human primordial germ cell fate. Cell. 2015;160(1–2):253–268. doi:S0092-8674(14)01583-9 [pii] 10.1016/j.cell.2014.12.013.
36. Crouau-Roy B, Amadou C, Bouissou C, Clayton J, Vernet C, Ribouchon MT, et al. Localization of the OTF3 gene within the human MHC class I region by physical and meiotic mapping. Genomics. 1994;21(1):241–243. doi:S0888-7543(84)71249-3 [pii] 10.1006/geno.1994.1249.
37. Schepers GE, Teasdale RD, Koopman P. Twenty pairs of sox: extent, homology, and nomenclature of the mouse and human sox transcription factor gene families. Dev Cell. 2002;3(2):167–170. doi:S153458070200223X [pii].
38. Clark AT, Rodriguez RT, Bodnar MS, Abeyta MJ, Cedars MI, Turek PJ, et al. Human STELLAR, NANOG, and GDF3 genes are expressed in pluripotent cells and map to chromosome 12p13, a hotspot for teratocarcinoma. Stem Cells. 2004;22(2):169–79. doi:10.1634/stemcells.22-2-169.
39. Le Bin GC, Munoz-Descalzo S, Kurowski A, Leitch H, Lou X, Mansfield W, et al. Oct4 is required for lineage priming in the developing inner cell mass of the mouse blastocyst. Development. 2014;141(5):1001–1010. doi:dev.096875 [pii] 10.1242/dev.096875.
40. Wu FR, Zhang Y, Ding B, Lei XH, Huang JC, Wang CH, et al. H3K27me3 may be associated with Oct4 and Sox2 in mouse preimplantation embryos. Genet Mol Res. 2014;13(4):10121–10129. doi:gmr4191 [pii] 10.4238/2014.December.4.6.
41. Toyooka Y, Oka S, Fujimori T. Early preimplantation cells expressing Cdx2 exhibit plasticity of specification to TE and ICM lineages through positional changes. Dev Biol. 2016;411(1):50–60. doi:S0012-1606(16)30026-4 [pii] 10.1016/j.ydbio.2016.01.011.
42. Goolam M, Scialdone A, Graham SJ, Macaulay IC, Jedrusik A, Hupalowska A, et al. Heterogeneity in Oct4 and Sox2 Targets Biases Cell Fate in 4-Cell Mouse Embryos. Cell. 2016;165(1):61–74. doi:S0092-8674(16)30061-7 [pii] 10.1016/j.cell.2016.01.047.
43. Niakan KK, Ji H, Maehr R, Vokes SA, Rodolfa KT, Sherwood RI, et al. Sox17 promotes differentiation in mouse embryonic stem cells by directly regulating extraembryonic gene expression and indirectly antagonizing self-renewal. Genes Dev. 2010;24(3):312–326. doi:24/3/312 [pii] 10.1101/gad.1833510.
44. Seguin CA, Draper JS, Nagy A, Rossant J. Establishment of endoderm progenitors by SOX transcription factor expression in human embryonic stem cells. Cell Stem Cell. 2008;3(2):182–195. doi:S1934-5909(08)00327-5 [pii] 10.1016/j.stem.2008.06.018.
45. Weidgang CE, Russell R, Tata PR, Kuhl SJ, Illing A, Muller M, et al. TBX3 Directs Cell-Fate Decision toward Mesendoderm. Stem Cell Reports. 2013;1(3):248–265. doi:10.1016/j.stemcr.2013.08.002 S2213-6711(13)00068-4 [pii].
46. Saitou M, Barton SC, Surani MA. A molecular programme for the specification of germ cell fate in mice. Nature. 2002;418(6895):293–300. doi:10.1038/nature00927 nature00927 [pii].
47. Hogan B. Developmental biology: decisions, decisions! Nature. 2002;418(6895):282–283. doi:10.1038/418282a 418282a [pii].

48. Matsui Y, Mochizuki K. A current view of the epigenome in mouse primordial germ cells. Mol Reprod Dev. 2014;81(2):160–70. doi:10.1002/mrd.22214.
49. Ly L, Chan D, Trasler JM. Developmental windows of susceptibility for epigenetic inheritance through the male germline. Semin Cell Dev Biol. 2015;43:96–105. doi:S1084-9521(15)00133-0 [pii] 10.1016/j.semcdb.2015.07.006.
50. Hackett JA, Sengupta R, Zylicz JJ, Murakami K, Lee C, Down TA, et al. Germline DNA demethylation dynamics and imprint erasure through 5-hydroxymethylcytosine. Science. 2013;339(6118):448–452. doi:science.1229277 [pii] 10.1126/science.1229277.
51. Freeman B. The active migration of germ cells in the embryos of mice and men is a myth. Reproduction. 2003;125(5):635–43.
52. Mamsen LS, Brochner CB, Byskov AG, Mollgard K. The migration and loss of human primordial germ stem cells from the hind gut epithelium towards the gonadal ridge. Int J Dev Biol. 2012;56(10–12):771–778. doi:120202lm [pii] 10.1387/ijdb.120202lm.
53. Mollgard K, Jespersen A, Lutterodt MC, Yding Andersen C, Hoyer PE, Byskov AG. Human primordial germ cells migrate along nerve fibers and Schwann cells from the dorsal hind gut mesentery to the gonadal ridge. Mol Hum Reprod. 2010;16(9):621–631. doi:gaq052 [pii] 10.1093/molehr/gaq052.
54. Teilum G. Special tumors of ovary and testis and related extragonadal lesions: comparative pathology and histological identification. Copenhagen: Munksgaard; 1976.
55. Gonzalez-Crussi F. Extragonadal teratomas. atlas of tumor pathology, Second Series. Armed Forces Institute of Pathology: Washington, D.C; 1982.
56. Oosterhuis JW, Stoop H, Honecker F, Looijenga LH. Why human extragonadal germ cell tumours occur in the midline of the body: old concepts, new perspectives. Int J Androl. 2007;30(4):256–63; discussion 263–4. doi:IJA793 [pii] 10.1111/j.1365-2605.2007.00793.x.
57. Runyan C, Gu Y, Shoemaker A, Looijenga L, Wylie C. The distribution and behavior of extragonadal primordial germ cells in Bax mutant mice suggest a novel origin for sacrococcygeal germ cell tumors. Int J Dev Biol. 2008;52(4):333–344. doi:072486cr [pii] 10.1387/ijdb.072486cr.
58. Upadhyay S, Zamboni L. Ectopic germ cells: natural model for the study of germ cell sexual differentiation. Proc Natl Acad Sci U S A. 1982;79(21):6584–8.
59. McLaren A. Germ cells and germ cell sex. Philos Trans R Soc Lond Ser B Biol Sci. 1995;350(1333):229–33. doi:10.1098/rstb.1995.0156.
60. Matsui Y, Zsebo K, Hogan BL. Derivation of pluripotential embryonic stem cells from murine primordial germ cells in culture. Cell. 1992;70(5):841–847. doi:0092-8674(92)90317-6 [pii].
61. Donovan PJ, de Miguel MP. Turning germ cells into stem cells. Curr Opin Genet Dev. 2003;13(5):463–471. doi:S0959437X03001217 [pii].
62. Hackett JA, Zylicz JJ, Surani MA. Parallel mechanisms of epigenetic reprogramming in the germline. Trends Genet. 2012;28(4):164–174. doi:S0168-9525(12)00014-5 [pii] 10.1016/j.tig.2012.01.005.
63. Sterneckert J, Hoing S, Scholer HR. Concise review: Oct4 and more: the reprogramming expressway. Stem Cells. 2012;30(1):15–21. doi:10.1002/stem.765.
64. Irie N, Tang WW, Azim Surani M. Germ cell specification and pluripotency in mammals: a perspective from early embryogenesis. Reprod Med Biol. 2014;13(4):203–215. doi:10.1007/s12522-014-0184-2 184 [pii].
65. Ferrari F, Apostolou E, Park PJ, Hochedlinger K. Rearranging the chromatin for pluripotency. Cell Cycle. 2014;13(2):167–168. doi:27028 [pii] 10.4161/cc.27028.
66. Magnusdottir E, Surani MA. How to make a primordial germ cell. Development. 2014;141(2):245–252. doi:141/2/245 [pii] 10.1242/dev.098269.
67. Festuccia N, Osorno R, Halbritter F, Karwacki-Neisius V, Navarro P, Colby D, et al. Esrrb is a direct Nanog target gene that can substitute for Nanog function in pluripotent cells. Cell Stem Cell. 2012;11(4):477–490. doi:S1934-5909(12)00480-8 [pii] 10.1016/j.stem.2012.08.002.
68. Kojima Y, Kaufman-Francis K, Studdert JB, Steiner KA, Power MD, Loebel DA, et al. The transcriptional and functional properties of mouse epiblast stem cells resemble the anterior primitive streak. Cell Stem Cell. 2014;14(1):107–120. doi:S1934-5909(13)00414-1 [pii] 10.1016/j.stem.2013.09.014.
69. Takahashi K, Yamanaka S. Induction of pluripotent stem cells from mouse embryonic and adult fibroblast cultures by defined factors. Cell. 2006;126(4):663–676. doi:S0092-8674(06)00976-7 [pii] 10.1016/j.cell.2006.07.024.
70. Takahashi K, Tanabe K, Ohnuki M, Narita M, Ichisaka T, Tomoda K, et al. Induction of pluripotent stem cells from adult human fibroblasts by defined factors. Cell. 2007;131(5):861–872. doi:S0092-8674(07)01471-7 [pii] 10.1016/j.cell.2007.11.019.
71. Kim JB, Sebastiano V, Wu G, Arauzo-Bravo MJ, Sasse P, Gentile L, et al. Oct4-induced pluripotency in adult neural stem cells. Cell. 2009;136(3):411–419. doi:S0092-8674(09)00071-3 [pii] 10.1016/j.cell.2009.01.023.
72. Kim JB, Greber B, Arauzo-Bravo MJ, Meyer J, Park KI, Zaehres H, et al. Direct reprogramming of human neural stem cells by OCT4. Nature. 2009;461(7264):649–643. doi:nature08436 [pii] 10.1038/nature08436.
73. Ramalho-Santos M, Yoon S, Matsuzaki Y, Mulligan RC, Melton DA. "Stemness": transcriptional profiling of embryonic and adult stem cells. Science. 2002;298(5593):597–600. doi:10.1126/science.1072530 1072530 [pii].
74. Miyoshi N, Ishii H, Nagano H, Haraguchi N, Dewi DL, Kano Y, et al. Reprogramming of mouse and human cells to pluripotency using mature microRNAs. Cell Stem Cell. 2011;8(6):633–8. doi:10.1016/j.stem.2011.05.001.
75. Hiura H, Toyoda M, Okae H, Sakurai M, Miyauchi N, Sato A, et al. Stability of genomic imprinting in human induced pluripotent stem cells. BMC Genet. 2013;14:32. doi:10.1186/1471-2156-14-32.

76. Stang A, Trabert B, Wentzensen N, Cook MB, Rusner C, Oosterhuis JW, et al. Gonadal and extragonadal germ cell tumours in the United States, 1973–2007. Int J Androl. 2012;35(4):616–25. doi:10.1111/j.1365-2605.2011.01245.x.
77. De Backer A. Surgical treatment and outcome of extracranial germ cell tumors in childhood [Doctoral Thesis]. Rotterdam: Erasmus University Medical Center; 2006, ISBN 90-8559-166-X.
78. Ozgur T, Atik E, Silfeler DB, Toprak S. Mature cystic teratomas in our series with review of the literature and retrospective analysis. Arch Gynecol Obstet. 2012;285(4):1099–101. doi:10.1007/s00404-011-2171-8.
79. Znaor A, Lortet-Tieulent J, Laversanne M, Jemal A, Bray F. International testicular cancer incidence trends: generational transitions in 38 countries 1900–1990. Cancer Causes Control. 2015;26(1):151–8. doi:10.1007/s10552-014-0486-z.
80. Stevens LC, Little CC. Spontaneous testicular teratomas in an inbred strain of mice. Proc Natl Acad Sci U S A. 1954;40(11):1080–7.
81. Stevens LC. A new inbred subline of mice (129-terSv) with a high incidence of spontaneous congenital testicular teratomas. J Natl Cancer Inst. 1973;50(1):235–42.
82. Slye M. Primary spontaneous tumors of the ovary in mice. Studies on the incidence and heritability of spontaneous tumors in mice. J Cancer Res. 1920;5:205–26.
83. Phillips RJ, Hulse EV. Two extragonadal teratomas in a mouse, with discussion of possible histogenesis. J Comp Pathol. 1982;92(2):273–284. doi:0021-9975(82)90086-X [pii].
84. Kehler J, Tolkunova E, Koschorz B, Pesce M, Gentile L, Boiani M, et al. Oct4 is required for primordial germ cell survival. EMBO Rep. 2004;5(11):1078–1083. doi:7400279 [pii] 10.1038/sj.embor.7400279.
85. Yamaguchi S, Kurimoto K, Yabuta Y, Sasaki H, Nakatsuji N, Saitou M, et al. Conditional knockdown of Nanog induces apoptotic cell death in mouse migrating primordial germ cells. Development. 2009;136(23):4011–4020. doi:136/23/4011 [pii] 10.1242/dev.041160.
86. Oosterhuis JW, Looijenga LH. Testicular germ-cell tumours in a broader perspective. Nat Rev Cancer. 2005;5(3):210–222. doi:nrc1568 [pii] 10.1038/nrc1568.
87. Rijlaarsdam MA, Tax DM, Gillis AJ, Dorssers LC, Koestler DC, de Ridder J, et al. Genome wide DNA methylation profiles provide clues to the origin and pathogenesis of germ cell tumors. PLoS One. 2015;10(4):e0122146. doi:10.1371/journal.pone.0122146 PONE-D-14-41158 [pii].
88. Grant P, Pearn JH. Foetus-in-foetu. Med J Aust. 1969;1(20):1016–9.
89. Willis RA. The borderland of embryology and pathology. Bull N Y Acad Med. 1950;26(7):440–60.
90. Griscom NT. The Roentgenology of Neonatal Abdominal Masses. Am J Roentgenol Radium Therapy, Nucl Med. 1965;93:447–63.
91. Du Plessis JP, Winship WS, Kirstein JD. Fetus in fetu and teratoma. A case report and review. S Afr Med J. 1974;48(50):2119–22.
92. Hopkins KL, Dickson PK, Ball TI, Ricketts RR, O'Shea PA, Abramowsky CR. Fetus-in-fetu with malignant recurrence. J Pediatr Surg. 1997;32(10):1476–1479. doi:S0022-3468(97)90567-4 [pii].
93. Khadaroo RG, Evans MG, Honore LH, Bhargava R, Phillipos E. Fetus-in-fetu presenting as cystic meconium peritonitis: diagnosis, pathology, and surgical management. J Pediatr Surg. 2000;35(5):721–723. doi:S0022-3468(00)48690-2 [pii] 10.1053/jpsu.2000.6037.
94. Isaacs H, Jr. Perinatal (fetal and neonatal) germ cell tumors. J Pediatr Surg. 2004;39(7):1003–1013. doi:S002234680400199X [pii].
95. Hoeffel CC, Nguyen KQ, Phan HT, Truong NH, Nguyen TS, Tran TT, et al. Fetus in fetu: a case report and literature review. Pediatrics. 2000;105(6):1335–44.
96. Spencer R. Parasitic conjoined twins: external, internal (fetuses in fetu and teratomas), and detached (acardiacs). Clin Anat. 2001;14(6):428–444. doi:10.1002/ca.1079 [pii] 10.1002/ca.1079.
97. Brand A, Alves MC, Saraiva C, Loio P, Goulao J, Malta J, et al. Fetus in fetu – diagnostic criteria and differential diagnosis – a case report and literature review. J Pediatr Surg. 2004;39(4):616–618. doi:S002234680300959X [pii].
98. de Lagausie P, de Napoli Cocci S, Stempfle N, Truong QD, Vuillard E, Ferkadji L, et al. Highly differentiated teratoma and fetus-in-fetu: a single pathology? J Pediatr Surg. 1997;32(1):115–116. doi:S0022-3468(97)90112-3 [pii].
99. Heifetz SA, Alrabeeah A, Brown BS, Lau H. Fetus in fetu: a fetiform teratoma. Pediatr Pathol. 1988;8(2):215–26.
100. Plattner G, Oxorn H. Familial incidence of ovarian dermoid cysts. Can Med Assoc J. 1973;108(7):892–3.
101. Feld D, Labes J, Nathanson M. Bilateral ovarian dermoid cysts in triplets. Obstet Gynecol. 1966;27(4):525–8.
102. Beutel K, Partsch CJ, Janig U, Nikischin W, Suttorp M. Oral mature teratoma containing epididymal tissue in a female neonate. Lancet. 2001;357(9252):283–284. doi:S0140-6736(00)03620-5 [pii] 10.1016/S0140-6736(00)03620-5.
103. Dehner LP. Gonadal and extragonadal germ cell neoplasia of childhood. Hum Pathol. 1983;14(6):493–511.
104. Tapper D, Lack EE. Teratomas in infancy and childhood. A 54-year experience at the Children's Hospital Medical Center. Ann Surg. 1983;198(3):398–410.
105. Hawkins EP. Pathology of germ cell tumors in children. Crit Rev Oncol Hematol. 1990;10(2):165–179. doi:1040-8428(90)90005-D [pii].
106. Heifetz SA, Cushing B, Giller R, Shuster JJ, Stolar CJ, Vinocur CD, et al. Immature teratomas in children: pathologic considerations: a report from the combined Pediatric Oncology Group/Children's Cancer Group. Am J Surg Pathol. 1998;22(9):1115–24.
107. Schneider DT, Calaminus G, Koch S, Teske C, Schmidt P, Haas RJ, et al. Epidemiologic analysis

of 1,442 children and adolescents registered in the German germ cell tumor protocols. Pediatr Blood Cancer. 2004;42(2):169–75. doi:10.1002/pbc.10321.
108. McKenney JK, Heerema-McKenney A, Rouse RV. Extragonadal germ cell tumors: a review with emphasis on pathologic features, clinical prognostic variables, and differential diagnostic considerations. Adv Anat Pathol. 2007;14(2):69–92. doi:10.1097/PAP.0b013e31803240e6 00125480-200703000-00002 [pii].
109. De Backer A, Madern GC, Pieters R, Haentjens P, Hakvoort-Cammel FG, Oosterhuis JW, et al. Influence of tumor site and histology on long-term survival in 193 children with extracranial germ cell tumors. Eur J Pediatr Surg. 2008;18(1):1–6. doi:10.1055/s-2007-989399.
110. Hawkins E, Issacs H, Cushing B, Rogers P. Occult malignancy in neonatal sacrococcygeal teratomas. A report from a Combined Pediatric Oncology Group and Children's Cancer Group study. Am J Pediatr Hematol Oncol. 1993;15(4):406–9.
111. Baker BA, Frickey L, Yu IT, Hawkins EP, Cushing B, Perlman EJ. DNA content of ovarian immature teratomas and malignant germ cell tumors. Gynecol Oncol. 1998;71(1):14–18. doi:S0090-8258(98)95102-2 [pii] 10.1006/gyno.1998.5102.
112. Rescorla FJ, Sawin RS, Coran AG, Dillon PW, Azizkhan RG. Long-term outcome for infants and children with sacrococcygeal teratoma: a report from the Childrens Cancer Group. J Pediatr Surg. 1998;33(2):171–176. doi:S0022-3468(98)90426-2 [pii].
113. Heerema-McKenney A, Harrison MR, Bratton B, Farrell J, Zaloudek C. Congenital teratoma: a clinicopathologic study of 22 fetal and neonatal tumors. Am J Surg Pathol. 2005;29(1):29–38. doi:00000478-200501000-00004 [pii].
114. Frazier AL, Weldon C, Amatruda J. Fetal and neonatal germ cell tumors. Semin Fetal Neonatal Med. 2012;17(4):222–230. doi:S1744-165X(12)00059-5 [pii] 10.1016/j.siny.2012.05.004.
115. Isaacs H, Jr. I. Perinatal brain tumors: a review of 250 cases. Pediatr Neurol. 2002;27(4):249–261. doi:S0887899402004721 [pii].
116. Blohm ME, Gobel U. Unexplained anaemia and failure to thrive as initial symptoms of infantile choriocarcinoma: a review. Eur J Pediatr. 2004;163(1):1–6. doi:10.1007/s00431-003-1361-1.
117. McNally OM, Tran M, Fortune D, Quinn MA. Successful treatment of mother and baby with metastatic choriocarcinoma. Int J Gynecol Cancer. 2002;12(4):394–398. doi:ijg01125 [pii].
118. Ngan KW, Jung SM, Lee LY, Chuang WY, Yeh CJ, Hsieh YY. Immunohistochemical expression of OCT4 in primary central nervous system germ cell tumours. J Clin Neurosci. 2008;15(2):149–152. doi:S0967-5868(07)00017-3 [pii] 10.1016/j.jocn.2006.08.013.
119. Abiko K, Mandai M, Hamanishi J, Matsumura N, Baba T, Horiuchi A, et al. Oct4 expression in immature teratoma of the ovary: relevance to histologic grade and degree of differentiation. Am J Surg Pathol. 2010;34(12):1842–1848. doi:10.1097/PAS.0b013e3181fcd707 00000478-201012000-00012 [pii].
120. Oosterhuis JW, Stoop JA, Rijlaarsdam MA, Biermann K, Smit VT, Hersmus R, et al. Pediatric germ cell tumors presenting beyond childhood? Andrology. 2015;3(1):70–7. doi:10.1111/andr.305.
121. Herszfeld D, Wolvetang E, Langton-Bunker E, Chung TL, Filipczyk AA, Houssami S, et al. CD30 is a survival factor and a biomarker for transformed human pluripotent stem cells. Nat Biotechnol. 2006;24(3):351–357. doi:nbt1197 [pii] 10.1038/nbt1197.
122. Chang MC, Vargas SO, Hornick JL, Hirsch MS, Crum CP, Nucci MR. Embryonic stem cell transcription factors and D2-40 (podoplanin) as diagnostic immunohistochemical markers in ovarian germ cell tumors. Int J Gynecol Pathol. 2009;28(4):347–55. doi:10.1097/PGP.0b013e318195da86.
123. Sasaki H, Matsui Y. Epigenetic events in mammalian germ-cell development: reprogramming and beyond. Nat Rev Genet. 2008;9(2):129–140. doi:nrg2295 [pii] 10.1038/nrg2295.
124. Yokomizo S, Tsujimura A, Nonomura N, Kirime S, Takahara S, Okuyama A. Metachronous bilateral testicular tumors in a child. J Urol. 2001;166(6):2341. doi:S0022-5347(05)65584-X [pii].
125. Abell MR, Holtz F. Testicular neoplasms in infants and children. I. Tumors of germ cell origin. Cancer. 1963;16:965–81.
126. Carney JA, Kelalis PP, Lynn HB. Bilateral teratoma of testis in an infant. J Pediatr Surg. 1973;8(1):49–54.
127. Gustavson KH, Gamstorp I, Meurling S. Bilateral teratoma of testis in two brothers with 47, XXY Klinefelter's syndrome. Clin Genet. 1975;8(1):5–10.
128. Kurman RJ, Norris HJ. Endodermal sinus tumor of the ovary: a clinical and pathologic analysis of 71 cases. Cancer. 1976;38(6):2404–19.
129. Gershenson DM. Management of ovarian germ cell tumors. J Clin Oncol. 2007;25(20):2938–2943. doi:25/20/2938 [pii] 10.1200/JCO.2007.10.8738.
130. Yanai-Inbar I, Scully RE. Relation of ovarian dermoid cysts and immature teratomas: an analysis of 350 cases of immature teratoma and 10 cases of dermoid cyst with microscopic foci of immature tissue. Int J Gynecol Pathol. 1987;6(3):203–12.
131. Mahdi H, Kumar S, Seward S, Semaan A, Batchu R, Lockhart D, et al. Prognostic impact of laterality in malignant ovarian germ cell tumors. Int J Gynecol Cancer. 2011;21(2):257–262. doi:10.1097/IGC.0b013e31820581e5 00009577-201102000-00011 [pii].
132. Isaacs Jr H. Fetal intracranial teratoma. A review. Fetal Pediatr Pathol. 2014;33(5–6):289–92. doi:10.3109/15513815.2014.969558.
133. Anteby EY, Ron M, Revel A, Shimonovitz S, Ariel I, Hurwitz A. Germ cell tumors of the ovary arising after dermoid cyst resection: a long-term follow-up study. Obstet Gynecol. 1994;83(4):605–8.
134. Kim R, Bohm-Velez M. Familial ovarian dermoids. J Ultrasound Med. 1994;13(3):225–8.

135. Nezhat C, Kotikela S, Mann A, Hajhosseini B, Veeraswamy A, Lewis M. Familial cystic teratomas: four case reports and review of the literature. J Minim Invasive Gynecol. 2010;17(6):782–786. doi:S1553-4650(10)00333-X [pii] 10.1016/j.jmig.2010.06.006.
136. Poremba C, Dockhorn-Dworniczak B, Merritt V, Li CY, Heidl G, Tauber PF, et al. Immature teratomas of different origin carried by a pregnant mother and her fetus. Diagn Mol Pathol. 1993;2(2):131–6.
137. Giambartolomei C, Mueller CM, Greene MH, Korde LA. A mini-review of familial ovarian germ cell tumors: an additional manifestation of the familial testicular germ cell tumor syndrome. Cancer Epidemiol. 2009;33(1):31–36. doi:S1877-7821(09)00038-1 [pii] 10.1016/j.canep.2009.04.015.
138. Barksdale Jr EM, Obokhare I. Teratomas in infants and children. Curr Opin Pediatr. 2009;21(3):344–9. doi:10.1097/MOP.0b013e32832b41ee.
139. Tan C, Scotting PJ. Stem cell research points the way to the cell of origin for intracranial germ cell tumours. J Pathol. 2013;229(1):4–11. doi:10.1002/path.4098.
140. Perlman EJ, Valentine MB, Griffin CA, Look AT. Deletion of 1p36 in childhood endodermal sinus tumors by two-color fluorescence in situ hybridization: a pediatric oncology group study. Genes Chromosomes Cancer. 1996;16(1):15–20. doi:10.1002/(SICI)1098-2264(199605)16:1<15::AID-GCC2>3.0.CO;2-6 [pii] 10.1002/(SICI)1098-2264(199605)16:1<15::AID-GCC2>3.0.CO;2-6.
141. Bussey KJ, Lawce HJ, Olson SB, Arthur DC, Kalousek DK, Krailo M, et al. Chromosome abnormalities of eighty-one pediatric germ cell tumors: sex-, age-, site-, and histopathology-related differences – a Children's Cancer Group study. Genes Chromosomes Cancer. 1999;25(2):134–146. doi:10.1002/(SICI)1098-2264(199906)25:2<134::AID-GCC9>3.0.CO;2-Y [pii].
142. Perlman EJ, Hu J, Ho D, Cushing B, Lauer S, Castleberry RP. Genetic analysis of childhood endodermal sinus tumors by comparative genomic hybridization. J Pediatr Hematol Oncol. 2000;22(2):100–5.
143. Mostert M, Rosenberg C, Stoop H, Schuyer M, Timmer A, Oosterhuis W, et al. Comparative genomic and in situ hybridization of germ cell tumors of the infantile testis. Lab Investig. 2000;80(7):1055–64.
144. Schneider DT, Schuster AE, Fritsch MK, Hu J, Olson T, Lauer S, et al. Multipoint imprinting analysis indicates a common precursor cell for gonadal and nongonadal pediatric germ cell tumors. Cancer Res. 2001;61(19):7268–76.
145. van Echten J, Timmer A, van der Veen AY, Molenaar WM, de Jong B. Infantile and adult testicular germ cell tumors. a different pathogenesis? Cancer Genet Cytogenet. 2002;135(1):57–62. doi:S0165460801006434 [pii].
146. Veltman IM, Schepens MT, Looijenga LH, Strong LC, van Kessel AG. Germ cell tumours in neonates and infants: a distinct subgroup? APMIS. 2003;111(1):152–160; discussion 60. doi:apm1110119 [pii].
147. Veltman I, Veltman J, Janssen I, Hulsbergen-van de Kaa C, Oosterhuis W, Schneider D, et al. Identification of recurrent chromosomal aberrations in germ cell tumors of neonates and infants using genomewide array-based comparative genomic hybridization. Genes Chromosomes Cancer. 2005;43(4):367–376. doi:10.1002/gcc.20208.
148. Zahn S, Sievers S, Alemazkour K, Orb S, Harms D, Schulz WA, et al. Imbalances of chromosome arm 1p in pediatric and adult germ cell tumors are caused by true allelic loss: a combined comparative genomic hybridization and microsatellite analysis. Genes Chromosomes Cancer. 2006;45(11):995–1006. doi:10.1002/gcc.20363.
149. Harms D, Zahn S, Gobel U, Schneider DT. Pathology and molecular biology of teratomas in childhood and adolescence. Klin Padiatr. 2006;218(6):296–302. doi:10.1055/s-2006-942271.
150. Palmer RD, Foster NA, Vowler SL, Roberts I, Thornton CM, Hale JP, et al. Malignant germ cell tumours of childhood: new associations of genomic imbalance. Br J Cancer. 2007;96(4):667–676. doi:6603602 [pii] 10.1038/sj.bjc.6603602.
151. Mosbech CH, Rechnitzer C, Brok JS, Rajpert-De Meyts E, Hoei-Hansen CE. Recent advances in understanding the etiology and pathogenesis of pediatric germ cell tumors. J Pediatr Hematol Oncol. 2014;36(4):263–70. doi:10.1097/MPH.0000000000000125.
152. Looijenga LH, Rosenberg C, van Gurp RJ, Geelen E, van Echten-Arends J, de Jong B, et al. Comparative genomic hybridization of microdissected samples from different stages in the development of a seminoma and a non-seminoma. J Pathol. 2000;191(2):187–192. doi:10.1002/(SICI)1096-9896(200006)191:2<187::AID-PATH584>3.0.CO;2-T [pii] 10.1002/(SICI)1096-9896(200006)191:2<187::AID-PATH584>3.0.CO;2-T.
153. Fujita T, Igarashi J, Okawa ER, Gotoh T, Manne J, Kolla V, et al. CHD5, a tumor suppressor gene deleted from 1p36.31 in neuroblastomas. J Natl Cancer Inst. 2008;100(13):940–949. doi:djn176 [pii] 10.1093/jnci/djn176.
154. Veltman I, van Asseldonk M, Schepens M, Stoop H, Looijenga L, Wouters C, et al. A novel case of infantile sacral teratoma and a constitutional t(12;15)(q13;q25) pat. Cancer Genet Cytogenet. 2002;136(1):17–22. doi:S0165460801006665 [pii].
155. Veltman IM, Vreede LA, Cheng J, Looijenga LH, Janssen B, Schoenmakers EF, et al. Fusion of the SUMO/Sentrin-specific protease 1 gene SENP1 and the embryonic polarity-related mesoderm development gene MESDC2 in a patient with an infantile teratoma and a constitutional t(12;15)(q13;q25). Hum Mol Genet. 2005;14(14):1955–1963. doi:ddi200 [pii] 10.1093/hmg/ddi200.
156. Jin Y, Mertens F, Kullendorff CM, Panagopoulos I. Fusion of the tumor-suppressor gene CHEK2 and the gene for the regulatory subunit B of protein phosphatase 2 PPP2R2A in childhood teratoma. Neoplasia. 2006;8(5):413–8. doi:10.1593/neo.06139.

157. Kiefer MC, Brauer MJ, Powers VC, Wu JJ, Umansky SR, Tomei LD, et al. Modulation of apoptosis by the widely distributed Bcl-2 homologue Bak. Nature. 1995;374(6524):736–9. doi:10.1038/374736a0.
158. Poynter JN, Hooten AJ, Frazier AL, Ross JA. Associations between variants in KITLG, SPRY4, BAK1, and DMRT1 and pediatric germ cell tumors. Genes Chromosomes Cancer. 2012;51(3):266–71. doi:10.1002/gcc.20951.
159. Fritsch MK, Schneider DT, Schuster AE, Murdoch FE, Perlman EJ. Activation of Wnt/beta-catenin signaling in distinct histologic subtypes of human germ cell tumors. Pediatr Dev Pathol. 2006;9(2):115–31. doi:10.2350/08-05-0097.1.
160. Fustino N, Rakheja D, Ateek CS, Neumann JC, Amatruda JF. Bone morphogenetic protein signalling activity distinguishes histological subsets of paediatric germ cell tumours. Int J Androl. 2011;34(4 Pt 2):e218–33. doi:10.1111/j.1365-2605.2011.01186.x.
161. Alagaratnam S, Lind GE, Kraggerud SM, Lothe RA, Skotheim RI. The testicular germ cell tumour transcriptome. Int J Androl. 2011;34(4 Pt 2):e133–e150; discussion e50–1. doi:10.1111/j.1365-2605.2011.01169.x.
162. Okpanyi V, Schneider DT, Zahn S, Sievers S, Calaminus G, Nicholson JC, et al. Analysis of the adenomatous polyposis coli (APC) gene in childhood and adolescent germ cell tumors. Pediatr Blood Cancer. 2011;56(3):384–91. doi:10.1002/pbc.22669.
163. Palmer RD, Murray MJ, Saini HK, van Dongen S, Abreu-Goodger C, Muralidhar B, et al. Malignant germ cell tumors display common microRNA profiles resulting in global changes in expression of messenger RNA targets. Cancer Res. 2010;70(7):2911–2923. doi:0008-5472.CAN-09-3301 [pii] 10.1158/0008-5472.CAN-09-3301.
164. Bussey KJ, Lawce HJ, Himoe E, Shu XO, Heerema NA, Perlman EJ, et al. SNRPN methylation patterns in germ cell tumors as a reflection of primordial germ cell development. Genes Chromosomes Cancer. 2001;32(4):342–352. doi:10.1002/gcc.1199 [pii].
165. Schneider DT, Schuster AE, Fritsch MK, Calaminus G, Gobel U, Harms D, et al. Genetic analysis of mediastinal nonseminomatous germ cell tumors in children and adolescents. Genes Chromosomes Cancer. 2002;34(1):115–125. doi:10.1002/gcc.10053 [pii].
166. Sievers S, Alemazkour K, Zahn S, Perlman EJ, Gillis AJ, Looijenga LH, et al. IGF2/H19 imprinting analysis of human germ cell tumors (GCTs) using the methylation-sensitive single-nucleotide primer extension method reflects the origin of GCTs in different stages of primordial germ cell development. Genes Chromosomes Cancer. 2005;44(3):256–64. doi:10.1002/gcc.20237.
167. Rijlaarsdam MA, Looijenga LH. An oncofetal and developmental perspective on testicular germ cell cancer. Semin Cancer Biol. 2014;29:59–74. doi:S1044-579X(14)00091-1 [pii] 10.1016/j.semcancer.2014.07.003.
168. Scotting PJ. Are cranial germ cell tumours really tumours of germ cells? Neuropathol Appl Neurobiol. 2006;32(6):569–574. doi:NAN797 [pii] 10.1111/j.1365-2990.2006.00797.x.
169. Lee SH, Appleby V, Jeyapalan JN, Palmer RD, Nicholson JC, Sottile V, et al. Variable methylation of the imprinted gene, SNRPN, supports a relationship between intracranial germ cell tumours and neural stem cells. J Neuro Oncol. 2011;101(3):419–28. doi:10.1007/s11060-010-0275-9.
170. Yang J, Cai J, Zhang Y, Wang X, Li W, Xu J, et al. Induced pluripotent stem cells can be used to model the genomic imprinting disorder Prader-Willi syndrome. J Biol Chem. 2010;285(51):40303–40311. doi:M110.183392 [pii] 10.1074/jbc.M110.183392.
171. Stevens LC. Experimental production of testicular teratomas in mice. Proc Natl Acad Sci U S A. 1964;52:654–61.
172. Stevens LC, Varnum DS. The development of teratomas from parthenogenetically activated ovarian mouse eggs. Dev Biol. 1974;37(2):369–380. doi:0012-1606(74)90155-9 [pii].
173. Stevens LC. The development of transplantable teratocarcinomas from intratesticular grafts of pre- and postimplantation mouse embryos. Dev Biol. 1970;21(3):364–382. doi:0012-1606(70)90130-2 [pii].
174. Damjanov I, Solter D. Experimental teratoma. Curr Top Pathol. 1974;59:69–130.
175. Dunn TB, Andervont HB. Histology of some neoplasms and non-neoplastic lesions found in wild mice maintained under laboratory conditions. J Natl Cancer Inst. 1963;31:873–901.
176. Artzt K, Damjanov I. Spontaneous extragonadal teratocarcinoma in a mouse. Lab Anim Sci. 1978;28(5):584–6.
177. van Berlo RJ, Oosterhuis JW, Schrijnemakers E, Schoots CJ, de Jong B, Damjanov I. Yolk-sac carcinoma develops spontaneously as a late occurrence in slow-growing teratoid tumors produced from transplanted 7-day mouse embryos. Int J Cancer. 1990;45(1):153–5.
178. van Berlo RJ, de Jong B, Oosterhuis JW, Dijkhuizen T, Buist J, Dam A. Cytogenetic analysis of murine embryo-derived tumors. Cancer Res. 1990;50(11):3416–21.
179. Eakin GS, Behringer RR. Tetraploid development in the mouse. Dev Dyn. 2003;228(4):751–66. doi:10.1002/dvdy.10363.
180. Eakin GS, Hadjantonakis AK, Papaioannou VE, Behringer RR. Developmental potential and behavior of tetraploid cells in the mouse embryo. Dev Biol. 2005;288(1):150–159. doi:S0012-1606(05)00635-4 [pii] 10.1016/j.ydbio.2005.09.028.
181. Walt H, Oosterhuis JW, Stevens LC. Experimental testicular germ cell tumorigenesis in mouse strains with and without spontaneous tumours differs from development of germ cell tumours of the adult human testis. Int J Androl. 1993;16(4):267–71.
182. Gu KL, Zhang Q, Yan Y, Li TT, Duan FF, Hao J, et al. Pluripotency-associated miR-290/302 family of microRNAs promote the dismantling of naive pluripotency. Cell Res. 2016;26(3):350–366. doi:cr20162 [pii] 10.1038/cr.2016.2.

183. Stevens LC. Testicular teratomas in fetal mice. J Natl Cancer Inst. 1962;28:247–67.
184. Damjanov I, Solter D. Animal model of human disease: teratoma and teratocarcinoma. Am J Pathol. 1976;83(1):241–4.
185. Heaney JD, Lam MY, Michelson MV, Nadeau JH. Loss of the transmembrane but not the soluble kit ligand isoform increases testicular germ cell tumor susceptibility in mice. Cancer Res. 2008;68(13):5193–5197. doi:68/13/5193 [pii] 10.1158/0008-5472.CAN-08-0779.
186. Ohinata Y, Payer B, O'Carroll D, Ancelin K, Ono Y, Sano M, et al. Blimp1 is a critical determinant of the germ cell lineage in mice. Nature. 2005;436(7048):207–213. doi:nature03813 [pii] 10.1038/nature03813.
187. Magnusdottir E, Dietmann S, Murakami K, Gunesdogan U, Tang F, Bao S, et al. A tripartite transcription factor network regulates primordial germ cell specification in mice. Nat Cell Biol. 2013;15(8):905–915. doi:ncb2798 [pii] 10.1038/ncb2798.
188. Nakaki F, Hayashi K, Ohta H, Kurimoto K, Yabuta Y, Saitou M. Induction of mouse germ-cell fate by transcription factors in vitro. Nature. 2013;501(7466):222–226. doi:nature12417 [pii] 10.1038/nature12417.
189. Krentz AD, Murphy MW, Kim S, Cook MS, Capel B, Zhu R, et al. The DM domain protein DMRT1 is a dose-sensitive regulator of fetal germ cell proliferation and pluripotency. Proc Natl Acad Sci U S A. 2009;106(52):22323–22328. doi:0905431106 [pii] 10.1073/pnas.0905431106.
190. Youngren KK, Coveney D, Peng X, Bhattacharya C, Schmidt LS, Nickerson ML, et al. The Ter mutation in the dead end gene causes germ cell loss and testicular germ cell tumours. Nature. 2005;435(7040):360–364. doi:nature03595 [pii] 10.1038/nature03595.
191. Papaioannou VE, McBurney MW, Gardner RL, Evans MJ. Fate of teratocarcinoma cells injected into early mouse embryos. Nature. 1975;258(5530):70–3.
192. Yoshinaga K, Nishikawa S, Ogawa M, Hayashi S, Kunisada T, Fujimoto T, et al. Role of c-kit in mouse spermatogenesis: identification of spermatogonia as a specific site of c-kit expression and function. Development. 1991;113(2):689–99.
193. Noguchi T, Noguchi M. A recessive mutation (ter) causing germ cell deficiency and a high incidence of congenital testicular teratomas in 129/Sv-ter mice. J Natl Cancer Inst. 1985;75(2):385–92.
194. Bhattacharya C, Aggarwal S, Zhu R, Kumar M, Zhao M, Meistrich ML, et al. The mouse dead-end gene isoform alpha is necessary for germ cell and embryonic viability. Biochem Biophys Res Commun. 2007;355(1):194–199. doi:S0006-291X(07)00186-6 [pii] 10.1016/j.bbrc.2007.01.138.
195. Rapley EA, Turnbull C, Al Olama AA, Dermitzakis ET, Linger R, Huddart RA, et al. A genome-wide association study of testicular germ cell tumor. Nat Genet. 2009;41(7):807–810. doi:ng.394 [pii] 10.1038/ng.394.
196. Kanetsky PA, Mitra N, Vardhanabhuti S, Li M, Vaughn DJ, Letrero R, et al. Common variation in KITLG and at 5q31.3 predisposes to testicular germ cell cancer. Nat Genet. 2009;41(7):811–815. doi:ng.393 [pii] 10.1038/ng.393.
197. Wang L, Yamaguchi S, Burstein MD, Terashima K, Chang K, Ng HK, et al. Novel somatic and germline mutations in intracranial germ cell tumours. Nature. 2014;511(7508):241–245. doi:nature13296 [pii] 10.1038/nature13296.
198. De Backer A, Madern GC, Hakvoort-Cammel FG, Haentjens P, Oosterhuis JW, Hazebroek FW. Study of the factors associated with recurrence in children with sacrococcygeal teratoma. J Pediatr Surg. 2006;41(1):173–181; discussion -81. doi:S0022-3468(05)00763-3 [pii] 10.1016/j.jpedsurg.2005.10.022.
199. Derikx JP, De Backer A, van de Schoot L, Aronson DC, de Langen ZJ, van den Hoonaard TL, et al. Factors associated with recurrence and metastasis in sacrococcygeal teratoma. Br J Surg. 2006;93(12):1543–8. doi:10.1002/bjs.5379.
200. Fraumeni Jr JF, Li FP, Dalager N. Teratomas in children: epidemiologic features. J Natl Cancer Inst. 1973;51(5):1425–30.
201. Altman RP, Randolph JG, Lilly JR. Sacrococcygeal teratoma: American Academy of Pediatrics Surgical Section Survey-1973. J Pediatr Surg. 1974;9(3):389–98.
202. Noseworthy J, Lack EE, Kozakewich HP, Vawter GF, Welch KJ. Sacrococcygeal germ cell tumors in childhood: an updated experience with 118 patients. J Pediatr Surg. 1981;16(3):358–364. doi:S0022346881000626 [pii].
203. Ward SP, Dehner LP. Sacrococcygeal teratoma with nephroblastoma (Wilm's tumor): a variant of extragonadal teratoma in childhood. A histologic and ultrastructural study. Cancer. 1974; 33(5):1355–63.
204. Biskup W, Calaminus G, Schneider DT, Leuschner I, Gobel U. Teratoma with malignant transformation: experiences of the cooperative GPOH protocols MAKEI 83/86/89/96. Klin Padiatr. 2006;218(6):303–8. doi:10.1055/s-2006-942272.
205. Miles RM, Stewart Jr GS. Sacrococcygeal teratomas in adult. Ann Surg. 1974;179(5):676–83.
206. Willis R. The borderland of embryology and pathology. 2 ed. London: Butterworths; 1962.
207. Oosterhuis JW, Van Berlo RJ, De Jong B, Dam A, Buist J, Tamminga R, et al. Sacral teratoma with late recurrence of yolk sac tumor: human counterpart of embryo or yolk sac derived teratoma? J Urol Pathol. 1993;1:257–67.
208. Chen YH, Chang CH, Chen KC, Diau GY, Loh IW, Chu CC. Malignant transformation of a well-organized sacrococcygeal fetiform teratoma in a newborn male. J Formos Med Assoc. 2007;106(5):400–402. doi:S0929-6646(09)60326-0 [pii] 10.1016/S0929-6646(09)60326-0.

209. Billmire DF, Grosfeld JL. Teratomas in childhood: analysis of 142 cases. J Pediatr Surg. 1986;21(6):548–551. doi:S0022346886002014 [pii].
210. De Backer A, Madern GC, Hazebroek FW. Retroperitoneal germ cell tumors: a clinical study of 12 patients. J Pediatr Surg. 2005;40(9):1475–1481. doi:S0022-3468(05)00424-0 [pii] 10.1016/j.jpedsurg.2005.05.048.
211. Daugaard G, von der Maase H, Olsen J, Rorth M, Skakkebaek NE. Carcinoma-in-situ testis in patients with assumed extragonadal germ-cell tumours. Lancet. 1987;2(8558):528–530. doi:S0140-6736(87)92922-9 [pii].
212. Hailemariam S, Engeler DS, Bannwart F, Amin MB. Primary mediastinal germ cell tumor with intratubular germ cell neoplasia of the testis – further support for germ cell origin of these tumors: a case report. Cancer. 1997;79(5):1031–1036. doi:10.1002/(SICI)1097-0142(19970301)79:5<1031::AID-CNCR21>3.0.CO;2-1 [pii].
213. Fossa SD, Aass N, Heilo A, Daugaard G, Skakkebaek NE, Stenwig AE, et al. Testicular carcinoma in situ in patients with extragonadal germ-cell tumours: the clinical role of pretreatment biopsy. Ann Oncol. 2003; 14(9):1412–8.
214. De Backer A, Madern GC, Hakvoort-Cammel FG, Oosterhuis JW, Hazebroek FW. Mediastinal germ cell tumors: clinical aspects and outcomes in 7 children. Eur J Pediatr Surg. 2006;16(5):318–22. doi:10.1055/s-2006-924647.
215. Lack EE. Extragonadal germ cell tumors of the head and neck region: review of 16 cases. Hum Pathol. 1985;16(1):56–64.
216. Dehner LP, Mills A, Talerman A, Billman GF, Krous HF, Platz CE. Germ cell neoplasms of head and neck soft tissues: a pathologic spectrum of teratomatous and endodermal sinus tumors. Hum Pathol. 1990;21(3):309–18.
217. De Backer A, Madern GC, van de Ven CP, Tibboel D, Hazebroek FW. Strategy for management of newborns with cervical teratoma. J Perinat Med. 2004;32(6):500–8. doi:10.1515/JPM.2004.122.
218. Azizkhan RG, Haase GM, Applebaum H, Dillon PW, Coran AG, King PA, et al. Diagnosis, management, and outcome of cervicofacial teratomas in neonates: a Childrens Cancer Group study. J Pediatr Surg. 1995;30(2):312–316. doi:0022-3468(95)90580-4 [pii].
219. Kuhn JJ, Schoem SR, Warnock GR. Squamous cell carcinoma arising in a benign teratoma of the maxilla. Otolaryngol Head Neck Surg. 1996;114(3):447–452. doi:S019459989600109X [pii].
220. Heffner DK, Hyams VJ. Teratocarcinosarcoma (malignant teratoma?) of the nasal cavity and paranasal sinuses A clinicopathologic study of 20 cases. Cancer. 1984;53(10):2140–54.
221. Houri T, Hashimoto N, Ibayashi N, Mori T, Fujimoto M, Ueda S, et al. Chromosomal translocation, t(1;11)(q12;p15), in an extragonadal immature teratoma. Cancer Genet Cytogenet. 1997;97(1):79–80. doi:S0165460897001969 [pii].
222. Pai SA, Naresh KN, Masih K, Ramarao C, Borges AM. Teratocarcinosarcoma of the paranasal sinuses: a clinicopathologic and immunohistochemical study. Hum Pathol. 1998;29(7):718–22.
223. De Backer A, Madern GC, Wolffenbuttel KP, Oosterhuis JW, Hakvoort-Cammel FG, Hazebroek FW. Testicular germ cell tumors in children: management and outcome in a series of 20 patients. J Pediatr Urol. 2006;2(3):197–201. doi:S1477-5131(05)00139-7 [pii] 10.1016/j.jpurol.2005.08.001.
224. Metcalfe PD, Farivar-Mohseni H, Farhat W, McLorie G, Khoury A, Bagli DJ. Pediatric testicular tumors: contemporary incidence and efficacy of testicular preserving surgery. J Urol. 2003;170(6 Pt 1):2412–2415; discussion 5–6. doi:10.1097/01.ju.0000097383.09743.f9.
225. Lee SD, Korean Society of Pediatric Urology. Epidemiological and clinical behavior of prepubertal testicular tumors in Korea. J Urol. 2004;172(2):674–678. doi:S0022-5347(05)61714-4 [pii] 10.1097/01.ju.0000129571.13955.6b.
226. Fan R, Zhang J, Cheng L, Lin J. Testicular and paratesticular pathology in the pediatric population: a 20 year experience at Riley hospital for children. Pathol Res Pract. 2013;209(7):404–408. doi:S0344-0338(13)00086-1 [pii] 10.1016/j.prp.2013.04.002.
227. Kato K, Ijiri R, Tanaka Y, Toyoda Y, Chiba K, Kitami K. Testicular immature teratoma with primitive neuroectodermal tumor in early childhood. J Urol. 2000;164(6):2068–2069. doi:S0022-5347(05)66968-6 [pii].
228. Manivel JC, Reinberg Y, Niehans GA, Fraley EE. Intratubular germ cell neoplasia in testicular teratomas and epidermoid cysts. Correlation with prognosis and possible biologic significance. Cancer. 1989;64(3):715–20.
229. Skakkebaek NE, Rajpert-De Meyts E, Main KM. Testicular dysgenesis syndrome: an increasingly common developmental disorder with environmental aspects. Hum Reprod. 2001;16(5):972–8.
230. Shu XO, Nesbit ME, Buckley JD, Krailo MD, Robinson LL. An exploratory analysis of risk factors for childhood malignant germ-cell tumors: report from the Childrens Cancer Group (Canada, United States). Cancer Causes Control. 1995;6(3):187–98.
231. Poynter JN, Radzom AH, Spector LG, Puumala S, Robison LL, Chen Z, et al. Family history of cancer and malignant germ cell tumors in children: a report from the Children's Oncology Group. Cancer Causes Control. 2010;21(2):181–9. doi:10.1007/s10552-009-9448-2.
232. Gordon A, Lipton D, Woodruff JD. Dysgerminoma: a review of 158 cases from the Emil Novak Ovarian Tumor Registry. Obstet Gynecol. 1981;58(4):497–504.
233. Comerci Jr JT, Licciardi F, Bergh PA, Gregori C, Breen JL. Mature cystic teratoma: a clinicopathologic evaluation of 517 cases and review of the literature. Obstet Gynecol. 1994;84(1):22–8.
234. De Backer A, Madern GC, Oosterhuis JW, Hakvoort-Cammel FG, Hazebroek FW. Ovarian

germ cell tumors in children: a clinical study of 66 patients. Pediatr Blood Cancer. 2006;46(4):459–64. doi:10.1002/pbc.20633.
235. Hartshorne GM, Lyrakou S, Hamoda H, Oloto E, Ghafari F. Oogenesis and cell death in human prenatal ovaries: what are the criteria for oocyte selection? Mol Hum Reprod. 2009;15(12):805–819. doi:gap055 [pii] 10.1093/molehr/gap055.
236. Anderson RA, Fulton N, Cowan G, Coutts S, Saunders PT. Conserved and divergent patterns of expression of DAZL, VASA and OCT4 in the germ cells of the human fetal ovary and testis. BMC Dev Biol. 2007;7:136. doi:1471-213X-7-136 [pii] 10.1186/1471-213X-7-136.
237. Byskov AG, Hoyer PE, Yding Andersen C, Kristensen SG, Jespersen A, Mollgard K. No evidence for the presence of oogonia in the human ovary after their final clearance during the first two years of life. Hum Reprod. 2011;26(8):2129–2139. doi:der145 [pii] 10.1093/humrep/der145.
238. Zhang C, Berney DM, Hirsch MS, Cheng L, Ulbright TM. Evidence supporting the existence of benign teratomas of the postpubertal testis: a clinical, histopathologic, and molecular genetic analysis of 25 cases. Am J Surg Pathol. 2013;37(6):827–35. doi:10.1097/PAS.0b013e31827dcc4c.
239. Semjen D, Kalman E, Tornoczky T, Szuhai K. Further evidence of the existence of benign teratomas of the postpubertal testis. Am J Surg Pathol. 2014;38(4):580–581. doi:10.1097/PAS.0000000000000186 00000478-201404000-00019 [pii].
240. Lewis BD, Hurt RD, Payne WS, Farrow GM, Knapp RH, Muhm JR. Benign teratomas of the mediastinum. J Thorac Cardiovasc Surg. 1983;86(5):727–31.
241. Williamson SR. Germ cell tumors of the mediastinum. In: Marchevsky AM, editor. Pathology of the mediastinum. Cambridge: Cambridge University Press; 2014. p. 146–68.
242. Stendel R, Pietila TA, Lehmann K, Kurth R, Suess O, Brock M. Ruptured intracranial dermoid cysts. Surg Neurol. 2002;57(6):391–398; discussion 8. doi:S0090301902007231 [pii].
243. Rosenblum MK. Germ Cell Tumours. In: Louis DN, editor. World Health Organization classification of tumours of the central nervous system. 4th ed. Lyon: IARC Press; 2007. p. 197–204.
244. Ulbright TM, Amin MB, Balzer B, Berney DM, Epstein JI, Guo C, et al. Germ cell tumours. In: Moch H, Humphrey PA, Ulbright TM, Reuter VE, editors. WHO classification of tumours of the urinary system and male genital organs. 4th ed. Lyon: IARC Press; 2016. p. 189–226.
245. Oosterhuis JW, Looijenga LH. The biology of human germ cell tumours: retrospective speculations and new prospectives. Eur Urol. 1993;23(1):245–50.
246. Kleinsmith LJ, Pierce Jr GB. Multipotentiality of single embryonal carcinoma cells. Cancer Res. 1964;24:1544–51.
247. Honecker F, Stoop H, Mayer F, Bokemeyer C, Castrillon DH, Lau YF, et al. Germ cell lineage differentiation in non-seminomatous germ cell tumours. J Pathol. 2006;208(3):395–400. doi:10.1002/path.1872.
248. Looijenga LH, Gillis AJ, Stoop H, Hersmus R, Oosterhuis JW. Relevance of microRNAs in normal and malignant development, including human testicular germ cell tumours. Int J Androl. 2007;30(4):304–314; discussion 14–5. doi:IJA765 [pii] 10.1111/j.1365-2605.2007.00765.x.
249. Poynter JN, Bestrashniy JR, Silverstein KA, Hooten AJ, Lees C, Ross JA, et al. Cross platform analysis of methylation, miRNA and stem cell gene expression data in germ cell tumors highlights characteristic differences by tumor histology. BMC Cancer. 2015;15:769. doi:10.1186/s12885-015-1796-6 10.1186/s12885-015-1796-6 [pii].
250. Hoozemans DA, Schats R, Lambalk CB, Homburg R, Hompes PG. Human embryo implantation: current knowledge and clinical implications in assisted reproductive technology. Reprod BioMed Online. 2004;9(6):692–715.
251. Morgani SM, Canham MA, Nichols J, Sharov AA, Migueles RP, Ko MS, et al. Totipotent embryonic stem cells arise in ground-state culture conditions. Cell Rep. 2013;3(6):1945–1957. doi:S2211-1247(13)00216-7 [pii] 10.1016/j.celrep.2013.04.034.
252. Batata MA, Whitmore Jr WF, Hilaris BS, Tokita N, Grabstald H. Cancer of the undescended or maldescended testis. AJR Am J Roentgenol. 1976;126(2):302–12. doi:10.2214/ajr.126.2.302.
253. Batata MA, Chu FC, Hilaris BS, Whitmore WF, Golbey RB. Testicular cancer in cryptorchids. Cancer. 1982;49(5):1023–30.
254. Halme A, Kellokumpu-Lehtinen P, Lehtonen T, Teppo L. Morphology of testicular germ cell tumours in treated and untreated cryptorchidism. Br J Urol. 1989;64(1):78–83.
255. Nettersheim D, Jostes S, Sharma R, Schneider S, Hofmann A, Ferreira HJ, et al. BMP inhibition in seminomas initiates acquisition of pluripotency via NODAL signaling resulting in reprogramming to an embryonal carcinoma. PLoS Genet. 2015;11(7):e1005415. doi:10.1371/journal.pgen.1005415 PGENETICS-D-15-00545 [pii].
256. Spiller CM, Feng CW, Jackson A, Gillis AJ, Rolland AD, Looijenga LH, et al. Endogenous Nodal signaling regulates germ cell potency during mammalian testis development. Development. 2012;139(22):4123–32. doi:10.1242/dev.083006.
257. Spiller CM, Gillis AJ, Burnet G, Stoop H, Koopman P, Bowles J, et al. Cripto: Expression, epigenetic regulation and potential diagnostic use in testicular germ cell tumors. Mol Oncol. 2016;10(4):526–37. doi:10.1016/j.molonc.2015.11.003.
258. Bredael JJ, Vugrin D, Whitmore Jr WF. Autopsy findings in 154 patients with germ cell tumors of the testis. Cancer. 1982;50(3):548–51.
259. Zaloudek CJ, Tavassoli FA, Norris HJ. Dysgerminoma with syncytiotrophoblastic giant cells. A histologically and clinically distinctive

subtype of dysgerminoma. Am J Surg Pathol. 1981; 5(4):361–7.
260. Moreira A. Germ cell tumours of the mediastinum. In: Travis W, editor. World Health Organization classification of tumours of the lung, pleura, thymus and heart. 4th ed. Lyon: IARC Press; 2015. p. 244–66.
261. Pugh R. Testicular tumours – introduction. In: Pugh R, editor. Pathology of the testis. Oxford: Blackwell Scientific; 1976. p. 139–59.
262. Oosterhuis JW, Kersemaekers AM, Jacobsen GK, Timmer A, Steyerberg EW, Molier M, et al. Morphology of testicular parenchyma adjacent to germ cell tumours. An interim report. APMIS. 2003;111(1):32–40; discussion 1–2.
263. Matsutani M, Sano K, Takakura K, Fujimaki T, Nakamura O, Funata N, et al. Primary intracranial germ cell tumors: a clinical analysis of 153 histologically verified cases. J Neurosurg. 1997;86(3):446–55. doi:10.3171/jns.1997.86.3.0446.
264. Moran CA, Suster S. Primary germ cell tumors of the mediastinum: I. Analysis of 322 cases with special emphasis on teratomatous lesions and a proposal for histopathologic classification and clinical staging. Cancer. 1997;80(4):681–90.
265. Nichols CR, Saxman S, Williams SD, Loehrer PJ, Miller ME, Wright C, et al. Primary mediastinal nonseminomatous germ cell tumors. A modern single institution experience. Cancer. 1990;65(7):1641–6.
266. Vicus D, Beiner ME, Klachook S, Le LW, Laframboise S, Mackay H. Pure dysgerminoma of the ovary 35 years on: a single institutional experience. Gynecol Oncol. 2010;117(1):23–6. doi:10.1016/j.ygyno.2009.12.024.
267. Kurman RJ, Norris HJ. Embryonal carcinoma of the ovary: a clinicopathologic entity distinct from endodermal sinus tumor resembling embryonal carcinoma of the adult testis. Cancer. 1976;38(6):2420–33.
268. Oosterhuis JW, Peeters SH, Smit VT, Stoop H, Looijenga LH, Elzevier HW, et al. Patient with two secondary somatic-type malignancies in a late recurrence of a testicular non-seminoma: illustration of potential and flaw of the cancer stem cell therapy concept. Int J Dev Biol. 2013;57(2–4):153–7. doi:10.1387/ijdb.130141jo.
269. Hanahan D, Weinberg RA. Hallmarks of cancer: the next generation. Cell. 2011;144(5):646–74. doi:10.1016/j.cell.2011.02.013.
270. Oosterhuis JW. The metastasis of human teratomas. In: Damjanov I, Solter D, Knowles BB, editors. The human teratomas. Clifton: Humana Press; 1983. p. 137–71.
271. Oosterhuis JW, Suurmeyer AJ, Sleyfer DT, Koops HS, Oldhoff J, Fleuren G. Effects of multiple-drug chemotherapy (cis-diammine-dichloroplatinum, bleomycin, and vinblastine) on the maturation of retroperitoneal lymph node metastases of nonseminomatous germ cell tumors of the testis. No evidence for De Novo induction of differentiation. Cancer. 1983;51(3):408–16.
272. De Graaff WE, Oosterhuis JW, Van der Linden S, Homan van der Heide JN, Schraffordt Koops H, Sleijfer DTh. Residual mature teratoma after chemotherapy for non seminomatous germ cell tumors of the testis occurs significantly less often in lung, than in retroperitoneal lymph node metastases. J Urol Pathol. 1991;1:75–81.
273. Trabert B, Chen J, Devesa SS, Bray F, McGlynn KA. International patterns and trends in testicular cancer incidence, overall and by histologic subtype, 1973–2007. Andrology. 2015;3(1):4–12. doi:10.1111/andr.293.
274. Arora RS, Alston RD, Eden TO, Geraci M, Birch JM. Comparative incidence patterns and trends of gonadal and extragonadal germ cell tumors in England, 1979 to 2003. Cancer. 2012;118(17):4290–7. doi:10.1002/cncr.27403.
275. Cools M, Drop SL, Wolffenbuttel KP, Oosterhuis JW, Looijenga LH. Germ cell tumors in the intersex gonad: old paths, new directions, moving frontiers. Endocr Rev. 2006;27(5):468–84. doi:10.1210/er.2006-0005.
276. Satge D, Sasco AJ, Cure H, Leduc B, Sommelet D, Vekemans MJ. An excess of testicular germ cell tumors in Down's syndrome: three case reports and a review of the literature. Cancer. 1997;80(5):929–35.
277. Volkl TM, Langer T, Aigner T, Greess H, Beck JD, Rauch AM, et al. Klinefelter syndrome and mediastinal germ cell tumors. Am J Med Genet A. 2006;140(5):471–81. doi:10.1002/ajmg.a.31103.
278. Bray F, Richiardi L, Ekbom A, Pukkala E, Cuninkova M, Moller H. Trends in testicular cancer incidence and mortality in 22 European countries: continuing increases in incidence and declines in mortality. Int J Cancer. 2006;118(12):3099–111. doi:10.1002/ijc.21747.
279. Dieckmann KP, Kulejewski M, Pichlmeier U, Loy V. Diagnosis of contralateral testicular intraepithelial neoplasia (TIN) in patients with testicular germ cell cancer: systematic two-site biopsies are more sensitive than a single random biopsy. Eur Urol. 2007;51(1):175–183; discussion 83–5. doi:10.1016/j.eururo.2006.05.051.
280. Dieckmann KP, Anheuser P, Sattler F, Von Kugelgen T, Matthies C, Ruf C. Sequential bilateral testicular tumours presenting with intervals of 20 years and more. BMC Urol. 2013;13:71. doi:10.1186/1471-2490-13-71.
281. Thomas GM, Dembo AJ, Hacker NF, DePetrillo AD. Current therapy for dysgerminoma of the ovary. Obstet Gynecol. 1987;70(2):268–75.
282. De Palo G, Pilotti S, Kenda R, Ratti E, Musumeci R, Mangioni C, et al. Natural history of dysgerminoma. Am J Obstet Gynecol. 1982;143(7):799–807.
283. Coffey J, Linger R, Pugh J, Dudakia D, Sokal M, Easton DF, et al. Somatic KIT mutations occur predominantly in seminoma germ cell tumors and are not predictive of bilateral disease: report of 220 tumors and review of literature. Genes Chromosomes Cancer. 2008;47(1):34–42. doi:10.1002/gcc.20503.
284. Heimdal K, Olsson H, Tretli S, Flodgren P, Borresen AL, Fossa SD. Familial testicular cancer in Norway and southern Sweden. Br J Cancer. 1996;73(7):964–9.

285. Czene K, Lichtenstein P, Hemminki K. Environmental and heritable causes of cancer among 9.6 million individuals in the Swedish Family-Cancer Database. Int J Cancer. 2002;99(2):260–6. doi:10.1002/ijc.10332.
286. Hemminki K, Li X. Familial risk in testicular cancer as a clue to a heritable and environmental aetiology. Br J Cancer. 2004;90(9):1765–70. doi:10.1038/sj.bjc.6601714.
287. Akyuz C, Koseoglu V, Gogus S, Balci S, Buyukpamukcu M. Germ cell tumours in a brother and sister. Acta Paediatr. 1997;86(6):668–9.
288. Greene MH, Mai PL, Loud JT, Pathak A, Peters JA, Mirabello L, et al. Familial testicular germ cell tumors (FTGCT) – overview of a multidisciplinary etiologic study. Andrology. 2015;3(1):47–58. doi:10.1111/andr.294.
289. Hartmann JT, Fossa SD, Nichols CR, Droz JP, Horwich A, Gerl A, et al. Incidence of metachronous testicular cancer in patients with extragonadal germ cell tumors. J Natl Cancer Inst. 2001;93(22):1733–8.
290. Kersemaekers AM, Honecker F, Stoop H, Cools M, Molier M, Wolffenbuttel K, et al. Identification of germ cells at risk for neoplastic transformation in gonadoblastoma: an immunohistochemical study for OCT3/4 and TSPY. Hum Pathol. 2005;36(5):512–21. doi:10.1016/j.humpath.2005.02.016.
291. Cools M, van Aerde K, Kersemaekers AM, Boter M, Drop SL, Wolffenbuttel KP, et al. Morphological and immunohistochemical differences between gonadal maturation delay and early germ cell neoplasia in patients with undervirilization syndromes. J Clin Endocrinol Metab. 2005;90(9):5295–303. doi:10.1210/jc.2005-0139.
292. Stoop H, Honecker F, van de Geijn GJ, Gillis AJ, Cools MC, de Boer M, et al. Stem cell factor as a novel diagnostic marker for early malignant germ cells. J Pathol. 2008;216(1):43–54. doi:10.1002/path.2378.
293. Lin Y, Yang Y, Li W, Chen Q, Li J, Pan X, et al. Reciprocal regulation of Akt and Oct4 promotes the self-renewal and survival of embryonal carcinoma cells. Mol Cell. 2012;48(4):627–40. doi:10.1016/j.molcel.2012.08.030.
294. Oosterhuis JW, Castedo SM, de Jong B, Cornelisse CJ, Dam A, Sleijfer DT, et al. Ploidy of primary germ cell tumors of the testis. Pathogenetic and clinical relevance. Lab Investig. 1989;60(1):14–21.
295. Litchfield K, Summersgill B, Yost S, Sultana R, Labreche K, Dudakia D, et al. Whole-exome sequencing reveals the mutational spectrum of testicular germ cell tumours. Nat Commun. 2015;6:5973. doi:10.1038/ncomms6973.
296. Kraggerud SM, Hoei-Hansen CE, Alagaratnam S, Skotheim RI, Abeler VM, Rajpert-De Meyts E, et al. Molecular characteristics of malignant ovarian germ cell tumors and comparison with testicular counterparts: implications for pathogenesis. Endocr Rev. 2013;34(3):339–76. doi:10.1210/er.2012-1045.
297. Oosterhuis JW, Rammeloo RH, Cornelisse CJ, De Jong B, Dam A, Sleijfer DT. Ploidy of malignant mediastinal germ-cell tumors. Hum Pathol. 1990;21(7):729–32.
298. Terashima K, Yu A, Chow WY, Hsu WC, Chen P, Wong S, et al. Genome-wide analysis of DNA copy number alterations and loss of heterozygosity in intracranial germ cell tumors. Pediatr Blood Cancer. 2014;61(4):593–600. doi:10.1002/pbc.24833.
299. Atkin NB, Baker MC. Specific chromosome change, i(12p), in testicular tumours? Lancet. 1982;2(8311):1349.
300. Gibas Z, Prout Jr GR, Sandberg AA. Malignant teratoma of the testis with an isochromosome no. 12, i(12p), as the sole structural cytogenetic abnormality. J Urol. 1984;131(4):762–3.
301. Castedo SM, de Jong B, Oosterhuis JW, Seruca R, Idenburg VJ, Dam A, et al. Chromosomal changes in human primary testicular nonseminomatous germ cell tumors. Cancer Res. 1989;49(20):5696–701.
302. Castedo SM, de Jong B, Oosterhuis JW, Seruca R, te Meerman GJ, Dam A, et al. Cytogenetic analysis of ten human seminomas. Cancer Res. 1989;49(2):439–43.
303. van Echten J, Oosterhuis JW, Looijenga LH, van de Pol M, Wiersema J, te Meerman GJ, et al. No recurrent structural abnormalities apart from i(12p) in primary germ cell tumors of the adult testis. Genes Chromosomes Cancer. 1995;14(2):133–44.
304. Looijenga LH, Zafarana G, Grygalewicz B, Summersgill B, Debiec-Rychter M, Veltman J, et al. Role of gain of 12p in germ cell tumour development. APMIS. 2003;111(1):161–171; discussion 72–3.
305. Speleman F, De Potter C, Dal Cin P, Mangelschots K, Ingelaere H, Laureys G, et al. i(12p) in a malignant ovarian tumor. Cancer Genet Cytogenet. 1990;45(1):49–53.
306. Hamers A, de Jong B, Suijkerbuijk RF, Geurts van Kessel A, Oosterhuis JW, van Echten J, et al. A 46,XY female with mixed gonadal dysgenesis and a 48,XY, +7, +i(12p) chromosome pattern in a primary gonadal tumor. Cancer Genet Cytogenet. 1991;57(2):219–24.
307. Dal Cin P, Drochmans A, Moerman P, Van den Berghe H. Isochromosome 12p in mediastinal germ cell tumor. Cancer Genet Cytogenet. 1989;42(2):243–51.
308. Sung MT, Maclennan GT, Lopez-Beltran A, Zhang S, Montironi R, Cheng L. Primary mediastinal seminoma: a comprehensive assessment integrated with histology, immunohistochemistry, and fluorescence in situ hybridization for chromosome 12p abnormalities in 23 cases. Am J Surg Pathol. 2008;32(1):146–55. doi:10.1097/PAS.0b013e3181379edf.
309. de Bruin TW, Slater RM, Defferrari R, Geurts van Kessel A, Suijkerbuijk RF, Jansen G, et al. Isochromosome 12p-positive pineal germ cell tumor. Cancer Res. 1994;54(6):1542–4.
310. Schneider DT, Zahn S, Sievers S, Alemazkour K, Reifenberger G, Wiestler OD, et al. Molecular genetic analysis of central nervous system germ cell tumors with comparative genomic hybridization. Mod Pathol Off J U S Can Acad Pathol Inc. 2006;19(6):864–73. doi:10.1038/modpathol.3800607.

311. Roelofs H, Mostert MC, Pompe K, Zafarana G, van Oorschot M, van Gurp RJ, et al. Restricted 12p amplification and RAS mutation in human germ cell tumors of the adult testis. Am J Pathol. 2000;157(4):1155–66. doi:10.1016/S0002-9440(10)64631-7.
312. Zafarana G, Gillis AJ, van Gurp RJ, Olsson PG, Elstrodt F, Stoop H, et al. Coamplification of DAD-R, SOX5, and EKI1 in human testicular seminomas, with specific overexpression of DAD-R, correlates with reduced levels of apoptosis and earlier clinical manifestation. Cancer Res. 2002;62(6):1822–31.
313. Rodriguez S, Jafer O, Goker H, Summersgill BM, Zafarana G, Gillis AJ, et al. Expression profile of genes from 12p in testicular germ cell tumors of adolescents and adults associated with i(12p) and amplification at 12p11.2–p12.1. Oncogene. 2003;22(12):1880–91. doi:10.1038/sj.onc.1206302.
314. Zafarana G, Grygalewicz B, Gillis AJ, Vissers LE, van de Vliet W, van Gurp RJ, et al. 12p-amplicon structure analysis in testicular germ cell tumors of adolescents and adults by array CGH. Oncogene. 2003;22(48):7695–701. doi:10.1038/sj.onc.1207011.
315. Geurts van Kessel A, van Drunen E, de Jong B, Oosterhuis JW, Langeveld A, Mulder MP. Chromosome 12q heterozygosity is retained in i(12p)-positive testicular germ cell tumor cells. Cancer Genet Cytogenet. 1989;40(1):129–34.
316. Sinke RJ, Suijkerbuijk RF, de Jong B, Oosterhuis JW. Geurts van Kessel A. Uniparental origin of i(12p) in human germ cell tumors. Genes Chromosomes Cancer. 1993;6(3):161–5.
317. Looijenga LH, Gillis AJ, Stoop HJ, Hersmus R, Oosterhuis JW. Chromosomes and expression in human testicular germ-cell tumors: insight into their cell of origin and pathogenesis. Ann N Y Acad Sci. 2007;1120:187–214. doi:10.1196/annals.1411.000.
318. Korkola JE, Houldsworth J, Bosl GJ, Chaganti RS. Molecular events in germ cell tumours: linking chromosome-12 gain, acquisition of pluripotency and response to cisplatin. BJU International. 2009;104(9 Pt B):1334–8. doi:10.1111/j.1464-410X.2009.08855.x.
319. Adamah DJ, Gokhale PJ, Eastwood DJ, Rajpert De-Meyts E, Goepel J, Walsh JR, et al. Dysfunction of the mitotic:meiotic switch as a potential cause of neoplastic conversion of primordial germ cells. Int J Androl. 2006;29(1):219–27. doi:10.1111/j.1365-2605.2005.00569.x.
320. de Jong B, Oosterhuis JW, Castedo SM, Vos A, te Meerman GJ. Pathogenesis of adult testicular germ cell tumors. A cytogenetic model. Cancer Genet Cytogenet. 1990;48(2):143–67.
321. Lee AS, Tang C, Rao MS, Weissman IL, Wu JC. Tumorigenicity as a clinical hurdle for pluripotent stem cell therapies. Nat Med. 2013;19(8):998–1004. doi:10.1038/nm.3267.
322. McIntyre A, Summersgill B, Grygalewicz B, Gillis AJ, Stoop J, van Gurp RJ, et al. Amplification and overexpression of the KIT gene is associated with progression in the seminoma subtype of testicular germ cell tumors of adolescents and adults. Cancer Res. 2005;65(18):8085–9. doi:10.1158/0008-5472.CAN-05-0471.
323. Samaniego F, Rodriguez E, Houldsworth J, Murty VV, Ladanyi M, Lele KP, et al. Cytogenetic and molecular analysis of human male germ cell tumors: chromosome 12 abnormalities and gene amplification. Genes Chromosomes Cancer. 1990;1(4):289–300.
324. Cutcutache I, Suzuki Y, Tan IB, Ramgopal S, Zhang S, Ramnarayanan K, et al. Exome-wide sequencing shows low mutation rates and identifies novel mutated genes in seminomas. Eur Urol. 2015;68(1):77–83. doi:10.1016/j.eururo.2014.12.040.
325. Mulder MP, Keijzer W, Verkerk A, Boot AJ, Prins ME, Splinter TA, et al. Activated ras genes in human seminoma: evidence for tumor heterogeneity. Oncogene. 1989;4(11):1345–51.
326. Tian Q, Frierson Jr HF, Krystal GW, Moskaluk CA. Activating c-kit gene mutations in human germ cell tumors. Am J Pathol. 1999;154(6):1643–7. doi:10.1016/S0002-9440(10)65419-3.
327. Kemmer K, Corless CL, Fletcher JA, McGreevey L, Haley A, Griffith D, et al. KIT mutations are common in testicular seminomas. Am J Pathol. 2004;164(1):305–13. doi:10.1016/S0002-9440(10)63120-3.
328. Bignell G, Smith R, Hunter C, Stephens P, Davies H, Greenman C, et al. Sequence analysis of the protein kinase gene family in human testicular germ-cell tumors of adolescents and adults. Genes Chromosomes Cancer. 2006;45(1):42–6. doi:10.1002/gcc.20265.
329. Goddard NC, McIntyre A, Summersgill B, Gilbert D, Kitazawa S, Shipley J. KIT and RAS signalling pathways in testicular germ cell tumours: new data and a review of the literature. Int J Androl. 2007;30(4):337–348; discussion 49. doi:10.1111/j.1365-2605.2007.00769.x.
330. Hoei-Hansen CE, Kraggerud SM, Abeler VM, Kaern J, Rajpert-De Meyts E, Lothe RA. Ovarian dysgerminomas are characterised by frequent KIT mutations and abundant expression of pluripotency markers. Mol Cancer. 2007;6:12. doi:10.1186/1476-4598-6-12.
331. Cheng L, Roth LM, Zhang S, Wang M, Morton MJ, Zheng W, et al. KIT gene mutation and amplification in dysgerminoma of the ovary. Cancer. 2011;117(10):2096–103. doi:10.1002/cncr.25794.
332. Hersmus R, Stoop H, van de Geijn GJ, Eini R, Biermann K, Oosterhuis JW, et al. Prevalence of c-KIT mutations in gonadoblastoma and dysgerminomas of patients with disorders of sex development (DSD) and ovarian dysgerminomas. PLoS One. 2012;7(8):e43952. doi:10.1371/journal.pone.0043952.
333. Przygodzki RM, Hubbs AE, Zhao FQ, O'Leary TJ. Primary mediastinal seminomas: evidence of single and multiple KIT mutations. Lab Investig. 2002;82(10):1369–75.
334. Fukushima S, Otsuka A, Suzuki T, Yanagisawa T, Mishima K, Mukasa A, et al. Mutually exclusive mutations of KIT and RAS are associated with KIT mRNA expression and chromosomal instability in primary intracranial pure germinomas. Acta Neuropathol. 2014;127(6):911–25. doi:10.1007/s00401-014-1247-5.

335. Mol CD, Dougan DR, Schneider TR, Skene RJ, Kraus ML, Scheibe DN, et al. Structural basis for the autoinhibition and STI-571 inhibition of c-Kit tyrosine kinase. J Biol Chem. 2004;279(30):31655–63. doi:10.1074/jbc.M403319200.
336. Przygodzki RM, Moran CA, Suster S, Khan MA, Swalsky PA, Bakker A, et al. Primary mediastinal and testicular seminomas: a comparison of K-ras-2 gene sequence and p53 immunoperoxidase analysis of 26 cases. Hum Pathol. 1996;27(9): 975–9.
337. Tate G, Suzuki T, Kishimoto K, Mitsuya T. A c-KIT codon 816 mutation, D816H, in the testicular germ cell tumor: case report of a Japanese patient with bilateral testicular seminomas. Acta Med Okayama. 2005;59(1):33–6.
338. Looijenga LH, de Leeuw H, van Oorschot M, van Gurp RJ, Stoop H, Gillis AJ, et al. Stem cell factor receptor (c-KIT) codon 816 mutations predict development of bilateral testicular germ-cell tumors. Cancer Res. 2003;63(22):7674–8.
339. Biermann K, Goke F, Nettersheim D, Eckert D, Zhou H, Kahl P, et al. c-KIT is frequently mutated in bilateral germ cell tumours and down-regulated during progression from intratubular germ cell neoplasia to seminoma. J Pathol. 2007;213(3):311–8. doi:10.1002/path.2225.
340. Rapley EA, Hockley S, Warren W, Johnson L, Huddart R, Crockford G, et al. Somatic mutations of KIT in familial testicular germ cell tumours. Br J Cancer. 2004;90(12):2397–2401. doi:10.1038/sj.bjc.6601880 6601880 [pii].
341. Caruana G, Cambareri AC, Gonda TJ, Ashman LK. Transformation of NIH3T3 fibroblasts by the c-Kit receptor tyrosine kinase: effect of receptor density and ligand-requirement. Oncogene. 1998;16(2):179–90. doi:10.1038/sj.onc.1201494.
342. Chian R, Young S, Danilkovitch-Miagkova A, Ronnstrand L, Leonard E, Ferrao P, et al. Phosphatidylinositol 3 kinase contributes to the transformation of hematopoietic cells by the D816V c-Kit mutant. Blood. 2001;98(5):1365–73.
343. Lind GE, Skotheim RI, Lothe RA. The epigenome of testicular germ cell tumors. APMIS. 2007;115(10):1147–1160. doi:APMapm_660.xml [pii] 10.1111/j.1600-0463.2007.apm_660.xml.x.
344. Okamoto K. Epigenetics: a way to understand the origin and biology of testicular germ cell tumors. Int J Urol. 2012;19(6):504–11. doi:10.1111/j.1442-2042.2012.02986.x.
345. Tsuchiya K, Reijo R, Page DC, Disteche CM. Gonadoblastoma: molecular definition of the susceptibility region on the Y chromosome. Am J Hum Genet. 1995;57(6):1400–7.
346. Lau Y, Chou P, Iezzoni J, Alonzo J, Komuves L. Expression of a candidate gene for the gonadoblastoma locus in gonadoblastoma and testicular seminoma. Cytogenet Cell Genet. 2000;91(1–4):160–164. doi:56838 [pii] 56838.
347. Lau YF, Li Y, Kido T. Gonadoblastoma locus and the TSPY gene on the human Y chromosome. Birth Defects Res C Embryo Today. 2009;87(1):114–22. doi:10.1002/bdrc.20144.
348. Oosterhuis JW, Stoop H, Dohle G, Boellaard W, van Casteren N, Wolffenbuttel K, et al. A pathologist's view on the testis biopsy. Int J Androl. 2011;34(4 Pt 2):e14–e19; discussion e20. doi:10.1111/j.1365-2605.2011.01204.x.
349. Kaprova-Pleskacova J, Stoop H, Bruggenwirth H, Cools M, Wolffenbuttel KP, Drop SL, et al. Complete androgen insensitivity syndrome: factors influencing gonadal histology including germ cell pathology. Mod Pathol Off J U S Can Acad Pathol Inc. 2014;27(5):721–730. doi:modpathol2013193 [pii] 10.1038/modpathol.2013.193.
350. Skaletsky H, Kuroda-Kawaguchi T, Minx PJ, Cordum HS, Hillier L, Brown LG, et al. The male-specific region of the human Y chromosome is a mosaic of discrete sequence classes. Nature. 2003;423(6942):825–837. doi:10.1038/nature01722 nature01722 [pii].
351. Page DC. Hypothesis: a Y-chromosomal gene causes gonadoblastoma in dysgenetic gonads. Development. 1987;101(Suppl):151–5.
352. Hoei-Hansen CE, Sehested A, Juhler M, Lau YF, Skakkebaek NE, Laursen H, et al. New evidence for the origin of intracranial germ cell tumours from primordial germ cells: expression of pluripotency and cell differentiation markers. J Pathol. 2006;209(1):25–33. doi:10.1002/path.1948.
353. Shahsiah R, Jahanbin B, Rabiei R, Ardalan FA, Sarhadi B, Izadi-Mood N. Malignant ovarian germ cell tumours in gonadal Y chromosome mosaicism. J Clin Pathol. 2011;64(11):973–976. doi:jcp.2011.090738 [pii] 10.1136/jcp.2011.090738.
354. Jacobsen C, Honecker F. Cisplatin resistance in germ cell tumours: models and mechanisms. Andrology. 2015;3(1):111–21. doi:10.1111/andr.299.
355. Chaganti RS, Houldsworth J. Genetics and biology of adult human male germ cell tumors. Cancer Res. 2000;60(6):1475–82.
356. Voorhoeve PM, le Sage C, Schrier M, Gillis AJ, Stoop H, Nagel R, et al. A genetic screen implicates miRNA-372 and miRNA-373 as oncogenes in testicular germ cell tumors. Cell. 2006;124(6):1169–1181. doi:S0092-8674(06)00291-1 [pii] 10.1016/j.cell.2006.02.037.
357. van Gurp RJ, Oosterhuis JW, Kalscheuer V, Mariman EC, Looijenga LH. Biallelic expression of the H19 and IGF2 genes in human testicular germ cell tumors. J Natl Cancer Inst. 1994;86(14):1070–5.
358. Netto GJ, Nakai Y, Nakayama M, Jadallah S, Toubaji A, Nonomura N, et al. Global DNA hypomethylation in intratubular germ cell neoplasia and seminoma, but not in nonseminomatous male germ cell tumors. Mod Pathol Off J U S Can Acad Pathol Inc. 2008;21(11):1337–1344. doi:modpathol2008127 [pii] 10.1038/modpathol.2008.127.
359. Eckert D, Biermann K, Nettersheim D, Gillis AJ, Steger K, Jack HM, et al. Expression of BLIMP1/PRMT5 and concurrent histone H2A/H4 arginine 3 dimethylation in fetal germ cells, CIS/IGCNU and germ

cell tumors. BMC Dev Biol. 2008;8:106. doi:1471-213X-8-106 [pii] 10.1186/1471-213X-8-106.
360. Wermann H, Stoop H, Gillis AJ, Honecker F, van Gurp RJ, Ammerpohl O, et al. Global DNA methylation in fetal human germ cells and germ cell tumours: association with differentiation and cisplatin resistance. J Pathol. 2010;221(4):433–42. doi:10.1002/path.2725.
361. Almstrup K, Nielsen JE, Mlynarska O, Jansen MT, Jorgensen A, Skakkebaek NE, et al. Carcinoma in situ testis displays permissive chromatin modifications similar to immature foetal germ cells. Br J Cancer. 2010;103(8):1269–1276. doi:6605880 [pii] 10.1038/sj.bjc.6605880.
362. Brait M, Maldonado L, Begum S, Loyo M, Wehle D, Tavora FF, et al. DNA methylation profiles delineate epigenetic heterogeneity in seminoma and non-seminoma. Br J Cancer. 2012;106(2):414–423. doi:bjc2011468 [pii] 10.1038/bjc.2011.468.
363. Kristensen DG, Nielsen JE, Jorgensen A, Skakkebaek NE, Rajpert-De Meyts E, Almstrup K. Evidence that active demethylation mechanisms maintain the genome of carcinoma in situ cells hypomethylated in the adult testis. Br J Cancer. 2014;110(3):668–678. doi:bjc2013727 [pii] 10.1038/bjc.2013.727.
364. Killian JK, Dorssers LCJ, Trabert B, Gillis AJM, Cook MB, Wang Y, et al. Malignant testicular germ cell tumor methylomes demonstrate differentiation-dependent somatic reprogramming but persistent erasure of imprinted loci and the pluripotency regulator DPPA3/STELLA. Genome Res. 2016;26: 1490–1504.
365. Gillis AJ, Stoop HJ, Hersmus R, Oosterhuis JW, Sun Y, Chen C, et al. High-throughput microRNAome analysis in human germ cell tumours. J Pathol. 2007;213(3):319–28. doi:10.1002/path.2230.
366. Novotny GW, Belling KC, Bramsen JB, Nielsen JE, Bork-Jensen J, Almstrup K, et al. MicroRNA expression profiling of carcinoma in situ cells of the testis. Endocr Relat Cancer. 2012;19(3):365–379. doi:ERC-11-0271 [pii] 10.1530/ERC-11-0271.
367. Rajpert-De Meyts E, Bartkova J, Samson M, Hoei-Hansen CE, Frydelund-Larsen L, Bartek J, et al. The emerging phenotype of the testicular carcinoma in situ germ cell. APMIS. 2003;111(1):267–278; discussion 78–9. doi:apm1110130 [pii].
368. Almstrup K, Hoei-Hansen CE, Wirkner U, Blake J, Schwager C, Ansorge W, et al. Embryonic stem cell-like features of testicular carcinoma in situ revealed by genome-wide gene expression profiling. Cancer Res. 2004;64(14):4736–4743. doi:10.1158/0008-5472.CAN-04-0679 64/14/4736 [pii].
369. Almstrup K, Hoei-Hansen CE, Nielsen JE, Wirkner U, Ansorge W, Skakkebaek NE, et al. Genome-wide gene expression profiling of testicular carcinoma in situ progression into overt tumours. Br J Cancer. 2005;92(10):1934–1941. doi:6602560 [pii] 10.1038/sj.bjc.6602560.
370. Rajpert-De Meyts E. Developmental model for the pathogenesis of testicular carcinoma in situ: genetic and environmental aspects. Hum Reprod Update. 2006;12(3):303–323. doi:dmk006 [pii] 10.1093/humupd/dmk006.
371. International Germ Cell Consensus Classification: a prognostic factor-based staging system for metastatic germ cell cancers. International Germ Cell Cancer Collaborative Group. J Clin Oncol. 1997;15(2): 594–603.
372. Oosterhuis JW, Andrews PW, Knowles BB, Damjanov I. Effects of cis-platinum on embryonal carcinoma cell lines in vitro. Int J Cancer. 1984;34(1):133–9.
373. Stampfer MR, Yaswen P. Human epithelial cell immortalization as a step in carcinogenesis. Cancer Lett. 2003;194(2):199–208. doi:S0304383502007073 [pii].
374. Wright WE, Piatyszek MA, Rainey WE, Byrd W, Shay JW. Telomerase activity in human germline and embryonic tissues and cells. Dev Genet. 1996;18(2):173–179. doi:10.1002/(SICI)1520-6408(1996)18:2<173::AID-DVG10>3.0.CO;2-3 [pii] 10.1002/(SICI)1520-6408(1996)18:2<173::AID-DVG10>3.0.CO;2-3.
375. Albanell J, Bosl GJ, Reuter VE, Engelhardt M, Franco S, Moore MA, et al. Telomerase activity in germ cell cancers and mature teratomas. J Natl Cancer Inst. 1999;91(15):1321–6.
376. Turnbull C, Rapley EA, Seal S, Pernet D, Renwick A, Hughes D, et al. Variants near DMRT1, TERT and ATF7IP are associated with testicular germ cell cancer. Nat Genet. 2010;42(7):604–607. doi:ng.607 [pii] 10.1038/ng.607.
377. Gilbert D, Rapley E, Shipley J. Testicular germ cell tumours: predisposition genes and the male germ cell niche. Nat Rev Cancer. 2011;11(4):278–288. doi:nrc3021 [pii] 10.1038/nrc3021.
378. Boublikova L, Buchler T, Stary J, Abrahamova J, Trka J. Molecular biology of testicular germ cell tumors: unique features awaiting clinical application. Crit Rev Oncol Hematol. 2014;89(3):366–385. doi:S1040-8428(13)00211-4 [pii] 10.1016/j.critrevonc.2013.10.001.
379. Koster R, di Pietro A, Timmer-Bosscha H, Gibcus JH, van den Berg A, Suurmeijer AJ, et al. Cytoplasmic p21 expression levels determine cisplatin resistance in human testicular cancer. J Clin Invest. 2010;120(10):3594–3605. doi:41939 [pii] 10.1172/JCI41939.
380. Bauer S, Mühlenberg T, Leahy M, Hoiczyk M, Gauler T, Schuler M, et al. Therapeutic potential of Mdm2 inhibition in malignant germ cell tumours. Eur Urol. 2010;57:679–687.
381. Koster R, Timmer-Bosscha H, Bischoff R, Gietema JA, De Jong S. Disruption of the MDM2-p53 interaction strongly potentiates p53-dependent apoptosis in cisplatin-resistant human testicular carcinoma cells via the Fas/FasL pathway. Cell death and disease. 2011;2:e148. doi:10.1038/cddis.2011.33.
382. Biswal BK, Beyrouthy MJ, Hever-Jardine MP, Amstrong D, Tomlinson CR, Christensen BC, et al. Acute hypersensitivity of pluripotent testicular cancer-derived embryonal carcinoma to low-dose 5-Aza

Deoxycytidine is associated with global DNA damage-associated p53 activation, anti-pluripotency and DNA demethylation. PLOS ONE. 2012;7:e53003. doi:10.1371/journal.pone.0053003.

383. Oosterhuis JW, Damjanov I. Treatment of primary embryo-derived teratocarcinomas in mice with cis-diamminedichloroplatinum. Eur J Cancer Clin Oncol. 1983;19(5):695–9.

384. Honecker F, Wermann H, Mayer F, Gillis AJ, Stoop H, van Gurp RJ, et al. Microsatellite instability, mismatch repair deficiency, and BRAF mutation in treatment-resistant germ cell tumors. J Clin Oncol. 2009;27(13):2129–2136. doi:JCO.2008.18.8623 [pii] 10.1200/JCO.2008.18.8623.

385. Mayer F, Wermann H, Albers P, Stoop H, Gillis AJ, Hartmann JT, et al. Histopathological and molecular features of late relapses in non-seminomas. BJU Int. 2011;107(6):936–43. doi:10.1111/j.1464-410X.2010.09631.x.

386. Michael H, Lucia J, Foster RS, Ulbright TM. The pathology of late recurrence of testicular germ cell tumors. Am J Surg Pathol. 2000;24(2):257–73.

387. Gillis AJ, Oosterhuis JW, Schipper ME, Barten EJ, van Berlo R, van Gurp RJ, et al. Origin and biology of a testicular Wilms' tumor. Genes Chromosomes Cancer. 1994;11(2):126–35.

388. Motzer RJ, Amsterdam A, Prieto V, Sheinfeld J, Murty VV, Mazumdar M, et al. Teratoma with malignant transformation: diverse malignant histologies arising in men with germ cell tumors. J Urol. 1998;159(1):133–138. doi:S0022-5347(01)64035-7 [pii].

389. Kum JB, Ulbright TM, Williamson SR, Wang M, Zhang S, Foster RS, et al. Molecular genetic evidence supporting the origin of somatic-type malignancy and teratoma from the same progenitor cell. Am J Surg Pathol. 2012;36(12):1849–1856. doi:10.1097/PAS.0b013e31826df1ab 00000478-201212000-00013 [pii].

390. Guo CC, Punar M, Contreras AL, Tu SM, Pisters L, Tamboli P, et al. Testicular germ cell tumors with sarcomatous components: an analysis of 33 cases. Am J Surg Pathol. 2009;33(8):1173–8. doi:10.1097/PAS.0b013e3181adb9d7.

391. Cunningham JJ, Ulbright TM, Pera MF, Looijenga LH. Lessons from human teratomas to guide development of safe stem cell therapies. Nat Biotechnol. 2012;30(9):849–857. doi:nbt.2329 [pii] 10.1038/nbt.2329.

392. Devouassoux-Shisheboran M, Mauduit C, Tabone E, Droz JP, Benahmed M. Growth regulatory factors and signalling proteins in testicular germ cell tumours. APMIS. 2003;111(1):212–224; discussion 24. doi:apm1110125 [pii].

393. Neumann JC, Chandler GL, Damoulis VA, Fustino NJ, Lillard K, Looijenga L, et al. Mutation in the type IB bone morphogenetic protein receptor Alk6b impairs germ-cell differentiation and causes germ-cell tumors in zebrafish. Proc Natl Acad Sci U S A. 2011;108(32):13153–13158. doi:1102311108 [pii] 10.1073/pnas.1102311108.

394. Neumann JC, Lillard K, Damoulis V, Amatruda JF. Zebrafish models of germ cell tumor. Methods Cell Biol. 2011;105:3–24. doi:B978-0-12-381320-6.00001-1 [pii] 10.1016/B978-0-12-381320-6.00001-1.

395. Basten SG, Davis EE, Gillis AJ, van Rooijen E, Stoop H, Babala N, et al. Mutations in LRRC50 predispose zebrafish and humans to seminomas. PLoS Genet. 2013;9(4):e1003384. doi:10.1371/journal.pgen.1003384 PGENETICS-D-12-02494 [pii].

396. Mizuno Y, Gotoh A, Kamidono S, Kitazawa S. Establishment and characterization of a new human testicular germ cell tumor cell line (TCam-2). Nihon Hinyokika Gakkai Zasshi. 1993;84(7):1211–8.

397. Berends JC, Schutte SE, van Dissel-Emiliani FM, de Rooij DG, Looijenga LH, Oosterhuis JW. Significant improvement of the survival of seminoma cells in vitro by use of a rat Sertoli cell feeder layer and serum-free medium. J Natl Cancer Inst. 1991;83(19):1400–3.

398. de Jong J, Stoop H, Gillis AJ, Hersmus R, van Gurp RJ, van de Geijn GJ, et al. Further characterization of the first seminoma cell line TCam-2. Genes Chromosomes Cancer. 2008;47(3):185–96. doi:10.1002/gcc.20520.

399. de Jong J, Stoop H, Gillis AJ, van Gurp RJ, van de Geijn GJ, Boer M, et al. Differential expression of SOX17 and SOX2 in germ cells and stem cells has biological and clinical implications. J Pathol. 2008;215(1):21–30. doi:10.1002/path.2332.

400. Nettersheim D, Gillis AJ, Looijenga LH, Schorle H. TGF-beta1, EGF and FGF4 synergistically induce differentiation of the seminoma cell line TCam-2 into a cell type resembling mixed non-seminoma. Int J Androl. 2011;34(4 Pt 2):e189–203. doi:10.1111/j.1365-2605.2011.01172.x.

401. Nettersheim D, Westernstroer B, Haas N, Leinhaas A, Brustle O, Schlatt S, et al. Establishment of a versatile seminoma model indicates cellular plasticity of germ cell tumor cells. Genes Chromosomes Cancer. 2012;51(7):717–26. doi:10.1002/gcc.21958.

402. Andrews PW, Damjanov I, Simon D, Banting GS, Carlin C, Dracopoli NC, et al. Pluripotent embryonal carcinoma clones derived from the human teratocarcinoma cell line Tera-2. Differentiation in vivo and in vitro. Lab Investig. 1984;50(2):147–62.

403. Andrews PW, Fenderson B, Hakomori S. Human embryonal carcinoma cells and their differentiation in culture. Int J Androl. 1987;10(1):95–104.

404. Andrews PW, Casper J, Damjanov I, Duggan-Keen M, Giwercman A, Hata J, et al. Comparative analysis of cell surface antigens expressed by cell lines derived from human germ cell tumours. Int J Cancer. 1996;66(6):806–816. doi:10.1002/(SICI)1097-0215(19960611)66:6<806::AID-IJC17>3.0.CO;2-0 [pii] 10.1002/(SICI)1097-0215(19960611)66:6<806::AID-IJC17>3.0.CO;2-0.

405. Scully RE. Gonadoblastoma. A review of 74 cases. Cancer. 1970;25(6):1340–56.

406. Ulbright TM, Young RH. Gonadoblastoma and selected other aspects of gonadal pathology in young patients with disorders of sex development. Semin

Diagn Pathol. 2014;31(5):427–440. doi:S0740-2570(14)00064-1 [pii] 10.1053/j.semdp.2014.07.001.
407. Houk CP, Hughes IA, Ahmed SF, Lee PA, Writing Committee for the International Intersex Consensus Conference Participants. Summary of consensus statement on intersex disorders and their management. International Intersex Consensus Conference. Pediatrics. 2006;118(2):753–757. doi:118/2/753 [pii] 10.1542/peds.2006-0737.
408. Cools M, Looijenga LH. Tumor risk and clinical follow-up in patients with disorders of sex development. Pediatr Endocrinol Rev. 2011;9(Suppl 1):519–24.
409. Hughes IA. Disorders of sex development: a new definition and classification. Best Pract Res Clin Endocrinol Metab. 2008;22(1):119–134. doi:S1521-690X(07)00105-4 [pii] 10.1016/j.beem.2007.11.001.
410. Subbiah V, Huff V, Wolff JE, Ketonen L, Lang Jr FF, Stewart J, et al. Bilateral gonadoblastoma with dysgerminoma and pilocytic astrocytoma with WT1 GT-IVS9 mutation: A 46 XY phenotypic female with Frasier syndrome. Pediatr Blood Cancer. 2009;53(7):1349–51. doi:10.1002/pbc.22152.
411. Looijenga LH, Hersmus R, de Leeuw BH, Stoop H, Cools M, Oosterhuis JW, et al. Gonadal tumours and DSD. Best Pract Res Clin Endocrinol Metab. 2010;24(2):291–310. doi:S1521-690X(09)00139-0 [pii] 10.1016/j.beem.2009.10.002.
412. Tanteles GA, Oakley S, Christian M, O'Neill D, Suri M. Denys-Drash syndrome and gonadoblastoma in a patient with Klinefelter syndrome. Clin Dysmorphol. 2011;20(3):131–5. doi:10.1097/MCD.0b013e328346f6dc.
413. Looijenga LH. Testicular germ cell tumors. Pediatr Endocrinol Rev. 2014;11(Suppl 2):251–62.
414. Achermann J. Disorders of sex development. In: Melmed S, editor. Williams textbook of endocrinology. 12th ed. Philadelphia: Elsevier; 2011. p. 868–934.
415. van der Zwan YG, Biermann K, Wolffenbuttel KP, Cools M, Looijenga LH. Gonadal maldevelopment as risk factor for germ cell cancer: towards a clinical decision model. Eur Urol. 2015;67(4):692–701. doi:S0302-2838(14)00651-4 [pii] 10.1016/j.eururo.2014.07.011.
416. Davidoff F, Federman DD. Mixed gonadal dysgenesis. Pediatrics. 1973;52(5):725–42.
417. Sohval AR. "Mixed" gonadal dysgenesis: a variety of hermaphroditism. Am J Hum Genet. 1963;15:155–8.
418. Ropke A, Pelz AF, Volleth M, Schlosser HW, Morlot S, Wieacker PF. Sex chromosomal mosaicism in the gonads of patients with gonadal dysgenesis, but normal female or male karyotypes in lymphocytes. Am J Obstet Gynecol. 2004;190(4):1059–1062. doi:10.1016/j.ajog.2003.09.053 S0002937803017794 [pii].
419. Cools M, Boter M, van Gurp R, Stoop H, Poddighe P, Lau YF, et al. Impact of the Y-containing cell line on histological differentiation patterns in dysgenetic gonads. Clin Endocrinol. 2007;67(2):184–192. doi:CEN2859 [pii] 10.1111/j.1365-2265.2007.02859.x.
420. Maier EM, Leitner C, Lohrs U, Kuhnle U. True hermaphroditism in an XY individual due to a familial point mutation of the SRY gene. J Pediatr Endocrinol Metab. 2003;16(4):575–80.
421. Isidor B, Capito C, Paris F, Baron S, Corradini N, Cabaret B, et al. Familial frameshift SRY mutation inherited from a mosaic father with testicular dysgenesis syndrome. J Clin Endocrinol Metab. 2009;94(9):3467–3471. doi:jc.2009-0226 [pii] 10.1210/jc.2009-0226.
422. Shahid M, Dhillon VS, Khalil HS, Haque S, Batra S, Husain SA, et al. A SRY-HMG box frame shift mutation inherited from a mosaic father with a mild form of testicular dysgenesis syndrome in Turner syndrome patient. BMC Med Genet. 2010;11:131. doi:1471-2350-11-131 [pii] 10.1186/1471-2350-11-131.
423. Hersmus R, Kalfa N, de Leeuw B, Stoop H, Oosterhuis JW, de Krijger R, et al. FOXL2 and SOX9 as parameters of female and male gonadal differentiation in patients with various forms of disorders of sex development (DSD). J Pathol. 2008;215(1):31–8. doi:10.1002/path.2335.
424. Buell-Gutbrod R, Ivanovic M, Montag A, Lengyel E, Fadare O, Gwin K. FOXL2 and SOX9 distinguish the lineage of the sex cord-stromal cells in gonadoblastomas. Pediatr Dev Pathol. 2011;14(5):391–5. doi:10.2350/10-12-0943-OA.1.
425. Hersmus R, van der Zwan YG, Stoop H, Bernard P, Sreenivasan R, Oosterhuis JW, et al. A 46,XY female DSD patient with bilateral gonadoblastoma, a novel SRY missense mutation combined with a WT1 KTS splice-site mutation. PLoS One. 2012; 7(7):e40858. doi:10.1371/journal.pone.0040858 PONE-D-12-13207 [pii].
426. Hertel JD, Huettner PC, Dehner LP, Pfeifer JD. The chromosome Y-linked testis-specific protein locus TSPY1 is characteristically present in gonadoblastoma. Hum Pathol. 2010;41(11):1544–1549. doi:S0046-8177(10)00123-1 [pii] 10.1016/j.humpath.2010.04.007.
427. Hersmus R, Stoop H, White SJ, Drop SL, Oosterhuis JW, Incrocci L, et al. Delayed recognition of Disorders of Sex Development (DSD): a missed opportunity for early diagnosis of malignant germ cell tumors. Int J Endocrinol. 2012;2012:671209. doi:10.1155/2012/671209.
428. Li Y, Vilain E, Conte F, Rajpert-De Meyts E, Lau YF. Testis-specific protein Y-encoded gene is expressed in early and late stages of gonadoblastoma and testicular carcinoma in situ. Urol Oncol. 2007;25(2):141–146. doi:S1078-1439(06)00166-9 [pii] 10.1016/j.urolonc.2006.08.002.
429. Carver BS, Shayegan B, Serio A, Motzer RJ, Bosl GJ, Sheinfeld J. Long-term clinical outcome after postchemotherapy retroperitoneal lymph node dissection in men with residual teratoma. J Clin Oncol. 2007;25(9):1033–1037. doi:JCO.2005.05.4791 [pii] 10.1200/JCO.2005.05.4791.
430. Malagon HD, Valdez AM, Moran CA, Suster S. Germ cell tumors with sarcomatous components: a clinicopathologic and immunohistochemical study of 46 cases. Am J Surg Pathol. 2007;31(9):1356–1362. doi:10.1097/PAS.0b013e318033c7c4 00000478-200709000-00008 [pii].

431. Comiter CV, Kibel AS, Richie JP, Nucci MR, Renshaw AA. Prognostic features of teratomas with malignant transformation: a clinicopathological study of 21 cases. J Urol. 1998;159(3):859–863. doi:S0022-5347(01)63754-6 [pii].
432. Ganjoo KN, Foster RS, Michael H, Donohue JP, Einhorn LH. Germ cell tumor associated primitive neuroectodermal tumors. J Urol. 2001;165(5):1514–1516. doi:S0022-5347(05)66339-2 [pii].
433. Michael H, Hull MT, Ulbright TM, Foster RS, Miller KD. Primitive neuroectodermal tumors arising in testicular germ cell neoplasms. Am J Surg Pathol. 1997;21(8):896–904.
434. Magers MJ, Kao CS, Cole CD, Rice KR, Foster RS, Einhorn LH, et al. "Somatic-type" malignancies arising from testicular germ cell tumors: a clinicopathologic study of 124 cases with emphasis on glandular tumors supporting frequent yolk sac tumor origin. Am J Surg Pathol. 2014;38(10):1396–409. doi:10.1097/PAS.0000000000000262.
435. Howitt BE, Magers MJ, Rice KR, Cole CD, Ulbright TM. Many postchemotherapy sarcomatous tumors in patients with testicular germ cell tumors are sarcomatoid yolk sac tumors: a study of 33 cases. Am J Surg Pathol. 2015;39(2):251–9. doi:10.1097/PAS.0000000000000322.
436. Moul JW, Schanne FJ, Thompson IM, Frazier HA, Peretsman SA, Wettlaufer JN, et al. Testicular cancer in blacks. A multicenter experience. Cancer. 1994;73(2):388–93.
437. Rosen A, Jayram G, Drazer M, Eggener SE. Global trends in testicular cancer incidence and mortality. Eur Urol. 2011;60(2):374–379. doi:S0302-2838(11)00493-3 [pii] 10.1016/j.eururo.2011.05.004.
438. Shanmugalingam T, Soultati A, Chowdhury S, Rudman S, Van Hemelrijck M. Global incidence and outcome of testicular cancer. Clin Epidemiol. 2013;5:417–427. doi:10.2147/CLEP.S34430 clep-5-417 [pii].
439. McGlynn KA, Devesa SS, Graubard BI, Castle PE. Increasing incidence of testicular germ cell tumors among black men in the United States. J Clin Oncol. 2005;23(24):5757–5761. doi:23/24/5757 [pii] 10.1200/JCO.2005.08.227.
440. Ekbom A, Richiardi L, Akre O, Montgomery SM, Sparen P. Age at immigration and duration of stay in relation to risk for testicular cancer among Finnish immigrants in Sweden. J Natl Cancer Inst. 2003;95(16):1238–40.
441. Montgomery SM, Granath F, Ehlin A, Sparen P, Ekbom A. Germ-cell testicular cancer in offspring of Finnish immigrants to Sweden. Cancer Epidemiol Biomarkers Prev. 2005;14(1):280–282. doi:14/1/280 [pii].
442. Levine H, Afek A, Shamiss A, Derazne E, Tzur D, Zavdy O, et al. Risk of germ cell testicular cancer according to origin: a migrant cohort study in 1,100,000 Israeli men. Int J Cancer. 2013;132(8):1878–85. doi:10.1002/ijc.27825.
443. McIver SC, Roman SD, Nixon B, Loveland KL, McLaughlin EA. The rise of testicular germ cell tumours: the search for causes, risk factors and novel therapeutic targets. F1000Res. 2013;2:55. doi:10.12688/f1000research.2-55.v1.
444. Moller H. Decreased testicular cancer risk in men born in wartime. J Natl Cancer Inst. 1989;81(21):1668–9.
445. Rajpert-De Meyts E, McGlynn KA, Okamoto K, Jewett MA, Bokemeyer C. Testicular germ cell tumours. Lancet. 2016;387(10029):1762–1774. doi:S0140-6736(15)00991-5 [pii] 10.1016/S0140-6736(15)00991-5.
446. Cook MB, Akre O, Forman D, Madigan MP, Richiardi L, McGlynn KA. A systematic review and meta-analysis of perinatal variables in relation to the risk of testicular cancer – experiences of the mother. Int J Epidemiol. 2009;38(6):1532–1542. doi:dyp287 [pii] 10.1093/ije/dyp287.
447. Trabert B, Zugna D, Richiardi L, McGlynn KA, Akre O. Congenital malformations and testicular germ cell tumors. Int J Cancer. 2013;133(8):1900–4. doi:10.1002/ijc.28207.
448. Fossa SD, Chen J, Schonfeld SJ, McGlynn KA, McMaster ML, Gail MH, et al. Risk of contralateral testicular cancer: a population-based study of 29,515 U.S. men. J Natl Cancer Inst. 2005;97(14):1056–1066. doi:97/14/1056 [pii] 10.1093/jnci/dji185.
449. Moller H, Skakkebaek NE. Risk of testicular cancer in subfertile men: case-control study. BMJ. 1999;318(7183):559–62.
450. Tollerud DJ, Blattner WA, Fraser MC, Brown LM, Pottern L, Shapiro E, et al. Familial testicular cancer and urogenital developmental anomalies. Cancer. 1985;55(8):1849–54.
451. Ferlin A, Ganz F, Pengo M, Selice R, Frigo AC, Foresta C. Association of testicular germ cell tumor with polymorphisms in estrogen receptor and steroid metabolism genes. Endocr Relat Cancer. 2010;17(1):17–25. doi:ERC-09-0176 [pii] 10.1677/ERC-09-0176.
452. Kristiansen W, Andreassen KE, Karlsson R, Aschim EL, Bremnes RM, Dahl O, et al. Gene variations in sex hormone pathways and the risk of testicular germ cell tumour: a case-parent triad study in a Norwegian-Swedish population. Hum Reprod. 2012;27(5):1525–1535. doi:des075 [pii] 10.1093/humrep/des075.
453. Cook MB, Akre O, Forman D, Madigan MP, Richiardi L, McGlynn KA. A systematic review and meta-analysis of perinatal variables in relation to the risk of testicular cancer – experiences of the son. Int J Epidemiol. 2010;39(6):1605–1618. doi:dyq120 [pii] 10.1093/ije/dyq120.
454. Cook MB, Trabert B, McGlynn KA. Organochlorine compounds and testicular dysgenesis syndrome: human data. Int J Androl. 2011;34(4 Pt2):e68–e84; discussion e–5. doi:10.1111/j.1365-2605.2011.01171.x.
455. McGlynn KA, Trabert B. Adolescent and adult risk factors for testicular cancer. Nat Rev Urol. 2012;9(6):339–349. doi:nrurol.2012.61 [pii] 10.1038/nrurol.2012.61.
456. Daling JR, Doody DR, Sun X, Trabert BL, Weiss NS, Chen C, et al. Association of marijuana use and

the incidence of testicular germ cell tumors. Cancer. 2009;115(6):1215–23. doi:10.1002/cncr.24159.

457. Swerdlow AJ, De Stavola BL, Swanwick MA, Maconochie NE. Risks of breast and testicular cancers in young adult twins in England and Wales: evidence on prenatal and genetic aetiology. Lancet. 1997;350(9093):1723–1728. doi:S0140673697055268 [pii].

458. Kharazmi E, Hemminki K, Pukkala E, Sundquist K, Tryggvadottir L, Tretli S, et al. Cancer risk in relatives of testicular cancer patients by histology type and age at diagnosis: a joint study from five nordic countries. Eur Urol. 2015;68(2):283–289. doi:S0302-2838(14)01387-6 [pii] 10.1016/j.eururo.2014.12.031.

459. Nathanson KL, Kanetsky PA, Hawes R, Vaughn DJ, Letrero R, Tucker K, et al. The Y deletion gr/gr and susceptibility to testicular germ cell tumor. Am J Hum Genet. 2005;77(6):1034–1043. doi:S0002-9297(07)63387-4 [pii] 10.1086/498455.

460. Pathak A, Stewart DR, Faucz FR, Xekouki P, Bass S, Vogt A, et al. Rare inactivating PDE11A variants associated with testicular germ cell tumors. Endocr Relat Cancer. 2015;22(6):909–917. doi:ERC-15-0034 [pii] 10.1530/ERC-15-0034.

461. Heimdal K, Olsson H, Tretli S, Fossa SD, Borresen AL, Bishop DT. A segregation analysis of testicular cancer based on Norwegian and Swedish families. Br J Cancer. 1997;75(7):1084–7.

462. Hemminki K, Vaittinen P, Dong C, Easton D. Sibling risks in cancer: clues to recessive or X-linked genes? Br J Cancer. 2001;84(3):388–391. doi:10.1054/bjoc.2000.1585 S0007092000915854 [pii].

463. Valberg M, Grotmol T, Tretli S, Veierod MB, Moger TA, Aalen OO. A hierarchical frailty model for familial testicular germ-cell tumors. Am J Epidemiol. 2014;179(4):499–506. doi:kwt267 [pii] 10.1093/aje/kwt267.

464. Kanetsky PA, Mitra N, Vardhanabhuti S, Vaughn DJ, Li M, Ciosek SL, et al. A second independent locus within DMRT1 is associated with testicular germ cell tumor susceptibility. Hum Mol Genet. 2011;20(15):3109–3117. doi:ddr207 [pii] 10.1093/hmg/ddr207.

465. Chung CC, Kanetsky PA, Wang Z, Hildebrandt MA, Koster R, Skotheim RI, et al. Meta-analysis identifies four new loci associated with testicular germ cell tumor. Nat Genet. 2013;45(6):680–685. doi:ng.2634 [pii] 10.1038/ng.2634.

466. Schumacher FR, Wang Z, Skotheim RI, Koster R, Chung CC, Hildebrandt MA, et al. Testicular germ cell tumor susceptibility associated with the UCK2 locus on chromosome 1q23. Hum Mol Genet. 2013;22(13):2748–2753. doi:ddt109 [pii] 10.1093/hmg/ddt109.

467. Koster R, Mitra N, D'Andrea K, Vardhanabhuti S, Chung CC, Wang Z, et al. Pathway-based analysis of GWAs data identifies association of sex determination genes with susceptibility to testicular germ cell tumors. Hum Mol Genet. 2014;23(22):6061–6068. doi:ddu305 [pii] 10.1093/hmg/ddu305.

468. Ruark E, Seal S, McDonald H, Zhang F, Elliot A, Lau K, et al. Identification of nine new susceptibility loci for testicular cancer, including variants near DAZL and PRDM14. Nat Genet. 2013;45(6):686–689. doi:ng.2635 [pii] 10.1038/ng.2635.

469. Kratz CP, Han SS, Rosenberg PS, Berndt SI, Burdett L, Yeager M, et al. Variants in or near KITLG, BAK1, DMRT1, and TERT-CLPTM1L predispose to familial testicular germ cell tumour. J Med Genet. 2011;48(7):473–476. doi:jmedgenet-2011-100001 [pii] 10.1136/jmedgenet-2011-100001.

470. Dalgaard MD, Weinhold N, Edsgard D, Silver JD, Pers TH, Nielsen JE, et al. A genome-wide association study of men with symptoms of testicular dysgenesis syndrome and its network biology interpretation. J Med Genet. 2012;49(1):58–65. doi:jmedgenet-2011-100174 [pii] 10.1136/jmedgenet-2011-100174.

471. Skakkebaek NE, Holm M, Hoei-Hansen C, Jorgensen N, Rajpert-De Meyts E. Association between testicular dysgenesis syndrome (TDS) and testicular neoplasia: evidence from 20 adult patients with signs of maldevelopment of the testis. APMIS. 2003;111(1):1–9; discussion -11. doi:apm1110103_1 [pii].

472. Jorgensen A, Lindhardt Johansen M, Juul A, Skakkebaek NE, Main KM, Rajpert-De Meyts E. Pathogenesis of germ cell neoplasia in testicular dysgenesis and disorders of sex development. Semin Cell Dev Biol. 2015;45:124–137. doi:S1084-9521(15)00173-1 [pii] 10.1016/j.semcdb.2015.09.013.

473. Lottrup G, Jorgensen A, Nielsen JE, Jorgensen N, Duno M, Vinggaard AM, et al. Identification of a novel androgen receptor mutation in a family with multiple components compatible with the testicular dysgenesis syndrome. J Clin Endocrinol Metab. 2013;98(6):2223–2229. doi:jc.2013-1278 [pii] 10.1210/jc.2013-1278.

474. Looijenga LH, Van Agthoven T, Biermann K. Development of malignant germ cells – the genvironmental hypothesis. Int J Dev Biol. 2013;57(2–4):241–253. doi:130026ll [pii] 10.1387/ijdb.130026ll.

475. Matson CK, Murphy MW, Sarver AL, Griswold MD, Bardwell VJ, Zarkower D. DMRT1 prevents female reprogramming in the postnatal mammalian testis. Nature. 2011;476(7358):101–104. doi:nature10239 [pii] 10.1038/nature10239.

476. Matson CK, Zarkower D. Sex and the singular DM domain: insights into sexual regulation, evolution and plasticity. Nat Rev Genet. 2012;13(3):163–174. doi:nrg3161 [pii] 10.1038/nrg3161.

477. Oram SW, Liu XX, Lee TL, Chan WY, Lau YF. TSPY potentiates cell proliferation and tumorigenesis by promoting cell cycle progression in HeLa and NIH3T3 cells. BMC Cancer. 2006;6:154. doi:1471-2407-6-154 [pii] 10.1186/1471-2407-6-154.

478. Zeron-Medina J, Wang X, Repapi E, Campbell MR, Su D, Castro-Giner F, et al. A polymorphic p53 response element in KIT ligand influences can-

cer risk and has undergone natural selection. Cell. 2013;155(2):410–422. doi:S0092-8674(13)01154-9 [pii] 10.1016/j.cell.2013.09.017.
479. Gill ME, Hu YC, Lin Y, Page DC. Licensing of gametogenesis, dependent on RNA binding protein DAZL, as a gateway to sexual differentiation of fetal germ cells. Proc Natl Acad Sci U S A. 2011;108(18):7443–7448. doi:1104501108 [pii] 10.1073/pnas.1104501108.
480. Skakkebaek NE. Possible carcinoma-in-situ of the testis. Lancet. 1972;2(7776):516–517. doi:S0140-6736(72)91909-5 [pii].
481. Skakkebaek NE, Berthelsen JG, Giwercman A, Muller J. Carcinoma-in-situ of the testis: possible origin from gonocytes and precursor of all types of germ cell tumours except spermatocytoma. Int J Androl. 1987;10(1):19–28.
482. Berthelsen JG, Skakkebaek NE, von der Maase H, Sorensen BL, Mogensen P. Screening for carcinoma in situ of the contralateral testis in patients with germinal testicular cancer. Br Med J (Clin Res Ed). 1982;285(6356):1683–6.
483. Dieckmann KP, Loy V. Prevalence of contralateral testicular intraepithelial neoplasia in patients with testicular germ cell neoplasms. J Clin Oncol. 1996;14(12):3126–32.
484. Rajpert-De Meyts E, Kvist M, Skakkebaek NE. Heterogeneity of expression of immunohistochemical tumour markers in testicular carcinoma in situ: pathogenetic relevance. Virchows Arch. 1996;428(3):133–9.
485. Sonne SB, Perrett RM, Nielsen JE, Baxter MA, Kristensen DM, Leffers H, et al. Analysis of SOX2 expression in developing human testis and germ cell neoplasia. Int J Dev Biol. 2010;54(4):755–760. doi:082668ss [pii] 10.1387/ijdb.082668ss.
486. Jorgensen A, Nielsen JE, Almstrup K, Toft BG, Petersen BL, Rajpert-De Meyts E. Dysregulation of the mitosis-meiosis switch in testicular carcinoma in situ. J Pathol. 2013;229(4):588–98. doi:10.1002/path.4154.
487. Mitchell RT, E Camacho-Moll M, Macdonald J, Anderson RA, Kelnar CJ, O'Donnell M, et al. Intratubular germ cell neoplasia of the human testis: heterogeneous protein expression and relation to invasive potential. Mod Pathol Off J U S Can Acad Pathol Inc. 2014;27(9):1255–1266. doi:modpathol2013246 [pii] 10.1038/modpathol.2013.246.
488. Gueler B, Sonne SB, Zimmer J, Hilscher B, Hilscher W, Graem N, et al. AZFa protein DDX3Y is differentially expressed in human male germ cells during development and in testicular tumours: new evidence for phenotypic plasticity of germ cells. Hum Reprod. 2012;27(6):1547–1555. doi:des047 [pii] 10.1093/humrep/des047.
489. Hoei-Hansen CE, Rajpert-De Meyts E, Daugaard G, Skakkebaek NE. Carcinoma in situ testis, the progenitor of testicular germ cell tumours: a clinical review. Ann Oncol. 2005;16(6):863–868. doi:mdi175 [pii] 10.1093/annonc/mdi175.
490. Sakuma Y, Sakurai S, Oguni S, Hironaka M, Saito K. Alterations of the c-kit gene in testicular germ cell tumors. Cancer Sci. 2003;94(6):486–91.
491. McIntyre A, Gilbert D, Goddard N, Looijenga L, Shipley J. Genes, chromosomes and the development of testicular germ cell tumors of adolescents and adults. Genes Chromosomes Cancer. 2008;47(7):547–57. doi:10.1002/gcc.20562.
492. Rajpert-De Meyts E. The role of puberty for testicular cancer. In: Bourguignon JP, editor. The Onset of puberty in perspective. Amsterdam: Elsevier Science; 2000. p. 201–13.
493. Ottesen AM, Skakkebaek NE, Lundsteen C, Leffers H, Larsen J, Rajpert-De Meyts E. High-resolution comparative genomic hybridization detects extra chromosome arm 12p material in most cases of carcinoma in situ adjacent to overt germ cell tumors, but not before the invasive tumor development. Genes Chromosomes Cancer. 2003;38(2):117–25. doi:10.1002/gcc.10244.
494. von Eyben FE, Jacobsen GK, Rorth M, Von Der Maase H. Microinvasive germ cell tumour (MGCT) adjacent to testicular germ cell tumours. Histopathology. 2004;44(6):547–554. doi:10.1111/j.1365-2559.2004.01889.x HIS1889 [pii].
495. von Eyben FE, Jacobsen GK, Skotheim RI. Microinvasive germ cell tumor of the testis. Virchows Arch. 2005;447(3):610–25. doi:10.1007/s00428-005-1257-8.
496. Browne TJ, Richie JP, Gilligan TD, Rubin MA. Intertubular growth in pure seminomas: associations with poor prognostic parameters. Hum Pathol. 2005;36(6):640–645. doi:S0046817705001620 [pii] 10.1016/j.humpath.2005.03.011.
497. Hvarness T, Nielsen JE, Almstrup K, Skakkebaek NE, Rajpert-De Meyts E, Claesson MH. Phenotypic characterisation of immune cell infiltrates in testicular germ cell neoplasia. J Reprod Immunol. 2013;100(2):135–145. doi:S0165-0378(13)00126-5 [pii] 10.1016/j.jri.2013.10.005.
498. Hadrup SR, Braendstrup O, Jacobsen GK, Mortensen S, Pedersen LO, Seremet T, et al. Tumor infiltrating lymphocytes in seminoma lesions comprise clonally expanded cytotoxic T cells. Int J Cancer. 2006;119(4):831–8. doi:10.1002/ijc.21894.
499. Li N, Wang T, Han D. Structural, cellular and molecular aspects of immune privilege in the testis. Front Immunol. 2012;3:152. doi:10.3389/fimmu.2012.00152.
500. Goedert JJ, Purdue MP, McNeel TS, McGlynn KA, Engels EA. Risk of germ cell tumors among men with HIV/acquired immunodeficiency syndrome. Cancer Epidemiol Biomarkers Prev. 2007;16(6):1266–1269. doi:16/6/1266 [pii] 10.1158/1055-9965.EPI-07-0042.
501. Menezes RX, Boetzer M, Sieswerda M, van Ommen GJ, Boer JM. Integrated analysis of DNA copy number and gene expression microarray data using gene sets. BMC Bioinformatics. 2009;10:203. doi:1471-2105-10-203 [pii] 10.1186/1471-2105-10-203.
502. Sparmann A, van Lohuizen M. Polycomb silencers control cell fate, development and cancer. Nat

Rev Cancer. 2006;6(11):846–856. doi:nrc1991 [pii] 10.1038/nrc1991.
503. Poulos C, Cheng L, Zhang S, Gersell DJ, Ulbright TM. Analysis of ovarian teratomas for isochromosome 12p: evidence supporting a dual histogenetic pathway for teratomatous elements. Mod Pathol Off J U S Can Acad Pathol Inc. 2006;19(6):766–771. doi:3800596 [pii] 10.1038/modpathol.3800596.
504. Smith HO, Berwick M, Verschraegen CF, Wiggins C, Lansing L, Muller CY, et al. Incidence and survival rates for female malignant germ cell tumors. Obstet Gynecol. 2006;107(5):1075–1085. doi:107/5/1075 [pii] 10.1097/01.AOG.0000216004.22588.ce.
505. Lifschitz-Mercer B, Walt H, Kushnir I, Jacob N, Diener PA, Moll R, et al. Differentiation potential of ovarian dysgerminoma: an immunohistochemical study of 15 cases. Hum Pathol. 1995;26(1):62–6.
506. Parkash V, Carcangiu ML. Transformation of ovarian dysgerminoma to yolk sac tumor: evidence for a histogenetic continuum. Mod Pathol Off J U S Can Acad Pathol Inc. 1995;8(8):881–7.
507. Aoki Y, Kase H, Fujita K, Tanaka K. Dysgerminoma with a slightly elevated alpha-fetoprotein level diagnosed as a mixed germ cell tumor after recurrence. Gynecol Obstet Invest. 2003;55(1):58–60. doi:68949 68949 [pii].
508. Cossu-Rocca P, Jones TD, Roth LM, Eble JN, Zheng W, Karim FW, et al. Cytokeratin and CD30 expression in dysgerminoma. Hum Pathol. 2006;37(8):1015–1021. doi:S0046-8177(06)00128-6 [pii] 10.1016/j.humpath.2006.02.018.
509. Moller H, Evans H. Epidemiology of gonadal germ cell cancer in males and females. APMIS. 2003;111(1):43–46; discussion 6–8. doi:apm1110107 [pii].
510. Stang A, Trabert B, Wentzensen N, Cook MB, Rusner C, Oosterhuis JW, et al. Burden of extragonadal germ cell tumours in Europe and the United States. Eur J Cancer. 2012;48(7):1116–1117. doi:S0959-8049(12)00215-8 [pii] 10.1016/j.ejca.2012.02.061.
511. dos Santos SI, Swerdlow AJ. Ovarian germ cell malignancies in England: epidemiological parallels with testicular cancer. Br J Cancer. 1991;63(5):814–8.
512. Walker AH, Ross RK, Pike MC, Henderson BE. A possible rising incidence of malignant germ cell tumours in young women. Br J Cancer. 1984;49(5):669–72.
513. Cools M, Stoop H, Kersemaekers AM, Drop SL, Wolffenbuttel KP, Bourguignon JP, et al. Gonadoblastoma arising in undifferentiated gonadal tissue within dysgenetic gonads. J Clin Endocrinol Metab. 2006;91(6):2404–2413. doi:jc.2005-2554 [pii] 10.1210/jc.2005-2554.
514. Koo YJ, Chun YK, Kwon YS, Lee IH, Kim TJ, Lee KH, et al. Ovarian gonadoblastoma with dysgerminoma in a woman with 46XX karyotype. Pathol Int. 2011;61(3):171–3. doi:10.1111/j.1440-1827.2010.02636.x.
515. Zhao S, Kato N, Endoh Y, Jin Z, Ajioka Y, Motoyama T. Ovarian gonadoblastoma with mixed germ cell tumor in a woman with 46, XX karyotype and successful pregnancies. Pathol Int. 2000;50(4):332–335. doi:pin1041 [pii].
516. Talerman A. Germ cell tumors of the ovary. In: Kurman RJ, editor. Blaustein's pathology of the female genital tract. New York: Springer-Verlag; 1987. p. 659–721.
517. A L Husaini H, Soudy H, El Din Darwish A, Ahmed M, Eltigani A, A L Mubarak M, et al. Pure dysgerminoma of the ovary: a single institutional experience of 65 patients. Med Oncol. 2012;29(4):2944–8. doi:10.1007/s12032-012-0194-z.
518. Nogales F. Germ cell tumours. In: Tavassoli FA, editor. World Health Organization classification of tumours: pathology and genetics of tumours of the breast and female genital organs. Lyon: IARC Press; 2003. p. 163–75.
519. Guillem V, Poveda A. Germ cell tumours of the ovary. Clin Transl Oncol. 2007;9(4):237–243. doi:1027 [pii].
520. Nogales FF, Dulcey I, Preda O. Germ cell tumors of the ovary: an update. Arch Pathol Lab Med. 2014;138(3):351–62. doi:10.5858/arpa.2012-0547-RA.
521. Sigismondi C, Scollo P, Ferrandina G, Candiani M, Angioli R, Vigano R, et al. Management of bilateral malignant ovarian germ cell tumors: a MITO-9 retrospective study. Int J Gynecol Cancer. 2015;25(2):203–7. doi:10.1097/IGC.0000000000000358.
522. Stettner AR, Hartenbach EM, Schink JC, Huddart R, Becker J, Pauli R, et al. Familial ovarian germ cell cancer: report and review. Am J Med Genet. 1999;84(1):43–46. doi:10.1002/(SICI)1096-8628(19990507)84:1<43::AID-AJMG9>3.0.CO;2-2 [pii].
523. Cyriac S, Rajendranath R, Louis AR, Sagar TG. Familial germ cell tumor. Indian J Hum Genet. 2012;18(1):119–121. doi:10.4103/0971-6866.96679 IJHG-18-119 [pii].
524. Huddart RA, Thompson C, Houlston R, Huddart RA, Nicholls EJ, Horwich A. Familial predisposition to both male and female germ cell tumours? J Med Genet. 1996;33(1):86.
525. Sharpe RM, Skakkebaek NE. Testicular dysgenesis syndrome: mechanistic insights and potential new downstream effects. Fertil Steril. 2008;89(2 Suppl):e33–e38. doi:S0015-0282(07)04302-6 [pii] 10.1016/j.fertnstert.2007.12.026.
526. Walker AH, Ross RK, Haile RW, Henderson BE. Hormonal factors and risk of ovarian germ cell cancer in young women. Br J Cancer. 1988;57(4):418–22.
527. Horn-Ross PL, Whittemore AS, Harris R, Itnyre J. Characteristics relating to ovarian cancer risk: collaborative analysis of 12 U.S. case-control studies. VI. Nonepithelial cancers among adults. Collaborative Ovarian Cancer Group. Epidemiology. 1992;3(6):490–5.
528. Adami HO, Hsieh CC, Lambe M, Trichopoulos D, Leon D, Persson I, et al. Parity, age at first childbirth, and risk of ovarian cancer. Lancet. 1994;344(8932):1250–1254. doi:S0140-6736(94)90749-8 [pii].

529. Albrektsen G, Heuch I, Kvale G. Full-term pregnancies and incidence of ovarian cancer of stromal and germ cell origin: a Norwegian prospective study. Br J Cancer. 1997;75(5):767–70.
530. Chen T, Surcel HM, Lundin E, Kaasila M, Lakso HA, Schock H, et al. Circulating sex steroids during pregnancy and maternal risk of non-epithelial ovarian cancer. Cancer Epidemiol Biomark Prev. 2011;20(2):324–336. doi:1055-9965.EPI-10-0857 [pii] 10.1158/1055-9965.EPI-10-0857.
531. Gobbi D, Fascetti Leon F, Aquino A, Melchionda F, Lima M. Metachronous bilateral ovarian teratoma: a germ-line familial disorder and review of surgical management options. J Pediatr Adolesc Gynecol. 2013;26(5):e105–e107. doi:S1083-3188(13)00102-2 [pii] 10.1016/j.jpag.2013.02.006.
532. Ayhan A, Bukulmez O, Genc C, Karamursel BS, Ayhan A. Mature cystic teratomas of the ovary: case series from one institution over 34 years. Eur J Obstet Gynecol Reprod Biol. 2000;88(2):153–157. doi:S0301211599001414 [pii].
533. Brown EH, Jr. Identical twins with twisted benign cystic teratoma of the ovary. Am J Obstet Gynecol. 1979;134(8):879–880. doi:0002-9378(79)90860-3 [pii].
534. Oud PS, Soeters RP, Pahlplatz MM, Hermkens HG, Beck HL, Reubsaet-Veldhuizen J, et al. DNA cytometry of pure dysgerminomas of the ovary. Int J Gynecol Pathol. 1988;7(3):258–67.
535. Kraggerud SM, Szymanska J, Abeler VM, Kaern J, Eknaes M, Heim S, et al. DNA copy number changes in malignant ovarian germ cell tumors. Cancer Res. 2000;60(11):3025–30.
536. Ichikawa Y, Yoshida S, Koyama Y, Hirai M, Ishikawa T, Nishida M, et al. Inactivation of p16/CDKN2 and p15/MTS2 genes in different histological types and clinical stages of primary ovarian tumors. Int J Cancer. 1996;69(6):466–470. doi:10.1002/(SICI)1097-0215(19961220)69:6<466::AID-IJC8>3.0.CO;2-2 [pii] 10.1002/(SICI)1097-0215(19961220)69:6<466::AID-IJC8>3.0.CO;2-2.
537. Lindeman RE, Gearhart MD, Minkina A, Krentz AD, Bardwell VJ, Zarkower D. Sexual cell-fate reprogramming in the ovary by DMRT1. Curr Biol. 2015;25(6):764–771. doi:S0960-9822(15)00066-4 [pii] 10.1016/j.cub.2015.01.034.
538. DeFalco T. DMRT1 keeps masculinity intact. Dev Cell. 2014;29(5):503–504. doi:S1534-5807(14)00335-9 [pii] 10.1016/j.devcel.2014.05.015.
539. McCarthy BJ, Shibui S, Kayama T, Miyaoka E, Narita Y, Murakami M, et al. Primary CNS germ cell tumors in Japan and the United States: an analysis of 4 tumor registries. Neuro Oncology. 2012;14(9):1194–1200. doi:nos155 [pii] 10.1093/neuonc/nos155.
540. Nichols CR, Heerema NA, Palmer C, Loehrer Sr PJ, Williams SD, Einhorn LH. Klinefelter's syndrome associated with mediastinal germ cell neoplasms. J Clin Oncol. 1987;5(8):1290–4.
541. Dexeus FH, Logothetis CJ, Chong C, Sella A, Ogden S. Genetic abnormalities in men with germ cell tumors. J Urol. 1988;140(1):80–4.
542. Kolodziejski L, Duda K, Niezabitowski A, Dyczek S, Staniec B. Occurrence of malignant non-germ cell components in primary mediastinal germ cell tumours. Eur J Surg Oncol J Eur Soc Surg Oncol Br Assoc Surg Oncol. 1999;25(1):54–60. doi:10.1053/ejso.1998.0600.
543. Neiman RS, Orazi A. Mediastinal non-seminomatous germ cell tumours: their association with non-germ cell malignancies. Pathol Res Pract. 1999;195(8):589–94. doi:10.1016/S0344-0338(99)80010-7.
544. Contreras AL, Punar M, Tamboli P, Tu SM, Pisters L, Moran C, et al. Mediastinal germ cell tumors with an angiosarcomatous component: a report of 12 cases. Hum Pathol. 2010;41(6):832–7. doi:10.1016/j.humpath.2009.11.008.
545. Dulmet EM, Macchiarini P, Suc B, Verley JM. Germ cell tumors of the mediastinum. A 30-year experience. Cancer. 1993;72(6):1894–901.
546. Hartmann JT, Nichols CR, Droz JP, Horwich A, Gerl A, Fossa SD, et al. Hematologic disorders associated with primary mediastinal nonseminomatous germ cell tumors. J Natl Cancer Inst. 2000;92(1):54–61.
547. Zhao GQ, Dowell JE. Hematologic malignancies associated with germ cell tumors. Expert Rev Hematol. 2012;5(4):427–37. doi:10.1586/ehm.12.24.
548. Curry WA, McKay CE, Richardson RL, Greco FA. Klinefelter's syndrome and mediastinal germ cell neoplasms. J Urol. 1981;125(1):127–9.
549. Nichols CR, Roth BJ, Heerema N, Griep J, Tricot G. Hematologic neoplasia associated with primary mediastinal germ-cell tumors. N Engl J Med. 1990;322(20):1425–9. doi:10.1056/NEJM199005173222004.
550. Krywicki R, Bowen K, Anderson L, Garland D, Cobb P, Jenkins T, et al. Mixed-lineage acute myeloid leukemia associated with a suprasellar dysgerminoma. Am J Clin Oncol. 1995;18(1):83–6.
551. Ulbright TM, Loehrer PJ, Roth LM, Einhorn LH, Williams SD, Clark SA. The development of non-germ cell malignancies within germ cell tumors. A clinicopathologic study of 11 cases. Cancer. 1984;54(9):1824–33.
552. Nichols CR. Mediastinal germ cell tumors. Clinical features and biologic correlates. Chest. 1991;99(2):472–9.
553. Hasle H, Jacobsen BB, Asschenfeldt P, Andersen K. Mediastinal germ cell tumour associated with Klinefelter syndrome. A report of case and review of the literature. Eur J Pediatr. 1992;151(10):735–9.
554. Hasle H, Mellemgaard A, Nielsen J, Hansen J. Cancer incidence in men with Klinefelter syndrome. Br J Cancer. 1995;71(2):416–20.
555. Kaido T, Sasaoka Y, Hashimoto H, Taira K. De novo germinoma in the brain in association with Klinefelter's syndrome: case report and review of the literature. Surg Neurol. 2003;60(6):553–558; discussion 9.
556. Queipo G, Aguirre D, Nieto K, Pena YR, Palma I, Olvera J, et al. Intracranial germ cell tumors: association with Klinefelter syndrome and sex chromosome

aneuploidies. Cytogenet Genome Res. 2008;121(3–4):211–4. doi:10.1159/000138887.
557. Satge D, Sommelet D, Geneix A, Nishi M, Malet P, Vekemans M. A tumor profile in Down syndrome. Am J Med Genet. 1998;78(3):207–216. doi:10.1002/(SICI)1096-8628(19980707)78:3<207::AID-AJMG1>3.0.CO;2-M [pii].
558. Hasle H. Pattern of malignant disorders in individuals with Down's syndrome. Lancet Oncol. 2001;2(7):429–436. doi:S1470-2045(00)00435-6 [pii] 10.1016/S1470-2045(00)00435-6.
559. Matsui I, Tanimura M, Kobayashi N, Sawada T, Nagahara N, Akatsuka J. Neurofibromatosis type 1 and childhood cancer. Cancer. 1993;72(9):2746–54.
560. Walker L, Thompson D, Easton D, Ponder B, Ponder M, Frayling I, et al. A prospective study of neurofibromatosis type 1 cancer incidence in the UK. Br J Cancer. 2006;95(2):233–238. doi:6603227 [pii] 10.1038/sj.bjc.6603227.
561. Hartley AL, Birch JM, Kelsey AM, Marsden HB, Harris M, Teare MD. Are germ cell tumors part of the Li-Fraumeni cancer family syndrome? Cancer Genet Cytogenet. 1989;42(2):221–226. doi:0165-4608(89)90090-3 [pii].
562. Hisada M, Garber JE, Fung CY, Fraumeni Jr JF, Li FP. Multiple primary cancers in families with Li-Fraumeni syndrome. J Natl Cancer Inst. 1998;90(8):606–11.
563. Daugaard G, Rorth M, von der Maase H, Skakkebaek NE. Management of extragonadal germ-cell tumors and the significance of bilateral testicular biopsies. Ann Oncol. 1992;3(4):283–9.
564. Konig R, Schonberger W, Grimm W. [Mediastinal teratocarcinoma and hypophyseal stalk germinoma in a patient with Klinefelter syndrome] Mediastinales Terato-Carzinom und Hypophysenstielgerminom bei einem Patienten mit Klinefelter Syndrom. Klin Padiatr. 1990;202(1):53–6. doi:10.1055/s-2007-1025486.
565. Moran CA, Suster S, Przygodzki RM, Koss MN. Primary germ cell tumors of the mediastinum: II. Mediastinal seminomas – a clinicopathologic and immunohistochemical study of 120 cases. Cancer. 1997;80(4):691–698. doi:10.1002/(SICI)1097-0142(19970815)80:4<691::AID-CNCR7>3.0.CO;2-Q [pii].
566. Moran CA, Suster S. Mediastinal seminomas with prominent cystic changes. A clinicopathologic study of 10 cases. Am J Surg Pathol. 1995;19(9):1047–53.
567. Moran CA, Suster S. Mediastinal yolk sac tumors associated with prominent multilocular cystic changes of thymic epithelium: a clinicopathologic and immunohistochemical study of five cases. Mod Pathol Off J U S Can Acad Pathol Inc. 1997;10(8):800–3.
568. Kesler KA, Brooks JA, Rieger KM, Fineberg NS, Einhorn LH, Brown JW. Mediastinal metastases from testicular nonseminomatous germ cell tumors: patterns of dissemination and predictors of long-term survival with surgery. J Thorac Cardiovasc Surg. 2003;125(4):913–923. doi:10.1067/mtc.2003.407 S0022522303000990 [pii].
569. de Graaff WE, Oosterhuis JW, de Jong B, Dam A, van Putten WL, Castedo SM, et al. Ploidy of testicular carcinoma in situ. Lab Investig. 1992;66(2):166–8.
570. Sandberg AA, Abe S, Kowalczyk JR, Zedgenidze A, Takeuchi J, Kakati S. Chromosomes and causation of human cancer and leukemia. L. Cytogenetics of leukemias complicating other diseases. Cancer Genet Cytogenet. 1982;7(2):95–136. doi:0165-4608(82)90009-7 [pii].
571. Mann BD, Sparkes RS, Kern DH, Morton DL. Chromosomal abnormalities of a mediastinal embryonal cell carcinoma in a patient with 47,XXY Klinefelter syndrome: evidence for the premeiotic origin of a germ cell tumor. Cancer Genet Cytogenet. 1983;8(3):191–196. doi:0165-4608(83)90134-6 [pii].
572. Rodriguez E, Mathew S, Reuter V, Ilson DH, Bosl GJ, Chaganti RS. Cytogenetic analysis of 124 prospectively ascertained male germ cell tumors. Cancer Res. 1992;52(8):2285–91.
573. Kaplan CG, Askin FB, Benirschke K. Cytogenetics of extragonadal tumors. Teratology. 1979;19(2):261–6. doi:10.1002/tera.1420190217.
574. Oosterhuis JW, van den Berg E, de Jong B, Timens W, Castedo SM, Rammeloo RH, et al. Mediastinal germ cell tumor with secondary nongerm cell malignancy, and extensive hematopoietic activity. Pathology, DNA-ploidy, and karyotyping. Cancer Genet Cytogenet. 1991;54(2):183–195. doi:0165-4608(91)90206-A [pii].
575. Ladanyi M, Samaniego F, Reuter VE, Motzer RJ, Jhanwar SC, Bosl GJ, et al. Cytogenetic and immunohistochemical evidence for the germ cell origin of a subset of acute leukemias associated with mediastinal germ cell tumors. J Natl Cancer Inst. 1990;82(3):221–7.
576. Orazi A, Neiman RS, Ulbright TM, Heerema NA, John K, Nichols CR. Hematopoietic precursor cells within the yolk sac tumor component are the source of secondary hematopoietic malignancies in patients with mediastinal germ cell tumors. Cancer. 1993;71(12):3873–81.
577. Weidner N. Germ-cell tumors of the mediastinum. Semin Diagn Pathol. 1999;16(1):42–50.
578. Chaganti RS, Rodriguez E, Mathew S. Origin of adult male mediastinal germ-cell tumours. Lancet. 1994;343(8906):1130–1132. doi:S0140-6736(94)90235-6 [pii].
579. Hasle H, Jacobsen BB. Origin of male mediastinal germ-cell tumours. Lancet. 1995;345(8956):1046.
580. Wikstrom AM, Hoei-Hansen CE, Dunkel L, Rajpert-De Meyts E. Immunoexpression of androgen receptor and nine markers of maturation in the testes of adolescent boys with Klinefelter syndrome: evidence for degeneration of germ cells at the onset of meiosis. J Clin Endocrinol Metab. 2007;92(2):714–719. doi:jc.2006-1892 [pii] 10.1210/jc.2006-1892.
581. Beattie LM. Testicular dysgenesis: report of two cases. Can Serv Med J. 1957;8(8):469–79.

582. Isurugi K, Imao S, Hirose K, Aoki H. Seminoma in Klinefelter's syndrome with 47, XXY, 15s+ karyotype. Cancer. 1977;39(5):2041–7.
583. Carroll PR, Morse MJ, Koduru PP, Chaganti RS. Testicular germ cell tumor in patient with Klinefelter syndrome. Urology. 1988;31(1):72–4.
584. Nakata Y, Yagishita A, Arai N. Two patients with intraspinal germinoma associated with Klinefelter syndrome: case report and review of the literature. AJNR Am J Neuroradiol. 2006;27(6):1204–1210. doi:27/6/1204 [pii].
585. Rapley EA, Crockford GP, Teare D, Biggs P, Seal S, Barfoot R, et al. Localization to Xq27 of a susceptibility gene for testicular germ-cell tumours. Nat Genet. 2000;24(2):197–200. doi:10.1038/72877.
586. Crockford GP, Linger R, Hockley S, Dudakia D, Johnson L, Huddart R, et al. Genome-wide linkage screen for testicular germ cell tumour susceptibility loci. Hum Mol Genet. 2006;15(3):443–451. doi:ddi459 [pii] 10.1093/hmg/ddi459.
587. North TE, de Bruijn MF, Stacy T, Talebian L, Lind E, Robin C, et al. Runx1 expression marks long-term repopulating hematopoietic stem cells in the midgestation mouse embryo. Immunity. 2002;16(5):661–672. doi:S1074761302002960 [pii].
588. de Bruijn MF, Ma X, Robin C, Ottersbach K, Sanchez MJ, Dzierzak E. Hematopoietic stem cells localize to the endothelial cell layer in the midgestation mouse aorta. Immunity. 2002;16(5):673–683. doi:S1074761302003138 [pii].
589. Le PT, Kurtzberg J, Brandt SJ, Niedel JE, Haynes BF, Singer KH. Human thymic epithelial cells produce granulocyte and macrophage colony-stimulating factors. J Immunol. 1988;141(4):1211–7.
590. Back MR, Hu B, Rutgers J, French S, Moore TC. Metastasis of an intracranial germinoma through a ventriculoperitoneal shunt: recurrence as a yolk-sac tumor. Pediatr Surg Int. 1997;12(1):24–7. doi:10.1007/BF01194796.
591. Bi WL, Bannykh SI, Baehring J. The growing teratoma syndrome after subtotal resection of an intracranial nongerminomatous germ cell tumor in an adult: case report. Neurosurgery 2005;56:E191–E194.
592. Czirjak S, Pasztor E, Slowik F, Szeifert G. Third ventricle germinoma after total removal of intrasellar teratoma. Case report. J Neurosurg. 1992;77(4):643–7. doi:10.3171/jns.1992.77.4.0643.
593. Ikeda J, Sawamura Y, Kato T, Abe H. Metachronous neurohypophyseal germinoma occurring 8 years after total resection of pineal mature teratoma. Surg Neurol. 1998;49(2):205–208; discussion 8–9. doi:S0090-3019(97)00158-4 [pii].
594. Sugimoto K, Nakahara I, Nishikawa M. Bilateral metachronous germinoma of the basal ganglia occurring long after total removal of a mature pineal teratoma: case report. Neurosurgery. 2002;50(3):613–616; discussion 6–7.
595. Janzarik WG, Muller K, Lubbert M, Spreer J, Trippel M, Uhlig I, et al. Occurrence of a germinoma 22 years after resection of a mature cerebral teratoma. J Neuro Oncol. 2008;88(2):217–9. doi:10.1007/s11060-008-9554-0.
596. Mao Q, Ma L, Pang Z, Liu J. Germinoma occurring 2 years after total resection of an intracranial epidermoid cyst in the pineal region. J Neuro Oncol. 2012;106(2):437–9. doi:10.1007/s11060-011-0683-5.
597. Maeda Y, Yoshikawa K, Kajiwara K, Ideguchi M, Amano T, Saka M, et al. Intracranial yolk sac tumor in a patient with Down syndrome. J Neurosurg Pediatr. 2011;7(6):604–8. doi:10.3171/2011.3.PEDS10500.
598. Araki C, Matsumoto S. Statistical reevaluation of pinealoma and related tumors in Japan. J Neurosurg. 1969;30(2):146–9. doi:10.3171/jns.1969.30.2.0146.
599. Jennings MT, Gelman R, Hochberg F. Intracranial germ-cell tumors: natural history and pathogenesis. J Neurosurg. 1985;63(2):155–67. doi:10.3171/jns.1985.63.2.0155.
600. Packer RJ, Cohen BH, Cooney K. Intracranial germ cell tumors. Oncologist. 2000;5(4):312–20.
601. Kuratsu J, Ushio Y. Epidemiological study of primary intracranial tumors: a regional survey in Kumamoto Prefecture in the southern part of Japan. J Neurosurg. 1996;84(6):946–50. doi:10.3171/jns.1996.84.6.0946.
602. Committee of Brain Tumor Registry of Japan. Report of brain tumor registry of Japan (1969–1996). Neurol Med Chir (Tokyo). 2003;43(Suppl):i–vii.
603. Kaneko S, Nomura K, Yoshimura T, Yamaguchi N. Trend of brain tumor incidence by histological subtypes in Japan: estimation from the Brain Tumor Registry of Japan, 1973–1993. J Neuro Oncol. 2002;60(1):61–9.
604. Wong TT, Chen YW, Guo WY, Chang KP, Ho DM, Yen SH. Germinoma involving the basal ganglia in children. Childs Nerv Syst. 2008;24(1):71–8. doi:10.1007/s00381-007-0495-2.
605. Villani A, Bouffet E, Blaser S, Millar BA, Hawkins C, Bartels U. Inherent diagnostic and treatment challenges in germinoma of the basal ganglia: a case report and review of the literature. J Neuro Oncol. 2008;88(3):309–14. doi:10.1007/s11060-008-9568-7.
606. Trentini GP, Maiorana A, De Benedittis A. Metachronous seminoma of the pineal region and right testis. Case report. Appl Pathol. 1985;3(3):129–33.
607. Peat DS, Trowell JE. Testicular seminoma in a patient with pineal germinoma. J Clin Pathol. 1994;47(8):771–2.
608. Daniel C, Fizazi K, Culine S, Zelek L, Wibault P, Theodore C. Metachronous gonadal and extragonadal primaries, or late relapse of germ cell tumor? Urol Oncol. 2001;6(2):49–52. doi:S1078-1439(00)00091-0 [pii].
609. Iwata H, Mori Y, Takagi H, Shirahashi K, Shinoda J, Shimokawa K, et al. Mediastinal growing teratoma syndrome after cisplatin-based chemotherapy and radiotherapy for intracranial germinoma. J Thorac Cardiovasc Surg. 2004;127(1):291–293. doi:10.1016/S0022 S0022-5223(03)01300-X [pii].

610. Bedano PM, Bonnin J, Einhorn LH. Metachronous intracranial germinoma in a patient with a previous primary mediastinal seminoma. J Clin Oncol. 2006;24(15):2386–2387. doi:24/15/2386 [pii] 10.1200/JCO.2005.02.1576.
611. Benesch M, Schreibmayer N, Ratschek M, Hollwarth M, Lackner H, Urban C. Mediastinal yolk sac tumor ten years after treatment of intracranial germinoma. Med Pediatr Oncol. 2003;40(1):54–6. doi:10.1002/mpo.10031.
612. Rubinstein LJ. Tumors of the central nervous system. Atlas of tumor pathology, Series 2. Washington D.C.: Armed Forces Institute of Pathology; 1972.
613. Hashimoto T, Sasagawa I, Ishigooka M, Kubota Y, Nakada T, Fujita T, et al. Down's syndrome associated with intracranial germinoma and testicular embryonal carcinoma. Urol Int. 1995;55(2):120–2.
614. Wakai S, Segawa H, Kitahara S, Asano T, Sano K, Ogihara R, et al. Teratoma in the pineal region in two brothers. Case reports. J Neurosurg. 1980;53(2):239–43. doi:10.3171/jns.1980.53.2.0239.
615. Kido G, Takeuchi T, Tsukiyama T, Nakamura S, Tsubokawa T, Henmi A. Tumor of the pineal region in three brothers. No Shinkei Geka. 1984;12(8):975–80.
616. Aoyama I, Kondo A, Ogawa H, Ikai Y. Germinoma in siblings: case reports. Surg Neurol. 1994;41(4):313–7.
617. Baek KH, Zaslavsky A, Lynch RC, Britt C, Okada Y, Siarey RJ, et al. Down's syndrome suppression of tumour growth and the role of the calcineurin inhibitor DSCR1. Nature. 2009;459(7250):1126–1130. doi:nature08062 [pii] 10.1038/nature08062.
618. Wong TT, Ho DM, Chang TK, Yang DD, Lee LS. Familial neurofibromatosis 1 with germinoma involving the basal ganglion and thalamus. Childs Nerv Syst. 1995;11(8):456–8.
619. Aker FV, Berkman ZM, Aydingoz I, Hakan T, Toksoy G. Pineal germinoma associated with multiple congenital melanocytic nevi: a unique presentation. Neuropathology. 2005;25(4):336–40.
620. Heery DM, Kalkhoven E, Hoare S, Parker MG. A signature motif in transcriptional co-activators mediates binding to nuclear receptors. Nature. 1997; 387(6634):733–6. doi:10.1038/42750.
621. Wolf SS, Patchev VK, Obendorf M. A novel variant of the putative demethylase gene, s-JMJD1C, is a coactivator of the AR. Arch Biochem Biophys. 2007;460(1):56–66. doi:S0003-9861(07)00031-8 [pii] 10.1016/j.abb.2007.01.017.
622. Kuroki S, Akiyoshi M, Tokura M, Miyachi H, Nakai Y, Kimura H, et al. JMJD1C, a JmjC domain-containing protein, is required for long-term maintenance of male germ cells in mice. Biol Reprod. 2013;89(4):93. doi:biolreprod.113.108597 [pii] 10.1095/biolreprod.113.108597.
623. Bartkova J, Hoei-Hansen CE, Krizova K, Hamerlik P, Skakkebaek NE, Rajpert-De Meyts E, et al. Patterns of DNA damage response in intracranial germ cell tumors versus glioblastomas reflect cell of origin rather than brain environment: implications for the anti-tumor barrier concept and treatment. Mol Oncol. 2014;8(8):1667–1678. doi:S1574-7891(14)00156-2 [pii] 10.1016/j.molonc.2014.07.001.
624. Sano K. Pathogenesis of intracranial germ cell tumors reconsidered. J Neurosurg. 1999;90(2):258–64. doi:10.3171/jns.1999.90.2.0258.
625. Ben-David U, Benvenisty N. The tumorigenicity of human embryonic and induced pluripotent stem cells. Nat Rev Cancer. 2011;11(4):268–277. doi:nrc3034 [pii] 10.1038/nrc3034.
626. Matsaniotis N, Karpouzas J, Economou-Mavrou C. Hypothyroidism and seminoma in association with Down's syndrome. J Pediatr. 1967;70(5):810–2.
627. Floret D, Renaud H, Monnet P. Sexual precocity and thoracic polyembryoma: Klinefelter syndrome? J Pediatr. 1979;94(1):163.
628. Masson P. [Not available] Etude sur le seminome. Rev Can Biol. 1946;5(4):361–87.
629. Scully RE. Spermatocytic seminoma of the testis. A report of 3 cases and review of the literature. Cancer. 1961;14:788–94.
630. Eble JN. Spermatocytic seminoma. Hum Pathol. 1994;25(10):1035–42.
631. Waheeb R, Hofmann MC. Human spermatogonial stem cells: a possible origin for spermatocytic seminoma. Int J Androl. 2011;34(4 Pt 2):e296–e305; discussion e. doi:10.1111/j.1365-2605.2011.01199.x.
632. Lim J, Goriely A, Turner GD, Ewen KA, Jacobsen GK, Graem N, et al. OCT2, SSX and SAGE1 reveal the phenotypic heterogeneity of spermatocytic seminoma reflecting distinct subpopulations of spermatogonia. J Pathol. 2011;224(4):473–83. doi:10.1002/path.2919.
633. Lombardi M, Valli M, Brisigotti M, Rosai J. Spermatocytic seminoma: review of the literature and description of a new case of the anaplastic variant. Int J Surg Pathol. 2011;19(1):5–10. doi:1066896910388645 [pii] 10.1177/1066896910388645.
634. Mikuz G, Bohm GW, Behrend M, Schafer G, Colecchia M, Verdorfer I. Therapy-resistant metastasizing anaplastic spermatocytic seminoma: a cytogenetic hybrid: a case report. Anal Quant Cytol Histol. 2014;36(3):177–82.
635. Floyd C, Ayala AG, Logothetis CJ, Silva EG. Spermatocytic seminoma with associated sarcoma of the testis. Cancer. 1988;61(2):409–14.
636. True LD, Otis CN, Delprado W, Scully RE, Rosai J. Spermatocytic seminoma of testis with sarcomatous transformation. A report of five cases. Am J Surg Pathol. 1988;12(2):75–82.
637. True LD, Otis CN, Rosai J, Scully RE, Delprado W. Spermatocytic seminoma of testis with sarcomatous transformation. Am J Surg Pathol. 1988;12(10):806.
638. Carriere P, Baade P, Fritschi L. Population based incidence and age distribution of spermatocytic seminoma. J Urol. 2007;178(1):125–128. doi:S0022-5347(07)00532-0 [pii] 10.1016/j.juro.2007.03.024.

639. Stevens MJ, Gildersleve J, Jameson CF, Horwich A. Spermatocytic seminoma in a maldescended testis. Br J Urol. 1993;72(5 Pt 1):657–9.
640. Muller J, Skakkebaek NE, Parkinson MC. The spermatocytic seminoma: views on pathogenesis. Int J Androl. 1987;10(1):147–56.
641. Dekker I, Rozeboom T, Delemarre J, Dam A, Oosterhuis JW. Placental-like alkaline phosphatase and DNA flow cytometry in spermatocytic seminoma. Cancer. 1992;69(4):993–6.
642. Kraggerud SM, Berner A, Bryne M, Pettersen EO, Fossa SD. Spermatocytic seminoma as compared to classical seminoma: an immunohistochemical and DNA flow cytometric study. APMIS. 1999;107(3):297–302.
643. Rosenberg C, Mostert MC, Schut TB, van de Pol M, van Echten J, de Jong B, et al. Chromosomal constitution of human spermatocytic seminomas: comparative genomic hybridization supported by conventional and interphase cytogenetics. Genes Chromosomes Cancer. 1998;23(4):286–291. doi:10.1002/(SICI)1098-2264(199812)23:4<286::AID-GCC2>3.0.CO;2-6 [pii].
644. Verdorfer I, Rogatsch H, Tzankov A, Steiner H, Mikuz G. Molecular cytogenetic analysis of human spermatocytic seminomas. J Pathol. 2004;204(3):277–81. doi:10.1002/path.1634.
645. Looijenga LH, Hersmus R, Gillis AJ, Pfundt R, Stoop HJ, van Gurp RJ, et al. Genomic and expression profiling of human spermatocytic seminomas: primary spermatocyte as tumorigenic precursor and DMRT1 as candidate chromosome 9 gene. Cancer Res. 2006;66(1):290–302. doi:66/1/290 [pii] 10.1158/0008-5472.CAN-05-2936.
646. Goriely A, Hansen RM, Taylor IB, Olesen IA, Jacobsen GK, McGowan SJ, et al. Activating mutations in FGFR3 and HRAS reveal a shared genetic origin for congenital disorders and testicular tumors. Nat Genet. 2009;41(11):1247–1252. doi:ng.470 [pii] 10.1038/ng.470.
647. Rajpert-De Meyts E, Jacobsen GK, Bartkova J, Aubry F, Samson M, Bartek J, et al. The immunohistochemical expression pattern of Chk2, p53, p19INK4d, MAGE-A4 and other selected antigens provides new evidence for the premeiotic origin of spermatocytic seminoma. Histopathology. 2003;42(3):217–226. doi:1587 [pii].
648. Kristensen DG, Mlynarska O, Nielsen JE, Jacobsen GK, Rajpert-De Meyts E, Almstrup K. Heterogeneity of chromatin modifications in testicular spermatocytic seminoma point toward an epigenetically unstable phenotype. Cancer Genet. 2012;205(9):425–431. doi:S2210-7762(12)00134-2 [pii] 10.1016/j.cancergen.2012.05.003.
649. Meng X, Lindahl M, Hyvonen ME, Parvinen M, de Rooij DG, Hess MW, et al. Regulation of cell fate decision of undifferentiated spermatogonia by GDNF. Science. 2000;287(5457):1489–1493. doi:8297 [pii].
650. Meng X, de Rooij DG, Westerdahl K, Saarma M, Sariola H. Promotion of seminomatous tumors by targeted overexpression of glial cell line-derived neurotrophic factor in mouse testis. Cancer Res. 2001;61(8):3267–71.
651. Sariola H, Meng X. GDNF-induced seminomatous tumours in mouse – an experimental model for human seminomas? APMIS. 2003;111(1):192–196; discussion 6. doi:apm1110123 [pii].
652. Matson CK, Murphy MW, Griswold MD, Yoshida S, Bardwell VJ, Zarkower D. The mammalian doublesex homolog DMRT1 is a transcriptional gatekeeper that controls the mitosis versus meiosis decision in male germ cells. Dev Cell. 2010;19(4):612–624. doi:S1534-5807(10)00428-4 [pii] 10.1016/j.devcel.2010.09.010.
653. Looijenga LH, Olie RA, van der Gaag I, van Sluijs FJ, Matoska J, Ploem-Zaaijer J, et al. Seminomas of the canine testis. Counterpart of spermatocytic seminoma of men? Lab Investig. 1994;71(4):490–6.
654. Portas TJ, Hermes R, Bryant BR, Goritz F, Ladds P, Hildebrant TB. Seminoma in a southern white rhinoceros (Ceratotherium simum simum). Vet Rec. 2005;157(18):556–558. doi:157/18/556 [pii].
655. Subramaniam K, Seydoux G. Dedifferentiation of primary spermatocytes into germ cell tumors in C. elegans lacking the pumilio-like protein PUF-8. Curr Biol. 2003;13(2):134–139. doi:S0960982203000058 [pii].
656. Bush JM, Gardiner DW, Palmer JS, Rajpert-De Meyts E, Veeramachaneni DN. Testicular germ cell tumours in dogs are predominantly of spermatocytic seminoma type and are frequently associated with somatic cell tumours. Int J Androl. 2011;34(4 Pt 2):e288–e295; discussion e95. doi:10.1111/j.1365-2605.2011.01166.x.
657. Kaku H, Usui H, Qu J, Shozu M. Mature cystic teratomas arise from meiotic oocytes, but not from pre-meiotic oogonia. Genes Chromosomes Cancer. 2016;55(4):355–64. doi:10.1002/gcc.22339.
658. Dick HM, Honore LH. Dental structures in benign ovarian cystic teratomas (dermoid cysts). A study of ten cases with a review of the literature. Oral Surg Oral Med Oral Pathol. 1985;60(3):299–307.
659. Kondi-Pafiti A, Filippidou-Giannopoulou A, Papakonstantinou E, Iavazzo C, Grigoriadis C. Epidermoid or dermoid cysts of the ovary? Clinicopathological characteristics of 28 cases and a new pathologic classification of an old entity. Eur J Gynaecol Oncol. 2012;33(6):617–9.
660. Roth LM, Talerman A. Recent advances in the pathology and classification of ovarian germ cell tumors. Int J Gynecol Pathol. 2006;25(4):305–20. doi:10.1097/01.pgp.0000225844.59621.9d.
661. Nogales FF, Preda O, Dulcey I. Gliomatosis peritonei as a natural experiment in tissue differentiation. Int J Dev Biol. 2012;56(10–12):969–74. doi:10.1387/ijdb.120172fn.

662. Blanchet J, Fouquette B. Case of malignant ovarian cystic teratoma with bilateral immature differentiation. Can Med Assoc J. 1979;120(11):1396–8.
663. Koonings PP, Campbell K, Mishell Jr DR, Grimes DA. Relative frequency of primary ovarian neoplasms: a 10-year review. Obstet Gynecol. 1989;74(6): 921–6.
664. Ferlay J, Soerjomataram I, Dikshit R, Eser S, Mathers C, Rebelo M, et al. Cancer incidence and mortality worldwide: sources, methods and major patterns in GLOBOCAN 2012. Int J Cancer. 2015;136(5):E359–86. doi:10.1002/ijc.29210.
665. Hegde P. Extragonadal omental teratoma: a case report. J Obstet Gynaecol Res. 2014;40(2):618–21. doi:10.1111/jog.12198.
666. Mazzarella P, Okagaki T, Richart RM. Teratoma of the uterine tube. A case report and review of the literature. Obstet Gynecol. 1972;39(3):381–8.
667. Ohshima K, Umeda A, Hosoi A, Yamamoto T, Munakata S. Mature cystic teratoma in Douglas' pouch. Case Rep Pathol. 2015;2015:202853. doi:10.1155/2015/202853.
668. Surti U, Hoffner L, Chakravarti A, Ferrell RE. Genetics and biology of human ovarian teratomas. I. Cytogenetic analysis and mechanism of origin. Am J Hum Genet. 1990;47(4):635–43.
669. Devouassoux-Shisheboran M, Vortmeyer AO, Silver SA, Zhuang Z, Tavassoli FA. Teratomatous genotype detected in malignancies of a non-germ cell phenotype. Lab Investig. 2000;80(1):81–6.
670. Schmidt J, Derr V, Heinrich MC, Crum CP, Fletcher JA, Corless CL, et al. BRAF in papillary thyroid carcinoma of ovary (struma ovarii). Am J Surg Pathol. 2007;31(9):1337–43. doi:10.1097/PAS.0b013e31802f5404.
671. Wolff EF, Hughes M, Merino MJ, Reynolds JC, Davis JL, Cochran CS, et al. Expression of benign and malignant thyroid tissue in ovarian teratomas and the importance of multimodal management as illustrated by a BRAF-positive follicular variant of papillary thyroid cancer. Thyroid Off J Am Thyroid Assoc. 2010;20(9):981–7. doi:10.1089/thy.2009.0458.
672. Miura K, Obama M, Yun K, Masuzaki H, Ikeda Y, Yoshimura S, et al. Methylation imprinting of H19 and SNRPN genes in human benign ovarian teratomas. Am J Hum Genet. 1999;65(5):1359–67. doi:10.1086/302615.
673. Linder D. Gene loss in human teratomas. Proc Natl Acad Sci U S A. 1969;63(3):699–704.
674. Linder D, Power J. Further evidence for post-meiotic origin of teratomas in the human female. Ann Hum Genet. 1970;34(1):21–30.
675. Linder D, Hecht F, McCaw BK, Campbell JR. Origin of extragonadal teratomas and endodermal sinus tumours. Nature. 1975;254(5501):597–8.
676. Linder D, McCaw BK, Hecht F. Parthenogenic origin of benign ovarian teratomas. N Engl J Med. 1975;292(2): 63–6. doi:10.1056/NEJM197501092920202.
677. Carritt B, Parrington JM, Welch HM, Povey S. Diverse origins of multiple ovarian teratomas in a single individual. Proc Natl Acad Sci U S A. 1982;79(23):7400–4.
678. Parrington JM, West LF, Povey S. The origin of ovarian teratomas. J Med Genet. 1984;21(1):4–12.
679. Ohama K, Nomura K, Okamoto E, Fukuda Y, Ihara T, Fujiwara A. Origin of immature teratoma of the ovary. Am J Obstet Gynecol. 1985;152(7 Pt 1):896–900.
680. Ohama K. Androgenesis and parthenogenesis in humans. In: Human genetics. Berlin: Springer-Verlag; 1987. p. 245–9.
681. Deka R, Chakravarti A, Surti U, Hauselman E, Reefer J, Majumder PP, et al. Genetics and biology of human ovarian teratomas. II. Molecular analysis of origin of nondisjunction and gene-centromere mapping of chromosome I markers. Am J Hum Genet. 1990;47(4):644–55.
682. Pangas SA. Regulation of the ovarian reserve by members of the transforming growth factor beta family. Mol Reprod Dev. 2012;79(10):666–79. doi:10.1002/mrd.22076.
683. Zhang M, Su YQ, Sugiura K, Wigglesworth K, Xia G, Eppig JJ. Estradiol promotes and maintains cumulus cell expression of natriuretic peptide receptor 2 (NPR2) and meiotic arrest in mouse oocytes in vitro. Endocrinology. 2011;152(11):4377–4385. doi:en.2011-1118 [pii] 10.1210/en.2011-1118.
684. Azoury RS, Jubayli NW, Barakat BY. Dermoid cyst of ovary containing fetus-like structure. Obstet Gynecol. 1973;42(6):887–91.
685. Lee YH, Kim SG, Choi SH, Kim IS, Kim SH. Ovarian mature cystic teratoma containing homunculus: a case report. J Korean Med Sci. 2003;18(6):905–907. doi:200312905 [pii] 10.3346/jkms.2003.18.6.905.
686. Kuno N, Kadomatsu K, Nakamura M, Miwa-Fukuchi T, Hirabayashi N, Ishizuka T. Mature ovarian cystic teratoma with a highly differentiated homunculus: a case report. Birth Defects Res A Clin Mol Teratol. 2004;70(1):40–6. doi:10.1002/bdra.10133.
687. Seckl MJ, Sebire NJ, Berkowitz RS. Gestational trophoblastic disease. Lancet. 2010;376(9742):717–729. doi:S0140-6736(10)60280-2 [pii] 10.1016/S0140-6736(10)60280-2.
688. Lurain JR. Gestational trophoblastic disease II: classification and management of gestational trophoblastic neoplasia. Am J Obstet Gynecol. 2011;204(1):11–18. doi:S0002-9378(10)00852-5 [pii] 10.1016/j.ajog.2010.06.072.
689. Hui P. Molar pregnancies. In: Kurman RJ, editor. World Health Organization classification of tumours of female reproductive organs. Lyon: IARC Press; 2014. p. 163–6.
690. Yoshida K, Nagasaka T, Nakashima N, Nishida Y, Saito M, Tomomitsu O. Elucidation of vascular structure of molar villi in complete hydatidiform mole by CD-34 antibody. Int J Gynecol Pathol. 2000;19(3):212–8.
691. Kim KR, Park BH, Hong YO, Kwon HC, Robboy SJ. The villous stromal constituents of complete hydatidiform mole differ histologically in very early pregnancy from the normally developing placenta.

Am J Surg Pathol. 2009;33(2):176–85. doi:10.1097/PAS.0b013e31817fada1.

692. Ngan HY, Kohorn EI, Cole LA, Kurman RJ, Kim SJ, Lurain JR, et al. Trophoblastic disease. Int J Gynaecol Obstet. 2012;119 Suppl 2:S130–S136. doi:S0020-7292(12)60026-5 [pii] 10.1016/S0020-7292(12)60026-5.
693. Eden A, Gaudet F, Waghmare A, Jaenisch R. Chromosomal instability and tumors promoted by DNA hypomethylation. Science. 2003;300(5618):455. doi:10.1126/science.1083557 300/5618/455 [pii].
694. Novakovic B, Saffery R. DNA methylation profiling highlights the unique nature of the human placental epigenome. Epigenomics. 2010;2(5):627–38. doi:10.2217/epi.10.45.
695. Xue WC, Chan KY, Feng HC, Chiu PM, Ngan HY, Tsao SW, et al. Promoter hypermethylation of multiple genes in hydatidiform mole and choriocarcinoma. J Mol Diagn. 2004;6(4):326–334. doi:S1525-1578(10)60528-4 [pii] 10.1016/S1525-1578(10)60528-4.
696. Hertig AT. Tumors of the female sex organs. Part I hydatidiform mole and choriocarcinoma. Armed Forces Institute of Pathology: Washington, D.C; 1956.
697. Murdoch S, Djuric U, Mazhar B, Seoud M, Khan R, Kuick R, et al. Mutations in NALP7 cause recurrent hydatidiform moles and reproductive wastage in humans. Nat Genet. 2006;38(3):300–302. doi:ng1740 [pii] 10.1038/ng1740.
698. Sebire NJ, Lindsay I, Fisher RA, Savage P, Seckl MJ. Overdiagnosis of complete and partial hydatidiform mole in tubal ectopic pregnancies. Int J Gynecol Pathol. 2005;24(3):260–264. doi:00004347-200507000-00010 [pii].
699. Kajii T, Ohama K. Androgenetic origin of hydatidiform mole. Nature. 1977;268(5621):633–4.
700. Lipata F, Parkash V, Talmor M, Bell S, Chen S, Maric V, et al. Precise DNA genotyping diagnosis of hydatidiform mole. Obstet Gynecol. 2010;115(4):784–794. doi:10.1097/AOG.0b013e3181d489ec 00006250-201004000-00018 [pii].
701. Fukunaga M, Endo Y, Ushigome S. Clinicopathologic study of tetraploid hydropic villous tissues. Arch Pathol Lab Med. 1996;120(6):569–72.
702. Wake N, Takagi N, Sasaki M. Androgenesis as a cause of hydatidiform mole. J Natl Cancer Inst. 1978;60(1):51–7.
703. Jacobs PA, Szulman AE, Funkhouser J, Matsuura JS, Wilson CC. Human triploidy: relationship between parental origin of the additional haploid complement and development of partial hydatidiform mole. Ann Hum Genet. 1982;46(Pt 3):223–31.
704. Lawler SD, Fisher RA, Pickthall VJ, Povey S, Evans MW. Genetic studies on hydatidiform moles. I. The origin of partial moles. Cancer Genet Cytogenet. 1982;5(4):309–20.
705. Genest DR. Partial hydatidiform mole: clinicopathological features, differential diagnosis, ploidy and molecular studies, and gold standards for diagnosis. Int J Gynecol Pathol. 2001;20(4):315–22.
706. Blum B, Benvenisty N. The tumorigenicity of diploid and aneuploid human pluripotent stem cells. Cell Cycle. 2009;8(23):3822–3830. doi:10067 [pii] 10.4161/cc.8.23.10067.
707. Stadtfeld M, Hochedlinger K. Induced pluripotency: history, mechanisms, and applications. Genes Dev. 2010;24(20):2239–2263. doi:24/20/2239 [pii] 10.1101/gad.1963910.
708. Baker DE, Harrison NJ, Maltby E, Smith K, Moore HD, Shaw PJ, et al. Adaptation to culture of human embryonic stem cells and oncogenesis in vivo. Nat Biotechnol. 2007;25(2):207–215. doi:nbt1285 [pii] 10.1038/nbt1285.
709. Draper JS, Moore HD, Ruban LN, Gokhale PJ, Andrews PW. Culture and characterization of human embryonic stem cells. Stem Cells Dev. 2004;13(4):325–36. doi:10.1089/scd.2004.13.325.
710. Maitra A, Arking DE, Shivapurkar N, Ikeda M, Stastny V, Kassauei K, et al. Genomic alterations in cultured human embryonic stem cells. Nat Genet. 2005;37(10):1099–1103. doi:ng1631 [pii] 10.1038/ng1631.
711. Werbowetski-Ogilvie TE, Bosse M, Stewart M, Schnerch A, Ramos-Mejia V, Rouleau A, et al. Characterization of human embryonic stem cells with features of neoplastic progression. Nat Biotechnol. 2009;27(1):91–97. doi:nbt.1516 [pii] 10.1038/nbt.1516.
712. Lefort N, Feyeux M, Bas C, Feraud O, Bennaceur-Griscelli A, Tachdjian G, et al. Human embryonic stem cells reveal recurrent genomic instability at 20q11.21 Nat Biotechnol. 2008;26(12):1364–1366. doi:nbt.1509 [pii] 10.1038/nbt.1509.
713. Spits C, Mateizel I, Geens M, Mertzanidou A, Staessen C, Vandeskelde Y, et al. Recurrent chromosomal abnormalities in human embryonic stem cells. Nat Biotechnol. 2008;26(12):1361–1363. doi:nbt.1510 [pii] 10.1038/nbt.1510.
714. Inzunza J, Sahlen S, Holmberg K, Stromberg AM, Teerijoki H, Blennow E, et al. Comparative genomic hybridization and karyotyping of human embryonic stem cells reveals the occurrence of an isodicentric X chromosome after long-term cultivation. Mol Hum Reprod. 2004;10(6):461–466. doi:10.1093/molehr/gah051 gah051 [pii].
715. Ohnishi K, Semi K, Yamamoto T, Shimizu M, Tanaka A, Mitsunaga K, et al. Premature termination of reprogramming in vivo leads to cancer development through altered epigenetic regulation. Cell. 2014;156(4):663–677. doi:S0092-8674(14)00015-4 [pii] 10.1016/j.cell.2014.01.005.
716. Shih CC, Forman SJ, Chu P, Slovak M. Human embryonic stem cells are prone to generate primitive, undifferentiated tumors in engrafted human fetal tissues in severe combined immunodeficient mice. Stem Cells Dev. 2007;16(6):893–902. doi:10.1089/scd.2007.0070.
717. Unzu C, Friedli M, Bosman A, Jaconi ME, Wildhaber BE, Rougemont AL. Human Hepatocyte-Derived Induced Pluripotent Stem Cells: MYC Expression, Similarities to Human Germ Cell Tumors, and

Safety Issues. Stem Cells Int. 2016;2016:4370142. doi:10.1155/2016/4370142.
718. Misra P, Husain Q, Svider PF, Sanghvi S, Liu JK, Eloy JA. Management of sinonasal teratocarcinosarcoma: a systematic review. Am J Otolaryngol. 2014;35(1):5–11. doi:S0196-0709(13)00117-8 [pii] 10.1016/j.amjoto.2013.04.010.
719. Kinjo T, Taniguchi H, Kushima R, Sekine S, Oda I, Saka M, et al. Histologic and immunohistochemical analyses of alpha-fetoprotein – producing cancer of the stomach. Am J Surg Pathol. 2012;36(1):56–65. doi:10.1097/PAS.0b013e31823aafec 00000478-201201000-00008 [pii].
720. Preda O, Dema A, Iacob M, Goyenaga P, Dulcey I, Aneiros Fernandez J, et al. Urothelial carcinoma of the renal pelvis with simultaneous trophoblastic and malignant clear cell endodermal-type differentiation. Virchows Arch. 2012;460(3):353–6. doi:10.1007/s00428-012-1211-5.
721. Nogales FF, Bergeron C, Carvia RE, Alvaro T, Fulwood HR. Ovarian endometrioid tumors with yolk sac tumor component, an unusual form of ovarian neoplasm. Analysis of six cases. Am J Surg Pathol. 1996;20(9):1056–66.
722. Liang L, Zhang Y, Malpica A, Ramalingam P, Euscher ED, Fuller GN, et al. Gliomatosis peritonei: a clinicopathologic and immunohistochemical study of 21 cases. Mod Pathol Off J U S Can Acad Pathol Inc. 2015;28(12):1613–1620. doi:modpathol2015116 [pii] 10.1038/modpathol.2015.116.
723. Alexander M, Cope N, Renninson J, Hong A, Simpson RH, Hirschowitz L. Relationship between endometriosis, endometrioid adenocarcinoma, gliomatosis peritonei, and carcinoid tumor in a patient with recurrent ovarian teratoma. Int J Gynecol Pathol. 2011;30(2):151–7. doi:10.1097/PGP.0b013e3181f6bcd4.
724. Virant-Klun I, Zech N, Rozman P, Vogler A, Cvjeticanin B, Klemenc P, et al. Putative stem cells with an embryonic character isolated from the ovarian surface epithelium of women with no naturally present follicles and oocytes. Differentiation. 2008;76(8):843–856. doi:S0301-4681(09)60024-4 [pii] 10.1111/j.1432-0436.2008.00268.x.
725. Virant-Klun I, Rozman P, Cvjeticanin B, Vrtacnik-Bokal E, Novakovic S, Rulicke T, et al. Parthenogenetic embryo-like structures in the human ovarian surface epithelium cell culture in postmenopausal women with no naturally present follicles and oocytes. Stem Cells Dev. 2009;18(1):137–49. doi:10.1089/scd.2007.0238.
726. van Echten J, de Jong B, Sinke RJ, Weghuis DO, Sleijfer DT, Oosterhuis JW. Definition of a new entity of malignant extragonadal germ cell tumors. Genes Chromosomes Cancer. 1995;12(1):8–15.
727. Noguera R, Navarro S, Carda C, Peydro-Olaya A, Llombart-Bosch A. Near-haploidy in a malignant sacrococcygeal teratoma. Cancer Genet Cytogenet. 1999;108(1):70–74. doi:S0165460898001150 [pii].
728. Chinoy RF, Soman CS, Swaroop D, Badwar RA. Extragonadal malignant teratoma of the foot. Indian J Cancer. 1992;29(2):96–9.
729. Ait Benali H, Lalya L, Allaoui M, Benmansour A, Elkhanoussi B, Benjelloun S, et al. Extragonadal mixed germ cell tumor of the right arm: description of the first case in the literature. World J Surg Oncol. 2012;10:69. doi:1477-7819-10-69 [pii] 10.1186/1477-7819-10-69.
730. Garcia-Galvis OF, Cabrera-Ozoria C, Fernandez JA, Stolnicu S, Nogales FF. Malignant Mullerian mixed tumor of the ovary associated with yolk sac tumor, neuroepithelial and trophoblastic differentiation (teratoid carcinosarcoma). Int J Gynecol Pathol. 2008;27(4):515–20. doi:10.1097/PGP.0b013e31817b06c7.
731. Salem F, Rosenblum MK, Jhanwar SC, Kancherla P, Ghossein RA, Carlson DL. Teratocarcinosarcoma of the nasal cavity and paranasal sinuses: report of 3 cases with assessment for chromosome 12p status. Hum Pathol. 2008;39(4):605–609. doi:S0046-8177(07)00486-8 [pii] 10.1016/j.humpath.2007.09.002.
732. Thomas J, Adegboyega P, Iloabachie K, Mooring JW, Lian T. Sinonasal teratocarcinosarcoma with yolk sac elements: a neoplasm of somatic or germ cell origin? Ann Diagn Pathol. 2011;15(2):135–139. doi:S1092-9134(10)00009-2 [pii] 10.1016/j.anndiagpath.2010.01.004.
733. Sinke RJ, Weghuis DO, Suijkerbuijk RF, Tanigami A, Nakamura Y, Larsson C, et al. Molecular characterization of a recurring complex chromosomal translocation in two human extragonadal germ cell tumors. Cancer Genet Cytogenet. 1994;73(1):11–6.
734. Egger G, Liang G, Aparicio A, Jones PA. Epigenetics in human disease and prospects for epigenetic therapy. Nature. 2004;429(6990):457–463. doi:10.1038/nature02625 nature02625 [pii].
735. Feinberg AP. Phenotypic plasticity and the epigenetics of human disease. Nature. 2007;447(7143):433–440. doi:nature05919 [pii] 10.1038/nature05919.
736. Jelinic P, Shaw P. Loss of imprinting and cancer. J Pathol. 2007;211(3):261–8. doi:10.1002/path.2116.
737. Cui H, Onyango P, Brandenburg S, Wu Y, Hsieh CL, Feinberg AP. Loss of imprinting in colorectal cancer linked to hypomethylation of H19 and IGF2. Cancer Res. 2002;62(22):6442–6.
738. Cui H, Cruz-Correa M, Giardiello FM, Hutcheon DF, Kafonek DR, Brandenburg S, et al. Loss of IGF2 imprinting: a potential marker of colorectal cancer risk. Science. 2003;299(5613):1753–1755. doi:10.1126/science.1080902 299/5613/1753 [pii].
739. Kim HT, Choi BH, Niikawa N, Lee TS, Chang SI. Frequent loss of imprinting of the H19 and IGF-II genes in ovarian tumors. Am J Med Genet. 1998;80(4):391–395. doi:10.1002/(SICI)1096-8628(19981204)80:4<391::AID-AJMG16>3.0.CO;2-H [pii].
740. Ben-Porath I, Thomson MW, Carey VJ, Ge R, Bell GW, Regev A, et al. An embryonic stem cell-like gene expression signature in poorly differentiated aggressive human tumors. Nat Genet. 2008;40(5):499–507. doi:ng.127 [pii] 10.1038/ng.127.
741. Kim J, Woo AJ, Chu J, Snow JW, Fujiwara Y, Kim CG, et al. A Myc network accounts for similarities between embryonic stem and cancer cell transcription

programs. Cell. 2010;143(2):313–324. doi:S0092-8674(10)01058-5 [pii] 10.1016/j.cell.2010.09.010.
742. Yun K. A new marker for rhabdomyosarcoma. Insulin-like growth factor II. Lab Investig. 1992; 67(5):653–64.
743. Baccarini P, Fiorentino M, D'Errico A, Mancini AM, Grigioni WF. Detection of anti-sense transcripts of the insulin-like growth factor-2 gene in Wilms' tumor. Am J Pathol. 1993;143(6):1535–42.
744. Akmal SN, Yun K, MacLay J, Higami Y, Ikeda T. Insulin-like growth factor 2 and insulin-like growth factor binding protein 2 expression in hepatoblastoma. Hum Pathol. 1995;26(8):846–51.
745. Korkola JE, Houldsworth J, Chadalavada RS, Olshen AB, Dobrzynski D, Reuter VE, et al. Down-regulation of stem cell genes, including those in a 200-kb gene cluster at 12p13.31, is associated with in vivo differentiation of human male germ cell tumors. Cancer Res. 2006;66(2):820–827. doi:66/2/820 [pii] 10.1158/0008-5472.CAN-05-2445.
746. Hirasawa R, Feil R. Genomic imprinting and human disease. Essays Biochem 2010;48:187–200.

4 Diagnostic Immunopathology of Germ Cell Tumors

Ovidiu Preda and Francisco F. Nogales

Immunodetection of various proteins has proved to be a useful tool in the diagnosis of germ cell tumors (GCT), and, curiously, these neoplasms have played an important role in the birth of modern immunohistochemistry. Indeed, the technique of horseradish peroxidase labeling, which represented the breakthrough in applied immunohistochemical technique, was originally developed by Paul Nakane and G Barry Pierce [1, 2] as a tool to label anti-basement membrane antibodies of the murine parietal yolk sac carcinoma.

Since its inception, immunohistochemistry has been successfully employed in the diagnosis of GCT by using the expression of proteins produced during a given developmental stage and related with a tumor type; for example, placental-like alkaline phosphatase as a marker of primitive germ cells and seminoma, alpha fetoprotein as a marker of early endoderm and yolk sac tumors, etc. However, since GCT are caricatures of normal embryogenesis, the recent advances in embryonal stem cells and their pluripotency markers have broadened our understanding of early developmental stages and their equivalent neoplasms. Although histology should always be taken as the diagnostic gold standard [3], the evaluation of the sequential expression of pluripotency markers in GCT has resulted in a better understanding of the relationship between tumor differentiation and diagnosis.

Diagnosis has been further enhanced by the analysis of tissue-specific markers that permit the identification of tissues such as those of neural or endodermal lineage whose presence may bear a prognostic significance in tumor grading. Thus, the joint recognition of pluripotency and tissue-specific marker has (a) improved histogenetic classification, (b) fine-tuned the diagnosis among the different varieties of GCT and their frequent overlap, (c) allowed assessment of the quantity and quality of immature tissues reproduced in the GCT and conditioned grading of immature teratomas, and (d) identified GCT patterns in somatic neoplasms.

In this chapter, we will review the antibodies used in the diagnosis of GCT, both those already used in practical diagnostic routine and others less frequently employed. We will describe their genes and chromosomal location, developmental implications and protein expression in the various types of GCT, and corresponding differential diagnosis with other tumor types.

O. Preda, MD, PhD
R&D. Department of Immunohistochemistry, Master Diagnostica SL, 18016 Granada, Spain
e-mail: ovidiu.preda@vitro.bio

F.F. Nogales, MD, PhD (✉)
Department of Pathology, University of Granada Medical School, 18016 Granada, Spain
e-mail: fnogales@ugr.es

F.F. Nogales, R.E. Jimenez (eds.), *Pathology and Biology of Human Germ Cell Tumors*,
DOI 10.1007/978-3-662-53775-6_4

4.1 Pluripotentiality, Stemness Markers

4.1.1 CD117

4.1.1.1 Nomenclature

CD117 or c-kit is also known as V-KIT Hardy-Zuckerman 4 feline sarcoma viral oncogene homolog, *KIT* oncogene, or mast cell growth factor receptor. However, it is better defined as a type III tyrosine kinase receptor [4].

4.1.1.2 Gene, Location, and Function

Using pulsed-field gel electrophoresis, the *KIT* gene was mapped in chromosome 4q12, close to its homologous *PDGFRA* gene [5, 6]. The gene works as a regulator of cell survival and proliferation, hematopoiesis, stem cell maintenance, gametogenesis, melanogenesis, and mast cell development, migration, and function [4]. It is present in an inactive form and is only activated in the presence of its ligand which is the real stem cell factor expressed by mastocytes and some fibroblasts [7]. *KIT* deficiencies due to gene mutations are responsible for the Piebald syndrome, an autosomal dominant genetic developmental abnormality of pigmentation, characterized by congenital patches of white skin and hair that lack melanocytes [8]. Also a deficient *KIT* gene is responsible for deficiencies in erythropoiesis, fertility, and gut motility.

4.1.1.3 Recommended Clones and Practical Considerations

Antibodies Diagnostic kits employ polyclonal antibodies (Dako c-Kit pharmDx®), while other antibodies using clone YR145 also perform well on paraffin-embedded formalin-fixed tissues and are the main products used in c-kit protein detection assays. Alkaline heat-induced antigen retrieval (HIER) is recommended.

Controls A well-calibrated protocol should produce strong staining in the appendiceal Cajal cells, while smooth muscle cells should be negative [9].

Staining patterns The antibody may have a membranous, cytoplasmic, or perinuclear staining.

4.1.1.4 Expression

C-kit is normally expressed by mast cells, some hematopoietic stem cells, germ cells, melanocytes, Cajal cells of the gastrointestinal tract, and also by some cells of the skin adnexa, breast epithelium, and cerebellar neurons [6].

Since more than 70 % of gastrointestinal stromal tumors (GIST) demonstrate activating mutations and protein overexpression [10], *C-Kit* is the main target in Imatinib mesylate (Gleevec®) therapy, an intracellular tyrosine kinase blocker preventing phosphorylation and the subsequent activation of growth receptors and their downstream signal transduction pathways [11]. In the treatment of GCT, different reports and clinical trials on chemotherapy refractory cases have only reported partial remission in primary and metastatic seminomas [12, 13].

In the development and differentiation of the testis, *KIT* intervenes in germ cell migration, proliferation, and apoptosis [14], maturation of undifferentiated spermatogonia or spermatogonial progenitor cells, and the formation of A1 spermatogonia, which subsequently undergo a series of rapid mitotic divisions prior to meiosis [15]. In the ovary, only the oocytes of the primordial and primary follicles are positive, while diffuse and weak cytoplasmic staining in the granulosa, theca, and interstitial cells is seen in all stages of follicle maturation [16].

KIT is overexpressed in seminomas/dysgerminomas/germinomas (for simplicity, throughout this chapter we will use the generic term of seminoma) (Fig. 4.1) and their precursor lesions: germ cell neoplasia in situ (GCNIS) and gonadoblastoma; it is used as a tool in their differentiation from embryonal carcinoma (EC) [17–19]. In spermatocytic tumors (ST), which do not express

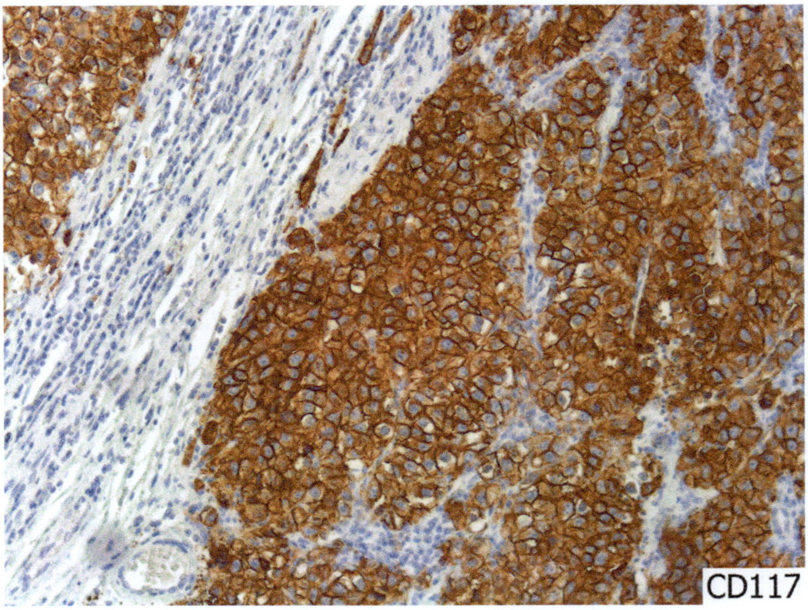

Fig. 4.1 CD117 membrane and cytoplasmic staining in classic seminoma

other GCT characteristic markers [20], CD117 is positive in up to 40 % of the cases [21].

	Positive	Negative
Germ cell tumors	Seminoma Spermatocytic tumor	Embryonal carcinomas Yolk sac tumors Choriocarcinoma Teratomas (epithelial elements)
Non germ cell tumors	GISTs, angiomyolipoma, mastocytosis, thymic carcinomas, lung carcinomas, etc.	Solitary fibrous tumor, alveolar soft part sarcoma, desmoplastic small round cell tumor, glomus tumor, leiomyoma, etc.

4.1.2 OCT4

4.1.2.1 Nomenclature

OCT4 is also known as POU5F1, OCT3, or OTF3 and has been used previously as OCT3/4 but is more recently known as OCT4 (octamer-binding transcription factor 4).

4.1.2.2 Gene Function and Chromosomal Location

OCT4 represents a mammalian nuclear transcription factor belonging to the POU domain encoded by a gene located in the human chromosome 6p21.3, being essential in blastocyst differentiation [22].

OCT4 protein expression is the sum of different isoforms (and pseudogenes) of which only the OCT4A can be directly linked functionally to pluripotency. For this reason, care should be taken as to the type of antibody used when stem cells of somatic tumors are targeted. In these cases, mRNA levels should be detected in order to confirm their stemness [23]. As a key protein and component in the regulatory network that maintains pluripotency in normal development and in GCT pathogenesis, OCT4 is extensively reviewed in Chap. 3.

4.1.2.3 Recommended Clones and Practical Considerations

Antibody The most widely used is a mouse monoclonal antibody (clone C-10) raised against amino acids 1–134 of OCT4 of human origin that specifically recognizes the OCT4A isoform. Alkaline HIER is recommended.

Controls Well-fixed sections of testicular parenchyma with unequivocal changes of GCNIS are recommended as positive controls. Due to the thick fibrous tunica albuginea and unexpected delays in sectioning of orchidectomy specimens, generally the most superficial tumor cells will show more intense staining, while those located in deeper areas of the sample will show weaker nuclear staining and more cytoplasmic background. Fresh sections should be stained as soon as possible to prevent oxidation and loss of antigenicity.

Staining pattern OCT4 predominantly produces a strong nuclear staining with only minimal cytoplasmic staining.

4.1.2.4 Expression

Focal OCT4 staining can be demonstrated in various normal tissues due to the presence of a population of progenitor cells [24]. Its expression is absent in the majority of somatic tumors [25, 26], but recent studies have shown it in high-grade carcinomas of the thyroid, gastrointestinal tract, or lung, being generally associated with distant metastases and poor response to conventional treatment [27–32]. Granular cytoplasmic staining in both normal tissue and neuroendocrine neoplasms, such as Merkel cell carcinomas of the skin, has been noted. This is probably related to cross-reactivity of some antibodies with the OCT4B isomer. The staining weakens as the differentiation grade increases, and it is inversely related with the Ki67 index [33].

In the first trimester of gestation, OCT4 is expressed in embryonal germ cells, while in the second trimester, secondary to the formation of oocytes and spermatogonia, its expression disappears [34, 35]. However, in newborns, it can be identified in testicular germ cells located within the luminal space as well as in gonads showing germ cell maturation delay [36].

In pathological conditions, it is re-expressed in the preinvasive germ cell lesions such as GCNIS and gonadoblastoma [37–39], as well as in seminoma (Fig. 4.2a) and EC [37, 40, 41–46]. It may also be expressed in isolated epithelial cells of ovarian immature teratomas [47], especially in high-grade tumors [48, 49]. OCT4 is expressed by epithelial areas in cases of GCT patterns originating from somatic neoplasms (see Chap. 6).

Albeit a more sensitive and specific marker for both primary and metastatic seminomas and EC [44, 50] than their classical markers [43, 45, 51], it must be borne in mind that rare cases of OCT4-negative, chemoresistant EC have been reported [52].

OCT4 does not differentiate seminomas from EC, and it is not expressed in solid areas of choriocarcinoma, yolk sac tumor (YST) (Fig. 4.2b), or ST. Positive staining in cells delineating papillary projections of EC, helps to differentiate

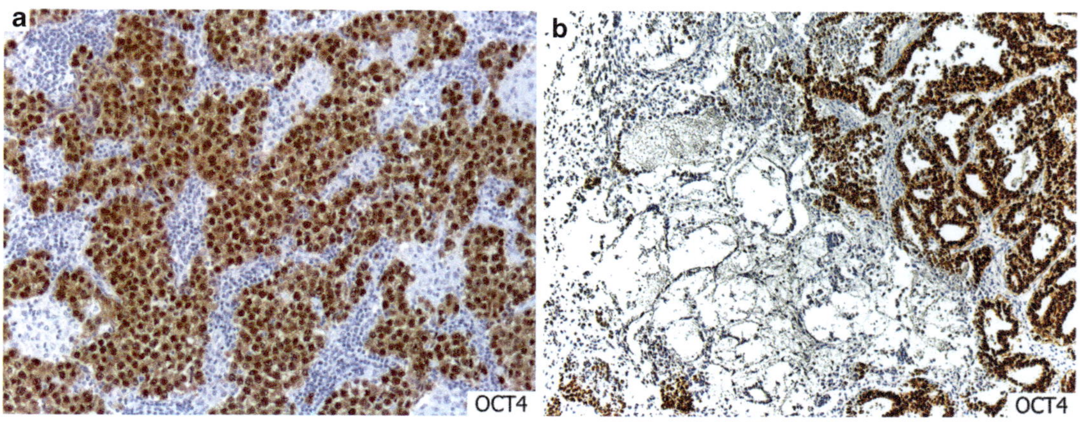

Fig. 4.2 (**a**) OCT4 nuclear staining with cytoplasmic background in classic seminoma. (**b**) OCT4 expresses in embryonal carcinoma, while areas of microcystic yolk sac tumor are negative

them from the Schiller-Duval perivascular structures of YST [53]. In summary, OCT4 is an antibody that differentiates more primitive, pluripotential GCT from differentiated ones.

	Positive	Negative
Germ cell tumors	Seminoma Embryonal carcinoma Primitive GCT areas in somatic tumors	Yolk sac tumor Choriocarcinoma Teratomas Spermatocytic tumor
Non-germ cell tumors	Virtually none	Generally weak staining and in isolated cells of undifferentiated carcinomas

4.1.3 SALL4

4.1.3.1 Nomenclature

Sal-like transcription factor 4

4.1.3.2 Gene Function and Chromosomal Location

SALL4 is a nuclear zinc finger transcription factor, part of the *Spalt-like* family genes, encoded by a gene located on the chromosomal region 20q13.2 and expressed in early developmental stages [54]. Together with *OCT4*, *NANOG*, and *SOX2*, it is part of a transcriptional core network of genes that maintains the pluripotent properties and self-renewal capacities of ESC [55–58]. *SALL4* regulates transcription of *OCT4* [57]. In comparison with *Oct4* and *Sox2,* it is the only one required for the development of the murine primitive endoderm from the inner cell mass in addition to the epiblast, while the development of trophoblastic lineage is not impaired by the lack of *Sall4* [59].

4.1.3.3 Recommended Clones and Practical Considerations

Antibody Mouse monoclonal antibodies EE-30 or 6E3, developed against recombinant proteins similar to the whole or just a fragment of SALL4 human protein, are both suitable for its detection. Alkaline HIER is recommended.

Controls Spermatogonia of normal testis (weak–moderate staining) or GCNIS (intense staining) might be used as a positive control.

Staining pattern As a nuclear transcription factor, SALL4 highlights the nuclei of the tumor cells.

4.1.3.4 Expression

SALL4 is detected in the primordial germ cells and interacts with other transcription factors in the development of the anorectal region, kidney, heart, limbs, and brain. SALL4 nuclear staining has been reported in the proximal renal tubules of the renal cortex, some neural tube elements, intestine, and hepatocytes [60]. Only weak to moderate expression can be seen in the oocytes and spermatogonia [61, 62].

As a master controller of pluripotency maintenance, its nuclear expression can be detected in all primitive and immature GCT with variable results in choriocarcinoma and immature teratoma areas. In trophoblastic tumors, the staining is generally restricted to more immature trophoblastic cells, while the differentiated syncytiotrophoblasts are negative (Fig. 4.3a). Pseudoglandular (Fig. 4.3b) and tubular-neuroectodermal elements of teratomas (Fig. 4.3c) show moderate to weak staining [53, 61–65].

Compared with the weak–moderate positivity of spermatogonia of the normal seminal epithelium, the atypical cell of GCNIS, as well as the medium and small cells of spermatocytic tumors, presents a stronger nuclear stain (Fig. 4.3d) [62]. Dysregulation of *SALL4* might be a common factor in the pathogenesis of all genetic groups of testicular GCT [62, 66]; however, positivity can occur in generally high-grade, poorly differentiated non-GCT, such as ovarian serous carcinomas and undifferentiated urothelial carcinomas [67, 68]. In nonepithelial malignant tumors, SALL4 may be positive in rhabdoid tumors, Wilms' tumors, melanomas, and rhabdomyosarcomas, among others [68]. The expression in lymphoid neoplasms is controversial, and only isolated cases of precursor B-cell lymphoblastic

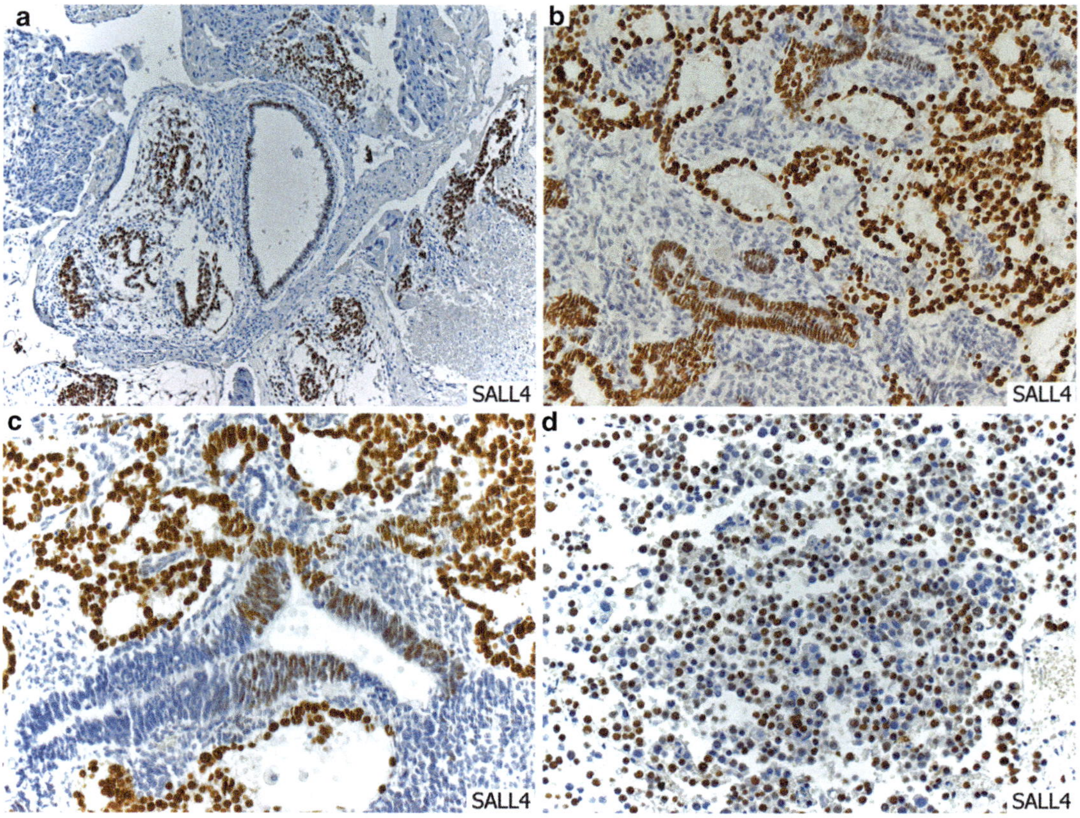

Fig. 4.3 (**a**) SALL4 shows diffuse and strong staining in embryonal carcinoma and yolk sac tumor, while a weak, focal staining in teratoma is seen. Areas of trophoblastic differentiation are negative in this case. (**b**) Moderate staining in immature glands of teratoma while epithelial cells of yolk sac tumor are strongly stained. (**c**) Immature neuroectodermal cells show partial and weak nuclear staining. (**d**) Moderate positivity in small and intermediate cells of spermatocytic tumor

leukemia/lymphomas showed elevated mRNA levels [67].

	Positive	Negative
Germ cell tumors	Seminoma Embryonal carcinoma Yolk sac tumors Choriocarcinoma (cytotrophoblast) Immature teratomas (immature elements) Spermatocytic tumor	Mature teratomas (epithelial elements)
Non-germ cell tumors	Ovarian serous carcinomas, urothelial carcinomas, hepatoid AFP-secreting and intestinal-type gastric adenocarcinomas, etc.	All with exceptions (see left)

4.1.4 SOX2

4.1.4.1 Nomenclature, Gene Function, and Chromosomal Location

SRY (sex-determining region Y-box 2), also known as SOX2, is one of the three members of SOXB1 subgroup of transcription factors encoded by a gene located in chromosome 3q26.33 [69].

In early implantation stages, it has a major role in trophectoderm differentiation; knockout embryos arrest at the morular stage and fail to cavitate [70]. Moreover, a reduction of *SOX2* expression in human ESC induces trophectodermal and partial endodermal differentiation [71]. The correct balance between *SOX2* and *OCT4* might be a key of pluripotency. On murine models, it has been demonstrated that together with *Sox3, Sox2* is expressed in the epiblast and extraembryonic ectoderm and is restricted to the forthcoming neural plate and chorion at gastrulation [72]. During early somitogenesis, all three genes are expressed in the neuroectoderm, and *Sox2* and *Sox3* are also expressed in the primitive streak ectoderm, gut endoderm, and prospective sensory placodes [72]. Consecutively, they may play an important role in maintaining neural crest stem cell multipotency [73, 74] and neuronal formation [75], while its upregulation may contribute to development of supratentorial PNET [76].

4.1.4.2 Recommended Clones and Practical Considerations

Antibody Rabbit polyclonal antibodies (such as AF2018 or AB5603) or mouse monoclonal E-4 have been the most frequently used for recent assays. From our personal experience, we recommend rabbit clone SP76; it works well with a pH 8 buffer for HIER, 10–20 minutes incubation of the antibody, and with a two-step polymer-based detection system.

Controls In our assay, embryonal neural tissue was used as positive control.

Staining pattern The antibody stains the nucleus with little or no cytoplasmic staining.

4.1.4.3 Expression

SOX2 positive staining is mainly reported in squamous cell carcinomas, correlating with amplifications of the 3q26 chromosomal region [30, 77–80]. Other pulmonary non-small cell carcinomas, excluding neuroendocrine ones, also frequently express SOX2 [81–84] as do gastric carcinomas [83, 85], pancreatic and biliary tumors [86], high-grade gliomas [86], and primitive neuroectodermal tumors [87]. Breast carcinomas express SOX2 predominantly in high-grade, HER2-positive tumors and those with a basal-like phenotype. Expression has been reported in metastases of otherwise SOX2-negative primary breast carcinomas [88].

Up to 60.5 % of ovarian carcinomas express SOX2, with a higher percentage of positivity in high-grade tumors and in more advanced stages [89]. Interestingly, a high level of SOX2 is associated with a better prognosis, probably due to a better response to platinum-based therapies [89].

Testicular parenchyma does not express SOX2 during its development [90] but occurs in Sertoli cells associated with GCNIS lesions [91, 92]. Even if it is considered as another important transcription factor in the induction and maintenance of pluripotency [70, 93, 94], in contrast with OCT4 and especially SALL4, SOX2 positivity in GCT is restricted to ECs, immature neuroepithelium, and squamous epithelium of mature teratoma (Fig. 4.4a) [65, 91, 92, 95]. Seminomas, YST (Fig. 4.4b), choriocarcinoma, and spermatocytic tumors are constantly negative [65, 91, 92, 95]. Accordingly, SOX2 expression differentiates seminomas from solid EC, while papillary structures in EC are clearly differentiated from Schiller-Duval bodies in YST [53]. Its negativity in the isolated foci of solid and hepatic areas of YST differentiates these YST variants from EC.

In mediastinal lesions, SOX2 is expressed in EC, isolated teratomatous glands, and foci of mature neural tissue and differentiates them from SOX2-negative lymphomas and epithelial tumors of the thymus [65].

	Positive	Negative
Germ cell tumors	Embryonal carcinomas Immature teratoma (neuroectodermal and squamous component) Gliomatosis peritonei	Seminoma Yolk sac tumors Choriocarcinoma Teratomas Spermatocytic tumor
Non-germ cell tumors	Squamous cell carcinoma, neuroendocrine carcinomas, primitive neuroectodermal tumors, melanomas, breast, pancreatic and biliary duct carcinomas, gliomas, etc.	

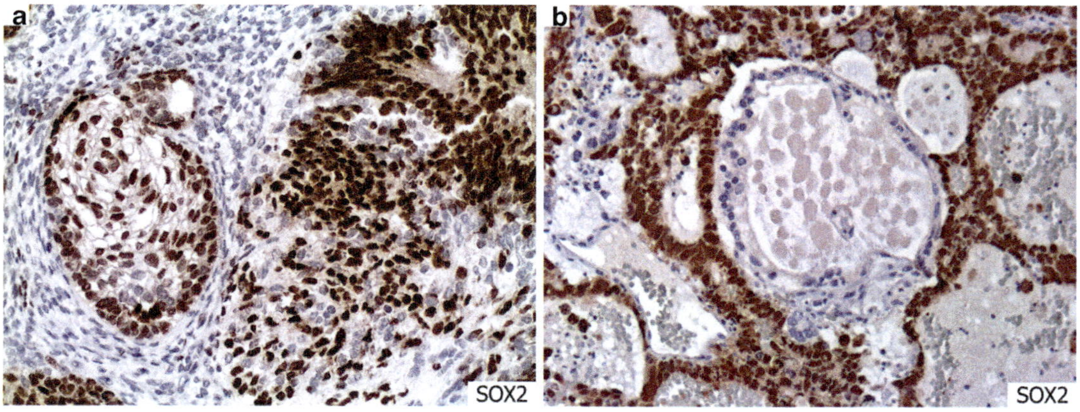

Fig. 4.4 (**a**) Nuclei of neuroectodermal and squamous epithelia of immature ovarian teratomas are stained by SOX2. (**b**) SOX2 expression is positive in embryonal carcinoma, while primitive endodermal elements of yolk sac tumor are negative

4.1.5 Lin28

4.1.5.1 Nomenclature

Protein lin-28 homolog, cell lineage abnormal 28, protein zinc finger CCHC domain-containing protein 1; ZCCHC1

4.1.5.2 Gene Function and Chromosomal Location

Lin28 was first identified in the nematode *Caenorhabditis elegans*, although their homologous genes exist in many species including mammals [96]. It is a monomeric cytoplasmic mRNA-binding protein that enhances the efficiency of protein synthesis. Additionally, Lin28 protein can be exported to the nucleus and nucleolus where it regulates the migration of mRNA to the cytoplasm [97].

Functionally, Lin28 plays an important role during embryonic development, particularly in skeletal muscle tissue, through upregulation of IGF2 mRNA, MYOD1, ARBP/36B4 ribosomal protein, and its own mRNA [98]. By contrast, Lin28 acts as a suppressor for the production of mature microRNA through its specific binding, followed by blocking and subsequent degradation of the let-7 precursor and thus contributing to the vital maintenance of embryonic stem cells [99]. Therefore, Lin28 may also promote protein synthesis through its direct association with the target mRNA, including that encoded by the transcription factor Oct4 [100]. For all these reasons, Lin28 is important for maintaining cell pluripotency during mammalian embryogenesis whereas, by contrast, it shows a very restricted expression in mature tissues. In fact, Lin28 together with OCT4, SOX2, and NANOG is used to reprogram human somatic cells into cells that exhibit the essential characteristics of ESC [101]. An experimental study has shown that Lin28 is essential in the development of mouse primordial germ cells and may be involved in their malignant transformation [102]. Likewise, this protein also plays an important role in the in vitro differentiation of germ cells from ESC and is expressed in the primordial germ cells up to the premeiotic stage [102, 103].

4.1.5.3 Recommended Clones and Practical Considerations

Antibody A rabbit polyclonal antibody has been used in all the studies on gonadal and extragonadal GCT [104–106]. We recommend rabbit clone EP150 that works well with a pH 8 buffer for HIER.

Controls For a positive control, seminoma or GCNIS can be used.

Staining pattern Cytoplasmic with occasional membranous accentuation.

4.1.5.4 Expression

Recent immunohistochemical studies have demonstrated a 100 % positive staining of Lin28 in all precursor lesions (GCNIS and gonadoblastoma), seminoma (Fig. 4.5), EC, and YST, some primary and extragonadal teratomas (especially immature neuroepithelium), and choriocarcinoma (mononucleated trophoblastic cells). Isolated cases of spermatocytic tumor show weak staining [104–106]. In addition to its value in the diagnosis of GCT, Lin28 stains a higher proportion of YST cells as compared with SALL4 [106].

Lin28 is variably expressed in isolated cases of breast, lung, ovary, colorectal, and liver carcinomas, where high expression is an indicator of poor prognosis [104, 106, 107]. The presence of Lin28 in colorectal cancer is significantly associated with lymph node metastasis [108]. In contrast with the expression of OCT4, which is only sporadically present in medulloblastoma and is significantly associated with a poor prognosis, Lin28 protein is more frequently detected but does not influence prognosis [109].

	Positive	Negative
Germ cell tumors	Seminoma Embryonal carcinoma Yolk sac tumors Immature teratoma (neuroepithelium) Choriocarcinoma (cytotrophoblast) Spermatocytic tumor	Teratomas (epithelial elements)
Non-germ cell tumors	Breast, lung, ovary, colorectal, or liver carcinomas	

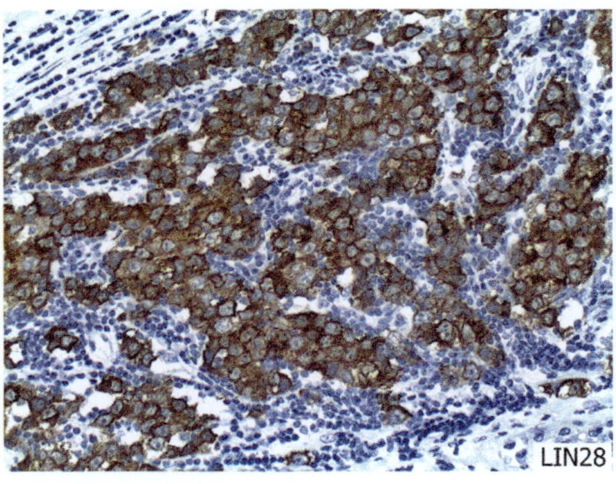

Fig. 4.5 Lin28 presents with a cytoplasmic and membrane staining pattern in a classic seminoma

4.1.6 NANOG

4.1.6.1 Nomenclature

NANOG is a homeobox transcription factor also known as FLJ12581.

4.1.6.2 Gene Function and Chromosomal Location

NANOG is another member of the network of pluripotency-maintaining factors, encoded by the human *NANOG* gene located on the chromosome region 12p13.31, a region frequently evoked in chromosomal studies of GCT. As compared with the murine model, where the *Oct4*, *Nanog*, and *Sox2* genes co-regulate cohorts of downstream genes [110], in humans, they directly repress or activate small sets of downstream transcriptional regulators, controlling a broader range of cellular processes. Due to this uncoupled regulatory loop, these three genes do not work as pan-repressors of differentiation, but each one controls specific cell fates [111]. In human ESC, *NANOG* seems to have a lineage-restricted function to repress the neuroectodermal and neural crest commitment but, together with *OCT4*, it is required for their self-renewal. In a similar way to OCT4, NANOG is expressed by the cells of the inner cell mass of the blastocyst and in the pregastrulation epiblast but is not expressed by unfertilized oocytes, 2–16-cell embryos, or early morulae; it is only restricted to the proximal epiblast, while OCT4 has a diffuse expression [112]. NANOG has also been detected in the nuclei of the murine and human germline stem cells in the embryonic testis but is lost as the gonocytes mature to form spermatogonia in the adult testis [113].

4.1.6.3 Recommended Clones and Practical Considerations

Antibody Since NANOG has a similar, but weaker, expression than the widely used OCT4 and SALL4, few studies have used it as an alternative antibody. AF1997 polyclonal goat unconjugated immunoglobulin and its biotinylated version BAF1197 are the most commonly used in these assays [65, 113].

Controls Similar to SALL4, GCNIS represents an optimal control.

Staining pattern The antibody produces a nuclear staining that is by comparison less defined than SALL4.

4.1.6.4 Expression

In GCT, NANOG parallels OCT4 and stains the neoplastic cells of precursor lesions, seminoma, and EC, while YST, teratoma, and choriocarcinoma are consistently negative [65, 113]. Levels of NANOG mRNA are 2 to 12 times higher in seminoma compared with EC [113]; this is to be expected since seminoma represents a more primitive cell stage than EC.

NANOG is positive in other malignancies, including carcinomas of the prostate, lung adenocarcinomas, gliomas, rectal and gastric carcinomas, and oral squamous or nasopharyngeal carcinomas [28, 114]. An epithelial-mesenchymal transdifferentiation has been described in some of these cases, with the expression of NANOG and other stemness markers at the invasive front of the tumor. In these neoplasms, the expression is correlated with a worse prognosis, distant metastasis, and recurrence after conventional therapies [115].

A comparative study of NANOG expression between pre- and postmenopausal normal ovaries and those harboring ovarian epithelial malignancies demonstrated a high percentage of NANOG positive cells in the lining of ovarian inclusion cysts in patients with carcinomas, while the surface epithelium was consistently negative [116]. NANOG-positive cells were absent in the ovarian mucinous tumors, while in the serous carcinomas, there was a positive population that increased with tumor grade [117]. Nevertheless, other studies have demonstrated the presence of NANOG-positive stem cells on the surface epithelium of the ovary and also in the epithelium of the distal part of the fimbriae, results that may lend some support to recent theories on ovarian serous carcinomas pathogenesis [118].

	Positive	Negative
Germ cell tumors	Seminoma Embryonal carcinoma	Yolk sac tumors Choriocarcinoma Teratomas Spermatocytic tumor
Non-germ cell tumors	Carcinomas of the prostate, lung adenocarcinomas, gliomas, rectal cancer, gastric carcinomas, and oral squamous or nasopharyngeal carcinomas Epithelial-mesenchymal transdifferentiation areas	

4.1.7 UTF-1

4.1.7.1 Nomenclature

UTF-1 from undifferentiated embryonic cell transcription factor 1

UTF-1 is another transcription factor expressed in undifferentiated ESC and downregulated during their differentiation [119].

4.1.7.2 Gene Function and Chromosomal Location

UTF-1 is located on the chromosomal region 10q26.3 [120]. Its expression in ESC is regulated by a dimer composed of *OCT4A* and *SOX2*, and it is specifically attached to the chromatin, being excluded from the nucleoli during all phases of cell progression. Similar to OCT4 and NANOG, the UTF-1 is expressed by cells of the inner cell mass of the blastocyst and in the epiblast but is soon downregulated during development. It is maintained in primordial germ cells and, unlike OCT4 and SOX2 but similar to SALL4, is detected in the spermatogonia of the adult testis [121, 122].

4.1.7.3 Recommended Clones and Practical Considerations

Antibody Most studies have used the mouse monoclonal antibody MAB4337 clone 5G10.2 which was raised against a recombinant GST fusion protein, human UTF1.

Controls GCNIS or seminoma is optimal.

Staining patterns With the clone 5G10.2, nuclear staining was obtained on various platforms. Alkaline pH is recommended for HIER.

4.1.7.4 Expression

The few available publications, mainly focused on GCT and their metastases, characterized UTF-1 as a marker for GCNIS and seminoma with variable staining intensity, while EC constantly shows strong and diffuse staining. A weak staining is detected in the YST epithelial component, while teratomas and choriocarcinomas are negative [122].

In other gonadal specific tumors, weak nuclear staining is also detected in a limited number of Sertoli and Leydig cell tumors. Similarly, carcinomas of the breast, stomach, and kidney and some soft tissue tumors such as alveolar soft part sarcoma, angiosarcoma, or epithelioid Ewing sarcoma may also be positive [122]. Semi-quantitative analysis of UTF-1 staining demonstrates a higher intensity in prostate and endometrial cancer, while lower levels are seen in colon and renal clear cell carcinomas [123].

	Positive	Negative
Germ cell tumors	Seminoma Embryonal carcinoma Yolk sac tumors	Choriocarcinoma Teratomas Spermatocytic tumor
Non-germ cell tumors	Sertoli and Leydig tumors Breast, stomach, and kidney carcinomas Soft tissue tumors like alveolar sarcoma, angiosarcoma, or epithelioid Ewing sarcoma, etc.	

4.1.8 TCL-1

4.1.8.1 Nomenclature

TCL-1 is also known as protein 1A of T-cell leukemia/lymphoma.

4.1.8.2 Gene Function and Chromosomal Location

TCL-1A is part of the TCL1 protein family and is encoded by a gene on chromosome region 14q32.13. This protein is involved in several chromosome translocations and gene inversions that characterize human prolymphocytic T-cell leukemia and some B-cell lymphomas, mainly through an increase in the phosphorylation and activation of the *AKT1*, *AKT2*, and *AKT3* genes resulting in an antiapoptotic response. Also, after the induction of the nuclear translocation of *AKT1*, the TCL1A protein increases proliferation, stabilizes mitochondrial membrane potential, and promotes cell survival [124].

4.1.8.3 Recommended Clones and Practical Considerations

Antibody Polyclonal antibodies and mouse monoclonal clone 27D6/20 have been used in the majority of the studies, both on lymphoid and GCT. In our limited experience, the rabbit clone EP105 was able to reproduce these results in all major types of GCT. An alkaline pH buffer should be used for HIER. Aside from the aforementioned clone EP105, mouse monoclonal antibodies have been developed against the whole TCL-1 human protein or just a fragment.

Controls The strongest nuclear and cytoplasm staining of the T lymphocytes from tonsils with no background on the B lymphocytes should be used as a positive control.

Staining patterns All developed antibodies exhibit nuclear and cytoplasmic staining.

4.1.8.4 Expression

Overexpression of human *TCL1* gene has been implicated in the development of T-cell prolymphocytic leukemia. In the B-cell neoplasms, *TCL1* gene

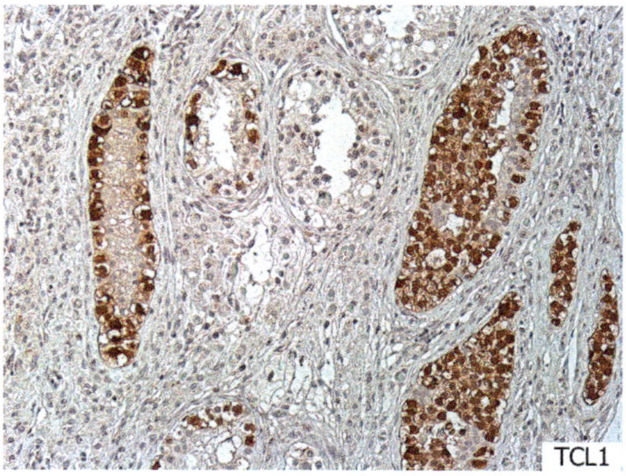

Fig. 4.6 TCL1 nuclear and cytoplasmic staining in germ cell neoplasia in situ and intratubular seminoma

	Positive	Negative
Germ cell tumors	Seminoma Focal in embryonal carcinoma and spermatocytic tumor	Yolk sac tumors Choriocarcinoma Teratomas
Non-germ cell tumors	Most B-cell lymphomas Isolated cases of breast, esophagus, colon, gastric, urothelium, thyroid, biliary duct, and renal carcinomas	Hodgkin/Reed-Sternberg cells, multiple myeloma, marginal cell lymphoma, anaplastic CD30-positive lymphoblastic T-cell lymphoma, peripheral T-cell lymphoma, or mycosis fungoides

acts as an oncogene, and the antibody is expressed in both the cytoplasmic and nuclear compartment of most B-cell lymphomas, including lymphoblastic lymphoma, chronic lymphocytic leukemia, mantle cell lymphoma, follicular lymphoma, diffuse large B-cell lymphoma (60 %), and primary cutaneous B-cell lymphoma (55 %). TCL1 also has a constant and intense staining in Burkitt lymphomas. TCL1 is not expressed by Hodgkin/Reed-Sternberg cells or multiple myeloma, marginal cell lymphoma, anaplastic CD30-positive lymphoblastic T-cell lymphoma, peripheral T-cell lymphoma, or mycosis fungoides. These data indicate that TCL1 is expressed in more differentiated B cells of both neoplastic and reactive conditions [125].

The TCL1 protein is overexpressed in primary seminomas and its precursors (Fig. 4.6), suggesting a role in the pathogenesis of GCT [65, 126, 127]. Indeed, a relation between OCT4 and TCL1 has been demonstrated in mouse ESC as OCT4 binds to Tcl1 promoter and activates its transcription. Furthermore, a role in the regulation of proliferation, but not differentiation, has been demonstrated [128], which is in accordance with the idea of limited pluripotency in seminomas. A decrease of TCL1 expression, both in number of cells and intensity, was observed in metastatic seminomas after radiotherapy, while chemotherapy seems to have little effect on its expression [129]. Focal nuclear staining has been demonstrated in some EC while nuclear and cytoplasmic positivity is observed in isolated cells of spermatocytic tumors [129].

4.1.9 KLF4

4.1.9.1 Nomenclature

KLF4 from the Krüppel-like factor 4 is alternatively known as epithelial/endothelial zinc finger protein EZF or Gut-enriched Krüppel-like factor (GKLF).

4.1.9.2 Gene Function and Chromosomal Location

KLF4 is a member of the KLF family of transcription factors that acts as an upstream regulator of *NANOG* and represents a direct downstream of LIF-Stat3 signaling. Due to LIF signaling, KLF4 levels increase, bind to the NANOG promoter, and induce NANOG overexpression. OCT4 and SOX2 expression is not correlated with NANOG expression [130]. However, it seems that KLF4, together with NANOG, OCT4, and SOX2, represents the main transcription factors involved in the reprogramming of the fibroblast into a pluripotent stem cell stage [131]. It can be both an activator and a repressor, depending on with which genes it interacts, and it is aberrantly expressed in breast and colon cancer [132]. It can bind to the promoter region of its own gene located on chromosomal region 9q31.2 and can activate its own transcription. It plays an important role in terminal differentiation of some epithelial cells of the digestive tract but also lung, genital tract, and vascular endothelium [133].

4.1.9.3 Recommended Clones and Practical Considerations

Antibody There is no standard regarding the clone to be used. Rabbit and mouse polyclonal antibodies are the most frequently employed, some of them requiring a 4 °C overnight incubation and thus have only limited use in routine diagnosis. Product H-180, a rabbit polyclonal raised against amino acids 1–180 of human KLF4, seems to provide a reasonable balance between nuclear staining and unspecific background.

Controls As a marker of undifferentiated GCT, GCNIS or seminoma could both be used as positive controls.

Staining patterns The antibody produces a nuclear staining similar to the other transcription factors.

4.1.9.4 Expression

Compared with NANOG, which highlights the nuclei of the spermatogonia in the human testis, KLF4 is strongly expressed in postmeiotic germ cells such as spermatids, where it plays an important role in spermiogenesis [133]. Limited data on KLF4 expression are available in GCT. Nuclear staining has been reported in GCNIS and seminoma [134].

In somatic tumors, KLF4 was proposed as yet another specific marker for monocytic differentiation in leukemia [135] and as a progression marker in nasopharyngeal [136] and prostate carcinomas [137].

	Positive	Negative
Germ cell tumors	Seminoma	Embryonal carcinoma Yolk sac tumors Choriocarcinoma Teratomas Spermatocytic tumor
Non-germ cell tumors	Monocytic leukemia, nasopharyngeal carcinomas, prostate adenocarcinomas	NA

4.1.10 SOX17

4.1.10.1 Nomenclature

SOX17 gene might be alternatively referred to as *SRY-related HMG-box gene 17*.

4.1.10.2 Gene Function and Chromosomal Location

SOX17 is another member of the family of 20 SOX transcription factors with key roles in embryogenesis, being a Wnt signaling pathway antagonist and a critical specifier in pluripotent germ cell fate, as mentioned in Chap. 3. The protein is coded by the *SOX17* gene located on the 8q11.23 chromosomal region, and its heterozygous mutations were detected in some congenital and sporadic urinary defects, generally grouped under the name of vesicoureteral reflux-3 (VUR-3). These are explained by the expression of SOX17 in the ureteric bud and metanephric mesenchyme of the developing kidney and urinary tract between 11.5–15.5 days after fertilization [138]. It is required also for both primitive mesoderm formation and normal looping of the embryonic heart tube as well as for the normal development of the definitive gut endoderm [139].

4.1.10.3 Recommended Clones and Practical Considerations

Antibody From the limited data available, it seems that a goat polyclonal antibody, raised against the recombinant human SOX17, gives a clean nuclear staining. Other commercially available goat polyclonal antibodies against N-terminus of Sox-17 of mouse origin and a mouse monoclonal raised against *E. coli*-derived recombinant human SOX17 are both prone to nonspecific cytoplasmic staining [91]. For these reasons, a pH 6 buffer is recommended for HIER.

Controls GCNIS and seminoma are used as positive controls.

Staining patterns SOX17 produces a nuclear staining.

4.1.10.4 Expression

SOX17 is expressed in male and female fetal gonocytes, but in contrast with male spermatogenesis, no expression is detected in oogenesis beyond the stage of gonocyte [91]. In spermatogenesis, SOX17 is expressed in the stages of spermatogonium, spermatocyte, and spermatid, showing no staining in the other cells of the seminal epithelium [91].

Limited data are available on the expression of SOX17 in testicular GCT, but in contrast with SOX2, it stains cells of GCNIS and seminoma, while EC is negative [91, 95]. Thus, it is useful in the differentiation of seminoma from EC. Nuclear staining is present in all tumor components of YST, while the expression in teratoma is variable but always occurs in glandular elements, as expected of its role in endodermal development. Choriocarcinomas are negative [95].

	Positive	Negative
Germ cell tumors	Seminoma Yolk sac tumors Teratomas (glands)	Embryonal carcinoma Choriocarcinoma Spermatocytic tumor
Non-germ cell tumors	NA	NA

4.1.11 AP-2γ

4.1.11.1 Nomenclature

AP-2γ, the activating enhancer-binding protein 2 gamma, is also called the transcription factor ERF-1 gene (estrogen receptor factor 1).

4.1.11.2 Gene Function and Chromosomal Location

The protein is encoded by the *TFAP2C* gene located on the 20q13.31 chromosome region. It is part of the AP2 family of transcription factors that interacts with various genes to control the ectoderm development, especially the skin and the neural crest [140]. In murine embryos, it has been shown to interact with transcription factors such as BLIMP1 and PRDM14 for the differentiation of epiblast-like cells into primordial germ cells [141]. It also cooperates with CDX2 in maintenance of the extraembryonic trophoblast. Together, they act in alternative pathways and are correlated with the repression of the pluripotency factor NANOG [142].

4.1.11.3 Recommended Clones and Practical Considerations

Antibody AP-2γ clone 6E4/4 is a mouse monoclonal antibody raised against bacterially produced AP-2 protein and specifically recognizes the C-terminus of AP-2γ. It is the most frequently used clone, and at a pH 6 for HIER, it produces a sharp staining on various platforms.

Controls GCNIS and seminomas can be used as positive control.

Staining patterns Nuclear staining is the accepted pattern for this antibody.

4.1.11.4 Expression

In embryonal tissue, it is expressed by gonocytes, reaching its highest expression at 12 weeks of gestation, followed by a gradual downregulation, which correlates with testicular maturation. It is detected until the 37th week of gestation, being absent in the adult testis [143]. It is overexpressed by GCNIS, gonadoblastoma, and seminoma. Only focal expression was observed in EC, choriocarcinomas, skin, and skin adnexa of teratomas. YST and spermatocytic tumor are negative [47, 143].

AP-2γ also stains the granulosa cell layer of ovarian follicles and is weakly expressed in isolated cells of immature testicular granulosa cell tumors [47]. There is an AP-2γ progressive downregulation and loss of staining in primary melanomas as compared with nevi [144]. It is upregulated in advanced-stage ovarian carcinoma compared with early-stage carcinomas, borderline tumors, and ovarian surface epithelium [145], suggesting its possible relationship with tumor progression.

	Positive	Negative
Germ cell tumors	Seminoma Focal in embryonal carcinoma, choriocarcinoma, and teratomas (epithelial elements)	Embryonal carcinoma Yolk sac tumors Choriocarcinoma Teratomas Spermatocytic tumor

	Positive	Negative
Non-germ cell tumors	Immature testicular granulosa cell tumors Nevi Advanced-stage ovarian carcinomas	Melanomas

4.1.12 IMP3

4.1.12.1 Nomenclature

IMP3 or IGF-II mRNA-binding protein 3 is also known as KH domain containing protein overexpressed in cancer, KOC1, L523S, or IGF2BP3.

4.1.12.2 Gene Function and Chromosomal Location

IMP3 is the third member of IGF-II mRNA-binding protein family, which contains a combination of two RNA recognition motifs and four hnRNP K homology domains. They are coded by the *IGF2BP3* gene located on the chromosomal region 7p15.3. They have a high affinity and multiple attachments to the IGF-II leader 3, an untranslated region of the insulin-like growth factor II, an important growth factor in embryogenesis [146].

In humans, *IGF2* gene expression is controlled by the imprinting control region 1 (ICR1) located in chromosome 11p15.5. DNA methylation defects involving the ICR1 are responsible for two growth disorders: (1) the Beckwith-Wiedemann syndrome (maternal ICR1 hypermethylation) characterized by a disproportionate overgrowth of the fetus, malformations, and a high risk of developing renal tumors, rhabdomyosarcomas, or hepatoblastoma and (2) the Silver-Russell syndrome (paternal ICR1 loss of methylation) which is generally characterized by growth restriction [147].

4.1.12.3 Recommended Clones and Practical Considerations

Antibody Clone 69.1 was raised against amino acids 2–580 of a recombinant protein similar to human IMP3 and has been used in the majority of the studies. Moreover, the IGF2BP3 (IMP3) mRNA is correlated with the status of protein expression by immunohistochemistry in ovarian clear cell carcinomas [148]. In our experience, the rabbit clone EP286 raised against a synthetic peptide corresponding to human IMP3 provides comparable results.

Controls Cytoplasmic and nuclear staining is observed in a large panel of fetal tissues, while ganglionic and granular layers of the cerebellar cortex or germ cells of the adult ovary or testis (see below) can be used as a positive control.

Staining patterns Cytoplasmic and nuclear staining is the accepted pattern.

4.1.12.4 Expression

IMP3 is expressed ubiquitously in fetal tissues, while in the gonads, it has been detected in resting and growing oocytes and granulosa cells of the ovary. In the adult testis, a cytoplasmic and nuclear staining is demonstrated in all spermatogonia, spermatocytes, and spermatozoa [149]. Extensive studies have demonstrated that IMP3 acts as an oncoprotein with no expression in benign tissues, whereas it is highly associated with aggressive and advanced cancer, promoting tumor cell proliferation, invasion, and metastasis [150, 151].

It has been suggested that IMP3 plays a role in GCT development, and its immunohistochemical expression has been demonstrated in the majority of these tumors, including spermatocytic tumor [149]. Its highest expression has been reported in EC, but what makes this antibody remarkable is the difference in its expression in male and female teratomas; it is positive in primary and metastatic testicular teratomas (Fig. 4.7) but negative in ovarian teratomas [151]. These data may further support reflect the malignant nature of testicular teratomas and the possibility of a different histogenesis from ovarian teratomas. This difference in expression might be explained by the parental imprinting of the *Igf-II* gene demonstrated in mice which shows that the difference in growth phenotype depends on the type of gamete contributing to the mutated allele. As the paternal

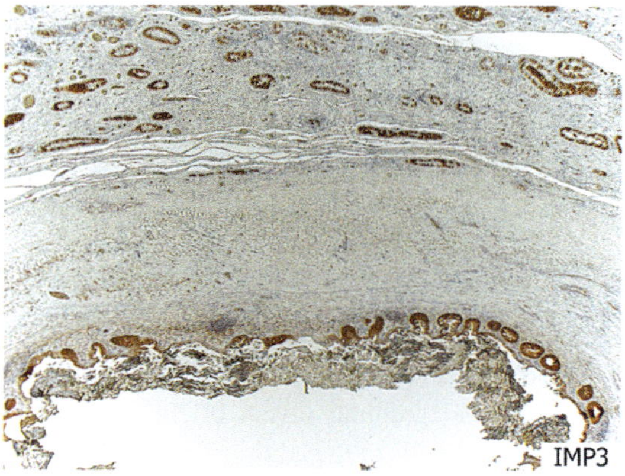

Fig. 4.7 Testicular teratoma and germ cell neoplasia in situ showing a nuclear and cytoplasmic expression for IMP3

allele is expressed in the majority of the tissues [152], the expression in only male teratomas might be simply sex related.

For primary and metastatic GCT diagnosis, the antibody is of little use as it cannot be used either as a marker for GCNIS, due to its expression in spermatogonia, or to differentiate ovarian GCT from somatic carcinomas as it stains mucinous, clear cell, serous, and faintly endometrioid carcinomas of the ovary [148].

	Positive	Negative
Germ cell tumors	Seminoma Embryonal carcinoma Yolk sac tumors Choriocarcinoma Spermatocytic tumor Testicular teratomas (epithelial elements)	Ovarian teratomas (epithelial elements)
Non-germ cell tumors	Mucinous, clear cell, and endometrioid carcinomas of the ovary GISTs, angiomyolipoma, mastocytosis, thymic carcinomas, lung carcinomas, etc.	Solitary fibrous tumor, alveolar soft part sarcoma, desmoplastic small round cell tumor, glomus tumor, leiomyoma, etc.

4.1.13 GATA3

4.1.13.1 Nomenclature

GATA3, also known as trans-acting T-cell-specific transcription factor GATA3 or GATA-binding factor 3, is one of the 6 members of GATA transcription factors with important roles in development. Indeed, except for GATA5, their inactivation is followed by embryonic death [153, 154].

4.1.13.2 Gene Function and Chromosomal Location

GATA3 is an essential enhancer of T-cell receptors, binding to its alpha and delta receptor genes via the 2 GATA-type zinc fingers. The human *GATA3* gene was mapped on the 10p15 chromosomal region, a critical region in the development of the parathyroid glands, inner ear, and kidneys [155]. Genetic mapping and subsequent functional studies indicate that deletion of chromosome 10 (del10p) or *GATA3* gene mutations induce its functional haplo-insufficiency and causes human HDR syndrome, also known as Barakat syndrome, characterized by hypoparathyroidism, sensorineural deafness, and renal disease [156]. In embryonic life, besides its importance in development and differentiation of the luminal breast epithelium, parathyroid gland, adipose tissue, and other non-hematopoietic organs [157], GATA3 is required for the differentiation of CD4-positive T-helper 2 (Th2) cells, a process relevant in immune and inflammatory responses [158].

4.1.13.3 Recommended Clones and Practical Considerations

Antibody With clone HG3-31 of GATA3, a mouse monoclonal antibody raised against the N-terminus of human recombinant GATA3, positivity in GCT has never been reported. However, the introduction of clone L50-823 raised against a peptide between trans-activation and DNA-binding domains of GATA3 has consistently demonstrated trophoblastic and endodermal expression [159]. Due to its reduced sensitivity, clone HG3-31 should not be used in routine diagnosis.

Controls Urothelial and breast normal epithelium and carcinomas can be used as positive controls.

Staining patterns Only nuclear pattern is considered positive.

4.1.13.4 Expression

Nuclear staining was detected in the epidermis, peritoneal mesothelium, and all trophoblastic cells of a 7-week-old embryo with some focal positivity in the caudal mesenchyme and endothelial cells [159]. Normal human secondary yolk sacs do not express GATA3, not even those from as early as the 5th and 6th weeks [160]. Meanwhile, the epidermis, epithelia of oral cavity, olfactory plate, periocular mesenchyme, skull base cartilage, isolated cells in the brain, and nerve glia as well as one-third of the tubular structures and glomerular mesangial cells were positive in a 10-week-old fetus [159].

In adult tissue, GATA3 staining is frequently encountered in many tissues, including the hematopoietic (blood, bone marrow, thymus, B, T, erythroid, and myeloid lineages), blood vessels (endothelial cells), adipocytes, adrenal gland, urothelium, mammary gland, brain, and hair follicles. In the kidney, the collecting ducts, distal tubules, and mesangial cells are positive. Variable staining in seminal vesicle and prostate basal cells is also observed. Terminal ducts of the parotid gland and the thymic cortex are also positive. In normal testis, no staining is observed. Decidual stromal change in the uterus and endometriosis may also be positive [159].

Initial immunohistochemical studies demonstrated a higher sensitivity of GATA3 nuclear staining for tumors of urothelial and breast origin, compared with classical markers [157]. An even lower expression is seen in triple negative, medullary, and metaplastic breast carcinomas [161]. Nevertheless, as more studies are published and clone L50-823 is used, GATA3 is seen to demonstrate an absence of specificity, being positive in many tissues such as parathyroid [162], salivary gland tumors [163]; mucinous and non-mucinous pancreatic adenocarcinomas; squamous cell carcinomas; basal cell carcinomas; chromophobe carcinomas of the kidney; mesotheliomas, poorly differentiated papillary, or mucinous adenocarcinomas of the lung; carcinomas of the stomach, colon, prostate, endometrium, thyroid, serous, and Brenner benign; and borderline ovarian tumors. Most extra-adrenal paragangliomas are positive, while the glandular component of synovial sarcomas may also be stained. Other types of sarcomas may also be stained [159]. It was also described as a sensitive marker of benign and malignant mesonephric lesions in the lower female genital tract [164].

Clone L50-823 also highlights the choriocarcinoma component of mixed GCT (both cyto- and syncytiotrophoblasts) (Fig. 4.8a) and the syncytiotrophoblastic cells of seminoma [159], which supports a GATA3 role in trophoblast differentiation [165]. Besides the staining in tumors with trophoblastic differentiation, endodermal elements are also strongly stained, while only focal elements of EC are identified [159]. This might be explained by cross-reactivity of the antibody with other GATA family members, as GATA-4, GATA-5, and GATA-6 are expressed in the endoderm in an overlapping manner [47, 158].

A study on YST using clone L50-823 demonstrated that GATA3 is only expressed by the primitive areas of YST (reticular-microcystic, endodermal sinus, polyvesicular, polyembryoma) (Fig. 4.8b), while the less frequent somatic patterns (like glandular, hepatoid, and solid) are consistently negative. In routine diagnostic procedures, most YST are identified by the presence of primitive areas which express GATA3; consequently it is a good marker for classic types of YST. However, hepatoid, glandular, and solid YST, which often constitute a diagnostic challenge, do not consistently express GATA3 [160].

	Positive	Negative
Germ cell tumors	Choriocarcinoma Syncytiotrophoblasts of seminomas Yolk sac tumors Squamous differentiation mature teratoma	Seminoma Embryonal carcinoma Teratomas Glandular and hepatoid YST Spermatocytic tumor
Non-germ cell tumors	Urothelial carcinomas, breast carcinomas, parathyroid carcinomas, mucinous and non-mucinous pancreatic adenocarcinomas, squamous and basal cell carcinomas, etc.	

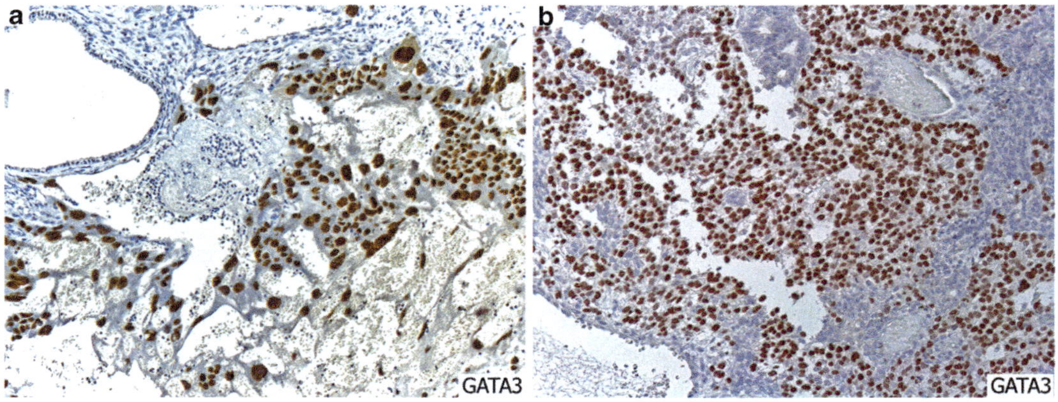

Fig. 4.8 (**a**) Area of trophoblastic differentiation expressing GATA3. (**b**) GATA3 nuclear staining highlights areas of primitive pattern of yolk sac tumor

Table 4.1 summarizes pluripotentiality, transcription factors, and developmental stage-related markers used in immunohistochemical diagnosis of GCT, while Table 4.2 analyzes antibodies published in research papers but less frequently used in diagnostic histopathology. Figure 4.9 depicts neoplastic differentiation pathways and its corresponding pluripotentiality markers in both type II GCT (associated with GCNIS or gonadoblastoma) and parthenogenetically activated Type IV GCT.

4.2 Classic Markers

4.2.1 PLAP

4.2.1.1 Nomenclature

Placental alkaline phosphatase or PLAP; it is also known as Regan isoenzyme.

4.2.1.2 Gene Function and Chromosomal Location

Alkaline phosphatase encompasses a family of four different genes that encodes hydrolase enzymes. The genes that encode the placental, the placental-like, and the intestinal form are located on chromosome 2q37, while the fourth one which encodes the hepatic, osseous, and renal type is located on chromosome 1p36.12 [166–168].

PLAP is normally secreted by the placental syncytiotrophoblasts and physiologically is involved in cellular transport, proliferation, and differentiation, regulation of metabolism, and gene transcription [169]. Even if its serum levels might be separately identified [170], due to its expression in normal conditions and several malignancies [53, 170–174], it is not considered a reliable diagnostic tool, although high levels in GCT are associated with seminomas [175, 176].

4.2.1.3 Recommended Clones and Practical Considerations

Antibody Mouse monoclonal antibodies such as 8A9 and NB-10 are the most frequently used. Clone SP15, a rabbit monoclonal antibody, can also be used with optimal results in the majority of the actual platforms. Alkaline buffer for the HIER is required.

Controls Placenta is the optimal positive control. Normal testicular and ovarian parenchyma should be negative. A cross-reactivity with skeletal and smooth muscle fibers occurs with clone 8A9 and should be accepted, as stated on various vendors' data sheets.

Staining patterns Membranous and cytoplasmic are the expected patterns.

4.2.1.4 Expression

A membranous and rarely cytoplasmic positivity was identified in different normal tissues and various tumors other than germ cells [177, 178].

Table 4.1 Pluripotentiality, developmental stage-related markers in GCT. The first five antibodies are the most useful ones in daily practice and recommended by the International Society of Urologic Pathology [3]

Antibody	EGC	GCNIS-GB	Sem	EC	YST	Choriocarcinoma	IT	MT	ST	Usefulness Availability
SALL4	+	+	+	+	+	±	± (glands and neuroepithelium)	–	±	+++
OCT4	+	+	+	+	–	–	–	–	–	+++
SOX2	–	–	–	+	–	–	± (neuroepithelium)	± (squamous differentiation)	–	+++
CD117	+	+	+	–	–	–	–	–	+	++
SOX17	+	+	+	–	+	–	–	± (glands)	–	+
GATA3		–	–	–	+	+ (syncytiotrophoblast)	–	+ (squamous differentiation)	–	±
Lin28	+	+	+	+	+	+ (cytotrophoblast)	± (epithelial component)	–	–	–
NANOG	+	+	+	+	–	–	–	–	–	–
AP-2γ	+	+	+	±	–	±	–	± (epithelial component)	–	–
IMP3	+	+	+	+	+	+	–	+M /– F	+	–
UTF-1	+	ND	+	+	+	–	–	–	–	–
TCL-1	ND	+	+	±	–	–	–	–	±	–
KLF4	ND	+	+	–	–	–	–	–	–	–

EGC embryonal germ cells, *GCNIS* germ cell neoplasia in situ, *GB* gonadoblastoma, *Sem* seminoma, *EC* embryonal carcinoma, *YST* yolk sac tumor, *IT* immature teratoma, *MT* mature teratoma, *ST* spermatocytic tumor, *GCT* germ cell tumor, *ND* not determined, *M* male, *F* female

Table 4.2 Antibodies published in research papers but less frequently used in diagnostic histopathology

Antibody	Pattern	Gene	Gene location	Expression in normal tissues	GCNIS-GB	Sem	EC	YST	ChoCa	T	ST	Somatic tumors
AuroraB [166–169]	Nuclear	*AURKB*	17p13.1	Spermatogonia ± spermatocytes	+	+	–	–	ND	–	ND	Poor prognostic marker in different carcinomas
Cyclin A2 [170, 171]	Nuclear	*CCNA2*	4q27	Spermatogonia	+	+	+	+	+	+	ND	Poor prognostic marker in stage I non-small cell lung cancer
ERβ [172, 173]	Nuclear	*ESR2*	14q23.2	Spermatogonia, primary and secondary spermatocytes, and round spermatids Sertoli cells	±	–/+	–/+	+	±	+	ND	ND
GATA4 [174–177]	Nuclear	*GATA4*	8p23.1	Granulosa and theca cells in both preantral and antral follicles, Sertoli and Leydig cells through fetal and postnatal development, fetal germ cells and prepubertal spermatogonia	ND	+	ND	+	ND	ND	ND	Sertoli and Leydig cell tumors
GDF3 [178, 179]	Cytoplasm	*GDF3*	12p13.1	Cerebral cortex, hippocampus, cerebellum	+	+	+	+	+	–	ND	ND
GLUT3 [180]	Membrane	*SLC2A3*	12p13.31	Vascular endothelium, spermatozoa	ND	+	+	+	–	+ mature T	–	–
GPR30 [181, 182]	Cytoplasm	*GPER1*	7p22.3	Sertoli cells, spermatogonia, and spermatocytes	+	+	+	–	–	–/+	ND	High-risk endometrial carcinomas with low hormone receptors
HMGA1 [183–185]	Cytoplasm	*HMGA1*	6p21	ND	+	+	+	–	ND	–	ND	Most malignant neoplasias

MAGEA4 [186–189]	Cytoplasm and nuclear	*MAGEA4*	Xq28	Pre-spermatogonia, spermatogonia, and early spermatocytes	±	+	–	–	–	–	+	Melanomas, certain carcinomas and sarcomas, uterine neoplasms, serous ovarian carcinomas
MAGEC2 [190, 191]	Nuclear	*MAGEC2*	Xq27.2	NA	+	+	NA	NA	NA	NA	+	Melanoma, urothelial carcinomas, etc.
NEK2 [192]	Nuclear	*NEK2*	1q32.3	–	+	–	–	–	–	–	ND	ND
NUT (ovarian GCT) [193, 194]	Cytoplasm and nuclear	*NUTM1*	15q13	NA	+	±	±	+ hepatoid and intestinal/glandular differentiation	ND	± immature hepatoid and intestinal/glandular differentiation	ND	Midline carcinomas
NY-ESO-1 [189, 195]	Cytoplasm	*CTAG1B*	Xq28	Gonocytes (from18 months with maximum level at 40 months); spermatogonia and in primary spermatocytes of the adult testes	+	–	–	–	–	–	+	Melanomas, certain carcinomas and sarcomas, uterine neoplasms, serous ovarian carcinomas
PATZ1 [196]	Cytoplasm	*PATZ1*	22q12.2	Sertoli, spermatogonia	+ cyt	+ cyt	+ cyt	+ cyt	+ cyt	+ cyt	NA	NA
RNF4 [197, 198]	Nuclear	*RNF4*	4p16.3	Spermatogonia, spermatocytes, spermatides	–	–	–	–	–	–	–	–
				Oocytes	+	+	+	+	+	+	–	–
SOX9 [199–201]	Nuclear	*SOX9*	17q24.3	NA	+ GCNIS but ± in gonado-blastoma	+	ND	ND	ND	ND	ND	Ductal pancreatic tumors
TSPY1 [202]	Nuclear	*TSPY1*	Yp11.2	NA	+	+		ND	ND	ND	ND	ND

GCNIS germ cell neoplasia in situ, *GB* gonadoblastoma, *Sem* seminoma, *EC* embryonal carcinoma, *YST* yolk sac tumor, *ChoCa* choriocarcinoma, *IT* immature teratoma, *MT* mature teratoma, *SpSem* spermatocytic seminoma

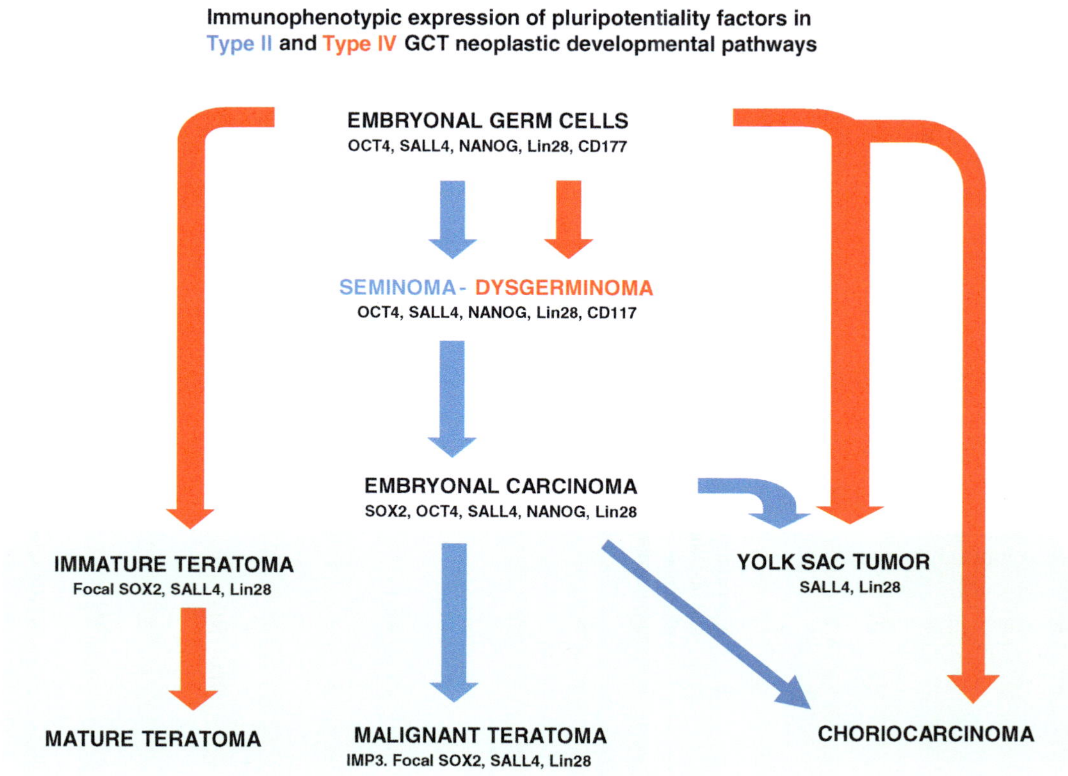

Fig. 4.9 Immunophenotypic expression of pluripotentiality factors in GCT developmental pathways. Type II tumors (postpubertal testicular and extragonadal), originated from germ cell neoplasia in situ (*blue arrows* and text), are different from ovarian parthenogenetically activated type IV (*red arrows* and text). Each tumor type exhibits a characteristic immunophenotype related to its degree of pluripotentiality, ranging from an unrestricted one in seminoma-dysgerminoma and embryonal carcinoma to absent in terminally differentiated tumors like choriocarcinoma and mature teratoma. Tumors of intermediate type of differentiation, such as yolk sac tumors, immature teratomas, and testicular postpubertal teratomas present only a partial, heterogeneous expression

In the fetal testis, distinct positivity is observed in the primitive germ cells [179], but normal adult testicular tissue is negative [180]. In our experience, significant loss of immunoreactivity occurs in poorly fixed or autolyzed material.

PLAP is re-expressed not only in precursor neoplastic germ cell lesions (GCNIS and gonadoblastoma) [38, 180, 181] but also in infiltrative gonadal and extragonadal, primary, or metastatic GCT [182–184]. Seminomas account for the strongest reactivity in up to 100 % of the cases (Fig. 4.10a), followed by a lower intensity of the staining in 97 % of EC (Fig. 4.10b) and 85 % of YST. Staining is focal in cytotrophoblasts of choriocarcinoma [180, 181], and areas of immature elements in teratoma may also be positive [180]176.

	Positive	Negative
Germ cell tumors	Seminoma Embryonal carcinoma Yolk sac tumors Choriocarcinoma Teratomas (epithelial elements)	Spermatocytic tumor
Non-germ cell tumors	Gastrointestinal, gynecologic, lung, breast, and urologic tumors	

4.2.2 CD30

4.2.2.1 Nomenclature

CD30 is recognized under various names such as Ki-1 or TNFRSF8 (tumor necrosis factor recep-

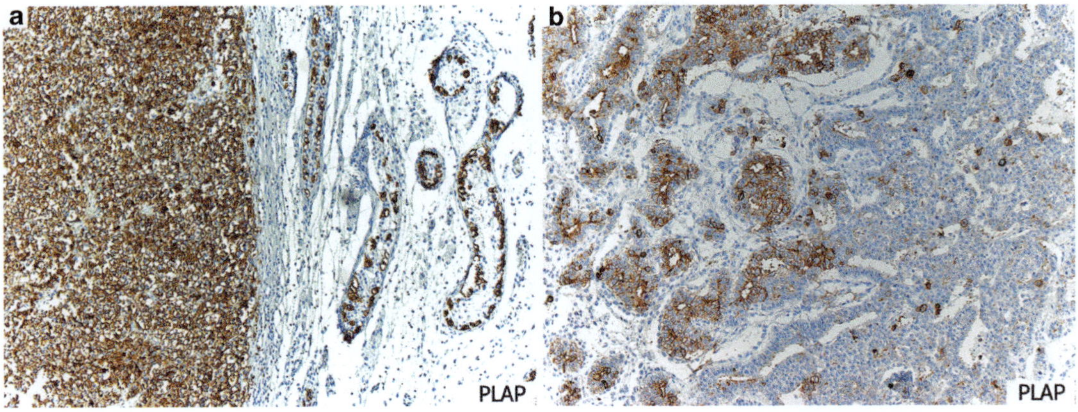

Fig. 4.10 (**a**) PLAP membranous expression in seminoma and germ cell neoplasia in situ while a similar pattern is seen in focal areas of embryonal carcinoma (**b**)

tor superfamily member 8) and is normally expressed by several embryonal tissues and consequently considered important in cell proliferation and development [185].

4.2.2.2 Gene Function and Chromosomal Location

CD30 is a cytokine receptor [186], part of the tumor necrosis factor superfamily with the encoding gene mapping to human chromosome 1p36 [187]. Functionally, it may play a role in the deletion of autoreactive T cells [188] and inhibition of apoptosis [189].

4.2.2.3 Recommended Clones and Practical Considerations

Antibody The monoclonal antibody generally used for its detection is Ber-H2, and it stains the cell membranes in alkaline HIER conditions.

Controls Tonsil-activated inter- and perifollicular B and T cells should be used as low-expressing positive controls; they must show a distinct membranous staining and focally a dot-like reaction in the Golgi area of the cytoplasm. Positive classic Hodgkin's lymphoma or EC could be used as positive controls. Staining should be as strong as possible and should not be accompanied by any unspecific staining or background.

Staining patterns In lymphoid neoplasia, membrane and Golgi area patterns are frequently demonstrable. In EC, only membrane staining is noted. The presence of cytoplasmic staining could be related to tissue fixation and/or autolysis.

4.2.2.4 Expression

It was first described in various lymphoproliferative disorders, particularly expressed on the surface of Reed-Sternberg cells of Hodgkin's lymphoma [190] and the cells of anaplastic large cell lymphoma (Ki-1 lymphoma) [191]. It is also overexpressed by the leukocytes of patients with chronic inflammatory diseases and has been suggested as a therapeutic target in autoimmune diseases [192].

In GCT, CD30 emerged as a highly specific antibody for EC (Fig. 4.11) [18, 43, 186, 193, 194]. Chemotherapy may modulate CD30 expression in EC, as it has been reported negative in a series of treated cases [195]. Additional EC markers may be necessary in these cases to reach a diagnosis [45]. As evidence of its close relationship with EC, minute foci of expression in classical seminoma would suggest early transformation into EC [196]. The accompanying undifferentiated GCNIS is constantly negative [17, 18, 43, 196], and, as expected, intratubular EC should stain positive. Generally, other GCT fail to express CD30 [17, 18, 43, 196, 197] and only isolated YST and mature teratoma may display focal positivity. This makes CD30 one of the

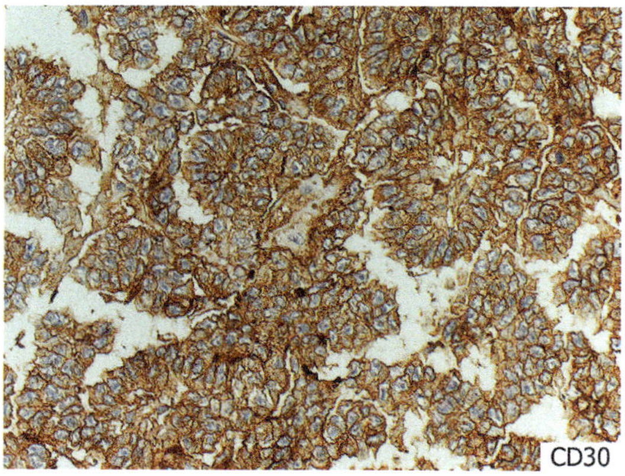

Fig. 4.11 CD30 specifically stains cell membranes of embryonal carcinoma

most specific antibodies in a GCT diagnostic panel.

	Positive	Negative
Germ cell tumors	Embryonal carcinoma	Seminoma Yolk sac tumor Choriocarcinoma Teratomas Spermatocytic tumor
Non-germ cell tumors	Hodgkin's lymphomas, anaplastic large cell lymphoma, anaplastic variant of diffuse large B-cell lymphoma, CD30-positive cutaneous lymphoproliferative disorders	

4.2.3 Alpha Fetoprotein

4.2.3.1 Gene Function and Chromosomal Location

Alpha-fetoprotein (AFP) is a member of the albuminoid gene superfamily, encoded by a gene localized on 4q11-q22 region [198]. This early protein is secreted by both primitive and secondary human yolk sac and is involved in the binding and transport of ligands. Among its many presumed functions, it has been proposed that it plays a role as growth regulator in fine-tuning the architectural interstitial growth patterns in developing organisms [199].

4.2.3.2 Recommended Clones and Practical Considerations

Antibody Polyclonal antibodies have been used to demonstrate the presence of the protein in different tissues. A recently introduced rabbit monoclonal antibody, clone EP209, gives crisp staining on endodermal elements of mixed GCT. Its specificity and sensitivity are still to be determined in ample studies.

Controls Well-fixed and non-autolysed embryonal liver is a reliable positive control.

Staining patterns Both normal and pathologic positive tissues give an often granular cytoplasmic staining that is particularly strong in intracytoplasmic lumina, as it occurs in the human yolk sac [200]. As a protein secreted in the serum, a strong nonspecific background reaction and positivity in necrotic areas and cystic contents is regularly seen.

4.2.3.3 Expression

AFP remains a highly specific immunohistochemical marker for YST, although, in the same way as β-hCG, it is also secreted by many non-GCT not only in the female genital tract [201] but also by tumors of other organs, usually those of endodermal origin with an associated hepatoid component [202].

Diagnostically relevant serum AFP isoforms are produced in non-neoplastic liver disease (L1) and liver cancers (L3), while L2 isoform is characteristic of YST [203]. Interestingly, both L2 and L3 isoforms can be elevated in pediatric YST reflecting both yolk sac and hepatic differentiations [204]. In the secondary human yolk sac, AFP positivity is limited to the endodermal layer where intracellular vesicles and endodermal tubules are highlighted by strong membrane AFP expression [200]. In primitive YST patterns, AFP produces a strong granular cytoplasmic positivity (Fig. 4.12a), especially in intracellular lumina. It may also be present in the hyaline globules, although most of them are not immunoreactive

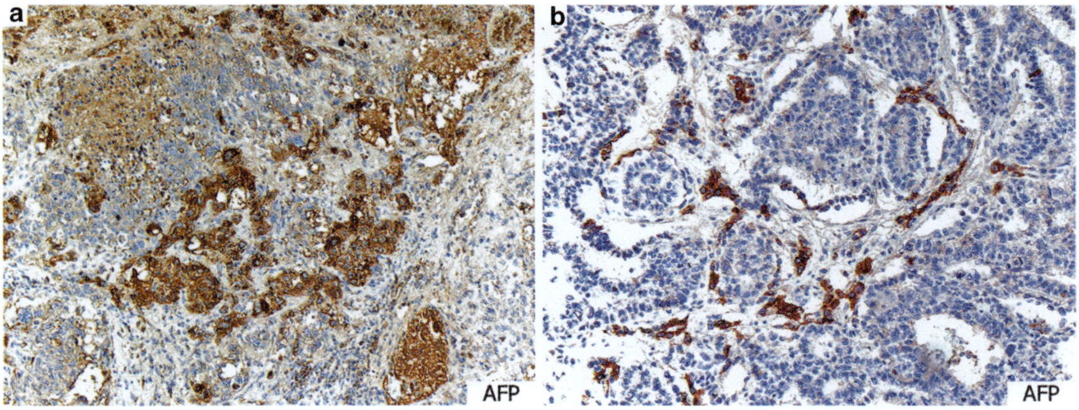

Fig. 4.12 (a) AFP granular cytoplasmic staining is seen in epithelial cells of yolk sac tumor. There is also a background in areas of necrosis and gland secretion. (b) Focal staining is seen in yolk sac tumor while embryonal cell carcinoma areas are negative

for this antigen [205]. Classical patterns have a heterogeneous but consistent AFP expression. However, in somatic glandular patterns, staining is either only focal or absent, often restricted to cytoplasm and the apex of isolated columnar cells, and difficult to identify at low power [206]. AFP-positive areas in embryoids (Fig. 4.12b) and EC reflect differentiation into YST [207]. GCNIS, gonadoblastoma, seminoma, EC, choriocarcinoma, and spermatocytic tumor are negative.

	Positive	Negative
Germ cell tumors	Yolk sac tumors Teratomas (some epithelial elements)	Seminoma Embryonal carcinoma Choriocarcinoma Teratomas Spermatocytic tumor
Non-germ cell tumors	Various AFP-secreting and hepatoid-differentiated tumors	

4.2.4 Podoplanin

4.2.4.1 Nomenclature

Podoplanin identifies a type 1 transmembrane sialoglycoprotein whose mRNA was first described in the murine osteoblastic cell line MC3Y3/E1 [208]. It was later recorded as antigen E.11 in the lymphatic endothelium and subsequently given other names, such as antigen M2A of the fetal gonocytes and Sertoli cell differentiation that is analog to T1α in the alveolar epithelial cells or gp36, a receptor for the influenza V virus or human Aggrus, a platelet aggregation inducing sialoprotein [209].

In 1997, Breitenede and Geleff described it in the podocytes and parietal cells of the Bowman's capsule of a murine nephrosis model [210, 211], and it became known as podoplanin.

4.2.4.2 Gene Function and Chromosomal Location

The protein is encoded by the gene *PDPN* that maps to human chromosome 1p36.21 [212]. Knockdown embryos die soon after birth due to the lack of type I lung alveolar cells [213] with associated generalized lymphedema [214], highlighting its role in the development and maturation of lung and lymphatic endothelium.

4.2.4.3 Recommended Clones and Practical Considerations

Antibody Clone D2-40 gives the highest rates of optimal results on the majority of the actual platforms. Alkaline antigen retrieval is indicated.

Controls Normal tonsil is an accessible and recommended positive control. Dendritic cells of the follicular centers, basal cells of the surface epithelium, and lymphatic endothelium should give intense staining.

Staining patterns Cytoplasmic and membranous staining is accepted in antibody evaluation. In seminoma, the antigen is even maintained in areas of necrosis.

4.2.4.4 Expression

D2-40 is expressed by various normal tissues, and it is mainly used to stain the lymph vessel endothelial cells [210, 215]. For this reason, it is frequently used to confirm the presence of endolymphatic emboli, lymph vessel malformations, or their neoplasms [216–220]. Its polarized positivity was noticed when it was first used to identify osteoblasts in the rat. These studies not only revealed the positivity of osteoblast cell membrane toward the osteoid, but also that type I lung alveolar cells were labeled only at their apical surface, and the epithelial cells of the thymus were stained at the surface facing the thymic parenchyma [210] It is also expressed in the yolk sac mesothelium. Mesotheliomas, especially the epithelioid variant, trichoepitheliomas, follicular dendritic cell tumors, meningiomas, and cartilage-derived tumors, are just some of the numerous neoplasms that express the antibody [209, 221–226].

D2-40 is normally expressed by the fetal gonocytes and the pubertal Sertoli cells. This double positivity is lost later with puberty, and it is re-expressed only by GCNIS [92, 221, 223]. D2-40 has strong membranous, but also cytoplasmic, paranuclear positivity in almost all seminomas and is considered by many as a specific marker for seminoma (Fig. 4.13a) [227–230].

EC is associated with the highest variability rates of staining, from negative to various percentages of positivity. Apical positivity is reported in up to 89 % of EC. In other studies, 11 % of cases present complete cytoplasmic positivity in less than 5 % of the cells [230–231]. In our experience, solid EC is negative, and only papillary and glandular areas show apical positivity in more than 50 % of the cases (Fig. 4.13b). Mesenchymal elements of teratoma may also show focal positivity.

	Positive	Negative
Germ cell tumors	Seminoma Embryonal carcinoma Teratomas (mesenchymal elements)	Yolk sac tumors Choriocarcinoma Teratomas (epithelial elements) Spermatocytic tumor
Non-germ cell tumors	Mesothelioma, lymph vessel tumors, trichoepitheliomas, follicular dendritic cell tumors, etc.	

4.2.5 Glypican-3

4.2.5.1 Nomenclature

Human glypican 3 protein was originally called OCI-5 or MXR-7.

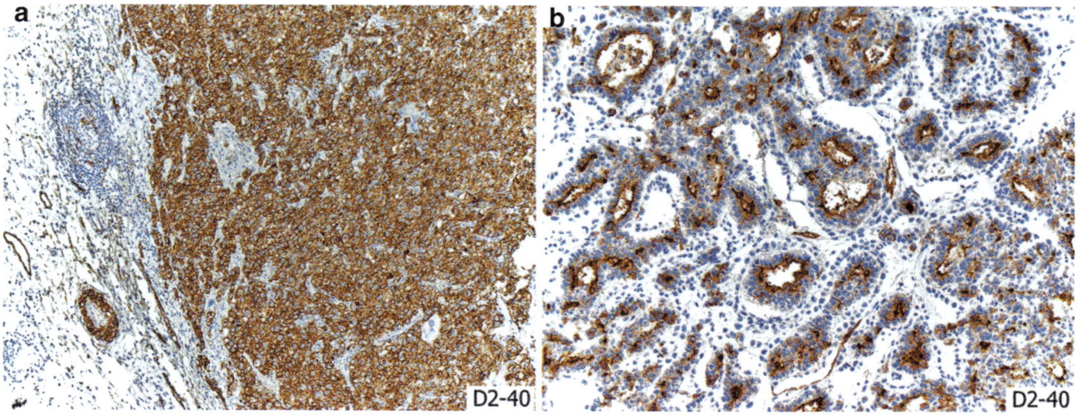

Fig. 4.13 (**a**) Strong and diffuse cytoplasmic D2-40/podoplanin expression in seminoma. Lymph vessel endothelium serves as a positive internal control. (**b**) Only apical D2-40 positivity is seen in areas of embryonal carcinoma

4.2.5.2 Gene Function and Chromosomal Location

Glypican 3 (GPC3) is a cell-surface heparan sulfate proteoglycan, one of the six glypican family members that have been identified in mammals [233]. It is encoded by a gene located on chromosomal region Xq26 and, similarly to AFP, seems to regulate cellular growth and apoptosis [234]. GPC3 is anchored to the cytoplasmic membrane by glycosylphosphatidylinositol and acts as a coreceptor for the growth factors that bind to heparin, such as fibroblast growth factor and insulin-like growth factor; this interaction has a role in cell growth and differentiation [235]. Gene mutations are responsible for the Simpson-Golabi-Behmel syndrome characterized by pre- and postnatal overgrowth [236].

4.2.5.3 Recommended Clones and Practical Considerations

Antibody The most consistent results are obtained with the use of clone 1G12, a mouse monoclonal antibody. Nevertheless, at least in GCT, we obtained similar results with clone SP86, a rabbit monoclonal antibody.

Controls For positive control, second-trimester placenta should be used; the membrane and strong cytoplasmic staining of the cyto- and syncytiotrophoblasts and some cells of the mesenchymal villous cores should also stain [237]. Fetal liver could also be used, but the results may be unreliable, due to frequent autolysis.

Staining patterns Membranous and cytoplasmic.

4.2.5.4 Expression

The antibody is expressed by the endodermal layer of the 8th to 11th week human yolk sac where, similar to AFP, it delineates inter- and intracellular tubules. Other endodermal-derived tissues like immature liver, lung, and pancreatic acinar cells are also positive. Moreover, the renal tubules of mesodermal origin and neuroectodermal elements are also positive [238]. Placental syncytiotrophoblasts are invariably positive, while results are varied in the cytotrophoblast [239–242].

GPC3 is expressed in normal tissues like the ovary, breast, and mesothelium, and its expression may be inactivated through a hypermethylation mechanism in their corresponding neoplasms, such as ovarian and breast carcinomas, cholangiocarcinomas, or mesotheliomas [243–245]. On the contrary, compared with the negativity in their supposed cells of origin, GPC3 is activated and highly expressed by hepatocellular carcinoma, hepatoblastoma, colonic adenocarcinoma, Wilms´ tumor, and other embryonal neoplasms [246–249]. All these data support the notion that in various locations, GPC3 acts like a suppressor or oncofetal protein [250]. One may imagine a similar mechanism for YST tumorigenesis, but this would be true only for the rare placental YST [251, 252], which may originate from the yolk sac itself, in the same way that experimental visceral YST develop from displaced yolk sac tissue after foetectomy [253, 254]. However, this theory is not applicable for the gonadal and midline tumors.

GPC3 is also positive in other various embryonal tumors such as neuroepithelial tubules in IT and medullo- and nephroblastoma [246]. 80 % of placental site trophoblastic tumors and 100 % of the placental site nodules with uterine location are also positive [255]. GPC3 has been described as a useful marker for liver neoplasia [249, 256, 257] where it has been introduced as a serum marker [258] used in conjunction with AFP levels to increase diagnostic sensitivity. It can also be positive in primary melanoma [259], ovarian clear cell carcinoma [260], or squamous cell carcinoma of the lung [261], among others. Furthermore, AFP-producing gastric adenocarcinomas of both clear cell and hepatoid variants also co-expresses GPC3 [262] and should be considered an important differential diagnosis, especially when evaluating lymph nodes, ovarian, or liver metastases.

When compared with AFP in gonadal and extragonadal GCT, GPC3 is consistently reported as a diffuse and more intense marker in almost 100 % of YST (Fig. 4.14a) [63, 241, 242, 263]. In our experience, GPC3 is expressed by the epithelial, reticular, or glandular-alveolar components and is particularly overexpressed in solid and hepatoid areas. The latter variant is rarely identi-

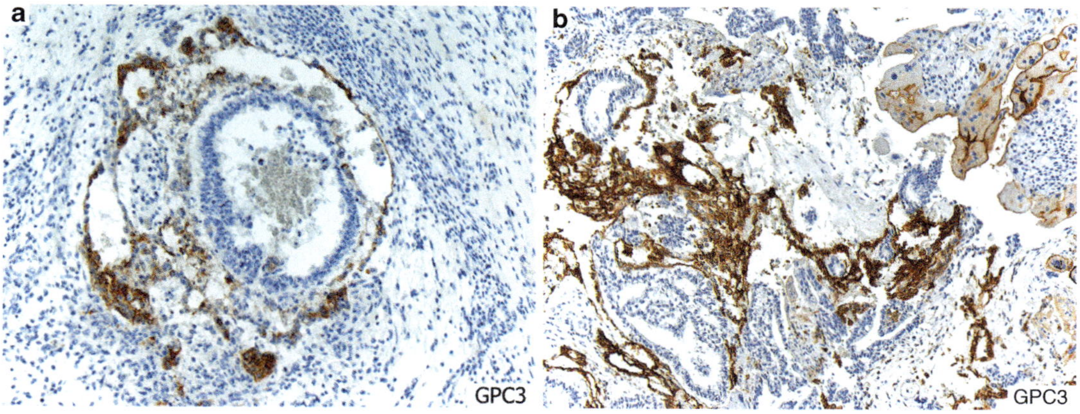

Fig. 4.14 (a) Strong cytoplasmic staining with a clean background characterizes Glypican 3 staining in yolk sac tumors and helps differentiate it from embryonal carcinoma in the majority of the cases. (b) Similar results with only weak to moderate staining in syncytiotrophoblast

fied and is easily misdiagnosed. In comparison with AFP, GPC3 highlights more epithelial areas of YST, presenting a cleaner and stronger staining than AFP. Moreover, rarely is GPC3 not expressed in YST, even highlighting minute epithelial areas missed on H&E examination and AFP immunostain [238].

EC are generally negative [49, 240, 241, 263], although contradictory results have been published of both gonadal and extragonadal GCT [241, 242, 264]. In our experience, almost 20 % of CD30- and OCT4-positive EC have a weak to moderate positive stain. These results might be interpreted as early endodermal differentiation of this stem cell neoplasm [238]. The importance of these results might become evident when evaluating small diagnostic biopsies in which GPC3 should not be automatically interpreted as an expression of only YST areas. In these cases, the concomitant evaluation of morphology and other antibodies is necessary to establish a correct diagnosis.

The positive staining in glands, stromal, and neuroepithelial elements of teratoma, together with the moderate positivity in the syncytiotrophoblasts (Fig. 4.14b), highlights the low specificity of GPC3 [238]. Surprisingly, in one study, one of four spermatocytic tumors showed strong cytoplasmic positivity in cells with large nuclei and a membranous pattern in intermediate ones [238], in contradiction to previous reports [263].

	Positive	Negative
Germ cell tumors	Yolk sac tumors Focal in Embryonal carcinoma Spermatocytic tumor Choriocarcinoma Teratomas	Seminoma Embryonal carcinoma Teratomas
Non-germ cell tumors	Hepatocarcinomas, neuro-, medullo-, and nephroblastoma, rhabdomyosarcoma, etc.	

4.2.6 Human Chorionic Gonadotropin

4.2.6.1 Nomenclature

Human chorionic gonadotropin (hCG), also known as beta polypeptide 3 (CGB3), represents one of the members of human glycoprotein hormone family.

4.2.6.2 Gene Function and Chromosomal Location

Together with hormones such as luteinizing hormone (LH), follicle-stimulating hormone (FSH), and thyroid-stimulating hormone (TSH), hCG is part of a family of glycoproteins that consists of a noncovalent dimer of alpha and beta subunits. The alpha subunit is shared by all the members of the family, while the beta subunit is responsible for the specific endocrine function and the one targeted in immunohistochemical staining [265]. The β-hCG is encoded by the *HCGB* gene which is a complex of seven *CGB* genes and, together with the *LHB* gene, map together as a cluster to the chromosomal region 19q13.33 [266]. The main role of hCG is to maintain the corpus luteum during pregnancy. It has also an important role in cell differentiation and proliferation [267].

4.2.6.3 Recommended Clones and Practical Considerations

Antibody Polyclonal antibodies have proved their sensitivity and specificity in hCG detection, by targeting the beta subunit. As β-hCG is secreted into the serum, a weak unspecific background may frequently accompany the strong staining of syncytiotrophoblasts. A weak diffusion of the staining into the cytoplasm of the cytotrophoblasts is also generally seen. All vendors recommend alkaline pH buffer for the HIER and high dilution when using a concentrated antibody.

Controls First-trimester placenta is an optimal positive control.

Staining patterns The cytoplasm of positive cells is intensely stained.

4.2.6.4 Expression

The GCT that is associated with β-hCG positivity, a marker of syncytiotrophoblast, is choriocarcinoma, often part of a mixed GCT. Isolated syncytiotrophoblasts in other GCT will also stain positively. Precursor GCT lesions, teratoma, YST, and spermatocytic tumor are constantly negative. EC may show weak to moderate staining in isolated cases [268]. Interestingly, some cases of classical seminoma show a diffuse or scattered cell staining pattern when a polyclonal antibody is used, correlating with elevated β-hCG serum levels and mRNA expression [269]. Other than gestational choriocarcinoma, which show similar aspects with the gonadal correspondent, placental site and epithelioid trophoblastic tumor may show β-hCG positivity in isolated, multinucleated cells; other tumors of cytotrophoblast and intermediate trophoblast origin are negative [270]. Complete hydatidiform mole shows strong expression of β-hCG and weak expression of PLAP, whereas weak hCG and strong PLAP expression is found in partial hydatidiform mole [271].

Carcinomas of various primary origins often show a trophoblastic differentiation. The majority are undifferentiated carcinomas of the urinary bladder and endometrium but also breast, pancreas, or stomach [272]. Theoretically, every undifferentiated carcinoma may present areas of trophoblastic differentiation which should be interpreted as an unspecific feature. Giant cell tumor of the bone has been also reported with β-hCG positivity and elevated serum levels [273].

	Positive	Negative
Germ cell tumors	Seminomas with syncytiotrophoblasts Choriocarcinoma	Embryonal carcinoma Yolk sac tumors Teratomas Spermatocytic tumor
Non-germ cell tumors	Trophoblastic differentiation of different somatic tumors, placental site, and epithelioid trophoblastic tumor	

4.3 Tissue Specific Markers Useful in GCT Diagnosis

4.3.1 Cytokeratins

Keratins (also known as cytokeratins) are intermediate filament-forming proteins that constitute the main cytoskeleton of epithelial cells. They are classified based on the Moll catalogue, which groups the basic-to-neutral type II keratins as K1–K8 and the acidic type I keratins as K9–K19 [274]. An updated classification including 24 types of keratins was later proposed in order to enable other mammalian keratins to be added [275]. Nevertheless, advances in sequencing of the human genome have divided keratin-related genes into the following four groups: (1) human epithelial keratins, (2) human hair keratins, (3) nonhuman epithelial/hair keratins, and (4) human keratin pseudogenes. The 54 actual human

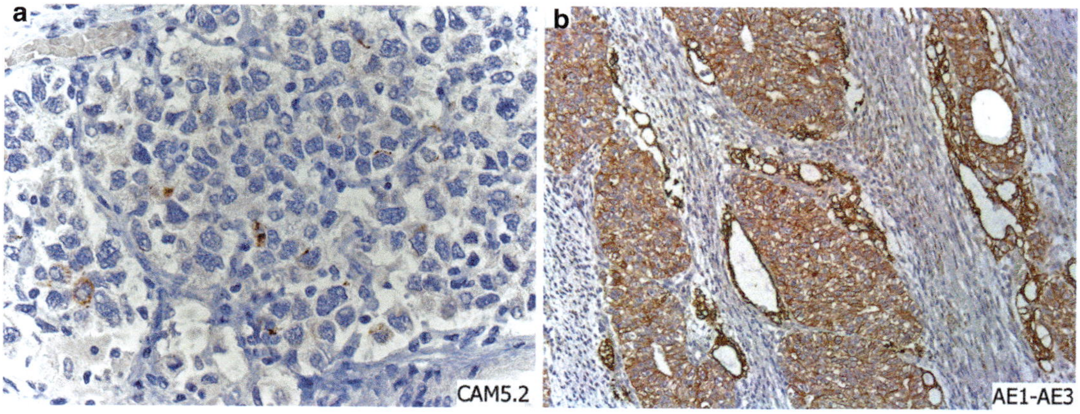

Fig. 4.15 (**a**) Focal, dot-like staining for cytokeratin CAM5.2 is seen in isolated cases of seminoma. (**b**) Cytokeratin AE1-AE1 exhibits a much stronger stain in areas of endodermal (yolk sac) differentiation than in embryonal carcinoma

keratins are grouped as 28 type I and 26 type II keratins and are coded by clusters of genes on 17q21.2 and 12q13.1 chromosomal regions, respectively [276]. Based on their molecular weight, they are also classified into low- and high-molecular-weight keratins. Due to their electric charge, they are generally expressed into heterotypic tetramers of type I and type II proteins, but there is no structural incompatibility between any of the members of the two subfamilies [277]. In this chapter, we will limit our description to the keratins relevant to GCT diagnosis.

Seminomas have cells with a poorly developed cytoskeleton. Consequently, the keratin expression is not always evident (Fig. 4.15a) [278] and is usually expressed in isolated cells together with scarce and loose filaments of vimentin [279]. This "immature," albeit complex, distribution of these intermediate filaments in seminoma, together with a more organized and higher expression especially of the high-molecular-weight keratins in non-seminomatous tumors, has been related with the multipotentiality and undifferentiated character of the initially transformed germ cell(s) [280]. Thus, they were initially used as a convenient diagnostic tool to differentiate seminomas from non-seminomas [278]. However, seminomas commonly express traces of keratin 8 and 18 which highlight isolated positive cells in a similar quantity as in EC. While keratin 19 is almost negative in EC, it is strongly expressed in YST, choriocarcinoma, and teratoma, in direct correlation with the maturity of the tissue [281]. These results agree with our personal observations that wide spectrum cytokeratins such as AE1-AE3 or CAM5.2 highlight the epithelial component of YST with a stronger and better definition compared with EC (Fig. 4.15b). Seminomas are cytokeratin-positive in up to one-third of cases [282]. Syncytiotrophoblastic differentiation of seminoma is also positive for wide-spectrum keratins.

Keratin 7 (clone OV-TL 12/30) and wide-spectrum keratins highlight more than 40 % of seminomas, while keratin 20 (clone Ks20.8) and high-molecular-weight keratins are invariably negative. EC does not stain for keratin 20 but expresses keratin 7 and wide-spectrum keratins [282]. Classical patterns of YST are generally negative for keratin 7, but areas of gut differentiation can be positive. Absence of both Keratin 7 and EMA expression is used in the differential diagnosis between classic forms of YST and clear cell carcinoma. [283].

4.3.2 CDX2

4.3.2.1 Nomenclature

CDX2 from caudal-type homeobox transcription factor 2 or CDX3 as a synonym

4.3.2.2 Gene Function and Chromosomal Location

CDX2 is a nuclear transcription factor, a member of the CDX family of homeobox genes located on chromosome 13q12.2.

It plays a key role in proliferation, differentiation, adhesion, and apoptosis of intestinal cells both in fetal and adult tissues [284]. A tumor suppression function was also described in murine models associating activation of the mTOR pathway and regulation of p21 and COX2 expression [285].

Murine models have demonstrated the important role of Cdx2 transcription factor that, together with TEAD4 and Notch activity, acts on the correct specification of trophectoderm lineage from the morular stage onwards, while both maternal and zygotic pools of *Cdx2* are required for correct preimplantation embryogenesis [286, 287]. In murine models, it has also been demonstrated that *Cdx2* is necessary for vasculogenesis in the yolk sac mesoderm, allantoic growth, chorioallantoic fusion, and the completion of the normal process of gastrulation [288].

CDX2 mRNA is expressed in normal colorectal epithelium and human small intestine. It is also expressed by ampullary epithelium, while weak staining is detected in the pancreatic ductal system and biliary tract. CDX2 is not expressed in normal epithelia of the stomach, esophagus, lung, breast, ovary, or uterus. Nevertheless, areas of intestinal-type metaplasia might be positive [289, 290].

4.3.2.3 Recommended Clones and Practical Considerations

Rabbit monoclonal antibody EPR2764 y and mouse DAK-CDX2 produce adequate results on various platforms. HIER is recommended in an alkaline pH buffer and a tissue array including the colon and pancreas should be used as a positive control for both high- and low-expressing tissues [291].

4.3.2.4 Expression

CDX2 expression has been demonstrated in colorectal adenocarcinomas and other gastrointestinal carcinomas, with minimal or no staining in adenocarcinomas from other organs, especially the breast and lung [292, 293], and consequently it is used as a marker of intestinal differentiation [294] and included in a panel of antibodies for malignancies of unknown origin. However, mucinous ovarian adenocarcinomas show immunoreactivity for CDX2, though heterogeneously and less intensely than in colorectal carcinomas (questioned by some studies) [292]. In endometrial and endocervical adenocarcinomas, expression of CDX2 is rare [290] and has only been related with the presence of morular metaplasia [290, 295]. Similarly, CDX2 is expressed by the morules of the cribriform variant of papillary thyroid carcinoma; gastric, duodenal, biliary, and colonic adenomas; polyps and endometrial hyperplasia; endometrial intestinal metaplasia; and endometrioid adenocarcinoma of the ovary [290]. Only 6 % of prostatic adenocarcinomas show focal expression, whereas staining is present in a higher percentage of cases with mucinous and signet ring cell differentiation [296]. 85 % of urachal adenocarcinomas present diffuse positivity, while similar results have been reported in cases of intestinal-type sinonasal adenocarcinoma [297, 298].

Its usefulness in the diagnosis of metastatic neuroendocrine tumors has been demonstrated by positivity in 47 % of cases, most of them of ileal, appendiceal, or colonic origin [299]. Together with TTF1, it helps differentiate them from their pulmonary equivalents. Similar results have been obtained in skin, ovarian, and thymic tumors. CDX2 is expressed more often in metastasis from small intestinal and appendiceal neuroendocrine tumors than in primary ovarian carcinoids [300].

Interestingly, in contrast with the lack of staining of stem cells and hematopoietic progenitors, 90 % of cases of acute myeloid leukemia show positivity for CDX-2 [301], while its presence in pediatric acute lymphoblastic leukemia correlates with persistence of residual disease [302].

We have recently demonstrated strong and diffuse nuclear expression of CDX2 in the cells of the endodermal layer of secondary human yolk sacs from the 7th to 8th week, becoming weaker and more focal in distribution during the involution period. In embryos showing gut structures, lining cells have a strong nuclear positivity, which is, however, absent in liver and other tissues [200].

In concordance with the consistent CDX2 expression in human yolk sac, indicative of its intestinal role, classical patterns of YST demonstrate focal positivity, while somatic glandular patterns show a stronger expression with a heterogeneous or diffuse distribution (Fig. 4.16). Therefore it would seem that CDX2 positivity highlights both areas of yolk sac and intestinal differentiation, the latter being more evident in somatic glandular patterns, especially those with vacuolated epithelia resembling the embryonal gut. Areas of endodermal-type differentiation of endometrioid adenocarcinomas and areas of morular metaplasia also show diffuse nuclear staining [206].

4.3.3 Villin

4.3.3.1 Nomenclature, Gene Function, and Chromosomal Location

Villin, also called villin-1 to differentiate it from its cytoplasmic correspondent villin-2 or ezrin, is a phospholipid-binding protein, part of a large family of actin-binding proteins that regulate actin dynamics, cell morphology, epithelial-to-mesenchymal transition, and cell migration [303]. It also protects cells from apoptosis by maintaining mitochondrial integrity, thus inhibiting the activation of caspase-9 and caspase-3 [304].

Villin is associated with the microvillar actin filaments and with the terminal web of epithelial cells and is expressed in significant amounts in renal, urogenital, and gastrointestinal epithelial cells [304]. In humans, it is coded by the *VIL1* gene which maps to the 2q35 chromosomal region [305]. Its genetic alterations are responsible for microvilli motility dysfunction which manifests as biliary atresia and progressive cholestasis and alteration of renal function [306].

4.3.3.2 Recommended Clones and Practical Considerations

The most frequently used antibody to detect villin in human tissues is the mouse monoclonal 1D2C3. As villin is regarded as an early marker of committed intestinal absorptive cells [303], its apical expression is seen in normal small intestine and proximal tubules of the nephron.

4.3.3.3 Expression

The apical expression, with or without cytoplasm staining [307], is normally reproduced in colonic, gastric, or pancreatic carcinomas. Only isolated cases of bronchioloalveolar, endometrial, ovarian, and kidney carcinomas have been reported [308]. As an early marker of endodermal cell lineage, villin is consistently expressed during early embryogenesis in both human yolk sac and early endoderm [200]. It is a highly sensitive endodermal epithelial marker in all cases of YST, both in classical and somatic glandular patterns (Fig. 4.17), being absent in pure EC and seminoma [206]. Its diffuse cytoplasmic expression parallels that of the human yolk sac and intestinal adenocarcinomas, where both diffuse cytoplasm and apical staining patterns occur [307, 309]. In our experience, villin is negative in ovarian clear cell carcinomas.

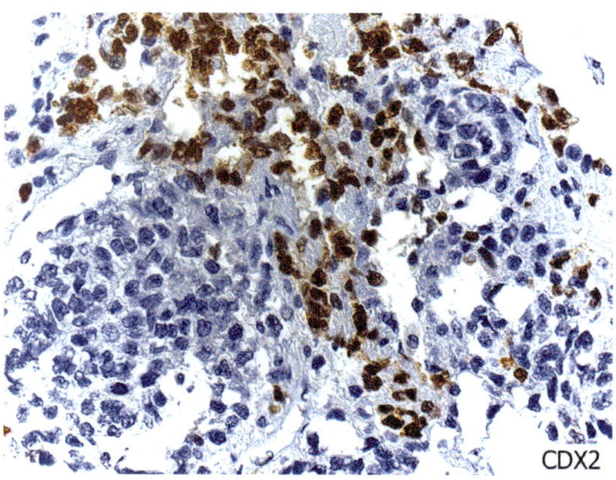

Fig. 4.16 CDX2 nuclear staining highlights minute foci of intestinal or yolk sac differentiation

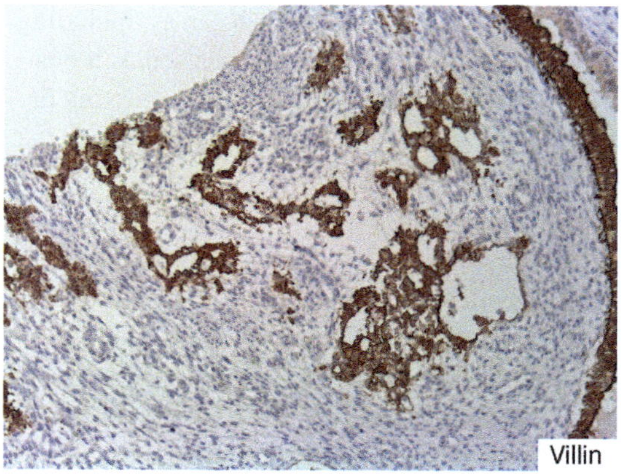

Fig. 4.17 Villin staining in glandular endoderm of yolk sac tumor coexists with similar staining in teratomatous glands of a mixed GCT

Villin is also considered as a marker of cells that arise from mesenchymal/epithelial conversion in the developing kidney [303], which would concur with its expression in some kidney carcinomas, especially of clear cell type [310].

4.3.4 HepPar-1

4.3.4.1 Nomenclature, Gene Function, and Chromosomal Location

Hepatocyte paraffin 1 (HepPar-1) is a murine monoclonal antibody (clone OCH1E5) developed in 1993 using a failed liver allograft as a source of the immunogen [311]; its antigen was not identified until recently. Using immunoprecipitation and mass spectrometry, it was identified as carbamoyl-phosphate synthetase 1 (CPS1), a rate-limiting enzyme in the urea cycle located in mitochondria [312]. Its granular pattern is due to mitochondrial antigen location as was demonstrated with immunoelectron microscopy using protein A–gold (pAg) for postembedding localization of antigens [313]. CPS1 catalyzes the first committed step of the hepatic urea cycle by synthesizing carbamoyl-phosphate from ammonia, bicarbonate, and two molecules of ATP. *CPS1* gene is located on the 2q34 chromosomal region, and its deficiency causes hyperammonemia ranging from lethal neonatal variants to environmentally induced adult-onset disease [314].

4.3.4.2 Recommended Clones and Practical Considerations

Clone OCH1E5 is the most frequently used. An alkaline HIER is always applied. The liver is the positive control of choice, always in conjunction with well-known negative controls such as lymphoid tissue.

4.3.4.3 Expression

Its specificity and sensitivity exceed 80 % for the detection of hepatic differentiated cells. It is still not totally specific as strong cytoplasmic staining has been demonstrated in cases of lung, gastric, and esophageal adenocarcinomas. Absorptive cells of the small intestine and intestinal metaplasia are also positive, while focal positivity has been reported in cholangio, urothelial, and pancreatic adenocarcinomas, among others [315–317]. Its sensitivity for hepatic tissue tumors is not complete, as high-grade hepatocarcinomas and their fibrolamellar variant tend to be negative [316]. The endodermal layer of the secondary human yolk sac is positive throughout its evolution, as would be expected with a hepatic and intestinal differentiation marker; the secondary human yolk sac appears to act as a temporary liver and intestine due to its absorptive and protein synthesis functions [200].

In YST, the antibody detects isolated epithelial cells of most classical and somatic glandular patterns (Fig. 4.18), as well as hepatoid histology [206, 317]. In our experience, it is negative in other GCT.

4.3.5 TTF1

4.3.5.1 Nomenclature, Gene Function, and Chromosomal Location

Thyroid transcription factor 1 (TTF1), also known as thyroid-specific enhancer-binding protein (TEBP), TITF1 or NKX2-1, is a homeodomain containing DNA-binding protein encoded by the *NKX2-1* gene located on 14q13.3 chromosomal region.

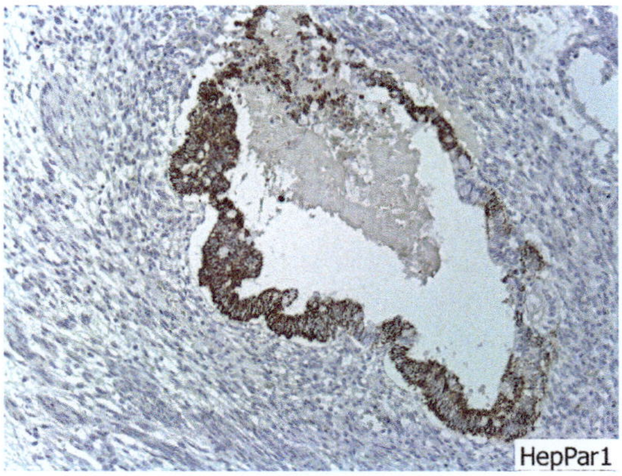

Fig. 4.18 HepPar-1 shows granular cytoplasmic staining in endodermal glandular areas in yolk sac tumor

Due to its major role in the development of the brain (forebrain regions, particularly the basal ganglia hypothalamus and pituitary gland), lung, and thyroid, germline mutations of the *NKX2-1* gene are associated with dysfunction and malformations of these organs [318]. The TTF1 protein binds and activates the promoter of thyroid-specific genes such as thyroglobulin, thyroperoxidase, and thyrotropin receptor, while gene mutations are associated with thyroid dysgenesis, generally characterized by congenital goiter with thyroglobulin synthesis defect [319]. TTF1 plays an important role in lung development, surfactant homeostasis, lung epithelial cell morphogenesis, and differentiation. The chromosomal region 14q13.3 is amplified in approximately 12 % of lung adenocarcinomas [320]. Neurological disorders are frequently characterized by benign chorea or are part of the brain-lung-thyroid syndrome [321, 318]. Ependymomas of the third ventricle seem to be the only brain tumor that expresses TTF1 [322].

4.3.5.2 Recommended Clones and Practical Considerations

Different published results are likely due to variations in the clone used for the protein detection. While clone 8G7G3/1 is the most frequently used, it has been demonstrated to be less sensitive than SPT24 in detecting lung neoplasms, especially carcinoids and squamous cell carcinomas. Nevertheless, SPT24 can be positive in carcinomas of primary origins other than the lung or thyroid, particularly colorectal and bladder urothelial carcinomas [323, 324]. Due to cross-reactivity with a liver protein, only clone 8G7G3/1 gives a cytoplasmic and specific staining in hepatocarcinomas [325].

4.3.5.3 Expression

In normal tissues, the expression of TTF1 is restricted to follicular epithelial cells and the C cells of the thyroid [326] and to type II pneumocytes, ciliated and nonciliated (Clara cells) cells in terminal and respiratory bronchioles, and basal cells of the distal bronchioles [327]. The tumors originated in these cells express the antibody in different percentages, generally related to the grade of differentiation [326, 327]. The staining tends to be stronger in thyroid than in lung parenchyma, and for this reason, the latter one is recommended for the fine-tuning of the antibody staining. In a recent study, clone 8G7G3/1 was negative in all but one case of classical YST patterns (polyvesicular) and had focal expression in three cases of somatic glandular pattern [206]. Similarly to HepPar-1, teratomatous elements reproducing normal positive cells are expected to express TTF1. This is particularly true for cases of struma ovarii.

4.4 Spermatocytic Tumor: A Special Germ Cell Tumor with a Particular Immunophenotype

As a special type of GCT, with a particular morphology, distinctive pathogenesis and exclusive to testicular parenchyma, spermatocytic tumor also presents a distinct immunoprofile. Spermatocytic tumor is extensively reviewed in Chapters 3 and 7. While the majority of the previously presented markers are generally not demonstrable or only stain-isolated cases of spermatocytic tumor, polyclonal CD117 antibodies stain the cell membrane of tumor cells in up to 40 % of cases [21], although results vary (Fig. 4.19a).

Due to their relative rarity and specific morphology, immunohistochemical studies on spermatocytic tumors often include isolated case reports that demonstrate expression of some markers for this entity. Indeed, focal cytoplasmic staining for low-molecular-weight cytokeratin (CAM 5.2) has been reported, and GPC3 positivity has been shown to have intense granular staining in its large-cell component, as well as membranous positivity in the intermediate cells [238]. Variable and contradictory results are also obtained with neuron-specific enolase (NSE) which normally stains normal spermatogonia [328, 329].

VASA or DEAD/H BOX 4 (DDX4), a specific germline lineage marker, is expressed in migratory primordial germ cells in the region of the gonadal ridge, normal spermatogonia, spermatocytes, and spermatids and is negative in Leydig cells, Sertoli cells, stromal cells, and spermatozoa [330, 331]. The encoding gene *DDX4* is located on the chro-

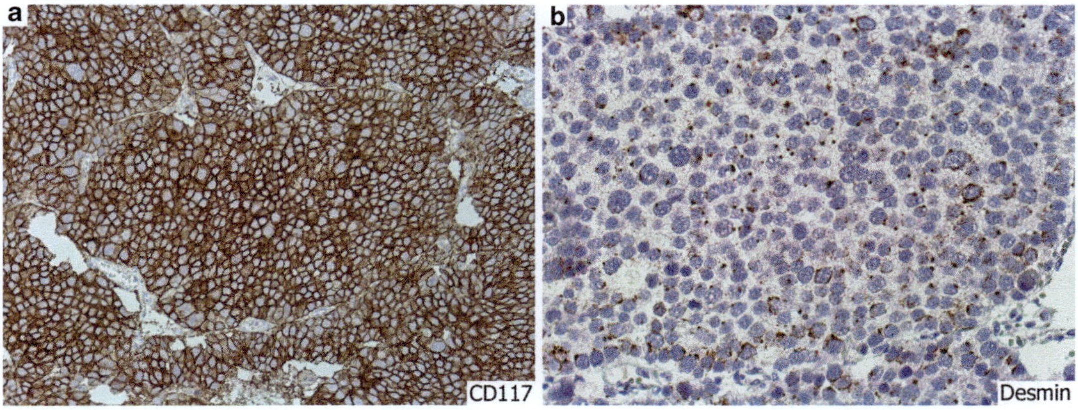

Fig. 4.19 Spermatocytic tumor. (**a**) High membrane expression of CD117 in all cell types. (**b**) Perinuclear, dot-like desmin stain

mosomal region 5q11.2 and encodes a member of the DEAD box family of ATP-dependent RNA helicases which play a major role in germ cell development [332]. In GCT, it is expressed with a higher intensity in spermatocytic tumor than in classical seminomas despite similar mRNA levels [330, 331], being negative in non-seminomas.

OCT2, SSX2–4, and SAGE1 have been studied in 36 spermatocytic tumors, some highlighting intratubular tumor. OCT2 (octamer-binding protein 2 coded by the gene *POU2F2* on chromosomal region 19q13.2) [333] is a transcription factor known as a pan B marker. *SSX 2–4* are part of the SSX gene family (synovial sarcoma X-chromosome breakpoint) which are all associated with the pathogenesis of synovial sarcoma and are coded by their correspondent *SSX* genes located on consecutive regions of the short arm of the X-chromosome [334]. SAGE1 is also known as sarcoma antigen 1 and is part of a class of tumor antigens recognized by cytolytic T lymphocytes expressed in tumors of different histologic types but not in normal tissues, except for spermatogenic cells and placenta [335]. It is encoded by the gene with the same name also located on the X-chromosome region q26.3 [335]. In spermatocytic tumor, the three markers failed to demonstrate an acceptable diagnostic sensitivity [336, 337].

Chk2 (a checkpoint kinase protein that is activated in response to DNA damage), involved in cell cycle arrest and encoded by a gene on the 22q12.1 chromosomal region, is recognized as a multiorgan cancer susceptibility gene [338]. MAGE-A4 is a member of the same family of VASA and SAGE1 proteins and is encoded by a gene on Xq28 chromosomal region [339]. These two markers are expressed by the gonocytes and spermatogonia and consistently by spermatocytic tumor [328].

p53 was expressed by 80 % of the cells in one study [328]; however, others have detected the protein only in a sarcomatous component, a phenomenon associated with less than 1 % of spermatocytic tumors, and correlated with tumor elevated apoptotic activity [340].

NY-ESO-1 (New York esophageal squamous cell carcinoma 1 or cancer/testis antigen 1 (CTAG1)) is encoded by the gene *CTAG1B* on Xq28 chromosomal region [341]. NY-ESO-1 is positive in 50 % of spermatocytic tumors, 13 % of classical seminomas, and 5 % of solid YST. Its sensitivity and specificity for the spermatocytic tumor diagnosis is about 82 % and 94 %, respectively. Neither the sarcomatous component of spermatocytic tumors nor any EC show reactivity with this antibody [342]. Similar results have been obtained with NUT (nuclear protein in testis) and GAGE7 (another cancer/testis related antigen) [342].

Desmin, other than its staining in any correspondent sarcomatous elements, produces a dot-like staining in some cases of spermatocytic seminoma [20]. (Fig. 4.19b). DMRT1 has been shown to be a immunohistochemically detectable, useful marker for diagnosing spermatocytic tumor (for a complete description of markers of spermatocytic tumor see Chap. 3).

References

1. Pierce Jr GB, Midgley Jr AR, Ram JS, Feldman JD. Pariental yolk sac carcinoma: clue to the histogenesis of Riechert's membrane of the mouse embryo. Am J Pathol. 1962;41:549–66.
2. Nakane PK, Pierce Jr GB. Enzyme-labeled antibodies for the light and electron microscopic localization of tissue antigens. J Cell Biol. 1967;33(2):307–18.
3. Ulbright TM, Tickoo SK, Berney DM, Srigley JR. Members of the IIiDUPG. Best practices recommendations in the application of immunohistochemistry in testicular tumors: report from the International Society of Urological Pathology consensus conference. Am J Surg Pathol. 2014;38(8):e50–9. doi:10.1097/PAS.0000000000000233.
4. Rothschild G, Sottas CM, Kissel H, Agosti V, Manova K, Hardy MP, et al. A role for kit receptor signaling in Leydig cell steroidogenesis. Biol Reprod. 2003;69(3):925–32. doi:10.1095/biolreprod.102.014548.
5. Vandenbark GR, deCastro CM, Taylor H, Dew-Knight S, Kaufman RE. Cloning and structural analysis of the human c-kit gene. Oncogene. 1992;7(7):1259–66.
6. Miettinen M, Lasota J. KIT (CD117): a review on expression in normal and neoplastic tissues, and mutations and their clinicopathologic correlation. Appl Immunohistochem Mol Morphol. 2005;13(3):205–20.
7. Ronnstrand L. Signal transduction via the stem cell factor receptor/c-Kit. Cell Mol Life Sci. 2004;61(19–20):2535–48. doi:10.1007/s00018-004-4189-6.
8. Syrris P, Malik NM, Murday VA, Patton MA, Carter ND, Hughes HE, et al. Three novel mutations of the proto-oncogene KIT cause human piebaldism. Am J Med Genet. 2000;95(1):79–81.
9. Søren Nielsen MV. NordiQC CD117 assessment run 26 2009. 2009. http://www.nordiqc.org/Run-26-B7/Assessment/assessment-26-CD117.htm.
10. Novelli M, Rossi S, Rodriguez-Justo M, Taniere P, Seddon B, Toffolatti L, et al. DOG1 and CD117 are the antibodies of choice in the diagnosis of gastrointestinal stromal tumours. Histopathology. 2010;57(2):259–70. doi:10.1111/j.1365-2559.2010.03624.x.
11. Demetri GD, von Mehren M, Blanke CD, Van den Abbeele AD, Eisenberg B, Roberts PJ, et al. Efficacy and safety of imatinib mesylate in advanced gastrointestinal stromal tumors. N Engl J Med. 2002;347(7):472–80. doi:10.1056/NEJMoa020461.
12. Einhorn LH, Brames MJ, Heinrich MC, Corless CL, Madani A. Phase II study of imatinib mesylate in chemotherapy refractory germ cell tumors expressing KIT. Am J Clin Oncol. 2006;29(1):12–3. doi:10.1097/01.coc.0000195086.47548.ef.
13. Pectasides D, Nikolaou M, Pectasides E, Koumarianou A, Valavanis C, Economopoulos T. Complete response after imatinib mesylate administration in a patient with chemoresistant stage IV seminoma. Anticancer Res. 2008;28(4C):2317–20.
14. Mauduit C, Hamamah S, Benahmed M. Stem cell factor/c-kit system in spermatogenesis. Hum Reprod Update. 1999;5(5):535–45.
15. Schrans-Stassen BH, van de Kant HJ, de Rooij DG, van Pelt AM. Differential expression of c-kit in mouse undifferentiated and differentiating type A spermatogonia. Endocrinology. 1999;140(12):5894–900. doi:10.1210/endo.140.12.7172.
16. Tuck AR, Robker RL, Norman RJ, Tilley WD, Hickey TE. Expression and localisation of c-kit and KITL in the adult human ovary. J Ovarian Res. 2015;8(1):31. doi:10.1186/s13048-015-0159-x.
17. Lau SK, Weiss LM, Chu PG. D2-40 immunohistochemistry in the differential diagnosis of seminoma and embryonal carcinoma: a comparative immunohistochemical study with KIT (CD117) and CD30. Mod Pathol. 2007;20(3):320–5. doi:10.1038/modpathol.3800749.
18. Leroy X, Augusto D, Leteurtre E, Gosselin B. CD30 and CD117 (c-kit) used in combination are useful for distinguishing embryonal carcinoma from seminoma. J Histochem Cytochem: Off J Histochem Soc. 2002;50(2):283–5.
19. Sever M, Jones TD, Roth LM, Karim FW, Zheng W, Michael H, et al. Expression of CD117 (c-kit) receptor in dysgerminoma of the ovary: diagnostic and therapeutic implications. Mod Pathol. 2005;18(11):1411–6. doi:10.1038/modpathol.3800463.
20. Lombardi M, Valli M, Brisigotti M, Rosai J. Spermatocytic seminoma: review of the literature and description of a new case of the anaplastic variant. Int J Surg Pathol. 2011;19(1):5–10. doi:10.1177/1066896910388645.
21. Kraggerud SM, Berner A, Bryne M, Pettersen EO, Fossa SD. Spermatocytic seminoma as compared to classical seminoma: an immunohistochemical and DNA flow cytometric study. APMIS : Acta Pathologica Microbiologica et Immunologica Scandinavica. 1999;107(3):297–302.
22. Guillaudeux T, Mattei MG, Depetris D. xx. In situ hybridization localizes the human OTF3 to chromosome 6p21.3→p22 and OTF3L to 12p13. Cytogenet Cell Genet. 1993;63:212–4.
23. Rijlaarsdam MA, van Herk HA, Gillis AJ, Stoop H, Jenster G, Martens J, et al. Specific detection of OCT3/4 isoform A/B/B1 expression in solid (germ cell) tumours and cell lines: confirmation of OCT3/4 specificity for germ cell tumours. Br J Cancer. 2011;105(6):854–63. doi:10.1038/bjc.2011.270.
24. Ono M, Kajitani T, Uchida H, Arase T, Oda H, Nishikawa-Uchida S et al. OCT4 expression in human uterine myometrial stem/progenitor cells. Hum Reprod. 2010;25(8):2059–2067. doi:deq163 [pii].
25. Cantz T, Key G, Bleidissel M, Gentile L, Han DW, Brenne A et al. Absence of OCT4 expression in somatic tumor cell lines. Stem Cells. 2008;26(3):692–697. doi:2007-0657 [pii] 1634/stemcells.2007-0657.
26. Mueller T, Luetzkendorf J, Nerger K, Schmoll HJ, Mueller LP. Analysis of OCT4 expression in an extended panel of human tumor cell lines from mul-

tiple entities and in human mesenchymal stem cells. Cell Mol Life Sci. 2009;66(3):495–503. doi:10.1007/s00018-008-8623-z.
27. Ruangpratheep C, Lohachittranond C, Poonpracha T, Punyarit P. OCT4 expression on a case of poorly differentiated (insular) carcinoma of the thyroid gland and minireview. J Med Assoc Thail. 2005; 88(Suppl 3):S281–9.
28. Chiou SH, Wang ML, Chou YT, Chen CJ, Hong CF, Hsieh WJ et al. Coexpression of Oct4 and Nanog enhances malignancy in lung adenocarcinoma by inducing cancer stem cell-like properties and epithelial-mesenchymal transdifferentiation. Cancer Res. 2010;70(24):10433–10444. doi: 10.1158/0008-5472.CAN-10-2638. 70/24/10433 [pii].
29. Chen Z, Xu WR, Qian H, Zhu W, Bu XF, Wang S, et al. Oct4, a novel marker for human gastric cancer. J Surg Oncol. 2009;99(7):414–9. doi:10.1002/jso.21270.
30. Saigusa S, Tanaka K, Toiyama Y, Yokoe T, Okugawa Y, Ioue Y, et al. Correlation of CD133, OCT4, and SOX2 in rectal cancer and their association with distant recurrence after chemoradiotherapy. Ann Surg Oncol. 2009;16(12):3488–98. doi:10.1245/s10434-009-0617-z.
31. Huang PZ, Lu CL, Li BK, Hong J, Huang L, Wang L et al. [OCT4 expression in hepatocellular carcinoma and its clinical significance]. Chin J Cancer. 2010;29(1):111–116. doi:1000-467X201001105 [pii].
32. Zhang X, Han B, Huang J, Zheng B, Geng Q, Aziz F et al. Prognostic significance of OCT4 expression in adenocarcinoma of the lung. Jpn J Clin Oncol. 2010;40(10):961–966. doi: 10.1093/jjco/hyq066. hyq066 [pii].
33. Alexander RE, Cheng L, Grignon DJ, Idrees MT. Cytoplasmic OCT4 staining is a sensitive marker of neuroendocrine differentiation. Hum Pathol. 2014; 45(1):27–32. doi:10.1016/j.humpath.2013.08.006.
34. Anderson RA, Fulton N, Cowan G, Coutts S, Saunders PT. Conserved and divergent patterns of expression of DAZL, VASA and OCT4 in the germ cells of the human fetal ovary and testis. BMC Dev Biol. 2007;7:136. doi:1471-213X-7-136.
35. Kerr CL, Hill CM, Blumenthal PD, Gearhart JD. Expression of pluripotent stem cell markers in the human fetal testis. Stem Cells. 2008;26(2):412–21. doi:10.1634/stemcells.2007-0605.
36. Cools M, van Aerde K, Kersemaekers AM, Boter M, Drop SL, Wolffenbuttel KP et al. Morphological and immunohistochemical differences between gonadal maturation delay and early germ cell neoplasia in patients with undervirilization syndromes. J Clin Endocrinol Metab. 2005;90(9):5295–5303. doi:jc.2005–0139.
37. Cheng L, Sung MT, Cossu-Rocca P, Jones TD, MacLennan GT, De Jong J, et al. OCT4: biological functions and clinical applications as a marker of germ cell neoplasia. J Pathol. 2007;211(1):1–9. doi:10.1002/path.2105.
38. Onida GA, Bosincu L, Dessole S, Nicolae A, Preda O, Cossu-Rocca P et al. Sertoli cell tumor with benign peritoneal implants associated with gonadoblastoma. Int J Gynecol Pathol. 2010;29(5):423–426.doi:10.1097/PGP.0b013e3181dc7c2e.00004347-201009000-00005 [pii].
39. Jones TD, Ulbright TM, Eble JN, Baldridge LA, Cheng L. OCT4 staining in testicular tumors: a sensitive and specific marker for seminoma and embryonal carcinoma. Am J Surg Pathol. 2004;28(7):935–940. doi:00000478--200407000-00014 [pii].
40. Jones TD, Ulbright TM, Eble JN, Cheng L. OCT4: a sensitive and specific biomarker for intratubular germ cell neoplasia of the testis. Clin Cancer Res. 2004;10(24):8544–8547. doi: 10.1158/1078-0432. CCR-04-0688. 10/24/8544 [pii].
41. Cheng L. Establishing a germ cell origin for metastatic tumors using OCT4 immunohistochemistry. Cancer. 2004;101(9):2006–10. doi:10.1002/cncr.20566.
42. Richie JP. OCT4 staining in testicular tumors. A sensitive and specific marker for seminoma and embryonal carcinoma. J Urol. 2005;174(2):569–570. doi:S0022–5347(01)68313–7 [pii].
43. Teng LH, Lu DH, Xu QZ, Fu YJ, Yang H, He ZL. Expression and diagnostic significance of OCT4, CD117 and CD30 in germ cell tumors. Zhonghua Bing Li Xue Za Zhi. 2005;34(11):711–5.
44. Lau SK, Chang KL. OCT4: a sensitive and specific immunohistochemical marker for metastatic germ cell tumors. Adv Anat Pathol. 2006;13(2):76–79. doi:10.1097/01.pap.0000213011.33232.72. 00125480-200603000-00002 [pii].
45. Sung MT, Jones TD, Beck SD, Foster RS, Cheng L. OCT4 is superior to CD30 in the diagnosis of metastatic embryonal carcinomas after chemotherapy. Hum Pathol. 2006;37(6):662–667. doi:10.1016/j.humpath.2006.01.019.
46. Looijenga LH, Stoop H, de Leeuw HP, de Gouveia Brazao CA, Gillis AJ, van Roozendaal KE, et al. POU5F1 (OCT3/4) identifies cells with pluripotent potential in human germ cell tumors. Cancer Res. 2003;63(9):2244–50.
47. Mosbech CH, Svingen T, Nielsen JE, Toft BG, Rechnitzer C, Petersen BL, et al. Expression pattern of clinically relevant markers in paediatric germ cell- and sex-cord stromal tumours is similar to adult testicular tumours. Virchows Arch. 2014;465(5):567–77. doi:10.1007/s00428-014-1635-1.
48. Abiko K, Mandai M, Hamanishi J, Matsumura N, Baba T, Horiuchi A, et al. Oct4 expression in immature teratoma of the ovary: relevance to histologic grade and degree of differentiation. Am J Surg Pathol. 2010;34(12):1842–8. doi:10.1097/PAS.0b013e3181fcd707.
49. Cheng L, Zhang S, Talerman A, Roth LM. Morphologic, immunohistochemical, and fluorescence in situ hybridization study of ovarian embryonal carcinoma with comparison to solid variant of yolk sac tumor and immature teratoma. Hum Pathol. 2010;41(5):716–23. doi:10.1016/j.humpath.2009.10.016.
50. Idrees MT, Williamson SR, Kieffer TW, Cheng L. The role of OCT4 immunohistochemistry in evalua-

tion of retroperitoneal lymph node dissections: a pilot study. Mod Pathol. 2013;26(12):1613–9. doi:10.1038/modpathol.2013.110.

51. Hattab EM, Tu PH, Wilson JD, Cheng L. OCT4 immunohistochemistry is superior to placental alkaline phosphatase (PLAP) in the diagnosis of central nervous system germinoma. Am J Surg Pathol. 2005;29(3):368–371. doi:00000478--200503000-00011 [pii].
52. Mueller T, Mueller LP, Holzhausen HJ, Witthuhn R, Albers P, Schmoll HJ. Histological evidence for the existence of germ cell tumor cells showing embryonal carcinoma morphology but lacking OCT4 expression and cisplatin sensitivity. Histochem Cell Biol. 2010;134(2):197–204. doi:10.1007/s00418-010-0710-1.
53. Nogales FF, Dulcey I, Preda O. Germ cell tumors of the ovary: an update. Arch Pathol Lab Med. 2014;138(3):351–62. doi:10.5858/arpa.2012-0547-RA.
54. Rao S, Zhen S, Roumiantsev S, McDonald LT, Yuan GC, Orkin SH. Differential roles of Sall4 isoforms in embryonic stem cell pluripotency. Mol Cell Biol. 2010;30(22):5364–5380. doi: 10.1128/MCB.00419–10. MCB.00419–10 [pii].
55. Wang J, Rao S, Chu J, Shen X, Levasseur DN, Theunissen TW et al. A protein interaction network for pluripotency of embryonic stem cells. Nature. 2006;444(7117):364–368. doi: 10.1038/nature05284. nature05284 [pii].
56. Wu Q, Chen X, Zhang J, Loh YH, Low TY, Zhang W et al. Sall4 interacts with Nanog and co-occupies Nanog genomic sites in embryonic stem cells. J Biol Chem. 2006;281(34):24090–24094. doi:10.1074/jbc.C600122200.
57. Zhang J, Tam WL, Tong GQ, Wu Q, Chan HY, Soh BS et al. Sall4 modulates embryonic stem cell pluripotency and early embryonic development by the transcriptional regulation of Pou5f1. Nat Cell Biol. 2006;8(10):1114–1123. doi: 10.1038/ncb1481.
58. Yang J, Gao C, Chai L, Ma Y. A novel SALL4/OCT4 transcriptional feedback network for pluripotency of embryonic stem cells. PLoS ONE. 2010;5(5):e10766. doi:10.1371/journal.pone.0010766.
59. Elling U, Klasen C, Eisenberger T, Anlag K, Treier M. Murine inner cell mass-derived lineages depend on Sall4 function. Proc Natl Acad Sci U S A. 2006;103(44):16319–16324. doi: 10.1073/pnas.0607884103.
60. Warren M, Wang W, Spiden S, Chen-Murchie D, Tannahill D, Steel KP, et al. A Sall4 mutant mouse model useful for studying the role of Sall4 in early embryonic development and organogenesis. Genesis. 2007;45(1):51–8. doi:10.1002/dvg.20264.
61. Cao D, Guo S, Allan RW, Molberg KH, Peng Y. SALL4 is a novel sensitive and specific marker of ovarian primitive germ cell tumors and is particularly useful in distinguishing yolk sac tumor from clear cell carcinoma. Am J Surg Pathol. 2009;33(6):894–904. doi:10.1097/PAS.0b013e318198177d.
62. Cao D, Li J, Guo CC, Allan RW, Humphrey PA. SALL4 is a novel diagnostic marker for testicular germ cell tumors. Am J Surg Pathol. 2009;33(7):1065–77. doi:10.1097/PAS.0b013e3181a13eef.
63. Wang F, Liu A, Peng Y, Rakheja D, Wei L, Xue D, et al. Diagnostic utility of SALL4 in extragonadal yolk sac tumors: an immunohistochemical study of 59 cases with comparison to placental-like alkaline phosphatase, alpha-fetoprotein, and glypican-3. Am J Surg Pathol. 2009;33(10):1529–39. doi:10.1097/PAS.0b013e3181ad25d5.
64. Cao D, Humphrey PA, Allan RW. SALL4 is a novel sensitive and specific marker for metastatic germ cell tumors, with particular utility in detection of metastatic yolk sac tumors. Cancer. 2009;115(12):2640–51. doi:10.1002/cncr.24308.
65. Liu A, Cheng L, Du J, Peng Y, Allan RW, Wei L, et al. Diagnostic utility of novel stem cell markers SALL4, OCT4, NANOG, SOX2, UTF1, and TCL1 in primary mediastinal germ cell tumors. Am J Surg Pathol. 2010;34(5):697–706. doi:10.1097/PAS.0b013e3181db84aa.
66. Looijenga LH, Oosterhuis JW. Pathogenesis of testicular germ cell tumours. Rev Reprod. 1999;4(2):90–100.
67. Cui W, Kong NR, Ma Y, Amin HM, Lai R, Chai L. Differential expression of the novel oncogene, SALL4, in lymphoma, plasma cell myeloma, and acute lymphoblastic leukemia. Mod Pathol. 2006;19(12):1585–1592. doi: 10.1038/modpathol.3800694. 3800694 [pii].
68. Miettinen M, Wang Z, McCue PA, Sarlomo-Rikala M, Rys J, Biernat W et al. SALL4 expression in germ cell and non-germ cell tumors: a systematic immunohistochemical study of 3215 cases. Am J Surg Pathol. 2014;38(3):410–420. doi:10.1097/PAS.0000000000000116. 00000478-201403000-00018 [pii].
69. Bass AJ, Watanabe H, Mermel CH, Yu S, Perner S, Verhaak RG et al. SOX2 is an amplified lineage-survival oncogene in lung and esophageal squamous cell carcinomas. Nat Genet. 2009;41(11):1238–1242. doi: ng.465 [pii] 10.1038/ng.465.
70. Keramari M, Razavi J, Ingman KA, Patsch C, Edenhofer F, Ward CM, et al. Sox2 is essential for formation of trophectoderm in the preimplantation embryo. PLoS ONE. 2010;5(11):e13952. doi:10.1371/journal.pone.0013952.
71. Adachi K, Suemori H, Yasuda SY, Nakatsuji N, Kawase E. Role of SOX2 in maintaining pluripotency of human embryonic stem cells. Genes Cells. 2010;15(5):455–470. doi:10.1111/j.1365-2443.2010.01400.x.
72. Wood HB, Episkopou V. Comparative expression of the mouse Sox1, Sox2 and Sox3 genes from pregastrulation to early somite stages. Mech Dev. 1999;86(1–2):197–201. doi: S0925477399001161 [pii].

73. Graham V, Khudyakov J, Ellis P, Pevny L. SOX2 functions to maintain neural progenitor identity. Neuron. 2003;39(5):749–765. doi:S0896627303004975 [pii].
74. Laga AC, Lai CY, Zhan Q, Huang SJ, Velazquez EF, Yang Q et al. Expression of the embryonic stem cell transcription factor SOX2 in human skin: relevance to melanocyte and merkel cell biology. Am J Pathol. 2010;176(2):903–913.doi:10.2353/ajpath.2010.090495. ajpath.2010.090495 [pii].
75. Puligilla C, Dabdoub A, Brenowitz SD, Kelley MW. Sox2 induces neuronal formation in the developing mammalian cochlea. J Neurosci. 2010;30(2):714–722. doi: 10.1523/JNEUROSCI.3852-09.2010. 30/2/714 [pii].
76. Phi JH, Kim JH, Eun KM, Wang KC, Park KH, Choi SA, et al. Upregulation of SOX2, NOTCH1, and ID1 in supratentorial primitive neuroectodermal tumors: a distinct differentiation pattern from that of medulloblastomas. J Neurosurg Pediatr. 2010;5(6):608–14. doi:10.3171/2010.2.PEDS1065.
77. Ji J, Zheng PS. Expression of Sox2 in human cervical carcinogenesis. Hum Pathol. 2010;41(10):1438–1447. doi: 10.1016/j.humpath.2009.11.021. S0046–8177(10)00113–9 [pii].
78. Wang S, Chandler-Militello D, Lu G, Roy NS, Zielke A, Auvergne R et al. Prospective identification, isolation, and profiling of a telomerase-expressing subpopulation of human neural stem cells, using sox2 enhancer-directed fluorescence-activated cell sorting. J Neurosci. 2010;30(44):14635–14648. doi:30/44/14635 [pii]. 10.1523/JNEUROSCI.1729-10.2010.
79. Ye F, Li Y, Hu Y, Zhou C, Chen H. Expression of Sox2 in human ovarian epithelial carcinoma. J Cancer Res Clin Oncol. 2011;137(1):131–7. doi:10.1007/s00432-010-0867-y.
80. Maier S, Wilbertz T, Braun M, Scheble V, Reischl M, Mikut R, et al. SOX2 amplification is a common event in squamous cell carcinomas of different organ sites. Hum Pathol. 2011;42(8):1078–88. doi:10.1016/j.humpath.2010.11.010.
81. Hussenet T, du Manoir S. SOX2 in squamous cell carcinoma: Amplifying a pleiotropic oncogene along carcinogenesis. Cell Cycle. 2010;9(8). doi:11203 [pii].
82. Lu Y, Futtner C, Rock JR, Xu X, Whitworth W, Hogan BL, et al. Evidence that SOX2 overexpression is oncogenic in the lung. PLoS ONE. 2010;5(6):e11022. doi:10.1371/journal.pone.0011022.
83. Mutoh H, Sashikawa M, Sugano K. Sox2 expression is maintained while gastric phenotype is completely lost in Cdx2-induced intestinal metaplastic mucosa. Differentiation. 2010. doi:10.1016/j.diff.2010.10.002. S0301-4681(10)00294-X [pii].
84. Sholl LM, Long KB, Hornick JL. Sox2 expression in pulmonary non-small cell and neuroendocrine carcinomas. Appl Immunohistochem Mol Morphol. 2010;18(1):55–61. doi:10.1097/PAI.0b013e3181b16b88.
85. Zhang X, Yu H, Yang Y, Zhu R, Bai J, Peng Z, et al. SOX2 in gastric carcinoma, but not Hath1, is related to patients' clinicopathological features and prognosis. J Gastrointest Surg. 2010;14(8):1220–6. doi:10.1007/s11605-010-1246-3.
86. Sanada Y, Yoshida K, Ohara M, Oeda M, Konishi K, Tsutani Y. Histopathologic evaluation of stepwise progression of pancreatic carcinoma with immunohistochemical analysis of gastric epithelial transcription factor SOX2: comparison of expression patterns between invasive components and cancerous or nonneoplastic intraductal components. Pancreas. 2006;32(2):164–70. doi:10.1097/01.mpa.0000202947.80117.a0.
87. Stolnicu S, Goyenaga P, Hincu M, Marian C, Murillo R, Nogales FF. Embryonal (botryoides) rhabdomyosarcoma of the uterus harboring a primitive neuroectodermal tumor component. Int J Gynecol Pathol. 2012;31(4):387–9. doi:10.1097/PGP.0b013e31823ff3e6.
88. Huang YH, Luo MH, Ni YB, Tsang JY, Chan SK, Lui PC, et al. Increased SOX2 expression in less differentiated breast carcinomas and their lymph node metastases. Histopathology. 2014;64(4):494–503. doi:10.1111/his.12257.
89. Pham DL, Scheble V, Bareiss P, Fischer A, Beschorner C, Adam A, et al. SOX2 expression and prognostic significance in ovarian carcinoma. Int J Gynecol Pathol. 2013;32(4):358–67. doi:10.1097/PGP.0b013e31826a642b.
90. Perrett RM, Turnpenny L, Eckert JJ, O'Shea M, Sonne SB, Cameron IT et al. The early human germ cell lineage does not express SOX2 during in vivo development or upon in vitro culture. Biol Reprod. 2008;78(5):852–858. doi:biolreprod.107.066175 [pii] 10.1095/biolreprod.107.066175.
91. de Jong J, Stoop H, Gillis AJ, van Gurp RJ, van de Geijn GJ, Boer M, et al. Differential expression of SOX17 and SOX2 in germ cells and stem cells has biological and clinical implications. J Pathol. 2008;215(1):21–30. doi:10.1002/path.2332.
92. Sonne SB, Perrett RM, Nielsen JE, Baxter MA, Kristensen DM, Leffers H et al. Analysis of SOX2 expression in developing human testis and germ cell neoplasia. Int J Dev Biol. 2010;54(4):755–760. doi:082668ss [pii] 10.1387/ijdb.082668ss.
93. Archer TC, Jin J, Casey ES. Interaction of Sox1, Sox2, Sox3 and Oct4 during primary neurogenesis. Dev Biol. 2010. doi: 10.1016/j.ydbio.2010.12.013.
94. Fong H, Hohenstein KA, Donovan PJ. Regulation of self-renewal and pluripotency by Sox2 in human embryonic stem cells. Stem Cells. 2008;26(8):1931–1938. doi: 10.1634/stemcells.2007-1002. 2007-1002 [pii]
95. Nonaka D. Differential expression of SOX2 and SOX17 in testicular germ cell tumors. Am J Clin Pathol. 2009;131(5):731–6. doi:10.1309/AJCP7MNCNBCRN8NO.
96. Moss EG, Tang L. Conservation of the heterochronic regulator Lin-28, its developmental expression and

microRNA complementary sites. Dev Biol. 2003; 258(2):432–42.
97. Shyh-Chang N, Daley GQ. Lin28: primal regulator of growth and metabolism in stem cells. Cell Stem Cell. 2013;12(4):395–406. doi:10.1016/j.stem.2013.03.005.
98. Polesskaya A, Cuvellier S, Naguibneva I, Duquet A, Moss EG, Harel-Bellan A. Lin-28 binds IGF-2 mRNA and participates in skeletal myogenesis by increasing translation efficiency. Genes Dev. 2007;21(9):1125–38. doi:10.1101/gad.415007.
99. Viswanathan SR, Daley GQ, Gregory RI. Selective blockade of microRNA processing by Lin28. Science. 2008;320(5872):97–100. doi:10.1126/science.1154040.
100. Qiu C, Ma Y, Wang J, Peng S, Huang Y. Lin28-mediated post-transcriptional regulation of Oct4 expression in human embryonic stem cells. Nucleic Acids Res. 2010;38(4):1240–8. doi:10.1093/nar/gkp1071.
101. Yu J, Vodyanik MA, Smuga-Otto K, Antosiewicz-Bourget J, Frane JL, Tian S et al. Induced pluripotent stem cell lines derived from human somatic cells. Science. 2007;318(5858):1917–1920. doi:10.1126/science.1151526.
102. West JA, Viswanathan SR, Yabuuchi A, Cunniff K, Takeuchi A, Park IH et al. A role for Lin28 in primordial germ-cell development and germ-cell malignancy. Nature. 2009;460(7257):909–913. doi:10.1038/nature0821. nature08210 [pii].
103. Childs AJ, Kinnell HL, He J, Anderson RA. LIN28 is selectively expressed by primordial and pre-meiotic germ cells in the human fetal ovary. Stem Cells Dev. 2012;21(13):2343–9. doi:10.1089/scd.2011.0730.
104. Xue D, Peng Y, Wang F, Allan RW, Cao D. RNA-binding protein LIN28 is a sensitive marker of ovarian primitive germ cell tumours. Histopathology. 2011;59(3):452–9. doi:10.1111/j.1365-2559.2011.03949.x.
105. Cao D, Liu A, Wang F, Allan RW, Mei K, Peng Y, et al. RNA-binding protein LIN28 is a marker for primary extragonadal germ cell tumors: an immunohistochemical study of 131 cases. Mod Pathol. 2011;24(2):288–96. doi:10.1038/modpathol.2010.195.
106. Cao D, Allan RW, Cheng L, Peng Y, Guo CC, Dahiya N, et al. RNA-binding protein LIN28 is a marker for testicular germ cell tumors. Hum Pathol. 2011;42(5):710–8. doi:10.1016/j.humpath.2010.09.007.
107. Xu C, Shen J, Xie S, Jiang Z, Huang L, Wang L. Positive expression of Lin28 is correlated with poor survival in gastric carcinoma. Med Oncol. 2013;30(1):382. doi:10.1007/s12032-012-0382-x.
108. King CE, Cuatrecasas M, Castells A, Sepulveda AR, Lee JS, Rustgi AK. LIN28B promotes colon cancer progression and metastasis. Cancer Res. 2011;71(12):4260–8. doi:10.1158/0008-5472.CAN-10-4637.
109. Rodini CO, Suzuki DE, Saba-Silva N, Cappellano A, de Souza JE, Cavalheiro S, et al. Expression analysis of stem cell-related genes reveal OCT4 as a predictor of poor clinical outcome in medulloblastoma. J Neuro-Oncol. 2012;106(1):71–9. doi:10.1007/s11060-011-0647-9.
110. Ivanova N, Dobrin R, Lu R, Kotenko I, Levorse J, DeCoste C, et al. Dissecting self-renewal in stem cells with RNA interference. Nature. 2006;442(7102):533–8. doi:10.1038/nature04915.
111. Wang Z, Oron E, Nelson B, Razis S, Ivanova N. Distinct lineage specification roles for NANOG, OCT4, and SOX2 in human embryonic stem cells. Cell Stem Cell. 2012;10(4):440–54. doi:10.1016/j.stem.2012.02.016.
112. Hyslop L, Stojkovic M, Armstrong L, Walter T, Stojkovic P, Przyborski S, et al. Downregulation of NANOG induces differentiation of human embryonic stem cells to extraembryonic lineages. Stem Cells. 2005;23(8):1035–43. doi:10.1634/stemcells.2005-0080.
113. Hart AH, Hartley L, Parker K, Ibrahim M, Looijenga LH, Pauchnik M, et al. The pluripotency homeobox gene NANOG is expressed in human germ cell tumors. Cancer. 2005;104(10):2092–8. doi:10.1002/cncr.21435.
114. Guo Y, Liu S, Wang P, Zhao S, Wang F, Bing L, et al. Expression profile of embryonic stem cell-associated genes Oct4, Sox2 and Nanog in human gliomas. Histopathology. 2011;59(4):763–75. doi:10.1111/j.1365-2559.2011.03993.x.
115. Luo W, Li S, Peng B, Ye Y, Deng X, Yao K. Embryonic stem cells markers SOX2, OCT4 and Nanog expression and their correlations with epithelial-mesenchymal transition in nasopharyngeal carcinoma. PLoS ONE. 2013;8(2):e56324. doi:10.1371/journal.pone.0056324.
116. He QZ, Luo XZ, Wang K, Zhou Q, Ao H, Yang Y, et al. Isolation and characterization of cancer stem cells from high-grade serous ovarian carcinomas. Cell Physiol Biochem: Inter J Exper Cell Physiol Biochem Pharmacol. 2014;33(1):173–84. doi:10.1159/000356660.
117. Schreiber L, Raanan C, Amsterdam A. CD24 and Nanog identify stem cells signature of ovarian epithelium and cysts that may develop to ovarian cancer. Acta Histochem. 2014;116(2):399–406. doi:10.1016/j.acthis.2013.09.007.
118. Auersperg N. The stem-cell profile of ovarian surface epithelium is reproduced in the oviductal fimbriae, with increased stem-cell marker density in distal parts of the fimbriae. Int J Gynecol Pathol. 2013;32(5):444–53. doi:10.1097/PGP.0b013e3182800ad5.
119. Nishimoto M, Fukushima A, Okuda A, Muramatsu M. The gene for the embryonic stem cell coactivator UTF1 carries a regulatory element which selectively interacts with a complex composed of Oct-3/4 and Sox-2. Mol Cell Biol. 1999;19(8):5453–65.
120. Fukushima A, Okuda A, Nishimoto M, Seki N, Hori TA, Muramatsu M. Characterization of functional domains of an embryonic stem cell coactivator UTF1 which are conserved and essential for potentiation of ATF-2 activity. J Biol Chem. 1998; 273(40):25840–9.

121. Kooistra SM, Thummer RP, Eggen BJ. Characterization of human UTF1, a chromatin-associated protein with repressor activity expressed in pluripotent cells. Stem Cell Res. 2009;2(3):211–8. doi:10.1016/j.scr.2009.02.001.
122. Wang P, Li J, Allan RW, Guo CC, Peng Y, Cao D. Expression of UTF1 in primary and metastatic testicular germ cell tumors. Am J Clin Pathol. 2010;134(4):604–12. doi:10.1309/AJCPB44HBKINJNYU.
123. Mouallif M, Albert A, Zeddou M, Ennaji MM, Delvenne P, Guenin S. Expression profile of undifferentiated cell transcription factor 1 in normal and cancerous human epithelia. Int J Exp Pathol. 2014;95(4):251–9. doi:10.1111/iep.12077.
124. Laine J, Kunstle G, Obata T, Sha M, Noguchi M. The protooncogene TCL1 is an Akt kinase coactivator. Mol Cell. 2000;6(2):395–407.
125. Aggarwal M, Villuendas R, Gomez G, Rodriguez-Pinilla SM, Sanchez-Beato M, Alvarez D, et al. TCL1A expression delineates biological and clinical variability in B-cell lymphoma. Mod Pathol. 2009;22(2):206–15. doi:10.1038/modpathol.2008.148.
126. Trinh DT, Shibata K, Hirosawa T, Umezu T, Mizuno M, Kajiyama H, et al. Diagnostic utility of CD117, CD133, SALL4, OCT4, TCL1 and glypican-3 in malignant germ cell tumors of the ovary. J Obstet Gynaecol Res. 2012;38(5):841–8. doi:10.1111/j.1447-0756.2011.01798.x.
127. Lau SK, Weiss LM, Chu PG. TCL1 protein expression in testicular germ cell tumors. Am J Clin Pathol. 2010;133(5):762–6. doi:10.1309/AJCPIPU1MPTBM2FQ.
128. Matoba R, Niwa H, Masui S, Ohtsuka S, Carter MG, Sharov AA, et al. Dissecting Oct3/4-regulated gene networks in embryonic stem cells by expression profiling. PLoS ONE. 2006;1:e26. doi:10.1371/journal.pone.0000026.
129. Cao D, Lane Z, Allan RW, Wang P, Guo CC, Peng Y, et al. TCL1 is a diagnostic marker for intratubular germ cell neoplasia and classic seminoma. Histopathology. 2010;57(1):152–7. doi:10.1111/j.1365-2559.2010.03583.x.
130. Zhang P, Andrianakos R, Yang Y, Liu C, Lu W. Kruppel-like factor 4 (Klf4) prevents embryonic stem (ES) cell differentiation by regulating Nanog gene expression. J Biol Chem. 2010;285(12):9180–9. doi:10.1074/jbc.M109.077958.
131. Takahashi K, Yamanaka S. Induction of pluripotent stem cells from mouse embryonic and adult fibroblast cultures by defined factors. Cell. 2006;126(4):663–76. doi:10.1016/j.cell.2006.07.024.
132. Rowland BD, Peeper DS. KLF4, p21 and context-dependent opposing forces in cancer. Nat Rev Cancer. 2006;6(1):11–23. doi:10.1038/nrc1780.
133. Godmann M, Gashaw I, Katz JP, Nagy A, Kaestner KH, Behr R. Kruppel-like factor 4, a "pluripotency transcription factor" highly expressed in male postmeiotic germ cells, is dispensable for spermatogenesis in the mouse. Mech Dev. 2009;126(8–9):650–64. doi:10.1016/j.mod.2009.06.1081.
134. Godmann M, Gashaw I, Eildermann K, Schweyer S, Bergmann M, Skotheim RI, et al. The pluripotency transcription factor Kruppel-like factor 4 is strongly expressed in intratubular germ cell neoplasia unclassified and seminoma. Mol Hum Reprod. 2009;15(8):479–88. doi:10.1093/molehr/gap040.
135. Klco JM, Kulkarni S, Kreisel FH, Nguyen TD, Hassan A, Frater JL. Immunohistochemical analysis of monocytic leukemias: usefulness of CD14 and Kruppel-like factor 4, a novel monocyte marker. Am J Clin Pathol. 2011;135(5):720–30. doi:10.1309/AJCPZ46PMMAWJROT.
136. Liu Z, Yang H, Luo W, Jiang Q, Mai C, Chen Y, et al. Loss of cytoplasmic KLF4 expression is correlated with the progression and poor prognosis of nasopharyngeal carcinoma. Histopathology. 2013;63(3):362–70. doi:10.1111/his.12176.
137. Le Magnen C, Bubendorf L, Ruiz C, Zlobec I, Bachmann A, Heberer M, et al. Klf4 transcription factor is expressed in the cytoplasm of prostate cancer cells. Eur J Cancer. 2013;49(4):955–63. doi:10.1016/j.ejca.2012.09.023.
138. Gimelli S, Caridi G, Beri S, McCracken K, Bocciardi R, Zordan P, et al. Mutations in SOX17 are associated with congenital anomalies of the kidney and the urinary tract. Hum Mutat. 2010;31(12):1352–9. doi:10.1002/humu.21378.
139. Katoh M. Molecular cloning and characterization of human SOX17. Int J Mol Med. 2002;9(2):153–7.
140. Williamson JA, Bosher JM, Skinner A, Sheer D, Williams T, Hurst HC. Chromosomal mapping of the human and mouse homologues of two new members of the AP-2 family of transcription factors. Genomics. 1996;35(1):262–4. doi:10.1006/geno.1996.0351.
141. Nakaki F, Hayashi K, Ohta H, Kurimoto K, Yabuta Y, Saitou M. Induction of mouse germ-cell fate by transcription factors in vitro. Nature. 2013;501(7466):222–6. doi:10.1038/nature12417.
142. Kuckenberg P, Buhl S, Woynecki T, van Furden B, Tolkunova E, Seiffe F, et al. The transcription factor TCFAP2C/AP-2gamma cooperates with CDX2 to maintain trophectoderm formation. Mol Cell Biol. 2010;30(13):3310–20. doi:10.1128/MCB.01215-09.
143. Pauls K, Jager R, Weber S, Wardelmann E, Koch A, Buttner R, et al. Transcription factor AP-2gamma, a novel marker of gonocytes and seminomatous germ cell tumors. Int J Cancer. 2005;115(3):470–7. doi:10.1002/ijc.20913.
144. Woenckhaus C, Giebel J, Failing K, Fenic I, Dittberner T, Poetsch M. Expression of AP-2alpha, c-kit, and cleaved caspase-6 and -3 in naevi and malignant melanomas of the skin. A possible role for caspases in melanoma progression? J Pathol. 2003;201(2):278–87. doi:10.1002/path.1424.
145. Odegaard E, Staff AC, Kaern J, Florenes VA, Kopolovic J, Trope CG, et al. The AP-2gamma transcription factor is upregulated in advanced-stage

ovarian carcinoma. Gynecol Oncol. 2006;100(3):462–8. doi:10.1016/j.ygyno.2005.09.022.
146. Monk D, Bentley L, Beechey C, Hitchins M, Peters J, Preece MA, et al. Characterisation of the growth regulating gene IMP3, a candidate for Silver-Russell syndrome. J Med Genet. 2002;39(8):575–81.
147. Poole RL, Leith DJ, Docherty LE, Shmela ME, Gicquel C, Splitt M, et al. Beckwith-Wiedemann syndrome caused by maternally inherited mutation of an OCT-binding motif in the IGF2/H19-imprinting control region, ICR1. Eur J Human Genet: EJHG. 2012;20(2):240–3. doi:10.1038/ejhg.2011.166.
148. Kobel M, Xu H, Bourne PA, Spaulding BO, Shih Ie M, Mao TL, et al. IGF2BP3 (IMP3) expression is a marker of unfavorable prognosis in ovarian carcinoma of clear cell subtype. Mod Pathol. 2009;22(3):469–75. doi:10.1038/modpathol.2008.206.
149. Hammer NA, Hansen T, Byskov AG, Rajpert-De Meyts E, Grondahl ML, Bredkjaer HE, et al. Expression of IGF-II mRNA-binding proteins (IMPs) in gonads and testicular cancer. Reproduction. 2005;130(2):203–12. doi:10.1530/rep.1.00664.
150. Shi M, Fraire AE, Chu P, Cornejo K, Woda BA, Dresser K, et al. Oncofetal protein IMP3, a new diagnostic biomarker to distinguish malignant mesothelioma from reactive mesothelial proliferation. Am J Surg Pathol. 2011;35(6):878–82. doi:10.1097/PAS.0b013e318218985b.
151. Goodman S, Zhang L, Cheng L, Jiang Z. Differential expression of IMP3 between male and female mature teratomas--immunohistochemical evidence of malignant nature. Histopathology. 2014;65(4):483–9. doi:10.1111/his.12409.
152. DeChiara TM, Robertson EJ, Efstratiadis A. Parental imprinting of the mouse insulin-like growth factor II gene. Cell. 1991;64(4):849–59.
153. Chou J, Provot S, Werb Z. GATA3 in development and cancer differentiation: cells GATA have it! J Cell Physiol. 2010;222(1):42–9. doi:10.1002/jcp.21943.
154. Pandolfi PP, Roth ME, Karis A, Leonard MW, Dzierzak E, Grosveld FG, et al. Targeted disruption of the GATA3 gene causes severe abnormalities in the nervous system and in fetal liver haematopoiesis. Nat Genet. 1995;11(1):40–4. doi:10.1038/ng0995-40.
155. Joulin V, Bories D, Eleouet JF, Labastie MC, Chretien S, Mattei MG, et al. A T-cell specific TCR delta DNA binding protein is a member of the human GATA family. EMBO J. 1991;10(7):1809–16.
156. Luo XJ, Deng M, Xie X, Huang L, Wang H, Jiang L et al. GATA3 controls the specification of prosensory domain and neuronal survival in the mouse cochlea. Hum Mol Genet. 2013;22(18):3609–3623. doi:10.1093/hmg/ddt212/ddt212. [pii].
157. Ordonez NG. Value of GATA3 immunostaining in tumor diagnosis: a review. Adv Anat Pathol. 2013; 20(5):352–360. doi:10.1097/PAP.0b013e3182a28a68. 00125480-201309000-00006 [pii].
158. Ranganath S, Murphy KM. Structure and specificity of GATA proteins in Th2 development. Mol Cell Biol. 2001;21(8):2716–25. doi:10.1128/MCB.21.8.2716-2725.2001.
159. Miettinen M, McCue PA, Sarlomo-Rikala M, Rys J, Czapiewski P, Wazny K, et al. GATA3: a multispecific but potentially useful marker in surgical pathology: a systematic analysis of 2500 epithelial and nonepithelial tumors. Am J Surg Pathol. 2014;38(1):13–22. doi:10.1097/PAS.0b013e3182a0218f.
160. Schuldt M, Rubio A, Preda O, Nogales FF. GATA3 expression is present in primitive patterns of Yolk Sac Tumors but is not expressed by differentiated variants. Histopathology. 2015;68(4):613–5. doi:10.1111/his.12776.
161. Wendroth SM, Mentrikoski MJ, Wick MR. GATA3 expression in morphologic subtypes of breast carcinoma: a comparison with gross cystic disease fluid protein 15 and mammaglobin. Ann Diagn Pathol. 2015; 19(1):6–9. doi:10.1016/j.anndiagpath.2014.12.001.
162. Ordonez NG. Value of GATA3 immunostaining in the diagnosis of parathyroid tumors. Appl Immunohistochem Mol Morphol. 2014;22(10):756–61. doi:10.1097/PAI.0000000000000007.
163. Schwartz LE, Begum S, Westra WH, Bishop JA. GATA3 immunohistochemical expression in salivary gland neoplasms. Head Neck Pathol. 2013; 7(4):311–5. doi:10.1007/s12105-013-0442-3.
164. Howitt BE, Emori MM, Drapkin R, Gaspar C, Barletta JA, Nucci MR, et al. GATA3 is a sensitive and specific marker of benign and malignant mesonephric lesions in the lower female genital tract. Am J Surg Pathol. 2015;39(10):1411–9. doi:10.1097/PAS.0000000000000471.
165. Ralston A, Cox BJ, Nishioka N, Sasaki H, Chea E, Rugg-Gunn P et al. Gata3 regulates trophoblast development downstream of Tead4 and in parallel to Cdx2. Development. 2010;137(3):395–403. doi:10.1242/dev.038828. 137/3/395 [pii].
166. Seargeant LE, Stinson RA. Evidence that three structural genes code for human alkaline phosphatases. Nature. 1979;281(5727):152–4.
167. McKenna MJ, Hamilton TA, Sussman HH. Comparison of human alkaline phosphatase isoenzymes. Structural evidence for three protein classes. Biochem J. 1979;181(1):67–73.
168. Henthorn PS, Raducha M, Kadesch T, Weiss MJ, Harris H. Sequence and characterization of the human intestinal alkaline phosphatase gene. J Biol Chem. 1988;263(24):12011–9.
169. Javadpour N. The role of biologic tumor markers in testicular cancer. Cancer. 1980;45(7 Suppl):1755–61.
170. Nielsen OS, Munro AJ, Duncan W, Sturgeon J, Gospodarowicz MK, Jewett MA, et al. Is placental alkaline phosphatase (PLAP) a useful marker for seminoma? Eur J Cancer. 1990;26(10):1049–54.
171. Fishman WH. Clinical and biological significance of an isozyme tumor marker--PLAP. Clin Biochem. 1987;20(6):387–392. doi:0009–9120(87)90003–8 [pii].
172. De Broe M, Pollet D. Multicenter evaluation of human placental alkaline phosphatase as a possible tumor-associated antigen in serum. Clin Chem. 1988;34(10):1995–9.

173. Muensch HA, Maslow WC, Azama F, Bertrand M, Dewhurst P, Hartman B. Placental-like alkaline phosphatase. Re-evaluation of the tumor marker with exclusion of smokers. Cancer. 1986;58(8): 1689–94.
174. Kommoss F, Oliva E, Bittinger F, Kirkpatrick CJ, Amin MB, Bhan AK, et al. Inhibin-alpha CD99, HEA125, PLAP, and chromogranin immunoreactivity in testicular neoplasms and the androgen insensitivity syndrome. Hum Pathol. 2000;31(9):1055–61. doi:10.1053/hupa.2000.16237.
175. Epenetos AA, Munro AJ, Tucker DF, Gregory W, Duncan W, MacDougall RH, et al. Monoclonal antibody assay of serum placental alkaline phosphatase in the monitoring of testicular tumours. Br J Cancer. 1985;51(5):641–4.
176. Koshida K, Uchibayashi T, Yamamoto H, Hirano K. Significance of placental alkaline phosphatase (PLAP) in the monitoring of patients with seminoma. Br J Urol. 1996;77(1):138–42.
177. Wick MR, Swanson PE, Manivel JC. Placental-like alkaline phosphatase reactivity in human tumors: an immunohistochemical study of 520 cases. Hum Pathol. 1987;18(9):946–54.
178. Durak H, Comunoglu NU, Comunoglu C, Guven A, Cam M, Dervisoglu S, et al. Specificity and sensitivity of differentiation antigens in superficial soft tissue tumors: comparison of SMA, calponin, H-caldesmon, C-kit. PLAP HPL Bratisl Lek Listy. 2010;111(8):432–8.
179. Hustin J, Collette J, Franchimont P. Immunohistochemical demonstration of placental alkaline phosphatase in various states of testicular development and in germ cell tumours. Int J Androl. 1987;10(1):29–35.
180. Koshida K, Wahren B. Placental-like alkaline phosphatase in seminoma. Urol Res. 1990;18(2):87–92.
181. Manivel JC, Jessurun J, Wick MR, Dehner LP. Placental alkaline phosphatase immunoreactivity in testicular germ-cell neoplasms. Am J Surg Pathol. 1987;11(1):21–9.
182. Uchida T, Shimoda T, Miyata H, Shikata T, Iino S, Suzuki H, et al. Immunoperoxidase study of alkaline phosphatase in testicular tumor. Cancer. 1981;48(6):1455–62.
183. Yoshida M, Koshiyama M, Konishi M, Fujii H, Nanno H, Hayashi M et al. Ovarian dysgerminoma showing high serum levels and positive immunostaining of placental alkaline phosphatase and neuron-specific enolase associated with elevation of serum prolactin level. Eur J Obstet Gynecol Reprod Biol. 1998;81(1):123–128. doi:S0301211598001626 [pii].
184. Schar BK, Otto VI, Hanseler E. Simultaneous detection of all four alkaline phosphatase isoenzymes in human germ cell tumors using reverse transcription-PCR. Cancer Res. 1997;57(17):3841–6.
185. Tamiolakis D, Papadopoulos N, Lambropoulou M, Venizelos J, Verettas D, Tsikouras P, et al. Ber-H2 (CD30) immunohistochemical staining of human fetal tissues. Int J Biol Sci. 2005;1(4):135–40.
186. Latza U, Foss HD, Durkop H, Eitelbach F, Dieckmann KP, Loy V, et al. CD30 antigen in embryonal carcinoma and embryogenesis and release of the soluble molecule. Am J Pathol. 1995;146(2):463–71.
187. Pera MF, Bennett W, Cerretti DP. Expression of CD30 and CD30 ligand in cultured cell lines from human germ-cell tumors. Lab Investig. 1997;76(4): 497–504.
188. Croager EJ, Gout AM, Abraham LJ. Involvement of Sp1 and microsatellite repressor sequences in the transcriptional control of the human CD30 gene. Am J Pathol. 2000;156(5):1723–31.
189. Herszfeld D, Wolvetang E, Langton-Bunker E, Chung TL, Filipczyk AA, Houssami S et al. CD30 is a survival factor and a biomarker for transformed human pluripotent stem cells. Nat Biotechnol. 2006;24(3):351–357. doi:10.1038/nbt1197.
190. Stein H, Gerdes J, Schwab U, Lemke H, Mason DY, Ziegler A, et al. Identification of Hodgkin and Sternberg-reed cells as a unique cell type derived from a newly-detected small-cell population. Int J Cancer. 1982;30(4):445–59.
191. Stein H, Mason DY, Gerdes J, O'Connor N, Wainscoat J, Pallesen G, et al. The expression of the Hodgkin's disease associated antigen Ki-1 in reactive and neoplastic lymphoid tissue: evidence that Reed-Sternberg cells and histiocytic malignancies are derived from activated lymphoid cells. Blood. 1985;66(4):848–58.
192. Oflazoglu E, Grewal IS, Gerber H. Targeting CD30/CD30L in oncology and autoimmune and inflammatory diseases. Adv Exp Med Biol. 2009;647:174–85. doi:10.1007/978-0-387-89520-8_12.
193. Pallesen G, Hamilton-Dutoit SJ. Ki-1 (CD30) antigen is regularly expressed by tumor cells of embryonal carcinoma. Am J Pathol. 1988;133(3):446–50.
194. Cossu-Rocca P, Jones TD, Roth LM, Eble JN, Zheng W, Karim FW et al. Cytokeratin and CD30 expression in dysgerminoma. Hum Pathol. 2006;37(8): 1015–1021. doi:10.1016/j.humpath.2006.02.018.
195. Berney DM, Shamash J, Pieroni K, Oliver RT. Loss of CD30 expression in metastatic embryonal carcinoma: the effects of chemotherapy? Histopathology. 2001;39(4):382–385. doi:1226 [pii].
196. Hittmair A, Rogatsch H, Hobisch A, Mikuz G, Feichtinger H. CD30 expression in seminoma. Hum Pathol. 1996;27(11):1166–71.
197. Ferreiro JA. Ber-H2 expression in testicular germ cell tumors. Hum Pathol. 1994;25:522–4.
198. Yang F, Luna VJ, Mcanelly RD, Naberhaus KH, Cupples RL, Bowman BH. Evolutionary and Structural Relationships among the Group-Specific Component. Albumin Alpha-Fetoprotein Nucleic Acids Res. 1985;13(22):8007–17.
199. Mizejewski GJ. Alpha-fetoprotein structure and function: relevance to isoforms, epitopes, and conformational variants. Exp Biol Med (Maywood). 2001;226(5):377–408.
200. Nogales FF, Dulcey I. The secondary human yolk sac has an immunophenotype indicative of both

hepatic and intestinal differentiation. Int J Dev Biol. 2012;56(9):755–60. doi:10.1387/ijdb.120080fn.
201. El-Bahrawy M. Alpha-fetoprotein-producing non-germ cell tumours of the female genital tract. Eur J Cancer. 2010;46(8):1317–1322. doi: 10.1016/j.ejca.2010.01.028.
202. Kishimoto T, Yano T, Hiroshima K, Inayama Y, Kawachi K, Nakatani Y. A case of *-fetoprotein-producing pulmonary carcinoma with restricted expression of hepatocyte nuclear factor-4* in hepatoid foci: a case report with studies of previous cases. Hum Pathol. 2008;39(7):1115–1120. doi: 10.1016/j.humpath.2007.12.013.
203. Tsuchida Y, Kaneko M, Fukui M, Sakaguchi H, Ishiguro T. Three different types of alpha-fetoprotein in the diagnosis of malignant solid tumors: use of a sensitive lectin-affinity immunoelectrophoresis. J Pediatr Surg. 1989;24(4):350–355. doi:S0022346889001740.
204. Kinoshita Y, Tajiri T, Souzaki R, Tatsuta K, Higashi M, Izaki T et al. Diagnostic value of lectin reactive alpha-fetoprotein for neoinfantile hepatic tumors and malignant germ cell tumors: preliminary study. J Pediatr Hematol Oncol. 2008;30(6):447–450. doi:10.1097/MPH.0b013e31816916ad.00043426-200806000-00006 [pii].
205. Nogales F, Fernandez PL, Alvaro T. Alfa-fetoprotein-positive globules in involuting human yolk sac. Hum Pathol. 1988;19(8):995.
206. Nogales FF, Quinonez E, Lopez-Marin L, Dulcey I, Preda O. A diagnostic immunohistochemical panel for yolk sac (primitive endodermal) tumours based on an immunohistochemical comparison with the human yolk sac. Histopathology. 2014;65(1):51–9. doi:10.1111/his.12373.
207. Jacobsen GK, Jacobsen M. Alpha-fetoprotein (AFP) and human chorionic gonadotropin (HCG) in testicular germ cell tumours. A prospective immunohistochemical study. Acta Pathol Microbiol Immunol Scand A. 1983;91(3):165–76.
208. Nose K, Saito H, Kuroki T. Isolation of a gene sequence induced later by tumor-promoting 12-O-tetradecanoylphorbol-13-acetate in mouse osteoblastic cells (MC3T3-E1) and expressed constitutively in ras-transformed cells. Cell Growth Diff: Mol Biol J Am Ass Cancer Res. 1990;1(11):511–8.
209. Ordonez NG. Podoplanin: a novel diagnostic immunohistochemical marker. Adv Anat Pathol. 2006;13(2): 83–88.doi:10.1097/01.pap.0000213007.48479.94.00125480-200603000-00004 [pii].
210. Breiteneder-Geleff S, Matsui K, Soleiman A, Meraner P, Poczewski H, Kalt R, et al. Podoplanin, novel 43-kd membrane protein of glomerular epithelial cells, is down-regulated in puromycin nephrosis. Am J Pathol. 1997;151(4):1141–52.
211. Wetterwald A, Hoffstetter W, Cecchini MG, Lanske B, Wagner C, Fleisch H et al. Characterization and cloning of the E11 antigen, a marker expressed by rat osteoblasts and osteocytes. Bone. 1996;18(2): 125–132. doi:8756328295004572 [pii].
212. Al-Rawi MAA, Mansel RE, Jiang WG. Molecular and cellular mechanisms of lymphangiogenesis. Eur J Surg Oncol. 2005;31(2):117–21. doi:10.1016/j.ejso.2004.08.015.
213. Ramirez MI, Millien G, Hinds A, Cao Y, Seldin DC, Williams MC. T1alpha, a lung type I cell differentiation gene, is required for normal lung cell proliferation and alveolus formation at birth. Dev Biol. 2003;256(1):61–72.
214. Schacht V, Ramirez MI, Hong YK, Hirakawa S, Feng D, Harvey N, et al. T1alpha/podoplanin deficiency disrupts normal lymphatic vasculature formation and causes lymphedema. EMBO J. 2003;22(14):3546–56. doi:10.1093/emboj/cdg342.
215. Evangelou E, Kyzas PA, Trikalinos TA. Comparison of the diagnostic accuracy of lymphatic endothelium markers: bayesian approach. Mod Pathol. 2005;18(11):1490–1497. doi: 10.1038/modpathol.3800457.
216. Kahn HJ, Marks A. A new monoclonal antibody, D2-40, for detection of lymphatic invasion in primary tumors. Lab Investig. 2002;82(9):1255–7.
217. Dadras SS, Paul T, Bertoncini J, Brown LF, Muzikansky A, Jackson DG et al. Tumor lymphangiogenesis: a novel prognostic indicator for cutaneous melanoma metastasis and survival. Am J Pathol. 2003;162(6):1951–1960. doi: 10.1016/S0002-9440(10)64328-3.
218. Dumoff KL, Chu C, Xu X, Pasha T, Zhang PJ, Acs G. Low D2-40 immunoreactivity correlates with lymphatic invasion and nodal metastasis in early-stage squamous cell carcinoma of the uterine cervix. Mod Pathol. 2005;18(1):97–104. doi: 10.1038/modpathol.3800269.
219. Schmid K, Birner P, Gravenhorst V, End A, Geleff S. Prognostic value of lymphatic and blood vessel invasion in neuroendocrine tumors of the lung. Am J Surg Pathol. 2005;29(3):324–328. doi:00000478-200503000-00005 [pii].
220. Sinzelle E, Duong Van Huyen JP, Breiteneder-Geleff S, Braunberger E, Deloche A, Kerjaschki D, et al. Intrapericardial lymphangioma with podoplanin immunohistochemical characterization of lymphatic endothelial cells. Histopathology. 2000;37(1):93–94. doi:his955-8 [pii].
221. Schacht V, Dadras SS, Johnson LA, Jackson DG, Hong YK, Detmar M. Up-regulation of the lymphatic marker podoplanin, a mucin-type transmembrane glycoprotein, in human squamous cell carcinomas and germ cell tumors. Am J Pathol. 2005;166(3):913–21. doi:10.1016/S0002-9440(10)62311-5.
222. Kahn HJ, Bailey D, Marks A. Monoclonal antibody D2-40, a new marker of lymphatic endothelium, reacts with Kaposi's sarcoma and a subset of angiosarcomas. Mod Pathol. 2002;15(4):434–40. doi:10.1038/modpathol.3880543.
223. Roy S, Chu A, Trojanowski JQ, Zhang PJ. D2-40, a novel monoclonal antibody against the M2A antigen as a marker to distinguish hemangioblastomas from renal cell carcinomas. Acta Neuropathol. 2005;109(5):497–502. doi:10.1007/s00401-005-0999-3.

224. Chu AY, Litzky LA, Pasha TL, Acs G, Zhang PJ. Utility of D2-40, a novel mesothelial marker, in the diagnosis of malignant mesothelioma. Mod Pathol. 2005;18(1): 105–110. doi:10.1038/modpathol. 3800259 [pii].
225. Ordonez NG. The diagnostic utility of immunohistochemistry in distinguishing between epithelioid mesotheliomas and squamous carcinomas of the lung: a comparative study. Mod Pathol. 2006;19(3):417–428. doi: 10.1038/modpathol.3800544.
226. Padgett DM, Cathro HP, Wick MR, Mills SE. Podoplanin is a better immunohistochemical marker for sarcomatoid mesothelioma than calretinin. Am J Surg Pathol. 2008;32(1):123–127. doi:10.1097/PAS.0b013e31814faacf. 00000478-200801000-00017 [pii].
227. Biermann K, Klingmuller D, Koch A, Pietsch T, Schorle H, Buttner R et al. Diagnostic value of markers M2A, OCT3/4, AP-2gamma, PLAP and c-KIT in the detection of extragonadal seminomas. Histopathology. 2006;49(3):290–297. doi: 10.1111/j.1365-2559.2006.02496.x.
228. Bailey D, Marks A, Ştratis M, Baumal R. Immunohistochemical staining of germ cell tumors and intratubular malignant germ cells of the testis using antibody to placental alkaline phosphatase and a monoclonal anti-seminoma antibody. Mod Pathol. 1991;4(2):167–71.
229. Iczkowski KA, Butler SL, Shanks JH, Hossain D, Schall A, Meiers I et al. Trials of new germ cell immunohistochemical stains in 93 extragonadal and metastatic germ cell tumors. Hum Pathol. 2008;39(2):275–281. doi: 10.1016/j.humpath.2007.07.002.
230. Idrees M, Saxena R, Cheng L, Ulbright TM, Badve S. Podoplanin, a novel marker for seminoma: a comparison study evaluating immunohistochemical expression of podoplanin and OCT3/4. Ann Diagn Pathol. 2010;14(5):331–336. doi: 10.1016/j.anndiagpath.2010.05.008.
231. Yu H, Pinkus GS, Hornick JL. Diffuse membranous immunoreactivity for podoplanin (D2-40) distinguishes primary and metastatic seminomas from other germ cell tumors and metastatic neoplasms. Am J Clin Pathol. 2007;128(5):767–775. doi: 10.1309/4GMREAULY257R3AY. 0618U45P37311296 [pii].
232. Marks A, Sutherland DR, Bailey D, Iglesias J, Law J, Lei M, et al. Characterization and distribution of an oncofetal antigen (M2A antigen) expressed on testicular germ cell tumours. Br J Cancer. 1999;80(3–4):569–78. doi:10.1038/sj.bjc.6690393.
233. De Cat B, David G. Developmental roles of the glypicans. Semin Cell Dev Biol. 2001;12(2):117–125. doi:10.1006/scdb.2000.0240. S1084-9521(00)90240-4 [pii].
234. De Cat B, Muyldermans SY, Coomans C, Degeest G, Vanderschueren B, Creemers J, et al. Processing by proprotein convertases is required for glypican-3 modulation of cell survival, Wnt signaling, and gastrulation movements. J Cell Biol. 2003;163(3):625–35. doi:10.1083/jcb.200302152.
235. Maeda D, Ota S, Takazawa Y, Aburatani H, Nakagawa S, Yano T et al. Glypican-3 expression in clear cell adenocarcinoma of the ovary. Mod Pathol. 2009;22(6):824–832. doi: 10.1038/modpathol.2009.40.
236. Gonzalez AD, Kaya M, Shi W, Song H, Testa JR, Penn LZ, et al. OCI-5/GPC3, a glypican encoded by a gene that is mutated in the Simpson-Golabi-Behmel overgrowth syndrome, induces apoptosis in a cell line-specific manner. J Cell Biol. 1998;141(6): 1407–14.
237. Søren Nielsen RR. NordiQC Glypican3 Assessment Run 42 2014. NordiQC, http://www.nordiqc.org/Run-42-B18-H6/Assessment/Run42_GLP3.pdf. 2014.
238. Preda O, Nicolae A, Aneiros-Fernandez J, Borda A, Nogales FF. Glypican 3 is a sensitive, but not a specific, marker for the diagnosis of yolk sac tumours. Histopathology.2011;58(2):312–4.doi:10.1111/j.1365-2559.2010.03735.x.
239. Khan S, Blackburn M, Mao DL, Huber R, Schlessinger D, Fant M. Glypican-3 (GPC3) expression in human placenta: localization to the differentiated syncytiotrophoblast. Histol Histopathol. 2001;16(1):71–8.
240. Esheba GE, Pate LL, Longacre TA. Oncofetal protein glypican-3 distinguishes yolk sac tumor from clear cell carcinoma of the ovary. Am J Surg Pathol. 2008;32(4):600–7. doi:10.1097/PAS.0b013e31815a565a.
241. Zynger DL, Dimov ND, Luan C, Teh BT, Yang XJ. Glypican 3: a novel marker in testicular germ cell tumors. Am J Surg Pathol. 2006;30(12):1570–1575. doi:10.1097/01.pas.0000213322.89670.48. 00000478-200612000-00010 [pii].
242. Zynger DL, Everton MJ, Dimov ND, Chou PM, Yang XJ. Expression of glypican 3 in ovarian and extragonadal germ cell tumors. Am J Clin Pathol. 2008;130(2):224–230. doi: 10.1309/8DN7DQRDFB4QNH3N.
243. Murthy SS, Shen T, De Rienzo A, Lee WC, Ferriola PC, Jhanwar SC, et al. Expression of GPC3, an X-linked recessive overgrowth gene, is silenced in malignant mesothelioma. Oncogene. 2000;19(3): 410–6. doi:10.1038/sj.onc.1203322.
244. Lin H, Huber R, Schlessinger D, Morin PJ. Frequent silencing of the GPC3 gene in ovarian cancer cell lines. Cancer Res. 1999;59(4):807–10.
245. Xiang YY, Ladeda V, Filmus J. Glypican-3 expression is silenced in human breast cancer. Oncogene. 2001;20(50):7408–12. doi:10.1038/sj.onc.1204925.
246. Saikali Z, Sinnett D. Expression of glypican 3 (GPC3) in embryonal tumors. Int J Cancer. 2000;89(5):418–422. doi:10.1002/1097-0215(20000920)89:5<418::AID-IJC4>3.0.CO;2-I. [pii].
247. Lage H, Dietel M, Froschle G, Reymann A. Expression of the novel mitoxantrone resistance associated gene MXR7 in colorectal malignancies. Int J Clin Pharmacol Ther. 1998;36(1):58–60.

248. Hsu HC, Cheng W, Lai PL. Cloning and expression of a developmentally regulated transcript MXR7 in hepatocellular carcinoma: biological significance and temporospatial distribution. Cancer Res. 1997;57(22):5179–84.
249. Zhu ZW, Friess H, Wang L, Abou-Shady M, Zimmermann A, Lander AD, et al. Enhanced glypican-3 expression differentiates the majority of hepatocellular carcinomas from benign hepatic disorders. Gut. 2001;48(4):558–64.
250. Baumhoer D, Tornillo L, Stadlmann S, Roncalli M, Diamantis EK, Terracciano LM. Glypican 3 expression in human nonneoplastic, preneoplastic, and neoplastic tissues: a tissue microarray analysis of 4,387 tissue samples. Am J Clin Pathol. 2008;129(6):899–906. doi:10.1309/HCQWPWD50XHD2DW6.
251. Matsuyama TA, Kushima M, Yamochi-Onizuka T, Ota H. Placental yolk sac tumor with divergent endodermal differentiation. Int J Gynecol Pathol. 2004;23(4):398–402. doi:00004347-200410000-00014 [pii].
252. Drut R, Mortera M, Drut RM. Yolk sac tumor of the placenta in Wiedemann-Beckwith syndrome. Pediatr Dev Pathol. 1998;1(6):534–537. doi:PD-97-32 [pii].
253. Vandeputte M, Sobis H, Billiau A, Van de Maele B, Leyten R. In utero tumor induction by murine sarcoma virus (Moloney) in the rat. I Biological characteristics. Int J Cancer. 1973;11(3):536–42.
254. Sobis H, Vandeputte M. In utero tumor induction by murine sarcoma virus (Moloney) in the rat. II. Histological and ultrastructural characteristics. Int J Cancer. 1973;11(3):543–54.
255. Ou-Yang RJ, Hui P, Yang XJ, Zynger DL. Expression of glypican 3 in placental site trophoblastic tumor. Diagn Pathol. 2010;5:64. doi: 10.1186/1746-1596-5-64. 1746-1596-5-64 [pii].
256. Capurro M, Wanless IR, Sherman M, Deboer G, Shi W, Miyoshi E et al. Glypican-3: a novel serum and histochemical marker for hepatocellular carcinoma. Gastroenterology. 2003;125(1):89–97. doi:S0016508503006899 [pii].
257. Yamauchi N, Watanabe A, Hishinuma M, Ohashi K, Midorikawa Y, Morishita Y et al. The glypican 3 oncofetal protein is a promising diagnostic marker for hepatocellular carcinoma. Mod Pathol. 2005;18(12):1591–1598. doi: 10.1038/modpathol.3800436.
258. Liu H, Li P, Zhai Y, Qu CF, Zhang LJ, Tan YF, et al. Diagnostic value of glypican-3 in serum and liver for primary hepatocellular carcinoma. World J Gastroenterol. 2010;16(35):4410–5.
259. Nakatsura T, Kageshita T, Ito S, Wakamatsu K, Monji M, Ikuta Y et al. Identification of glypican-3 as a novel tumor marker for melanoma. Clin Cancer Res. 2004;10(19):6612–6621. doi: 10.1158/1078-0432.CCR-04-0348. 10/19/6612 [pii].
260. Umezu T, Shibata K, Kajiyama H, Yamamoto E, Nawa A, Kikkawa F. Glypican-3 expression predicts poor clinical outcome of patients with early-stage clear cell carcinoma of the ovary. J Clin Pathol. 2010;63(11):962–966. doi: 10.1136/jcp.2010.080234.
261. Aviel-Ronen S, Lau SK, Pintilie M, Lau D, Liu N, Tsao MS et al. Glypican-3 is overexpressed in lung squamous cell carcinoma, but not in adenocarcinoma. Mod Pathol. 2008;21(7):817–825. doi: 10.1038/modpathol.2008.37.
262. Ushiku T, Uozaki H, Shinozaki A, Ota S, Matsuzaka K, Nomura S et al. Glypican 3-expressing gastric carcinoma: distinct subgroup unifying hepatoid, clear-cell, and alpha-fetoprotein-producing gastric carcinomas. Cancer Sci. 2009;100(4):626–632. doi: 10.1111/j.1349-7006.2009.01108.x.
263. Ota S, Hishinuma M, Yamauchi N, Goto A, Morikawa T, Fujimura T, et al. Oncofetal protein glypican-3 in testicular germ-cell tumor. Virchows Arch. 2006;449(3):308–14. doi:10.1007/s00428-006-0238-x.
264. Zynger DL, McCallum JC, Luan C, Chou PM, Yang XJ. Glypican 3 has a higher sensitivity than alpha-fetoprotein for testicular and ovarian yolk sac tumour: immunohistochemical investigation with analysis of histological growth patterns. Histopathology. 2010;56(6):750–757. doi: 10.1111/j.1365-2559.2010.03553.x.
265. Cahoreau C, Klett D, Combarnous Y. Structure-function relationships of glycoprotein hormones and their subunits' ancestors. Front Endocrinol. 2015;6:26. doi:10.3389/fendo.2015.00026.
266. Talmadge K, Boorstein WR, Fiddes JC. The human genome contains seven genes for the beta-subunit of chorionic gonadotropin but only one gene for the beta-subunit of luteinizing hormone. DNA. 1983;2(4):281–9.
267. Cole LA. Biological functions of hCG and hCG-related molecules. Reprod biol Endocrinol: RBE. 2010;8:102. doi:10.1186/1477-7827-8-102.
268. Lempiainen A, Sankila A, Hotakainen K, Haglund C, Blomqvist C, Stenman UH. Expression of human chorionic gonadotropin in testicular germ cell tumors. Urol Oncol. 2014;32(5):727–34. doi:10.1016/j.urolonc.2013.11.007.
269. Hes O, Pivovarcikova K, Stehlik J, Martinek P, Vanecek T, Bauleth K, et al. Choriogonadotropin positive seminoma-a clinicopathological and molecular genetic study of 15 cases. Ann Diagn Pathol. 2014;18(2):89–94. doi:10.1016/j.anndiagpath.2013.12.004.
270. Shih IM. Trophogram, an immunohistochemistry-based algorithmic approach, in the differential diagnosis of trophoblastic tumors and tumorlike lesions. Ann Diagn Pathol. 2007;11(3):228–34. doi:10.1016/j.anndiagpath.2007.04.001.
271. Losch A, Kainz C. Immunohistochemistry in the diagnosis of the gestational trophoblastic disease. Acta Obstet Gynecol Scand. 1996;75(8):753–6.
272. Stenman UH, Alfthan H, Hotakainen K. Human chorionic gonadotropin in cancer. Clin Biochem. 2004;37(7):549–61. doi:10.1016/j.clinbiochem.2004.05.008.
273. Lawless ME, Jour G, Hoch BL, Rendi MH. Beta-human chorionic gonadotropin expression in recurrent and metastatic giant cell tumors of bone: a potential mimicker of germ cell tumor. Int J Surg Pathol. 2014;22(7):617–22. doi:10.1177/1066896914534466.

274. Moll R, Franke WW, Schiller DL, Geiger B, Krepler R. The catalog of human cytokeratins: patterns of expression in normal epithelia, tumors and cultured cells. Cell. 1982;31(1):11–24.
275. Rogers GE, Powell BC. Organization and expression of hair follicle genes. J Invest Dermatol. 1993; 101(1 Suppl):50S–5S.
276. Schweizer J, Bowden PE, Coulombe PA, Langbein L, Lane EB, Magin TM et al. New consensus nomenclature for mammalian keratins. J Cell Biol. 2006;174(2):169–74. doi:10.1083/jcb.200603161.
277. Hatzfeld M, Franke WW. Pair formation and promiscuity of cytokeratins: formation in vitro of heterotypic complexes and intermediate-sized filaments by homologous and heterologous recombinations of purified polypeptides. J Cell Biol. 1985;101(5 Pt 1): 1826–41.
278. Battifora H, Sheibani K, Tubbs RR, Kopinski MI, Sun TT. Antikeratin antibodies in tumor diagnosis. Distinction between seminoma and embryonal carcinoma. Cancer. 1984;54(5):843–8.
279. Miettinen M, Virtanen I, Talerman A. Intermediate filament proteins in human testis and testicular germ-cell tumors. Am J Pathol. 1985;120(3):402–10.
280. Denk H, Moll R, Weybora W, Lackinger E, Vennigerholz F, Beham A, et al. Intermediate filaments and desmosomal plaque proteins in testicular seminomas and non-seminomatous germ cell tumours as revealed by immunohistochemistry. Virchows Archiv A Pathol Anat Histopathol. 1987;410(4):295–307.
281. Lifschitz-Mercer B, Fogel M, Moll R, Jacob N, Kushnir I, Livoff A, et al. Intermediate filament protein profiles of human testicular non-seminomatous germ cell tumors: correlation of cytokeratin synthesis to cell differentiation. Differentiation. 1991;48(3):191–8.
282. Cheville JC, Rao S, Iczkowski KA, Lohse CM, Pankratz VS. Cytokeratin expression in seminoma of the human testis. Am J Clin Pathol. 2000;113(4):583–8. doi:10.1309/3QLC-5MF1-JYXU-A5XX.
283. Ramalingam P, Malpica A, Silva EG, Gershenson DM, Liu JL, Deavers MT. The use of cytokeratin 7 and EMA in differentiating ovarian yolk sac tumors from endometrioid and clear cell carcinomas. Am J Surg Pathol. 2004;28(11):1499–505.
284. Walters JR, Howard A, Rumble HE, Prathalingam SR, Shaw-Smith CJ, Legon S. Differences in expression of homeobox transcription factors in proximal and distal human small intestine. Gastroenterology. 1997;113(2):472–7.
285. Suh N, Yang XJ, Tretiakova MS, Humphrey PA, Wang HL. Value of CDX2, villin, and alpha-methylacyl coenzyme A racemase immunostains in the distinction between primary adenocarcinoma of the bladder and secondary colorectal adenocarcinoma. Mod Pathol. 2005;18(9):1217–22. doi:10.1038/modpathol.3800407.
286. Jedrusik A, Cox A, Wicher K, Glover DM, Zernicka-Goetz M. Maternal-zygotic knockout reveals a critical role of Cdx2 in the morula to blastocyst transition. Dev Biol. 2015;398(2):147–52. doi:10.1016/j.ydbio.2014.12.004.
287. Rayon T, Menchero S, Nieto A, Xenopoulos P, Crespo M, Cockburn K, et al. Notch and hippo converge on Cdx2 to specify the trophectoderm lineage in the mouse blastocyst. Dev Cell. 2014;30(4):410–22. doi:10.1016/j.devcel.2014.06.019.
288. Chawengsaksophak K, de Graaff W, Rossant J, Deschamps J, Beck F. Cdx2 is essential for axial elongation in mouse development. Proc Natl Acad Sci U S A. 2004;101(20):7641–5. doi:10.1073/pnas.0401654101.
289. Moskaluk CA, Zhang H, Powell SM, Cerilli LA, Hampton GM, Frierson HF. Cdx2 protein expression in normal and malignant human tissues: an immunohistochemical survey using tissue microarrays. Mod Pathol: Off J US Canadian Acad Pathol Inc. 2003; 16(9):913–9.
290. Nicolae A, Goyenaga P, McCluggage WG, Preda O, Nogales FF. Endometrial intestinal metaplasia: a report of two cases, including one associated with cervical intestinal and pyloric metaplasia. Int J Gynecol Pathol. 2011;30(5):492–6. doi:10.1097/PGP.0b013e318211d586.
291. Søren Nielsen RR, Mogens Vyberg. NordiQC CDX2 Assessment Run 38 2013. NordiQC, http://www.nordiqc.org/Run-38/Assessment/assessment-38-CDX2.htm. 2013.
292. Werling RW, Yaziji H, Bacchi CE, Gown AM. CDX2, a highly sensitive and specific marker of adenocarcinomas of intestinal origin: an immunohistochemical survey of 476 primary and metastatic carcinomas. Am J Surg Pathol. 2003;27(3):303–10.
293. Groisman GM, Bernheim J, Halpern M, Brazowsky E, Meir A. Expression of the intestinal marker Cdx2 in secondary adenocarcinomas of the colorectum. Arch Pathol Lab Med. 2005;129(7):920–3. doi:10.1043/1543-2165(2005)129[920:EOTIMC]2.0.CO;2.
294. Saad RS, Ghorab Z, Khalifa MA, Xu M. CDX2 as a marker for intestinal differentiation: Its utility and limitations. World J Gastrointest Surg. 2011;3(11): 159–66. doi:10.4240/wjgs.v3.i11.159.
295. Houghton O, Connolly LE, McCluggage WG. Morules in endometrioid proliferations of the uterus and ovary consistently express the intestinal transcription factor CDX2. Histopathology. 2008;53(2):156–65. doi:10.1111/j.1365-2559.2008.03083.x.
296. Osunkoya AO, Epstein JI. Primary mucin-producing urothelial-type adenocarcinoma of prostate: report of 15 cases. Am J Surg Pathol. 2007;31(9):1323–9. doi:10.1097/PAS.0b013e31802ff7c4.
297. Paner GP, McKenney JK, Barkan GA, Yao JL, Frankel WL, Sebo TJ, et al. Immunohistochemical analysis in a morphologic spectrum of urachal epithelial neoplasms: diagnostic implications and pitfalls. Am J Surg Pathol. 2011;35(6):787–98. doi:10.1097/PAS.0b013e3182189c11.
298. Franchi A, Palomba A, Miligi L, Ranucci V, Innocenti DR, Simoni A, et al. Intestinal metaplasia of the sino-

nasal mucosa adjacent to intestinal-type adenocarcinoma. A morphologic, immunohistochemical, and molecular study. Virchows Arch. 2015;466(2):161–8. doi:10.1007/s00428-014-1696-1.
299. Saqi A, Alexis D, Remotti F, Bhagat G. Usefulness of CDX2 and TTF-1 in differentiating gastrointestinal from pulmonary carcinoids. Am J Clin Pathol. 2005;123(3):394–404. doi:10.1309/UKN6-PVRK-XHG4-22DA.
300. Desouki MM, Lioyd J, Xu H, Cao D, Barner R, Zhao C. CDX2 may be a useful marker to distinguish primary ovarian carcinoid from gastrointestinal metastatic carcinoids to the ovary. Hum Pathol. 2013;44(11):2536–41. doi:10.1016/j.humpath.2013.06.014.
301. Scholl C, Bansal D, Dohner K, Eiwen K, Huntly BJ, Lee BH, et al. The homeobox gene CDX2 is aberrantly expressed in most cases of acute myeloid leukemia and promotes leukemogenesis. J Clin Invest. 2007;117(4):1037–48. doi:10.1172/JCI30182.
302. Riedt T, Ebinger M, Salih HR, Tomiuk J, Handgretinger R, Kanz L, et al. Aberrant expression of the homeobox gene CDX2 in pediatric acute lymphoblastic leukemia. Blood. 2009;113(17):4049–51. doi:10.1182/blood-2008-12-196634.
303. Khurana S, George SP. Regulation of cell structure and function by actin-binding proteins: villin's perspective. FEBS Lett. 2008;582(14):2128–39. doi:10.1016/j.febslet.2008.02.040.
304. Wang Y, Srinivasan K, Siddiqui MR, George SP, Tomar A, Khurana S. A novel role for villin in intestinal epithelial cell survival and homeostasis. J Biol Chem. 2008;283(14):9454–64. doi:10.1074/jbc.M707962200.
305. Rousseau-Merck MF, Simon-Chazottes D, Arpin M, Pringault E, Louvard D, Guenet JL, et al. Localization of the villin gene on human chromosome 2q35-q36 and on mouse chromosome 1. Hum Genet. 1988; 78(2):130–3.
306. Phillips MJ, Azuma T, Meredith SL, Squire JA, Ackerley CA, Pluthero FG, et al. Abnormalities in villin gene expression and canalicular microvillus structure in progressive cholestatic liver disease of childhood. Lancet. 2003;362(9390):1112–9. doi:10.1016/S0140-6736(03)14467-4.
307. Wong HH, Chu P. Immunohistochemical features of the gastrointestinal tract tumors. J Gastrointest Oncol. 2012;3(3):262–84.
308. Bacchi CE, Gown AM. Distribution and pattern of expression of villin, a gastrointestinal-associated cytoskeletal protein, in human carcinomas: a study employing paraffin-embedded tissue. Lab Investig. 1991;64(3):418–24.
309. West AB, Isaac CA, Carboni JM, Morrow JS, Mooseker MS, Barwick KW. Localization of villin, a cytoskeletal protein specific to microvilli, in human ileum and colon and in colonic neoplasms. Gastroenterology. 1988;94(2):343–52.
310. Moll R, Robine S, Dudouet B, Louvard D. Villin: a cytoskeletal protein and a differentiation marker expressed in some human adenocarcinomas. Virchows Archiv B Cell Pathol Includ Mol Pathol. 1987;54(3):155–69.
311. Wennerberg AE, Nalesnik MA, Coleman WB. Hepatocyte paraffin 1: a monoclonal antibody that reacts with hepatocytes and can be used for differential diagnosis of hepatic tumors. Am J Pathol. 1993;143(4):1050–4.
312. Butler SL, Dong H, Cardona D, Jia M, Zheng R, Zhu H, et al. The antigen for Hep Par 1 antibody is the urea cycle enzyme carbamoyl phosphate synthetase 1. Lab Investig. 2008;88(1):78–88. doi:10.1038/labinvest.3700699.
313. Pang Y, von Turkovich M, Wu H, Mitchell J, Mount S, Taatjes D, et al. The binding of thyroid transcription factor-1 and hepatocyte paraffin 1 to mitochondrial proteins in hepatocytes: a molecular and immunoelectron microscopic study. Am J Clin Pathol. 2006;125(5):722–6. doi:10.1309/EBCB-6H54-K1N2-P9QL.
314. Haberle J, Shchelochkov OA, Wang J, Katsonis P, Hall L, Reiss S, et al. Molecular defects in human carbamoyl phosphate synthetase I: mutational spectrum, diagnostic and protein structure considerations. Hum Mutat. 2011;32(6):579–89. doi:10.1002/humu.21406.
315. Jeung JA, Coran JJ, Liu C, Cardona DM. Hepatocyte paraffin 1 antigen as a biomarker for early diagnosis of Barrett esophagus. Am J Clin Pathol. 2012;137(1):111–20. doi:10.1309/AJCPYOBVGS4CGA8Y.
316. Kakar S, Muir T, Murphy LM, Lloyd RV, Burgart LJ. Immunoreactivity of Hep Par 1 in hepatic and extrahepatic tumors and its correlation with albumin in situ hybridization in hepatocellular carcinoma. Am J Clin Pathol. 2003;119(3):361–6.
317. Fan Z, van de Rijn M, Montgomery K, Rouse RV. Hep par 1 antibody stain for the differential diagnosis of hepatocellular carcinoma: 676 tumors tested using tissue microarrays and conventional tissue sections. Mod Pathol. 2003;16(2):137–44. doi:10.1097/01.MP.0000052103.13730.20.
318. Thorwarth A, Schnittert-Hubener S, Schrumpf P, Muller I, Jyrch S, Dame C, et al. Comprehensive genotyping and clinical characterisation reveal 27 novel NKX2-1 mutations and expand the phenotypic spectrum. J Med Genet. 2014;51(6):375–87. doi:10.1136/jmedgenet-2013-102248.
319. Trueba SS, Auge J, Mattei G, Etchevers H, Martinovic J, Czernichow P, et al. PAX8, TITF1, and FOXE1 gene expression patterns during human development: new insights into human thyroid development and thyroid dysgenesis-associated malformations. J Clin Endocrinol Metab. 2005;90(1):455–62. doi:10.1210/jc.2004-1358.
320. Weir BA, Woo MS, Getz G, Perner S, Ding L, Beroukhim R, et al. Characterizing the cancer genome in lung adenocarcinoma. Nature. 2007;450(7171):893–8. doi:10.1038/nature06358.
321. Shetty VB, Kiraly-Borri C, Lamont P, Bikker H, Choong CS. NKX2-1 mutations in brain-lung-

thyroid syndrome: a case series of four patients. J Pediat Endocrinol Metabol: JPEM. 2014;27(3–4):373–8. doi:10.1515/jpem-2013-0109.
322. Zamecnik J, Chanova M, Kodet R. Expression of thyroid transcription factor 1 in primary brain tumours. J Clin Pathol. 2004;57(10):1111–3. doi:10.1136/jcp.2004.017467.
323. Matoso A, Singh K, Jacob R, Greaves WO, Tavares R, Noble L, et al. Comparison of thyroid transcription factor-1 expression by 2 monoclonal antibodies in pulmonary and nonpulmonary primary tumors. Appl Immunohistochem Mol Morphol. 2010;18(2):142–9. doi:10.1097/PAI.0b013e3181bdf4e7.
324. Comperat E, Zhang F, Perrotin C, Molina T, Magdeleinat P, Marmey B, et al. Variable sensitivity and specificity of TTF-1 antibodies in lung metastatic adenocarcinoma of colorectal origin. Mod Pathol. 2005;18(10):1371–6. doi:10.1038/modpathol.3800422.
325. Gu K, Shah V, Ma C, Zhang L, Yang M. Cytoplasmic immunoreactivity of thyroid transcription factor-1 (clone 8G7G3/1) in hepatocytes: true positivity or cross-reaction? Am J Clin Pathol. 2007;128(3):382–8. doi:10.1309/CADCVWHR2QF6JMVN.
326. Fischer S, Asa SL. Application of immunohistochemistry to thyroid neoplasms. Arch Pathol Lab Med. 2008;132(3):359–72. doi:10.1043/1543-2165(2008)132[359:AOITTN]2.0.CO;2.
327. Nakamura N, Miyagi E, Murata S, Kawaoi A, Katoh R. Expression of thyroid transcription factor-1 in normal and neoplastic lung tissues. Mod Pathol: Off J US Canadian Acad Pathol Inc. 2002;15(10):1058–67.
328. Rajpert-De Meyts E, Jacobsen GK, Bartkova J, Aubry F, Samson M, Bartek J, et al. The immunohistochemical expression pattern of Chk2, p53, p19INK4d, MAGE-A4 and other selected antigens provides new evidence for the premeiotic origin of spermatocytic seminoma. Histopathology. 2003; 42(3):217–26.
329. Kang JL, Rajpert-De Meyts E, Skakkebaek NE. Immunoreactive neuron-specific enolase (NSE) is expressed in testicular carcinoma-in-situ. J Pathol. 1996;178(2):161–5. doi:10.1002/(SICI)1096-9896(199602)178:2<161::AID-PATH452>3.0.CO;2-H.
330. Zeeman AM, Stoop H, Boter M, Gillis AJ, Castrillon DH, Oosterhuis JW, et al. VASA is a specific marker for both normal and malignant human germ cells. Lab Investig. 2002;82(2):159–66.
331. Aggarwal N, Parwani AV. Spermatocytic seminoma. Arch Pathol Lab Med. 2009;133(12):1985–8. doi:10.1043/1543-2165-133.12.1985.
332. Castrillon DH, Quade BJ, Wang TY, Quigley C, Crum CP. The human VASA gene is specifically expressed in the germ cell lineage. Proc Natl Acad Sci U S A. 2000;97(17):9585–90. doi:10.1073/pnas.160274797.
333. Ko HS, Fast P, McBride W, Staudt LM. A human protein specific for the immunoglobulin octamer DNA motif contains a functional homeobox domain. Cell. 1988;55(1):135–44.
334. Clark J, Rocques PJ, Crew AJ, Gill S, Shipley J, Chan AM, et al. Identification of novel genes, SYT and SSX, involved in the t(X;18)(p11.2;q11.2) translocation found in human synovial sarcoma. Nat Genet. 1994;7(4):502–8. doi:10.1038/ng0894-502.
335. Martelange V, De Smet C, De Plaen E, Lurquin C, Boon T. Identification on a human sarcoma of two new genes with tumor-specific expression. Cancer Res. 2000;60(14):3848–55.
336. Lim J, Goriely A, Turner GD, Ewen KA, Jacobsen GK, Graem N, et al. OCT2, SSX and SAGE1 reveal the phenotypic heterogeneity of spermatocytic seminoma reflecting distinct subpopulations of spermatogonia. J Pathol. 2011;224(4):473–83. doi:10.1002/path.2919.
337. Looijenga LH. Spermatocytic seminoma: toward further understanding of pathogenesis. J Pathol. 2011;224(4):431–3. doi:10.1002/path.2939.
338. Cybulski C, Gorski B, Huzarski T, Masojc B, Mierzejewski M, Debniak T, et al. CHEK2 is a multiorgan cancer susceptibility gene. Am J Hum Genet. 2004;75(6):1131–5. doi:10.1086/426403.
339. Rogner UC, Wilke K, Steck E, Korn B, Poustka A. The melanoma antigen gene (MAGE) family is clustered in the chromosomal band Xq28. Genomics. 1995;29(3):725–31. doi:10.1006/geno.1995.9945.
340. Bishop EF, Badve S, Morimiya A, Saxena R, Ulbright TM. Apoptosis in spermatocytic and usual seminomas: a light microscopic and immunohistochemical study. Mod Pathol. 2007;20(10):1036–44. doi:10.1038/modpathol.3800933.
341. Chen YT, Boyer AD, Viars CS, Tsang S, Old LJ, Arden KC. Genomic cloning and localization of CTAG, a gene encoding an autoimmunogenic cancer-testis antigen NY-ESO-1, to human chromosome Xq28. Cytogenet Cell Genet. 1997;79(3–4):237–40.
342. Kao CS, Badve SS, Ulbright TM. The utility of immunostaining for NUT, GAGE7 and NY-ESO-1 in the diagnosis of spermatocytic seminoma. Histopathology. 2014;65(1):35–44. doi:10.1111/his.12365.

Management of Germ Cell Tumors

5

Manish Kohli, Ben Y. Zhang, and Brian A. Costello

5.1 Introduction

Germ cell tumors are one of the few solid tumors in which long-term cure has been achieved in an overwhelming proportion of patients, even when the initial presenting occurs with widespread metastasis. Testicular germ cell tumors form the bulk of all germ cell tumors with 95 % of tumors originating in the testis being germ cell tumors. Fewer than 10 % of germ cell tumors arise from extra-gonadal primary sites with the most common sites being mediastinum and retroperitoneum in both males and females. In females, additionally ovarian-based germ cell tumors are observed. In 2015, the estimated new cases in the USA of by far the most common type of germ cell tumor, testicular germ cell tumors, were 8430 with an estimated death of 380 cases [1]. Over the past four decades, the cure rates for this cancer have dramatically increased from 65 % during the 1970s [2] to greater than 95 % 5-year disease-free survival [3] by using either single or integrating combined modality treatments which include surgery, radiation, and/or chemotherapy. This makes testicular germ cell tumors a role model tumor type to follow. The two outstanding reasons for this success have been the availability of effective treatment interventions and the value of integrating highly sensitive and specific tumor markers in the diagnosis, staging, monitoring, and management of germ cell tumors. This chapter will summarize the management of germ cell tumors, with an overwhelming emphasis on the management of testicular germ cell tumors since it is clinically the most frequent observed and treated type of germ cell tumor.

5.2 Clinical Presentations of Testicular Germ Cell Tumors

The initial presentation of most testicular germ cell tumors is with a nodule or swelling of one testicle accompanied with heaviness and discomfort with or without a dragging sensation in the lower abdomen. The median age for testicular germ cell tumors is between 15 and 35 years of age. Occasionally patients may also present with symptoms of orchitis or epididymo-orchitis which are accompanied with acute and abrupt episode of pain with or without accompanied swelling. Symptoms of metastatic disease on initial presentation can include a growing lump in

M. Kohli, MD (✉) • B.A. Costello, MD
Mayo Clinic, 200, 1st St. SW,
Rochester, MN 55905, USA
e-mail: Kohli.manish@mayo.edu; Costello.Brian@mayo.edu

B.Y. Zhang, MD
Division of Medical Oncology, Mayo Clinic,
200, 1st St. SW, Rochester, MN 55905, USA
e-mail: Zhang.Ben@mayo.edu

F.F. Nogales, R.E. Jimenez (eds.), *Pathology and Biology of Human Germ Cell Tumors*,
DOI 10.1007/978-3-662-53775-6_5

the left neck region, cough and shortness of breathing, loss of weight and appetite, neurological symptoms similar to stroke or peripheral neuropathies, recent onset of breast tenderness and gynecomastia, or bone pains [4]. Examinations of a testicular swelling with these symptoms raise the suspicion of testicular cancer prompting appropriate further work-up.

General physical examination of suspicious testicular no-fluid-filled swelling followed by diagnostic scrotal ultrasound evaluation can diagnose with a high degree of accuracy malignant from nonmalignant testicular tumors. Ultrasonography-based findings for seminomas may include hypoechoic masses without cystic areas, while non-seminoma ultrasound findings can include inhomogeneous masses with cystic areas, calcifications, and masses with indistinct margins. Such findings are followed by measuring blood-based germ cell tumor-specific markers including alpha-fetoprotein (AFP), beta-human chorionic gonadotropin (β-hCG), and imaging scans including a high-resolution computed tomography (CT) scan of the abdomen and pelvis [3]. Based on test results that suggest a germ cell tumor, a unilateral radical inguinal orchiectomy should be entertained, as an orchiectomy provides histological confirmation, accurate pathological subtyping, and staging of tumor apart from local control and riddance of potential sanctuary sites.

5.3 Pathology of Testicular Germ Cell Tumor

The orchiectomy specimen provides the histological classification and immunohistochemistry characteristics based on which testicular germ cell tumors are divided into seminoma and non-seminoma. Seminomas are further classified as seminoma and seminoma with syncytiotrophoblastic cells. Spermatocytic tumor is a rare category of GCT [5] that typically present in the seventh decade of life in males and are typically treated with surgical resection alone. Rarely spermatocytic tumors can present with sarcoma as well on initial presentations which carries a poorer prognosis.

Non-seminomatous testicular germ cell tumors include embryonal carcinoma, teratoma, trophoblastic (choriocarcinoma), yolk sac tumors, and mixed germ cell tumors. Both seminoma and non-seminoma represent nearly 50 % each of all testicular germ cell tumors at initial presentation, but with the non-seminomas which are generally faster-growing tumors than seminoma, it is more likely that initial presentations occur in more advanced stages compared to seminomas which present more frequently with localized stage disease. In fact 80 % of all seminomas present in the localized stages, while greater than 50 % of non-seminomas present with advanced stage disease.

Pathological classification of the tumor specimen for lympho-vascular invasion (LVI) is a key initial staging feature that has impact on subsequent management strategies. Accurate initial staging includes work-up based on integrating clinical, pathological, and radiographic features; based on the tumor extent (T), as assessed by pathology, and lymph nodal (N) and metastatic involvement (M), as assessed by CT scan of the abdomen and pelvis and radiography of the chest or other parts of the body; and based on the pre-orchiectomy symptoms and signs in each patient and post orchiectomy serum tumor marker levels. Additional scans such as magnetic resonance imaging (MRI) scans of the testes are performed not routinely but as needed for clarifying the initial work-up. The TNM staging system developed by the American Joint Committee on Cancer (AJCC) and the International Union Against Cancer (UICC) applies to all germ cell tumor staging. It is the only staging system in solid tumors which has formally incorporated levels of tumor-specific serum (S) markers into TNM staging. For testicular germ cell tumors, the marker levels are measured *after* an orchiectomy (Table 5.1). The post-orchiectomy level of the serum markers, AFP, β-HCG, and LDH, is graded between S0 and S3 and then taken together with TNM stage to form a final composite tumor stage (Tables 5.1 and 5.2). Assessing the risk for relapse and future prognosis has been extremely well categorized in the past four decades and is also stage dependent. In 1997, a consensus was reached on a uniform and validated prognostic model by the International Germ Cell Cancer Collaborative Group

Table 5.1 Staging for testicular germ cell tumors

Taken from the American Cancer Society Website: http://www.cancer.org/cancer/testicularcancer/detailedguide/testicular-cancer-staging
The TNM staging system
A staging system is a standard way for your cancer care team to sum up the extent of your cancer. Testicular cancer is staged using the TNM system created by the American Joint Committee on Cancer (AJCC). It's based on four key pieces of information:
T refers to how much the main (primary) *tumor* has spread to tissues next to the testicle
N describes how much the cancer has spread to regional (nearby) lymph *nodes*
M indicates whether the cancer has *metastasized* (spread to distant lymph nodes or other organs of the body)
S indicates the *serum* (blood) levels of tumor markers that are made by some testicular cancers
Letters or numbers appear after T, N, M, and S to provide more details about each piece of information. The numbers 0 through 4 indicate increasing severity. The letters "IS" after the T stand for in situ, which means the tumor is contained in one place and has not yet penetrated to a deeper layer of tissue. The letter X after T, N, M, or S means "cannot be assessed" because the information is not known
Primary tumor (T)
TX: The primary tumor cannot be assessed
T0: There is no evidence of primary tumor
Tis: Carcinoma in situ (noninvasive cancer cells)
T1: The tumor has not spread beyond the testicle and epididymis (the tubes next to the testicles where sperm mature). The cancer has not reached nearby blood vessels or lymph vessels. The cancer might have grown through the inner layer surrounding the testicle (tunica albuginea), but it has not reached the outer layer covering the testicle (tunica vaginalis)
T2: Similar to T1 except that the cancer has spread to blood or lymph vessels near the tumor or the tunica vaginalis
T3: The tumor is growing into the spermatic cord (which contains blood vessels, lymph vessels, nerves, and the vas deferens)
T4: The tumor is growing into the skin surrounding the testicles (scrotum)
Regional lymph nodes (N)
NX: Regional (nearby) lymph nodes cannot be assessed
N0: No spread to regional lymph nodes is seen on imaging tests
N1: The cancer has spread to at least one lymph node, but no lymph node is larger than 2 cm (about ¾ inch) across
N2: The cancer has spread to at least one lymph node that is larger than 2 cm but is not bigger than 5 cm (2 inches) across
N3: The cancer has spread to at least one lymph node that is larger than 5 cm across
If the lymph nodes were taken out during surgery, there is a slightly different classification:
pNX: Regional (nearby) lymph nodes cannot be assessed
pN0: Examination of regional lymph nodes removed with surgery reveals no cancer spread
pN1: Examination of regional lymph nodes removed with surgery reveals cancer spread in one to five lymph nodes, but no lymph node is larger than 2 cm (about ¾ inch) across
pN2: Examination of regional lymph nodes removed with surgery reveals cancer spread in at least one lymph node that is bigger than 2 cm but not larger than 5 cm across *or* spread to more than five lymph nodes that aren't bigger than 5 cm, *or* the cancer is growing out the side of a lymph node
pN3: Examination of regional lymph nodes removed with surgery reveals cancer spread in at least one lymph node that is bigger than 5 cm across
Distant metastasis (M)
M0: There is no distant metastasis (no spread to lymph nodes outside the area of the tumor or other organs, such as the lungs)
M1: Distant metastasis is present
M1a: The tumor has metastasized to distant lymph nodes or to the lung
M1b: The tumor has metastasized to other organs, such as the liver, brain, or bone

(continued)

Table 5.1 (continued)

Serum tumor markers (S)

For staging, serum (blood) levels of tumor markers are measured *after* the testicle containing the cancer has been removed with surgery

	LDH (U/l)	HCG (mIU/ml)	AFP (ng/ml)
SX	Marker studies not available or not done		
S0	Normal	Normal	Normal
S1*	<1.5 × Normal	<5000	<1000
S2+	1.5–10 × Normal	5000–50,000	1000–10,000
S3+	>10 × Normal	>50,000	>10,000

Stage grouping

Once the T, N, M, and S categories have been determined, they are combined in a process called *stage grouping* to assign an overall stage (using Roman numerals and letters)

Stage	T	N	M	S
Stage 0	Tis (in situ)	N0	M0	S0
Stage I	T1–T4	N0	M0	SX
Stage IA	T1	N0	M0	S0
Stage IB	T2–T4	N0	M0	S0
Stage IS	Any T	N0	M0	S1–S3
Stage II	Any T	N1–N3	M0	SX
Stage IIA	Any T	N1	M0	S0–S1
Stage IIB	Any T	N2	M0	S0–S1
Stage IIC	Any T	N3	M0	S0–S1
Stage III	Any T	Any N	M1	SX
Stage IIIA	Any T	Any N	M1a	S0–S1
Stage IIIB	Any T	N1–N3	M0	S2
	Any T	Any N	M1a	S2
Stage IIIC	Any T	N1–N3	M0	S3
	Any T	Any N	M1a	S3
	Any T	Any N	M1b	Any S

Note: Normal values vary among laboratories

LDH lactate dehydrogenase (measured in units per liter [U/l]), *HCG* human chorionic gonadotropin (measured in milli-international units per milliliter [mIU/ml]), *AFP* alpha-fetoprotein (measured in nanograms per milliliter [ng/ml]), < means less than, > means more than, * all the markers must be in the stated range to be considered S1, + only one marker needs to be in the stated range to be considered S2 or S3

(IGCCCG) [6] which is based on this TNM staging and forms the current basis for prognostication (Table 5.3) and also for defining definitive treatment strategies based on prognosis. Table 5.2 highlights the 5-year prognosis of good, intermediate, and poor prognosis categories.

This initial approach sets the stage for further treatment decisions that can range from surveillance to combination treatments in patients presenting with localized stage disease based on prognosis and regardless of pathological subtype.

Apart from using a pathological classification for testicular germ cell tumors, they can also be classified on anatomical location as gonadal and extra-gonadal in origin. The vast majority of germ cell tumors are of gonadal origin. In general, extra-gonadal germ cell tumors carry a poorer prognosis. In order to diagnose suspected extra-gonadal germ cell tumors, it is necessary to perform a biopsy of a presenting mediastinal or retroperitoneal or pelvic area mass for establishing a histological diagnosis and not an orchiectomy. In such presentations, additional serum tumor markers are also evaluated in aiding the initial evaluation of suspected extra-gonadal germ cell tumors.

Table 5.2 Survival rates per clinical staging in Testicular germ cell tumors

The SEER database does not divide survival rates by AJCC TNM stage. Instead, it divides cancers into summary stages: localized, regional, and distant

Localized means that the cancer is still only in the testicle. This includes most AJCC stage I tumors (stage 0 cancers are not included in these statistics)

Regional means that the cancer has spread to nearby lymph nodes or tissues. This includes T4 tumors and cancers with lymph node spread (all stage II cancers and some stage IIIB and IIIC cancers)

Distant means that the cancer has spread to organs or lymph nodes away from the tumor, such as all M1 cancers (which can be stage IIIA, IIIB, or IIIC)

Stage	5-year relative survival rate
Localized	99 %
Regional	96 %
Distant	73 %

Taken from the American Cancer Society Website: http://www.cancer.org/cancer/testicularcancer/detailedguide/testicular-cancer-staging

Table 5.3 IGCCCG prognostic classification of *germ* cell tumors

	Non-seminoma	Seminoma
	Good risk	
Location of primary tumor	Testis/retroperitoneal primary *AND*	Any *AND*
Non-pulmonary visceral mets	No *AND*	No *AND*
AFP (ng/mL)	<1000 *AND*	Normal *AND*
β-hCG (IU/L)	<5000 *AND*	Any *AND*
LDH (N × upper limit of normal)	< 1.5	Any
	Intermediate risk	
Location of primary tumor	Testis/retroperitoneal primary *AND*	Any *AND*
Non-pulmonary visceral mets	No *AND*	Yes *AND*
AFP (ng/mL)	≥ 1000, ≤10,000 *OR*	Normal *AND*
β-hCG (IU/L)	≥ 5000, ≤50,000 *OR*	Any *AND*
LDH (N × upper limit of normal)	≥ 1.5, ≤10	Any
	Poor risk	
Location of primary tumor	Mediastinal primary *OR*	N/A
Non-pulmonary visceral mets	Yes *OR*	
AFP (ng/mL)	>10,000 *OR*	
β-hCG (IU/L)	>50,000 *OR*	
LDH (N × upper limit of normal)	> 10	

5.4 Serum Tumor Markers in Germ Cell Cancers

As mentioned, serum alpha-fetoprotein (AFP), beta-human chorionic gonadotropin (β-hCG), and lactate dehydrogenase (LDH) play an important role in the diagnosis, prognosis, and follow-up assessment of patients in germ cell cancers [7, 8]. The interpretation of tumor marker levels in germ cell cancer management [9] has been refined for clinical application based on the results of multiple randomized clinical trials and will be discussed further in this section.

5.5 Alpha-Fetoprotein (AFP)

AFP is normally produced by the fetal yolk sac and therefore is essentially undetectable in normal adult males (normal levels in males 10–15 μg/l). As a tumor marker in germ cell cancers, it is observed to be elevated in yolk sac testicular tumors and embryonal carcinoma. Seminoma cells do not produce AFP, and therefore by definition, patients with germ cell tumors presenting with elevated AFP levels are treated as non-seminomatous germ cell tumors [9, 10].

In non-seminomas, elevated AFP levels can vary with stage of presentation as it is increased in 10–20 % with stage I disease, 20–40 % with low-volume stage II disease, and 40–60 % with advanced disseminated disease. The half-life of AFP is approximately 5–7 days [9], and, therefore, serum AFP should normalize 25–35 days (five half-lives) following effective treatment for a non-seminoma [9, 11, 12]. Failure to normalize can raise the suspicion of inadequately treated or ineffectively treated disease. There can be misleading false-positive reasons for elevated AFP levels in males. Benign liver conditions such as hepatitis, fatty liver disease, hepatocellular carcinoma, and other gastrointestinal malignancies can produce an elevation in AFP, and these should be kept in mind especially after patients have undergone localized or systemic treatments. In addition, tumor lysis during early chemotherapy could also produce a falsely positive result [9].

5.6 Beta-Human Chorionic Gonadotropin (β-hCG)

β-hCG is a common serum tumor marker for germ cell cancers. The serum concentration of β-hCG is elevated in approximately 15–20 % of patients with advanced pure testicular seminoma. In addition, 10–20 % of stage I, 20–30 % of low-volume stage II, and 40 % of advanced disseminated testicular non-seminomas present with an elevated serum β-hCG [9, 12]. The biologic half-life of β-hCG is approximately 1.5–3 days [9]. Therefore, during monitoring of treatment effect, serum β-hCG should normalize 1–2 weeks following effective treatment for a β-hCG-producing testicular cancer [9].

As with AFP, there can be false-positive reasons for elevated β-hCG levels other than germ cell tumors including in one report with the use of marijuana [13]. Other malignancies including neuroendocrine tumors, non-germ cell genitourinary cancers, lung cancer, and cancers of the female genital tract are also occasionally associated with elevated β-hCG levels. In addition, hypogonadism can also cause increased production of LH and hCG by the pituitary gland via the pituitary–gonadal axis feedback mechanism, thus leading to a falsely elevated serum β-hCG. Of note, this scenario could be observed after primary treatment of testicular cancer such as orchiectomy. Similar to the tumor lysis effect on AFP, β-hCG could also be falsely elevated during the first cycle of chemotherapy [9].

5.7 Lactate Dehydrogenase (LDH)

Approximately 40–60 % of patients with testicular cancer have an elevated serum LDH [8, 9]. As a germ cell marker and specifically testicular germ cell marker, LDH is less sensitive and specific than AFP and β-hCG, in diagnosis, and during monitoring of treatment effect or following the completion of therapy, although sometimes it may be the only positive tumor marker associated with a seminoma presentation.

5.8 Use of Tumor Markers in Testicular Germ Cell Tumors

5.8.1 As Diagnostic Markers

AFP, β-hCG, and LDH checked prior to orchiectomy for all patients with suspected testicular germ cell cancer help in establishing a diagnosis and for proceeding to performing an orchiectomy. These markers should be repeated *after* orchiectomy as the expected outcome postsurgery is for markers to decrease and normalize. Since not all patients with seminomatous and non-seminomatous germ cell tumors actually present with elevated tumor markers, these markers are not taken alone to determine the need for a diagnostic orchiectomy.

In rare circumstances, when a reproductive age male patient presents with a primary testicular tumor with symptomatic widespread metastases, along with a markedly elevated or rapidly increasing AFP and/or β-hCG, these markers' levels aid in starting chemotherapy empirically in the absence of a diagnostic orchiectomy as it avoids delay of treatments.

5.8.2 As Prognostic Markers

As mentioned before, the International Germ Cell Consensus Classification (IGCCCG) prognostic classification is used universally and is integral in the practice of germ cell tumor management. This was developed in 1997 after a long-term follow-up of 660 patients with pure seminoma and 5202 patients with non-seminoma. The IGCCCG created clinically based prognostic classification system based on post-orchiectomy serum tumor marker levels, the site of primary tumor, and the presence or absence of non-pulmonary visceral metastases (Table 5.1) [6] .

Patients with good-, intermediate-, and poor-risk non-seminomatous germ cell tumors were shown to have 5-year survivals of 92 %, 80 %, and 48 %, respectively. In comparison, patients with good-risk and intermediate-risk seminomatous germ cell tumors had 5-year survivals of 86 % and 72 %, respectively [6].

5.8.3 For Disease Monitoring and Follow-Up

Serum tumor markers are used to monitor response following surgical resection and radiation therapy, as well as after each cycle of systemic chemotherapy. Tumor marker levels that either do not return to normal range or initially return to baseline but subsequently become elevated indicate residual or relapsed disease. Early detection of relapsed disease allows prompt initiation of salvage therapy.

For patients with stage I and II non-seminoma, post-orchiectomy serum tumor marker levels are crucial in determining the next treatment step. Patients with persistently elevated tumor markers have a high risk of relapse and therefore undergo systemic chemotherapy as their primary treatment [14–16]. For patients with normal post-orchiectomy serum tumor markers, those with stage IIA disease are considered for either systemic chemotherapy or retroperitoneal lymphadenectomy (RPLND). On the other hand, patients with stage IIB disease and normal post-orchiectomy tumor markers are offered chemotherapy followed by RPLND or surveillance if the sites of metastatic disease are not confined to the lymphatic drainage in the retroperitoneum [14].

After stage-specific treatments, tumor markers are also used for assessing the risk of recurrence, which is much higher in the first 2–3 years. Relapsed disease after 5 years is infrequent [17, 18]. Therefore, the National Comprehensive Cancer Network (NCCN) recommends an intensive early surveillance schedule including monitoring tumor markers every 3 months in the first 1–2 years. The frequency of surveillance typically becomes spaced out to every 6–12 months in years 3–5 and annually in years 5–10 [14].

5.9 Principles for the Treatment of Testicular Seminoma and Non-seminomatous Tumors

Following the clinical presentation, radiological and tumor marker evaluation of a testicular mass, the first intervention for a testicular cancer begins with a radical orchiectomy. This provides pathological staging and confirmation of histological subtypes. In rare cases a delayed radical orchiectomy is reserved in those advanced stage patients who are in need of immediate systemic therapy for rapidly growing tumors diagnosed as testicular germ cell tumors based on results of a metastatic site biopsy. Fertility and sperm banking using cryopreservation prior to initiating any treatment for germ cell tumors are also discussed during the initial treatment planning, as testicular cancers are associated with gonadal dysgenesis and nearly 50 % of men have impaired spermatogenesis [19–21] which increases after orchiectomy or other treatments [22].

In general, seminomatous testicular tumors differ from non-seminomas in several ways. Seminomas are slower growing than non-seminomas and present more often initially in either localized or locoregional stages with only 5 % presenting in advanced stages with spread beyond retroperitoneal lymph nodes. Non-seminomatous germ cell tumors are in comparison faster to grow, and on initial presentation, nearly 33 % of all non-seminomatous testicular tumors present with stages I, II, and III. Seminomas are extremely radiosensitive compared to non-seminomas and therefore radiation therapy is a part of the treatment preamble for seminomas. Seminomas rarely spread hematogenously compared to non-seminomas, which is known to occur more commonly with non-seminomas (pulmonary, liver, brain, and bones), and so advanced stage seminomas are never considered to have poor prognosis as defined by the IGCCCG prognostic stratification system in use. Stage- and pathology-specific treatment guidelines have been well established [23] and are summarized in the following sections.

5.10 Management of Seminomas

5.10.1 Stage I

For stage I testicular seminomas, a radical orchiectomy is usually curative. Afterwards, the preferred management option is to offer a structured schedule of surveillance for at least 5 years. The surveillance schedule suggested includes a history and physical examination along with tumor marker measurements every 3–4 months for 2 years, 6–12 months for years 3–4, and then annually. Additionally imaging scans of the abdomen and pelvis are recommended every 6 months for the first 2 years, every 6–12 months for year 3, and then annually for years 4 and 5 along with a chest X-ray as needed for years 1–5 [23]. In the largest series of 1954 men who underwent a surveillance strategy for stage I seminomas, relapse was observed in 369 (18.9 %) patients after a median time of 13.7 months (range, 2.3–173.6 months) [24]. Of the 369 patients who relapsed, 16 relapsed after a 5-year follow-up. At the time of relapse, over 230 of 369 patients were salvaged with radiation therapy, while 136 underwent systemic chemotherapy, and three underwent surgery for the relapse. The overall survival after 5, 10, and 15 years for this cohort was 98.1 %, 95.5 %, and 91.6 %, respectively, and the disease-specific survival at 5, 10, and 15 years was 99.6 %, 99.4 %, and 99.3 %, respectively.

For men who are unable to comply or refuse surveillance as a management option or may harbor higher-risk features, such as lympho-vascular invasion, post-orchiectomy radiation or systemic chemotherapy with one or two cycles of carboplatin offers similar excellent long-term outcomes [25].

The choice to offer an individual patient any of the three options involves balancing the risk–benefit profile for the individual patient [23]. A shared decision with the patient is taken after explaining the pros and cons of each option. Typically, this involves highlighting the risk of overtreatment if systemic chemotherapy or radiation is chosen by way of acute and long-term toxicities over surveillance for stage I seminoma patients, as the vast majority of patients will not need treatments. This is balanced by including in the discussion the risk of relapse for stage I seminoma post-orchiectomy and the excellent salvage options available at the time of relapse.

5.10.2 Stage II

By definition, stage II disease involves lymphadenopathy below the diaphragm. Following orchiectomy and resolution of tumor markers, a radiological evaluation for bulky (>5 cm) versus non-bulky retroperitoneal lymphadenopathy is usually performed for deciding the optimal management of stage II disease. Non-bulky stage II disease patients have excellent long-term outcomes with radiation therapy alone, which includes radiation fields extended to para-aortic and ipsilateral lymph nodes to a total dose of 30–36 Gy. On the other hand for treating stage II bulky retroperitoneal lymphadenopathy or patients with symptomatic (such as back pain) non-bulky retroperitoneal lymphadenopathy, primary chemotherapy with four cycles of cisplatin and etoposide or three cycles of cisplatin, etoposide, and bleomycin is preferred. Post treatment, patients are again followed every 3 months in year 1 with examination, tumor markers, and biannual CT imaging of the abdomen and pelvis. In years 2–5 follow-up for recurrence is every 6 months and then annually for years 6–10.

5.10.3 Stage III/Advanced Stage

Patients with advanced stage present with either lymphadenopathy above and below the diaphragm or meet the definition of advanced stage disease on the basis of non-pulmonary visceral metastasis or persistently elevated tumor markers. The approach to treatment planning starts with an initial evaluation for prognostic risk based on the IGCCCG risk stratification system (Table 5.3). By definition no seminomas are graded as "poor prognosis" and so stage III disease is either "good" prognosis or "intermediate" prognosis. This relevance for prognostication is important to establish as it has a direct impact on treatment planning. Seminomas with good prognosis receive systemic chemotherapy with either four treatments of cisplatin and etoposide or three of cisplatin, etoposide, and bleomycin, while intermediate-risk seminomas receive four treatments with cisplatin, etoposide, and bleomycin.

Posttreatment management for bulky stage II and advanced stage (III) seminoma treated with systemic chemotherapy is complex and depends on the response to chemotherapy [23]. Immediately following systemic therapy, patients should undergo imaging with CT scans of the chest, abdomen, and pelvis along with tumor markers. If there are no residual masses, or for residual masses less than 3 cm in size and normal tumor marker levels, patients are offered surveillance with exams, tumor markers, and a chest X-ray every 2 months in year 1, every 3 months in year 2, every 6 months in year 3, and then annually. CT of the abdomen and pelvis should be performed 3–6 months in the first 2 years and then as clinically indicated thereafter. If post-chemotherapy residual masses on imaging reveal a size greater than 3 cm and normal markers, a positron emission tomography (PET) scan is obtained, at least 6 weeks following treatments. If results of the scan are positive for uptake (considered to have a standardized uptake value (SUV) of >4.0 units), then a surgical resection of the mass (or lymph-nodal mass) is considered, or second-line systemic chemotherapy may be advocated if resection is considered to be risky. If the PET scan is negative, then patients are followed up as for no residual masses after chemotherapy as most of these masses resolve over time. Finally, if after completing chemotherapy, imaging reveals progressive disease or rising tumor markers, then second-line systemic chemotherapy is considered. The overall 5-year survival with the above management for patients with good-risk stage III seminomas is 92 and 72 % for the intermediate-risk group.

In addition to monitoring for treatment effects, long-term side effects of systemic chemotherapy are also monitored in the follow-up. In particular bleomycin [26, 27] can have long-term side effects in causing pulmonary toxicity and peripheral vascular disease, while cisplatin can increase the risk of nerve damage and renal insufficiency [28]. As mentioned previously, gonadal effects of treatments can lead to infertility in young men, and this potential complication of therapy can be mitigated by offering cryopreservation to men before initiating therapy.

5.11 Management of Non-seminomas

5.11.1 Stage 1

The rate of cure for non-seminoma testicular germ cell tumors like seminoma depends on the stage and prognosis at the time of initial presentation. For stage I disease patients, the 5-year survival exceeds 95 % with appropriate management

[3, 23]. Treatment planning begins following orchiectomy, and after obtaining pathological "T" staging, depending on the presence or absence of lympho-vascular invasion (LVI) for staging T1 versus T2 tumors. The absence of LVI with no radiological evidence of pelvic, abdominal, or distant metastasis (clinical stage IA) carries the most favorable prognosis. Such patients are excellent candidates for a strategy of surveillance as was observed in one large retrospective case series of 1139 clinical stage I (CSI) nonseminoma patients who were offered and followed with active surveillance between 1998 and 2010. Relapse was observed in 221 (19 %) with a median time to relapse of 4 months (range, 2–61 months) for LVI positive (CSIB) and 8 months (range, 2–77 months) for LVI negative (CSIA). The majority of relapses occurred within the first 2–3 years after orchiectomy for CSI nonseminoma (90 %) [29]. Active surveillance involves strict adherence to a structured follow-up schedule that involves repeat examinations with tumor marker measurements every 1–2 months in year one (after orchiectomy), every 2 months in year two, 3 months in year three, 4 months in year four, 6 months in year five, and then annually. Additionally imaging with CT abdomen and pelvis is performed every 3–4 months in year one, 4–6 months in year 2, 6–12 months in years 3–5, and then annually.

For those patients who are either unable to follow this schedule, nerve-sparing retroperitoneal lymphadenopathy (RPLND) for CSI or primary chemotherapy with one to two treatment cycles of cisplatin, etoposide, and bleomycin is offered as an option which produces excellent outcomes. RPLND dissection is both diagnostic to stage microscopic pathological nodal involvement and performed with curative intent. However, these options can have complications such as retrograde ejaculation and infertility with RPLND, or long-term complications of systemic chemotherapy. An individual assessment of risk–benefit ratio has to be made while offering any of the three choices, all of which have excellent long-term survival results.

A specific substage classified in CSI is CSIS, which stands for persistent tumor marker(s) elevation post-orchiectomy. Such patients are best treated with three to four treatment cycles of systemic chemotherapy as they usually harbor disease outside the retroperitoneum.

5.11.2 Stage II

Imaging of retroperitoneal lymph nodes in patients presenting with stage IIA disease is by definition less than 2 cm and between 2 and 5 cm for stage IIB. CSIIA patients with normalized post-orchiectomy tumor markers are best treated with nerve-sparing RPLND which confirms pathological nodal involvement. Following a RPLND adjuvant systemic chemotherapy is typically reserved for men if the pathological nodal involvement is detected to be greater than 2 cm as the relapse risk is highest for this category. For clinical stage IIB and IIC (lymph node size greater than 5 cm), primary systemic chemotherapy with either four treatments of cisplatin and etoposide or three of cisplatin, etoposide, and bleomycin is offered.

For all stage II patients treated with primary systemic chemotherapy, posttreatment tumor markers and imaging are performed 3–4 weeks after the last treatment. Negative markers with resolution of lymph nodal masses are typically followed with surveillance, while if markers have normalized, but imaging reveals remnant nodal masses, a RPLND is offered in such cases. If markers have not normalized, such carry a poor prognosis and are generally offered salvage chemotherapy with ifosfamide- and taxane-based regimens.

5.11.3 Stage III

Advanced or stage III disease is treated with primary systemic chemotherapy following orchiectomy and tumor marker and radiographic imaging. The number of treatments is based on the results of these investigations and the final prognostic category. Good prognosis advanced stage disease is treated with either four treatments of cisplatin and etoposide or three of cisplatin, etoposide, and bleomycin, while intermediate to poor prognosis is treated with four treatments of cisplatin, etoposide, and bleomycin. Between 20 and 25 % of patients with advanced germ cell tumors present with hepatic, bone, or brain metastases or with mediastinal masses or with extremely high tumor markers. The success of cure in such poor prognosis is low.

After completion of primary systemic chemotherapy for patients with complete response surveillance with examination, tumor marker and radiographic assessments are used to guide further management. Patients with normalized tumor markers who have radiographic residual disease undergo resection. The presence of necrotic debris or teratoma requires no further treatments. However, if viable germ cell tumor is detected pathologically, in order to maximize the rate of cure, additional two cycles of systemic chemotherapy are offered. Further follow-up should include history and examination with tumor markers every 2–3 months for the first 2 years, every 3–6 months in year 3, every 6 months in years 4 and 5, and then annually. Radiography with CT imaging of the abdomen and pelvis with chest X-rays is performed every 6 months in the first 2 years and then annually until year 5 and as clinically indicated after that.

During follow-up after chemotherapy, detection of masses that appear to increase in size in the presence of normal tumor markers should raise the suspicion of a growing teratoma syndrome in such presentations. The treatment for these chemo-insensitive masses is surgical resection without chemotherapy.

5.12 Resistant of Recurrent Disease

For advanced stage patients who recur after primary systemic chemotherapy or who fail to show a complete response to therapy, second-line systemic chemotherapy with combinations of ifosfamide, taxane, and platinum agents or gemcitabine is preferred as it offers a chance of cure in at least 25 % of such cases [30, 31]. High-dose or autologous stem cell rescue is offered to patients who are not cured with the above combinations [32–34].

5.13 Ovarian Germ Cell Tumors

Germ cell tumors may also arise from the ovary accounting for approximately 2–3 % of ovarian malignancies in distinction to the more common epithelial type [35, 36]. Ovarian germ cell tumors most commonly occur in women in their 20s. The WHO histological classification for ovarian germ cell tumors differs from testicular germ cell cancers [37, 38]. Broadly, ovarian germ cell tumors are divided into embryo-like (teratomas and dysgerminomas) or extraembryonic (placenta-like) cell populations or a mixture of both. Teratomas can have several subtypes including benign cystic mature teratoma (dermoid cyst) which is the most common ovarian germ cell tumor and immature teratoma. Mature teratomas can also develop a somatic malignant neoplasm (mature teratoma with malignant degeneration). Dysgerminomas are the female equivalent to seminomas in males. On the other hand, yolk sac tumors are ovarian carcinomas that differentiate toward primitive endodermal structures, while the rare mixed germ cell tumors can have a mixture of all of the above elements. They are clinically treated based on the most aggressive histological component. Among malignant ovarian germ cell tumors, dysgerminomas, yolk sac tumors, and immature teratoma account for greater than 90 % of all cases. Embryonal carcinoma, mixed germ cell tumors, non-gestational choriocarcinoma, and polyembryoma are far more rare in occurrence. For ovarian germ cell tumors based on clinical TNM classification, there are four stages as opposed to testicular cancers, which have three. Overall, although there are a number of similarities between ovarian germ cell tumors and testicular cancers, there are some key differentiating features of both which influence which treatment modalities are used. Both diseases share in common significant chemosensitivity of disease and, therefore, overall favorable prognosis.

Approximately 60–70 % of ovarian germ cell tumors present in stage I when diagnosed, which means that cancer is found in one or both ovaries. Most of these are confined to a single ovary (stage IA disease).

Clinically, the majority of ovarian germ cell tumor patients present with abdominal pain and a palpable mass. Since many ovarian germ cell tumors produce hormones including β-hCG, symptoms of pregnancy can often be mimicked in the initial presentation. Other less common symptoms include abdominal distention, fever, urinary symptoms, and vaginal bleeding. When an ovarian germ cell tumor is suspected, for evaluating these nonspecific symptoms, additional testing with β-hCG and AFP tumor markers is often performed. β-hCG is commonly produced by embryonal carcinomas with trophoblastic component, ovarian carcinomas, mixed germ cell tumors, and some dysgerminomas, while elevated AFP levels are suspected in yolk sac tumors, embryonal carcinoma, mixed germ cell tumors with yolk sac tumor component, and some immature teratomas. As with testicular germ cell tumors, these markers aid in diagnosis and also for monitoring response and recurrence following treatments. Imaging with transvaginal ultrasound or MRI in the initial work-up also plays a key role.

Surgery is typically the first step in treatment for ovarian germ cell tumors, both for diagnosis and therapy [39]. Preoperative elevated levels of β-hCG in children, teens, or young women can indicate specific subtypes of germ cell tumors as mentioned and apart from providing diagnostic clues can also aid in preparing for surgery while preserving fertility. Since most of these tumors are unilateral and occur in young women, fertility preservation is an integral part in treatment planning along with the surgical approach which typically involves offering a unilateral salpingo-oophorectomy with preservation of the uterus if these organs appear normal during surgical resection. For bilateral disease however, a bilateral salpingo-oophorectomy is done, though, in both cases, the uterus is typically not removed.

For patients diagnosed with stage IA dysgerminoma following surgery, surveillance is offered. Surgical principles are heavily dependent on maximal cytoreduction if disease is advanced. The recurrence rate is 15–20 % in early-stage presentations, and adjuvant combination chemotherapy postsurgery or at the time of recurrence offers a high chance of cure. Dysgerminomas are also very radiosensitive like testicular seminoma, though use of adjuvant radiotherapy for stage IA disease has fallen out of favor due to long-term side effects. For patients with dysgerminoma and stage greater than IA, combination adjuvant chemotherapy in the form of bleomycin, etoposide, and cisplatin (BEP) is given usually with three treatment cycles for fully resected disease and four cycles for those with microscopic residual disease [40]. Surveillance post treatment is critical and should include serial history, physical examinations, tumor markers, and radiographical surveillance to be performed for at least 2 years if the tumor markers were not elevated initially. For dysgerminomas late relapses can occur and annual surveillance up to 10 years is a clinical consideration as well.

For patients with non-dysgerminomas, typically combination chemotherapy after resection of disease is offered regardless of final stage. As with dysgerminomas, the most common regimen is BEP. Patients with stage IA, grade 1 immature teratoma which is fully resected can be observed without postoperative chemotherapy especially in the case of a motivated patient. For patients with immature teratoma of similar stage, but grade 2 or 3, there is a higher risk of recurrent disease, and, therefore, postoperative chemotherapy is considered. All patients with non-dysgerminomas should receive active and serial follow-ups on nearly similar schedules as male patients with non-seminomatous germ cell testicular tumors.

References

1. Siegel RL, Miller KD, Jemal A. Cancer statistics, 2015. CA Cancer J Clin. 2015;65(1):5–29. doi:10.3322/caac.21254.
2. Einhorn LH. Treatment of testicular cancer: a new and improved model. J Clin Oncol. 1990;8(11):1777–81.
3. Bosl GJ, Motzer RJ. Testicular germ-cell cancer. N Engl J Med. 1997;337(4):242–53. doi:10.1056/NEJM199707243370406.

4. Tseng Jr A, Horning SJ, Freiha FS, Resser KJ, Hannigan Jr JF, Torti FM. Gynecomastia in testicular cancer patients. Prognostic and therapeutic implications. Cancer. 1985;56(10):2534–8.
5. Muller J, Skakkebaek NE, Parkinson MC. The spermatocytic seminoma: views on pathogenesis. Int J Androl. 1987;10(1):147–56.
6. International Germ Cell Consensus Classification: a prognostic factor-based staging system for metastatic germ cell cancers. International Germ Cell Cancer Collaborative Group. J Clin Oncol. 1997;15(2):594–603.
7. Lange PH, McIntire KR, Waldmann TA, Hakala TR, Fraley EE. Serum alpha fetoprotein and human chorionic gonadotropin in the diagnosis and management of nonseminomatous germ-cell testicular cancer. N Engl J Med. 1976;295(22):1237–40. doi:10.1056/NEJM197611252952207.
8. Lippert MC, Javadpour N. Lactic dehydrogenase in the monitoring and prognosis of testicular cancer. Cancer. 1981;48(10):2274–8.
9. Gilligan TD, Seidenfeld J, Basch EM, Einhorn LH, Fancher T, Smith DC, et al. American Society of Clinical Oncology Clinical Practice Guideline on uses of serum tumor markers in adult males with germ cell tumors. J Clin Oncol. 2010;28(20):3388–404. doi:10.1200/JCO.2009.26.4481.
10. Nazeer T, Ro JY, Amato RJ, Park YW, Ordonez NG, Ayala AG. Histologically pure seminoma with elevated alpha-fetoprotein: a clinicopathologic study of ten cases. Oncol Rep. 1998;5(6):1425–9.
11. Klein EA. Tumor markers in testis cancer. Urol Clin North Am. 1993;20(1):67–73.
12. Weissbach L, Bussar-Maatz R, Mann K. The value of tumor markers in testicular seminomas. Results of a prospective multicenter study. Eur Urol. 1997;32(1):16–22.
13. Garnick MB. Spurious rise in human chorionic gonadotropin induced by marihuana in patients with testicular cancer. N Engl J Med. 1980;303(20):1177. doi:10.1056/NEJM198011133032014.
14. Motzer RJ, Agarwal N, Beard C, Bolger GB, Boston B, Carducci MA, et al. NCCN clinical practice guidelines in oncology: testicular cancer. J Natl Compr Cancer Netw. 2009;7(6):672–93.
15. Rabbani F, Sheinfeld J, Farivar-Mohseni H, Leon A, Rentzepis MJ, Reuter VE, et al. Low-volume nodal metastases detected at retroperitoneal lymphadenectomy for testicular cancer: pattern and prognostic factors for relapse. J Clin Oncol. 2001;19(7):2020–5.
16. Stephenson AJ, Bosl GJ, Bajorin DF, Stasi J, Motzer RJ, Sheinfeld J. Retroperitoneal lymph node dissection in patients with low stage testicular cancer with embryonal carcinoma predominance and/or lymphovascular invasion. J Urol. 2005;174(2):557–60. discussion 60 doi:10.1097/01.ju.0000165163.03805.37.
17. Oldenburg J, Alfsen GC, Waehre H, Fossa SD. Late recurrences of germ cell malignancies: a population-based experience over three decades. Br J Cancer. 2006;94(6):820–7. doi:10.1038/sj.bjc.6603014.
18. Geldart TR, Gale J, McKendrick J, Kirby J, Mead G. Late relapse of metastatic testicular nonseminomatous germ cell cancer: surgery is needed for cure. BJU Int. 2006;98(2):353–8. doi:10.1111/j.1464-410X.2006.06250.x.
19. Fossa SD, Kravdal O. Fertility in Norwegian testicular cancer patients. Br J Cancer. 2000;82(3):737–41. doi:10.1054/bjoc.1999.0989.
20. Ohl DA, Sonksen J. What are the chances of infertility and should sperm be banked? Semin Urol Oncol. 1996;14(1):36–44.
21. Panidis D, Rousso D, Stergiopoulos K, Papathanasiou K, Delkos D, Papaletsos M. The effect of testicular seminoma in semen quality. Eur J Obstet Gynecol Reprod Biol. 1999;83(2):219–22.
22. Gordon Jr W, Siegmund K, Stanisic TH, McKnight B, Harris IT, Carroll PR, et al. A study of reproductive function in patients with seminoma treated with radiotherapy and orchidectomy: (SWOG-8711). Southwest Oncology Group. Int J Radiat Oncol Biol Phys. 1997;38(1):83–94.
23. Motzer RJ, Agarwal N, Beard C, Bhayani S, Bolger GB, Buyyounouski MK, et al. Testicular cancer. J Natl Compr Cancer Netw. 2012;10(4):502–35.
24. Mortensen MS, Lauritsen J, Gundgaard MG, Agerbaek M, Holm NV, Christensen IJ, et al. A nationwide cohort study of stage I seminoma patients followed on a surveillance program. Eur Urol. 2014;66(6):1172–8. doi:10.1016/j.eururo.2014.07.001.
25. Oliver RT, Mason MD, Mead GM, von der Maase H, Rustin GJ, Joffe JK, et al. Radiotherapy versus single-dose carboplatin in adjuvant treatment of stage I seminoma: a randomised trial. Lancet. 2005;366(9482):293–300. doi:10.1016/S0140-6736(05)66984-X.
26. Williams SD, Birch R, Einhorn LH, Irwin L, Greco FA, Loehrer PJ. Treatment of disseminated germ-cell tumors with cisplatin, bleomycin, and either vinblastine or etoposide. N Engl J Med. 1987;316(23):1435–40. doi:10.1056/NEJM198706043162302.
27. Jules-Elysee K, White DA. Bleomycin-induced pulmonary toxicity. Clin Chest Med. 1990;11(1):1–20.
28. Lauritsen J, Mortensen MS, Kier MG, Christensen IJ, Agerbaek M, Gupta R, et al. Renal impairment and late toxicity in germ-cell cancer survivors. Ann Oncol. 2015;26(1):173–8. doi:10.1093/annonc/mdu506.
29. Kollmannsberger C, Tandstad T, Bedard PL, Cohn-Cedermark G, Chung PW, Jewett MA, et al. Patterns of relapse in patients with clinical stage I testicular cancer managed with active surveillance. J Clin Oncol. 2015;33(1):51–7. doi:10.1200/JCO.2014.56.2116.
30. Culine S, Kramar A, Theodore C, Geoffrois L, Chevreau C, Biron P, et al. Randomized trial comparing bleomycin/etoposide/cisplatin with alternating cisplatin/cyclophosphamide/doxorubicin and vinblastine/bleomycin regimens of chemotherapy for patients with intermediate- and poor-risk metastatic nonseminomatous germ cell tumors: Genito-Urinary Group of the French Federation of Cancer Centers

Trial T93MP. J Clin Oncol. 2008;26(3):421–7. doi:10.1200/JCO.2007.13.8461.
31. Motzer RJ, Nichols CJ, Margolin KA, Bacik J, Richardson PG, Vogelzang NJ, et al. Phase III randomized trial of conventional-dose chemotherapy with or without high-dose chemotherapy and autologous hematopoietic stem-cell rescue as first-line treatment for patients with poor-prognosis metastatic germ cell tumors. J Clin Oncol. 2007;25(3):247–56. doi:10.1200/JCO.2005.05.4528.
32. Einhorn LH, Williams SD, Chamness A, Brames MJ, Perkins SM, Abonour R. High-dose chemotherapy and stem-cell rescue for metastatic germ-cell tumors. N Engl J Med. 2007;357(4):340–8. doi:10.1056/NEJMoa067749.
33. Bhatia S, Abonour R, Porcu P, Seshadri R, Nichols CR, Cornetta K, et al. High-dose chemotherapy as initial salvage chemotherapy in patients with relapsed testicular cancer. J Clin Oncol. 2000;18(19):3346–51.
34. Bokemeyer C, Harstrick A, Beyer J, Metzner B, Ruther U, Hartmann JT, et al. The use of dose-intensified chemotherapy in the treatment of metastatic nonseminomatous testicular germ cell tumors. German Testicular Cancer Study Group. Semin Oncol. 1998;25(2 Suppl 4):24–32. discussion 45–8
35. Brown J, Friedlander M, Backes FJ, Harter P, O'Connor DM, de la Motte RT, et al. Gynecologic Cancer Intergroup (GCIG) consensus review for ovarian germ cell tumors. Int J Gynecol Cancer. 2014;24(9 Suppl 3):S48–54. doi:10.1097/IGC.0000000000000223.
36. Gershenson DM. Management of ovarian germ cell tumors. J Clin Oncol. 2007;25(20):2938–43. doi:10.1200/JCO.2007.10.8738.
37. Palenzuela G, Martin E, Meunier A, Beuzeboc P, Laurence V, Orbach D, et al. Comprehensive staging allows for excellent outcome in patients with localized malignant germ cell tumor of the ovary. Ann Surg. 2008;248(5):836–41. doi:10.1097/SLA.0b013e31818447cd.
38. Billmire D, Vinocur C, Rescorla F, Cushing B, London W, Schlatter M et al. Outcome and staging evaluation in malignant germ cell tumors of the ovary in children and adolescents: an intergroup study. J Pediatr Surg. 2004;39(3):424–9; discussion 429.
39. Gershenson DM. Update on malignant ovarian germ cell tumors. Cancer. 1993;71(4 Suppl):1581–90.
40. Gershenson DM, Morris M, Cangir A, Kavanagh JJ, Stringer CA, Edwards CL, et al. Treatment of malignant germ cell tumors of the ovary with bleomycin, etoposide, and cisplatin. J Clin Oncol. 1990;8(4):715–20.

6 Germ Cell Tumors of the Female Genital Tract

Francisco F. Nogales and Maollly Schuldt

6.1 Introduction and General Considerations

In the female genital tract, germ cell tumors (GCT) are found mainly in the ovary, being merely anecdotal in other locations. They occur twice as often in the ovary than in the testis in both children and adults [1]. Most are benign teratomas [2]; malignant GCT are infrequent. Since epidemiologic data from tumor registries deal almost exclusively with malignant tumors, it is difficult to establish a general incidence for benign forms despite their being considered the most frequent cause of ovarian mass in adolescents and young adults [3]. GCT are rarely congenital, and their incidence begins to rise during late infancy and early adulthood [4, 5].

As analyzed in Chap. 3, the origin of female genital tract GCT is far from univocal. Most ovarian teratomas originate from parthenogenetically activated oocytes in various stages of meiosis [6, 7], corresponding to type IV pluripotential tumors in Oosterhuis and Looijenga's classification, discussed in Chap. 3 and followed here. This is reflected in the ability to differentiate into practically any embryonal or adult tissue and an outstanding capacity of organization, reproducing complex organs such as cerebellum and conforming tissue relationships that may even have a certain axial symmetry, caricaturizing fetal appearances [8].

The parthenogenetic origin of ovarian type IV GCT is, however, different from most testicular and extragonadal GCT, which arise from totipotent-state, probably type II, primordial germ cells. However, the exception to this would be GCT in female fetuses, neonates, and infants, where ovarian GCT such as yolk sac tumors and immature teratoma probably originate from pluripotent primed embryonal stem cells derived from germ cells prior to maternal imprinting, corresponding to type I tumors in the Oosterhuis and Looijenga's classification. Nevertheless, distinction between parthenogenetically originated (type IV) and infantile type I GCT is by no means clear-cut.

Finally, in elderly women, GCT are found in association with somatic tumors of Müllerian nature and possibly arise from malignant stem cells analogous to induced pluripotential stem cells (iPSC) corresponding to type VI GCT.

As a consequence of their different origin, type IV ovarian teratomas and type I tumors rarely exhibit genetic markers such as 12p isochromosome and chromosome 12 overrepresentation [9].

F.F. Nogales, MD, PhD (✉) • M. Schuldt, MD
Department of Pathology, University of Granada Medical School, 18016 Granada, Spain
e-mail: fnogales@ugr.es; maollyschuldt@hotmail.com

F.F. Nogales, R.E. Jimenez (eds.), *Pathology and Biology of Human Germ Cell Tumors*, DOI 10.1007/978-3-662-53775-6_6

For this reason, although equating ovarian and testicular germ cell tumors is not biologically correct, their morphology and most of their diagnostic immunohistochemical features are similar, the exception being the expression of some proteins in male but not in female teratomas [10]. Interestingly, some tumor types, such as strumal carcinoid, are unique to the ovary.

In the ovary, GCT are only bilateral in 5 % [11] and lack a familial distribution [12] with some rare exceptions [13, 14]. When they do occur in multiple members of the same family or in families with testicular GCT, a gene conferring susceptibility to GCT may exist [15].

In further contrast with testicular tumors, usual parthenogenetic ovarian GCT do not have a precursor or in situ lesion. Indeed, no parthenotes, similar to those found in the oocytes from the LT/Sv mice strain which spontaneously develop teratoma [16], are found in the human ovary. However, gonadoblastoma, a characteristic tumor of disorders of sex development (DSD), behaves as the equivalent of the germ cell neoplasia in situ (GCNIS), the precursor lesion of most testicular tumors that is almost invariably associated with a background of gonadal Y-chromosome mosaicism; consequently, it should not be considered as true ovarian tissue. Indeed, tumors originated from gonadoblastoma are similar to type II testicular GCT in their histologic types and their relative proportions.

The WHO classification (Table 6.1) [17] of ovarian GCT has remained unchanged for more than a quarter of a century (see Chap. 1) and in many ways overlaps with that of older classifications of testicular tumors. It is a descriptive, histopathologic classification that does not take into account the different cellular developmental origins of GCT during different ages of life that are relevant to their behavior. Indeed, it does not include distinct bio- and clinicopathologic categories such as GCT associated with disorders of sex differentiation (DSD) or those originating from somatic neoplasms. It lists together both common tumors and extremely unusual lesions in the ovary (i.e., embryonal carcinoma, sebaceous tumors). Moreover, the term monodermal teratoma comprises a mixed bag of ectopic tissue

Table 6.1 WHO classification of germ cell tumors of the ovary [19]

Germ cell tumors
Dysgerminoma
Yolk sac tumor
Embryonal carcinoma
Non-gestational choriocarcinoma
Mature teratoma
Immature teratoma
Mixed germ cell tumor
Monodermal teratoma and somatic-type tumors arising from a dermoid cyst
Struma ovarii, benign
Struma ovarii, malignant
Carcinoid
Strumal carcinoid
Mucinous carcinoid
Neuroectodermal-type tumors
Sebaceous tumors
Sebaceous adenoma
Sebaceous carcinoma
Other rare monodermal teratomas
Carcinomas
Squamous cell carcinoma
Others
Germ cell-sex cord-stromal tumors
Gonadoblastoma, including gonadoblastoma with malignant germ cell tumor
Mixed germ cell-sex cord-stromal tumor, unclassified

lesions of debatable histogenesis that may have either germ- or stem cell origins. However, the introduction of new stem cell pluripotency antibodies and genetic markers provides a new understanding of the identification and taxonomy of malignant GCT types [18] that may help their reclassification according to other criteria than mere histologic ones (see Chap. 3).

6.2 Primitive Germ Cell Tumors of the Ovary

Primitive GCT are defined as malignant neoplasms that caricaturize early embryonal developmental stages. Primitive ovarian GCT represent only 2 to 5 % of all ovarian cancers [20, 21], but they constitute the most common malignant ovarian tumor in women under the age of 20 [5]. Recent epidemiologic data [22, 23] have demonstrated a trimodal age distribution of malignant GCT in women: a high incidence peak at the end of the second decade of life and two of lower amplitude (in the immediate postmenopause and at age 70). These peaks are possibly related to different pathogeneses of ovarian GCT in each age group: the first one is represented by parthenogenetic tumors, the second includes malignant degenerations of benign teratomas, and the third is comprised of GCT patterns originated from somatic tumors.

Malignant GCT represent a model of progressive neoplastic differentiation from pluripotent stem cells where every stage of development and tissue is caricaturized by a well-defined tumor type [24]. Thus, dysgerminomas reproduce the morphology of primordial germ cells, embryonal carcinoma mimics inner mass embryonic stem cells, polyembryoma duplicates the trilaminar embryo, yolk sac tumors the primitive endoderm and its differentiations, choriocarcinoma the early placenta, and both immature and mature teratoma, the embryonal somatic organs. However, malignancy is not excluded by full differentiation, as mature tissues may ultimately undergo malignant change, giving rise to a malignant neoplasm within a mature teratoma (see Chap. 12) [25].

There are over 23 instances of malignant testicular GCT for every malignant ovarian one [22]. As a reflection of their different pathogeneses, the proportion of histologic types of germ cell tumors varies widely between testis and ovary. Compared with the testis, the ovary shows a substantially lower proportion of seminoma/dysgerminomas (2:50) and mixed GCT (33:<1) and a negligible presence of embryonal carcinoma compared to the testis, while yolk sac tumors (YST) and choriocarcinomas share similar proportions in each organ (1:1 and 0.3: <0.1, respectively) [26]. An unknown, but significant, proportion of malignant tumors in phenotypic females originates in underdiagnosed DSD cases with Y-chromosome mosaicism, which further narrows the relative number of malignant GCT originating from ovarian tissue.

In the ovary, the proportion of the different histologic types of GCT is variable. For some [27, 28], dysgerminoma is the more common malignant GCT. However, immature teratoma seems to be the more frequent GCT in various statistics [20, 29–31]. Since parthenogenetic benign teratomas are the most frequent ovarian GCT, the same would be true for immature teratomas, some of which may share a similar origin.

Malignant ovarian GCT have highly aggressive behavior. Before the introduction of modern combination chemotherapy in the 1970s and 1980s, the prognosis for patients with malignant ovarian germ cell tumors was extremely poor [32]. In non-dysgerminomas, the survival rate for patients with apparent stage I disease was only 5–20 %, and the survival for those with advanced-stage disease was insignificant [33]. Nowadays, GCT are fortunately a paradigm of tumor response, with a 5-year cause-specific survival of 97 % for dysgerminoma and 92 % for non-dysgerminoma [34]. Adverse prognostic factors are represented by a stage higher than the International Federation of Gynecology and Obstetrics (FIGO) IA, the presence of metastases, and an age at diagnosis greater than 40–45 years [34, 35]. Since malignant GCT often involve young females, conservative surgery with preservation of fertility is paramount [32]. A high proportion of women treated with this type of surgery and chemotherapy retain menstrual function and fertility [36]; this may be related to the extent of chemotherapy administered [37]. Monitoring anti-Müllerian hormone levels as a biomarker of gonadal function and ovarian oocyte reserve may be helpful [37].

Although malignant ovarian GCT incidence seems to be stable or decreasing in some statistics [20], others show an increasing trend [22] (see Chap. 2).

6.2.1 Dysgerminoma

This tumor of the ovary and dysgenetic gonads is the histologic equivalent of the more common testicular seminoma and extragonadal germinomas, with which it shares a high rate of 12p abnormalities [38] and KIT mutations and amplifications [39]. Due to its frequent association to DSD, it was named *disgerminoma* (sic) in Robert Meyer's 1930 seminal paper on ovarian tumors and sexuality [40]. Its frequency in phenotypic females is less than that of teratomas and yolk sac tumors combined, comprising 5–10 % of malignant ovarian tumors occurring in the first two decades of life and being less frequent than immature teratoma [20, 29, 34]. Its well-known precursor lesion is gonadoblastoma, although in an unknown number of cases, it may arise *de novo*.

6.2.1.1 Clinical Features and Treatment

Practically, dysgerminoma involves only adolescents and young adults, being rare in children and in older patients. A coexisting subclinical DSD, such as androgen insensitivity syndrome, should be unmasked by sex-determining region Y (SRY) analysis [41].

Although 30–50 % of dysgerminoma patients are fertile [42, 43], few series studied karyotypic features [44] of malignant ovarian GCT patients, thus not differentiating the relative numbers of true 46, XX females from those with a Y-containing karyotype. Dysgerminomas originated from gonadoblastoma have a different set of chromosomal abnormalities than de novo dysgerminomas in true 46, XX females. However, it must be borne in mind that rarely gonadoblastomas may occur in fertile females [45, 46] and in true hermaphrodites [47], where a mosaicism cannot be excluded. Occasionally, DSD-associated dysgerminomas can be familial [48].

Since dysgerminoma involves young patients, abdominal mass may be associated with pregnancy and can cause fetal death and dystocia. Rare cases of pseudopregnancy have been reported [49]. Dysgerminoma biomarkers include human chorionic gonadotropin (hCG), which may be present in dysgerminomas with focal syncytiotrophoblastic differentiation [50]. AFP is usually secreted when dysgerminoma is part of a mixed GCT with primitive endodermal components of YST. Among paraneoplastic manifestations, there are many reports of hypercalcemia [51] that should be differentiated, especially in the older literature, from ovarian small cell carcinoma with hypercalcemia, which is a histological look-alike. Rarely, it may be associated with systemic lupus erythematosus [52] and cholestasis [53].

In over half the cases, tumors are found at FIGO stage I [54, 55]. Nevertheless, there is a high rate of undetected deposits in para-aortic lymph nodes that may recur [54], implying that clinical stage I tumors are often understaged [35]. Combination platin-based chemotherapy cure rates approach 98 % in early stages and 75 % in advanced stages [31]. No adjuvant chemotherapy is recommended for true FIGO IA stage. Consequently, surgery should be fertility sparing with complete staging procedures, including omentectomy, peritoneal washings, and peritoneal biopsies. Only when karyotypic features or abdominal findings of gonadal dysgenesis are present, bilateral adnexectomy is indicated in order to avoid possible development of a metachronic GCT in the contralateral gonad.

6.2.1.2 Pathology

Macroscopy

Although it is the GCT with the highest rate of bilaterality, this only occurs in 10 % of cases [28]. Tumors are usually solid and well encapsulated; on cut section they are homogeneous or lobulated (Fig. 6.1) but rarely cystic. They are usually white or tan in color, except in congestive, torsioned tumors. Heterogeneous and gritty areas should be sampled in order to demonstrate other histologic components. Punctate red, hemorrhagic foci are often associated with syncytiotrophoblastic differentiation. Marked necrosis should prompt an extensive sampling of the specimen.

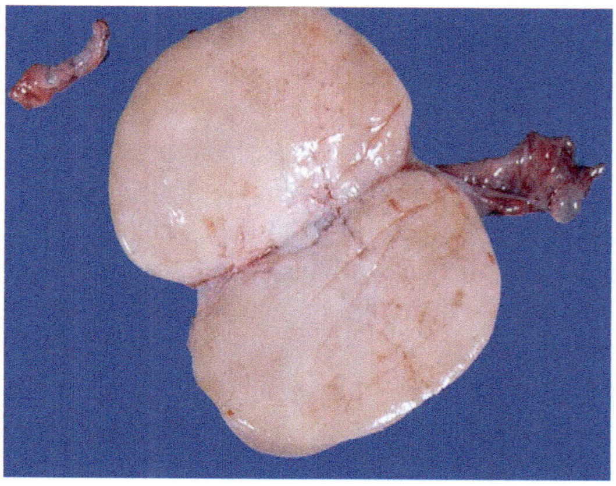

Fig. 6.1 Dysgerminoma in Swyer syndrome. Tumor is attached to a fallopian tube and is homogeneous and rubbery. A contralateral gonadal streak was present

Microscopy

An extensive microscopic description of its testicular counterpart, seminoma, is found in Chap. 7. Here we will review histopathologic features present in dysgerminomas in females and dysgenetic gonads that are relevant to differential diagnosis with other ovarian tumors.

At low power, it is necessary to search for coarse or morular microcalcifications that may remain from a preexisting gonadoblastoma, overgrown by the malignant germ cells of dysgerminoma. Dysgerminoma cells are arranged in a variety of patterns: diffuse or solid, lobular, trabecular, etc. They have uniform nuclei with prominent nucleoli and well-defined membranes. A variable degree of atypia is present but lacks any prognostic significance.

Cells of dysgerminoma have a labile structure due to a poorly developed cytoskeleton, and consequently dysgerminoma histology is often subject to rapid autolytic changes being very sensitive to frozen section artifacts. Retraction of cytoplasmic gels during fixation is often reflected in the formation of clear cells. Invariably, the intervening septa show lymphocytes and, less often, histiocytic, epithelioid non-caseating granulomas, the consequence of an enhanced T-lymphocytic response induced by neoplastic germ cells [56]. This inflammatory response is an invaluable diagnostic aid in poorly fixed cases where identification of the large, clear cells is difficult. Only in rare instances may extensive chronic inflammatory infiltrates complicate diagnosis by effacing the dysgerminoma architecture.

Distortion due to autolysis may produce large empty tubular or pseudofollicular colloid-filled spaces that may mimic other ovarian tumors in the young, such as small cell carcinoma, hypercalcemic type (SCCH), and even struma ovarii (Figs. 6.2a, b). Differential diagnoses also include ovarian lymphoma and clear cell carcinoma (CCC). Dysgerminoma shares many similarities with SCCH since it involves the same age group and may also present with serum hypercalcemia. Especially in poorly fixed cases that imitate pseudofollicular spaces, the diagnosis is made by the absence of lymphocytic infiltrate in SCCH and the characteristic immunophenotype of dysgerminoma. CCCs involve an older age group but can show solid areas with cytoplasmic clearing and even plasma cell stromal infiltrates [57]. Usually, a thorough sampling will reveal the tubulocystic areas of CCCs. Lymphoma in young patients is a more remote differential and when an adequate immunohistochemical panel is used, diagnosis should not present any problems, although in rare cases, ovarian lymphomas can be negative for B and T-lymphocyte (CD3) markers [58] and thus may mimic an undifferentiated tumor.

Syncytiotrophoblastic focal differentiation occurs in about 3–5 % of cases. Syncytiotrophoblasts are found focally either alone or accompanied by isolated mononuclear cells (Fig. 6.3a). Often they are surrounded by erythrocytes from burst capillaries.

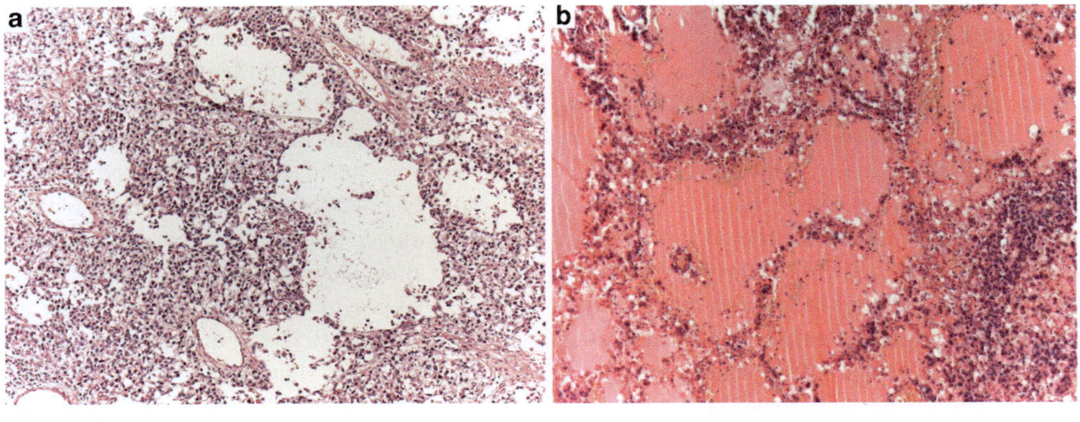

Fig. 6.2 Artifactual changes in dysgerminoma complicating diagnosis: (**a**) pseudofollicular change mimics small cell carcinoma with hypercalcemia. (**b**) Large pools of proteinaceous fluid separated by septa mimic struma ovarii

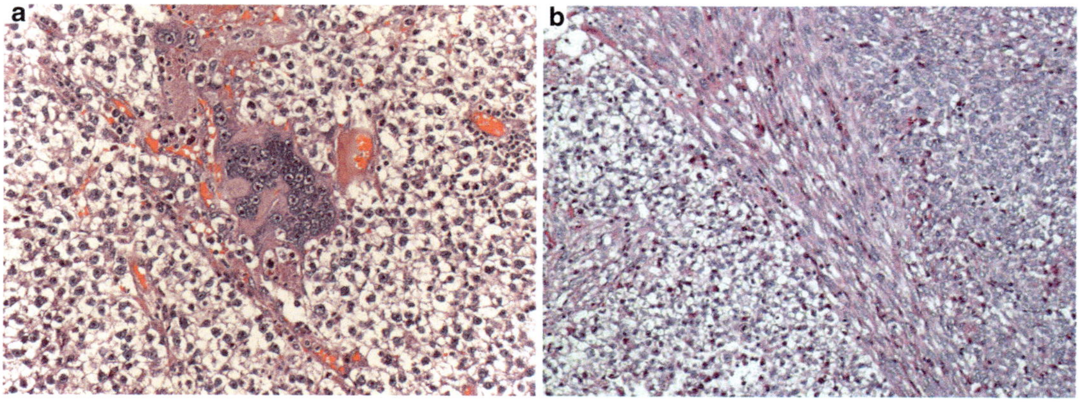

Fig. 6.3 Unusual forms of dysgerminoma. (**a**) Syncytiotrophoblastic differentiation. (**b**) Dysgerminoma displaying somatic malignant change (fibrosarcoma)

In the same way as germinomas of other locations (see Chap. 12), rarely dysgerminoma may be complicated by somatic malignancies (see Chap. 12). A case of coexisting fibrosarcoma [59] has been reported in the ovary (Fig. 6.3b).

Immunohistochemistry

In the past, diagnostic immunohistochemistry for dysgerminomas has relied on the classic membrane staining of placental-like alkaline phosphatase (PLAP). In our experience, however, PLAP is relatively unreliable, especially in poorly fixed material; furthermore, it may stain positively in a high percentage of cases of both embryonal carcinoma and YST [18]. CD117 (c-kit) [60], however, is a conspicuous membrane marker, and c-kit mutations have been shown to occur in a third of dysgerminoma cases [39]. Clone D2–40 of podoplanin is a stable cytoplasmic and membrane marker [61], working well in poorly fixed and necrotic material [62]. OCT4 is regularly expressed in dysgerminoma and in gonadoblastoma, often a precursor lesion, but also in embryonal carcinoma [63]. However, considering the exceptionality of embryonal carcinoma in the ovary, OCT4 can be considered as a selective marker of ovarian dysgerminoma [18], especially in cases of poorly fixed tissue with tubular structures, helping to differentiate them from SCCH and even struma ovarii. OCT4 also identifies isolated dysgerminoma cells engulfed by fibrosis or effaced by inflammation [18] (Fig. 6.4). SALL4 represents a sensitive marker for all malignant ovarian GCT being also expressed in dysgerminoma. Cytokeratin expression does not exclude the diagnosis of dysgerminoma, since positives ranging from strongly diffuse to focal and dot-like can be seen in up to a third of cases [18]. Frequently used cocktails such as CAM 5.2 and AE1/AE3 are positive in 19.2 % and 7.7 %, respectively [64]. Cytokeratin 7 is occasionally focally positive; consequently, it is not advisable to differentiate dysgerminoma from embryonal carcinoma by cytokeratin expression alone, as initially proposed [65]. Expression of cytokeratins has unknown significance, but it may be related to an eventual differentiation of primitive germ cells into somatic-type cells. This assump-

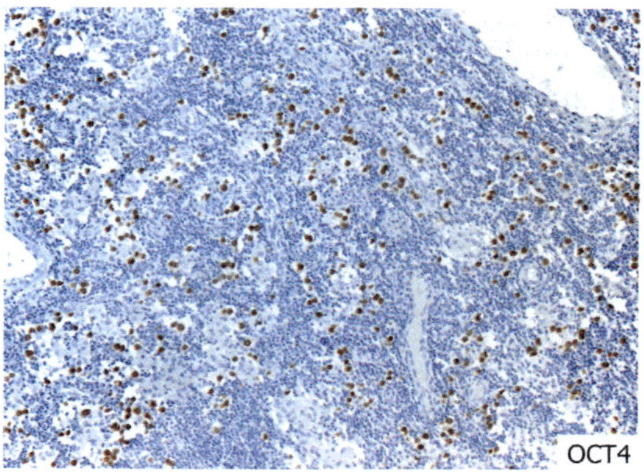

Fig. 6.4 OCT4 expression identifies scattered dysgerminoma cells among a dense lymphocytic inflammatory background

tion is partly supported by the focal positivity of blood group-related antigens in the cytoplasm of dysgerminomas [66], which may reflect a differentiation into somatic-type tissues or yolk sac tumor. Additionally, trophoblastic cell differentiation in dysgerminoma is always cytokeratin positive. Finally, rare cases may be positive for neuron-specific enolase (NSE) [67].

6.2.2 Yolk Sac Tumors: Primitive Endodermal Tumors

Yolk sac tumors (YST) do not represent a discrete histopathologic entity but rather a morphologically heterogeneous group of neoplasms capable of differentiating into various types of endodermal tissues, ranging from the primitive stages of primary and secondary yolk sac to endodermal somatic embryonal and mature adult tissues. Different developmental stages often coexist in the same neoplasm, although usually primitive-type ones predominate. As in any other tumor reproducing embryonic structures, YST are potentially able to differentiate into fully mature tissues [68] such as mucinous and hepatoid neoplasms. Although it is typically a true germ cell-derived tumor in the majority of cases, it can also be a heterogeneous component of somatic Müllerian neoplasms.

Due to its changing multifaceted histology and histogenesis, few tumors have attracted so much historical interest, as summarized in pertinent reviews [69–71].

YST are found at all ages in both the ovary and testis. It is possible that tumors have a different pathogenesis according to the age group in which they present: in neonates, infants, and young adults, they correspond to type I GCT, while in sexually mature women and the elderly, they would belong to type VI. In DSD with a female phenotype, they represent type II GCT.

The current terminology includes both endodermal sinus and yolk sac tumor. The former, which is not recommended [27], is used mostly by non-pathologists without a clear understanding of its meaning: the term resulted from comparative pathological findings between similar morphological structures (endodermal sinuses) in the murine placenta and human tumors. The human placenta and yolk sac, however, are totally different from their murine counterpart. This, together with the relatively rare presence of endodermal sinuses in human tumors, makes this term obsolete [71]. However, neither is the recommended term yolk sac tumor devoid of shortcomings, as it may imply that the actual yolk sac is the origin of the tumor. Furthermore, it is often translated into many other languages, with the name *Yolk* believed to be an eponym [71]. The recently proposed term *primitive endodermal tumors* [71, 72] defines more accurately the various endodermal and mesenchymal differentiations found in these neoplasms [27]. This broader term would also include some somatic clear cell, AFP-secreting neoplasms from the stomach, liver, bladder, and other areas which have an embryonal endodermal glandular appearance. This approach is analogous to the use of the term primitive neuroectodermal tumor (PNET), which includes all possible and complex types of differentiation from primitive neuroectoderm: neuronal, glial, melanotic, etc. [71].

6.2.2.1 Clinical Features and Treatment

Most YST are found at 16–19 years of age. Although only 10 % of cases occur before age 10, they are the most frequent malignant gonadal GCT at this early age [73]. Their frequency is almost identical to that of dysgerminoma. Although in adult and adolescents (possibly belonging to type II) YST share with dysgerminoma a similar rate of 12p abnormalities [44], in children (type I GCT), 12p is not represented [73]. However, YST and dysgerminomas present substantial clinicopathologic differences:

(a) While dysgerminoma shows a 10 % bilaterality, YST are invariably unilateral.
(b) YST occur mostly in 46, XX females, being less frequently originated in gonadoblastoma than dysgerminomas [74, 75].
(c) Dysgerminoma is rarely found in children and postmenopausal patients; [76] in contrast, YST may occur in older patients [77, 78].
(d) In the ovary, YST associate with dysgerminoma and less frequently with other malig-

Table 6.2 Comparison of clinicopathologic features between ovarian dysgerminoma and yolk sac tumor

	Dysgerminoma	Yolk sac tumor
GCT pathogenetic types	II, others (?)	I, II, IV(MCT-associated ?), and VI
12p [73]	Adolescents and adults	Adolescents and young adults, absent in children
Associated GB	Frequent	Rare
Bilaterality	10 %	<1 %
Age	Young and adults	Children, young adults and postmenopausal
Associated GCT	YST, EC, teratoma, CC	IT (children), MCT, dysgerminoma
Associated Müllerian tumors	No	Endometrioid carcinoma and CCC

Key: *GB* gonadoblastoma, *EC* embryonal carcinoma, *CC* choriocarcinoma, *MCT* mature cystic teratoma, *IT* immature teratoma, *CCC* clear cell carcinoma

nant GCT [28]. Indeed, some reports point out to a possible histogenetic continuum of transformation between dysgerminoma and YST [66]. However, their most frequent association with another GCT is with mature teratoma [79, 80]. All these data make difficult to ascribe adult ovarian YST to a particular pathogenetic GCT type.

(e) Type VI GCT arising in somatic Müllerian neoplasms are nearly always YST patterns; only infrequently are they teratomatous structures, such as neurogenic tissues or muscle, but are never dysgerminomas. These points are summarized in Table 6.2.

Tumors present as a rapidly growing abdominal mass. Clinical features include a relatively frequent association with pregnancy that may behave as hormonally functioning, due to stromal luteinization of the tumor [81]. Association with ataxia-telangiectasia syndrome, a condition associated with an abnormal production of alpha-fetoprotein (AFP) has been reported [82]. AFP elevation is usual [83], showing raised levels between 51,100 ± 12,400 ng/ml. Nevertheless, in a differential clinical diagnosis, AFP is also a marker for tumors harboring immature endodermal tissues such as mediastinal, hepatic, gastric, and urinary bladder tumors, hepatoid carcinoma, immature teratoma, and Sertoli-Leydig cell tumors with the liver as heterologous component [84], which exceptionally reach the high serum levels usually associated with YST.

Older series preceding modern chemotherapy [75] reveal their high grade of malignancy: 70 % of tumors were found at stage I at the time of diagnosis but presented subclinical metastases in 84 % [80]. Untreated cases spread rapidly, involving preferentially the liver, lymph nodes, and peritoneum: brain metastases usually occur in advanced stages of the disease [85].

6.2.2.2 Pathology

Macroscopy

Tumors are unilateral and large, with an average size of 15 cm; their weight may reach 5 kg [25]. They are usually well encapsulated, although large herniations and rupture may be present. On cut section, the tumor tissue is gray or yellow with numerous irregular zones of hemorrhage and liquefaction. Cysts are usually situated at the periphery and contain mucinous or gelatinous material. Rarely, the tumors are unicystic [86] or uniformly multicystic; the latter appearance correlates with a histological pattern of polyvesicular type [87, 88] (Fig. 6.5a). Solid tumors with a fibrous or elastic consistency correspond to YST with an abundant mesenchymal component (Fig. 6.5b). One-seventh of cases are associated with an otherwise typical mature cystic teratoma in the ipsi- or contralateral ovary [89].

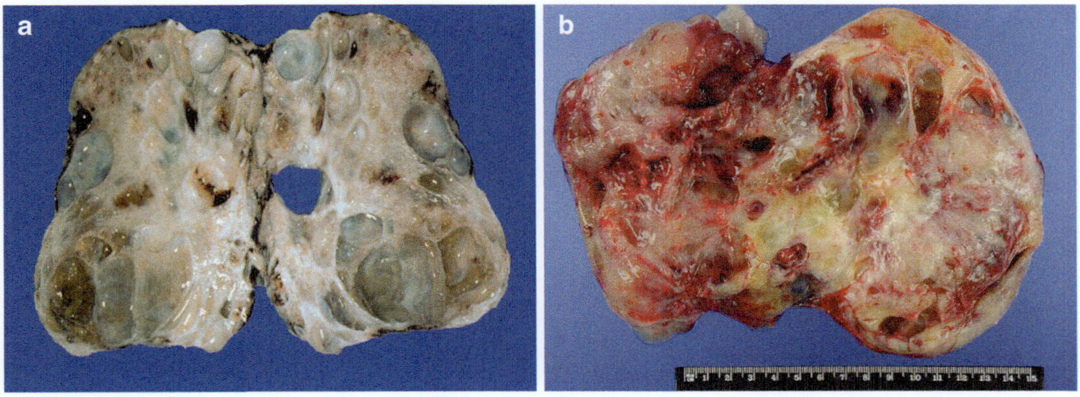

Fig. 6.5 Unusual macroscopic findings in YST. (**a**) Multicystic mass corresponding to a polyvesicular vitelline tumor. (**b**) Fibrous, elastic mass corresponding histologically to a mesenchymal overgrowth

Microscopy [71]

Tumors usually present an admixture of endodermal extraembryonal and somatic patterns in variable proportions.

1. *Extraembryonal endodermal differentiations.* The primitive extraembryonal areas represent their most characteristic diagnostic feature and their identification facilitates the diagnosis of YST in cases where other complex morphological patterns predominate. These patterns resemble visceral and parietal type of murine yolk sac carcinoma and represent the most primitive attempt of endodermal differentiation including primitive yolk sac cavity and mesenchyme. Broadly, the following histological patterns are described:
   (a) *Reticular-microcystic* (Fig. 6.6a). This is the most frequent and characteristic, having a basophilic loose mesenchymal background containing a labyrinth of anastomosing microcysts lined by a flattened epithelium, often presenting with hyaline globules which are also present within the lumina. Masses of amorphous eosinophilic basement membrane material are found in the stroma.
   (b) *Endodermal sinus.* They are tubulopapillary sinusoidal structures with vascular cores lined by cuboidal or columnar endodermal epithelium that protrude into a space that frequently communicates with the microcystic network (Fig. 6.6b). This structure, however, occurs in only 20 % of tumors and should be differentiated from papillary formations present in clear cell tumors. Experimentally, this visceral structure is only present in neoplasms derived from displaced rat visceral yolk sac [90]. Similar papillary structures are found in the endodermal component of teratomas developed from xenotransplanted human induced pluripotent stem cells [71].
   (c) *Parietal.* This highly unusual pattern closely resembles mouse parietal yolk sac carcinoma; AFP-negative tumor cells are embedded in an amorphous hyaline material [91] (Fig. 6.6c). This feature can occur as a post-chemotherapy conversion [92]. Certainly, this morphology is difficult to explain in humans, since the secondary human yolk sac only has a tenuous, transitory basement membrane [93], which does not resemble the thick Reichert's membrane of the rat yolk sac.
   (d) *Polyvesicular.* This rare variant is composed of numerous cystic spaces (Figs. 6.6d, e) lined by a mesothelial-like epithelium that merges with columnar, clear vacuolated cells. It is more frequent in the ovary, where it may rarely represent the

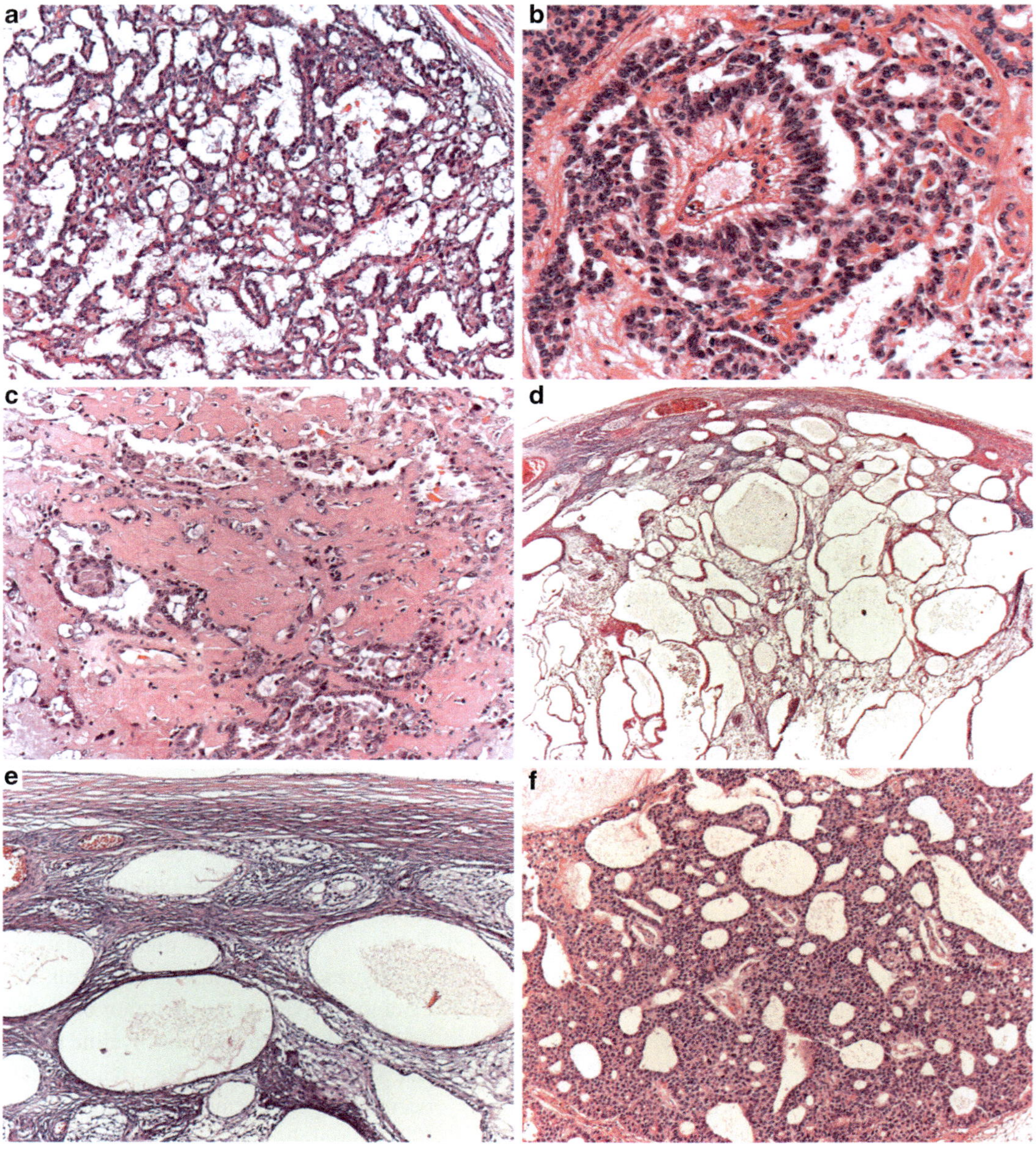

Fig. 6.6 Heterogeneous histology of YST. (**a**) Characteristic microcystic pattern. (**b**) Endodermal sinuses. (**c**) Parietal variant. (**d**, **e**) Polyvesicular vitelline pattern showing a cystic pattern. Vesicles have a biphasic lining similar to allantois. (**f**, **g**) Unusual cribriform variant with tubular formations and hemopoiesis (**h**) reproducing human yolk sac histology. (**i**) Solid pattern with microcysts and hyaline globules. (**j**) Glandular papillary variant imitates intestinal tube formation and is lined by clear cells surrounding a space (**k**) filled with dense eosinophilic material. (**l**) Columnar cells in glandular pattern with characteristic apical and basal vacuolation, often surrounded by marked periglandular stromal rarefaction (**m**). In rare occasions glandular YST shows a tubular pattern lined by compact non-vacuolated cells (**n**). Hepatoid pattern (**o**) showing a trabecular arrangement. Hepatic cells have moderate atypia and exhibit numerous hyaline globules (**p**). Mesenchymal overgrowth of a YST (**q**), composed by a loose mesenchyme with only few isolated epithelial elements present. Glandular YST (*top*) coexisting with an insular type carcinoid (**r**). Mucinous tumor of the ovary (**s**) presenting with an aggressive clinical course. Immunohistochemistry revealed a full YST immunophenotype

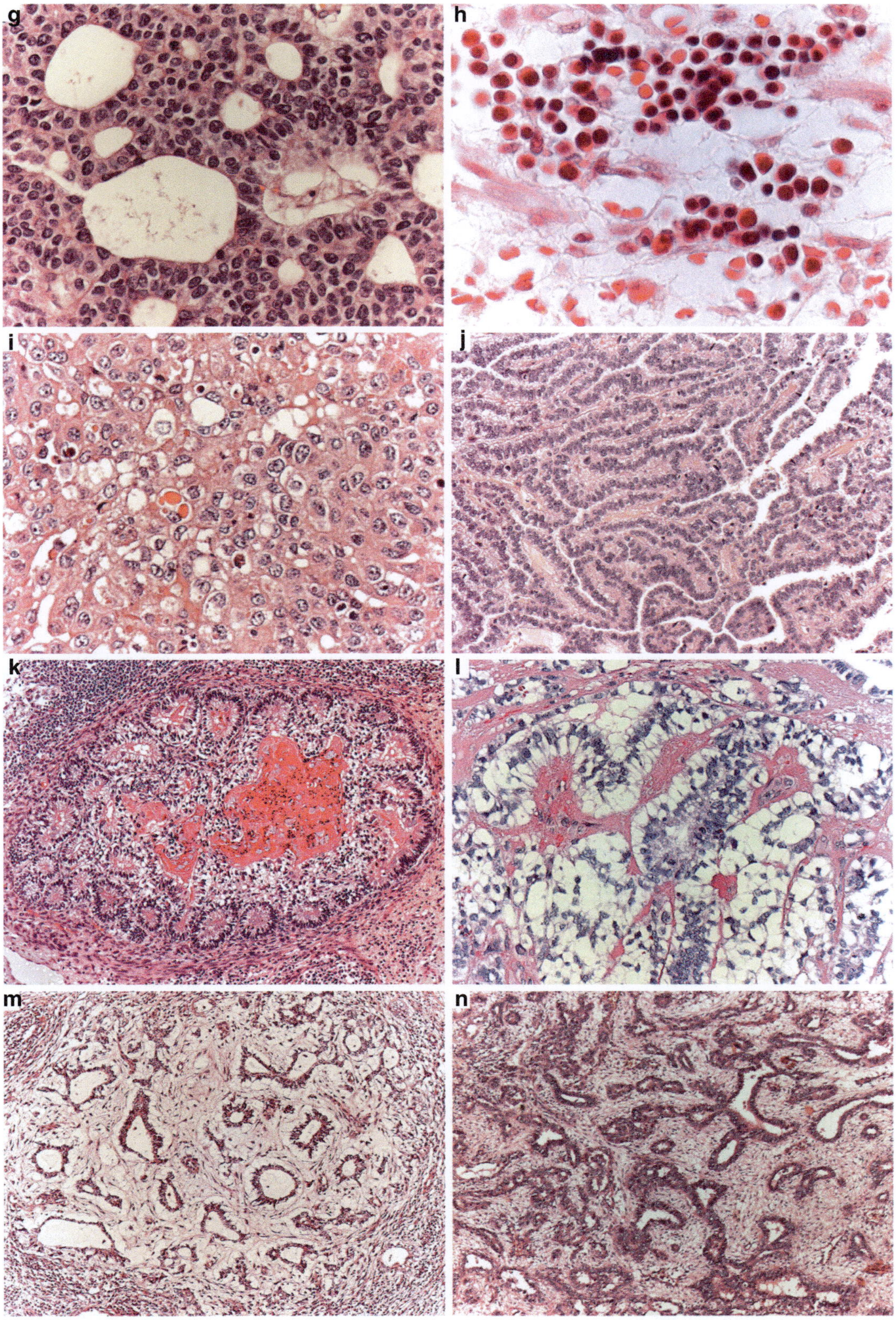

Fig. 6.6 (continued)

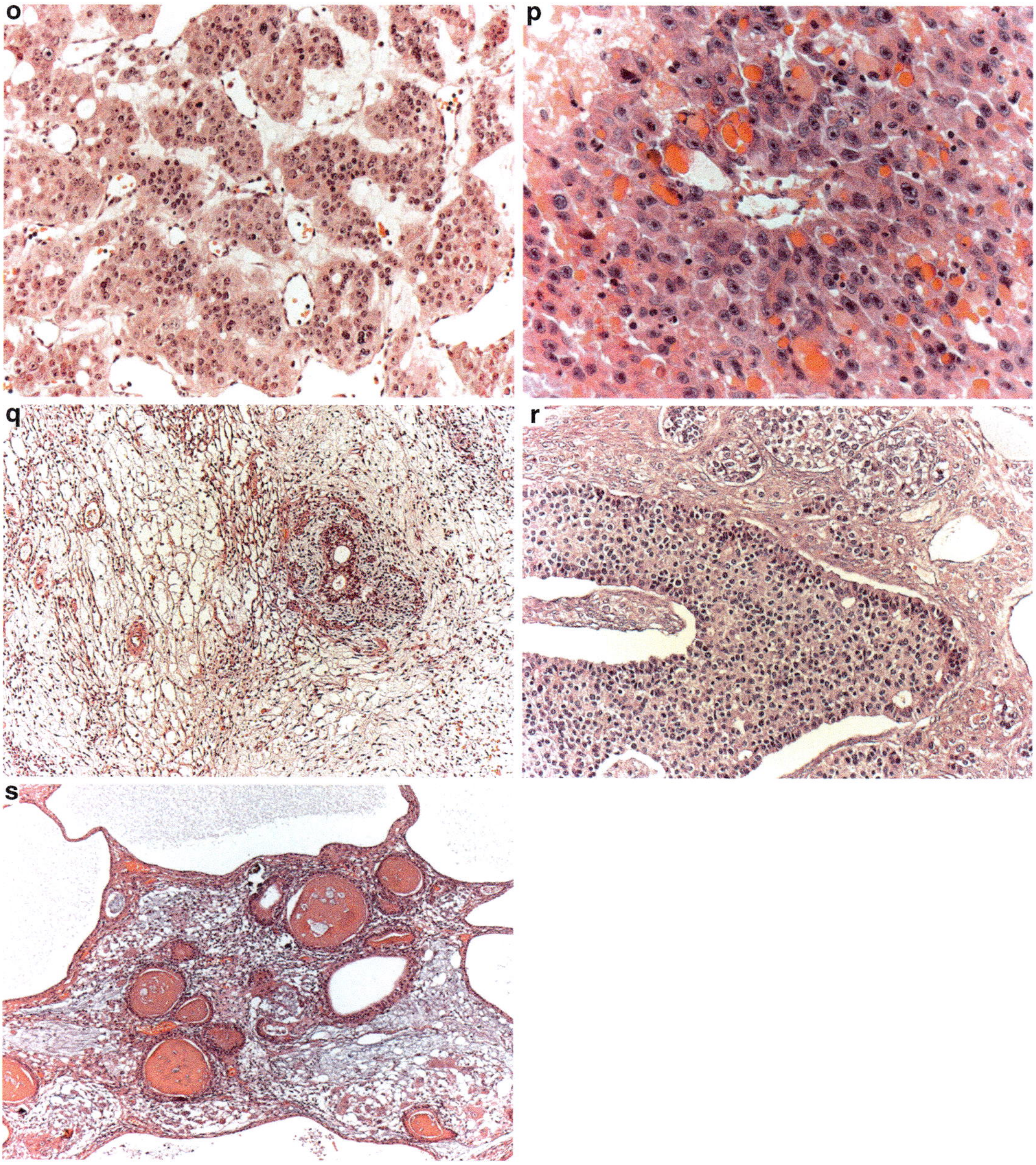

Fig. 6.6 (continued)

predominant pattern [87, 88], being readily confused with other multicystic tumors (clear cell, endometrioid adenofibromas, cystic mesotheliomas, etc.). We believe that these vesicles reproduce the morphology of the allantois, which has similar lining of a biphasic epithelium.

(e) *Cribriform-tubular.* This exceptional YST growth pattern is the only one that caricaturizes the histology of a 6–7-week human yolk sac (also see Fig. 6.7a). Histologically, it has polygonal cells arranged in tubular formations (Figs. 6.6f, g) that, in a similar way to the secondary human yolk sac, may even display hematopoiesis (Fig. 6.6h). In mediastinal and lung YST, these hematopoietic cells may act as precursors of some hematological malignancies (see Chap. 12). Hematopoietic cells should be differentiated from apoptotic epithelial cells.

(f) *Solid.* This pattern [94] has no correspondence with any embryonal structures and consists of a solid sheet of polygonal, often vacuolated, cells (Fig. 6.6i) with frequent hyaline globules. They have eosinophilic or clear cytoplasm, often negative for AFP [95], and may also present abortive tubular formations. Solid YST lacks the plump cytoplasm and characteristic nuclei of embryonal carcinomas and the consistent chronic lymphocytic infiltrates of dysgerminoma.

2. *Somatic endodermal differentiations.* They reproduce endodermal somatic derivatives such as respiratory, intestinal, and liver tissues. These morphological variations can mimic many other tumors among which glandular patterns are the most frequent.

(a) *Glandular patterns* [96]. We prefer to use this broader term rather than others such as enteroblastic, enteroid, intestinal, endometrioid-like, etc. They represent a differentiation into somatic endodermal epithelia of the embryonal gut. Glandular patterns consist of a complex network of cysts and glandular spaces that often have papillary projections (Fig. 6.6j). Occasionally, the glandular spaces are filled with a dense eosinophilic material (Fig. 6.6k) and lined by columnar epithelial cells with apical or subnuclear vacuolation (Fig. 6.6l), similar to those of the embryonal gut, that may differentiate into isolated goblet and neuroendocrine cells. Often, a marked periglandular stromal rarefaction is present (Fig. 6.6m). Unusually, somatic glandular patterns may have gland-like spaces with empty lumina lined by compact, closely packed columnar cells lacking vacuoles but set in a loose stroma (Fig. 6.6n). Due to their similarities with many adenocarcinomas, glandular patterns are frequently misinterpreted [97]. Similar areas may also occur in AFP-positive gastric clear cell carcinomas and fetal-type lung adenocarcinomas. Gastric tumors may metastasize to the ovary and may mimic a primary ovarian YST, especially if the metastasis is unilateral. In these cases, the absence of concurrent primitive YST patterns in the ovarian mass and the presence of angioinvasion may support the diagnosis of metastasis.

(b) *Hepatic differentiation* [98] is a relatively common focal differentiation. True hepatocytes are found in a solid or trabecular arrangement (Fig. 6.6o), often presenting with hyaline globules (Fig. 6.6p). The hepatic nature of these cells has been documented ultrastructurally [99] and may reveal bile secretion and even have associated hematopoiesis. This growth pattern is a frequent source of misinterpretation [97] as it can be confused with dysgerminomas, hepatoblastoma, hepatoid carcinoma, and clear cell carcinomas, but usually the presence of primitive YST foci provides the clue to their identity.

3. *Other secondary mesenchymal or endodermal epithelial tumor patterns in YST.* They represent either overgrowths or full differentiations of endodermal lineages present in YST.

(a) *Mesenchymal overgrowth* occurs [100] as a mesenchymal expansion of loose, myxoid tissue stimulated by endodermal epithelium, taking place in periglandular areas and thus mimicking a similar epithelial-mesenchymal transition seen in early embryogenesis. Primitive mesenchyme may differentiate into any derivatives such as smooth or striated muscle and cartilage. Rarely, mesenchymal tissues constitute the bulk of tumor with the presence of only a few engulfed epithelial foci (Fig. 6.6q). In these cases, the diagnosis of YST is only possible after extensive sampling.

(b) *Carcinoids.* Since the neuroendocrine cells of the gut have an endodermal origin, YST may, albeit rarely, differentiate into an extensive epithelial secondary pattern resembling *mucinous carcinoid (adenocarcinoid),* displaying prominent goblet cells and neuroendocrine components [101]. They should be differentiated from metastases of gastric carcinomas. Foci of *insular type of carcinoid* originate from YST, particularly from glandular, intestinal-type patterns [102] (Fig. 6.6r), and may produce insulin [103].

4. *Endodermal differentiations occurring in other ovarian tumors.* Immature teratomas may differentiate endodermal foci that have prognostic significance [97, 104].
   (a) *Immature endodermal teratomas* [68], (type I GCT) in children, reproduce endodermal and mesenchymal tissues in the absence of neuroectodermal differentiation and represent a neoplasm developmentally related to YST.
   (b) *Mucinous tumors of the ovary.* Some histologically well-differentiated mucinous tumors occurring in young patients rarely may have an aggressive clinical course and present a YST immunophenotype expressing SALL4, villin, AFP, and Glypican-3 (Fig. 6.6s) [105]. They could represent a mature mucinous differentiation of a YST. This hypothesis is partly supported by recent genetic findings proposing that some intestinal-type mucinous tumors may have a germ cell origin [106].

Immunohistochemistry [96, 107]

The heterogeneous immunophenotype of YST is a consequence of the various differentiations and developmental stages that coexist in these tumors. For this reason, the use of a broad immunohistochemical diagnostic panel covering various endodermal phenotypes is advised [96].

Comparative expression of various proteins in the human yolk sac (HYS) and YST has been attempted. As a referential point, the HYS immunophenotype has been recently studied by analysis of the expression of pluripotentiality and endodermal proteins [107]. The HYS (Fig. 6.7a) has a consistent expression of immunohistochemical markers associated with hepatic (HepPar-1 and GPC3) and intestinal (villin and CDX2) functions. In addition, it expresses SALL4 during all its functional period. In contrast, LIN28, another pluripotentiality protein, is only expressed in early yolk sacs up to the 5th week of development. Additionally, AFP is secreted by endodermal cells via a transient canalicular network that represents the substrate of a transport system functioning during the period of maximum activity of the HYS [107].

Alpha-fetoprotein (AFP) has been considered a gold standard for the diagnosis of YST, although it is known that this embryonal protein can be expressed, among others, in ovarian clear cell carcinoma [108] and in ovarian metastases of gastric carcinoma. In the normal yolk sac, AFP is present as early as the 5th week showing a granular cytoplasmic positivity, which is concentrated in the intra- and intercellular lumina and tubular surfaces that constitute a complex transfer system. In classic YST patterns, AFP expression is constant but heterogeneous in the epithelium, (Fig. 6.7b) where it also delineates occasional cellular lumina. In somatic glandular or solid [95] patterns, however, its distribution tends to be focal or absent [109]. Thus, it can be said that AFP negativity does not necessarily preclude a diagnosis of YST.

Glypican-3 (GPC3). In the normal yolk sac, GPC3 displays a cytoplasmic or membranous expression (Fig. 6.7c) that also highlights inter- and intracellular tubules. GPC3 is also extensively or patchily expressed in YST [110]. GPC3 is diffusely expressed in classical YST patterns but has a heterogeneous distribution, and it is even absent in somatic glandular variants. Similar in distribution to AFP, it is a more sensitive antibody, although not as specific as AFP [111], being also expressed in tumors of the female genital tract such as the primitive neural tubules of immature teratoma, hepatocellular carcinoma, squamous cell carcinomas, carcinosarcomas, and placental site trophoblastic tumors, among many others. Some clear cell carcinomas may also express GPC3, and consequently its demonstration may not be that useful in the differential diagnosis [112] of YST originating from somatic tumors.

Hepatocyte paraffin-1 (HepPar-1) is expressed throughout the lifespan of the human yolk sac, reflecting its vicarious hepatic role. HepPar-1 positivity has been reported in both the hepatoid areas of YST and in hepatoid carcinomas [113]. Our study suggests that the frequent, but focal, HepPar-1 positivity in YST does not necessarily mean that it identifies a hepatic tissue, but that it can also reflect a HYS differentiation. Well-differentiated glandular YST of intestinal type show an intense and diffuse HepPar-1 expression (Fig. 6.7d) similar to that found in the small intestine [114].

CDX2 expression is consistently present in the human yolk sac [107]. In classical patterns, CDX2

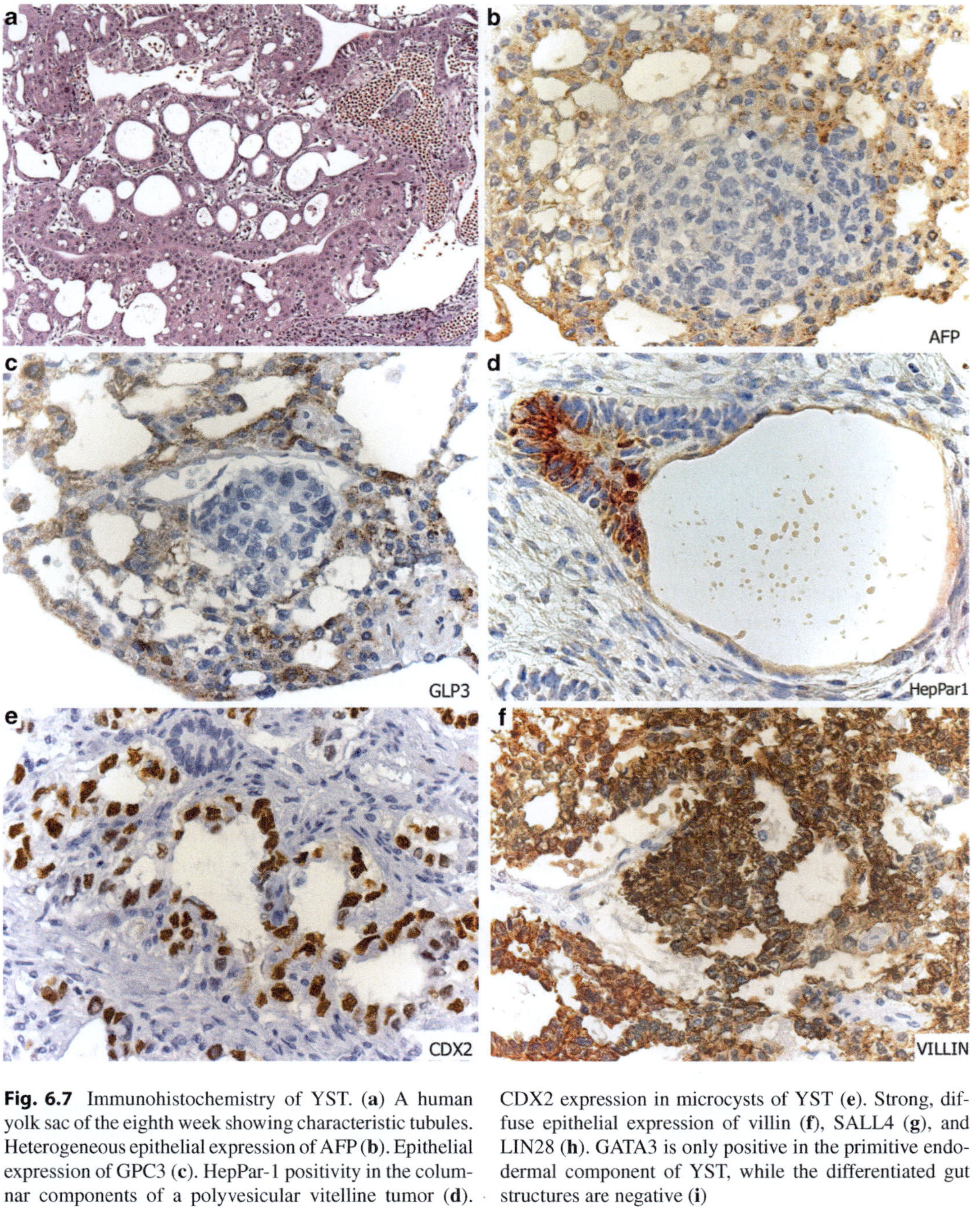

Fig. 6.7 Immunohistochemistry of YST. (**a**) A human yolk sac of the eighth week showing characteristic tubules. Heterogeneous epithelial expression of AFP (**b**). Epithelial expression of GPC3 (**c**). HepPar-1 positivity in the columnar components of a polyvesicular vitelline tumor (**d**). CDX2 expression in microcysts of YST (**e**). Strong, diffuse epithelial expression of villin (**f**), SALL4 (**g**), and LIN28 (**h**). GATA3 is only positive in the primitive endodermal component of YST, while the differentiated gut structures are negative (**i**)

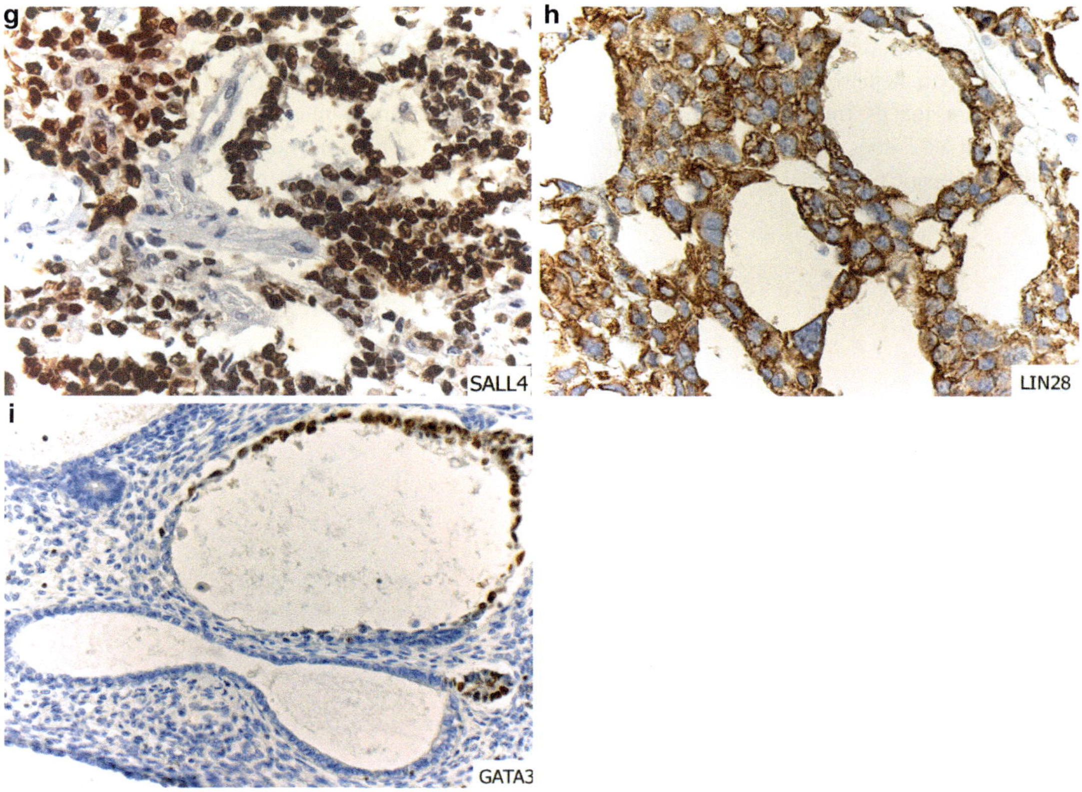

Fig. 6.7 (continued)

expression is focal [115, 116] (Fig. 6.7e), but in somatic glandular patterns, it displays a stronger, diffusely positive, staining [96]. Therefore, it would seem that CDX2 positivity will highlight both areas of HYS and intestinal differentiation, the latter being more evident in somatic glandular patterns, especially those with vacuolated epithelia resembling the embryonal gut.

Villin is consistently expressed during early embryogenesis in both HYS and early endoderm [107] and is a highly sensitive endodermal epithelial component marker in all cases of YST, both in classical and somatic glandular patterns (Fig. 6.7f), being absent in pure embryonal carcinoma and seminoma/dysgerminoma. Its diffuse cytoplasmic expression parallels that of the HYS [107] and intestinal adenocarcinomas, where both diffuse and apical staining patterns occur [117]. We believe that villin is an excellent marker of both primitive and differentiated YST variants. To our knowledge, villin has not been previously used as a marker for the epithelial components of YST.

SALL4, as a marker of cells retaining pluripotency, is a highly sensitive marker for YST (Fig. 6.7g), but with a low specificity, since it is also consistently expressed by all malignant primitive germ cell tumors of both gonadal and extragonadal locations [118–120]. It is also present throughout the life cycle of the HYS, reflecting that this temporary organ retains a degree of pluripotency in its endodermal cell component. This could explain the eventual differentiation of ectopic mature endodermal tissues such as liver cell and enteric tissue in the placenta, which may originate from displaced yolk sac remnants [121, 122].

LIN28 is also a highly sensitive marker for malignant GCT [108, 123–125] (Fig. 6.7h). However, its expression is not specific for YST, although it has been proposed that it might be useful for immunohistochemical detection of YST metastases and for differential diagnoses with clear cell carcinoma of the ovary [108, 124, 125], where both AFP and GPC3 can also be positive. Other stemness markers, such as OCT4 and SOX2, are not expressed in YST [126].

GATA3 [127] exhibits nuclear expression only in primitive patterns of YST being, however, often negative in glandular and hepatic variants [128]. Conversely, *NUT* (nuclear protein in the testis) expression is absent in primitive patterns but is positive in differentiated variants, as it seems to play a role in intestinal or hepatoid differentiations [129].

As a summary, a diagnostic antibody panel, including both markers of pluripotentiality (SALL4 and LIN28) and endodermal identity (AFP, GPC3, and villin), is useful in recognizing the multiple differentiations present in YST. The overlapping immunophenotypes of primitive and differentiated YST areas lend support to the newly proposed term of primitive endodermal tumors [71]. This diagnostic panel will also prove useful in the differential diagnosis of unusual histological variants of YST in uncommon locations [130], as well as in the absence of classical diagnostic patterns and AFP expression. Furthermore, it helps both to identify YST arising from somatic neoplasms [131] and differentiate them from clear cell carcinomas, especially in elderly patients. Additionally, the negativity of CK7 and EMA is characteristic of primitive YST and discriminates them from endometrioid and clear cell carcinomas [132, 131]. In contrast, glandular somatic patterns often show an incomplete YST immunophenotype, being potentially negative for AFP or GPC3 and unexpectedly positive for CK7, EMA [133], HepPar-1, and CDX2, especially when they occur in association with somatic neoplasms (type VI GCT). Consequently, AFP or GPC3 negativity in the presence of other markers such as villin, SALL4, or LIN28 should not preclude the diagnosis of YST.

6.2.3 Postpubertal-Type Testicular GCT (Type II) in Phenotypic Females

Apart from the relatively common dysgerminoma, embryonal carcinoma, polyembryoma, mixed germ cell tumors, and choriocarcinoma are highly unusual in the ovary but frequent in the testis where they are associated with germ cell neoplasia in situ (GCNIS) [134–136] and, consequently, are dealt with at length in Chap. 7. It is possible that many reports of these tumors in women, especially in the older literature, may correspond to phenotypic females with a Y-chromosome-containing DSD [137], where the precursor lesion would be gonadoblastoma and therefore analogous in pathogenesis to type II testicular tumors. Hence, such diagnoses in phenotypic females should prompt karyotypic or molecular genetic studies [18]. Their real incidence in true 46,XX females is difficult to assess, since many cases reported in the past do not indicate presumptive clinical or genetic data relating to DSDs such as pubertal, menstrual, and gestational status; negative sex chromatin; or karyotype, which may help to classify them as true female neoplasms. As stated in Chap. 3, it is possible that type II GCT of the ovary may have their origin in mild forms of ovarian dysgenesis, possibly transient, which leave no obvious final phenotypic traces.

6.2.3.1 Embryonal Carcinoma (EC)

It represents the archetypal stem cell neoplasm, where every cell is pluripotential [138, 139]. Due to their many histopathologic similarities, it is possible that some EC reported in older series were misinterpretations of solid [140] or hepatoid forms of YST [141] or even dysgerminomas. The characteristic immunophenotype of EC, expressing OCT4, CD30, and SOX2, should facilitate diagnosis. However, EC may be associated with yolk sac or trophoblastic differentiation and thus may reveal focal expression for AFP or GPC3. Exceptionally, we have seen EC in true female children and adolescents in association with YST and dysgerminoma, but never in a pure form. In the older literature, there is only one large histopathologic series [142] with limited clinical details, as well as a few isolated case reports, often only partially illustrated or supported by modern immunohistochemistry. One recent study analyzed six ovarian mixed GCT with a component of EC with characteristic immunophenotype of pluripotential markers and presence of 12p chromosome alteration, but the Y-chromosome or testis-specific protein Y-encoded-1 (TSPY-1) status was not mentioned [140].

6.2.3.2 Polyembryoma

This recently reviewed, fascinating GCT pattern [143] caricaturizes the morphology of the trilaminar embryo, forming blastocyst-like embryoid bodies scattered throughout the tumor. They are often associated with EC, YST, or trophoblast. Embryoids are surrounded by a circumferential halo of rarefied loose stroma. Their embryonic disks delineate a small amniotic cavity over the ectodermal plate and, more prominently, a cystic primitive yolk sac cavity from the endoderm (Fig. 6.8). Foci of liver cells or primitive embryonic gut are often related to the primitive yolk sac area. A rarefied, basophilic mesoblast surrounds the embryoid, which may present isolated syncytiotrophoblasts. The identity and position of polyembryoma among primitive GCT is still debated. For some, it represents the "most immature of teratomas" [28, 143] due to its organoid appearance and coexistence of teratoid tissues, while for others, it is a form of mixed GCT [144]. We believe, however, that considering its situation in the embryonal developmental pathway, where stem cells become embryonal germ cells in the naïve pluripotent state (see Chap. 3) and its frequent association with EC, polyembryoma represents an attempt of organoid differentiation of a stem cell proliferation.

6.2.3.3 Mixed Germ Cell Tumors

They consist of the admixture of various patterns of malignant GCT, analogous to those commonly found in the postpubertal testis. However, the rare mixed ovarian GCT have the following particular features that makes them different from testicular and some extragonadal neoplasms:

1. In the ovary, the most characteristic combinations of malignant GCT are YST and dysgerminoma [28], and immature teratoma and dysgerminoma (Fig. 6.9), as opposed to the usual combinations of testicular tumors, where EC is common, whereas it is exceptional in ovarian mixed GCT.
2. In the ovary, malignant GCT patterns can be associated to benign, differentiated ones. In particular, YST may associate with mature cystic teratoma in the ipsi- or contralateral ovary [79], both of them lacking 12p alteration. This possibly indicates a pathogenetic relationship in type I or IV GCT between both neoplasms [89]. This mixed form is characteristic in the ovary.
3. Another situation involving coexistence of tumor types is represented by the neoplastic transformation of mature components of cystic teratoma, a situation characteristic in the ovary but infrequent in other sites (see Chap. 12). This combination should not be considered a mixed GCT.

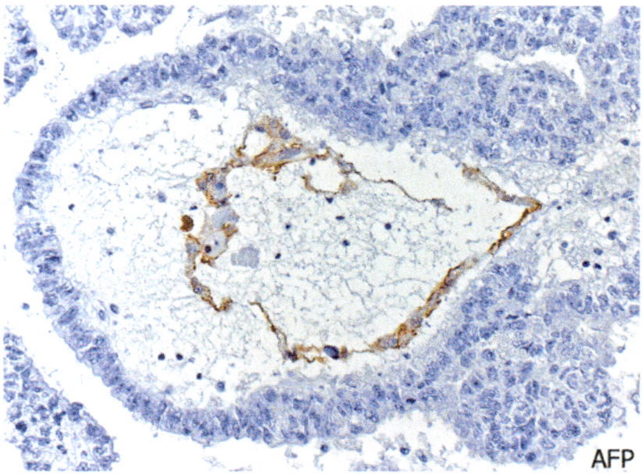

Fig. 6.8 An embryoid body showing a primitive yolk sac cavity, positive for AFP

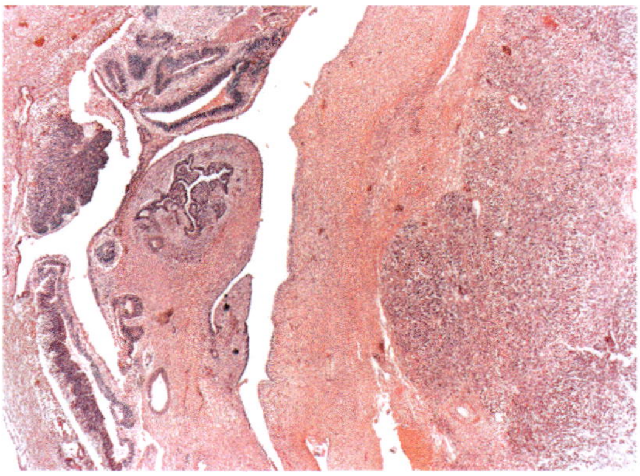

Fig. 6.9 A rare example of mixed GCT (dysgerminoma and immature teratoma) in a 46, XX female

4. In high-grade immature teratoma of the ovary, immature neural structures coexist with immature endodermal ones [97]. This should be interpreted as a manifestation of the pluripotentiality of the teratoma rather than an associated tumor.
5. Teratomas in mixed ovarian GCT have a pathogenesis akin to postpubertal testicular types (type II tumors), different to typical parthenogenetic (type IV) benign teratomas of the ovary [145]. Thus, this type of neoplasm is likely to occur in DSD with a Y-chromosome component. IMP3 expression, characteristic of male teratomas and absent in ovarian ones [10], has not been studied in ovarian mixed GCT.

6.2.3.4 Choriocarcinoma

Choriocarcinoma of non-gestational germ cell origin is also rare and truly exceptional as a pure neoplasm, being usually DSD associated and a component of a mixed GCT [146]. Rarely, it may develop in a mature cystic teratoma (Fig. 6.10). In fertile patients, it should be differentiated both clinically and histopathologically from gestational choriocarcinoma originated from an ovarian pregnancy [147], as both share similar histology, symptoms, image findings, and high serum levels of β-hCG. However, in cases where the differentiation between both types is difficult, genetic analysis can demonstrate either a monospermic complete mole as a precursor of gestational choriocarcinoma [148] or the paternal DNA sequences of an androgenetic mole.

Macroscopically they are solid or cystic hemorrhagic masses, and histologically they show extensive hemorrhage and necrosis that often makes identification of the trophoblast difficult. Both mononuclear cyto- and extravillous (intermediate) trophoblast are intimately admixed in a plexiform arrangement. Numerous tissue blocks should be taken in order to demonstrate other GCT patterns such as mature somatic tissues.

Although non-gestational choriocarcinoma has different chemosensitivity from gestational type, methotrexate-based regimens can be used [149]. However, it has to be taken into account that these neoplasms are likely to be associated with other GCT types and, consequently, cisplatin regimens are preferable.

6.3 Ovarian Teratomas

The usual definition of teratoma implies the differentiation of tissues derived from all three germ layers. Teratomas imitate somatic embryogenesis and a histologic correlation with developmental stages has been proposed [150]. However, this correlation is certainly more accurate when the sequential expression of pluripotentiality markers is analyzed [18]. There are exceptions, as some teratomas only develop one particular tissue foreign to those normally differentiated in any given organ, representing derivations from a single germ cell layer (monophyletic or monodermal teratomas), such as the endoderm in struma ovarii or ectoderm in an ovarian neurogenic cyst. Teratomas show a gradient of maturation and organoid arrangement that includes coexisting embryonal tissues.

Pathogenetically (see Chap. 3), mature, cystic ovarian teratomas, are GCT of parthenogenetic origin (type IV) and constitute the most frequent GCT in 46, XX females. However, there seems to exist a continuum between the type IV mature,

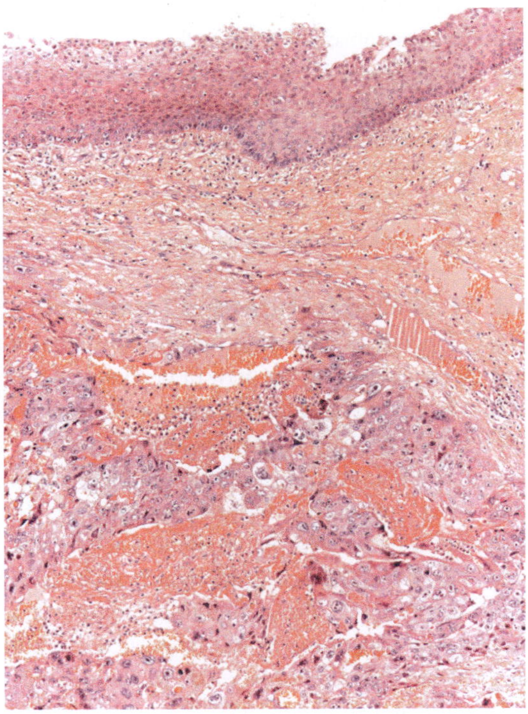

Fig. 6.10 Unusual coexistence of a mature cystic teratoma with non-gestational choriocarcinoma

cystic ovarian teratomas and type I immature teratomas. While the tissues of most teratomas are differentiated (mature teratomas, cystic or solid), a small proportion show varying amounts of immature tissues, which are responsible for their aggressive behavior. Some may show the presence of secondary malignancies originated from the terminally differentiated tissues. While in children the proportion of mature to immature teratomas is relatively similar (2:1) [3], in adult women mature teratomas show an overwhelming predominance over immature ones.

6.3.1 Immature Teratomas (IT)

Immature teratomas (*IT*) reproduce progressive stages of tissue organization reflected in variable amounts of immature tissues, usually of neural origin. Xenotransplant of human embryonal stem cells or iPSC produces experimental immature teratomas identical to spontaneous ones [151]. Most patients are 46, XX with the exception of rare bilateral IT, some of which may occur associated with a Y-chromosome genotype [137]. 12p alterations are absent, but gains from 1p, 16p, 19, and 22q are reported [44]. As mentioned in Chap. 3, immature teratomas of infancy belong to type I GCT; however, in adult women, boundaries between type I and IV tumors are not clear-cut, since both immature and mature (dermoid-like) areas may coexist in the same tumor [158].

6.3.1.1 Clinical Features and Treatment

Most IT occur in adolescent and young females and are exceptional in the peri- or postmenopause [152]. Symptoms are nonspecific, with an average duration of 3 months before surgery, and consist of abdominal mass, pain, vaginal bleeding, and fever in a quarter of patients [153]. Paraneoplastic limbic autoimmune encephalitis [154] may occur. Although the tumors are usually solid, preoperative ultrasonograms may reveal a multicystic neoplasm, in which case a careful surgical incision should be made to avoid surgical rupture and spillage [155].

A third of tumors can exhibit AFP elevation and less than 10 % may show increased serum hCG levels [156]. A unilateral stage I mass is found in up to 69 % of cases [155, 157]. Bilaterality is rare [11]; however, 10 % may be associated with a mature cystic teratoma (MCT) in the contralateral ovary, and, exceptionally, they can be preceded by a previous resection of a MCT in the same ovary [158].

Abdominal extension is present at the time of surgery in a third of cases. Rarely they may present with metastases in soft tissues [159] as well as in the brain, lung, and liver [160].

Surgery with preservation of fertility is the procedure of choice since IT involves young patients. In stage 1A, grade 1 tumors, adjuvant therapy is not necessary. Resection of stage 1A tumors has been reported as curative in pediatric IT [156]. More advanced stages and grades should be treated with current combination chemotherapy. Gliomatosis peritonei is not treated except for recurrent cases or in the development of growing teratoma syndrome [161]; it remains stable for many years and a possible complication is abdominal hemorrhage [162].

6.3.1.2 Pathology

Macroscopy

IT are usually bulky, solid tumors, rarely bilateral, that can be associated with MCT in a fourth of cases [158]. Three quarters show a smooth capsule, and only a third may present with rup-

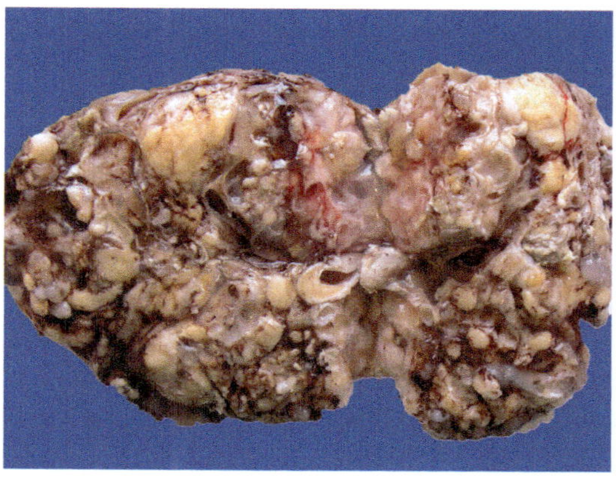

Fig. 6.11 Characteristic gross appearance of immature ovarian teratoma with abundant encephaloid nodular formations

ture or herniations. On cut section, most tumors are solid, white, encephaloid (Fig. 6.11) and may present small cysts. Small foci of bone or cartilage may be present. A multilocular appearance occurs in a third of cases. Only rarely are tumors parvilocular or pedunculated within a large, solitary cyst, and up to a fourth of cases may show embedded or peripheral dermoid-like cysts containing sebum and hairs [158]. Hemorrhage and necrosis are often found and are especially evident in cases of torsion. Sampling must include a tissue block for every centimeter in diameter; it is mandatory that the mass should be adequately sampled as this is crucial for grading the tumor.

Microscopy

A mixture of mature and immature tissues with predominance of neural ones characterizes this tumor. *Grading*, assessed in the relative amounts and degree of immaturity of the neural tissues, is prognostically related as demonstrated in prechemotherapy series of IT [155, 157]. *Grade 0*, or mature solid teratomas, do not show the usual dermoid features and are composed of mature tissues of predominant neural origin. *Grade 1* tumors contain rare foci of immature neural tissue (<1 low-power field (LPF) in any one slide), while *grade 2* and grade 3 tumors contain 2–3 LPFs or 4 or more LPFs of immature neural tissue, respectively [28]. This system can be simplified into a two-tier classification with *low grades* referring to grade 1 teratomas and *high grades* encompassing grades 2 and 3 [163].

Structures relevant to histologic grading. The characteristic *immature neural structures* evaluated in grading are neural blastematous nodules containing rosettes and neuroepithelial tubules (Fig. 6.12a) with a crowded cell lining with abundant mitoses that may be occasionally pigmented. Their sharp-edged apical borders are well delineated by an inner limiting membrane (Fig. 6.12b) which differentiates them from other immature tubules of nonneural origin. It is important to take into account some other immature tubular structures that can mimic neuroepithelial tubules. *Endodermal tubules* lined by tall columnar vacuolated epithelium (Fig. 6.12c), similar to glandular YST, are, in our experience, intimately admixed with immature neural tubules in grade 3 tumors, especially in pediatric IT (type I GCT), where they are found in a third of cases [156]. While some authors may regard these tubules in high-grade IT as microscopic foci of YST, we believe that the presence of these does not necessarily signify a mixed GCT but yet another immature embryologic structure differentiated in a high-grade tumor and thus may influence prognosis [97]. Microscopic areas of primitive, microcystic YST can be found, especially in high-grade IT in children (Fig. 6.12d). Endodermal tubules are often identified by marked periglandular stromal basophilic rarefaction (Fig. 6.12e). Hence, the finding of endodermal tubules should also be taken into account when grading a teratoma. Solid, undifferentiated areas, rarely forming embryoid bodies, may be present in high-grade tumors and resemble EC, even sharing a similar immunophenotype (see below).

Elements which are not relevant to grading include *metanephric tubules*. Nevertheless, these may mimic neuroepithelial tubules as they are lined by crowded cells with scanty cytoplasm. However, they lack the inner limiting membrane characteristic of neurotubules and are associated with abortive glomerular structures (Fig. 6.12f). Other areas showing developmental immature features which are not relevant to grading are *mesenchymal derivatives*, such as immature skeletal muscle, which should be differentiated from true overgrowth of sarcoma [28]. Immature mesenchymal areas often resemble myxoid liposarcoma with a characteristic vascular arrangement. Other immature areas such as tooth buds, lack any importance other than providing a beautiful microphotograph. Marked vascular hyperplasia with both endothelial and adventitial proliferations, similar to that occurring in central nervous system tumors, is frequently found in any grade. However, it may be quite prominent in high-grade IT, where it may be associated with a malignant overgrowth of primitive neuroectodermal tumors (PNET), including malignant retinal anlage-type tumors [164].

Only when IT elements are a component of a type II mixed GCT, they may it show 12p chromosome abnormalities [145]. Some teratomas

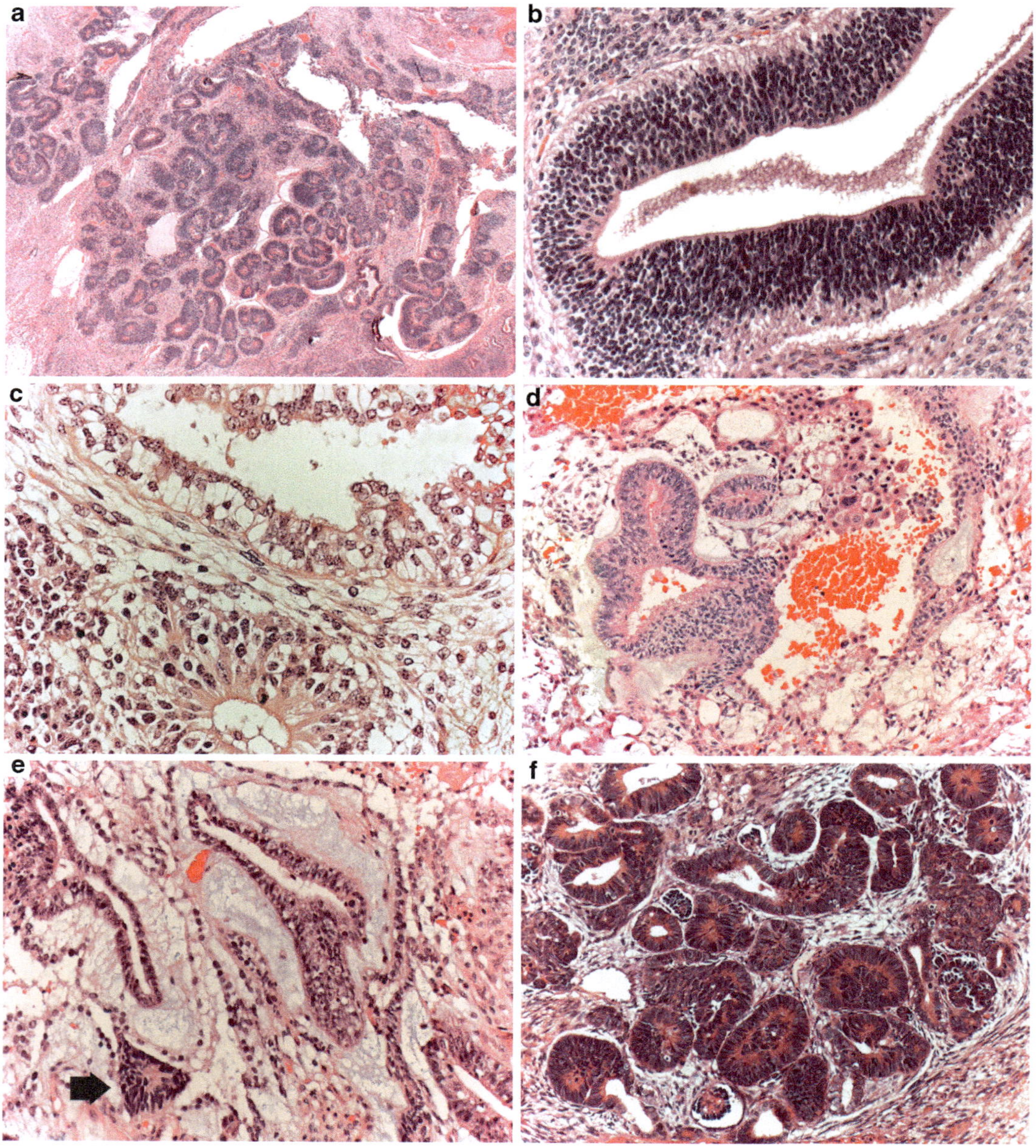

Fig. 6.12 Structures relevant to histologic grading in IT. Neural tubules in a high-grade tumor (**a**) with a basophilic cellular lining with numerous mitoses. The luminal edge is usually sharp (**b**). Coexistence of endodermal tubules (*top*) with a neural rosette (*bottom*) (**c**). Neural tubules admixed with a small focus of reticular YST in a grade III teratoma (**d**). Endodermal tubules with periglandular stromal rarefaction coexist with a small focus of neuroectoderm (*arrow*) (**e**). Neural tubules mimicry: metanephric tubules resemble neural structure but also display abortive glomeruli (**f**)

are almost exclusively composed of both mature and immature endodermal structures in various stages of differentiation. For these, the name immature endodermal teratoma has been used, and their relationship with YST is open to speculation [68].

6.3.1.3 Complications and Unusual Features

Malignant overgrowth of some immature tissue components in IT is a rare event, being less frequent than the development of secondary somatic malignancies in testicular GCT. Since neural tissues

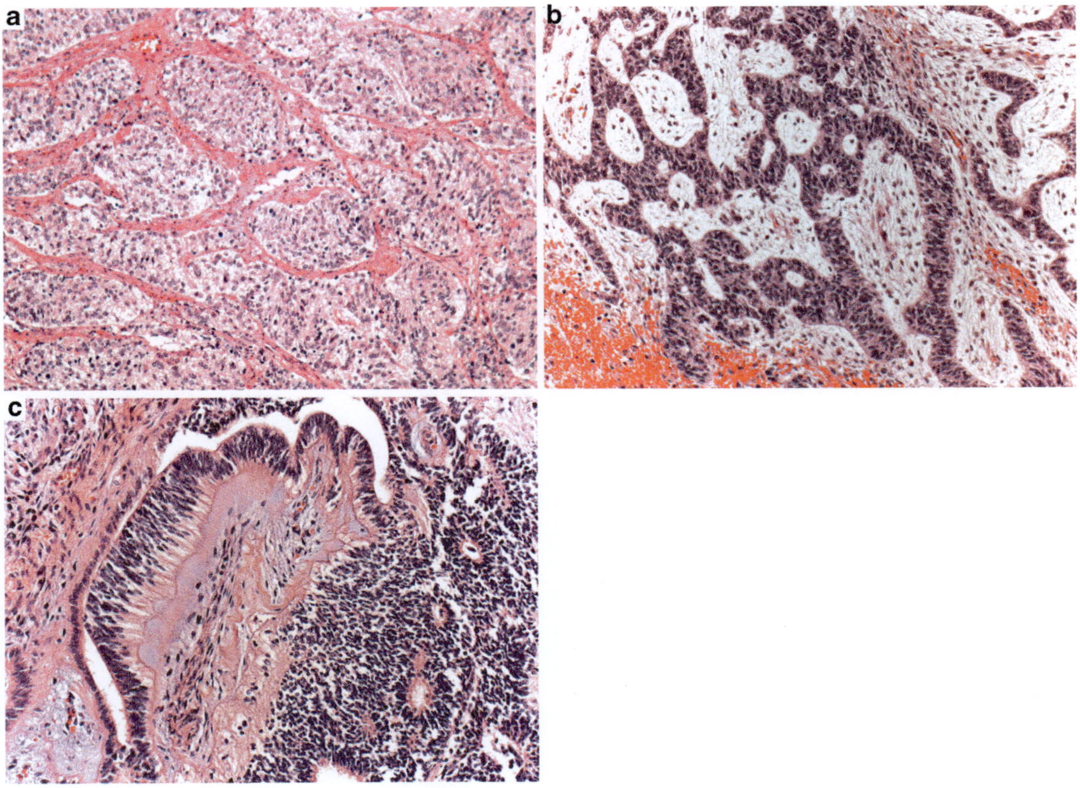

Fig. 6.13 PNET overgrowths in immature teratoma. Nested pattern with vascular hyperplasia (**a**). adamantiform pattern (**b**) and ependymal differentiation (**c**)

predominate in IT, overgrowth by astrocytoma [165] and highly malignant PNET (Fig. 6.13a, b) may occur. Cases reported as neuroblastoma, neurocytoma, etc., arising in IT, would belong to this category. GCT-associated PNET often have central genotype [166], lacking the ESWR-1 rearrangement, and consequently diagnosed as medulloblastoma-like [167, 168]. Ependymoma is only rarely associated with teratoma (Fig. 6.13c).

Rhabdomyosarcoma has also been reported as an overgrowth from IT presenting with peritoneal and retroperitoneal involvement [169].

Mature implants and metastases from IT. Immature tissues from ovarian teratoma can achieve full differentiation in abdominal and less frequently, extra-abdominal deposits, either spontaneously or subsequent to treatment. This phenomenon takes place in both *growing teratoma syndrome* and *gliomatosis peritonei* and may occur simultaneously [170].

Growing teratoma syndrome. The presence of expansile masses of mature teratoid tissues in the abdomen, pelvis, retroperitoneum, liver, lung, or pleura after chemotherapy for ovarian IT was initially reported by DiSaia [171] and later found in testicular tumors [172] (see Chap. 7). Only rarely may it also be present in untreated patients. Resected masses are usually nodular and comprised of cystic structures that expand by continued secretion from their serous or mucinous epithelia, causing compression of various organs and structures. Long-term follow-up [173] often shows no changes in the benign tissues, although eventually they may rarely develop endodermal type VI GCT, such as glandular YST and carcinoid [174, 173] or intestinal-type adenocarcinomas [170].

Gliomatosis peritonei (GP) [175] is a relatively frequent phenomenon associated to ovarian IT, being rare in teratomas of other organs. GP presents as widespread small solid nodules of mature glia in the peritoneum (Fig. 6.14a, b) and abdominal lymph nodes. Rarely, it may occur outside the abdomen [176]. Its behavior is benign, although it may recur [177]. Implants grow rapidly but remain

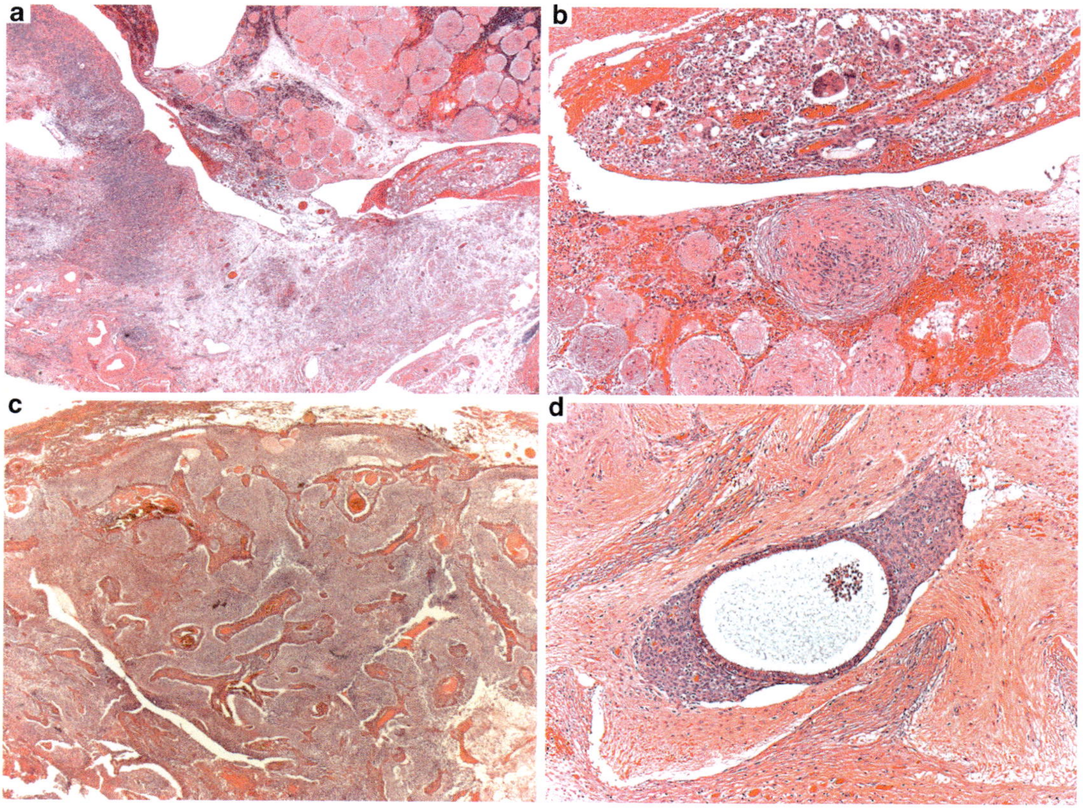

Fig. 6.14 Gliomatosis peritonei. Glial nodules in the ovarian surface (**a**) and peritoneum (**b**) coexisting with a marked granulomatous response to embedded hairs. Marked hyperplastic vascular change in gliomatosis peritonei (**c**) may be responsible for hemoperitoneum. Rarely glial nodules coexist with endometriosis (**d**) (*center*)

unchanged for life, undergoing only involutive phenomena of an ischemic nature. Astrocytes represent their main component, but other neural lineage elements, such as neurons and ependyma, and mesenchymal tissues such as cartilage can be found [178]. Secondary changes in glial nodules may include (a) degenerative astrocytic changes; (b) granulomatous and follicular chronic inflammatory response; (c) association with hormonally related changes, such as decidual peritoneal metaplasia; (d) endothelial and adventitial vascular hyperplasia [162] (Fig. 6.14c); (e) complete regression [179, 180]; and (f) development of glioma [181]. Association with endometriosis has been reported, where endometrium and glia are intimately admixed (Fig. 6.14d). This intriguing phenomenon may either imply a common origin for each component or a possible glial induction of endometrial differentiation [182].

Two pathogenetic mechanisms are considered in GP: (a) direct seeding of immature neural cells from a primary tumor with subsequent differentiation and (b) metaplasia from peritoneal stem cells.

An implantation mechanism is supported by the following clinicopathologic data:

(i) The GP nodules present with a wide cellular heterogeneity of neural and nonneural components, showing coexistence of mature astroglia with neural blastemal areas and other non-ectodermal tissues such as cartilage. This would favor an implantative origin from pluripotential teratoma cells that undergo multiple differentiations in the peritoneal environment.
(ii) The frequent presence of shed keratin and embedded hairs within the nodules

(Fig. 6.14b), possibly originating from the primary ovarian neoplasm, is a strong clue for an implantative origin.

(iii) The presence of lymphovascular involvement and even extraperitoneal deposits [176].

(iv) There are rare cases of GP associated with ventriculoperitoneal shunts which would constitute a natural experiment of the implantative capacity of glial cells present in the cerebrospinal fluid into the peritoneum [183].

However, a metaplastic origin is sustained by a heterozygosity pattern of GP nodules, identical to the normal tissue but different from the coexistent ovarian teratoma [184, 185]. SOX2 is expressed by gliomatosis [175] and is one of the key factors for the maintenance of pluripotency in stem cells. Its expression is required for inducing stem cells to differentiate toward the neural lineage, and, consequently, SOX2 may play a key role in the pathogenesis of gliomatosis peritonei [181]. Thus, GP would constitute a peritoneal change in response to growth factors secreted from teratoma or macrophages.

While an implantative origin from ovarian teratoma remains the more probable mechanism in most cases, metaplastic glial differentiation from peritoneal stem cells could explain cases of GP with a monomorphic astrocytic cell population, as well as those associated with endometriosis [186].

6.3.1.4 Immunohistochemistry

The main role of immunohistochemistry in these otherwise diagnostically straightforward tumors lies in identifying features relevant to grading that may be not conclusively recognized with hematoxylin eosin stains. This could be achieved by distinguishing various tissue components as well as their degree of immaturity [96]. Identification of neural areas can be complemented by characteristic neural makers such as glial fibrillary acidic protein, nestin, and others. In immaturity assessment, SALL4 (Fig. 6.15a) is expressed in immature components that may be present in grade 3 tumors, not being expressed in lower grades. On the other hand, SOX2 is only expressed in immature neural areas (Fig. 6.15b), while any of the endodermal markers used in diagnosis of YST (AFP, GPC3, villin, HepPar-1, CDX2, etc.) would be expressed in immature endodermal areas. It is worth remembering that GPC3 is also expressed in neuroepithelium and could highlight both endo- and neuroectodermal components. A combination of SALL4, SOX2, and markers such as villin, AFP, and GPC3 will manifest both immature endodermal and neural areas and facilitate semiquantitative grading. Finally, SOX2 marking is particularly useful in the demonstration of the rare PNET overgrowths of teratoma, including glioma arising from gliomatosis peritonei [181].

In some grade 3 tumors, we have identified [105] focal solid areas with an EC immunophenotype, coexpressing OCT4, SOX2, and CD30 (Fig. 6.16a,

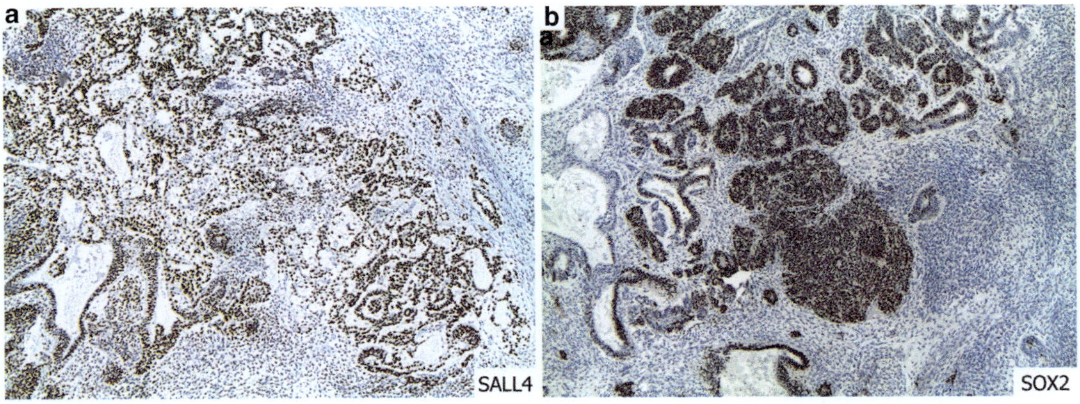

Fig. 6.15 Immunohistochemical support for diagnosis of immature areas of IT. SALL4 stains diffusely all immature elements (**a**), while SOX2 (**b**) only highlights immature neural areas

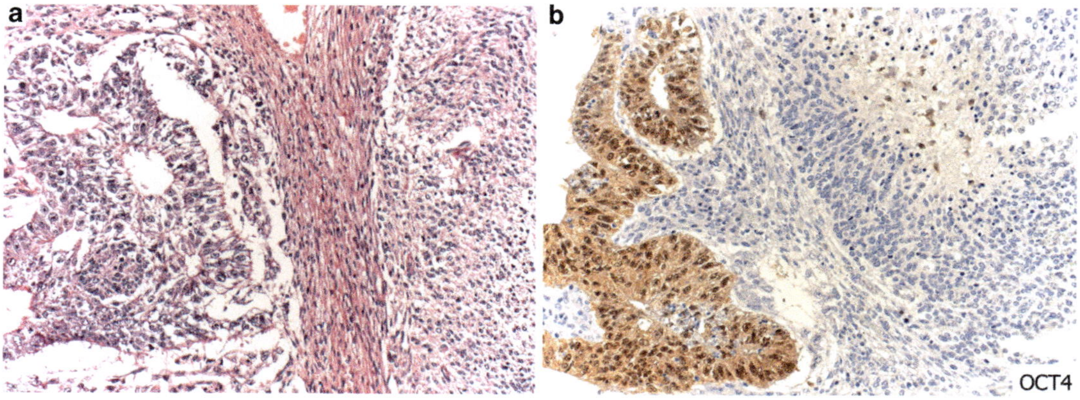

Fig. 6.16 Rare foci of embryonal carcinoma-like areas (**a**) in high-grade IT. The cellular, fenestrated areas that coexist with neural elements (*right*) coexpress OCT4 (**b**) as well as CD30 and SOX2

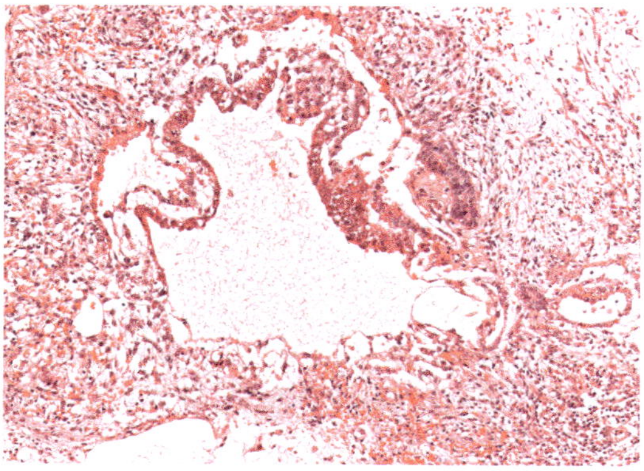

Fig. 6.17 Isolated embryoid found in a high-grade IT

b), which would represent the stem cell population of some IT. In these unusual cases, EC-like areas are not associated with the usual admixture of tumor types present in testicular-type mixed GCT but may develop into organoid structures of embryoid bodies (Fig. 6.17) [28]. This finding may explain reports of OCT4 positivity in IT [187].

6.3.2 Mature Teratomas

The most frequent GCT in females represent the final stages of multidifferentiation of a parthenogenetically induced germ cell [7, 188] after meiosis I [189]. These fascinating ovarian tumors are capable of reproducing practically any adult tissue and organ, even replicating a rudimentary external human form (fetiform teratoma).

Ovarian teratomas that can also originate from supernumerary ovaries [190] present with fully differentiated tissues with a cystic appearance (*dermoids*) and are only rarely solid or macroscopically similar to IT. Mature cystic teratomas (MCT) constitute the most common ovarian tumor in the young and, consequently, are prevalent in countries with a young average population. They can be diagnosed at any age but only a small proportion is found in postmenopausal women. Familial incidence, including presentation in identical twins, is rare [191–193].

6.3.2.1 Clinical Features and Treatment

They present with the usual symptoms of abdominal mass, mostly when they are associated with pregnancy. Preoperative diagnosis is usually performed due to their characteristic ultrasound and MRI appearances [193].

Torsion seems to be the most frequent complication of MCT. In rare cases it may lead to amputation and eventual parasitic implantation of tumor in the abdominal cavity. During pregnancy, MCT may be a mechanical obstacle to uterine expansion and delivery and cause ectopic pregnancies. Rupture, either spontaneous or iatrogenic, is relatively uncommon but may produce granulomatous serositis and hemoperitoneum. Rupture may be followed by fistulization into adjacent organs. Exceptional cases of peritoneal melanosis may be associated with chronic hemorrhagic spillage of pigmentary substances including melanin from the

Table 6.3 Clinical complications in mature teratoma

Complications	References	Frequency
Torsion	[153, 200]	3–16 %
Amputation	[201–203]	RR
Parasitic implantation	[204–207]	RR
Associated with pregnancy	[208–210]	10 %
Torsion	[208, 211–213]	FR
Ectopic pregnancy	[209, 210]	RR
Incarceration	[214]	RR
Prolapse	[214]	RR
Dystocia and rupture	[215]	RR
Recurrence after surgery	[216–218]	3–4 %
Neoplastic degeneration	see Table 6.5 [28, 219, 220]	0.2–2 %
Rupture and consequences (spontaneous or iatrogenic)	[153, 221, 222]	<1 %
Chemical peritonitis and granulomatous serositis	[223–225]	RR
Fistulization and perforation into adjacent organs (rectum, small and large bowel, urinary bladder)	[226–229]	RR
Hemoperitoneum	[230]	RR
Peritoneal melanosis (melanin) Peritoneal melanosis (iron compounds)	[231] [232]	
Infections		
Salmonella	[233, 234]	RR
Actinomyces	[235]	RR
Schistosoma	[236]	RR
Filaria	[237]	RR
Autoimmune disorders[(a)]		
Hemolytic anemia	[238, 239]	RR
Dermatomyositis	[240]	RR
Polyarthritis	[241]	
Generalized pruritus	[242]	RR
Exercise-induced urticarial vasculitis	[243]	RR
Limbic encephalitis with anti-NMDR antibodies	[194, 197, 198, 244, 245]	RR
Hashimoto thyroiditis	[246, 247]	RR
Hormonal secretion		
Hyperprolactinemia	[248, 249]	RR
Hypercalcemia	[250]	RR
Hyperthyroidism	[251, 252]	
Cushing's syndrome	[253]	RR
Virilization	[254–257]	RR
Hyperestrogenism	[258, 259]	RR
Other complications		
Migraine Peptic ulcer in gastric tissue Metastases to teratoma	[260] [232, 261, 262] [263]	RR RR RR

Key: *FR* frequently reported, *RR* rarely reported, ([a]) often with disappearance of symptoms after tumor removal

MCT. As any other tissue, it may harbor various infectious bacteria or parasites.

MCT may associate with generalized autoimmune disorders triggered by various components of the cyst; hemolytic anemia, dermatomyositis, polyarthritis, pruritus, urticarial vasculitis, etc. are occasionally reported. All symptoms disappear after removal of the teratoma. However, the most frequently reported autoimmune complication of MCT is limbic encephalitis with antibodies against N-methyl-D-aspartate receptor (NMDAR). This rare, often fatal, form of encephalitis has psychiatric symptoms, particularly psychosis, and was first reported in 2005 [194–196]. It has now become a clinically recognizable disease, diagnosed by the presence of anti-NMDAR antibodies in the cerebrospinal fluid, only treatable with removal of the ovarian teratoma and subsequent immunosuppressive treatment [197]. More than half of the ovarian teratomas reported in association with anti-NMDAR encephalitis are MCT, while a quarter of them correspond to IT. Although also associated with teratomas of other organs, the lower frequency of teratomas in extraovarian locations and the smaller proportion of neural tissues present in them would partly explain their rarity. Often, MCT in this syndrome are small and thus difficult to diagnose by ultrasound [198], making higher resolution MRI studies necessary. A similar autoimmune phenomenon, but without systemic consequences, occurs in teratoid thyroid tissue, where Hashimoto's thyroiditis has been reported.

Endocrine symptoms may be associated with the presence of functioning endocrine glands developed in teratomas. Examples include hypophysis able to develop pituitary adenomas, producing Cushing's syndrome or hyperprolactinemia. Hypersecretion of androgens is not unusual and is related, not to the teratoid tissues themselves but to the induced ovarian stromal cells at the periphery of the expanding mass (peripheral luteinization).

Other rare complications include metastases of breast tumors to teratoma.

Treatment must take into account the future fertility of the patient, with preservation of ovarian tissue and limitation of the risk of adhesion formation [199]. Laparoscopic treatment has less postoperative complications and morbidity than oophorectomy, but, on the other hand, it may produce spillage into the abdominal cavity and subsequent adhesions. Furthermore, incomplete resections of multifocal tumors may lead to recurrence, which can occur in up to 4 % of cases, especially in rare multiple tumors within the ovary. A comprehensive summary of reported complications (except malignant transformation) occurring in MCT is shown in Table 6.3.

6.3.3 Pathology

6.3.3.1 Macroscopy [153]

Bilaterality occurs in 15 % of cases and not infrequently contralateral ovaries harbor small, occult MCT. Rarely, tumors are multifocal in the same ovary [264]. Average size is 8 cm, although it may vary remarkably, the record size being awarded to a 42 kg tumor [265]. Tumors are round or ovoid with a prominent vascular network and have a smooth capsule with occasional adhesions. Their consistency immediately after excision is fluid becoming doughy at room temperature. Tumors are usually cystic and contain

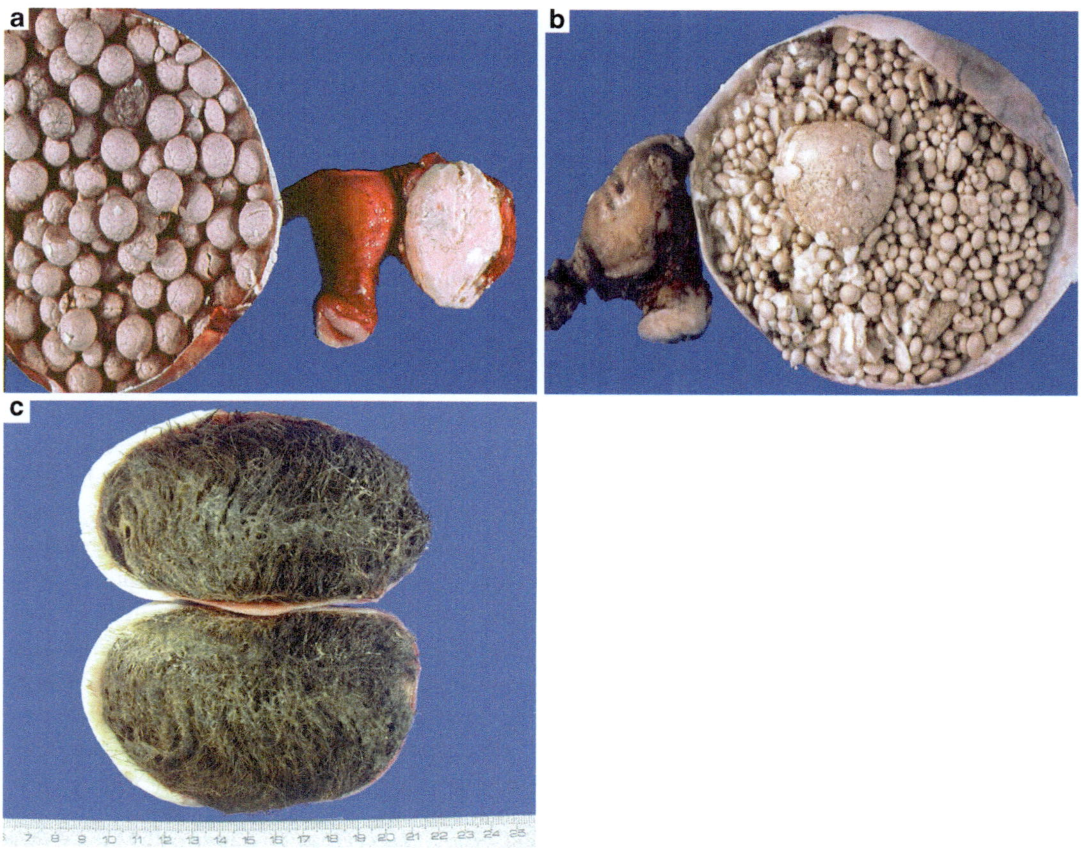

Fig. 6.18 Unusual macroscopic findings in mature cystic teratomas. Hair and sebum aggregate to form "golf balls" (**a**) or "jelly beans" (**b**). (**c**) This rare hairy teratoma replicates the illustration of the first description of a teratoma: *Johannes Sculteto: Trichiasis admiranda siue morbus pilaris mirabilis noribergae (Nürnberg, 1658)*

yellow sebum mixed with hairs in 98 % of cases; only on rare occasions are they predominantly solid. Sometimes the sebum becomes aggregated in sebum and hair spheroids ("golf balls" or "jelly beans") (Fig. 6.18a, b) that have a peculiar MRI appearance [266]. Infrequently, contents are exclusively fluid: serous, mucinous, or resembling cerebrospinal fluid. The wall is usually elastic, although irregular calcified plaques can be present in older patients.

A dermoid protuberance, often named in the older literature after Wilms or Rokitansky, is found in 87 % of cases. Although singular in most instances, multiple protuberances can occur. Hairs may constitute the main bulk of cystic contents (Fig. 6.18c), and although their color is often unrelated to that of the patient, they may become gray with age. Teeth are found embedded in the protuberance, usually in groups of two or three, but up to more than 300 teeth have been counted in a single specimen [267]. Only seldom are they set in a well-formed mandible surrounded by gingiva [268]. The teeth are of permanent type, usually canines, incisors, or molars, and are often dysplastic [269].

Unusual macroscopic findings include eversion of the inner lining of the cyst [270]. Highly organized structures resembling fetuses (*fetiform teratoma*) [271] may be found within cysts, and some may even have metameric, axial, and symmetrical organization [272] with nails, vertebral bodies, lung, limbs, and even external genitalia [273], thus deserving the name of *homunculus*, latin for little man (or perhaps better *muliercula*: little woman, taking into account the presence of female external genitalia in some tumors) [273, 274].

6.3.3.2 Microscopy [153]

Typically, the inner surface of the cyst is lined by epidermis and skin adnexa which desquamate into the lumen. Most tissues are found within the protuberance and mostly reproduce those corresponding to the rostral part of embryos. They are haphazardly mixed, although an attempt to maintain a tissue relationship similar to that found in adult organs (polarity of growth) is frequent: bronchi lined with respiratory epithelium with bronchial-type glands and cartilage, intestinal, or gastric mucosa with corresponding layers and even harboring Cajal's cells as pacemarkers in the muscle. Teeth buds imitate normal odontogenesis and thyroid tissue is often in proximity to thymus. Practically any type of tissue except perhaps gonads or adrenal has been reported. Relative percentages of tissues and various curiosities are shown in Table 6.4, and the presence of some unusual tissues is depicted in Fig. 6.19a–k.

Table 6.4 Tissues or organs found in mature teratomas

Tissues/organs	Reference	Frequency
Skin	[210]	98.9 %
Adnexa	[210]	92.9 %
Pituitary	[253, 275, 276]	RR
Eye structures	[210]	
Breast incl. lactation	[279, 280]	RR
Melanocytes	[210]	14.8 %
Langerhans cell	[210]	
Adipose tissue	[210]	29.9 %
Brown fat	Fig. 6.19d [210]	RR
Smooth muscle	[210]	30.3 %
Cartilage	[210]	30.3 %
Bone	[210]	0.7 %
Kidney	[282]	RR
Hematopoietic marrow	[210]	RR
Prostatic tissue	[283–286]	RR
Male sex glands	Fig. 6.19e	RR
Thymus	Fig. 6.19f [287]	RR
Corpora cavernosa	Fig. 6.19g	
Respiratory epithelium	[210]	31.0 %
Schneiderian epithelium	Fig. 6.19h	RR
Gastrointestinal, general	[210]	19.7 %
Intestinal segments	[288, 289]	RR
Stomach	[261, 262, 290]	RR
Cajal cells	[291, 292]	RR
Thyroid	[210]	8.1 %
Teeth	[267]	12.7 %
Glia	[210]	39.9 %
Microglia	[293]	RR
Ependyma	[210]	8.8 %
Neurons	[210]	7.0 %
Choroid plexus	[210]	4.9 %
Leptomeninges	[210]	3.8 %
Peripheral nerves	[210]	27.5 %
Sympathetic ganglia	[210]	9.5 %
Schwann cells	[210]	1.9 %
Cerebellum	[294–296]	RR
Neurohypophysis	[210]	0.6 %

Functioning luteinized stroma, occasionally associated with virilization [256], may occur in the ovarian tissue at the periphery of the cysts [257].

Ageing of the various tissue components is mainly reflected in degenerative phenomena of the nervous tissue, such as microcalcification, microcystic degeneration, gliosis, and gemistocytic change (Fig. 6.19i); Rosenthal fibers (Fig. 6.19j) may also occur [297] and are possibly related to an impaired vascular supply. Chronic lymphocytic inflammation of neural structures has been related to autoimmune anti-NMDAR encephalitis [28].

Reactive phenomena, such as a foreign body reaction to keratin and hair as well as an oleo-

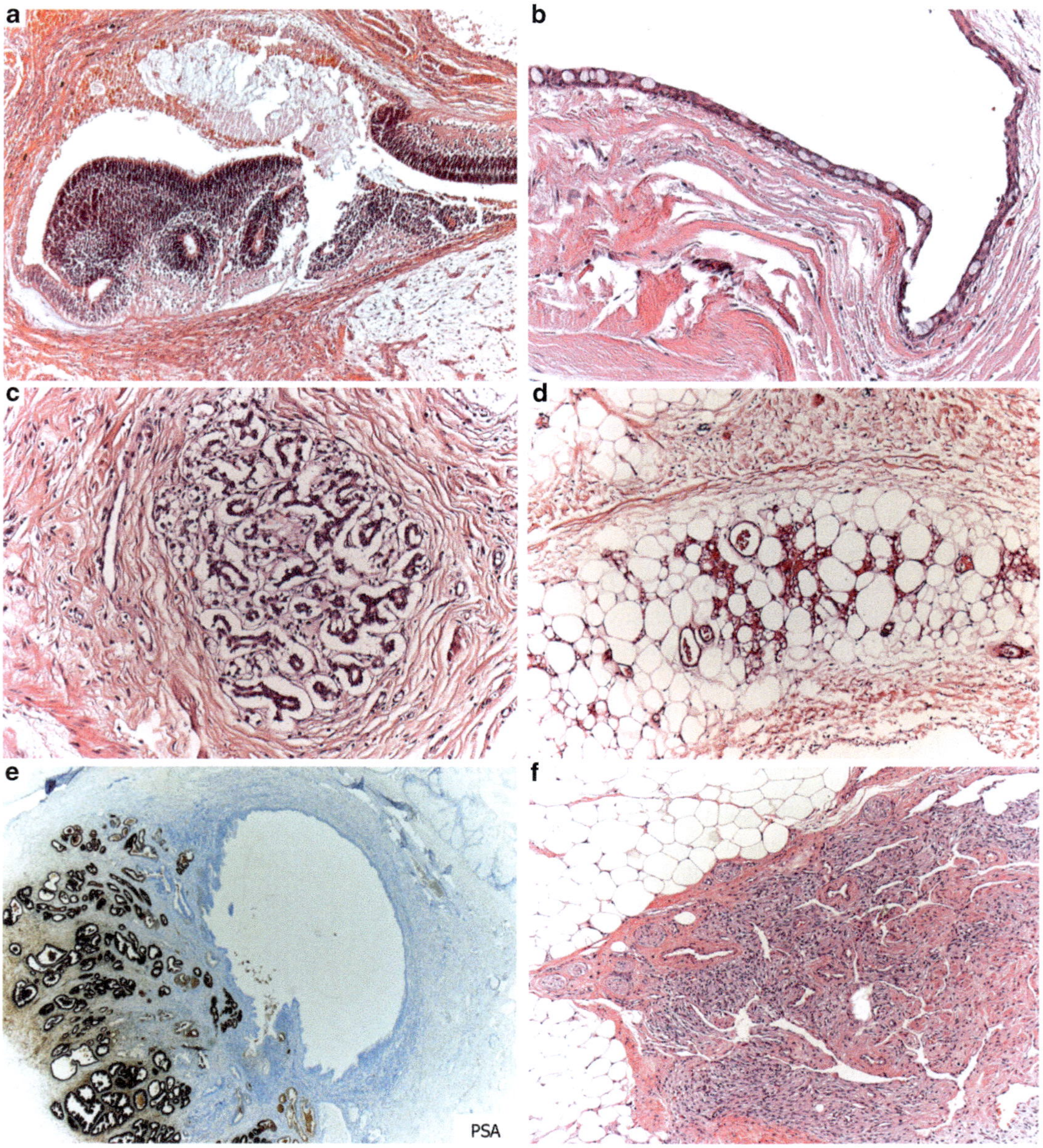

Fig. 6.19 Unusual tissues in mature cystic teratoma. (**a**) Retinal structures may simulate the immature neural tubules of IT. (**b**) Conjunctival epithelium, (**c**) breast lobules, (**d**) fetal fat, (**e**) PSA-positive prostate, (**f**) vascular structures similar to corpora cavernosa, (**g**) thymus, (**h**) Schneiderian epithelium, (**i**) gemistocytic transformation of astrocytes, (**j**) Rosenthal fibers, and (**k**) well-formed cerebellum

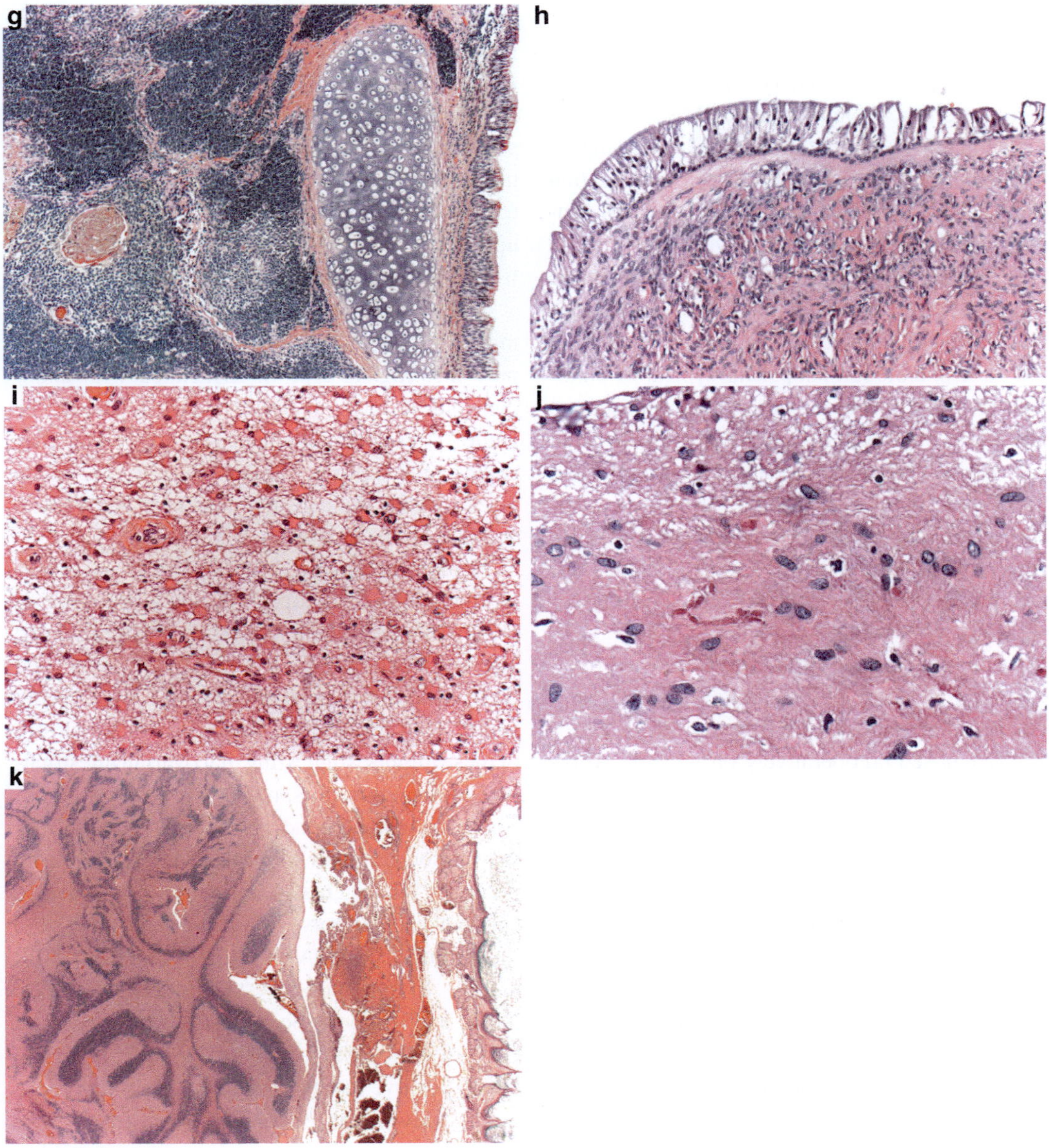

Fig. 6.19 (continued)

granulomatous response to sebum, are often frequent in older patients although they may occur at any age. When prominent, they may efface teratoid tissues and only stable structures such as hair or bones remain. In the tumor wall, mesovarium and mesosalpinx, cystoid oleogranulomas occur and represent an indirect, but pathognomonic, sign of the presence of a neighboring teratoma [153]. Coagulative necrosis, unrelated to torsion, is rare and in one instance [198] has occurred in relationship with autoimmune anti-NMDAR encephalitis. Myospherulosis, as a response to oleous contents, develops occasionally in surrounding tissues [298]. A summary of reported secondary lesions in MCTs is shown in Table 6.3.

6.3.3.3 Secondary Neoplasms Arising in Mature Teratomas

The presence of secondary tumors developing from the differentiated tissues of MCT is a phenomenon that occurs in the ovary in a much higher proportion than in teratomas of any other location (see Chap. 12). Mature teratoid tissues can potentially develop into any tumor type found in the adult. Malignant change occurs in patients in the fifth and sixth decades in 1.5–3 % of MCT [220, 299], often being related to senescence of the teratoid tissues, although neoplasms typical of childhood, such as primitive neuroectodermal tumors, may also occur. Local factors are possibly involved in tumorigenesis, as there are neither associated multicentric tumors in the rest of the female genital tract nor simultaneous malignant change in cases of contralateral mature cystic teratomas, the only exception being bilateral mucinous tumor associated with bilateral teratomas [300].

Squamous cell carcinomas constitute the overwhelming majority of secondary tumors in older patients. Due to their relative rarity, they are seldom diagnosed preoperatively on imaging workups. It has been proposed that elevated serum markers of SCC (squamous cell carcinoma antigen) and CA19.9 may be helpful in diagnosis and even in staging of the tumor [301]. At the time of diagnosis, the tumor often extends beyond the ovary adherent to pelvic organs in over half of cases and ascites is present [302]. Omental, lymph node, and distant metastases are frequent. Due to their advanced stage at the time of diagnosis and their resistance to chemo- or radiotherapy, survival is poor [220, 299].

Macroscopy

The usually unilateral cysts show irregular, thickened, or ulcerated plaques and/or fungating intracavitary growths. Since they involve postmenopausal women, any teratoma in this age group warrants extensive sampling, especially when necrosis, hemorrhage, or thickened or pigmented areas (Fig. 6.20a, b) are present.

Microscopy

Squamous cell carcinoma is the most common histologic type of secondary malignant tumor found in MCT. Most originate from the epidermis, as demonstrated by the presence of in situ lesions (Fig. 6.20c) [153], but are not related to human papilloma virus. An origin from bronchogenic structures is also contemplated. A squamous papillomatous pattern (Fig. 6.20d), reminiscent of a schneiderian papilloma of the upper respiratory tract, should be differentiated from a proliferative Brenner tumor.

Differential diagnosis from rare primary squamous cell ovarian carcinomas [303], either of pure histological type or associated with endometrioid lesions or Brenner tumor, should be made if teratoid structures are not found after extensive sampling. Squamous cell carcinoma often invades the capsule and lymphatic vessels and has a poor prognosis (Fig. 6.20e).

Benign, malignant, and borderline intestinal-type mucinous tumors [300, 304] originating from mature teratoma exhibit well-differentiated intestinal epithelium and are often associated with pseudomyxoma ovarii (Fig. 6.20f). A recent study [106] has demonstrated the homozygosity of these tumors, suggesting that ovarian intestinal-type mucinous tumors may have a teratoid origin. Those showing appendiceal-type histology [305] may associate with a form of pseudomyxoma peritonei less aggressive than that originated in appendiceal tumors. Immunohistochemically, these tumors show a variable expression of CK20 and 7. In mucinous cystadenomas, a CK7+/CK20− phenotype occurs as commonly as a CK7−/CK20+ phenotype [300], being diffusely CK20 positive in over a third of cases and, more often, only focal or partially positive [304]. Borderline or malignant mucinous tumors tend to have a lack of expression of cytokeratin 7 [300]. Mucin (Muc) family antibodies, although often nonspecific, may be helpful in establishing an origin from teratoid tissues [306].

Melanoma (Fig. 6.20b). Although most ovarian melanomas are metastatic in nature, origin from teratoid tissues may occur, representing one of the rarest origins of metastatic melanoma. The only series dealing with ovarian melanoma originating in teratoma [307] revealed a wide age range from 18–72 years. All were unilateral and histologically showed large epithelioid cells with eosinophilic cytoplasm, small cells, spindle-shaped cells, or a combination of these. Melanin pigment was not always present, but their melanocytic nature was shown by electron microscopic demonstration of melanosomes and positivity for one or more melanocytic markers.

Potentially, practically any tumor pattern found in adults and children can occur in the tissues of a mature teratoma. An updated list of reported sarcomas, cutaneous, neural, and miscellaneous tumors, and their rare combinations is presented in Table 6.5.

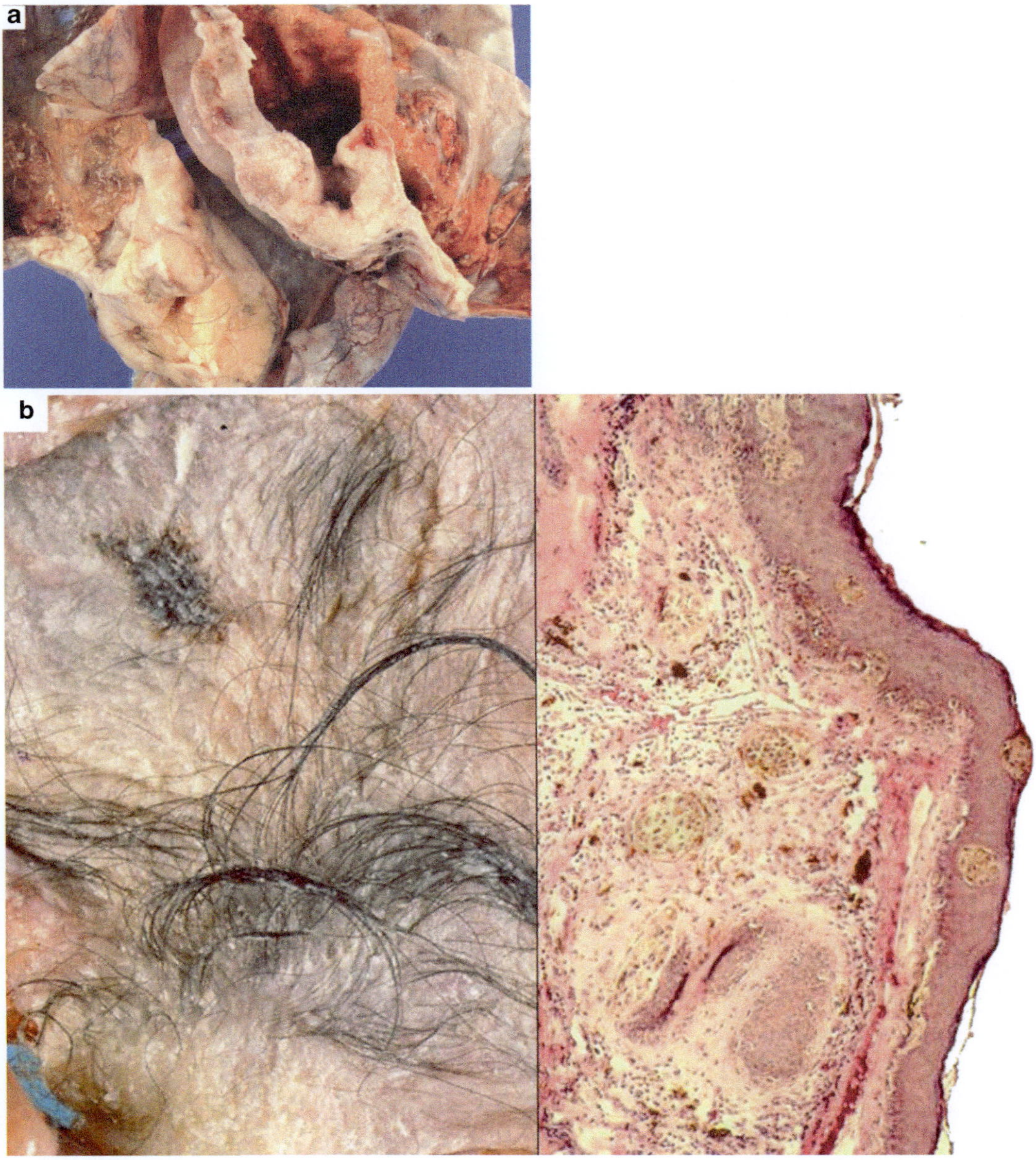

Fig. 6.20 Secondary tumors in teratoma. Squamous cell carcinoma presenting as a thickened wall in a cystic teratoma (**a**). Pigmented lesion in a teratoma that corresponded to a malignant melanoma (330) (**b**). Surface squamous cell carcinoma (**c**) in teratoma with early invasion (**d**). A squamous papillary growth simulating Schneiderian papilloma (**e**). Stromal and vascular invasion by squamous cell carcinoma (**f**). Intestinal mucinous overgrowth in teratoma with pseudomyxoma ovarii. Pilomatrixoma in an ovarian teratoma (**g**) with coexisting neural tissues (*right*). Sebaceous carcinoma (**h**) in teratoma, positive for adipophilin (**i**). A calcified growth corresponding to an odontoma (**j**)

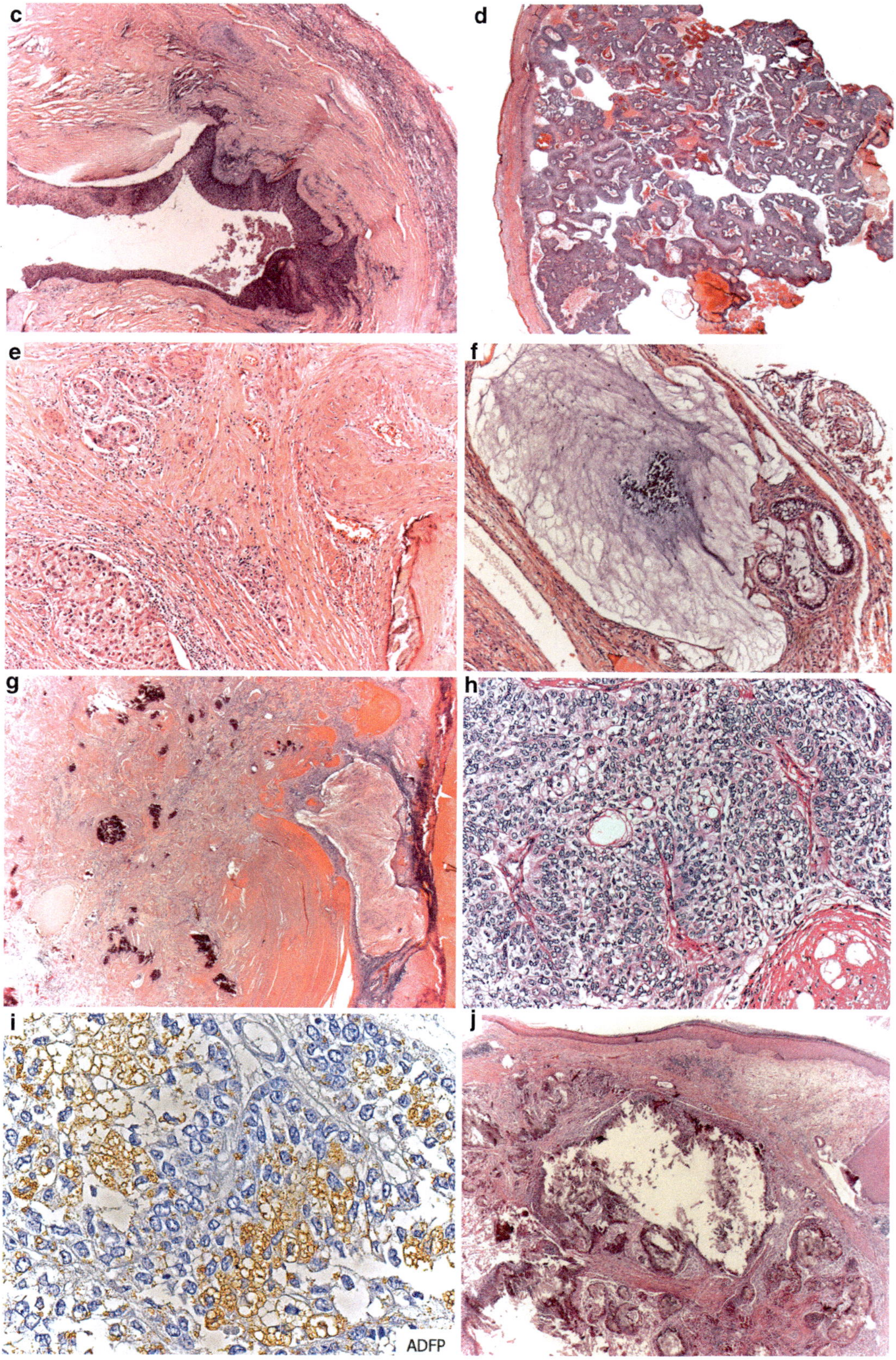

Fig. 6.20 (continued)

Table 6.5 Benign and malignant somatic tumors arising in mature cystic teratoma

Tumor type	References
Squamous cell carcinoma (80–97 %)	
Cutaneous (including in situ lesions)	[219, 220, 308, 309]
Small cell type squamous carcinoma	[220, 310–312]
Adenosquamous variants	[313, 314]
Bronchogenic	[315]
Adenocarcinomas (7 %)	[219, 316]
Intestinal type	[106, 304, 317–319]
Respiratory type	[316]
Prostatic	[320, 321]
Breast or adnexal	[322]
Other skin and adnexal tumors	
Extramammary Paget's disease	[323–326]
Malignant melanoma (0.2–1.8 %)	[307, 309, 327–330]
Basal cell carcinoma	[219, 331]
Pilomatrixoma (Fig. 6.20g)	[332]
Sebaceous adenoma and carcinoma (Fig. 6.20h, i)	[333, 334]
Sweat gland adenocarcinoma	[335, 336]
Blue nevus	[337]
Junctional activity or benign nevi	[338–340]
Sarcomas (7 %)	[219]
Fibrosarcoma	[341, 342]
Leiomyosarcoma	[343]
Osteosarcoma	[344, 345]
Chondrosarcoma	[346, 347]
Angiosarcoma	[348, 349]
Rhabdomyosarcoma	[350]
Malignant fibrous histiocytoma	[351]
Spindle cell sarcoma	[352]
Neurogenic tumors	
Meningioma	[353, 354]
Paraganglioma	[355, 356]
Glioblastoma	[357–359]
Ependymoma	[360, 361]
Oligodendroglioma	[362, 363]
Pilocytic astrocytoma	[364]
Medulloblastoma	[365, 366]
Neuroblastoma	[367, 368]
Neuroectodermal primitive tumor	[369–371]
Ganglioneuroma	[293]
Neurocytoma	[372]
Choroid plexus papilloma	[373]
Miscellaneous	
Low grade mucinous tumors	[374, 375]
Non-Hodgkin lymphoma	[376, 377]
Pulmonary papillary adenoma like	[378]
Small cell carcinoma pulmonary type	[379, 380]
Neuroendocrine carcinoma	[380–383]
Carcinoid	[372, 384–386]
Clear cell carcinoma NOS	[219]
Transitional cell carcinoma	[344]
Salivary gland type carcinoma	[210]
Undifferentiated carcinoma	[219, 342]
Schneiderian papilloma (Fig. 6.20d)	[220, 299, 387]
Odontoma (Fig. 6.20j)	
Chordoma	[388]
Glomus tumor	[389]
Combinations	
Squamous cell Ca and sarcoma (spindle cell squamous carcinoma)	[344, 390, 391]
Squamous cell Ca. and carcinoid	[390, 392]
Brenner tumor and struma ovarii	[393]

Monodermal (Monophyletic) Teratomas

This imprecise term refers to tumors revealing a one-sided differentiation of a tissue not native to the ovary, derived from one of the germ cell layers, usually ecto- or endoderm, since the ovary is an organ of mesodermal derivation. Thus, tumors originating from tissues of mesodermal origin such as chondroma [394] are not included. Neither are those of celomic origin such as epidermoid cysts [395] and some rare ectopic tumor patterns such as ovarian extrarenal Wilms' tumor [396], extra-axial ependymomas [361], hepatoid carcinomas [397], and PNET that may result from tumor stem cell univocal differentiation. The same would apply to rare heterologous differentiations such as hepatic or intestinal areas in granulosa and Sertoli-Leydig cell tumors [84, 398].

Struma Ovarii (SO)

SO is a term applied to teratomas which contain thyroid tissue as the unique or predominant component, usually 50 % or more. SO occurs in an older age group than simple MCT and presents as an abdominal mass that is unusually associated with hyperthyroidism or pseudo Meigs' syndrome [399, 400]. The great majority of cases are benign. Malignant change is uncommon but may represent a challenge in the understanding of its clinical behavior and treatment.

Ultrasonographic findings are nonspecific, usually revealing a heterogeneous, predominantly solid mass [401].

Macroscopy

Most cases are unilateral and associated with teratoma and can reach up to 20 cm in size, although a small SO is not an unusual incidental finding (Fig. 6.21a). Larger tumors often have a pure thyroidal histology and may be malignant. Tumors are well encapsulated and multicystic and their contents are usually brown or green colloid. Rarely, they may present as a unicystic tumor that is only diagnosed in the microscopic study (Fig. 6.21b, c).

Microscopy

Thyroid tissue is readily identified with colloid-filled cystic follicles grouped in nodules. The full range of microscopic changes seen in the eutopic thyroid including microfollicular, solid, tubular,

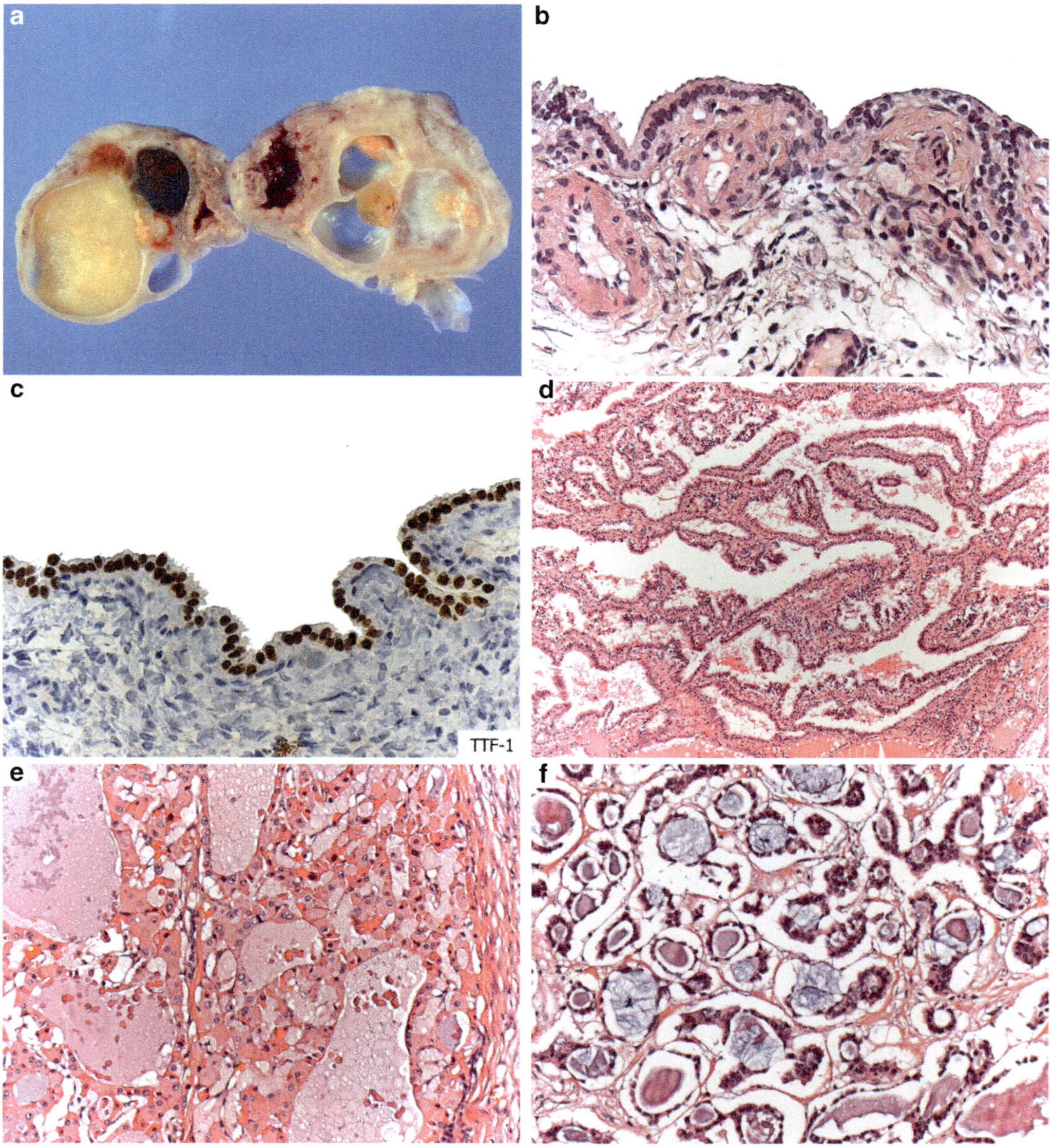

Fig. 6.21 Struma ovarii. (**a**) Cystic struma with colloid contents. Thyroidal lining is reduced to a single cell line (**b**) positive for TTF-1 (**c**). Pseudopapillary (**d**), Hürthle cell (**e**), and mucinous (**f**) changes

trabecular, Hürthle cell, pseudopapillary and mucinous [305] changes (Fig. 6.21d–f), hyperplasia, atrophy, lymphocytic infiltration, etc. can be present in the teratomatous thyroid. Cases exhibiting densely packed cellular follicles or pseudopapillary change have been termed proliferative struma [399], and although benign, they represent the main differential diagnosis with histologically malignant struma. Luteinized stromal cells are not uncommon in the periphery.

The enigmatic association with Brenner tumor (Fig. 6.22a, b) [402, 403] and mucinous cystadenoma suggests that, in these cases, the urothelial and mucinous components so intimately admixed with thyroid tissue would represent a combination of teratoid endodermal tissue differentiations rather than a coincidental association of a common epithelial neoplasm with teratoma.

The pathology of malignant SO parallels that of thyroid carcinomas [399, 404, 405]. Papillary carcinomas of both classic, tall cell, and follicular variants (Fig. 6.23a–e) are the most frequent, and they should be differentiated from proliferative struma using the same criteria applied in the eutopic thyroid to differentiate between papillary carcinomas of follicular variant and hyperplastic nodules (ground glass nuclei, vascular invasion (Fig. 6.23f), etc.) [399, 405]. True follicular carcinomas are unusual and poorly differentiated carcinomas (Fig. 6.23d, e) are even more infrequent. It must be borne in mind that true medullary carcinomas do not occur in the ovary, their closest histological relative possibly being represented by strumal carcinoid.

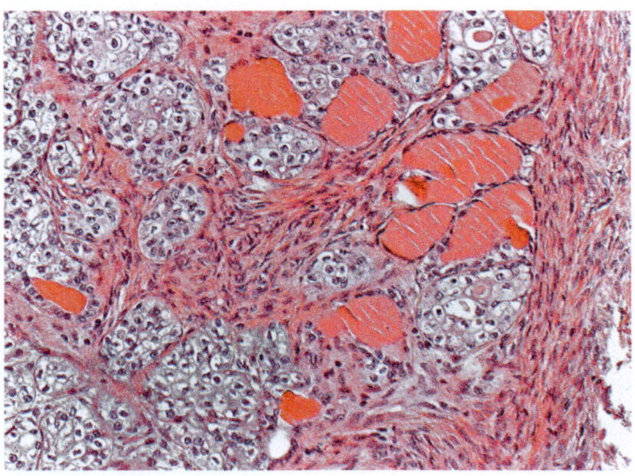

Fig. 6.22 Mixed Brenner tumor and struma. Brenner nests are intimately admixed with thyroid follicles

Malignant SO are usually large masses measuring over 10 cm in size, often with adhesions and ascites. Their behavior is indolent, with a good survival rate at both 10 (89 %) and 25 years (84 %) [406]; extraovarian spread or recurrences rarely develop. The hormonal milieu of pregnancy apparently may provoke more aggressive behavior [405, 407]. Slow growing peritoneal implants, previously called *benign strumosis,* are only examples of well-differentiated metastases. Histologically, both a papillary pattern and a poor degree of differentiation are associated with a more aggressive course [406].

The expression of specific markers such as thyroglobulin (Fig. 6.23e) and thyroid transcription factor-1 (TTF-1) is useful in identifying the thyroidal nature of the neoplastic tissue in cases when it is difficult to recognize, such as the thin cuboidal lining of the cavity of cystic SO (Fig. 6.21b, c) and in solid, poorly differentiated thyroid carcinomas with minimal follicular formation. These markers also help in establishing a thyroidal phenotype in the differential diagnosis with other ovarian tumors with tubular or trabecular arrangement, such as Sertoli-Leydig cell and carcinoid tumors.

Classic papillary carcinoma is a relatively straightforward diagnosis, but the more complex diagnosis of the follicular variant of papillary carcinomas may be facilitated by the expression of HMBE-1 and CK19 positivity [408]. BRAF point mutations of the type commonly observed in papillary carcinomas of the eutopic thyroid have also been reported in malignant struma [407, 409].

Histologic diagnosis of malignancy occurs in most cases without any evidence of extraovarian spread. In these, a conservative approach with salpingo-oophorectomy and analysis of thyroglobulin serum levels to exclude occult disease is advised [399, 404, 405]. Postoophorectomy treatment should follow similar protocols to those used in eutopic thyroid carcinoma, potentially including thyroidectomy or radioiodine ablation, and hormonal suppressive therapy. Long-term follow-up for both metastases and abdominal recurrences can be accomplished by serial serum thyroglobulin measurements and radioiodine scans [410].

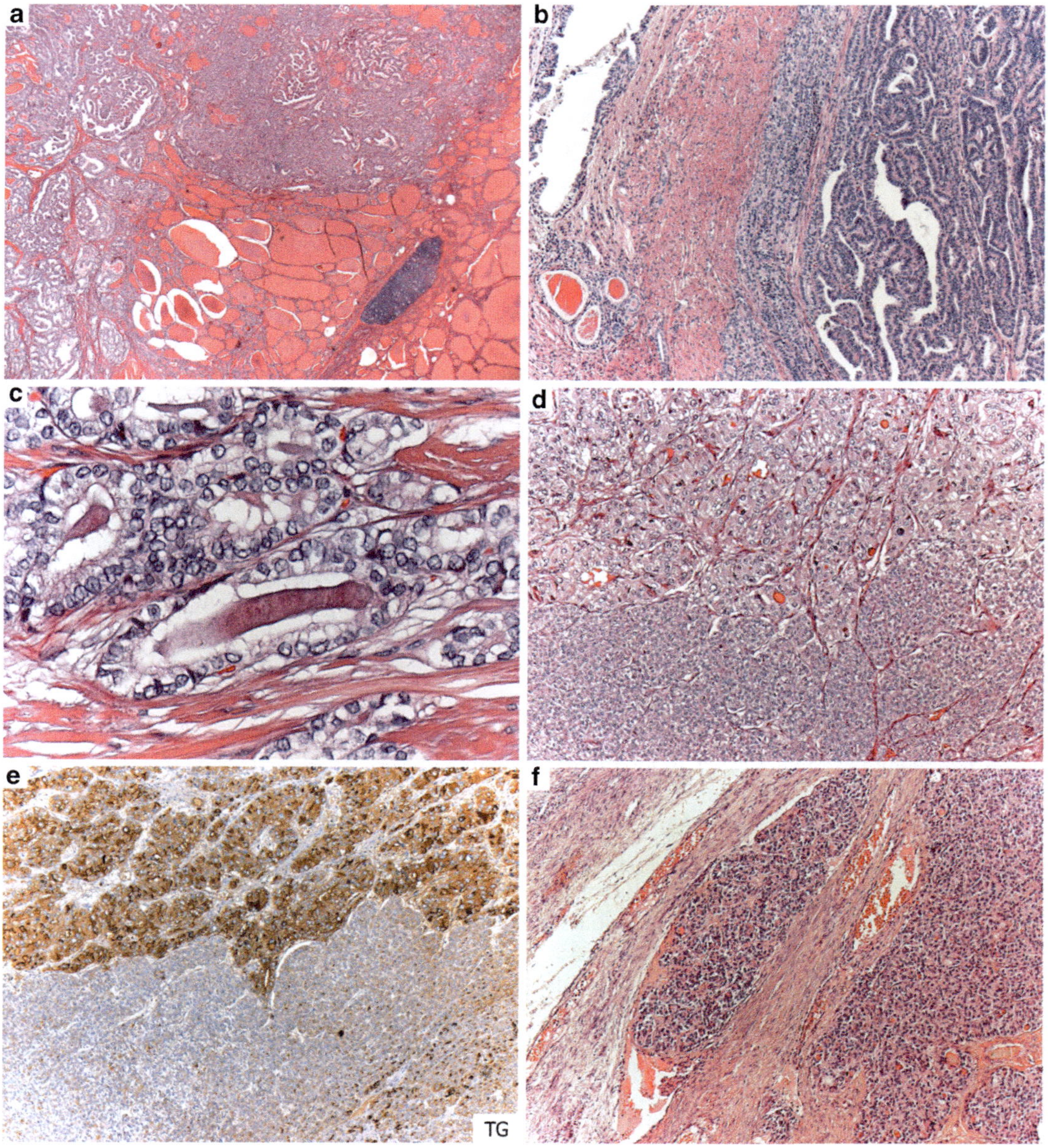

Fig. 6.23 Malignant struma ovarii. Papillary carcinoma in SO (**a**). Rare cases may exhibit a tall cell pattern (**b**). Follicular variant of papillary carcinoma with characteristic nuclei (**c**). Poorly differentiated thyroid carcinoma (**d**) with a patchy expression of thyroglobulin (**e**). Vascular invasion (**f**) in malignant struma

6.3.3.4 Carcinoid Tumors

Ovarian carcinoids are infrequent, well-differentiated neuroendocrine tumors with restricted malignant potential and a good long-term prognosis. They are often unilateral and small and coexist with mature cystic teratoma (57–80 %), although in 15–42 % of cases [411, 412], tumors are larger and not associated with teratoma. Neoplasms have a heterogeneous morphology, with various neuroendocrine differentiations of embryonal fore-, mid-, and hindgut, often coexisting in the same neoplasm [305]. Each predominant histologic subtype presents a relatively characteristic clinicopathologic profile.

On a histological basis alone, they are similar to carcinoids metastatic to the ovary, from which they should be differentiated, since they have a different prognosis.

Histologically, the following carcinoid patterns are present in the ovary:

(a) The most frequent *insular* patterns of midgut type.
(b) *Trabecular* patterns of fore- and hindgut type that have in some series a similar frequency to insular carcinoids [411] but in others [28] are considered less common than strumal carcinoid.
(c) Rare *mucinous* carcinoids, similar to appendiceal adenocarcinoids (goblet cell carcinoids).
(d) A combination of thyroid tissue with trabecular or insular variants: *strumal carcinoid.* This peculiar tumor is exclusive to the ovary, not found in the testis or in any other GCT location and different from medullary carcinoma of the thyroid.
(e) *Metastatic* carcinoids from the gastrointestinal tract differ from primary carcinoids by their monomorphic pattern; the absence of accompanying teratoid tissues; their bilateral, multinodular ovarian involvement; and frequent vascular and peritoneal invasion. Clinically, the age distribution of patients with ovarian carcinoid is similar to that of those with secondary neoplasms originating in mature teratomas, and, in most cases, symptoms are nonspecific. Rare functioning cases may present with Cushing's and hyperinsulinism syndromes secondary to ectopic ACTH or insulin secretion by ovarian carcinoid [103, 386] as well as an association with other neuroendocrine tumors in multiple endocrine neoplasia I syndrome [360].

Up to a third of cases are associated with carcinoid syndrome [413], which is more frequent in patients with large insular carcinoid tumors and/or metastatic disease [411]. In the absence of hepatic involvement, carcinoid syndrome may occur by a direct release of vasoactive substances into the systemic circulation bypassing the liver. Its symptoms include flushes, diarrhea, wheezing, cramping, pruritus, etc. Late-onset, right-sided congestive heart failure related to tricuspid fibrosis can occur, which in rare cases may resolve after tumor removal [414]. Massive medial hypertrophy of the ovarian drainage vein has also been reported [415]. Carcinoid syndrome occurs more frequently in association with insular, midgut-type carcinoids, less so with strumal carcinoid, where it may be associated with constipation [416] and rarely, if ever, with trabecular [411] and mucinous carcinoids [28].

Macroscopically, tumors are unilateral, although they may be associated with MCT or mucinous cystadenoma in the contralateral ovary. Often, carcinoids are found as nodular growths in a teratoma (Fig. 6.24a) or, less frequently, as uniform solid masses. They are only rarely discovered as incidental microscopic findings in a MCT. On cut section, tissue is firm and tan or yellow; mucinous microcysts may be present. An association with both benign and malignant ovarian mucinous tumors [417] has been reported.

Microscopically, insular carcinoid is a midgut carcinoid constituted by anastomosing sheets and nests of uniform, non-mitotic regular cells with salt and pepper nuclei arranged in gland-like or acinar formations (Fig. 6.24b) that may contain proteinaceous or calcified material. A granular, chromogranin positive, cytoplasm is usually evident in most cells and is prominent at the periphery (Figs. 6.24c, d). Luteinized stromal cells are often present in the pushing margins. Their differential diagnosis includes adult granulosa, Sertoli-Leydig cell, and Brenner tumors.

Trabecular carcinoids are arranged in bland uni-or bicellular ribbons of cylindrical cells with clear cytoplasm (Fig. 6.24e) in a background of a dense stroma. Granules are not prominent. Sertoli-Leydig cell tumors with a trabecular pattern should be considered in the differential diagnosis.

Primary mucinous (goblet cell) carcinoid is composed of small glands, cords, or nests populated by varying proportions of goblet and granular cells. Mucin pools are present in the stroma. When not associated with other mature teratoid components, they are indistinguishable from metastases from an appendiceal mucinous carcinoid [418]. A histologic spectrum, ranging from well-differentiated to atypical crowded or microcystic glands to openly malignant goblet cell carcinoma (Fig. 6.24f) with angioinvasive features, is found (Fig. 6.24g) [419]. Mucinous glands lined by goblet cells (Fig. 6.24h) are often focally present in various types of ovarian carcinoid and should not be considered as mucinous carcinoid [28].

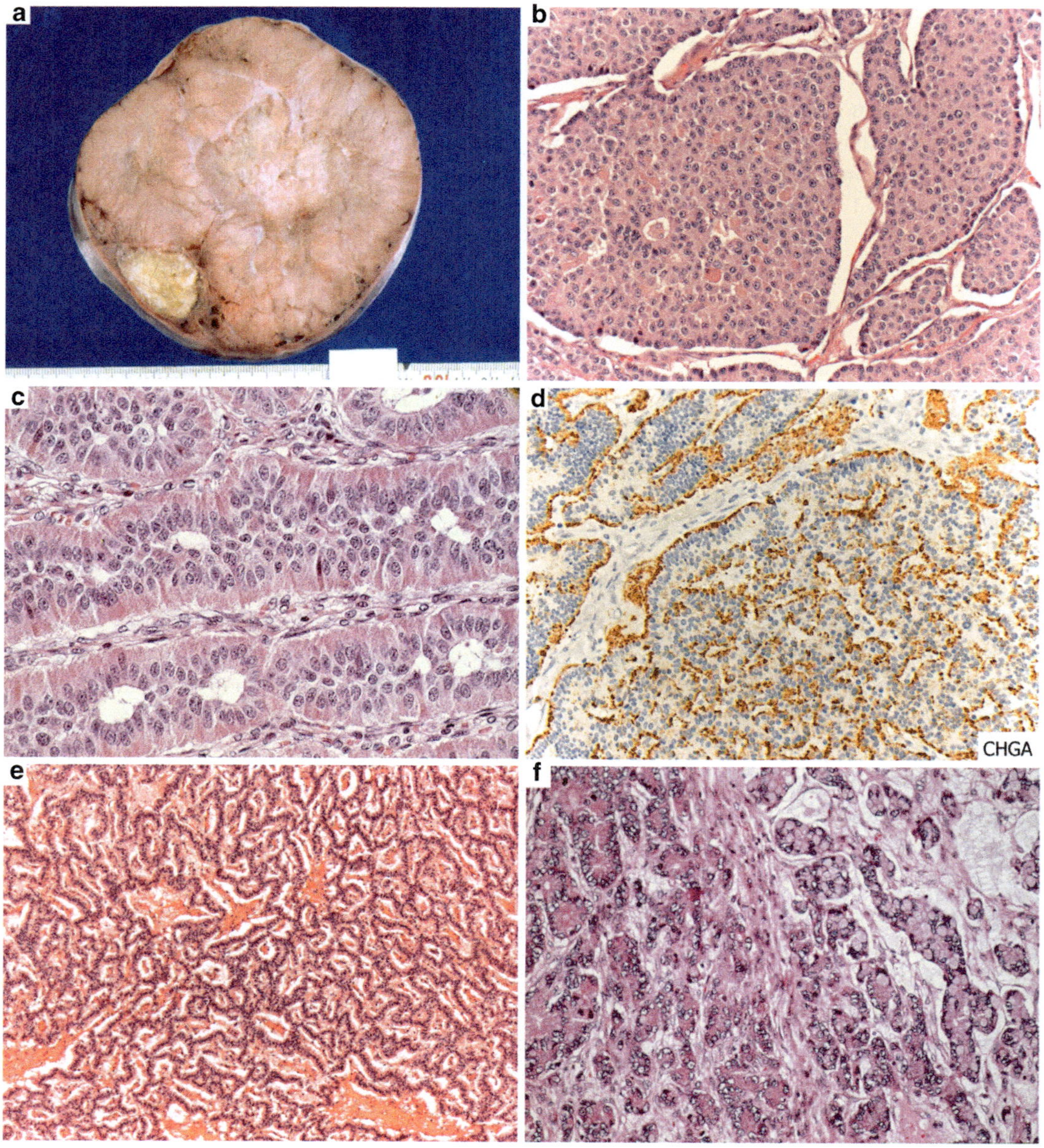

Fig. 6.24 Ovarian carcinoids. Characteristic gross appearance of carcinoid originating from teratoma (**a**). Tumor is tan colored. Insular pattern (**b**) with gland-like spaces. Insular carcinoid with prominent granular cytoplasm at the periphery (**c**), expressing chromogranin (**d**). Trabecular carcinoid (**e**). Adenocarcinoid with compact granular areas coexisting with chords of goblet cells (**f**). Vascular invasion in the lung (**g**) in a patient with ovarian mucinous carcinoid. Trabecular carcinoid with colonic-type mucinous glands (**h**). Strumal carcinoid (**i**) with thyroid tissue (*left*), trabecular carcinoid (*center*), and appendiceal-type mucinous epithelium (*bottom right*). Strumal carcinoid: trabecular areas coexist with abortive follicles (**j**)

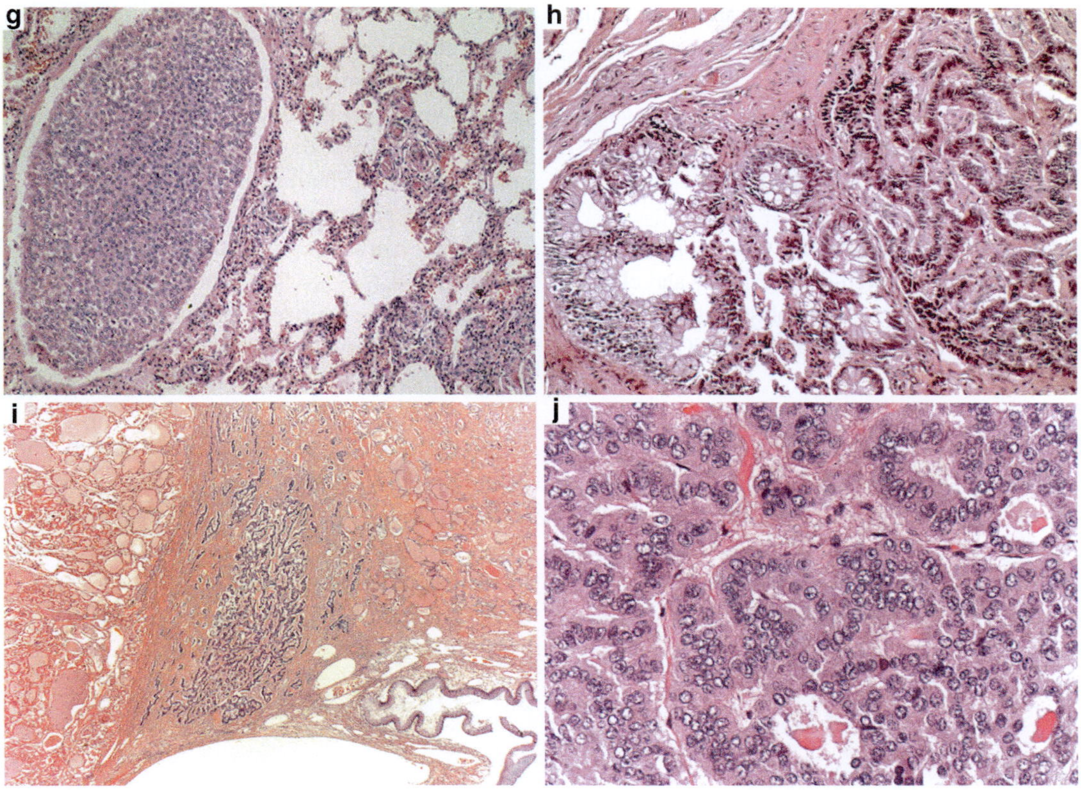

Fig. 6.24 (continued)

Strumal carcinoid shows an intimate admixture of thyroid colloid-containing follicles and trabecular carcinoid (Fig. 6.24i, j), although insular carcinoid nests and, more commonly, mucinous glands with interspersed neuroendocrine cells can also be present. Notable differences with medullary carcinoma of the thyroid include the fact that strumal carcinoid is almost invariably benign, rarely secretes calcitonin, or has an amyloid stroma. In the rare event of metastases, it may do so as carcinoid [258] and exceptionally as follicular carcinoma [420]. The frequent mucinous component can be appendiceal type and may be associated with pseudomyxoma peritonei [305].

Rare primary *neuroendocrine carcinomas of the ovary* are usually of large cell type and are not associated with teratoma. Their germ cell origin is questionable. They should be differentiated from small cell pulmonary type neuroendocrine carcinomas that originate from MCT [380].

6.3.3.5 Immunohistochemistry

Immunohistochemically, each histologic variant of ovarian carcinoid has a relatively characteristic ultrastructure and neuroendocrine immunophenotype. Insular carcinoid shows pleomorphic neurosecretory granules, which, in contrast, are rounded and uniform in trabecular carcinoid. Strumal carcinoid may show a hybrid (amphicrine) phenotype displaying thyroidal and neuroendocrine features [421] including the expression of TTF-1 in follicular areas. All carcinoids express chromogranin-A, synaptophysin, and CD56 in varying amounts, as well as, specifically, neuropeptides such as YY, calcitonin, substance P, insulin, etc. but rarely have any hormonally functioning activity. Prostate-specific acid phosphatase may be expressed by trabecular strumal carcinoids. CDX2 positivity has been considered a discriminating factor between metastases from gastrointestinal and primary carcinoids, being overexpressed in metastases and only weakly positive in ovarian, teratoma-associated primaries [412].

6.4 Mixed Germ and Sex Cord-Stromal Cell Tumors

Simultaneous growth of primitive sex cord and germ cells occurs in two well-defined histo- and clinicopathologic entities: gonadoblastoma and mixed germ cell-sex cord-stromal tumor (MGC-SCST).

6.4.1 Gonadoblastoma (GB)

GB [422] is a benign lesion that is, however, a precursor of primitive GCT. It characteristically involves gonads of patients with DSD which genetically contain the gonadoblastoma locus on the Y-chromosome (GBY), responsible for the testis-specific protein Y-encoded-1 (TSPY-1) [423]. In these patients, mutations or deletions of genes involved in the development of functional Sertoli cells disrupt their regulation of germ cell maturation, contributing to the maintenance of pluripotency features in the germ cells with a corresponding marker expression, being identical to testicular GCNIS [424]. Consequently, these germ cells may give rise to primitive GCT, more often dysgerminomas and yolk sac tumors, as well as other type II GCT such as EC and mixed GCT.

Since GB almost invariably occurs in Y-chromosome-containing cells, it is not *stricto sensu* an ovarian (female) tumor. GB also represents a curious crossroads between genetic malformation and neoplasia. Its genetic and clinicopathologic features are also considered in Chaps. 3, 7, and 10.

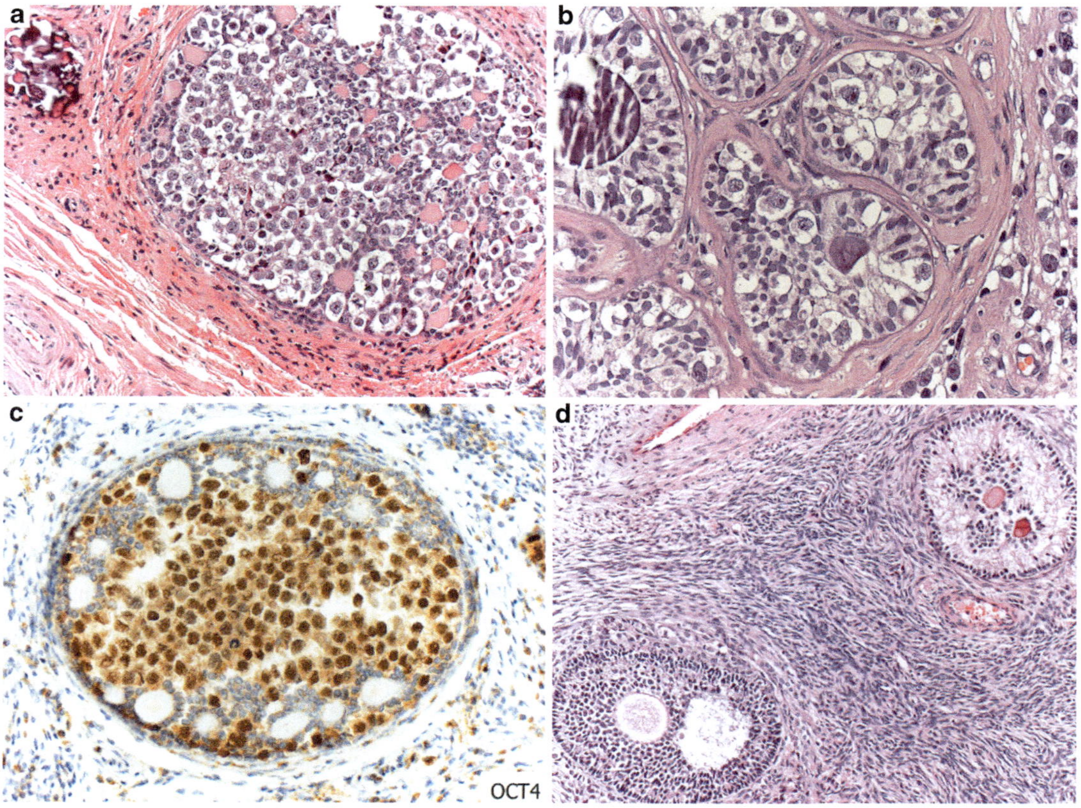

Fig. 6.25 Gonadoblastoma. (**a**) Partly calcified nodules of gonadoblastoma in a streak gonad. (**b**) Higher magnification of gonadoblastoma with large, atypical germ cells positive for OCT4 (**c**) coexisting with small sex cord cells. (**d**) Gonadoblastoma-like structures resembling sex cord tumor with annular tubules in a 15 year old girl

GB is diagnosed in all age groups with an average age of 18 years and may occur in identical twins. Many present in virilized, phenotypic females [425], frequently with Swyer syndrome. The bilateral gonadal lesions are often an incidental finding in amenorrheic patients presenting microcalcifications [426]. In phenotypic females, they occur in either streak or unknown nature gonads. They are often microscopic unless they are overgrown by primitive GCT, which may present in either gonad.

Microscopically, the characteristic appearance is that of well-delineated insular aggregates of intimately admixed germ and immature sex cord cells arranged in pseudofollicular spaces filled with an eosinophilic, hyaline material that often becomes calcified (Fig. 6.25a, b). The clear germ cell component presents features identical to GCNIS, expressing pluripotentiality markers such as OCT4 (Fig. 6.25c) and SALL4, which contrasts with their negativity in the otherwise α-inhibin-positive sex cord cells. In children, microscopic foci of cribriform, microfollicular structures with eosinophilic contents (Fig. 6.25d) resembling sex cord tumor with annular tubules, but lacking germ cells, have been called gonadoblastoma-like structures [427]. They merely represent examples of abnormal folliculogenesis that involute with sexual maturity.

Initial overgrowth by dysgerminoma can be seen in discrete areas and is usually revealed by the presence of lymphocytic infiltrates peripheral to GB.

6.4.2 Unclassified Mixed Germ Cell-Sex Cord-Stromal Tumor (UMGC-SCST)

This exceptionally rare tumor was initially reported in the French literature [428] and later in English by Talerman, who coined its current nomenclature [429]. Due to the few reported cases, there is no clear profile of age, distribution, or clinical presentation. UMGC-SCST have been considered in both reviews [430–432] and in a few well-documented case reports [433–437].

Although this tumor occurs in both female and male gonads, it is much more frequent in the ovary. Female patients lack any DSD features and have a normal 46,XX karyotype, some of them being able to reach term pregnancies [431]. It is diagnosed due to the presence of an abdominal mass in girls younger than 10 years, but may also involve any age and only rarely presents with isosexual precocity [434].

Macroscopically, tumors are large and unilateral [431] and only occasionally, bilateral [438].

Histologically, ovarian UMGC-SCST grow in a predominantly solid pattern (Figs. 6.26a, b), although tubular/cribriform, trabecular (Fig. 6.26c), or haphazard, often mixed, arrangements can be found. All patterns show an admixture of sex cord cells with scanty cytoplasm and rounded elongated nuclei that coexist with large, atypical, clear germ cells of variable size, some presenting mitoses [432] and granular (lead shot) chromatin reminiscent of spermatocytic tumors (Fig. 6.26d). Coexistence with gonadoblastoma areas has been reported [439, 440].

Retiform [435, 436] and heterologous [437] areas are occasionally present. Similar features are found in Sertoli-Leydig cell tumors which, in their retiform variant, are primitive tumors of sex cord lineage that mirror early ovarian organogenesis, showing sex cord blastema and its connection with the primitive rete cords [441]. It is therefore not surprising that the earliest descriptions of UMGC-SCST in the French literature used the term *Pflügeromes* [442] due to their resemblance to the trabecular embryonic sex cords described by Pflüger [443].

Immunohistochemically, their germ cells express pluripotentiality markers such as OCT4 and SALL4 [430, 432] as well as other germ cell markers such as PLAP and CD117. 12p amplification does not usually occur in the germ cell component, being present only in isolated instances [430]. Sex cord components characteristically express α-inhibin.

Both pluripotential germ cell elements and immature sex cord components are able to metastasize [434] and recur in the abdominal cavity [433]. Only rare cases give rise to malignant germ cell tumors [444].

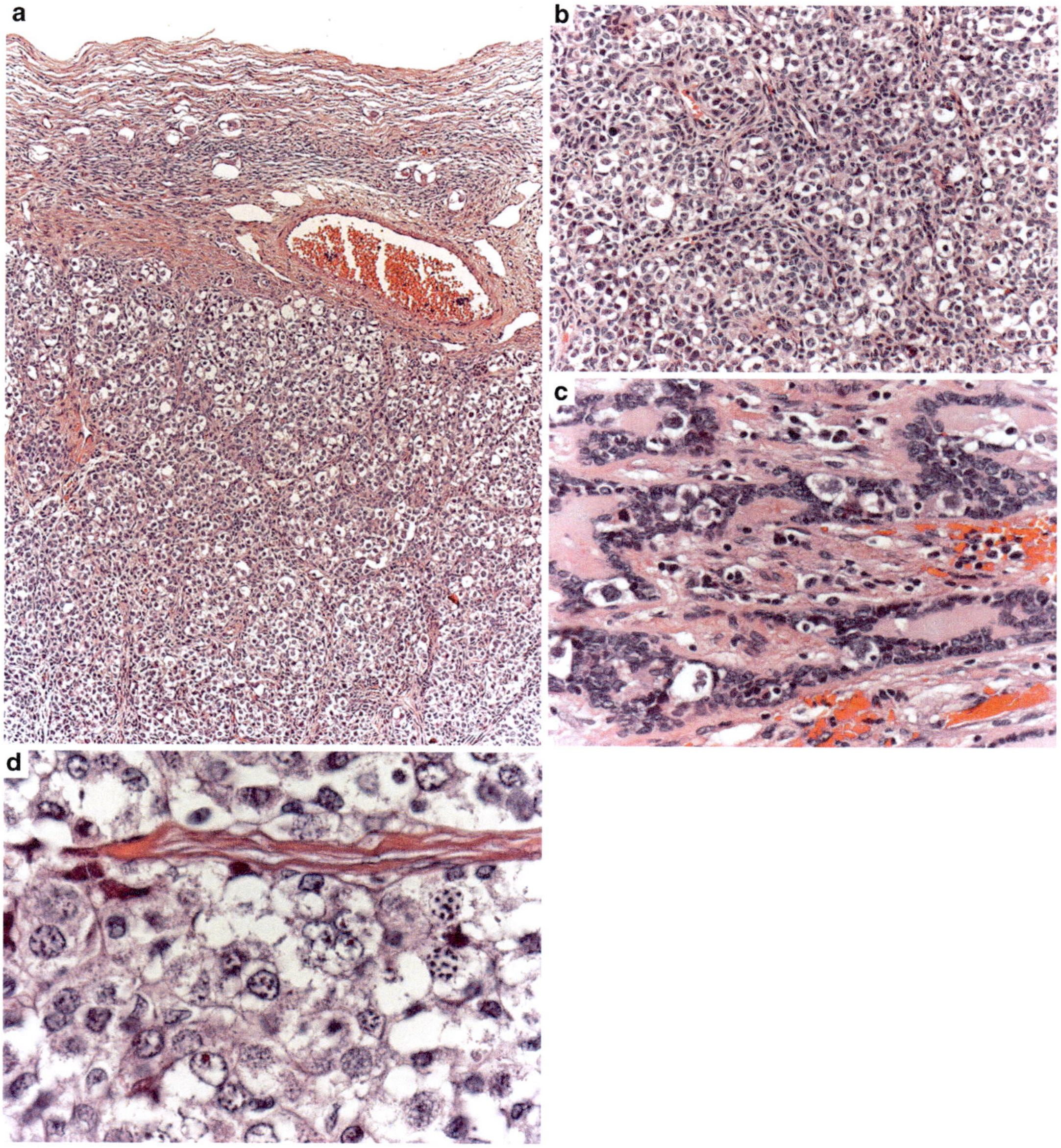

Fig. 6.26 Mixed germ cell-sex cord-stromal tumor in a young girl (**a**), exhibiting a diffusely mixed population (**b**) of clear germ cells and sex cord elements. This trabecular pattern reminiscent of embryonic sex cords (**c**) occurred in a 24-year-old phenotypic female. Higher magnification of solid area in the same case (**d**). Abundant germ cells with nuclei exhibiting a "lead shot"-chromatin coexist with smaller sex cord cells with fusiform nuclei

6.5 Germ Cell Tumors Originating from Somatic Müllerian Neoplasms (Type VI GCT)

This group of rare, type VI GCT includes secondary malignant GCT growth patterns arising from somatic-cell-derived tumors. They likely originate from tumor stem cells behaving as induced pluripotential cells (iPSC) (see Chap. 3) and occur in association with neoplasms of Müllerian derivation, including endometriosis, and are not related to chemotherapy. The GCT patterns reproduced are *exclusively non-dysgerminomas* and frequently, glandular YST. They occur predominantly in pre-and postmenopausal patients, reaching a peak in the seventh decade of life. Often they are diagnosed in advanced clinical stages and behave aggressively. Similar tumors occur with a lesser frequency in the testis (see Chap. 7) as well as in extragonadal locations such as the uterus, head and neck region [445], stomach [446], urinary tract [109], etc. (see Chaps. 11 and 12).

In the WHO classification of female genital tract tumors [447], they are not considered a discrete entity. However, although currently very unusual tumors, they mainly involve elderly women, an age group that is increasing in numbers due the progressive aging of the population in developed countries. This, coupled with their difficult diagnosis and poor prognosis, would merit considering them as a separate category in tumor classifications of ovarian and uterine tumors. In the literature, most publications are case reports with only few relevant series [77, 78].

Recent data from histopathologic-based registries from the European Union (see Chap. 2), the United Kingdom [22], and the United States [23] reveal an unusual GCT incidence peak at 65–75 that corresponds to GCT arising from somatic Müllerian neoplasms (GCT-SMN). Although they are not referred to as such in the various registries, it is possible that many such cases pass unrecognized due to their rarity and difficult diagnosis, often complicated by the malignant GCT overgrowth effacing the original somatic tumor [78]. European and US population-based statistics predict that almost 30 % of the population will be over 65 by 2035 and the population pyramid inverted by 2080, with a substantial increase of women over the age of 85 [448, 449]. This would mean that, in the not too distant future, age-related tumors such as these and type II endometrial cancers will experiment a remarkable increase in incidence.

Historically, GCT-SMN were described for the first time in the female genital tract associated with endometriosis [450] and, later, with endometrioid adenocarcinomas and carcinosarcomas of both ovary and endometrium [77, 451, 452] and much less frequently with serous [453] and mucinous carcinomas [454].

We have studied 25 cases of ovarian and uterine GCT-SMN arising from both endometrioid and clear cell carcinomas [133, 455]. Clinically,

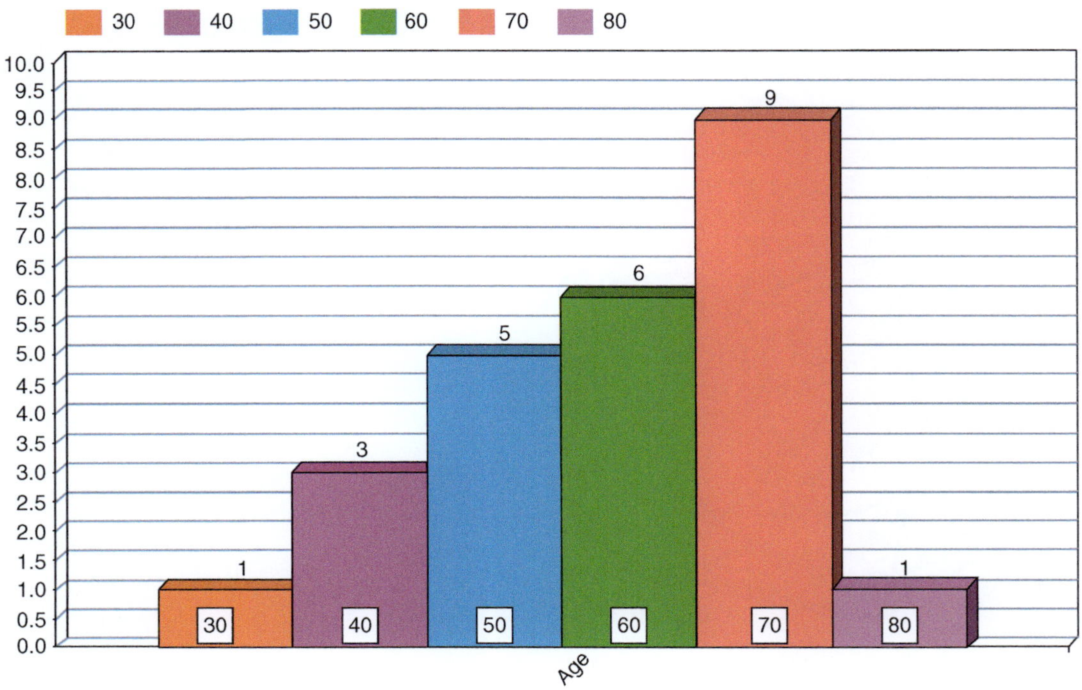

Fig. 6.27 Age distribution in 25 cases of somatic Müllerian malignancies of ovary and uterus associated with germ cell tumor

their ages ranged from 30 to 94 with the majority of cases in the sixth and seventh decades of life (Fig. 6.27). They presented with the usual nonspecific symptoms of ovarian and uterine tumors. At the time of surgery, they were often found in FIGO stages II–IV. Survival data indicate both poor prognosis and response to treatment [77]. A r.

Histologically, most tumors show diffuse intermingling of Müllerian somatic tumors with non-dysgerminomatous GCT.

The main histologic categories of GCT-SMN correspond to:

(a) Endometrioid neoplasms associated with YST
(b) Clear cell carcinoma (CCC) associated with YST
(c) Endometrioid or CCC associated with YST and other GCT patterns

6.5.1 Endometrioid Neoplasms Associated with YST

These somatically derived YST [456] represent the most numerous category of GCT-SMN. We have analyzed a series of 15 cases [455] corresponding to women aged between 30 and 80 years. Of these, most (11/15) were in the fifth and sixth decades of life. Tumors were in an advanced stage in over half of cases. Mortality was high with a poor response to chemotherapy. Macroscopically, the tumors were bulky with an average size of 15 cm and only once were bilateral. A third of cases were associated with endometriosis or endometrioid cystoadenofibromas. The endometrioid adenocarcinoma constituted the main neoplastic component and was well differentiated with morular formation in four cases, moderately differentiated in six and the remaining five were carcinosarcomas. A recently

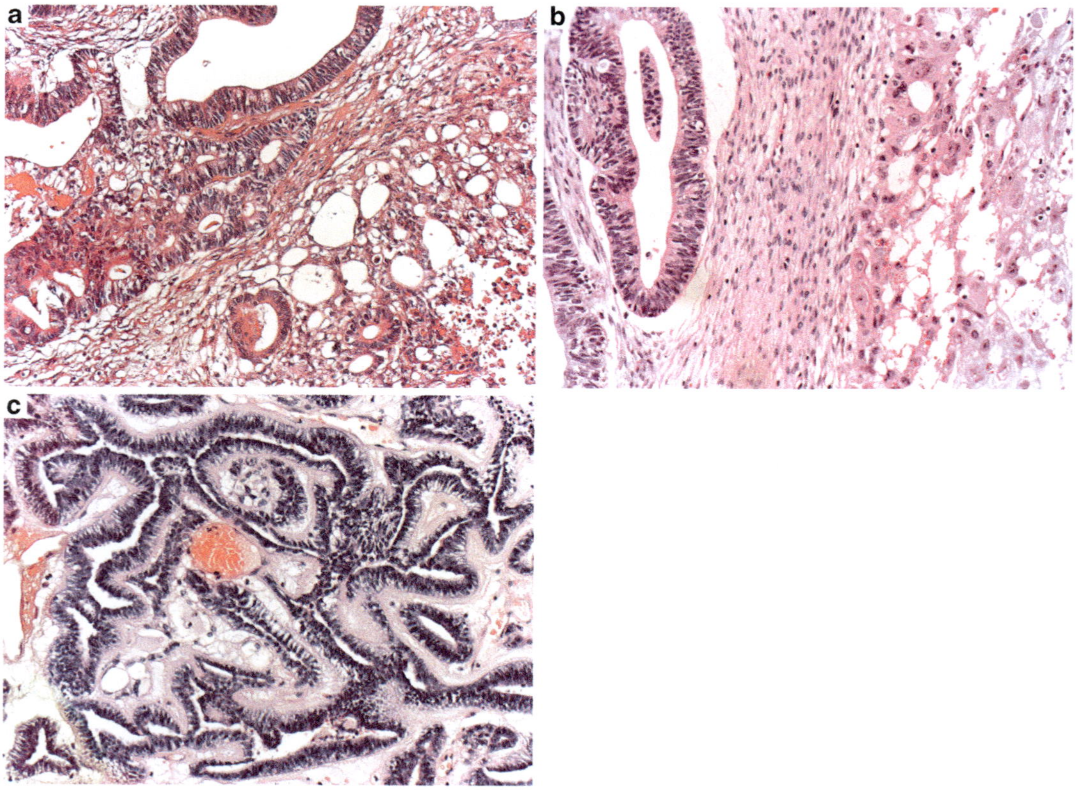

Fig. 6.28 Interphase of well-differentiated endometrioid adenocarcinoma with microcystic (**a**) and hepatoid (**b**) YST. Glandular YST (**c**) resembles endometrioid adenocarcinoma except for the diffuse basal vacuolation and periglandular rarefaction

published series [456] provides a similar a clinicopathologic profile. The YST element is characteristically heterogeneous, showing small foci of classic microcystic YST (Fig. 6.28a, b) in half of cases, coexisting with areas of glandular pattern that were the predominant or the exclusive pattern in three cases (Fig. 6.28c). The cylindrical cell lining of glandular spaces of both endometrioid tumor and glandular YST can be very similar, and, often, apical and basal

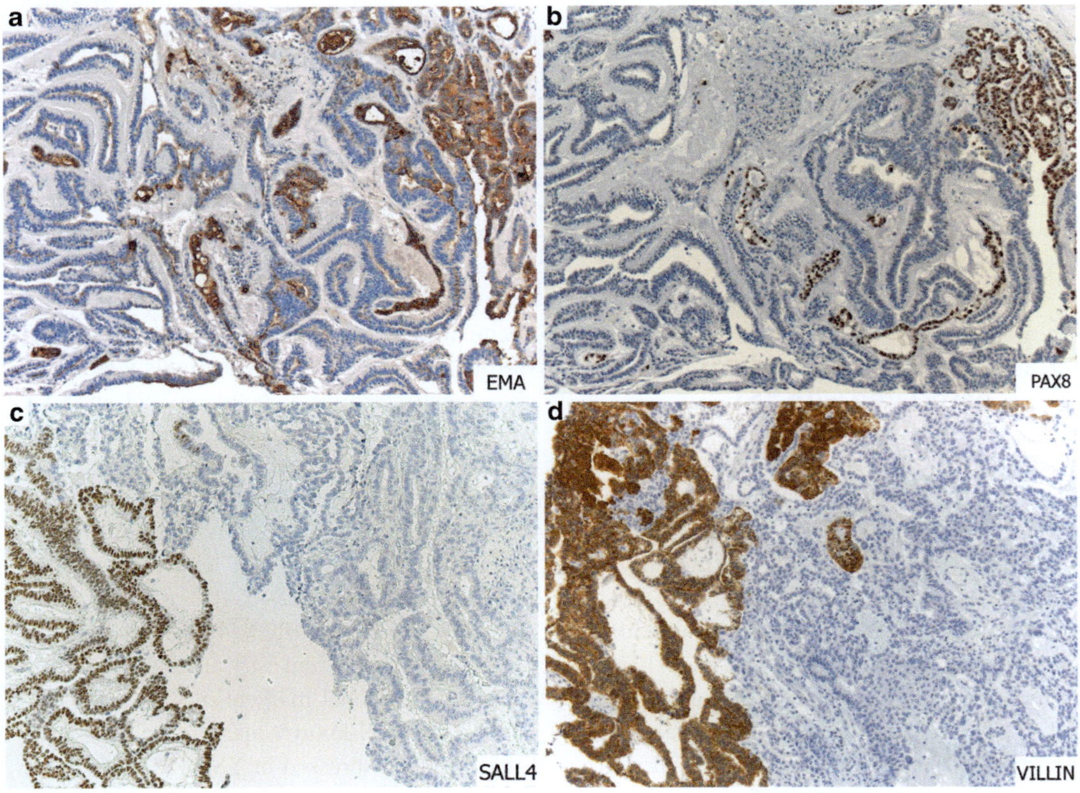

Fig. 6.29 Coexistence of glandular YST with endometrioid adenocarcinoma. (**a**, **b**) EMA and PAX8 are only expressed in the endometrioid component, while only glandular YST expresses SALL4 (**c**) and villin (**d**)

Table 6.6 Differential diagnosis by immunophenotype of endometrioid carcinoma, clear cell tumors (CCC), and YST [131, 133]

Antibodies	Endometrioid Ca	CCC	YST (preferently glandular variants)
HNF1-β	Negative or focal	Diffuse positive	Focal but intense in YST glandular
Napsin A	Negative or focal	Diffuse positive	Negative
CK7	Diffuse positive	Diffuse positive	Focal (gastropancreatic differentiation?)
PAX8	Diffuse positive	Diffuse positive	Negative
EMA	Diffuse positive	Diffuse positive	Focal but intense in YST glandular areas
AFP	Negative or focal	Negative or focal	Focal and irregular
SALL4	Negative	Negative	Diffuse positive
Villin	Negative	Negative	Diffuse positive
HepPar-1	Negative	Negative	Focal irregular
GPC3	Negative	Focal irregular	Focal irregular
GATA3	Negative	Focal irregular	Focal irregular

vacuolation of the YST intestinal-type glands is the only reliable histologic discriminating feature (Fig. 6.28c). For this reason, immunohistochemistry is necessary to identify the differential diagnostic phenotypes of each component. The endometrioid component expresses EMA (Fig. 6.29a), CK7 [132], and estrogen receptors [77] as well as PAX8 (Fig. 6.29b) [457]. In contrast, glandular YST can be focally positive for CK7 and EMA but expresses diffusely SALL4 and villin (Fig. 6.29c, d) as well as focal, patchy AFP and GPC3. Intestinal differentiation in YST reveals a strong CDX2 and HepPar-1 expression. Immunohistochemical differential diagnostic data are shown in Table 6.6.

6.5.2 Clear Cell Carcinoma (CCC) Associated with YST

We have analyzed twelve cases of this unusual association from several European consultation materials [133]. They involved a similar group than endometrioid tumors, ranging from 37 to 94 years, with an average of 62. They were also smaller in size, ranging from 2 to 20 cm in diameter, average 12 cm. Endometriosis was present in all but two cases and 50 % were in advanced clinical stages. Histologically, all corresponded to characteristic tubulopapillary patterns of clear cell carcinomas (CCC) except one case of borderline clear cell adenofibroma. The YST component was almost invariably of glandular type, being predominant in over half of cases. Clinically, eight patients died 16 months to three years after diagnosis. Four are alive and well with short follow ups of up to 4 years.

Since the initial descriptions of YST [458], CCC has been the main differential diagnosis with YST. Since glandular YST is the most frequently associated pattern with CCC [133], the difficulty in its differentiation from CCC lies in the fact that glandular YST may have a papillary growth pattern and extensive vacuolation, so closely resembling CCC that it can be taken for a hybrid phenotype (Fig. 6.30a). Histologically, glandular YST coexists and merges with CCC (Fig. 6.30b) but differs from it in their more regular nuclei, polarized vacuoles in apical or subnuclear position, subepithelial stromal rarefaction, and occasional presence of goblet cells and hyaline globules. These features should differentiate it from the usually complex CCC histology [459]. For these reasons, immunohistochemistry is necessary for differential diagnosis and should not be based on the expression of a single antibody [112, 460], since some otherwise "characteristic markers" can be expressed by both components. Therefore, diagnostic accuracy is improved by contrasting wider immunohistochemical panels for CCC [459] and YST [96]. Before these comprehensive panels were available, the differentiation between YST and CCC was problematic; thus it is difficult to know if previously reported cases of CCC with AFP expression were either pure CCCs with ectopic AFP secretion, papillary glandular YST resembling CCC exhibiting AFP expression, or an admixture of both [461–464].

Neither AFP nor GPC3 [460] fully discriminates between CCC and YST and the same is true for other traditional markers of CCC and YST: HNF1-β, a well-known marker for most CCC [465], is also expressed in YST [133, 464] (Fig. 6.30c), which limits its usefulness in differentiating these patterns. The absence of expression of CK7 and EMA [466] has also been considered characteristic of YST. However, it can also be focally present in glandular YST (Fig. 6.30d), where this expression may represent gastropancreatic differentiation [133].

PAX8 and Napsin A are more specific antibodies (Fig. 6.30e) expressed in CCC [467–469] but not in YST patterns even the glandular ones [133].

SALL4 is an excellent marker of cells retaining stemness, and consequently it is expressed in YST of both classic and glandular type where it represents, as in carcinoma of the stomach, fetal gut differentiation [470]. It is, however,

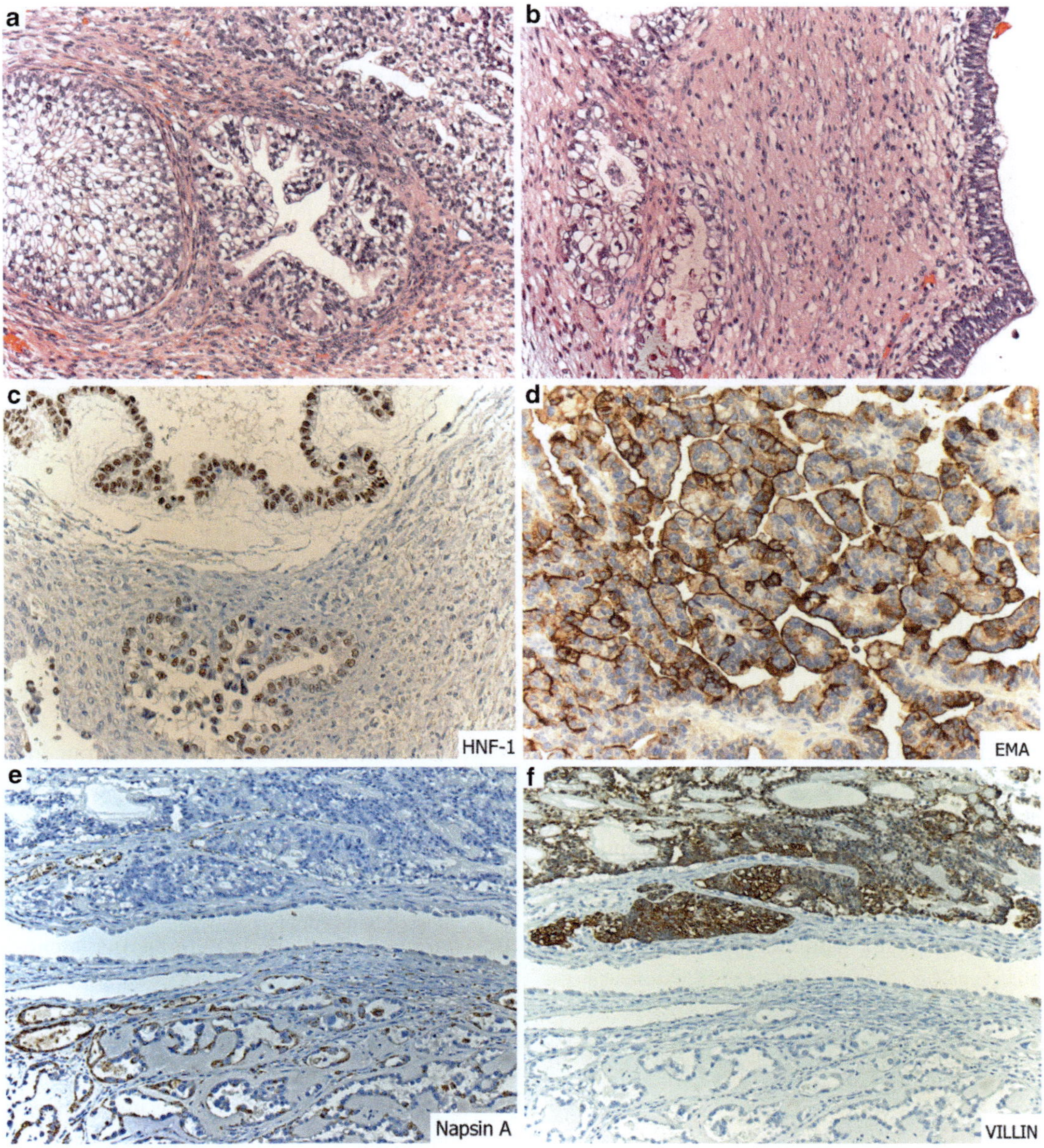

Fig. 6.30 (**a**) YST resembling clear cell carcinoma (CCC). (**b**) Interphase between CCC (*left*) and glandular YST (*right*). HNF1-ß (**c**) is expressed by both glandular YST (*top*) and CCC (*bottom*). EMA (**d**) can show a strong expression in glandular papillary YST. However, Napsin A (**e**) is only expressed in CCC, while villin (**f**) only stains glandular YST

negative in CCC [119]. Villin, although considered a relatively nonspecific antibody, is constantly expressed in YST (Fig. 6.30f) but not in CCC [96]. Finally, HepPar-1 is frequently focally positive in YST and invariably absent in CCC.

As a summary (Table 6.6), CCC and glandular YST may coexpress HNF1-β, CK7, EMA, and GPC3. Identification of CCC areas is based on expression of PAX8 and NapsinA, while YST is best identified by its AFP, SALL4, villin, and, eventually, focal HepPar-1 expression.

6.5.3 Endometrioid or CCC Associated with YST and Other Polydifferentiated GCT Patterns

This rare association occurs in neoplasms of the uterus and ovary as a combination of endometrioid or CCC, with glandular YST and other GCT growth patterns such as choriocarcinoma, mature teratoma, immature neural structures similar to those found in IT, as well as mimics of embryonal carcinoma and polyembryoma. It is possible that, in the past, some tumors in this category were reported as immature teratoma in extraovarian locations such as the uterus [471, 472]. Only isolated well-documented cases [473] have been reported in the literature.

We have studied five cases, associated with endometrioid carcinoma and CCC [474]. All had a glandular YST component. Coexistence of glandular YST and CCC was observed in some cysts which presented a biphasic lining of both clear cell and endoderm. Other components included, in three cases, an abundance of both neural blastema and neuroepithelial pigmented tubules of the type seen in immature teratoma (Fig. 6.31), one of them differentiating a glioblastoma multiforme pattern. In two instances, there were patterns resembling areas of polyembryoma with embryoids expressing a characteristic ecto- and endodermal immunophenotype (Fig. 6.32). Another included a choriocarcinoma and yet another, mature components such as squamous and respiratory epithelia, striated muscle and cartilage of the type seen on mature teratoma, but lacking a dermoid appearance. No areas of dysgerminoma were present.

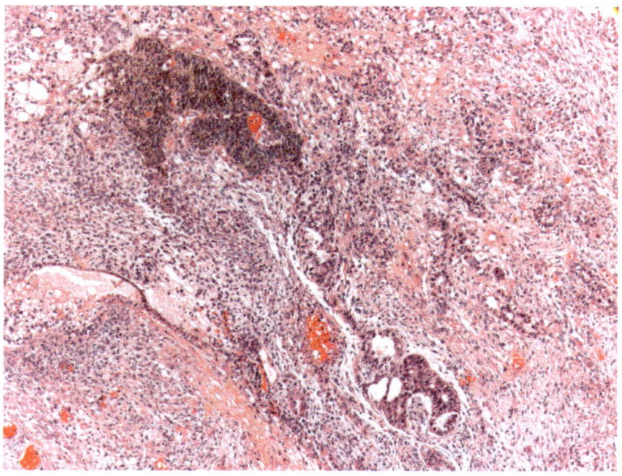

Fig. 6.31 Pigmented neuroepithelium in a mixed CCC-glandular YST

Immunohistochemically, all GCT components (glandular YST, neural components, choriocarcinoma, and embryoids) show a characteristic phenotype. Some areas may have an incomplete EC immunophenotype expressing OCT4, but weak and focal SOX2 and CD30. These incomplete EC features partly resemble type II testicular-type tumors and would exemplify the developmental capacities of iPSC. The existence in these tumors of cells retaining a high degree of stemness [452] is confirmed by the constant presence of foci of cells expressing OCT4 (see Fig. 3.34) [474] (Fig. 6.33), similar to iPSC. OCT4 expression has been reported in high-grade immature teratomas [187], but not in Müllerian-derived GCT.

Tumor stem cells have been demonstrated in neoplasms classically associated with polydifferentiations such as uterine carcinosarcomas [475], endometrioid carcinoma [476, 477], endometriosis [478], and CCC [479]. An intriguing paper [480] has demonstrated that the ovarian surface epithelium in the menopause is able to produce stem cells capable of giving rise to embryoid body-, oocyte-, and blastocyst-like structures but not teratomas when transplanted to mice. This could be exemplified by a recent case of mature uterine teratoma where the comparison of DNA profile of normal uterine tissue and a uterine teratoma suggested an origin from a pluripotential stem cell of the uterus [481].

As a summary, somatic-derived malignant stem cells behaving as an induced pluripotential stem cell (iPSC) can give rise to a malignant stem cell tumor with features of embryonal carcinoma, even with an organoid arrangement into embryoids. They may also differentiate into endodermal structures of YST and trophoblast as well as

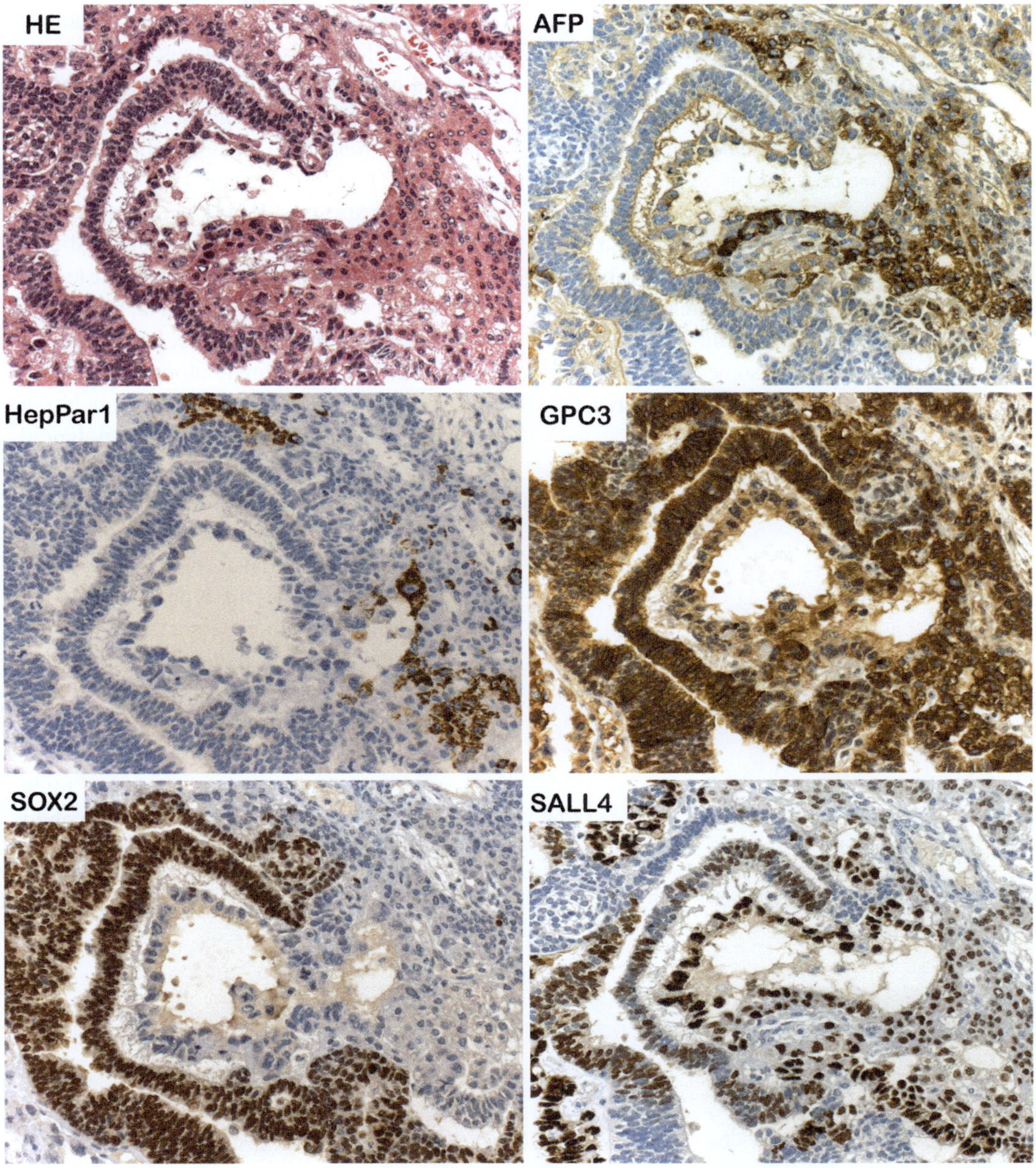

Fig. 6.32 Embryoid body in an endometrioid carcinoma associated with embryonal carcinoma-like patterns. The area corresponding to endodermal differentiation expresses AFP and only focally HepPar-1. GPC3 stains both ecto- and endodermal areas, while SOX2 is only positive in the ectoderm. SALL4 stains all immature areas

embryonal mesenchymal and neural structures present in teratoid carcinosarcoma. It is also speculated that rare "neometaplastic" monophyletic neoplasm such as extrarenal Wilms, extra-axial ependymomas, hepatoid tumors, and even intestinal mucinous tumors may also represent differentiations from induced pluripotential stem cells.

6.6 Extraovarian Germ Cell Tumors of the Female Genital Tract

Extraovarian GCT in the female genital tract are rare. Their true incidence is, as in other extragonadal teratomas, difficult to know, since most reports are clinically oriented and few have a well-documented histopathologic description. Likewise, the loose usage of the term teratoma in the older literature to describe some polydifferentiated neoplasms may have included some cases of mixed Müllerian tumors. Additionally, in the uterus, cervix, and perhaps the fallopian tube, iatrogenic fetal implants [482] may display various mature tissues embedded in the endometrium [483, 484] that can be erroneously diagnosed as mature teratoma.

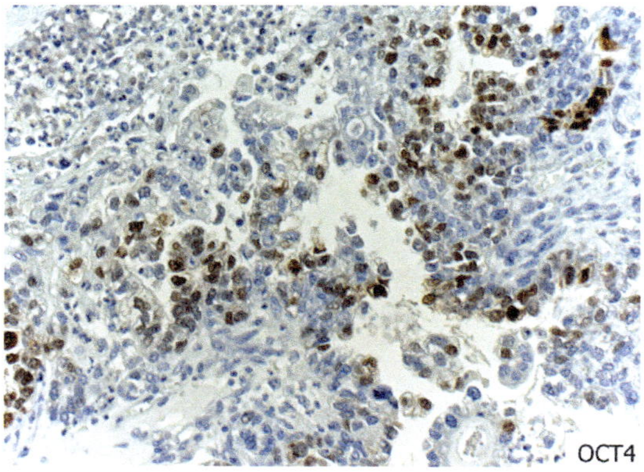

Fig. 6.33 Uterine carcinosarcoma associated with YST. Atypical glandular areas have a patchy expression of OCT4

Fallopian tube GCT are possibly the most common GCT in female genital tract outside the ovary and are also type IV tumors with a parthenogenetic origin. Most represent incidental findings and are small mature teratomas that tend to adopt either a solid [485] (Figs. 6.34a, b) or cystic [486, 487] appearance. There are, however, some reported cases of immature teratomas [488].

Uterine and cervical GCT are extremely rare lesions (see Chap. 2) and are often polypoid, mature teratomas [489–491] even showing the presence of thyroid tissue [492]. It is possible that cases reported as immature teratomas or those with malignant transformation [493, 472] may represent examples of type VI tumors of somatic Müllerian tumors associated with the polydifferentiated GCT patterns, as described above. In the cervix, both mature [494, 495] and immature [496] teratomas have been reported, as well as rare cases of YST [497, 498].

Vulvovaginal GCT include primary YST [499–501] that may correspond to type I GCT in childhood as well as rare immature teratoma [502].

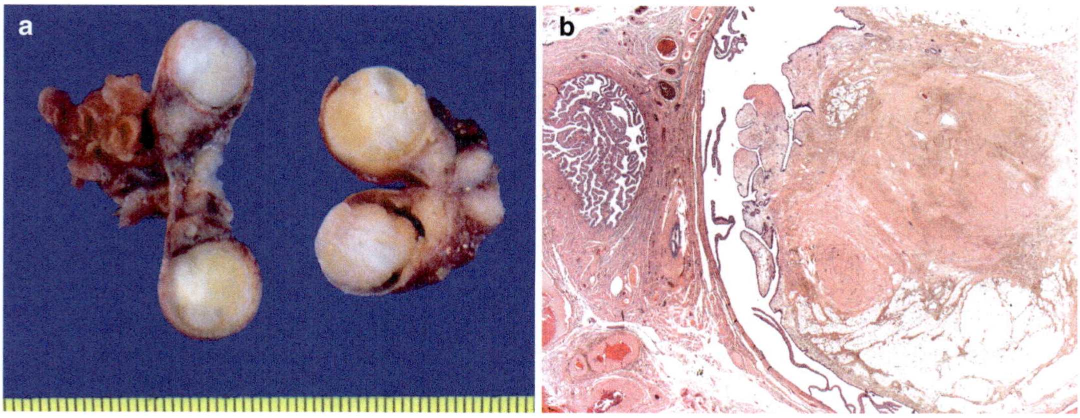

Fig. 6.34 A small solid mature teratoma in the lumen of a fallopian tube (**a**) is composed of fibroadipose tissue with some sebaceous glands (**b**)

Acknowledgments The authors are grateful to the following pathologists for kindly providing the cases illustrating the following figures:

Figure 6-3B. Dr. Isabel Alvarado-Cabrero, Hospital Siglo XXI, Mexico DF, Mexico

Figure 6-6B. Dr. Carmen Tornos, Stony Brook University, NY. USA

Figure 6-6R. Dr. Silvia Chiarelli, Università degli Studi di Padova. Italy

Figure 6-6S. Dr. Harry Hollema, University of Groningen. The Netherlands

Figure 6-20A. Dr. Miguel Marin-Padilla, Geisel School of Medicine at Dartmouth, Hanover, NH, USA

Figure 6-23B. Dr. Lara Alberte, CHUVI Vigo, Spain

Figure 6-30. Dr. Baljeet Kaur. Hammersmith Hospital, London, UK

Figure6-31. Dr. Maria J Pareja.Complejo Hospitalario Universitario de Huelva, Spain

Bibliography

1. Marsden HB, Birch JM, Swindell R. Germ cell tumours of childhood: a review of 137 cases. J Clin Pathol. 1981;34(8):879–83.
2. Katsube Y, Berg JW, Silverberg SG. Epidemiologic pathology of ovarian tumors: a histopathologic review of primary ovarian neoplasms diagnosed in the Denver Standard Metropolitan Statistical Area, 1 July-31 December 1969 and 1 July-31 December 1979. Int J Gynecol Pathol. 1982;1(1):3–16.
3. Lo Curto M, D'Angelo P, Cecchetto G, Klersy C, Dall'Igna P, Federico A, et al. Mature and immature teratomas: results of the first paediatric Italian study. Pediatr Surg Int. 2007;23(4):315–22. doi:10.1007/s00383-007-1890-1.
4. Quirk JT, Natarajan N, Mettlin CJ. Age-specific ovarian cancer incidence rate patterns in the United States. Gynecol Oncol 2005;99(1):248–50. doi:S0090-8258(05)00506-8 [pii] 10.1016/j.ygyno.2005.06.052.
5. Moller H, Evans H. Epidemiology of gonadal germ cell cancer in males and females. APMIS 2003;111(1):43–6; discussion 6–8. doi:apm1110107 [pii].
6. Lee ST, Choi MH, Lee EJ, Gong SP, Jang M, Park SH et al. Establishment of autologous embryonic stem cells derived from preantral follicle culture and oocyte parthenogenesis. Fertil Steril 2008;90(5):1910–20. doi:S0015-0282(07)00207-5 [pii] 10.1016/j.fertnstert.2007.01.099.
7. Oliveira FG, Dozortsev D, Diamond MP, Fracasso A, Abdelmassih S, Abdelmassih V et al. Evidence of parthenogenetic origin of ovarian teratoma: case report. Hum Reprod 2004;19(8):1867–70. doi:10.1093/humrep/deh345. deh345 [pii].
8. Weiss JR, Burgess JR, Kaplan KJ. Fetiform teratoma (homunculus). Arch Pathol Lab Med 2006;130(10):1552–6. doi:CR51040 [pii] 10.1043/1543-2165(2006)130[1552:FTH]2.0.CO;2.
9. Sheikine Y, Genega E, Melamed J, Lee P, Reuter VE, Ye H. Molecular genetics of testicular germ cell tumors. Am J Cancer Res. 2012;2(2):153–67.
10. Goodman S, Zhang L, Cheng L, Jiang Z. Differential expression of IMP3 between male and female mature teratomas–immunohistochemical evidence of malignant nature. Histopathology. 2014;65(4):483–9. doi:10.1111/his.12409.
11. Sigismondi C, Scollo P, Ferrandina G, Candiani M, Angioli R, Vigano R, et al. Management of bilateral malignant ovarian germ cell tumors: a MITO-9 Retrospective Study. Int J Gynecol Cancer. 2015;25(2):203–7. doi:10.1097/IGC.0000000000000358.
12. Horton Z, Schlatter M, Schultz S. Pediatric germ cell tumors. Surg Oncol. 2007;16(3) doi:10.1016/j.suronc.2007.07.005.
13. Gobbi D, Fascetti Leon F, Aquino A, Melchionda F, Lima M. Metachronous bilateral ovarian teratoma: a germ-line familial disorder and review of surgical management options. J Pediatr Adolesc Gynecol 2013;26(5):e105–7. doi:10.1016/j.jpag.2013.02.006. S1083-3188(13)00102-2 [pii].
14. Galani E, Alamanis C, Dimopoulos MA. Familial female and male germ cell cancer. A new syndrome? Gynecol Oncol 2005;96(1):254–5. doi:S0090-8258(04)00802-9 [pii] 10.1016/j.ygyno.2004.09.047.
15. Giambartolomei C, Mueller CM, Greene MH, Korde LA. A mini-review of familial ovarian germ cell tumors: an additional manifestation of the familial testicular germ cell tumor syndrome. Cancer Epidemiol 2009;33(1):31–6. doi:10.1016/j.canep.2009.04.015. S1877-7821(09)00038-1 [pii].
16. Stevens LC, Varnum DS. The development of teratomas from parthenogenetically activated ovarian mouse eggs. Dev Biol. 1974;37(2):369–80. doi:0012-1606(74)90155-9 [pii].
17. Prat J, Cao D, Carinelli SG, Nogales FF, Vang R, Zaloudek CJ. Germ cell tumours. In: WHO classification of tumours of female reproductive organs. 4th ed. Lyon: International Agency for Research on Cancer; 2014. p. 57–62.
18. Nogales FF, Dulcey I, Preda O. Germ cell tumors of the ovary: an update. Arch Pathol Lab Med. 2014;138(3):351–62. doi:10.5858/arpa.2012-0547-RA.
19. Prat J, Cao D, Carinelli SG, Nogales FF, Vang R, Zaloudek CJ. World health organization classification of tumours In: Kurman RY, Young RH. editor. Tumours of the female genital organs. 2nd ed. Lyon: IARCPress; 2014.
20. Smith HO, Berwick M, Verschraegen CF, Wiggins C, Lansing L, Muller CY et al. Incidence and survival rates for female malignant germ cell tumors. Obstet Gynecol. 2006;107(5):1075–85. doi:107/5/1075 [pii] 10.1097/01.AOG.0000216004.22588.ce.
21. Quirk JT, Natarajan N. Ovarian cancer incidence in the United States, 1992–1999. Gynecol Oncol 2005;97(2):519–23. doi: S0090-8258(05)00123-X [pii] 10.1016/j.ygyno.2005.02.007.
22. Arora RS, Alston RD, Eden TO, Geraci M, Birch JM. Comparative incidence patterns and trends

of gonadal and extragonadal germ cell tumors in England, 1979 to 2003. Cancer. 2012;118(17):4290–7. doi:10.1002/cncr.27403.
23. Stang A, Trabert B, Wentzensen N, Cook MB, Rusner C, Oosterhuis JW, et al. Gonadal and extragonadal germ cell tumours in the United States, 1973–2007. Int J Androl. 2012;35(4):616–25. doi:10.1111/j.1365-2605.2011.01245.x.
24. Pierce GB, Speers WC. Tumors as caricatures of the process of tissue renewal: prospects for therapy by directing differentiation. Cancer Res. 1988;48(8):1996–2004.
25. Nogales FF. Haines & Taylor obstetrical and gynaecological pathology. In: Fox H, Wells M, editors. Haines & Taylor obstetrical and gynaecological pathology. 5th ed. Edinburgh: Churchill Livingstone; 2003. p. 771–820.
26. Ulbright TM. Germ cell tumors of the gonads: a selective review emphasizing problems in differential diagnosis, newly appreciated, and controversial issues. Mod Pathol 2005;18 Suppl 2:S61–79. doi:3800310 [pii] 10.1038/modpathol.3800310.
27. Prat JC, D, Carinelli SG, Nogales FF, Vang R, Zaloudek CJ, International Agency for Research on Cancer., World Health Organization. WHO classification of tumours of female reproductive organs. World Health Organization classification of tumours. 4th ed. Lyon: International Agency for Research on Cancer; 2014. p. 57–62.
28. Oliva H, Young RH. Germ cell tumours of the ovary: selected topics. Diagnostic Histopathol. 2014;20(9):364–375.
29. Pectasides D, Pectasides E, Kassanos D. Germ cell tumors of the ovary. Cancer Treat Rev 2008;34(5):427–41. doi:10.1016/j.ctrv.2008.02.002. S0305-7372(08)00025-X [pii].
30. Lee KH, Lee IH, Kim BG, Nam JH, Kim WK, Kang SB et al. Clinicopathologic characteristics of malignant germ cell tumors in the ovaries of Korean women: a Korean Gynecologic Oncology Group Study. Int J Gynecol Cancer 2009;19(1):84–7. doi:10.1111/IGC.0b013e31819914c5. 00009577-200901000-00016 [pii].
31. Vazquez I, Rustin GJ. Current controversies in the management of germ cell ovarian tumours. Curr Opin Oncol 2013;25(5):539–45. doi:10.1097/01.cco.0000432609.39293.77. 00001622-201309000-00014 [pii].
32. Gershenson DM. Treatment of ovarian cancer in young women. Clin Obstet Gynecol. 2012;55(1):65–74. doi:10.1097/GRF.0b013e318248045b. 00003081-201203000-00008 [pii].
33. Gershenson DM. Management of ovarian germ cell tumors. J Clin Oncol. 2007;25(20):2938–43. doi:25/20/2938 [pii] 10.1200/JCO.2007.10.8738.
34. Solheim O, Gershenson DM, Trope CG, Rokkones E, Sun CC, Weedon-Fekjaer H et al. Prognostic factors in malignant ovarian germ cell tumours (The Surveillance, Epidemiology and End Results experience 1978–2010). Eur J Cancer 2014;50(11):1942–50. doi:10.1016/j.ejca.2014.03.288. S0959-8049(14)00573-5 [pii].
35. Mangili G, Sigismondi C, Lorusso D, Cormio G, Scollo P, Vigano R et al. Is surgical restaging indicated in apparent stage IA pure ovarian dysgerminoma? The MITO group retrospective experience. Gynecol Oncol 2011;121(2):280–4. doi:10.1016/j.ygyno.2011.01.006. S0090-8258(11)00018-7 [pii].
36. Gershenson DM, Miller AM, Champion VL, Monahan PO, Zhao Q, Cella D et al. Reproductive and sexual function after platinum-based chemotherapy in long-term ovarian germ cell tumor survivors: a Gynecologic Oncology Group Study. J Clin Oncol. 2007;25(19):2792–7. doi:25/19/2792 [pii] 10.1200/JCO.2006.08.4590.
37. Solheim O, Trope CG, Rokkones E, Kaern J, Paulsen T, Salvesen HB et al. Fertility and gonadal function after adjuvant therapy in women diagnosed with a malignant ovarian germ cell tumor (MOGCT) during the "cisplatin era". Gynecol Oncol 2015;136(2):224–9. doi:10.1016/j.ygyno.2014.12.010 S0090-8258(14)01565-0 [pii].
38. Cossu-Rocca P, Zhang S, Roth LM, Eble JN, Zheng W, Karim FW et al. Chromosome 12p abnormalities in dysgerminoma of the ovary: a FISH analysis. Mod Pathol 2006;19(4):611–5. doi:3800576 [pii] 10.1038/modpathol.3800576.
39. Cheng L, Roth LM, Zhang S, Wang M, Morton MJ, Zheng W, et al. KIT gene mutation and amplification in dysgerminoma of the ovary. Cancer. 2011;117(10):2096–103. doi:10.1002/cncr.25794.
40. Meyer R. Ovarialtumoren und Geschlechtlichkeit. Ein Beitrag zur funktionellen Betrachtung der Geschwülste: I."Disgerminome" beider Geschlechter bei Störung in der Entwicklung der Keimdrüsen, II. Granulosazelltumoren mit "Verweiblichung," III. Arrhenoblastome mit "Vermännlichung". Klin Wochenschr. 1930;9:2237–40.
41. Stewart CJ, Baker E, Beaton C, Crook M, Peverall J, Wallace S. Detection of Y-chromosome in gonadal tumours using fluorescence in situ hybridization: diagnostic value in intersex conditions including older patients with clinically unsuspected androgen insensitivity syndrome. Histopathology 2008;52(2):175–82. doi:10.1111/j.1365-2559.2007.02927.x. HIS2927 [pii].
42. Boyes DA, Pankratz E, Walliford BW, White GW, Fairey RN. Experience with dysgerminomas at the Cancer Control Agency of British Columbia. Gynecol Oncol. 1978;6(2):123–9.
43. Linton EB. Dysgerminoma of the ovary. J Reprod Med. 1981;26(5):255–60.
44. Kraggerud SM, Szymanska J, Abeler VM, Kaern J, Eknaes M, Heim S, et al. DNA copy number changes in malignant ovarian germ cell tumors. Cancer Res. 2000;60(11):3025–30.
45. Zhao S, Kato N, Endoh Y, Jin Z, Ajioka Y, Motoyama T. Ovarian gonadoblastoma with mixed germ cell tumor in a woman with 46, XX karyotype and successful pregnancies. Pathol Int. 2000;50(4):332–5. doi:pin1041 [pii]
46. Koo YJ, Chun YK, Kwon YS, Lee IH, Kim TJ, Lee KH, et al. Ovarian gonadoblastoma with dysgerminoma in a woman with 46XX karyotype. Pathol Int. 2011;61(3):171–3. doi:10.1111/j.1440-1827.2010.02636.x.

47. Tanaka Y, Fujiwara K, Yamauchi H, Mikami Y, Kohno I. Pregnancy in a woman with a Y chromosome after removal of an ovarian dysgerminoma. Gynecol Oncol 2000;79(3):519–21. doi:10.1006/gyno.2000.6004. S0090-8258(00)96004-9 [pii].
48. Paliwal P, Sharma A, Birla S, Kriplani A, Khadgawat R. Identification of novel SRY mutations and SF1 (NR5A1) changes in patients with pure gonadal dysgenesis and 46, XY karyotype. Mol Hum Reprod 2011;17(6):372–8. doi:10.1093/molehr/gar002 gar002 [pii].
49. Rozenholc A, Abdulcadir J, Pelte MF, Petignat P. A pelvic mass on ultrasonography and high human chorionic gonadotropin level: not always an ectopic pregnancy. BMJ Case Rep. 2012;2012. doi:10.1136/bcr.01.2012.5577 bcr0120125577 [pii] bcr.01.2012.5577 [pii].
50. Brettell JR, Miles PA, Herrera G, Greenberg H. Dysgerminoma with syncytiotrophoblastic giant cells presenting as a hydatidiform mole. Gynecol Oncol. 1984;18(3):393–401.
51. Nourani M, Manera RB. Pediatric ovarian dysgerminoma presenting with hypercalcemia and chronic constipation: a case report. J Pediatr Hematol Oncol. 2013;35(7):e272–3. doi:10.1097/MPH.0b013e31829bcaf1.
52. Rotman M, Dorfmann H, de Seze S, Kahn MF. Coexistence d'un lupus erythemateux disseminé et d'un seminome de l'ovarie. Guérison apparente rapide du lupus après ablation de la tumeur. Recul de 7 ans. Nouv Press Med. 1972;1(13):853–7.
53. Yeh KE, Marcus PS, Fong TL. Paraneoplastic cholestasis associated with ovarian dysgerminoma. Obstet Gynecol. 2015;126(2):431–4. doi:10.1097/AOG.0000000000000699.
54. Buskirk SJ, Schray MF, Podratz KC, Lee RA, Stanhope CR, Gaffey TA, et al. Ovarian dysgerminoma: a retrospective analysis of results of treatment, sites of treatment failure, and radiosensitivity. Mayo Clin Proc. 1987;62(12):1149–57.
55. De Palo G, Lattuada A, Kenda R, Musumeci R, Zanini M, Pilotti S, et al. Germ cell tumors of the ovary: the experience of the National Cancer Institute of Milan. I. Dysgerminoma. Int J Radiat Oncol Biol Phys. 1987;13(6):853–60. doi:0360-3016(87)90099-X [pii].
56. Schweyer S, Soruri A, Baumhoer D, Peters J, Cattaruzza M, Radzun HJ, et al. Expression of CXC chemokine IP-10 in testicular germ cell tumours. J Pathol. 2002;197(1):89–97. doi:10.1002/path.1094.
57. Kato N, Kurotaki H, Uchigasaki S, Fukase M, Kurose A. Ovarian clear cell carcinoma with plasma cell-rich inflammatory stroma: a clear cell carcinoma subgroup with distinct clinicopathological features. Histopathology. 2015; doi:10.1111/his.12783.
58. Sellami-Dhouib R, Nasfi A, Mejri NT, Doghri R, Charfi L, Sassi S, et al. Ovarian ALK+ diffuse large B-cell lymphoma: a case report and a review of the literature. Int J Gynecol Pathol. 2013;32(5):471–5. doi:10.1097/PGP.0b013e31826cbd6e.
59. Alvarado-Cabrero I, Valencia-Cedillo R, Mohs-Alfaro M, De Anda-Gonzalez J. Ovarian dysgerminoma associated with fibrosarcoma: a case report. Int J Gynecol Pathol. 2011;30(5):466–9. doi:10.1097/PGP.0b013e318217139e.
60. Iczkowski KA, Butler SL, Shanks JH, Hossain D, Schall A, Meiers I et al. Trials of new germ cell immunohistochemical stains in 93 extragonadal and metastatic germ cell tumors. Hum Pathol 2008;39(2):275–81. doi:S0046-8177(07)00357-7 [pii] 10.1016/j.humpath.2007.07.002.
61. Idrees M, Saxena R, Cheng L, Ulbright TM, Badve S. Podoplanin, a novel marker for seminoma: A comparison study evaluating immunohistochemical expression of podoplanin and OCT3/4. Ann Diagn Pathol 2010;14(5):331–6. doi:10.1016/j.anndiagpath.2010.05.008 S1092-9134(10)00095-X [pii].
62. Chang MC, Vargas SO, Hornick JL, Hirsch MS, Crum CP, Nucci MR. Embryonic stem cell transcription factors and D2-40 (podoplanin) as diagnostic immunohistochemical markers in ovarian germ cell tumors. Int J Gynecol Pathol. 2009;28(4):347–55. doi:10.1097/PGP.0b013e318195da86.
63. Rijlaarsdam MA, van Herk HA, Gillis AJ, Stoop H, Jenster G, Martens J et al. Specific detection of OCT3/4 isoform A/B/B1 expression in solid (germ cell) tumours and cell lines: confirmation of OCT3/4 specificity for germ cell tumours. Br J Cancer 2011;105(6):854–63. doi:10.1038/bjc.2011.270 bjc2011270 [pii].
64. Cossu-Rocca P, Jones TD, Roth LM, Eble JN, Zheng WX, Karim FWA, et al. Cytokeratin and CD30 expression in dysgerminoma. Hum Pathol. 2006;37(8):1015–21. doi:10.1016/j.humpath.2006.02.018.
65. Battifora H, Sheibani K, Tubbs RR, Kopinski MI, Sun TT. Antikeratin antibodies in tumor diagnosis. Distinction between seminoma and embryonal carcinoma. Cancer. 1984;54(5):843–8.
66. Parkash V, Carcangiu ML. Transformation of ovarian dysgerminoma to yolk sac tumor: evidence for a histogenetic continuum. Mod Pathol. 1995;8(8):881–7.
67. Tatekawa Y, Kemmotsu H, Mouri T, Joe K, Ohkawa H. A case of pediatric ovarian dysgerminoma associated with high serum levels and positive immunohistochemical staining of neuron-specific enolase. J Pediatr Surg. 2004;39(9):1437–9. doi:S0022346804003628 [pii].
68. Nogales FF, Ruiz Avila I, Concha A, del Moral E. Immature endodermal teratoma of the ovary: embryologic correlations and immunohistochemistry. Hum Pathol. 1993;24(4):364–70.
69. Young RH. A brief history of the pathology of the gonads. Mod Pathol. 2005;18:S3–S17. doi:10.1038/modpathol.3800305.
70. Young RH. The yolk sac tumor: reflections on a remarkable neoplasm and two of the many intrigued by It-Gunnar Teilum and Aleksander Talerman-and the bond it formed between them. Int J Surg Pathol 2014;22(8):677–87.doi:10.1177/1066896914558265 1066896914558265 [pii].

71. Nogales FF, Preda O, Nicolae A. Yolk sac tumours revisited. A review of their many faces and names. Histopathology. 2012;60(7):1023–33. doi:10.1111/j.1365-2559.2011.03889.x.
72. Nogales FF, Talerman A, Kubik-Huch RA, Tavassoli FA, Devousssoux-Shisheboran M. Germ cell tumours In: Tavassoli FA, P Devilee, Eds. Tumours of the breast and female genital tract. Lyon: IARC Press; 2003. p. 163–75.
73. Mosbech CH, Rechnitzer C, Brok JS, Rajpert-De Meyts E, Hoei-Hansen CE. Recent advances in understanding the etiology and pathogenesis of pediatric germ cell tumors. J Pediatr Hematol Oncol. 2014;36(4):263–70. doi:10.1097/MPH.0000000000000125.
74. Simon RA, Laughlin TS, Nuccie B, Wang N, Rothberg PG, Wang X. A 46 XY phenotypic female adolescent with bilateral gonadal tumors consisting of five different components. Int J Gynecol Pathol 2008;27(3):407–11. doi:10.1097/PGP.0b013e31815d05d4 00004347-200807000-00013 [pii].
75. Langley FA, Govan AD, Anderson MC, Gowing NF, Woodcock AS, Harilal KR. Yolk sac and allied tumours of the ovary. Histopathology. 1981;5(4):389–401.
76. Cooper C, Cooper M, Carter J, Russell P. Gonadoblastoma progressing to dysgerminoma in a 55-year-old woman with normal karyotype. Pathology 2007;39(2):284–5. doi:773354847 [pii] 10.1080/00313020701230708.
77. Nogales FF, Bergeron C, Carvia RE, Alvaro T, Fulwood HR. Ovarian endometrioid tumors with yolk sac tumor component, an unusual form of ovarian neoplasm. Analysis of six cases. Am J Surg Pathol. 1996;20(9):1056–66.
78. Roth LM, Talerman A, Levy T, Sukmanov O, Czernobilsky B. Ovarian yolk sac tumors in older women arising from epithelial ovarian tumors or with no detectable epithelial component. Int J Gynecol Pathol. 2011;30(5):442–51. doi:10.1097/PGP.0b013e3182164386.
79. Elhosseiny A, Gonzalez O, Hedjazi M, Pillari V, Howard R. Endodermal sinus tumor and benign cystic teratoma of the ovary occurring simultaneously. N Y State J Med. 1989;89(4):233–4.
80. Kurman RJ, Norris HJ. Endodermal sinus tumor of the ovary: a clinical and pathologic analysis of 71 cases. Cancer. 1976;38(6):2404–19.
81. Arima N, Tanimoto A, Hayashi R, Hamada T, Sasaguri Y. Ovarian yolk sac tumor with virilization during pregnancy: Immunohistochemical demonstration of Leydig cells as functioning stroma. Pathol Int. 2000;50(6):520–5. doi:10.1046/j.1440-1827.2000.01073.x.
82. Pecorelli S, Sartori E, Favalli G, Ugazio AG, Gastaldi A. Ataxia-telangiectasia and endodermal sinus tumor of the ovary: report of a case. Gynecol Oncol. 1988;29(2):240–4. doi:0090-8258(88)90219-3 [pii].
83. Billmire DF, Cullen JW, Rescorla FJ, Davis M, Schlatter MG, Olson TA et al. Surveillance after initial surgery for pediatric and adolescent girls with stage I ovarian germ cell tumors: report from the Children's Oncology Group. J Clin Oncol 2014;32(5):465–70. doi:10.1200/JCO.2013.51.1006 JCO.2013.51.1006 [pii].
84. Mooney EE, Nogales FF, Tavassoli FA. Hepatocytic differentiation in retiform Sertoli-Leydig cell tumors: distinguishing a heterologous element from Leydig cells. Hum Pathol. 1999;30(6):611–7.
85. Williams SD, Einhorn LH. Brain metastases in disseminated germinal neoplasms: incidence and clinical course. Cancer. 1979;44(4):1514–6.
86. Clement PB, Young RH, Scully RE. Endometrioid-like variant of ovarian yolk sac tumor. A clinicopathological analysis of eight cases. Am J Surg Pathol. 1987;11(10):767–78.
87. Nogales Jr FF, Matilla A, Nogales O, galera-Davidson HL. Yolk sac tumors with pure and mixed polyvesicular vitelline patterns. Hum Pathol. 1978;9(5):553–66.
88. Young RH, Ulbright TM, Policarpio-Nicolas MLC. Yolk sac tumor with a prominent polyvesicular vitelline pattern a report of three cases. Am J Surg Pathol. 2013;37(3):393–8. doi:10.1097/Pas.0b013e31827dcc2b.
89. Vos A, Oosterhuis JW, de Jong B, Castedo SM, Hollema H, Buist J, et al. Karyotyping and DNA flow cytometry of metastatic ovarian yolk sac tumor. Cancer Genet Cytogenet. 1990;44(2):223–8. doi:0165-4608(90)90051-B [pii].
90. Vandeputte M, Sobis H. Experimental rat model for human yolk sac tumor. Eur J Cancer Clin Oncol. 1988;24(3):551–8.
91. Ulbright TM, Roth LM, Brodhecker CA. Yolk sac differentiation in germ cell tumors. A morphologic study of 50 cases with emphasis on hepatic, enteric, and parietal yolk sac features. Am J Surg Pathol. 1986;10(3):151–64.
92. Damjanov I, Amenta PS, Zarghami F. Transformation of an AFP-positive yolk sac carcinoma into an AFP-negative neoplasm. Evidence for in vivo cloning of the human parietal yolk sac carcinoma. Cancer. 1984;53(9):1902–7.
93. Nogales-Fernandez F, Silverberg SG, Bloustein PA, Martinez-Hernandez A, Pierce GB. Yolk sac carcinoma (endodermal sinus tumor): ultrastructure and histogenesis of gonadal and extragonadal tumors in comparison with normal human yolk sac. Cancer. 1977;39(4):1462–74.
94. Kao CS, Idrees MT, Young RH, Ulbright TM. Solid pattern yolk sac tumor: a morphologic and immunohistochemical study of 52 cases. Am J Surg Pathol. 2012;36(3):360–7. doi:10.1097/PAS.0b013e31823c510b.
95. Kao CS, Idrees MT, Young RH, Ulbright TM. Solid pattern yolk sac tumor: a morphologic and immunohistochemical study of 52 cases. Am J Surg Pathol. 2012;36(3):360–7.
96. Nogales FF, Quinonez E, Lopez-Marin L, Dulcey I, Preda O. A diagnostic immunohistochemical panel for yolk sac (primitive endodermal) tumours based on an immunohistochemical comparison with the human yolk sac. Histopathology. 2014;65(1):51–9. doi:10.1111/his.12373.

97. Heifetz SA, Cushing B, Giller R, Shuster JJ, Stolar CJ, Vinocur CD, et al. Immature teratomas in children: pathologic considerations: a report from the combined Pediatric Oncology Group/Children's Cancer Group. Am J Surg Pathol. 1998;22(9):1115–24.
98. Prat J, Bhan AK, Dickersin GR, Robboy SJ, Scully RE. Hepatoid yolk sac tumor of the ovary (endodermal sinus tumor with hepatoid differentiation): a light microscopic, ultrastructural and immunohistochemical study of seven cases. Cancer. 1982;50(11):2355–68.
99. Salazar HK, Tobon, A. et al. Endodermal cell derivatives in embryonal carcinoma of the ovary: an electron microscopy study of two cases. Am J Pathol. 1974;74:108a.(abstr).
100. Michael H, Ulbright TM, Brodhecker CA. The pluripotential nature of the mesenchyme-like component of yolk sac tumor. Arch Pathol Lab Med. 1989;113(10):1115–9.
101. Nogales FF, Buritica C, Regauer S, Gonzalez T. Mucinous carcinoid as an unusual manifestation of endodermal differentiation in ovarian yolk sac tumors. Am J Surg Pathol. 2005;29(9):1247–51. doi:00000478-200509000-00017 [pii].
102. Dickersin GR, Oliva E, Young RH. Endometrioid-like variant of ovarian yolk sac tumor with foci of carcinoid: an ultrastructural study. Ultrastruct Pathol. 1995;19(5):421–9.
103. Battocchio M, Zatelli MC, Chiarelli S, Trento M, Ambrosio MR, Pasquali C, et al. Ovarian tumors secreting insulin. Endocrine. 2015;49(3):611–9. doi:10.1007/s12020-015-0605-y.
104. Mann JR, Gray ES, Thornton C, Raafat F, Robinson K, Collins GS, et al. Mature and immature extracranial teratomas in children: The UK Children's Cancer Study Group experience. J Clin Oncol. 2008;26(21):3590–7. doi:10.1200/jco.2008.16.0622.
105. Nogales FF. Unpublished observations.
106. Fujii K, Yamashita Y, Yamamoto T, Takahashi K, Hashimoto K, Miyata T et al. Ovarian mucinous tumors arising from mature cystic teratomas–a molecular genetic approach for understanding the cellular origin. Hum Pathol 2014;45(4):717–24. doi:10.1016/j.humpath.2013.10.031 S0046-8177(13)00478-4 [pii].
107. Nogales FF, Dulcey I. The secondary human yolk sac has an immunophenotype indicative of both hepatic and intestinal differentiation. Int J Dev Biol. 2012;56(9):755–760. doi:10.1387/ijdb.120080fn 120080fn [pii].
108. Xue D, Peng Y, Wang F, Allan RW, Cao D. RNA-binding protein LIN28 is a sensitive marker of ovarian primitive germ cell tumours. Histopathology. 2011;59(3):452–9. doi:10.1111/j.1365-2559.2011.03949.x.
109. Preda O, Dema A, Iacob M, Goyenaga P, Dulcey I, Aneiros Fernandez J, et al. Urothelial carcinoma of the renal pelvis with simultaneous trophoblastic and malignant clear cell endodermal-type differentiation. Virchows Arch. 2012;460(3):353–6. doi:10.1007/s00428-012-1211-5.
110. Zynger DL, McCallum JC, Luan C, Chou PM, Yang XJ. Glypican 3 has a higher sensitivity than alpha-fetoprotein for testicular and ovarian yolk sac tumour: immunohistochemical investigation with analysis of histological growth patterns. Histopathology 2010;56(6):750–7. doi:10.1111/j.1365-2559.2010.03553.x HIS3553 [pii].
111. Preda O, Nicolae A, Aneiros-Fernandez J, Borda A, Nogales FF. Glypican 3 is a sensitive, but not a specific, marker for the diagnosis of yolk sac tumours. Histopathology 2011;58(2):312–4; author reply 4–5. doi:10.1111/j.1365-2559.2010.03735.x.
112. Esheba GE, Pate LL, Longacre TA. Oncofetal protein glypican-3 distinguishes yolk sac tumor from clear cell carcinoma of the ovary. Am J Surg Pathol. 2008;32(4):600–7. doi:10.1097/PAS.0b013e31815a565a.
113. Pitman MB, Triratanachat S, Young RH, Oliva E. Hepatocyte paraffin 1 antibody does not distinguish primary ovarian tumors with hepatoid differentiation from metastatic hepatocellular carcinoma. Int J Gynecol Pathol. 2004;23(1):58–64. doi:10.1097/01.pgp.0000101141.31270.a1.
114. Gulluoglu MG, Karayigit E, Ozden I, Kapran Y, Dizdaroglu F. Does HepPar-1 immunoexpression have a role in differential diagnosis of periampullary cancer? Pathology. 2008;40(1):35–41. doi:10.1080/00313020701716391.
115. Pelosi G, Petrella F, Sandri MT, Spaggiari L, Galetta D, Viale G. A primary pure yolk sac tumor of the lung exhibiting CDX-2 immunoreactivity and increased serum levels of alkaline phosphatase intestinal isoenzyme. Int J Surg Pathol. 2006;14(3):247–51. doi:14/3/247 [pii] 1177/1066896906290657.
116. Bing Z, Pasha T, Tomaszewski JE, Zhang P. CDX2 expression in yolk sac component of testicular germ cell tumors. Int J Surg Pathol 2009;17(5):373–7. doi:10.1177/1066896909338598 1066896909338598 [pii].
117. Wong HH, Chu P. Immunohistochemical features of the gastrointestinal tract tumors. J Gastrointest Oncol. 2012;3(3):262–84. doi:10.3978/j.issn.2078-6891.2012.019.
118. Cao D, Li J, Guo CC, Allan RW, Humphrey PA. SALL4 is a novel diagnostic marker for testicular germ cell tumors. Am J Surg Pathol. 2009;33(7):1065–77. doi:10.1097/PAS.0b013e3181a13eef.
119. Cao D, Guo S, Allan RW, Molberg KH, Peng Y. SALL4 is a novel sensitive and specific marker of ovarian primitive germ cell tumors and is particularly useful in distinguishing yolk sac tumor from clear cell carcinoma. Am J Surg Pathol. 2009;33(6):894–904. doi:10.1097/PAS.0b013e318198177d.
120. Wang F, Liu A, Peng Y, Rakheja D, Wei L, Xue D, et al. Diagnostic utility of SALL4 in extragonadal yolk sac tumors: an immunohistochemical study of 59 cases with comparison to placental-like alkaline phosphatase, alpha-fetoprotein, and glypican-3. Am J Surg Pathol. 2009;33(10):1529–39. doi:10.1097/PAS.0b013e3181ad25d5.

121. Khalifa MA, Gersell DJ, Hansen CH, Lage JM. Hepatic (hepatocellular) adenoma of the placenta: a study of four cases. Int J Gynecol Pathol. 1998;17(3):241–4.
122. Matsuyama TA, Kushima M, Yamochi-Onizuka T, Ota H. Placental yolk sac tumor with divergent endodermal differentiation. Int J Gynecol Pathol. 2004;23(4):398–402. doi:00004347-200410000-00014 [pii].
123. West JA, Viswanathan SR, Yabuuchi A, Cunniff K, Takeuchi A, Park IH et al. A role for Lin28 in primordial germ-cell development and germ-cell malignancy. Nature 2009;460(7257):909–13. doi:10.1038/nature08210 nature08210 [pii].
124. Cao D, Allan RW, Cheng L, Peng Y, Guo CC, Dahiya N et al. RNA-binding protein LIN28 is a marker for testicular germ cell tumors. Hum Pathol 2011;42(5):710–8. doi:10.1016/j.humpath.2010.09.007 S0046-8177(10)00364-3 [pii].
125. Cao D, Liu A, Wang F, Allan RW, Mei K, Peng Y, et al. RNA-binding protein LIN28 is a marker for primary extragonadal germ cell tumors: an immunohistochemical study of 131 cases. Mod Pathol. 2011;24(2):288–96. doi:10.1038/modpathol.2010.195.
126. Nogales FF, Dulcey I, Preda O. Issues in Gynecologic Pathology. Germ Cell Tumors of the Ovary. An Update. Arch Pathol Lab Med. 2014.
127. Miettinen M, McCue PA, Sarlomo-Rikala M, Rys J, Czapiewski P, Wazny K, et al. GATA3: a multispecific but potentially useful marker in surgical pathology: a systematic analysis of 2500 epithelial and nonepithelial tumors. Am J Surg Pathol. 2014;38(1):13–22. doi:10.1097/PAS.0b013e3182a0218f.
128. Schuldt M, Rubio A, Preda O, Nogales FF. GATA3 expression is present in primitive patterns of Yolk Sac Tumors but is not expressed by differentiated variants. Histopathology. 2015; doi:10.1111/his.12776.
129. Iacobelli JF, Charles AK, Crook M, Stewart CJ. NUT protein immunoreactivity in ovarian germ cell tumours. Pathology 2015;47(2):118–122. doi:10.1097/PAT.0000000000000208 01268031-201502000-00005 [pii].
130. Damato S, Haldar K, McCluggage WG. Primary endometrial yolk sac tumor with endodermal-intestinal differentiation masquerading as metastatic colorectal adenocarcinoma. Int J Gynecol Pathol. 2015; doi:10.1097/PGP.0000000000000236.
131. Retamero JA, Schuldt M, Nogales FF. Yolk sac tumors in postmenopausal patients arising in endometrioid adenocarcinoma. AJSP Rev Rep. 2016;21:189–94.
132. Ramalingam P, Malpica A, Silva EG, Gershenson DM, Liu JL, Deavers MT. The use of cytokeratin 7 and EMA in differentiating ovarian yolk sac tumors from endometrioid and clear cell carcinomas. Am J Surg Pathol. 2004;28(11):1499–505.
133. Nogales FF, Schuldt M, Cruz BN, Oosterhuis JW, Kaur B, Matias-Guiu X et al. Germ cell tumor growth patterns originated from clear cell carcinoma of the ovary and endometrium. Comparative immunohistochemical study proposing a putative origin from neoplastic stem cells of somatic tumors. In preparation 2017.
134. De Backer A, Madern GC, Pieters R, Haentjens P, Hakvoort-Cammel FG, Oosterhuis JW, et al. Influence of tumor site and histology on long-term survival in 193 children with extracranial germ cell tumors. Eur J Pediatr Surg. 2008;18(1):1–6. doi:10.1055/s-2007-989399.
135. Topuz S, Iyibozkurt AC, Akhan SE, Keskin N, Yavuz E, Salihoglu Y, et al. Malignant germ cell tumors of the ovary: a review of 41 cases and risk factors for recurrence. Eur J Gynaecol Oncol. 2008;29(6):635–7.
136. Moch H, Humphrey P, Ubright TM, Reuter VE. WHO classification of tumours of the urinary system and male genital organs, World health organization classification of tumours. 4th ed. Lyon: IARC; 2016.
137. Hennes E, Zahn S, Lopes LF, Schonberger S, Leuschner I, Gobel U, et al. Molecular genetic analysis of bilateral ovarian germ cell tumors. Klin Padiatr. 2012;224(6):359–65. doi:10.1055/s-0032-1327606.
138. Kleinsmith LJ, Pierce Jr GB. Multipotentiality of single embryonal carcinoma cells. Cancer Res. 1964;24:1544–51.
139. Malecki M, Anderson M, Beauchaine M, Seo S, Tombokan X, Malecki R. TRA-1-60, SSEA-4, Oct4A, Nanog clones of pluripotent stem cells in the embryonal carcinomas of the ovaries. J Stem Cell Res Ther. 2012;2(5) doi:130 [pii].
140. Cheng L, Zhang SB, Talerman A, Roth LM. Morphologic, immunohistochemical, and fluorescence in situ hybridization study of ovarian embryonal carcinoma with comparison to solid variant of yolk sac tumor and immature teratoma. Hum Pathol. 2010;41(5):716–23. doi:10.1016/j.humpath.2009.10.016.
141. Prat J, Bhan AK, Dickersin GR, Robboy SJ, Scully RE. Hypothesis: when is a seminoma not a seminoma? J Clin Pathol. 1981;34(11):1308–9.
142. Kurman RJ, Norris HJ. Embryonal carcinoma of the ovary: a clinicopathologic entity distinct from endodermal sinus tumor resembling embryonal carcinoma of the adult testis. Cancer. 1976;38(6):2420–33.
143. Young RH, Stall JN, Sevestre H. The polyembryoma: one of the most intriguing human neoplasms, with comments on the investigator who brought it to light, Albert Peyron. Int J Gynecol Pathol. 2016;35(2):93–105. doi:10.1097/PGP.0000000000000282.
144. Ulbright TM, Young RH. Tumors of the tesis and adjacent structures. Atlas of tumor pathology (4th series). American Registry of Pathology: Silver Spring; 2013.
145. Poulos C, Cheng L, Zhang S, Gersell DJ, Ulbright TM. Analysis of ovarian teratomas for isochromosome 12p: evidence supporting a dual histogenetic pathway for teratomatous elements. Mod Pathol 2006;19(6):766–71. doi:3800596 [pii] 10.1038/modpathol.3800596.

146. Wang D, Hu Y, He Y, Xie C, Yin R. Pure ovarian choriocarinoma mimicking ectopic pregnancy in true hermaphroditism. Acta Obstet Gynecol Scand 2009;88(7):850–2.doi:10.1080/00016340902902883 910379889 [pii].
147. Joneborg U, Papadogiannakis N, Lindell G, Marions L. Choriocarcinoma following ovarian hydatidiform mole: a case report. J Reprod Med. 2011;56(11–12):511–4.
148. Sehn JK, Kuroki LM, Hopeman MM, Longman RE, McNicholas CP, Huettner PC. Ovarian complete hydatidiform mole: case study with molecular analysis and review of the literature. Hum Pathol 2013;44(12):2861–4. doi:10.1016/j.humpath.2013.07.027 S0046-8177(13)00316-X [pii].
149. Gardner JK, Konner JA. Germ cell and sex cord stromal tumors: clinical introduction. In: Soslow RA, Tornos C, Eds. Diagnostic Pathology of Ovarian Tumors. New York: Springer; 2011. p. 145–53.
150. Cho NH, Kim YT, Lee JH, Song C, Cho SW, Cho SH, et al. Diagnostic challenge of fetal ontogeny and its application on the ovarian teratomas. Int J Gynecol Pathol. 2005;24(2):173–82. doi:00004347-200504000-00010 [pii].
151. Gutierrez-Aranda I, Ramos-Mejia V, Bueno C, Munoz-Lopez M, Real PJ, Macia A, et al. Human induced pluripotent stem cells develop teratoma more efficiently and faster than human embryonic stem cells regardless the site of injection. Stem Cells. 2010;28(9):1568–70. doi:10.1002/stem.471.
152. Doss BJ, Jacques SM, Qureshi F, Chang CH, Christensen CW, Morris RT et al. Immature teratomas of the genital tract in older women. Gynecol Oncol 1999;73(3):433–8. doi:S0090-8258(99)95345-3 [pii] 10.1006/gyno.1999.5345.
153. Nogales FF. Germ cell tumours of the ovary. In: Fox H, Wells M, editors. Haines & Taylor obstetrical and gynaecological pathology. 5th ed. Edinburgh: Churchill Livingstone; 2003. p. 771–820.
154. Dabner M, McCluggage WG, Bundell C, Carr A, Leung Y, Sharma R, et al. Ovarian teratoma associated with anti-N-methyl D-aspartate receptor encephalitis: a report of 5 cases documenting prominent intratumoral lymphoid infiltrates. Int J Gynecol Pathol. 2012;31(5):429–37. doi:10.1097/PGP.0b013e31824a1de2.
155. Nogales Jr FF, Favara BE, Major FJ, Silverberg SG. Immature teratoma of the ovary with a neural component ("solid" teratoma). A clinicopathologic study of 20 cases. Hum Pathol. 1976;7(6):625–42.
156. Marina NM, Cushing B, Giller R, Cohen L, Lauer SJ, Ablin A, et al. Complete surgical excision is effective treatment for children with immature teratomas with or without malignant elements: A Pediatric Oncology Group/Children's Cancer Group Intergroup Study. J Clin Oncol. 1999;17(7):2137–43.
157. Norris HJ, Zirkin HJ, Benson WL. Immature (malignant) teratoma of the ovary: a clinical and pathologic study of 58 cases. Cancer. 1976;37(5):2359–72.
158. Yanai H, Matsuura H, Kawasaki M, Takada Y, Tabuchi Y, Yoshino T. Immature teratoma of the ovary with a minor rhabdomyosarcomatous component and fatal rhabdomyosarcomatous metastases: the first case in a child. Int J Gynecol Pathol. 2002;21(1):82–5.
159. Byun JC, Choi IJ, Han MS, Lee SC, Roh MS, Cha MS. Soft tissue metastasis of an immature teratoma of the ovary. J Obstet Gynaecol Res. 2011;37(11):1689–93. doi:10.1111/j.1447-0756.2011.01553.x.
160. Kurata A, Hirano K, Nagane M, Fujioka Y. Immature teratoma of the ovary with distant metastases: favorable prognosis and insights into chemotherapeutic retroconversion. Int J Gynecol Pathol. 2010;29(5):438–44. doi:10.1097/PGP.0b013e3181cef16b.
161. Merard R, Ganesan R, Hirschowitz L. Growing teratoma syndrome: a report of 2 cases and review of the literature. Int J Gynecol Pathol 2015;34(5):465–72. doi:10.1097/PGP.0000000000000180. 00004347-201509000-00009 [pii].
162. Nogales FF, Aguilar D. Florid vascular proliferation in grade 0 glial implants from ovarian immature teratoma. Int J Gynecol Pathol. 2002;21(3):305–7.
163. O'Connor DM, Norris HJ. The influence of grade on the outcome of stage I ovarian immature (malignant) teratomas and the reproducibility of grading. Int J Gynecol Pathol. 1994;13(4):283–9.
164. King ME, Micha JP, Allen SL, Mouradian JA, Chaganti RS. Immature teratoma of the ovary with predominant malignant retinal anlage component. A parthenogenically derived tumor. Am J Surg Pathol. 1985;9(3):221–31.
165. Liang L, Olar A, Niu N, Jiang Y, Cheng W, Bian XW, et al. Primary glial and neuronal tumors of the ovary or peritoneum: a clinicopathologic study of 11 cases. Am J Surg Pathol. 2016; doi:10.1097/PAS.0000000000000635.
166. Euscher ED, Deavers MT, Lopez-Terrada D, Lazar AJ, Silva EG, Malpica A. Uterine tumors with neuroectodermal differentiation: a series of 17 cases and review of the literature. Am J Surg Pathol 2008;32(2):219–28. doi:10.1097/PAS.0b013e318093e421 00000478-200802000-00006 [pii].
167. Clinkard DJ, Khalifa M, Osborned RJ, Bouffet E. Successful management of medulloblastoma arising in an immature ovarian teratoma in pregnancy. Gynecol Oncol 2011;120(2):311–2. doi:10.1016/j.ygyno.2010.10.022 S0090-8258(10)00768-7 [pii].
168. Morency EG, Lerner D, Garcia R, Kalir T. High-grade sarcoma masquerading as growing teratoma syndrome after resection of ovarian immature teratoma: report of a case. Int J Gynecol Pathol. 2012;31(3):276–9. doi:10.1097/PGP.0b013e31823ef912.
169. Malagon HD, Valdez AM, Moran CA, Suster S. Germ cell tumors with sarcomatous components: a clinicopathologic and immunohistochemical study of 46 cases. Am J Surg Pathol 2007;31(9):1356–62. doi:10.1097/PAS.0b013e318033c7c4 00000478-200709000-00008 [pii].

170. Mengshol SC, Demars LR, Schned AR. Gliomatosis peritonei and teratomatous implant with carcinomatous transformation presenting 54 years following oophorectomy for dermoid cyst. Gynecol Oncol. 2004;92(1):353–6.
171. DiSaia PJ, Saltz A, Kagan AR, Morrow CP. Chemotherapeutic retroconversion of immature teratoma of the ovary. Obstet Gynecol. 1977;49(3):346–50.
172. Logothetis CJ, Samuels ML, Trindade A, Johnson DE. The growing teratoma syndrome. Cancer. 1982;50(8):1629–35.
173. Kato N, Uchigasaki S, Fukase M. How does secondary neoplasm arise from mature teratomas in growing teratoma syndrome of the ovary? A report of two cases. Pathol Int. 2013;63(12):607–10. doi:10.1111/pin.12112.
174. Djordjevic B, Euscher ED, Malpica A. Growing teratoma syndrome of the ovary: review of literature and first report of a carcinoid tumor arising in a growing teratoma of the ovary. Am J Surg Pathol 2007;31(12):1913–8. doi:10.1097/PAS.0b013e318073cf44 00000478-200712000-00017 [pii].
175. Nogales FF, Preda O, Dulcey I. Gliomatosis peritonei as a natural experiment in tissue differentiation. Int J Dev Biol 2012;56(10–12):969–74. doi:10.1387/ijdb.120172fn 120172fn [pii].
176. Wu XH, Han LY, Xu X, Li ZT. Recurrent immature teratoma of the ovary: a case report of radical secondary cytoreduction with replacement of the aortic bifurcation. Gynecol Oncol. 2004;95(3):746–9. doi:10.1016/j.ygyno.2004.08.007.
177. Yoon NR, Lee JW, Kim BG, Bae DS, Sohn I, Sung CO, et al. Gliomatosis peritonei is associated with frequent recurrence, but does not affect overall survival in patients with ovarian immature teratoma. Virchows Arch. 2012;461(3):299–304. doi:10.1007/s00428-012-1285-0.
178. Gonzalez-Campora R, Nogales Jr FF, Davidson HG, Mendez JA. Case report: ultrastructure of mature neurogenic implants from ovarian immature teratoma. Histopathology. 1979;3(3):233–40.
179. Webman R, Talishinskiy T, Raetz E, Lala S, Tomita S. Spontaneous regression of thoracic and extraperitoneal glial implants in child with gliomatosis peritonei after resection of ovarian teratoma. J Pediatr Hematol Oncol. 2014; doi:10.1097/MPH.0000000000000230.
180. Webman R, Talishinskiy T, Raetz E, Lala S, Tomita S. Spontaneous regression of thoracic and extraperitoneal glial implants in child with gliomatosis peritonei after resection of ovarian teratoma. J Pediatr Hematol Oncol. 2015;37(3):230–1. doi:10.1097/MPH.0000000000000230.
181. Liang L, Zhang Y, Malpica A, Ramalingam P, Euscher ED, Fuller GN, et al. Gliomatosis peritonei: a clinicopathologic and immunohistochemical study of 21 cases. Mod Pathol. 2015;28(12):1613–20. doi:10.1038/modpathol.2015.116.
182. Kim NR, Lim S, Jeong J, Cho HY. Peritoneal and nodal gliomatosis with endometriosis, accompanied with ovarian immature teratoma: a case study and literature review. Korean J Pathol. 2013;47(6):587–91. doi:10.4132/KoreanJPathol.2013.47.6.587.
183. Lobotesis K, U-King-Im JM, Cross JJ, Gillard JH, Antoun NM. Gliomatosis peritonei associated with a ventriculo-peritoneal shunt. Clin Radiol 2009;64(1):95–9. doi:10.1016/j.crad.2008.08.015 S0009-9260(08)00374-7 [pii].
184. Ferguson AW, Katabuchi H, Ronnett BM, Cho KR. Glial implants in gliomatosis peritonei arise from normal tissue, not from the associated teratoma. Am J Pathol. 2001;159(1):51–5. doi:10.1016/s0002-9440(10)61672-0.
185. Kwan MY, Kalle W, Lau GT, Chan JK. Is gliomatosis peritonei derived from the associated ovarian teratoma? Hum Pathol. 2004;35(6):685–8. doi:S004681770400108X [pii].
186. Alexander M, Cope N, Renninson J, Hong A, Simpson RH, Hirschowitz L. Relationship between endometriosis, endometrioid adenocarcinoma, gliomatosis peritonei, and carcinoid tumor in a patient with recurrent ovarian teratoma. Int J Gynecol Pathol. 2011;30(2):151–7. doi:10.1097/PGP.0b013e3181f6bcd4.
187. Abiko K, Mandai M, Hamanishi J, Matsumura N, Baba T, Horiuchi A et al. Oct4 expression in immature teratoma of the ovary: relevance to histologic grade and degree of differentiation. Am J Surg Pathol 2010;34(12):1842–8. doi:10.1097/PAS.0b013e3181fcd707 00000478-201012000-00012 [pii].
188. Linder D, McCaw BK, Hecht F. Parthenogenic origin of benign ovarian teratomas. N Engl J Med. 1975;292(2):63–6. doi:10.1056/NEJM197501092920202.
189. Dahl N, Gustavson KH, Rune C, Gustavsson I, Pettersson U. Benign ovarian teratomas. An analysis of their cellular origin. Cancer Genet Cytogenet. 1990;46(1):115–23.
190. Besser MJ, Posey DM. Cystic teratoma in a supernumerary ovary of the greater omentum. A case report. J Reprod Med. 1992;37(2):189–93.
191. Nezhat C, Kotikela S, Mann A, Hajhosseini B, Veeraswamy A, Lewis M. Familial cystic teratomas: four case reports and review of the literature. J Minim Invasive Gynecol 2010;17(6):782–6. doi:10.1016/j.jmig.2010.06.006 S1553-4650(10)00333-X [pii].
192. Brown Jr EH. Identical twins with twisted benign cystic teratoma of the ovary. Am J Obstet Gynecol. 1979;134(8):879–80. doi:0002-9378(79)90860-3 [pii].
193. Saba L, Guerriero S, Sulcis R, Virgilio B, Melis G, Mallarini G. Mature and immature ovarian teratomas: CT, US and MR imaging characteristics. Eur J Radiol 2009;72(3):454–63. doi:10.1016/j.ejrad.2008.07.044 S0720-048X(08)00457-9 [pii].
194. Dalmau J, Tuzun E, Wu HY, Masjuan J, Rossi JE, Voloschin A, et al. Paraneoplastic anti-N-methyl-

D-aspartate receptor encephalitis associated with ovarian teratoma. Ann Neurol. 2007;61(1):25–36. doi:10.1002/ana.21050.
195. Dalmau J, Rosenfeld MR. Update on paraneoplastic neurologic disorders. Commun Oncol. 2010;7(5):219–24.
196. Vitaliani R, Mason W, Ances B, Zwerdling T, Jiang Z, Dalmau J. Paraneoplastic encephalitis, psychiatric symptoms, and hypoventilation in ovarian teratoma. Ann Neurol. 2005;58(4):594–604. doi:10.1002/ana.20614.
197. Kayser MS, Dalmau J. Anti-NMDA receptor encephalitis, autoimmunity, and psychosis. Schizophr Res 2014. doi:S0920-9964(14)00546-5 [pii] 10.1016/j.schres.2014.10.007.
198. Dulcey I, Cespedes MU, Ballesteros JL, Preda O, Aneiros-Fernandez J, Clavero PA et al. Necrotic mature ovarian teratoma associated with anti-N-methyl-D-aspartate receptor encephalitis. Pathol Res Pract 2012;208(8):497–500. doi:10.1016/j.prp.2012.05.018 S0344-0338(12)00156-2 [pii].
199. Templeman CL, Fallat ME, Lam AM, Perlman SE, Hertweck SP, O'Connor DM. Managing mature cystic teratomas of the ovary. Obstet Gynecol Surv. 2000;55(12):738–45.
200. Kim MJ, Kim NY, Lee DY, Yoon BK, Choi D. Clinical characteristics of ovarian teratoma: age-focused retrospective analysis of 580 cases. Am J Obstet Gynecol. 2011;205(1):32 e1–4. doi:10.1016/j.ajog.2011.02.044 S0002-9378(11)00234-1 [pii].
201. Peitsidou A, Peitsidis P, Goumalatsos N, Papaspyrou R, Mitropoulou G, Georgoulias N. Diagnosis of an autoamputated ovary with dermoid cyst during a Cesarean section. Fertil Steril. 2009;91(4):1294 e9–12. doi:10.1016/j.fertnstert.2008.12.029 S0015-0282(08)04725-0 [pii].
202. Kusaka M, Mikuni M. Ectopic ovary: a case of autoamputated ovary with mature cystic teratoma into the cul-de-sac. J Obstet Gynaecol Res. 2007;33(3):368–70. doi:JOG538 [pii] 10.1111/j.1447-0756.2007.00538.x.
203. Siegel MJ, McAlister WH, Shackelford GD. Radiographic findings in ovarian teratomas in children. AJR Am J Roentgenol. 1978;131(4):613–5. doi:10.2214/ajr.131.4.613.
204. Shetty NS, Vallabhaneni S, Patil A, Babu MM, Baig A. Unreported location and presentation for a parasitic ovarian dermoid cyst in an indirect inguinal hernia. Hernia. 2013;17(2):263–5. doi:10.1007/s10029-011-0876-z.
205. Sinha R, Sundaram M, Lakhotia S. Multiple intraabdominal parasitic cystic teratomas. J Minim Invasive Gynecol 2009;16(6):789–91. doi:10.1016/j.jmig.2009.08.003 S1553-4650(09)00443-9 [pii].
206. Takeda A, Imoto S, Mori M, Yamada J, Nakamura H. Early abdominal pregnancy complicated by parasitic dermoid cyst: diagnosis by diffusion-weighted magnetic resonance imaging and management by laparoendoscopic single-site surgery. J Minim Invasive Gynecol 2012;19(5):647–50. doi:10.1016/j.jmig.2012.03.024 S1553-4650(12)00127-6 [pii].
207. Cucinella G, Granese R, Venezia R, Mangione D, Calagna G, Perino A. Parasitic dermoid cyst coexisting with absence of an adnexa. Acta Obstet Gynecol Scand. 2011;90(6):677–8. doi:10.1111/j.1600-0412.2011.01125.x.
208. Walid MS, Boddy MG. Bilateral dermoid cysts of the ovary in a pregnant woman: case report and review of the literature. Arch Gynecol Obstet. 2009;279(2):105–8. doi:10.1007/s00404-008-0695-3.
209. Feltingoff M, Heller DS, Bleiweiss IJ. Ovarian pregnancy with ipsilateral mature cystic teratoma of the ovary: case report. Mt Sinai J Med. 1992;59(1):82–4.
210. Nogales FF, Fernandez-Sanz J, Fulwood H, Matilla A, Galera-Davidson H. Estudio clínico-patológico de 288 teratomas quísticos de ovario. Patologia (Madrid). 1979;16:12–25.
211. Goh W, Bohrer J, Zalud I. Management of the adnexal mass in pregnancy. Curr Opin Obstet Gynecol 2014;26(2):49–53. doi:10.1097/GCO.0000000000000048 00001703-201404000-00002 [pii].
212. Tariel O, Huissoud C, Rudigoz RC, Dubernard G. [Presumed benign ovarian tumors during pregnancy]. J Gynecol Obstet Biol Reprod (Paris). 2013;42(8):842–55. doi:10.1016/j.jgyn.2013.09.038 S0368-2315(13)00284-6 [pii].
213. Erdemoglu M, Kuyumcuoglu U, Kale A. Pregnancy and adnexal torsion: analysis of 20 cases. Clin Exp Obstet Gynecol. 2010;37(3):224–5.
214. Ulstein M, Rana G, Yangzom K, Gurung R. Prolapse of an ovarian tumor during labor. Acta Obstet Gynecol Scand. 1987;66(8):721–2.
215. Nader R, Thubert T, Deffieux X, de Laveaucoupet J, Ssi-Yan-Kai G. Delivery induced intraperitoneal rupture of a cystic ovarian teratoma and associated chronic chemical peritonitis. Case Rep Radiol. 2014;2014:189409. doi:10.1155/2014/189409.
216. Yoshikata R, Yamamoto T, Kobayashi M, Ota H. Immunohistochemical characteristics of mature ovarian cystic teratomas in patients with postoperative recurrence. Int J Gynecol Pathol. 2006;25(1):95–100. doi:00004347-200601000-00014 [pii].
217. Anteby EY, Ron M, Revel A, Shimonovitz S, Ariel I, Hurwitz A. Germ cell tumors of the ovary arising after dermoid cyst resection: a long-term follow-up study. Obstet Gynecol. 1994;83(4):605–8.
218. Chapron C, Dubuisson JB, Samouh N, Foulot H, Aubriot FX, Amsquer Y, et al. Treatment of ovarian dermoid cysts. Place and modalities of operative laparoscopy. Surg Endosc. 1994;8(9):1092–5.
219. Arora DS, Haldane S. Carcinosarcoma arising in a dermoid cyst of the ovary. J Clin Pathol. 1996;49(6):519–21.
220. Amerigo J, Nogales Jr FF, Fernandez-Sanz J, Oliva H, Velasco A. Squamous cell neoplasms arising from ovarian benign cystic teratoma. Gynecol Oncol. 1979;8(3):277–83.
221. Pansky M, Shade D, Moskovitch M, Halperin R, Ben-Ami I, Smorgick N. Inadvertent rupture of benign cystic

teratoma does not impair future fertility. Am J Obstet Gynecol. 2010;203(5):442 e1–4. doi:10.1016/j.ajog.2010.06.062 S0002-9378(10)00842-2 [pii].

222. Pantoja E, Noy MA, Axtmayer RW, Colon FE, Pelegrina I. Ovarian dermoids and their complications. Comprehensive historical review. Obstet Gynecol Surv. 1975;30(1):1–20.
223. Stuart GC, Smith JP. Ruptured benign cystic teratomas mimicking gynecologic malignancy. Gynecol Oncol. 1983;16(1):139–43.
224. Gendre J, Sebban-Rozot C, Regent D, Ranchoup Y, Ridereau-Zins C, Vullierme MP et al. [Peritoneal parasitic teratoma and chemical dermoid peritonitis]. J Radiol 2011;92(5):382–92. doi:10.1016/j.jradio.2011.03.005 S0221-0363(11)00117-X [pii].
225. Maiti S, Fatima Z, Anjum ZK, Hopkins RE. Ruptured ovarian cystic teratoma in pregnancy with diffuse peritoneal reaction mimicking advanced ovarian malignancy: a case report. J Med Case Rep 2008;2:203. doi:10.1186/1752-1947-2-203 1752-1947-2-203 [pii].
226. Sasaki H, Nagasako K, Harada M, Kobayashi S, Uetake K. Benign cystic teratoma of the ovary with rupture into the rectum: report of a unique rectal tumor. Dis Colon Rectum. 1979;22(4):248–51.
227. Suzuki M, Fukasawa M, Hara T. Dermoid cyst with thyroid follicle perforated into bladder and ileum: a case report. Nihon Hinyokika Gakkai Zasshi. 1999;90(5):586–9.
228. von-Walter AR, Nelken RS. Benign cystic ovarian teratoma with a fistula into the small and large bowel. Obstet Gynecol 2012;119(2 Pt 2):434–6. doi:10.1097/AOG.0b013e3182319185 00006250-201202001-00007 [pii].
229. Sardesai S, Raghoji V, Dabade R, Shaikh H. Benign cytic teratoma of ovary perforating into the urinary bladder: a rare case. J Obstet Gynaecol India 2012;62(Suppl 1):54–5. doi:10.1007/s13224-013-0354-x 354 [pii].
230. Candela G, Di Libero L, Varriale S, Manetta F, Napolitano S, Scetta G, et al. Hemoperitoneum caused by the rupture of a giant ovarian teratoma in a 9-year-old female. Case report and literature review. Ann Ital Chir. 2009;80(2):141–4.
231. Fukushima M, Sharpe L, Okagaki T. Peritoneal melanosis secondary to a benign dermoid cyst of the ovary: a case report with ultrastructural study. Int J Gynecol Pathol. 1984;2(4):403–9.
232. Jaworski RC, Boadle R, Greg J, Cocks P. Peritoneal "melanosis" associated with a ruptured ovarian dermoid cyst: report of a case with electron-probe energy dispersive X-ray analysis. Int J Gynecol Pathol. 2001;20(4):386–9.
233. Brelje MC, Garcia-Bunuel R. Salmonella infection in an ovarian dermoid cyst in pregnancy: report of a case. Obstet Gynecol. 1964;24:279–80.
234. Matsubayashi T, Hamajima T, Asano K, Mizukami A, Seguchi M, Kohno C, et al. Salmonella infection of an ovarian dermoid cyst. Pediatr Int. 2001;43(2):164–5. doi:ped1351 [pii].
235. Salehpour S, Akbari Sene A. Super infection of an ovarian dermoid cyst with actinomyces in an infertile woman. Int J Fertil Steril. 2013;7(2):134–7.
236. Melato M, Muuse MM, Hussein AM, Falconieri G. Schistosomiasis in a cystic teratoma of the ovary. Clin Exp Obstet Gynecol. 1987;14(1):57–9.
237. Sane SY, Patel CV. A filarial worm in the wall of a cystic teratoma of the ovary–(a case report). J Postgrad Med. 1989;35(4):217–8.
238. Cobo F, Pereira A, Nomdedeu B, Gallart T, Ordi J, Torne A, et al. Ovarian dermoid cyst-associated autoimmune hemolytic anemia: a case report with emphasis on pathogenic mechanisms. Am J Clin Pathol. 1996;105(5):567–71.
239. Payne D, Muss HB, Homesley HD, Jobson VW, Baird FG. Autoimmune hemolytic anemia and ovarian dermoid cysts: case report and review of the literature. Cancer. 1981;48(3):721–4.
240. Ibarra M, Chou P, Pachman LM. Ovarian teratoma mimicking features of juvenile dermatomyositis in a child. Pediatrics 2011;128(5):e1293–6. doi:10.1542/peds.2010-2115 peds.2010-2115 [pii].
241. Wiese W, Alansari H, Tranchida P, Madrid FF. Paraneoplastic polyarthritis in an ovarian teratoma. J Rheumatol. 2004;31(9):1854–7. doi:0315162X-31-1854 [pii].
242. Deen R, de Silva A, Wijesuriya R. Twisted benign ovarian teratoma presenting with pain and generalized pruritus: a case report. J Med Case Rep 2013;7(1):130. doi:10.1186/1752-1947-7-130 1752-1947-7-130 [pii].
243. Di Stefano F, Siriruttanapruk S, Di Gioacchino M. Exercise-induced urticarial vasculitis as a paraneoplastic manifestation of cystic teratoma. Rheumatology (Oxford). 2003;42(11):1418–9. doi:10.1093/rheumatology/keg346 42/11/1418 [pii].
244. Frawley KJ, Calvo-Garcia MA, Krueger DA, McMasters RL. 'Benign' ovarian teratoma and N-methyl-D-aspartate receptor (NMDAR) encephalitis in a child. Pediatr Radiol. 2012;42(1):120–3. doi:10.1007/s00247-011-2111-6.
245. McCarthy A, Dineen J, McKenna P, Keogan M, Sheehan J, Lynch T, et al. Anti-NMDA receptor encephalitis with associated catatonia during pregnancy. J Neurol. 2012;259(12):2632–5. doi:10.1007/s00415-012-6561-z.
246. Lupi I, Fessehatsion R, Manca A, Cossu-Rocca P, Martino E, Macchia E. Hashimoto's thyroiditis in a benign cystic teratoma of the ovary: case report and literature review. Gynecol Endocrinol. 2012;28(1):39–42. doi:10.3109/09513590.2011.579659.
247. Amareen VN, Haddad FH, Al-Kaisi NS. Hypothyroidism due to Hashimoto thyroiditis post struma ovarii excision. Saudi Med J. 2004;25(7):948–50. doi:20030587' [pii].
248. Elms AF, Carlan SJ, Rich AE, Cerezo L. Ovarian tumor-derived ectopic hyperprolactinemia. Pituitary. 2012;15(4):552–5. doi:10.1007/s11102-011-0366-4.
249. Klonoff DC, Kahn DG, Rosenzweig W, Wilson CB. Hyperprolactinemia in a patient with a pitu-

itary and an ovarian dermoid tumor: case report. Neurosurgery. 1990;26(2):335–9.
250. Takeuchi K, Murata K, Funaki K, Kitazawa S, Kitazawa R. Immunohistochemical detection of parathyroid hormone-related protein in a squamous cell carcinoma arising from mature cystic teratoma causing humoral hypercalcemia of malignancy. Gynecol Oncol 2000;79(3):504–7. doi:10.1006/gyno.2000.5945 S0090-8258(00)95945-6 [pii].
251. Gunasekaran S, Kopecka E, Maung KH, England RJ. Struma ovarii and the thyroid surgeon. J Laryngol Otol 2012;126(8):858–60. doi:10.1017/S0022215112000904 S0022215112000904 [pii].
252. Deffieux X, Thubert T, Huchon C, Demoulin G, Rivain AL, Faivre E et al. [Complications of presumed benign ovarian tumors]. J Gynecol Obstet Biol Reprod (Paris). 2013;42(8):816–32. doi:10.1016/j.jgyn.2013.09.036 S0368-2315(13)00282-2 [pii].
253. Axiotis CA, Lippes HA, Merino MJ, deLanerolle NC, Stewart AF, Kinder B. Corticotroph cell pituitary adenoma within an ovarian teratoma. A new cause of Cushing's syndrome. Am J Surg Pathol. 1987;11(3):218–24.
254. Tsujimoto T, Takaichi M, Endo H, Yasuda K, Kishimoto M, Noto H, et al. A patient with diabetes and breast cancer in whom virilization was caused by a testosterone-producing mature cystic teratoma containing a Brenner tumor. e. 2011;341(1):74–7. doi:10.1097/MAJ.0b013e3181f54e59.
255. Hoffman JG, Strickland JL, Yin J. Virilizing ovarian dermoid cyst with leydig cells. J Pediatr Adolesc Gynecol 2009;22(3):e39–40. doi:10.1016/j.jpag.2008.05.012 S1083-3188(08)00194-0 [pii].
256. Lopez-Beltran A, Calanas AS, Jimena P, Escudero AL, Campello TR, Munoz-Torres M, et al. Virilizing mature ovarian cystic teratomas. Virchows Arch. 1997;431(2):149–51.
257. Rutgers JL, Scully RE. Functioning ovarian tumors with peripheral steroid cell proliferation: a report of twenty-four cases. Int J Gynecol Pathol. 1986;5(4):319–37.
258. Robboy SJ, Scully RE. Strumal carcinoid of the ovary: an analysis of 50 cases of a distinctive tumor composed of thyroid tissue and carcinoid. Cancer. 1980;46(9):2019–34.
259. Ober WB. Solid ovarian teratoma with struma ovarii, theca lutein reaction, and endometrial hyperplasia. Report of a case. J Obstet Gynaecol Br Emp. 1960;67:451–4.
260. Murdoch W, Sadoski J, Rosin FC. Multisystem manifestations of benign ovarian teratomas. J Am Board Fam Med. 2014;27(3):421–3. doi:10.3122/jabfm.2014.03.130221 27/3/421 [pii].
261. Sahin AA, Ro JY, Chen J, Ayala AG. Spindle cell nodule and peptic ulcer arising in a fully developed gastric wall in a mature cystic teratoma. Arch Pathol Lab Med. 1990;114(5):529–31.
262. Tiltman AJ, Glass DR, Duffield MS. Peptic ulceration in a mature cystic teratoma of the ovary. Histopathology. 1995;26(4):375–7.
263. Kirova YM, Feuilhade F, de Baecque-Fontaine C, Le Bourgeois JP. Metastasis of a breast carcinoma in a mature teratoma of the ovary. Eur J Gynaecol Oncol. 1999;20(3):223–5.
264. Damewood M, Rosenshein NB, Woodruff JD. Multiple benign cystic teratomas of the ovary. Report of two cases and review of the literature. Diagn Gynecol Obstet. 1982;4(3):243–5.
265. Monk BE, Isaacs AJ, Bayliss R. Giant ovarian dermoid masked by obesity. Postgrad Med J. 1980;56(660):748–9.
266. Canda AE, Astarcioglu H, Obuz F, Canda MS. Cystic ovarian teratoma with intracystic floating globules. Abdom Imaging. 2005;30(3):369–71. doi:10.1007/s00261-004-0266-4.
267. Devoize L, Collangettes D, Le Bouedec G, Mishellany F, Orliaguet T, Dallel R et al. Giant mature ovarian cystic teratoma including more than 300 teeth. Oral Surg Oral Med Oral Pathol Oral Radiol Endod 2008;105(3):e76–9. doi:10.1016/j.tripleo.2007.10.001 S1079-2104(07)00819-0 [pii].
268. Chavan SS, Yenni VV. Mandible like structure with fourteen teeth in a benign cystic teratoma. Indian J Pathol Microbiol. 2009;52(4):595–6. doi:10.4103/0377-4929.56145 IndianJPathol Microbiol_2009_52_4_595_56145 [pii].
269. Ingale Y, Shankar AA, Routray S, Agrawal M, Kadam A, Patil T. Ectopic teeth in ovarian teratoma: a rare appearance. Case Rep Dent. 2013;2013:970464. doi:10.1155/2013/970464.
270. Willis RA. The spread of tumours in the human body. 3rd ed. London: Butterworths; 1973.
271. Kuno N, Kadomatsu K, Nakamura M, Miwa-Fukuchi T, Hirabayashi N, Ishizuka T. Mature ovarian cystic teratoma with a highly differentiated homunculus: a case report. Birth Defects Res A Clin Mol Teratol. 2004;70(1):40–6. doi:10.1002/bdra.10133.
272. Abbott TM, Hermann Jr WJ, Scully RE. Ovarian fetiform teratoma (homunculus) in a 9-year-old girl. Int J Gynecol Pathol. 1984;2(4):392–402.
273. Joutel N, Pirot N, Pop I, Khoutech K, Bisson V, Cingotti M et al. Homunculus: fetiform teratoma. Gynecol Obstet Fertil 2012;40(2):121–3. doi:10.1016/j.gyobfe.2011.07.046 S1297-9589(11)00273-6 [pii].
274. Weldon-Linne CM, Rushovich AM. Benign ovarian cystic teratomas with homunculi. Obstet Gynecol. 1983;61(3 Suppl):88S–94S.
275. Di Donato N, Zanello M, Corti B, Seracchioli R, Paradisi R. Incidental diagnosis of ectopic adenohypophyseal tissue in an ovarian dermoid cyst: a case report. Eur J Obstet Gynecol Reprod Biol. 2014;183:203–4. doi:10.1016/j.ejogrb.2014.10.044. S0301-2115(14)00568-5 [pii].
276. McKeel Jr DW, Askin FB. Ectopic hypophyseal hormonal cells in benign cystic teratoma of the ovary. Light microscopic histochemical dye staining and immunoperoxidase cytochemistry. Arch Pathol Lab Med. 1978;102(3):122–8.
277. Khan N, Sen Ray P, Hakim S, Ziyauddin F. Retinal tissue in mature cystic teratoma of ovary presenting

with full-term pregnancy. BMJ Case Rep. 2014;2014. doi:10.1136/bcr-2013-202914. bcr2013202914 [pii] bcr-2013-202914 [pii].
278. Mercur L, Farkas TG, Anker PM, Brill R, Mund ML. Developing eye in an ovarian teratoma. Arch Ophthalmol. 1976;94(4):597–8.
279. Oi RH, Dobbs M. Lactating breast tissue in benign cystic teratoma. Am J Obstet Gynecol. 1978;130(6):729–31. doi:0002-9378(78)90342-3 [pii].
280. Ulirsch RC, Goldman RL. An unusual teratoma of the ovary: neurogenic cyst with lactating breast tissue. Obstet Gynecol. 1982;60(3):400–2.
281. Muretto P. The relationship of Langerhans cells to melanocytes and Schwann cells in mature cystic teratomas of the ovary. Int J Surg Pathol. 2007;15(3):266–71. doi:15/3/266 [pii] 10.1177/1066896907302227.
282. Nogales Jr FF, Ortega I, Rivera F, Armas JR. Metanephrogenic tissue in immature ovarian teratoma. Am J Surg Pathol. 1980;4(3):297–9.
283. Lopez JI, de Santamaria IS, Claros IJ, Garmendia M, Gaafar A, Bilbao FJ. Prostatic remnants in mature cystic teratoma of the ovary. Ann Diagn Pathol 2008;12(5):378–80. doi:10.1016/j.anndiagpath.2007.04.005 S1092-9134(07)00054-8 [pii].
284. Nogales FF, Vergara E, Medina MT. Prostate in ovarian mature cystic teratoma. Histopathology. 1995;26(4):373–5.
285. Melniker LA, Slavutin LJ. Prostatic tissue in a benign cystic teratoma of the ovary. Diagn Gynecol Obstet. 1980;2(2):139–45.
286. Shelekhova KV, Hejda V, Kazakov DV, Michal M. Mature cystic teratoma of the ovary with male accessory sexual glands including seminal vesicles, prostatic tissue, and bulbo-urethral glands: a case report. Virchows Arch. 2008;452(1):109–11. doi:10.1007/s00428-007-0546-9.
287. Glavina-Durdov M, Forempoher G. Thymus tissue in ovarian cystic teratoma: possible clues to its histogenesis. Pathol Res Pract 2000;196(11):795–7; discussion 8–9. doi:S0344-0338(00)80116-8 [pii] 10.1016/S0344-0338(00)80116-8.
288. Agarwal K, Agarwal C, Agarwal S, Chaddha R. Mature teratoma with loops of colon: a case report. Indian J Pathol Microbiol. 2007;50(2):405–7.
289. Woodfield B, Katz DA, Cantrell CJ, Bogard PJ. A benign cystic teratoma with gastrointestinal tract development. Am J Clin Pathol. 1985;83(2):236–40.
290. Zaragoza Vargas PE, Dickter S, Baquera Heredia J, Ortiz Hidalgo C. Gastric wall in benign cystic teratoma of the ovary. Ginecol Obstet Mex. 2007;75(1):43–5.
291. Agaimy A, Lindner M, Wuensch PH. Interstitial cells of Cajal (ICC) in mature cystic teratoma of the ovary. Histopathology 2006;48(2):208–9. doi:HIS2211 [pii] 10.1111/j.1365-2559.2005.02211.x.
292. Agaimy A, Wunsch PH. Coexistence of interstitial cell of Cajal hyperplasia and microcarcinoidosis in a mature cystic teratoma of the ovary. Histopathology 2008;52(2):260–2. doi:HIS2910 [pii] 10.1111/j.1365-2559.2007.02910.x.
293. Marucci G, Collina G. Mature cystic teratoma of the ovary with a small ganglioneuroma. Pathologica. 2006;98(1):41–3.
294. Ishida M, Iwai M, Yoshida K, Kagotani A, Okabe H. Well-differentiated cerebellum within a mature cystic teratoma of the ovary. Int J Clin Exp Pathol. 2014;7(3):1255–7.
295. Dominguez-Rosado I, Michel J, Ortiz-Hidalgo C. Ovarian mature cystic teratoma exhibiting well-differentiated cerebellum in a 14-year-old female. Pediatr Dev Pathol. 2005;8(2):247–9. doi:10.1007/s10024-004-7083-1.
296. MacSween RN. Foetal cerebellar tissue in an ovarian teratoma. J Pathol Bacteriol. 1968;96(2):513–4. doi:10.1002/path.1700960235.
297. Nogales FF, Aguilar D. Neural tissue in human teratomas. In: The biology of teratomas. Clifton: Human Press; 1983.
298. PN V. Spontaneous myospherulosis developing at contact with a mature cystic dysembryoma of the ovary. Pathogenetic discussion apropos of a case diagnosed by laparoscopic biopsy. J Gynecol Obstet Biol Reprod (Paris). 1988;17(1):15–8.
299. Hirakawa T, Tsuneyoshi M, Enjoji M. Squamous cell carcinoma arising in mature cystic teratoma of the ovary. Clinicopathologic and topographic analysis. Am J Surg Pathol. 1989;13(5):397–405.
300. McKenney JK, Soslow RA, Longacre TA. Ovarian mature teratomas with mucinous epithelial neoplasms: morphologic heterogeneity and association with pseudomyxoma peritonei. Am J Surg Pathol. 2008;32(5):645–55. doi:10.1097/PAS.0b013e31815b486d.
301. Choi EJ, Koo YJ, Jeon JH, Kim TJ, Lee KH, Lim KT. Clinical experience in ovarian squamous cell carcinoma arising from mature cystic teratoma: A rare entity. Obstet Gynecol Sci. 2014;57(4):274–80. doi:10.5468/ogs.2014.57.4.274.
302. Park JY, Kim DY, Kim JH, Kim YM, Kim YT, Nam JH. Malignant transformation of mature cystic teratoma of the ovary: experience at a single institution. Eur J Obstet Gynecol Reprod Biol 2008;141(2):173–8. doi:10.1016/j.ejogrb.2008.07.032 S0301-2115(08)00298-4 [pii].
303. Park JY, Song JS, Choi G, Kim JH, Nam JH. Pure primary squamous cell carcinoma of the ovary: a report of two cases and review of the literature. Int J Gynecol Pathol 2010;29(4):328–34. doi:10.1097/PGP.0b013e3181c6d965 00004347-201007000-00005 [pii].
304. Vang R, Gown AM, Zhao C, Barry TS, Isacson C, Richardson MS et al. Ovarian mucinous tumors associated with mature cystic teratomas: morphologic and immunohistochemical analysis identifies a subset of potential teratomatous origin that shares features of lower gastrointestinal tract mucinous tumors more commonly encountered as secondary tumors in the ovary. Am J Surg Pathol 2007;31(6):854–69. doi:10.1097/PAS.0b013e31802efb45 00000478-200706000-00005 [pii].

305. Quinonez E, Schuldt M, Retamero JA, Nogales FF. Ovarian strumal carcinoid containing appendiceal-type mucinous tumor patterns presenting as pseudomyxoma peritonei. Int J Gynecol Pathol. 2015;34(3):293–7. doi:10.1097/PGP.0000000000000138.
306. Nakatsuka S, Wakimoto T, Ozaki K, Nagano T, Kimura H, Nakajo K, et al. Mucinous borderline-like tumor of the gastrointestinal type arising from mature cystic teratoma of the ovary and its immunohistochemical cytokeratin and mucin phenotype. J Obstet Gynaecol Res. 2012;38(2):471–5. doi:10.1111/j.1447-0756.2011.01729.x.
307. McCluggage WG, Bissonnette JP, Young RH. Primary malignant melanoma of the ovary: a report of 9 definite or probable cases with emphasis on their morphologic diversity and mimicry of other primary and secondary ovarian neoplasms. Int J Gynecol Pathol 2006;25(4):321–9. doi:10.1097/01.pgp.0000215301.39900.07 00004347-200610000-00002 [pii].
308. Hurwitz JL, Fenton A, McCluggage WG, McKenna S. Squamous cell carcinoma arising in a dermoid cyst of the ovary: a case series. BJOG 2007;114(10):1283–7. doi:BJO1478 [pii] 10.1111/j.1471-0528.2007.01478.x.
309. Pantoja E, Rodriguez-Ibanez I, Axtmayer RW, Noy MA, Pelegrina I. Complications of dermoid tumors of the ovary. Obstet Gynecol. 1975;45(1):89–94.
310. Marcial-Rojas RA, Medina R. Cystic teratomas of the ovary; a clinical and pathological analysis of two hundred sixty-eight tumors. AMA Arch Pathol. 1958;66(4):577–89.
311. Klionsky BL, Nickens OJ, Amortegui AJ. Squamous cell carcinoma in situ arising in adult cystic teratoma of the ovary. Arch Pathol. 1972;93(2):161–3.
312. Nogales FF, Luna MA, Vergara E, O'Valle F, Lopez-Cantarero M. Small-cell epidermoid carcinoma with melanocyte colonization arising in a mature ovarian cystic teratoma. Arch Pathol Lab Med. 1995;119(6):551–4.
313. Karateke A, Gurbuz A, Kir G, Haliloglu B, Kabaca C, Devranoglu B et al. Mucoepidermoid variant of adenosquamous carcinoma arising in ovarian dermoid cyst: a case report and review of the literature. Int J Gynecol Cancer 2006;16 Suppl 1:379–84. doi:IJG233 [pii] 10.1111/j.1525-1438.2006.00233.x.
314. Iwasa A, Oda Y, Kaneki E, Ohishi Y, Kurihara S, Yamada T et al. Squamous cell carcinoma arising in mature cystic teratoma of the ovary: an immunohistochemical analysis of its tumorigenesis. Histopathology 2007;51(1):98–104. doi:HIS2727 [pii] 10.1111/j.1365-2559.2007.02727.x.
315. Fox H. Bronchial carcinoma in an ovarian cystic teratoma (dermoid). J Clin Pathol. 1965;18:164–6.
316. Yahata T, Kawasaki T, Serikawa T, Suzuki M, Tanaka K. Adenocarcinoma arising from respiratory ciliated epithelium in benign cystic teratoma of the ovary: a case report with analyzes of the CT, MRI, and pathological findings. J Obstet Gynaecol Res. 2008;34(3):408–12. doi:10.1111/j.1447-0756.2008.00784.x JOG784 [pii].
317. Hershkovitz D, Vlodavsky E, Simon E, Ben-Izhak O. KRAS mutation positive mucinous adenocarcinoma originating in mature ovarian teratoma: Case report and review of literature. Pathol Int. 2013;63(12):611–4. doi:10.1111/pin.12123.
318. Fujiwara K, Ginzan S, Silverberg SG. Mature cystic teratomas of the ovary with intestinal wall structures harboring intestinal- type epithelial neoplasms. Gynecol Oncol 1995;56(1):97–101. doi:S0090-8258(85)71017-7 [pii] 10.1006/gyno.1995.1017.
319. Clark ME, Will MD. Intestinal-type adenocarcinoma arising in a mature cystic teratoma of the ovary. Int J Gynecol Pathol. 2016; doi:10.1097/PGP.0000000000000258.
320. Halabi M, Oliva E, Mazal PR, Breitenecker G, Young RH. Prostatic tissue in mature cystic teratomas of the ovary: a report of four cases, including one with features of prostatic adenocarcinoma, and cytogenetic studies. Int J Gynecol Pathol. 2002;21(3):261–7.
321. Stanhiser J, Mahdi H, Rosa G, Harper H, Shepard D, Rose PG, et al. Prostate-type adenocarcinoma in mature cystic ovarian teratoma. Int J Gynecol Pathol. 2016;35(2):185–90. doi:10.1097/PGP.0000000000000217.
322. Wheeler L, Westhoff GL, O'Keefe MC, Kong CS, Karam A. Adenocarcinoma with breast/adnexal and upper gastrointestinal differentiation arising in an ovarian mature cystic teratoma: a case report and review of the literature. Int J Gynecol Pathol. 2016;35(1):72–7. doi:10.1097/PGP.0000000000000218.
323. Monteagudo C, Torres JV, Llombart-Bosch A. Extramammary Paget's disease arising in a mature cystic teratoma of the ovary. Histopathology. 1999;35(6):582–4. doi:his833d [pii].
324. Shimizu S, Kobayashi H, Suchi T, Torii Y, Narita K, Aoki S. Extramammary Paget's disease arising in mature cystic teratoma of the ovary. Am J Surg Pathol. 1991;15(10):1002–6.
325. Randall BJ, Ritchie C, Hutchison RS. Paget's disease and invasive undifferentiated carcinoma occurring in a mature cystic teratoma of the ovary. Histopathology. 1991;18(5):469–70.
326. Shintaku M, Nakagawa E, Tsubura A. Extramammary Paget's disease arising in a mature cystic teratoma of the ovary: immunohistochemical expression of androgen receptor. Pathol Int. 2011;61(8):498–500. doi:10.1111/j.1440-1827.2011.02692.x.
327. Hyun HS, Mun ST. Primary malignant melanoma arising in a cystic teratoma. Obstet Gynecol Sci. 2013;56(3):201–4. doi:10.5468/ogs.2013.56.3.201.
328. Lee J, Magpayo J, Panchari C, Peng SK. Primary malignant melanoma of the ovary arising in a cystic teratoma; case report and review of the literature. Dermatol Online J. 2014;20(10). doi:13030/qt0tr2b2z9 [pii].
329. Godoy HE, Kesterson JP, Kasznica JM, Lele S. Malignant melanoma arising in mature cystic teratoma of the ovary. Gynecol Oncol Case Rep 2011;2(1):14–5. doi:10.1016/j.gynor.2011.10.007 S2211-338X(11)00016-0 [pii].

330. Boughton RS, Hughmanick S, Marin-Padilla M. Malignant melanoma arising in an ovarian cystic teratoma in pregnancy. J Am Acad Dermatol. 1987;17(5 Pt 2):871–5.
331. Mathios AJ, McCausland AM. Basal cell carcinoma originating in a benign cystic teratoma of the ovary. Obstet Gynecol. 1973;42(6):892–6.
332. Alfsen GC, Strom EH. Pilomatrixoma of the ovary: a rare variant of mature teratoma. Histopathology. 1998;32(2):182–3.
333. An HJ, Jung YH, Yoon HK, Jung SJ. Sebaceous carcinoma arising in mature cystic teratoma of ovary. Korean J Pathol. 2013;47(4):383–7. doi:10.4132/KoreanJPathol.2013.47.4.383.
334. Chumas JC, Scully RE. Sebaceous tumors arising in ovarian dermoid cysts. Int J Gynecol Pathol. 1991;10(4):356–63.
335. Yasui W, Kameda T, Harada D, Nakatani H, Hata J, Ito H, et al. A case of sweat gland carcinoma arising in a mature cystic teratoma of the ovary. Gan No Rinsho. 1987;33(7):875–80.
336. Morimitsu Y, Nakashima O, Nakashima Y, Kojiro M, Shimokobe T. Apocrine adenocarcinoma arising in cystic teratoma of the ovary. Arch Pathol Lab Med. 1993;117(6):647–9.
337. Kudo M. The nature of "blue nevus" in cystic teratomas of the ovary. An ultrastructural evidence for Schwann cell origin. Acta Pathol Jpn. 1985;35(3):693–8.
338. Kuroda N, Hirano K, Inui Y, Yamasaki Y, Toi M, Nakayama H, et al. Compound melanocytic nevus arising in a mature cystic teratoma of the ovary. Pathol Int. 2001;51(11):902–4. doi:1288 [pii]
339. Tsang P, Berman L, Kasznica J. Adnexal tumor and a pigmented nevoid lesion in a benign cystic ovarian teratoma. Arch Pathol Lab Med. 1993;117(8):846–7.
340. Amortegui AJ, Kanbour AI. Compound nevus in a cystic teratoma of the ovary. Arch Pathol Lab Med. 1981;105(2):115–6.
341. Kruger S, Schmidt H, Kupker W, Rath FW, Feller AC. Fibrosarcoma associated with a benign cystic teratoma of the ovary. Gynecol Oncol 2002;84(1):150–4. doi:10.1006/gyno.2001.6408 S009082580196408X [pii].
342. Stamp GW, McConnell EM. Malignancy arising in cystic ovarian teratomas. A report of 24 cases. Br J Obstet Gynaecol. 1983;90(7):671–5.
343. Kelley RR, Scully RE. Cancer developing in dermoid cysts of the ovary. A report of 8 cases, including a carcinoid and a leiomyosarcoma. Cancer. 1961;14:989–1000.
344. Allam-Nandyala P, Bui MM, Caracciolo JT, Hakam A. Squamous cell carcinoma and osteosarcoma arising from a dermoid cyst–a case report and review of literature. Int J Clin Exp Pathol. 2010;3(3):313–8.
345. Ngwalle KE, Hirakawa T, Tsuneyoshi M, Enjoji M. Osteosarcoma arising in a benign dermoid cyst of the ovary. Gynecol Oncol. 1990;37(1):143–7.
346. Yasunaga M, Saito T, Eto T, Okadome M, Ariyoshi K, Nishiyama K, et al. Dedifferentiated chondrosarcoma arising in a mature cystic teratoma of the ovary: a case report and review of the literature. Int J Gynecol Pathol. 2011;30(4):391–4. doi:10.1097/PGP.0b013e318211f56e.
347. Climie AR, Heath LP. Malignant degeneration of benign cystic teratomas of the ovary. Review of the literature and report of a chondrosarcoma and carcinoid tumor. Cancer. 1968;22(4):824–32.
348. Contreras AL, Malpica A. Angiosarcoma arising in mature cystic teratoma of the ovary: a case report and review of the literature. Int J Gynecol Pathol 2009;28(5):453–7. doi:10.1097/PGP.0b013e31819d4574 00004347-200909000-00006 [pii].
349. den Bakker MA, Ansink AC, Ewing-Graham PC. "Cutaneous-type" angiosarcoma arising in a mature cystic teratoma of the ovary. J Clin Pathol. 2006;59(6):658–60. doi:59/6/658 [pii] 10.1136/jcp.2005.029751.
350. Kefeli M, Kandemir B, Akpolat I, Yildirim A, Kokcu A. Rhabdomyosarcoma arising in a mature cystic teratoma with contralateral serous carcinoma: case report and review of the literature. Int J Gynecol Pathol. 2009;28(4):372–5. doi:10.1097/PGP.0b013e3181929269.
351. Ueda G, Sato Y, Yamasaki M, Inoue M, Hiramatsu K, Tanaka Y, et al. Malignant fibrous histiocytoma arising in a benign cystic teratoma of the ovary. Gynecol Oncol. 1977;5(3):313–22. doi:0090-8258(77)90041-5 [pii].
352. Richardson G, Robertson DI, O'Connor ME, Nation JG, Stuart GC. Malignant transformation occurring in mature cystic teratomas of the ovary. Can J Surg. 1990;33(6):499–503.
353. Adams H, Went P, Tolnay M, Sartorius G, Singer G. Meningothelial meningioma in a mature cystic teratoma of the ovary. Pathologe. 2007;28(4):278–80. doi:10.1007/s00292-006-0828-z.
354. Takeshima Y, Kaneko M, Furonaka O, Jeet AV, Inai K. Meningioma in mature cystic teratoma of the ovary. Pathol Int 2004;54(7):543–8. doi:10.1111/j.1440-1827.2004.01662.x PIN1662 [pii].
355. Elliot VJ, Shaw EC, Walker M, Jaynes E, Theaker JM. Ovarian paraganglioma arising from mature cystic teratoma. Int J Gynecol Pathol. 2012;31(6):545–6. doi:10.1097/PGP.0b013e31824d3564.
356. Mahdavi A, Silberberg B, Malviya VK, Braunstein AH, Shapiro J. Gangliocytic paraganglioma arising from mature cystic teratoma of the ovary. Gynecol Oncol. 2003;90(2):482–5. doi:S0090825803002725 [pii].
357. den Boon J, van Dijk CM, Helfferich M, Peterse HL. Glioblastoma multiforme in a dermoid cyst of the ovary. A case report. Eur J Gynaecol Oncol. 1999;20(3):187–8.
358. Bjersing L, Cajander S, Rogo K, Ottosson UB, Stendahl U. Glioblastoma multiform in a dermoid cyst of the ovary. Eur J Gynaecol Oncol. 1989;10(6):389–92.
359. Yadav A, Lellouch-Tubiana A, Fournet JC, Quazza JE, Kalifa C, Sainte-Rose C, et al. Glioblastoma

multiforme in a mature ovarian teratoma with recurring brain tumours. Histopathology. 1999;35(2): 170–3. doi:his695 [pii].
360. Spaulding R, Alatassi H, Stewart Metzinger D, Moghadamfalahi M. Ependymoma and carcinoid tumor associated with ovarian mature cystic teratoma in a patient with multiple endocrine neoplasia I. Case Rep Obstet Gynecol. 2014;2014:712657. doi:10.1155/2014/712657.
361. Stolnicu S, Furtado A, Sanches A, Nicolae A, Preda O, Hincu M et al. Ovarian ependymomas of extra-axial type or central immunophenotypes. Hum Pathol 2011;42(3):403–8. doi:10.1016/j.humpath.2010.07.017 S0046-8177(10)00311-4 [pii].
362. Ud Din N, Memon A, Aftab K, Ahmad Z, Ahmed R, Hassan S. Oligodendroglioma arising in the glial component of ovarian teratomas: a series of six cases and review of literature. J Clin Pathol 2012;65(7):631–4. doi:10.1136/jclinpath-2012-200714 jclinpath-2012-200714 [pii].
363. Opris I, Ducrotoy V, Bossut J, Lamy A, Sabourin JC. Oligodendroglioma arising in an ovarian mature cystic teratoma. Int J Gynecol Pathol. 2009;28(4):367–71. doi:10.1097/PGP.0b013e318196c4c0.
364. Skopelitou A, Mitselou A, Michail M, Mitselos V, Stefanou D. Pilocytic astrocytoma arising in a dermoid cyst of the ovary: a case presentation. Virchows Arch. 2002;440(1):105–6.
365. Wey SL, Chen CK, Chen TC, Chen TY. Desmoplastic medulloblastoma arising from an ovarian teratoma: a case report and review of the literature. Int J Surg Pathol 2013;21(4):427–31. doi:10.1177/1066896912464048 1066896912464048 [pii].
366. Olah KS, Needham PG, Jones B. Multiple neuroectodermal tumors arising in a mature cystic teratoma of the ovary. Gynecol Oncol. 1989;34(2):222–5. doi:0090-8258(89)90147-9 [pii].
367. Block M, Gilbert E, Davis C. Metastatic neuroblastoma arising in an ovarian teratoma with long-term survival. Case report and review of the literature. Cancer. 1984;54(3):590–5.
368. Reid HA, van der Walt JD, Fox H. Neuroblastoma arising in a mature cystic teratoma of the ovary. J Clin Pathol. 1983;36(1):68–73.
369. Kleinman GM, Young RH, Scully RE. Primary neuroectodermal tumors of the ovary. A report of 25 cases. Am J Surg Pathol. 1993;17(8):764–78.
370. Aguirre P, Scully RE. Malignant neuroectodermal tumor of the ovary, a distinctive form of monodermal teratoma: report of five cases. Am J Surg Pathol. 1982;6(4):283–92.
371. Boor PJ, Schoene WC. Fetal cerebellar tissue associated with a primitive neuro-epithelial tumor in an ovarian teratoma. Can J Neurol Sci. 1975;2(2):139–41.
372. Hirschowitz L, Ansari A, Cahill DJ, Bamford DS, Love S. Central neurocytoma arising within a mature cystic teratoma of the ovary. Int J Gynecol Pathol. 1997;16(2):176–9.
373. von Gunten M, Burger H, Vajtai I. Choroid plexus papilloma developing in a dermoid cyst of ovary. Histopathology 2006;49(2):204–5. doi:HIS2383 [pii] 10.1111/j.1365-2559.2006.02383.x.
374. Marquette S, Amant F, Vergote I, Moerman P. Pseudomyxoma peritonei associated with a mucinous ovarian tumor arising from a mature cystic teratoma. A case report. Int J Gynecol Pathol 2006;25(4):340–3. doi:10.1097/01.pgp.0000209571.16340.20 00004347-200610000-00005 [pii].
375. Ronnett BM, Seidman JD. Mucinous tumors arising in ovarian mature cystic teratomas: relationship to the clinical syndrome of pseudomyxoma peritonei. Am J Surg Pathol. 2003;27(5):650–7.
376. Valli R, Froio E, Alvarez de Celis MI, Mandato VD, Piana S. Diffuse large B-cell lymphoma occurring in an ovarian cystic teratoma: expanding the spectrum of large B-cell lymphoma associated with chronic inflammation. Hum Pathol 2014;45(12):2507–11. doi:10.1016/j.humpath.2014.09.002 S0046-8177(14)00368-2 [pii].
377. Gandhi N, Soomro IN, O'Connor S, Sovani V. Primary lymphoma arising in a mature cystic teratoma of the ovary. Histopathology. 2012;61(6):1238–40. doi:10.1111/j.1365-2559.2012.04343.x.
378. Damiani S. Pulmonary papillary adenoma-like tumour arising in ovarian teratoma. Virchows Arch. 2004;445(1):96–7. doi:10.1007/s00428-004-1022-4.
379. Ikota H, Kaneko K, Takahashi S, Kawarai M, Tanaka Y, Yokoo H, et al. Malignant transformation of ovarian mature cystic teratoma with a predominant pulmonary type small cell carcinoma component. Pathol Int. 2012;62(4):276–80. doi:10.1111/j.1440-1827.2012.02804.x.
380. Rubio A, Schuldt M, Chamorro C, Crespo-Lora V, Nogales FF. Ovarian small cell carcinoma of pulmonary type arising in mature cystic teratomas with metastases to the contralateral ovary. Int J Surg Pathol 2015;23(5):388–92.doi:10.1177/1066896915586254 1066896915586254 [pii].
381. Miyamoto M, Takano M, Goto T, Kato M, Sasaki N, Furuya K. Large cell neuroendocrine carcinoma arising in mature cystic teratoma: a case report and review of the literature. Eur J Gynaecol Oncol. 2012;33(4):414–8.
382. Chenevert J, Bessette P, Plante M, Tetu B, Dube V. Mixed ovarian large cell neuroendocrine carcinoma, mucinous adenocarcinoma, and teratoma: a report of two cases and review of the literature. Pathol Res Pract 2009;205(9):657–61. doi:10.1016/j.prp.2009.01.013 S0344-0338(09)00040-5 [pii].
383. Choi YD, Lee JS, Choi C, Park CS, Nam JH. Ovarian neuroendocrine carcinoma, non-small cell type, associated with serous carcinoma. Gynecol Oncol 2007;104(3):747–52. doi:S0090-8258(06)00900-0 [pii] 10.1016/j.ygyno.2006.11.008.
384. Ting WH, Hsiao SM, Lin HH, Wei MC. Primary carcinoid tumor of the ovary arising in a mature cystic teratoma: a case report. Eur J Gynaecol Oncol. 2014;35(1):100–2.
385. Guney N, Sayilgan T, Derin D, Ozcan D. Primary carcinoid tumor arising in a mature cystic teratoma

of the ovary: a case report. Eur J Gynaecol Oncol. 2009;30(2):223–5.
386. Huang B, Wu X, Zhou Q, Hu Y, Zhao H, Zhu H, et al. Cushing's syndrome secondary to ectopic ACTH secretion from carcinoid tumor within an ovarian mature teratoma: a case report and review of the literature. Gynecol Endocrinol. 2014;30(3):192–6. doi:10.3109/09513590.2013.871518.
387. Hsu CY, Yang CF, Chen WY, Chiang H. Adenosquamous carcinoma and schneiderian papilloma-like lesion in a mature cystic teratoma of the ovary: a case report. Zhonghua Yi Xue Za Zhi (Taipei). 1996;57(5):375–9.
388. Las Heras F, Pritzker KP, Colgan TJ. Chordoma arising in a mature cystic teratoma of the ovary: a case report. Pathol Res Pract 2007;203(6):467–71. doi:S0344-0338(07)00031-3 [pii] 10.1016/j.prp.2006.12.010.
389. Silver SA, Tavassoli FA. Glomus tumor arising in a mature teratoma of the ovary: report of a case simulating a metastasis from cervical squamous carcinoma. Arch Pathol Lab Med. 2000;124(9):1373–5. doi:10.1043/0003-9985(2000)124<1373:GTAIAM>2.0.CO;2.
390. Genadry R, Parmley T, Woodruff JD. Secondary malignancies in benign cystic teratomas. Gynecol Oncol. 1979;8(2):246–51. doi:0090-8258(79)90032-5 [pii].
391. Hanada M, Tsujimura T, Shimizu H. Multiple malignancies (squamous cell carcinoma and sarcoma) arising in a dermoid cyst of the ovary. Acta Pathol Jpn. 1981;31(4):681–8.
392. Leroy C, Gontier MF, Gontier F, Minh HN, Smadja A. Malignant dermoid cyst of the ovary with double tumor cell population, revealed by an umbilical metastasis. Ann Pathol. 1992;12(3):178–82.
393. Terada T, Tateoka K. Ovarian cystic tumor composed of Brenner tumor and struma ovarii. Int J Clin Exp Pathol. 2012;5(3):274–7.
394. Nogales FF. Primary chondrosarcoma of the ovary. Histopathology. 1982;6:376.
395. Nogales Jr FF, Silverberg SG. Epidermoid cysts of the ovary: a report of five cases with histogenetic considerations and ultrastructural findings. Am J Obstet Gynecol. 1976;124(5):523–8.
396. Isaac MA, Vijayalakshmi S, Madhu CS, Bosincu L, Nogales FF. Pure cystic nephroblastoma of the ovary with a review of extrarenal Wilms' tumors. Hum Pathol. 2000;31(6):761–4. doi:10.1053/hupa.2000.7627.
397. Stolnicu S, Preda O, Dohan M, Puscasiu L, Garcia-Galvis OF, Nogales FF. Pseudoglandular hepatoid differentiation in endometrioid carcinoma of the ovary simulates oxyphilic cell change. Int J Gynecol Pathol. 2008;27(4):521–5. doi:10.1097/PGP.0b013e318178e566.
398. Nogales FF, Concha A, Plata C, Ruiz-Avila I. Granulosa cell tumor of the ovary with diffuse true hepatic differentiation simulating stromal luteinization. Am J Surg Pathol. 1993;17(1):85–90.
399. Devaney K, Snyder R, Norris HJ, Tavassoli FA. Proliferative and histologically malignant struma ovarii: a clinicopathologic study of 54 cases. Int J Gynecol Pathol. 1993;12(4):333–43.
400. Jin C, Dong R, Bu H, Yuan M, Zhang Y, Kong B. Coexistence of benign struma ovarii, pseudo-Meigs' syndrome and elevated serum CA 125: case report and review of the literature. Oncol Lett 2015;9(4):1739–42. doi:10.3892/ol.2015.2927 ol-09-04-1739 [pii].
401. Outwater EK, Siegelman ES, Hunt JL. Ovarian teratomas: tumor types and imaging characteristics. Radiographics. 2001;21(2):475–90. doi:10.1148/radiographics.21.2.g01mr09475.
402. Moon S, Waxman M. Mixed ovarian tumor composed of Brenner and thyroid elements. Cancer. 1976;38(5):1997–2001.
403. Yoshida M, Obayashi C, Tachibana M, Minami R. Coexisting Brenner tumor and struma ovarii in the right ovary: case report and review of the literature. Pathol Int 2004;54(10):793–7. doi:PIN1757 [pii] 10.1111/j.1440-1827.2004.01757.x.
404. Shaco-Levy R, Peng RY, Snyder MJ, Osmond GW, Veras E, Bean SM, et al. Malignant struma ovarii: a blinded study of 86 cases assessing which histologic features correlate with aggressive clinical behavior. Arch Pathol Lab Med. 2012;136(2):172–8. doi:10.5858/arpa.2011-0092-OA.
405. Garg K, Soslow RA, Rivera M, Tuttle MR, Ghossein RA. Histologically bland "extremely well differentiated" thyroid carcinomas arising in struma ovarii can recur and metastasize. Int J Gynecol Pathol 2009;28(3):222–30. doi:10.1097/PGP.0b013e31818a2b99 00004347-200905000-00003 [pii].
406. Robboy SJ, Shaco-Levy R, Peng RY, Snyder MJ, Donahue J, Bentley RC et al. Malignant struma ovarii: an analysis of 88 cases, including 27 with extraovarian spread. Int J Gynecol Pathol 2009;28(5):405–22. doi:10.1097/PGP.0b013e3181a27777 00004347-200909000-00001 [pii].
407. Wolff EF, Hughes M, Merino MJ, Reynolds JC, Davis JL, Cochran CS, et al. Expression of benign and malignant thyroid tissue in ovarian teratomas and the importance of multimodal management as illustrated by a BRAF-positive follicular variant of papillary thyroid cancer. Thyroid. 2010;20(9):981–7. doi:10.1089/thy.2009.0458.
408. Boutross-Tadross O, Saleh R, Asa SL. Follicular variant papillary thyroid carcinoma arising in struma ovarii. Endocr Pathol. 2007;18(3):182–6. doi:10.1007/s12022-007-0022-8.
409. Schmidt J, Derr V, Heinrich MC, Crum CP, Fletcher JA, Corless CL et al. BRAF in papillary thyroid carcinoma of ovary (struma ovarii). Am J Surg Pathol 2007;31(9):1337–43. doi:10.1097/PAS.0b013e31802f5404 00000478-200709000-00005 [pii].
410. Reed NS, Gomez-Garcia E, Gallardo-Rincon D, Barrette B, Baumann K, Friedlander M et al. Gynecologic Cancer InterGroup (GCIG) con-

sensus review for carcinoid tumors of the ovary. Int J Gynecol Cancer 2014;24(9 Suppl 3):S35–41. doi:10.1097/IGC.0000000000000265 00009577-201411003-00009 [pii].
411. Soga J, Osaka M, Yakuwa Y. Carcinoids of the ovary: an analysis of 329 reported cases. J Exp Clin Cancer Res. 2000;19(3):271–80.
412. Desouki MM, Lioyd J, Xu H, Cao D, Barner R, Zhao C. CDX2 may be a useful marker to distinguish primary ovarian carcinoid from gastrointestinal metastatic carcinoids to the ovary. Hum Pathol. 2013;44(11):2536–41. doi:10.1016/j.humpath.2013.06.014.
413. Davis KP, Hartmann LK, Keeney GL, Shapiro H. Primary ovarian carcinoid tumors. Gynecol Oncol 1996;61(2):259–65. doi:S0090-8258(96)90136-5 [pii] 10.1006/gyno.1996.0136.
414. Goldman T, Adamson K, Yang E. Resolution of right-sided heart failure symptoms after resection of a primary ovarian carcinoid tumor. Tex Heart Inst J. 2014;41(5):533–6. doi:10.14503/THIJ-13-3314.
415. Dessauvagie BF, Lai PH, Oost E, Thomas A, Stewart CJ. Medial hypertrophy of the ovarian vein: a novel type of vascular pathology associated with a primary ovarian carcinoid tumor. Int J Gynecol Pathol. 2015;34(1):36–9. doi:10.1097/PGP.0000000000000114.
416. Muller KE, Tafe LJ, Gonzalez JL, West LA, Schned AR. Ovarian strumal carcinoid producing peptide YY associated with severe constipation: a case report and review of the literature. Int J Gynecol Pathol. 2015;34(1):30–5. doi:10.1097/PGP.0000000000000117.
417. Robboy SJ. Insular carcinoid of ovary associated with malignant mucinous tumors. Cancer. 1984;54(10):2273–6.
418. Mandai M, Konishi I, Tsuruta Y, Suginami N, Kusakari T, Iwasaki T, et al. Krukenberg tumor from an occult appendiceal adenocarcinoid: a case report and review of the literature. Eur J Obstet Gynecol Reprod Biol. 2001;97(1):90–5.
419. Baker PM, Oliva E, Young RH, Talerman A, Scully RE. Ovarian mucinous carcinoids including some with a carcinomatous component: a report of 17 cases. Am J Surg Pathol. 2001;25(5):557–68.
420. Armes JE, Ostor AG. A case of malignant strumal carcinoid. Gynecol Oncol 1993;51(3):419–23. doi:S0090-8258(83)71316-8 [pii] 10.1006/gyno.1993.1316.
421. Kimura N, Sasano N, Namiki T. Evidence of hybrid cell of thyroid follicular cell and carcinoid cell in strumal carcinoid. Int J Gynecol Pathol. 1986;5(3):269–77.
422. Scully RE. Gonadoblastoma; a gonadal tumor related to the dysgerminoma (seminoma) and capable of sex-hormone production. Cancer. 1953;6(3):455–63.
423. Hertel JD, Huettner PC, Dehner LP, Pfeifer JD. The chromosome Y-linked testis-specific protein locus TSPY1 is characteristically present in gonadoblastoma. Hum Pathol. 2010;41(11):1544–9. doi:10.1016/j.humpath.2010.04.007.
424. Ulbright TM, Young RH. Gonadoblastoma and selected other aspects of gonadal pathology in young patients with disorders of sex development. Semin Diagn Pathol 2014;31(5):427–40. doi:10.1053/j.semdp.2014.07.001 S0740-2570(14)00064-1 [pii].
425. Scully RE. Gonadoblastoma. A review of 74 cases. Cancer. 1970;25(6):1340–56.
426. Schiller PJ, Tagatz GE. Gonadal calcification. Indication for gonadectomy in gonadal dysgenesis. Obstet Gynecol. 1976;48(4):486–90.
427. Manivel JC, Dehner LP, Burke B. Ovarian tumorlike structures, biovular follicles, and binucleated oocytes in children: their frequency and possible pathologic significance. Pediatr Pathol. 1988;8(3):283–92.
428. Serment H, Laffargue P, Piana L, Blanc B. Tumeurs endocrines de l'ovaire chez l'enfant. Rev Fr Gynecol Obstet. 1970;65(9):465–77.
429. Talerman A. A distinctive gonadal neoplasm related to gonadoblastoma. Cancer. 1972;30(5):1219–24.
430. Michal M, Vanecek T, Sima R, Mukensnabl P, Hes O, Kazakov DV, et al. Mixed germ cell sex cord-stromal tumors of the testis and ovary. Morphological, immunohistochemical, and molecular genetic study of seven cases. Virchows Arch. 2006;448(5):612–22. doi:10.1007/s00428-006-0155-z.
431. Talerman A, Roth LM. Recent advances in the pathology and classification of gonadal neoplasms composed of germ cells and sex cord derivatives. Int J Gynecol Pathol. 2007;26(3):313–21. doi:10.1097/01.pgp.0000250148.52215.ce.
432. Roth LM, Cheng L. Expression of transcription factors and nuclear receptors in mixed germ cell-sex cord stromal tumor and related tumors of the gonads. Int J Gynecol Pathol. 2015; doi:10.1097/PGP.0000000000000192.
433. Arroyo JG, Harris W, Laden SA. Recurrent mixed germ cell-sex cord-stromal tumor of the ovary in an adult. Int J Gynecol Pathol. 1998;17(3):281–3.
434. Lacson AG, Gillis DA, Shawwa A. Malignant mixed germ-cell-sex cord-stromal tumors of the ovary associated with isosexual precocious puberty. Cancer. 1988;61(10):2122–33.
435. Tokuoka S, Aoki Y, Hayashi Y, Yokoyama T, Ishii T. A mixed germ cell-sex cord-stromal tumor of the ovary with retiform tubular structure: a case report. Int J Gynecol Pathol. 1985;4(2):161–70.
436. Tavassoli FA. A combined germ cell–gonadal stromal–epithelial tumor of the ovary. Am J Surg Pathol. 1983;7(1):73–84.
437. Zuntova A, Motlik K, Horejsi J, Eckschlager T. Mixed germ cell-sex cord stromal tumor with heterologous structures. Int J Gynecol Pathol. 1992;11(3):227–33.
438. Jacobsen GK, Braendstrup O, Talerman A. Bilateral mixed germ cell sex-cord stroma tumour in a young adult woman. Case Rep APMIS Suppl. 1991;23:132–7.
439. Bhathena D, Haning Jr RV, Shapiro S, Hafez GR. Coexistence of a gonadoblastoma and mixed germ cell-sex cord stroma tumor. Pathol Res Pract.

1985;180(2):203–8. doi:10.1016/S0344-0338(85)80171-0.
440. Colafranceschi M, Massi D. Gonadoblastoma with coexistent features of mixed germ cell-sex cord stroma tumor: a case report. Tumori. 1995;81(3):215–8.
441. Mooney EE, Nogales FF, Bergeron C, Tavassoli FA. Retiform Sertoli-Leydig cell tumours: clinical, morphological and immunohistochemical findings. Histopathology. 2002;41(2):110–7.
442. Cabanne F. Gonadoblastomes et tumeurs de l 'ebauche gonadique. Ann Anat Pathol (Paris). 1971;16(4):387–404.
443. Pflüger E. Ueber die Bewegungen der Ovarien. Archiv für Anatomie, Physiologie und wissenschaftliche Medicin. 1859;S:30–2.
444. Pang S, Zhang L, Shi Y, Liu Y. Unclassified mixed germ cell-sex cord-stromal tumor with multiple malignant cellular elements in a young woman: a case report and review of the literature. Int J Clin Exp Pathol. 2014;7(8):5259–66.
445. Shanmugaratnam K, Kunaratnam N, Chia KB, Chiang GS, Sinniah R. Teratoid carcinosarcoma of the paranasal sinuses. Pathology. 1983;15(4):413–9.
446. Kinjo T, Taniguchi H, Kushima R, Sekine S, Oda I, Saka M, et al. Histologic and immunohistochemical analyses of alpha-fetoprotein–producing cancer of the stomach. Am J Surg Pathol. 2012;36(1):56–65. doi:10.1097/PAS.0b013e31823aafec.
447. Kurman RJ, Carcangiu ML, Herrington CS, Young RH. WHO classification of tumours of female reproductive organs. 4th ed. Lyon: International Agency for Research on Cancer; 2014.
448. http://ec.europa.eu/eurostat/statistics-explained/index.php/Population_structure_and_ageing. Accessed: December 2016.
449. Praxis Strategy Group Analysis of U.S. Census Population Projections, https://www.census.gov/population/projections/data/national/2014.html.
450. Lankerani MR, Aubrey RW, Reid JD. Endometriosis of the colon with mixed "germ cell" tumor. Am J Clin Pathol. 1982;78(4):555–9.
451. Rutgers JL, Young RH, Scully RE. Ovarian yolk sac tumor arising from an endometrioid carcinoma. Hum Pathol. 1987;18(12):1296–9.
452. OF G-G, Cabrera-Ozoria C, Fernandez JA, Stolnicu S, Nogales FF. Malignant mullerian mixed tumor of the ovary associated with yolk sac tumor, neuroepithelial and trophoblastic differentiation (teratoid carcinosarcoma). Int J Gynecol Pathol. 2008;27(4):515–20. doi:10.1097/PGP.0b013e31817b06c7.
453. Varia M, McCluggage WG, Oommen R. High grade serous carcinoma of the ovary with a yolk sac tumour component in a postmenopausal woman: report of an extremely rare phenomenon. J Clin Pathol. 2012;65(9):853–4. doi:10.1136/jclinpath-2011-200606.
454. Lopez JM, Malpica A, Deavers MT, Ayala AG. Ovarian yolk sac tumor associated with endometrioid carcinoma and mucinous cystadenoma of the ovary. Ann Diagn Pathol. 2003;7(5):300–5. doi:S1092913403000819 [pii].
455. Retamero JA, Schuldt M, Nogales FF. Yolk Sac tumours arising from Endometrioid carcinoma of the ovary and endometrium. A review of 14 cases. AJSP Reviews Reports. 2016;(in press).
456. McNamee T, Damato S, McCluggage WG. Yolk sac tumours of the female genital tract in older adults derive commonly from somatic epithelial neoplasms: somatically derived yolk sac tumours. Histopathology. 2016;69(5):739–51. doi:10.1111/his.13021.
457. Yemelyanova A, Gown AM, Wu LS, Holmes BJ, Ronnett BM, Vang R. PAX8 expression in uterine adenocarcinomas and mesonephric proliferations. Int J Gynecol Pathol. 2014;33(5):492–9. doi:10.1097/PGP.0b013e3182a54afa.
458. Teilum G. Endodermal sinus tumors of the ovary and testis. Comparative morphogenesis of the so-called mesoephroma ovarii (Schiller) and extraembryonic (yolk sac-allantoic) structures of the rat's placenta. Cancer. 1959;12:1092–105.
459. Offman SL, Longacre TA. Clear cell carcinoma of the female genital tract (not everything is as clear as it seems). Adv Anat Pathol. 2012;19(5):296–312. doi:10.1097/PAP.0b013e31826663b1.
460. Maeda D, Ota S, Takazawa Y, Aburatani H, Nakagawa S, Yano T, et al. Glypican-3 expression in clear cell adenocarcinoma of the ovary. Mod Pathol. 2009;22(6):824–32. doi:10.1038/modpathol.2009.40.
461. Ballotta MR, Bianchini E, Borghi L, Fortini RM. Clear cell carcinoma simulating the "endometrioid-like variant" of yolk sac tumor. Pathologica. 1995;87(1):87–90.
462. Cetin A, Bahat Z, Cilesiz P, Demirbag N, Yavuz E. Ovarian clear cell adenocarcinoma producing alpha-fetoprotein: case report. Eur J Gynaecol Oncol. 2007;28(3):241–4.
463. El-Bahrawy M. Alpha-fetoprotein-producing non-germ cell tumours of the female genital tract. Eur J Cancer. 2010;46(8):1317–22. doi:10.1016/j.ejca.2010.01.028.
464. Meguro S, Yasuda M. alpha-Fetoprotein-producing ovarian tumor in a postmenopausal woman with germ cell differentiation. Ann Diagn Pathol. 2013;17(1):140–4. doi:10.1016/j.anndiagpath.2011.07.010.
465. Hoang LN, Han G, McConechy M, Lau S, Chow C, Gilks CB, et al. Immunohistochemical characterization of prototypical endometrial clear cell carcinoma–diagnostic utility of HNF-1beta and oestrogen receptor. Histopathology. 2014;64(4):585–96. doi:10.1111/his.12286.
466. Ramalingam P, Malpica A, Silva EG, Liu JL, Gershenson DM, Deavers MT. The use of cytokeratin 7 in differentiating yolk sac tumors from endometrioid and clear cell carcinomas of the ovary. Lab Investig. 2003;83(1):207A–A.
467. Iwamoto M, Nakatani Y, Fugo K, Kishimoto T, Kiyokawa T. Napsin A is frequently expressed in

clear cell carcinoma of the ovary and endometrium. Hum Pathol. 2015;46(7):957–62. doi:10.1016/j.humpath.2015.03.008.
468. Nonaka D, Chiriboga L, Soslow RA. Expression of pax8 as a useful marker in distinguishing ovarian carcinomas from mammary carcinomas. Am J Surg Pathol. 2008;32(10):1566–71. doi:10.1097/PAS.0b013e31816d71ad.
469. Schaner ME, Ross DT, Ciaravino G, Sorlie T, Troyanskaya O, Diehn M, et al. Gene expression patterns in ovarian carcinomas. Mol Biol Cell. 2003;14(11):4376–86. doi:10.1091/mbc.E03-05-0279.
470. Ushiku T, Shinozaki A, Shibahara J, Iwasaki Y, Tateishi Y, Funata N, et al. SALL4 represents fetal gut differentiation of gastric cancer, and is diagnostically useful in distinguishing hepatoid gastric carcinoma from hepatocellular carcinoma. Am J Surg Pathol. 2010;34(4):533–40. doi:10.1097/PAS.0b013e3181d1dcdd.
471. Iwanaga S, Shimada A, Hasuo Y, Yoh S, Miyajima S, Nishimura H, et al. Immature teratoma of the uterine fundus. Kurume Med J. 1993;40(3):153–8.
472. Ansah-Boateng Y, Wells M, Poole DR. Coexistent immature teratoma of the uterus and endometrial adenocarcinoma complicated by gliomatosis peritonei. Gynecol Oncol. 1985;21(1):106–10.
473. Yu XJ, Zhang L, Liu ZP, Shi YQ, Liu YX. Ovarian malignant mixed germ cell tumor with clear cell carcinoma in a postmenopausal woman. Int J Clin Exp Pathol. 2014;7(12):8996–9001.
474. Nogales FF, Schuldt M, Prat J. Germ cell tumour patterns occurring in somatic malignacies of the female genital tract. 2016.
475. Choijamts B, Jimi S, Kondo T, Naganuma Y, Matsumoto T, Kuroki M, et al. CD133+ cancer stem cell-like cells derived from uterine carcinosarcoma (malignant mixed Mullerian tumor). Stem Cells. 2011;29(10):1485–95. doi:10.1002/stem.711.
476. Rutella S, Bonanno G, Procoli A, Mariotti A, Corallo M, Prisco MG, et al. Cells with characteristics of cancer stem/progenitor cells express the CD133 antigen in human endometrial tumors. Clin Cancer Res. 2009;15(13):4299–311. doi:10.1158/1078-0432.CCR-08-1883.
477. Chang SJ, Wang TY, Tsai CY, Hu TF, Chang MD, Wang HW. Increased epithelial stem cell traits in advanced endometrial endometrioid carcinoma. BMC Genomics. 2009;10:613. doi:10.1186/1471-2164-10-613.
478. Gotte M, Wolf M, Staebler A, Buchweitz O, Kelsch R, Schuring AN, et al. Increased expression of the adult stem cell marker Musashi-1 in endometriosis and endometrial carcinoma. J Pathol. 2008;215(3):317–29. doi:10.1002/path.2364.
479. Mizuno T, Suzuki N, Makino H, Furui T, Morii E, Aoki H, et al. Cancer stem-like cells of ovarian clear cell carcinoma are enriched in the ALDH-high population associated with an accelerated scavenging system in reactive oxygen species. Gynecol Oncol. 2015;137(2):299–305. doi:10.1016/j.ygyno.2014.12.005.
480. Virant-Klun I, Rozman P, Cvjeticanin B, Vrtacnik-Bokal E, Novakovic S, Rulicke T, et al. Parthenogenetic embryo-like structures in the human ovarian surface epithelium cell culture in postmenopausal women with no naturally present follicles and oocytes. Stem Cells Dev. 2009;18(1):137–49. doi:10.1089/scd.2007.0238.
481. Wang WC, Lee MS, Ko JL, Lai YC. Origin of uterine teratoma differs from that of ovarian teratoma: a case of uterine mature cystic teratoma. Int J Gynecol Pathol 2011;30(6):544–8. doi:10.1097/PGP.0b013e31821c3205 00004347-201111000-00006 [pii].
482. Siddon A, Hui P. Glial heterotopia of the uterine cervix: DNA genotyping confirmation of its fetal origin. Int J Gynecol Pathol. 2010;29(4):394–7. doi:10.1097/PGP.0b013e3181c5a7e8.
483. Newton 3rd CW, Abell MR. Iatrogenic fetal implants. Obstet Gynecol. 1972;40(5):686–91.
484. Russell P, de Costa C, Yeoh G. Fetal glial allograft in the endometrium: case report of a recurrent pseudotumor. Pathology. 1993;25(3):247–9.
485. Chao TJ, Chao J, Kuan LJ, Li YT, Kuo TC, Chang YC, et al. Mature solid teratoma of the fallopian tube associated with uterine leiomyomas. J Chinese Med Assoc. 2008;71(8):425–7. doi:10.1016/s1726-4901(08)70095-9.
486. Kutteh WH, Albert T. Mature cystic teratoma of the fallopian tube associated with an ectopic pregnancy. Obstet Gynecol. 1991;78(5 Pt 2):984–6.
487. Hurd Jr JK. Benign cystic teratoma of the fallopian tube. Obstet Gynecol. 1978;52(3):362–4.
488. Frost RG, Roongpisuthipong A, Cheek BH, Majmudar BN. Immature teratoma of the fallopian tube. A case report. J Reprod Med. 1989;34(1):62–4.
489. Papadia A, Rutigliani M, Gerbaldo D, Fulcheri E, Ragni N. Mature cystic teratoma of the uterus presenting as an endometrial polyp. Ultrasound Obstet Gynecol. 2007;29(4):477–8. doi:10.1002/uog.3969.
490. Takahashi O, Shibata S, Hatazawa J, Takisawa J, Sato H, Ota H, et al. Mature cystic teratoma of the uterine corpus. Acta Obstet Gynecol Scand. 1998;77(9):936–8.
491. Kamgobe E, Massinde A, Matovelo D, Ndaboine E, Rambau P, Chaula T. Uterine myometrial mature teratoma presenting as a uterine mass: a review of literature. BMC Clin Pathol. 2016;16:5. doi:10.1186/s12907-016-0026-8.
492. Akai M, Isoda H, Sawada S, Matsuo I, Kanzaki H, Sakaida N, et al. A case of struma uteri. AJR Am J Roentgenol. 2005;185(1):216–8. doi:10.2214/ajr.185.1.01850216.
493. Newsom-Davis T, Poulter D, Gray R, Ameen M, Lindsay I, Papanikolaou K et al. Case report: malignant teratoma of the uterine corpus. BMC Cancer 2009;9:195. doi:10.1186/1471-2407-9-195 1471-2407-9-195 [pii].

494. Lim SC, Kim YS, Lee YH, Lee MS, Lim JY. Mature teratoma of the uterine cervix with lymphoid hyperplasia. Pathol Int. 2003;53(5):327–31. doi:1468 [pii].
495. Iwanaga S, Ishii H, Nagano H, Shimizu M, Nishida T, Yakushiji M. Mature cystic teratoma of the uterine cervix. Asia Oceania J Obstet Gynaecol. 1990;16(4):363–6.
496. Panesar NK, Sidhu JS. Uterine cervical teratoma with divergent neuroepithelial differentiation and development of an oligodendroglioma: report of a case and review of the literature. Ann Diagn Pathol. 2007;11(4):293–6. doi:10.1016/j.anndiagpath.2006.03.019.
497. Spatz A, Bouron D, Pautier P, Castaigne D, Duvillard P. Primary yolk sac tumor of the endometrium: a case report and review of the literature. Gynecol Oncol 1998;70(2):285–8. doi:S0090-8258(98)95036-3 [pii] 10.1006/gyno.1998.5036.
498. Wang C, Li G, Xi L, Gu M, Ma D. Primary yolk sac tumor of the endometrium. Int J Gynaecol Obstet 2011;114(3):291–3. doi:10.1016/j.ijgo.2011.03.020 S0020-7292(11)00259-1 [pii].
499. Mardi K, Gupta N, Bindra R. Primary yolk sac tumor of cervix and vagina in an adult female: a rare case report. Indian J Cancer. 2011;48(4):515–6. doi:10.4103/0019-509X.92249 IndianJournalof Cancer_2011_48_4_515_92249 [pii].
500. Mochizuki K, Obatake M, Taura Y, Inamura Y, Takatsuki M, Nagayasu T, et al. Yolk sac tumor of the vulva: a case report with recurrence after long-term follow-up. Pediatr Surg Int. 2012;28(9):931–4. doi:10.1007/s00383-012-3153-z.
501. Flanagan CW, Parker JR, Mannel RS, Min KW, Kida M. Primary endodermal sinus tumor of the vulva: a case report and review of the literature. Gynecol Oncol 1997;66(3):515–8. doi:S0090-8258(97)94759-4 [pii] 10.1006/gyno.1997.4759.
502. Mamoon N, Mushtaq S, Akhter N, Aslam A, Chaudary A, Rashid M. Immature teratoma of the vulva with an inguinal lymph node metastasis: report of a case and review of literature. Int J Gynecol Pathol 2010;29(2):197–200. doi:10.1097/PGP.0b013e3181b8e73e 00004347-201003000-00015 [pii].

7 Testicular Germ Cell Tumors

Rafael E. Jimenez, Sounak Gupta,
Loren P. Herrera-Hernandez, and Thomas J. Sebo[1]

7.1 General Concepts

Aside from benign teratomas of the ovary, germ cell tumors of the testis are the most common germ cell tumors (GCT). They also correspond to the vast majority of testicular tumors, encompassing about 98 % of the neoplasms arising in the male gonad [1–3].

Testicular GCT share the basic characteristics with GCT at other sites, but possess several unique features. Contrary to the ovarian counterpart, most are malignant. This is particularly the case in testicular teratomas, specifically the postpubertal type, which is malignant independently of the degree of maturation of its components. While for the most part, the histologic spectrum is similar to other sites, a specific type, spermatocytic tumor (ST), is unique to the male gonad. Most testicular GCT are characterized by the existence of an in situ lesion, germ cell neoplasia in situ (GCNIS), which is pivotal in their histogenesis. Finally, the phenomenon of regression, in which primary GCT may involute and frequently disappear, is characteristic of the testis.

This chapter deals preferentially with postpubertal testicular GCT; while on occasion pediatric tumors will be mentioned, detailed discussion of these may be found in Chap. 6.

7.2 Epidemiology

Testicular cancers correspond to approximately 1–2 % of tumors in men; however, they are the most common malignancy in men between the ages of 15 and 34 [1–3]. Approximately 95 % of testicular tumors correspond to GCT, with the majority of non-GCT histologies occurring in men over 50 years of age [4]. There is an increasing incidence of testicular GCT, particularly in those countries with an already high incidence [5]. Marked variations among countries and geographical regions have been described, with a higher incidence in Scandinavian and northern European countries and lowest in Middle Eastern and Asian countries [5]. Racial differences have also been widely reported within racially diverse countries. In the USA, the incidence per 100,000 varies from 1.1 in African-Americans to 6.3 cases in whites [3]. A more extensive review of GCT epidemiology can be found in Chap. 2.

[1]Dr. Sebo sadly passed away during the preparation of this manuscript. We will always be grateful for his collaboration in this project. He will be missed for his teaching, his mentorship, and his friendship.

R.E. Jimenez, MD, MHA (✉)
L.P. Herrera-Hernandez, MD • T.J. Sebo, MD, PhD
Laboratory Medicine and Pathology, Mayo Clinic College of Medicine, Rochester, MN, USA
e-mail: jimenez.rafael@mayo.edu; herrerahernandez.loren@mayo.edu

S. Gupta, MBBS, PhD
Mayo Clinic, Rochester, MN, USA
e-mail: gupta.sounak@mayo.edu

F.F. Nogales, R.E. Jimenez (eds.), *Pathology and Biology of Human Germ Cell Tumors*,
DOI 10.1007/978-3-662-53775-6_7

Statistical studies have suggested several risk factors that may increase the risk of developing testicular cancer. These include higher economic status, professional workers, technicians, and other occupational categories [6–9], immunodeficiency [10], history of sexually transmitted disease [11], past mumps orchitis [12], in utero exposure to estrogen [13], cannabis use [14, 15], testicular trauma [11], Down syndrome [16], Marfan syndrome [17], and Klinefelter syndrome [18]. Early male patterned baldness and severe acne have a negative association [19]. Most of these associations are weak, or are not demonstrated consistently in different studies. However, the following risk factors have been repeatedly linked with an increased risk for the development of testicular GCT:

1. *Cryptorchidism*: A maldescended or undescended testis is the most consistently associated risk factor for testicular cancer, conferring 3.7–7.5 times higher risk of developing this malignancy [20]. Conversely, 5–10 % of patients with testicular cancer have a history of cryptorchidism [21]. The mechanism is unclear; their association could be the result of a common cause for both pathologies, but, alternatively, it is more likely that it is related to the adverse environmental conditions to which the undescended testis is subjected. The varying relative risks depending on the location of the testis (abdominal vs. inguinal), and the impact early orchidopexy has in reducing the risk, would favor the latter possibility. Orchidopexy before age 13 is associated with a 2.23-fold relative risk, compared to 5.4 for patients who had orchidopexy after that age [22]. However, not all series have demonstrated a risk reduction dependent on the age at orchidopexy [23]. The GCT most commonly observed in patients with cryptorchidism is seminoma. Cryptorchidic patients have a higher proportion of seminoma vs. non-seminomatous GCT, compared to the general population [24].
2. *Contralateral GCT*: A previous history of testicular GCT in the contralateral testis confers approximately a 24.5–27.5-fold relative risk of developing testicular cancer [25, 26]. The metachronous tumor has a highest risk of developing in the first 5 years after the initial diagnosis [27]. The risk is higher if the testis is or was cryptorchidic or atrophic.
3. *Family history of testicular cancer*: Sons of fathers with testicular cancer have approximately 4 times the risk of the general population of developing a testicular tumor. This risk is almost doubled if the affected first-degree relative is a brother [28–30]. Migration studies have demonstrated that immigrants to a region with lower or higher incidence of testicular cancer carry the risk of the region from where they originate. Familial testicular cancer has been associated with several candidate genes, which include among others *TGCT1* – which may also predispose to cryptorchidism – located in chromosome Xq27 [31], *KITLG* in 12q22 [32, 33], *SPRY4* in 5q31.3 [34], *BAK1* in 6p21.3 [35], *DMRT1* in 9p24.3 [36, 37], *TERT* in 5p15.33, and *ATF7IP* in 2p13.1 [37]. Most of these genes have been associated to the KITLG-KIT pathway, implicated in primordial germ cell development [38].
4. *Disorders of sex development (DSD)*: Abnormally differentiated or maldeveloped gonads, with our without maldeveloped sex organs, are linked to a higher risk of development of testicular cancer [21]. DSD include gonadal dysgenesis with a Y chromosome, true hermaphroditism, and pseudohermaphroditism due to androgen insensitivity syndrome [39]. Approximately 25 to 30 % of patients with gonadal dysgenesis and a Y chromosome develop a GCT [40], usually gonadoblastoma, while those with an abnormal androgen function have a risk of around 5–10% [39]. This association is discussed at length in Chap. 10 and further below in the section on gonadoblastoma.
5. *Male infertility*: An increase in testicular cancer rates has paralleled a decrease in fertility and semen quality [41]. It is unclear if there is a causal relationship between both conditions or the association is a manifestation of a com-

mon cause [42–44]. Given the connection of both infertility and testicular cancer with cryptorchidism, the latter is not an unlikely possibility [38]. Similarly, gonadal dysgenesis, or other DSD, account for a significant portion of cases of infertility [38]. Patients with infertility have a higher incidence of testicular cancer than those that have undergone vasectomy [45].

It has been proposed that patients with an identified risk of developing GCT may benefit from a testicular biopsy. The two best scenarios in which this potential benefit has been studied are in the setting of cryptorchidism, and contralateral GCT. Patients with a history of cryptorchidism have a 2–8 percent risk of having GCNIS, the precursor lesion of GCT [26]. Of 1500 cryptorchidic patients whose biopsies did not reveal GCNIS and were followed for up to 8 years, none developed an invasive GCT [46]. Conversely, 50 % of patients with a positive biopsy for GCNIS will develop an invasive GCT in the same timeframe [47]. Timing, frequency, location, and management of a positive biopsy are still debated in the literature [46, 48–50]. Similarly, testicular biopsy of the contralateral testis on patients with a GCT may identify a second primary in earlier stages. In this setting, approximately 5 % of contralateral testis biopsies will be positive for GCNIS. Of these, approximately 30 % will have an invasive tumor upon orchiectomy, and 50 % more will develop an invasive tumor within 5 years [47]. As in the cryptorchidism scenario, patients with a negative biopsy rarely develop a second GCT [51]. Radiation therapy can then be offered to these patients to reduce the risk of a second neoplasm [52, 53]. Arguments against the widespread use of contralateral biopsy include the need for close follow-up on patients with negative biopsy (given the rate of approximately 1 % false positive), the need of radiation therapy resulting in irreversible infertility and impairment of endocrine Leydig cell function, a questionable outcome advantage over patients handled with only surveillance, and the availability of local resection as therapy for small tumors [54].

7.3 Histogenesis

A detailed explanation of the histogenesis of germ cell tumors has already been presented in Chap. 3. Herein we will summarize those aspects specific to testicular GCT. Testicular tumors fall mainly within three of the types of the Oosterhuis and Looijenga classification [55], described in Chap. 3 . They include prepubertal-type tumors (type I GCT), including teratomas and yolk sac tumors (YST), postpubertal tumors (type II GCT, seminomatous and non-seminomatous GCT), by far the most common ones, and ST (type III GCT) [55, 56]. Each of these categories has a characteristic epidemiological, biological, and clinical profile, as summarized in Table 7.1. While our understanding of the molecular pathology of types I and III is limited, most of the current knowledge is related to the most common form, type II or postpubertal GCT.

The different histologic types of type II GCT are intimately related. The precursor lesion of GCT, GCNIS, shares many phenotypical and morphologic features with seminoma, suggesting a precursor relationship [57–59]. However, molecular abnormalities more characteristic of other forms of invasive GCT, such as embryonal carcinoma (EC) or YST, can also be seen in GCNIS, suggesting that the in situ lesion may also play a role of precursor to other forms of GCT as well [57]. Further, the existence of "specialized" forms of intratubular tumors such as intratubular EC or seminoma suggests that transformation to other subtypes can occur within the tubular compartment [60–62], at least in a subset of cases, with later events inducing an invasive counterpart. Plasticity among different forms of invasive GCT subtypes has been documented experimentally and clinically and is likely more common than intratubular transformation. Patients with testicular seminomas can present with non-seminomatous metastases [63], and it is not uncommon that seminomas show isolated syncytiotrophoblastic giant cells, "early" epithelial differentiation [64], and expression of markers associated with other histologies, such as alpha-fetoprotein, CD30, or cytokeratin [65–67]. Similarly, patients with

Table 7.1 Pathogenetic types of GCT in the testis

Type	Histologic spectrum	Association with GCNIS	Other anatomical sites	Age of presentation	Incidence (per 100,000)	Originating cell	Cytogenetic abnormality	Ploidy
I	Prepubertal-type teratoma Dermoid cysts Possibly epidermoid cysts Prepubertal yolk sac tumor	No	Ovary Sacrum Retroperitoneum Mediastinum Neck Midline brain	Neonates Children Occasionally postpubertal	0.12	Early primordial germ cell/gonocyte	Gains: 1q, 12(p13), 20q Losses: 1p, 4, 6q	Teratoma: Diploid Yolk sac tumor: Aneuploid
II	Seminoma Embryonal carcinoma Yolk sac tumor Choriocarcinoma and other trophoblastic tumors Postpubertal teratoma Mixed germ cell tumor	Yes	Ovary Dysgenetic gonad Mediastinum Midline brain	Postpubertal Median age: Seminoma 35 years Non-seminoma: 25 years	6.0	Primordial germ cell/gonocyte	Gains: X, 7, 8, *12p*, 21 Losses: Y, 1p, 11, 13, 18	Aneuploid (± triploid)
III	Spermatocytic tumor	No	None	>45	0.2	Spermatogonium/spermatocyte	Gains: 9	Aneuploid

Modified from Oosterhuis and Looijenga [55] and Reuter [56]

pure EC of the testis may present with teratomatous metastases [68, 69], and EC cells transplanted into peritoneal cavity of mice can differentiate into teratoma [70, 71]. YST may differentiate into mesenchymal tissues and may even show malignant somatic transformation (see Chap. 12), blurring the line between this histologic type and teratomas.

A common cytogenetic abnormality found in invasive types of type II GCT is the presence of isochromosome 12p (i12p) [56, 72, 73]. The isochromosome results in an overrepresentation of genes from the short arm of chromosome 12. Even cases lacking i12p have other forms of overrepresentation of these genes, such as duplication of the 12p11.21 region. Polysomy 12, while also seen in GCT, is less specific (Fig. 7.1) [73]. The i12p thus serves as a "common denominator" for the different subtypes of invasive type II GCT, providing further evidence of their intimate relationship at the molecular level. Interestingly, i12p is not found in GCNIS [74, 75], suggesting that its role is more important in the development of invasive capabilities, rather than in the original malignant transformation of germ cells.

In summary, these observations suggest a close association between the different histologies of GCT and suggest that different molecular events translate into morphologic transitions. Further analysis of the relationships between the subtypes of type II GCT is presented in Chap. 3.

In contrast with these suggested pathways, type I (pediatric GCT) and type III GCT (i.e., ST) have different histogeneses. Pediatric or prepubertal-type teratomas and YST are not associated with GCNIS and lack association with i12p. As explained in Chap. 3, it has been theorized that the cell of origin for this type of GCT is an earlier form of germ cell, such as a primordial

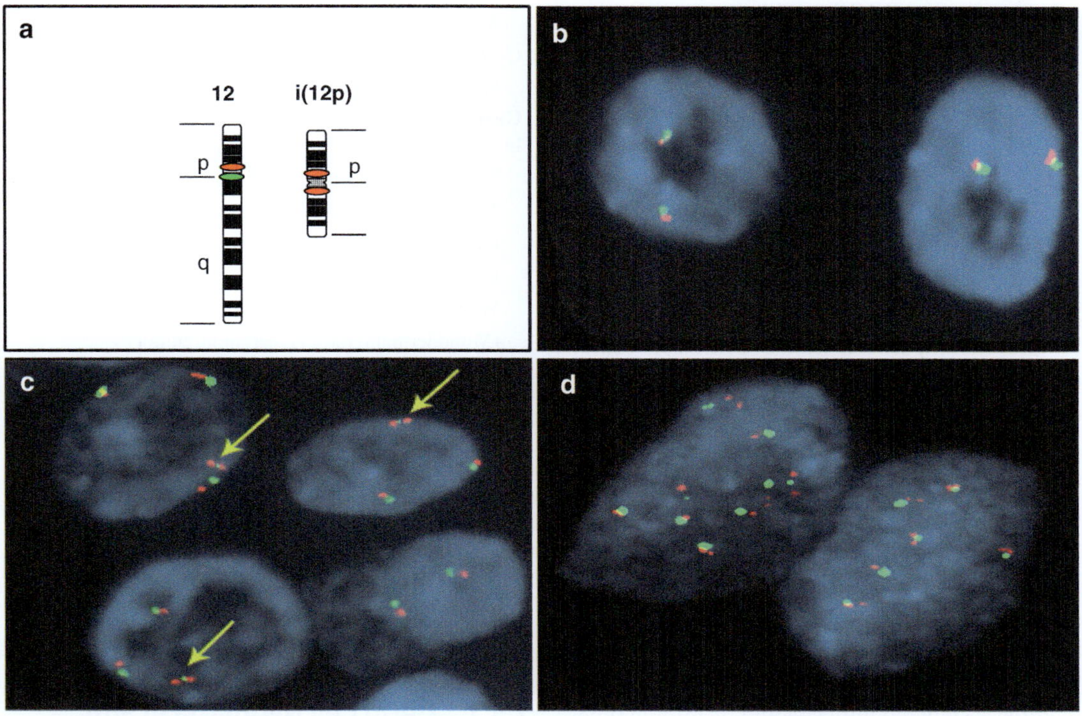

Fig. 7.1 Cytogenetic abnormalities of chromosome 12. The short arm of chromosome 12 is consistently abnormal in type II testicular GCT and can be detected with fluorescent in situ hybridization. Schematic representation of probes used to detect isochromosome 12p are shown in (**a**) when compared to normal hybridization pattern (**b**), abnormalities include isochromosome 12p (**c**), and gain of 12p (**d**)

or embryonic stem cell, compared to that of type II tumors. Similarly, ST lacks association with GCNIS and i12p and shows frequent cytogenetic abnormalities at the level of chromosome 9. Based on genomic imprinting analysis, the cell of origin for ST is believed to be a more mature germ cell than the one that gives rise to type II or type I tumors [55].

7.4 Classification of GCT

The pathogenetic mechanisms described above are relevant to the classification of GCT. Type II tumors, which constitute the majority of testicular GCT, are histologically characterized by their association with GCNIS. In fact, since GCNIS is an excellent surrogate for the molecular and biologic events in the pathogenesis of type II tumors, testicular GCT may be divided according to the presence or absence of GCNIS. Thus, tumors associated with GCNIS would correspond to type II GCT, while those not associated with GCNIS would encompass both type I and III neoplasms. The WHO classification of testicular neoplasms has incorporated this approach in its 2016 edition (see Table 7.2) [76].

For management purposes, testicular GCT are usually classified into seminomatous and non-seminomatous tumors. Seminomatous tumors include only pure seminomas. Non-seminomatous tumors encompass all mixed germ cell tumors (including those with seminoma as one of the components, even if predominant), as well as the less common pure forms of the other type II histologic types. The rationale for this classification resides in the completely different therapeutic approach for both categories, largely explained by the high radioensitivity of seminoma. However, the categorization into seminomatous and non-seminomatous tumors is overly simplistic and carries the risk of overlooking important features associated with specific histologic types. Additionally, it should be applied exclusively to type II tumors, and attention should be paid at not including within the non-seminomatous categories tumors with unique behavior and prognosis, such as ST or prepubertal-type teratomas.

Table 7.2 2016 WHO classification of GCT of the testis

Germ cell tumors derived from germ cell neoplasia in situ
Noninvasive germ cell neoplasia
Germ cell neoplasia in situ
Specific forms of intratubular germ cell neoplasia
Tumors of a single histological type (pure tumors)
Seminoma
Seminoma with syncytiotrophoblast cells
Non-seminomatous germ cell tumors
Embryonal carcinoma
Yolk sac tumor, postpubertal type
Trophoblastic tumors
Choriocarcinoma
Non-choriocarcinomatous trophoblastic tumors
Placental site trophoblastic tumor
Epitheliod trophoblastic tumor
Cystic trophoblastic tumor
Teratoma, postpubertal type
Teratoma with somatic-type malignancy
Non-seminomatous germ cell tumors of more than one histological type
Mixed germ cell tumors
Germ cell tumors of unknown type
Regressed germ cell tumors
Germ cell tumors unrelated to germ cell neoplasia in situ
Spermatocytic tumor
Teratoma, prepubertal type
Dermoid cyst
Epidermoid cyst
Well-differentiated neuroendocrine tumor
(Monodermal teratoma, carcinoid tumor)
Mixed teratoma and yolk sac tumor, prepubertal type
Yolk sac tumor, prepubertal type
Tumors containing both germ cell and sex cord-stromal elements
Gonadoblastoma

7.5 Tumors Associated with Germ Cell Neoplasia In Situ

7.5.1 Germ Cell Neoplasia In Situ

7.5.1.1 General Aspects

GCNIS is defined as the presence of malignant germ cells within the seminiferous tubules of the testis. Classically, these neoplastic germ cells are characterized by having abundant clear cytoplasm and a prominent enlarged nucleus, and they are usually located in the basal layers of the tubule (Fig. 7.2). It was originally described by Skakkebaek in 1972 [77]. Because of its undifferentiated nature, the term *intratubular germ cell neoplasia unclassified type* has been historically preferred in the USA over the term carcinoma in situ, more popular in European countries, as the neoplastic cells do not show epithelial differentiation nor necessarily give rise to invasive carcinomas. Recently, the term germ cell neoplasia in situ has been proposed for this lesion and was incorporated in the 2016 WHO classification [76]. This terminology will be used throughout this chapter.

7.5.1.2 Incidence and GCT Risk

The neoplastic nature of GCNIS has been demonstrated by epidemiological, morphological, and biological studies. Fifty percent of patients with GCNIS will develop an invasive form of germ cell tumor within 5 years [47]. Similarly, more than 90 % of patients with an invasive form of GCT show GCNIS in the adjacent testicular parenchyma [78–80]. GCNIS is found in patients with a high risk of developing invasive GCT, including patients with cryptorchidism, gonadal dysgenesis, contralateral GCT, infertility, and androgen insensitivity syndrome [46, 48, 49]. Morphologically, the cells of GCNIS are markedly similar to the individual cells of seminoma and share a similar immunophenotype, including the expression of c-KIT, OCT4, SALL4, and placental-like alkaline phosphatase (PLAP) [81–83]. Molecularly, they are aneuploid and share allelic losses with invasive GCT [57].

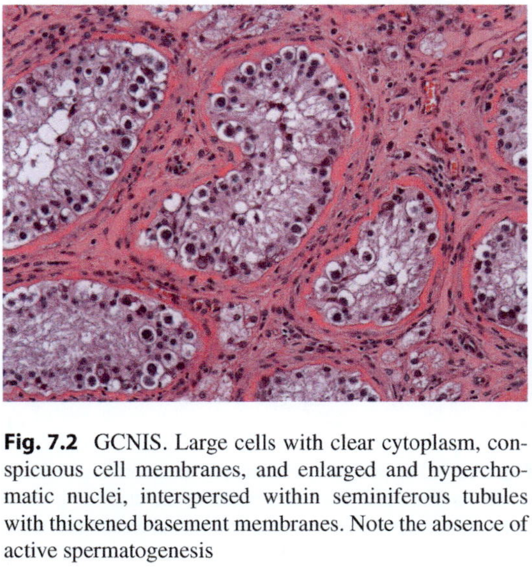

Fig. 7.2 GCNIS. Large cells with clear cytoplasm, conspicuous cell membranes, and enlarged and hyperchromatic nuclei, interspersed within seminiferous tubules with thickened basement membranes. Note the absence of active spermatogenesis

7.5.1.3 Morphology

No specific findings are grossly detected in cases of GCNIS that are associated with an invasive tumor. Given that it is commonly associated with infertility, changes of atrophy are not uncommon. Microscopically, GCNIS is characterized by the presence of large cells with clear cytoplasm, conspicuous cell membranes, enlarged and hyperchromatic nuclei, and prominent nucleoli (Fig. 7.2) [58]. These cells stand out in low power and are usually found in the basal layer of seminiferous tubules with atrophic features. In fact, it is extremely rare to find it in tubules with ongoing spermatogenesis. The involved tubules tend to be small and have a thickened basement membrane. Aside from the GCNIS cells, they usually contain only Sertoli cells. The distribution of GCNIS is usually patchy, and affected tubules may be located immediately adjacent to completely unremarkable ones. Peritubular lymphoid and granulomatous inflammation may also be present. GCNIS may show pagetoid spread into the rete testis (Fig. 7.3) [84] and less frequently into the epididymis and vas deferens [85].

PAS stains demonstrate abundant intracytoplasmic glycogen and may be useful as a screening stain. GCNIS is almost always positive for PLAP with a membranous/cytoplasmic pattern

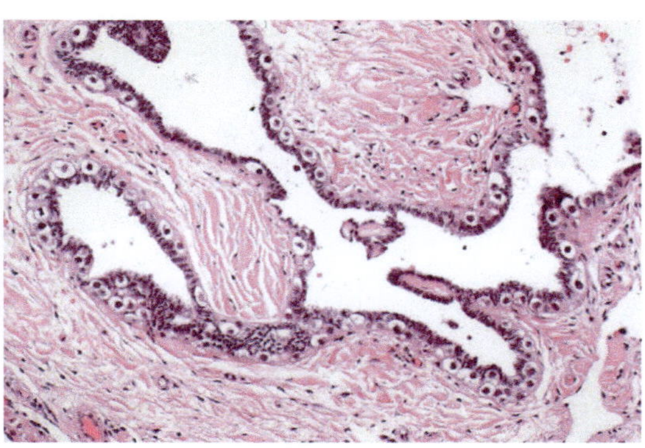

Fig. 7.3 GCNIS. GCNIS with pagetoid involvement of the rete testis

(Fig. 7.4a) and is quite useful, as normal spermatogonia are rarely positive for this marker. OCT4 is also highly specific, displaying nuclear staining only in neoplastic germ cells (Fig. 7.4b). c-KIT is highly sensitive, but its specificity is rather low, as nonneoplastic germ cells may react with this marker. The same can be said for SALL4 [82].

While GCNIS shares most of its molecular profile with invasive GCT, particularly seminoma, 12p amplification, usually in the form of i(12p), is rarely present in GCNIS, contrary to the case in invasive GCT. It is thus believed that the presence of i(12p) is probably involved in the ability of GCNIS to invade the stroma (see below) [82].

7.5.1.4 Differential Diagnosis

GCNIS must be distinguished from other intratubular forms of GCT, such as intratubular seminoma (Fig. 7.5a). These result for the most part from intratubular spread of invasive GCT, as they are rarely seen in the absence of an invasive component. Intratubular seminomas share many of the features of GCNIS; however, they fill and distend the involved tubules. They may be seen without an invasive seminoma counterpart, but usually they are accompanied by some kind of invasive GCT [61, 62]. Whether they represent an exaggerated form of GCNIS or an intratubular spread of an invasive seminoma is unresolved. Intratubular EC shows the epithelial differentiation and high-grade cytologic features of its invasive counterpart, including necrosis (Fig. 7.5b) [60]. Intratubular teratoma [86] and YST [87] have also been reported. Intratubular ST shows the three characteristic types of cells of the invasive counterpart (see below) and is always associated with an invasive component [58]. Metastatic carcinomas to the testis may show involvement of seminiferous tubules resembling GCNIS [88].

Prepubertal testes tend to show enlarged, slightly atypical germ cells with abnormal chromatin. In this setting, particularly in young

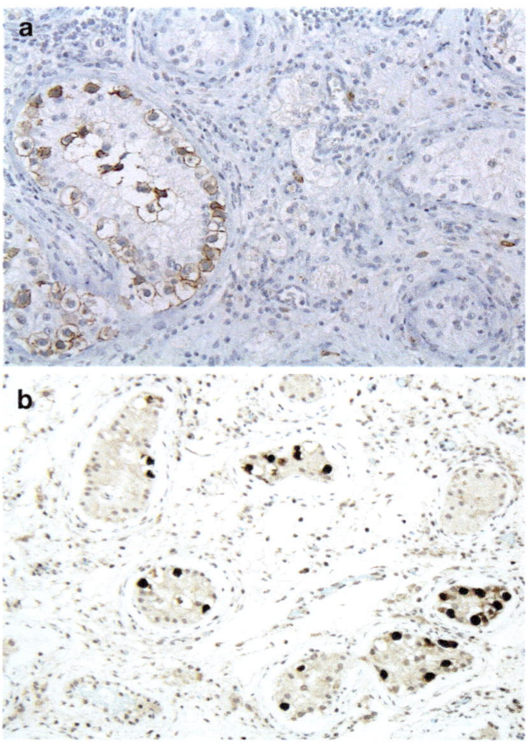

Fig. 7.4 (**a**, **b**) *GCNIS*, immunohistochemistry. GCNIS with characteristic membranous expression of PLAP (**a**) and nuclear expression of OCT4 (**b**)

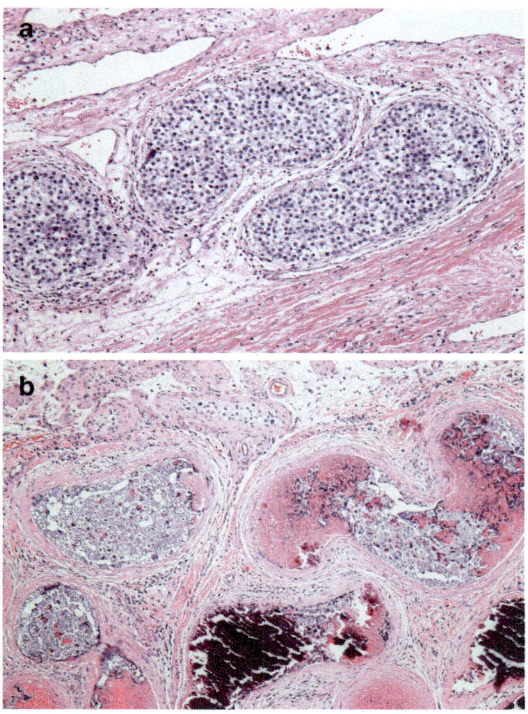

Fig. 7.5 (**a**, **b**) *GCNIS*, differential diagnostic considerations. (**a**) Intratubular seminoma filling and distending seminiferous tubules (**b**) Intratubular embryonal carcinoma, with intraluminal pleomorphic cells, necrosis, and calcifications

patients (less than 2 years), the atypical cells may share an immunophenotype with GCNIS. They are thought to represent delayed maturation of germ cells, frequently associated with DSD [89]. However, they are likely to be present diffusely throughout the parenchyma and are not limited to the basal layer within the tubules. Distinction between these two settings is clinically relevant, as these abnormal germ cells in prepubertal patients do not convey a high risk to progression to an invasive neoplasm. OCT4 is the preferred marker to use in the workup of this differential diagnosis [82].

Adult testis may also harbor occasional atypical germ cells, which usually show hyperchromatic nuclei, multinucleation, or enlarged size [90]. They lack the classic prominent nucleoli and typical distribution of GCNIS and do not share their immunophenotype (Table 7.3). While they may represent a manifestation of testicular dysgenesis, and thus may be present in the background of GCT, by themselves, they do not convey the high-risk implications of GCNIS.

7.5.1.5 Pathogenesis

During embryological development, germ cells, once located in the gonadal ridges where they are surrounded by mesenchymal cells, are termed gonocytes and are characterized by the expression of stem cell markers, including PLAP, NANOG, c-KIT, SOX2, and OCT4. Gonadal stromal cells express transcription factor SRY, which results in early development of Sertoli cells through the activation of SOX9 [58]. Sertoli cells create a microenvironment that allows differentiation of gonocytes into spermatogonia. This process marked by the gradual loss of expression of the stem markers mentioned above, the acquisition of expression of germ cell-specific proteins MAGE4A, VASA, TSPY, OCT2, and SSX2, and the migration toward the basement membrane of the seminiferous tubule. Disturbances of the microenvironment result in arrest of fetal germ cell differentiation. It is in this arrested stage that mutations in oncogenes or tumor-suppressing genes are believed to occur, resulting in transformation into a neoplastic cell [58]. In fact, GCNIS widely shares the immunophenotype of gonocytes, suggesting an arrest at this stage [91, 92].

With the acquisition of additional mutational events, likely potentiated by the changes in hormonal milieu at puberty, the neoplastic cells eventually acquire the capacity to invade through the basement membrane. The 12p abnormalities are consistently present in invasive tumors and are absent in GCNIS [74, 75]. It is thus likely that the abnormal region on 12p harbors genes that enable the tumor cells to survive, proliferate, and develop invasive growth independent of signals from the intratubular Sertoli cells and the adjacent Leydig cells. Candidate genes include KITLG, NANOG, BCAT1, and CCND2, but

Table 7.3 Differential diagnosis of GCNIS and atypical germ cells in postpubertal patients

	GCNIS	Atypical germ cells
Morphologic features:		
Involved tubules	Atrophic, absent spermatogenesis	Atrophic and normal with ongoing spermatogenesis
Distribution	Segmental	Diffuse and scattered
Location	Exclusively basal	Basal or luminal
Nuclear features	"Squared off," regular	Irregular, polylobated, or multinucleation
Immunohistochemistry:		
OCT4	+	–
PLAP	+	–
Podoplanin	+	–
SALL4	+	+
SOX17	+	+
CD117	+	+

their exact definition remains elusive [58, 93]. The pathway to this may include intratubular transformation to either intratubular seminoma or intratubular EC (nonlinear progression) [94] or development first of an invasive seminoma and then transformation into EC (linear progression) [57]. YST, teratoma, and choriocarcinoma (CC) appear to evolve by differentiation from EC (Fig. 7.6).

7.5.2 Seminoma

7.5.2.1 General Aspects

Seminoma is the most common GCT comprising approximately 50 % of these tumors in the postpubertal setting. The average age at presentation is 40 years with most patients presenting between 35 and 45 years of age. Seminoma is unusual in childhood, and after the fifth decade [95], however, in a study by Berney et al. [96], seminoma accounted for 82 % of cases of GCT in the elderly. Bilateral involvement is seen in up to 5 % of cases. Seminoma has shown the highest incidence of bilaterality among GCT in several studies [97, 98]. Patients with seminoma often present with a painless testicular mass or dull aching sensation. Up to 2–3 % of patients present with symptoms related to metastatic disease in the retroperitoneum. Typically serum levels of AFP are normal; however mild elevations have been described in pure seminomas [99]. Significant elevations are regarded as evidence of nonseminomatous elements and should prompt a careful search for these components. Liver disease including metastasis of pure seminoma may explain the presence of mild to moderate elevations of AFP. Mild serum β-hCG elevation is

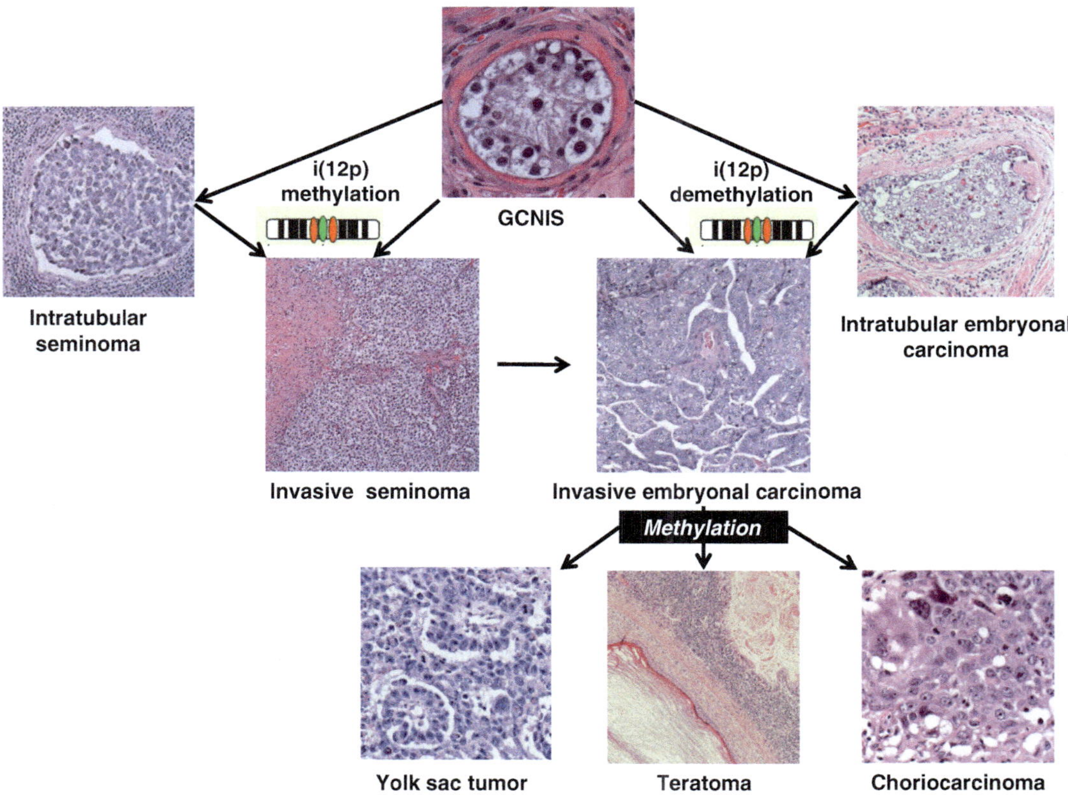

Fig. 7.6 Pathogenesis of germ cell neoplasia. Most forms of invasive GCT arise from GCNI, usually after the development of 12p abnormalities. GCNIS advances to seminoma or embryonal carcinoma, and from the latter, other forms arise, usually in association with DNA methylation

observed in seminomas with syncytiotrophoblasts and can be associated with the development of gynecomastia [100, 101].

7.5.2.2 Macroscopy

Grossly seminoma usually has a nodular configuration with well-circumscribed borders. The cut surface is lobulated, cream, tan, or white gray and shows variable consistency, reflecting the amount of fibrous tissue within the tumor (Fig. 7.7). Punctate foci of hemorrhage are often associated with the microscopic presence of syncytiotrophoblasts. Larger foci of hemorrhage and necrosis can be seen in tumors of large size. Invasion into the testis mediastinum and spermatic cord are uncommon with a reported frequency of 5–8 % [95].

7.5.2.3 Microscopy

Seminoma often has a diffuse sheetlike, nested, or trabecular growth. Characteristically the tumor is traversed by bands of connective tissue of variable thickness containing mature lymphocytes, predominantly of T cell type (Fig. 7.8a). These lymphocytes are also seen interspersed among tumor cells, but can be very prominent, occasionally obscuring the tumor cells. Cytologically the tumor cells show a uniform appearance with polygonal to round configuration and moderate amounts of clear, amphophilic, or eosinophilic cytoplasm with fairly distinct cell membranes (Fig. 7.8b). The latter attribute helps differentiate this tumor from EC, which exhibits a syncytial growth pattern. The nucleus is usually centrally placed and shows an evenly distributed chromatin with prominent nucleoli. The nuclear membrane is not perfectly round, with tendency to show angulated or flat contours, a term described by some as "squared-off" nuclei [102]. Some seminoma cells may have an eccentrically placed nucleus resulting in a plasmacytoid appearance. In contrast to plasma cells, a perinuclear hof is not observed. Mitotic activity is variable and may be quite brisk. Importantly, the degree of mitotic activity does not appear to correlate with tumor behavior. The term "anaplastic seminoma" was used in the past to designate seminomas with a mitotic rate equal or greater than three mitoses per high-power field and increased pleomorphism. This terminology has been discouraged due to insufficient data supporting a worsened prognosis in this subset of tumors [103, 104]. Similarly, necrosis, even when extensive, does not suggest a more aggressive behavior (Fig. 7.9). Prominent cytological atypia and pleomorphism, suggestive of early transition to EC, can be seen in seminoma [105]. In general most authors advice against designating these foci as EC, unless clear epithelial features such as glands or papillae are observed [106]. Up to 50 % of seminomas show granulomatous inflammation (Fig. 7.10a) [107]. This may range from scattered clusters of epithelioid histiocytes to an exuberant and diffuse reaction that may masquerade the tumor cells mimicking granulomatous orchitis. A subset of seminomas may exhibit ossification and calcification.

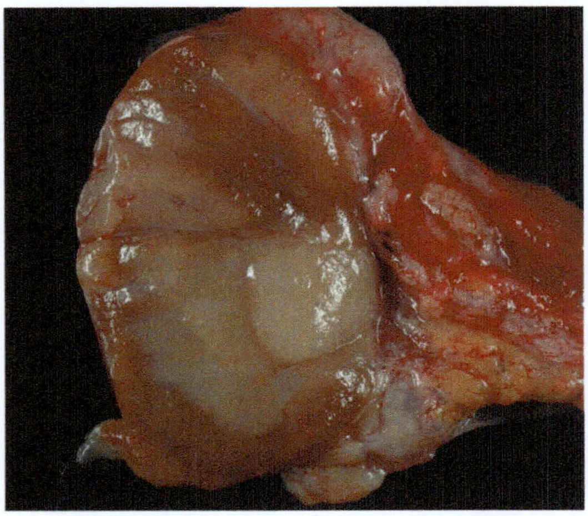

Fig. 7.7 Seminoma. Gross image of a seminoma showing a relatively well-circumscribed, lobulated, tan-white-colored bulging mass with a smooth surface

Scattered syncytiotrophoblasts can be seen in up to 20 % of seminomas (Fig. 7.10b) [108]. These are characterized by multinucleation, large size, and usually denser cytoplasm than the adjacent seminoma cells. Some may show prominent vacuolization of the cytoplasm. As mentioned above, these may be associated with small foci of hemorrhage. A distinction should be made with CC, which will require the presence of both syncytiotrophoblasts and cytotrophoblasts. The presence of syncytiotrophoblasts should thus be documented in the pathology report as it may explain the presence of mild elevations of β-hCG [109].

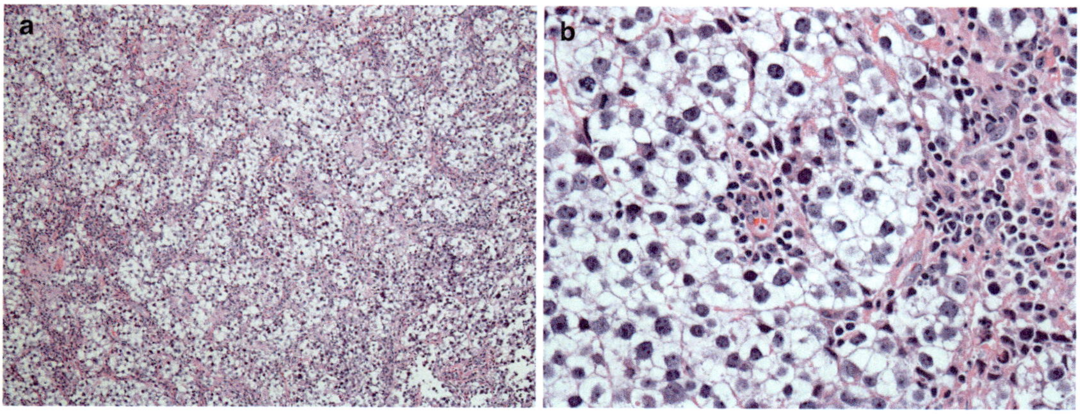

Fig. 7.8 (**a**, **b**) Seminoma. At low magnification a sheet-like pattern of growth with interspersed bands of connective tissue containing lymphocytes is seen (**a**). Higher magnification demonstrates uniform polygonal cells with clear cytoplasm, angulated or "squared-off" nuclei and distinct cell membranes (**b**)

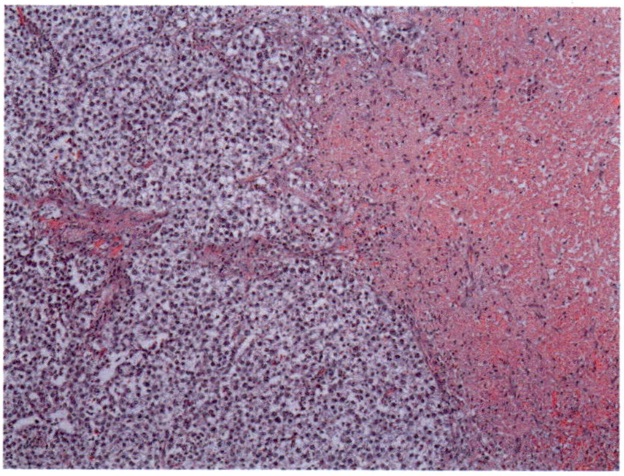

Fig. 7.9 Seminoma. Seminoma with prominent areas of necrosis

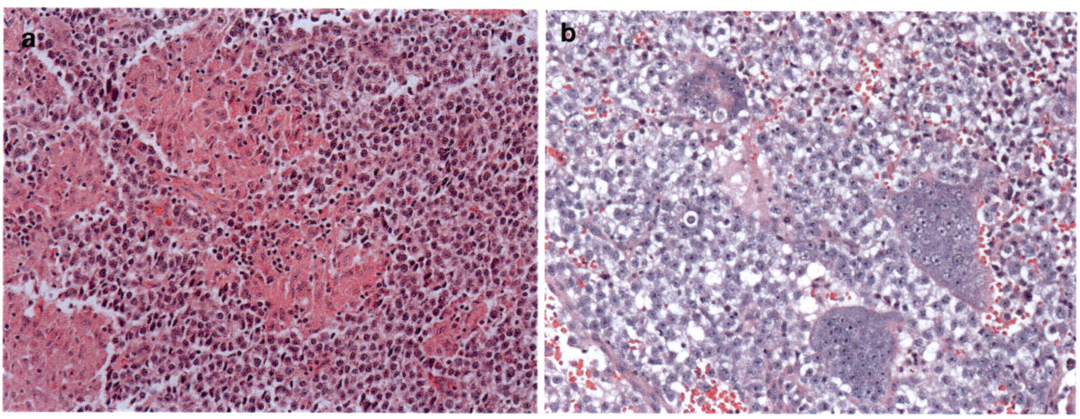

Fig. 7.10 (**a**, **b**) Seminoma. Seminomas are frequently associated with foci of granulomatous inflammation (**a**) and syncytiotrophoblastic giant cells (**b**)

Some seminomas show a tendency to form tubular, cribriform, or microcystic structures (Fig. 7.11), by virtue of developing spaces in between the solid growth of tumor cells [110–114]. These spaces are often associated with prominent intratumoral edema, but in other cases it may be artifactually induced by poor fixation or tumor degeneration. The result is a neoplasm that deviates from the classical appearance of seminoma and introduces the possibility of other diagnosis, such as YST or Sertoli cell tumor [112]. Usually typical areas of solid growth can be seen nearby. Difficult cases may require immunohistochemistry to resolve this differential diagnosis. Occasionally, seminomas may have prominent signet-ring cell change, a finding that may elicit the differential diagnosis with metastatic carcinoma (Fig. 7.11) [115]. Some seminomas may show exclusively or predominantly an intertubular growth pattern, without forming a discrete mass or nodule [116]. Intratubular seminoma was discussed in the section of GCNIS. It is not infrequently seen in association with invasive seminoma. Finally, some seminomas may have a very prominent lymphoid infiltrate that not only obscures the neoplastic cells but may even simulate a lymphoma involving the testis [117, 118]. Awareness of this phenomenon should prompt the careful search of typical seminoma cells and the use of appropriate immunohistochemical markers.

7.5.2.4 Immunohistochemistry

Seminoma is typically positive for OCT4, PLAP, c-KIT, podoplanin, SOX17, and SALL4 (Table 7.4). Keratin A1/AE3 expression is usually focal and weak, although, in our institutional experience, we have come across rare cases with diffuse reactivity. Seminoma lacks expression of CD30, SOX2, AFP, glypican-3, and ß-hCG [82].

7.5.2.5 Differential Diagnosis

The differential diagnosis of seminoma includes other forms of GCT, Sertoli cell tumors, and lymphoma. Solid variants of EC and YST are the most common types of GCT confused with seminomas. EC displays more pleomorphism than seminoma, and typically there is at least focal glandular differentiation. When both types of tumors are present, comparison of the nuclear characteristics may help in classifying an area of solid growth. Additionally, nuclei of EC frequently overlap, and cell borders are difficult to differentiate, imparting a "syncytium" appear-

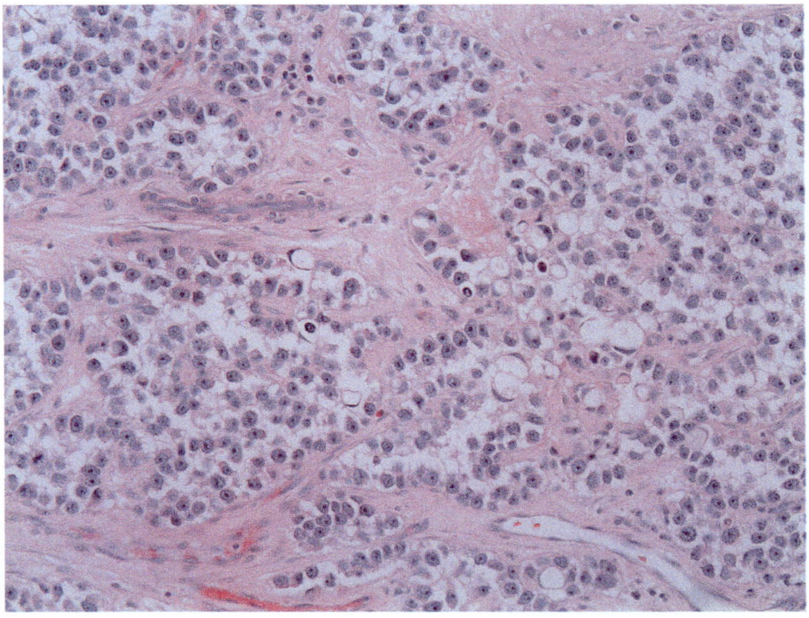

Fig. 7.11 Seminoma. Seminoma with tubular pattern of growth. Note also the focal signet-ring cell change

Table 7.4 Immunophenotype of GCT of the testis and their most common differential diagnosis considerations

	Seminoma	Embryonal carcinoma	Yolk sac tumor	Choriocarcinoma	Spermatocytic tumor	Sertoli cell tumor	Leydig cell tumor	Lymphoma	Adenomatoid tumor	Mesothelioma	Rete testis adenocarcinoma	Metastasis
SALL4	+	+	+/–	+	+	–	–	–	–	–	–	–
OCT4	+	+	–	–	–	–	–	–	–	–	–	–
PLAP	+	+	+	+	–	–	–	–	–	–	–	–
CD117	+	–	–	–	+/–	–	–	–/+	–	–	–	–/+
CD30	–	+	–	–	–	–	–	–/+	–	–	–	–
Pankeratin AE1/AE3	–	+	+	+	–	+/–	–/+	–	+	+	+	+/–
AFP	–	–	+	–	–	–	–	–	–	–	+	–
hCG	–	–	–	+	–	–	–	–	–	–	–	–
Glypican-3	–	–	+	+	–	–	–	–	–	–	–	–
Inhibin	– [a]	– [a]	– [a]	–	–	+	+	–	–	–	–	–
Melan-A	–	–	–	–	–	+	+	–	–	–	–	–/+
LCA/ CD20	–	–	–	–	–	–	–	+	–	–	–	–
Calretinin	–	–	–	–	–	–	+	–	+	+	+	–
WT1	–	–	–	–	–	+	+	–	+	+	–	–
Somatic markers (e.g., PSA, RCC, HMB45, CDX2, etc.)	–	–	–	–	–	–	–	–	–	–	–	Variable, depending on primary

Note: immunophenotype of teratoma is variable, depending on the somatic components present
[a]May be positive in scattered syncytiotrophoblastic giant cells

ance. Cytoplasm in EC tends to be more amphophilic and dense. Differential expression of cytokeratin, CD30, c-KIT, and SOX17 may help in difficult cases. YST may be confused with seminoma, specially its solid and microcystic growth patterns. Solid YST lacks the fibrous septae and lymphocytic infiltrate of seminoma and is rarely unaccompanied by more classical patterns, except in biopsy specimens. Cytologically, they show more variation in size and shape of the nuclei and are accompanied by hyaline globules and basement membrane material deposition. In tumors with microcystic architecture, attention should be paid to the cells lining the microcysts. In seminoma, they maintain their size and shape, while in YST they tend to be flattened, forming a lining [112]. A potentially critical misdiagnosis occurs when Sertoli cell tumors grow with a solid growth pattern mimicking seminomas, sometimes even with an associated lymphocytic infiltrate. Attention should be paid to the much lower nuclear grade of Sertoli cell tumors. The absence of associated GCNIS is an important diagnostic clue that should prompt appropriate immunohistochemical stains [119]. As stated above, some lymphoid stroma-rich seminomas may mimic lymphomas, while those associated with a prominent granulomatous reaction may lead to a misdiagnosis of granulomatous orchitis. Seminomas are the prototype of testicular neoplasms with a solid growth pattern. The differential diagnosis of this morphologic category is presented in Table 7.5.

7.5.2.6 Prognosis

Clinical stage I seminoma has a cure rate of 99 %. Relapse rates at this stage are 15–19 % [120], with most posttreatment relapses occurring within 3 years post-orchiectomy [121]. Metastases typically occur first in the retroperitoneal lymph nodes and subsequently at supradiaphragmatic sites. Metastatic seminoma within the good prognosis group as determined by the International Germ Cell Cancer Collaborative Group has an overall 95 % survival rate. Intermediate prognosis seminoma is rare; therefore survival rates reported are limited by low patient numbers [120]. There is contradictory data with regard to prognostic value of rete testis invasion, tumor size, and invasion into the tunica albuginea in stage I seminoma [122–126].

7.5.3 Embryonal Carcinoma

7.5.3.1 General Aspects

In its pure form, EC represents 3 % of testicular neoplasms, but nearly 40 % of testicular GCT have a component of EC. The peak incidence of this tumor is in the second to third decade with an average of 32 years, about a decade earlier than seminomas. EC is rare in the prepubertal setting and appears to be associated with DSD [102]. The most common presentation is that of a palpable testicular mass (80 %) followed by hormonal-related symptoms, such as gynecomastia (10 %) and symptomatic metastatic disease (10 %) [127]. Metastases occur via hematogenous and lymphatic pathways and primarily affect the periaortic lymph nodes, lung, and liver [63]. Although older reports have noted serum AFP elevations in patients with pure EC, current literature suggests this may be the result of unrecognized YST elements [128]. ß-hCG elevation is commonly seen in EC and reflects the presence of syncytiotrophoblasts.

7.5.3.2 Macroscopy

EC has a bulging soft, pale tan, pink, or dark-brown cut surface with areas of hemorrhage and necrosis (Fig. 7.12). The tumor borders are often ill defined. Tumor extension into the mediastinum testis and testicular adnexa is observed in up to 25 % of cases [106]. Compared to seminoma, EC are smaller at presentation, with average tumor size of 2.5 cm in diameter.

7.5.3.3 Microscopy

EC is composed of large pleomorphic cells with indistinct borders and nuclear crowding or overlap. Most tumor cells show vesicular chromatin and macronucleoli. The cytoplasm is predominantly basophilic although foci of amphophilic or clear cytoplasm are not unusual.

Table 7.5 Differential diagnosis of testicular tumors with a solid growth pattern

	Key features	Pearls for differential diagnosis
Seminoma	Solid sheets or nests Interstitial, tubular, trabecular, or sclerotic variants Fibrous septae dividing sheets of tumor cells Tumor cells evenly spread, no overlap Open chromatin with prominent nucleoli Squared-off nuclei	Clear to pale cytoplasm Variable lymphoplasmacytic infiltrate Granulomatous inflammation (30 % of cases) Fibrosis and sclerosis may be prominent Scattered syncytiotrophoblastic giant cells frequently present GCNIS in surrounding tubules
Sertoli cell tumor	At least focal tubular differentiation Uniform cuboidal or columnar cells Light eosinophilic to pale cytoplasm with vacuoles Round-ovoid nuclei, inconspicuous nucleoli, rare mitosis Fibrous septae and lymphoid infiltrates may be present	Tumors with solid growth, fibrous septae, and lymphoid infiltrates may strongly resemble seminoma Absence of GCNIS Smaller, more irregular nuclei, with less prominent nucleoli and less mitoses than seminoma
Yolk sac tumor (solid pattern)	Usually associated with other patterns of YST Solid sheets of polygonal cells with pale eosinophilic or clear cytoplasm Variable nuclear shape and size Hyaline globules Basement membrane deposition GCNIS present	More pleomorphism than what is seen in seminoma, but less than embryonal carcinoma Absence of fibrous septae and lymphocytic infiltrate
Embryonal carcinoma (solid pattern)	Areas of papillary and glandular pattern Large polygonal, highly pleomorphic cells with amphophilic or clear cytoplasm Necrosis and hemorrhage Frequent mitoses and apoptosis GCNIS present	Highest degree of pleomorphism Additional sampling may reveal areas more typical for embryonal carcinoma
Spermatocytic tumor	Solid growth pattern uninterrupted by fibrous septae Scant fibrous or edematous stroma Three distinct cell types: Small lymphocyte-like: 6–8 μm Intermediate cells: 15–20 μm Giant cells: 50–100 μm May have intratubular growth No GCNIS	Cells with pale cytoplasm, more pleomorphic on medium power than seminoma Lack of GCNIS, fibrous septae, lymphocytic infiltrate, or granulomatous inflammation differentiates it from classic seminoma Tends to occur in older patients, although age ranges may overlap
Leydig cell tumor	Solid sheets of oxyphilic cells Abundant eosinophilic cytoplasm, round nuclei, prominent nucleoli Cytoplasmic lipofuscin or Reinke crystals Fibrous, hyalinized, edematous, or myxoid stroma May show fatty metaplasia, spindle, clear cell, or microcystic changes	Dense eosinophilic cytoplasm separates them from the other entities; however, some may have cytoplasmic clearing Absence of GCNIS separates them from GCT
Lymphoma	More likely in older patients and bilateral Generally diffuse large B cell type Intertubular (interstitial) growth pattern Intratubular growth may be seen Variable sclerosis Large atypical cells with angulated nuclei and eosinophilic cytoplasm	Seminoma rarely shows a prominent interstitial growth pattern Cells in seminoma tend to have clearer and more abundant cytoplasm Absent GCNIS in lymphoma may aid in distinction
Metastasis	Metastatic clear cell renal cell carcinoma, prostatic adenocarcinoma, and melanoma may grow in sheets of clear/pale cells Morphologic features variable depending on primary origin of tumor	Usually known history of malignancy Metastases tend to occur in older patients, compared to germ cell tumors High level of suspicion in morphologic features that do not fit working diagnosis May require IHC to establish primary

YST yolk sac tumor, *GCNIS* germ cell neoplasia in situ, *GCT* germ cell tumors, *IHC* immunohistochemistry

Single-cell necrosis with frequent apoptotic bodies and mitosis are also readily encountered. Architecturally solid, papillary, and glandular patterns are the most common (Figs. 7.13, 7.14, and 7.15). The majority of EC show coexistence of two or more of the abovementioned patterns. Syncytiotrophoblasts are common in EC with a reported frequency of 46 % [129]. They are typically not associated with hemorrhage, in contrast with CC. Small amounts of undifferentiated spindled cellular stroma are accepted by some authors as part of EC, while others regard this as immature teratomatous mesenchyme [102, 127]. A granulomatous response of varied severity may be seen in EC [129]. Angiolymphatic invasion is commonly encountered at the periphery of the tumor. In fact, EC is the most common tumor identified in foci of lymphovascular invasion in mixed GCT. Intratubular EC is also similarly found at the edge of the tumor and is characterized by smudged hyperchromatic tumor cells admixed with abundant eosinophilic necrotic debris. Recognition of a thickened tubular basement membrane and residual Sertoli cells is helpful in avoiding misdiagnosis of intratubular EC as angiolymphatic invasion. Coarse calcifications are often seen in intratubular EC (Fig. 7.5b).

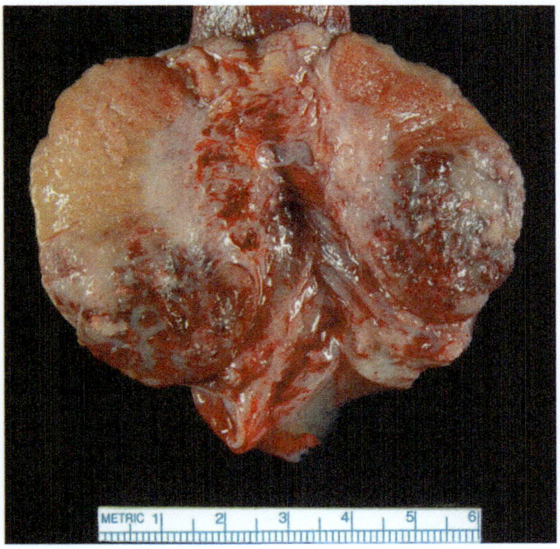

Fig. 7.12 Embryonal carcinoma. Gross image shows an ill-defined mass with a bulging, soft, pale-tan cut surface and prominent areas of hemorrhage and necrosis

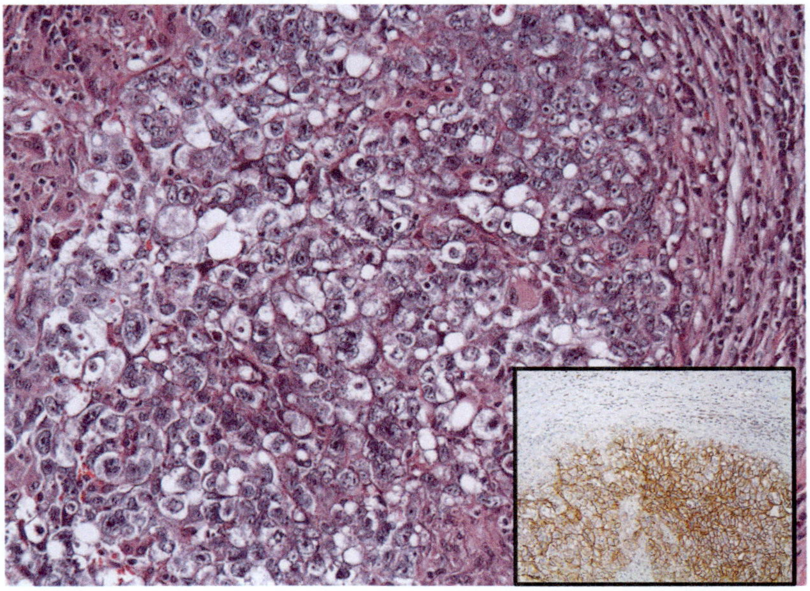

Fig. 7.13 *Embryonal carcinoma*, solid architectural pattern. These neoplasms show high-grade nuclear features, characterized by large pleomorphic cells with nuclear crowding and basophilic to clear cytoplasm. Foci of single-cell necrosis and apoptosis are easily appreciated. *Inset* shows strong membranous expression of CD30 by immunohistochemistry

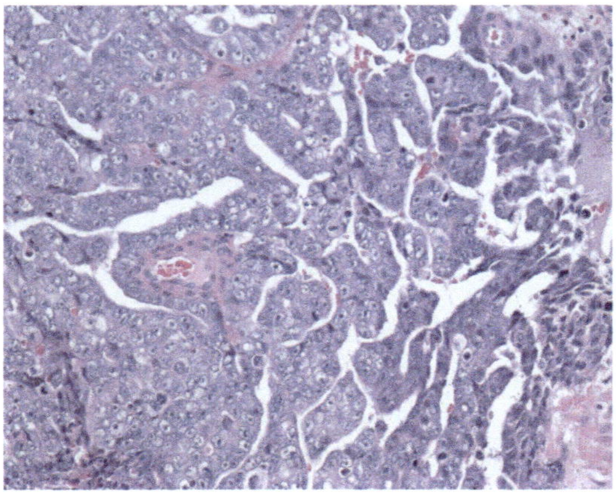

Fig. 7.14 Embryonal carcinoma. Glandular architectural pattern

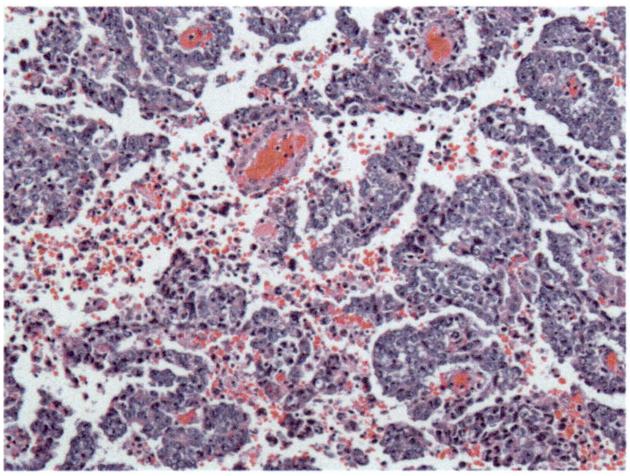

Fig. 7.15 Embryonal carcinoma. Papillary architectural pattern with prominent fibrovascular cores

The morphologic spectrum of EC is relatively broad [129]. It includes patterns commonly associated with YST (see below) or that may mimic seminoma or other neoplasms. An important pattern to recognize is the so-called appliqué pattern. This results in smudged degenerated cells located at the periphery of tumor nests. These cells closely resemble syncytiotrophoblasts and may lead to confusion with CC. Secretory-type change is characterized by subnuclear vacuoles similar to secretory endometrium. The pseudo-endodermal sinus pattern includes the presence of structures similar to the Schiller-Duval bodies of YST [130]. Sievelike pattern is reminiscent of the microcystic pattern of YST. In these patterns, differentiation from YST is based on the degree of pleomorphism of the tumor cells. In cases difficult to categorize, immunohistochemical staining with CD30 and OCT4 would favor EC. Solid forms may show tumor cells with prominent cell membranes and even mild lymphocytic infiltrate, resembling seminoma. Again, these EC variants will present higher degrees of pleomorphism than seminoma, and in difficult cases keratin and CD30 staining should resolve the problem. Other less common patterns include nested pattern, micropapillary pattern, anastomosing glands, necklace pattern, and a blastocyst-like pattern, characterized by large vesicles with edema fluid. Given the importance of reporting the percentage of EC in a mixed GCT (see below), accurate recognition of these patterns is mandatory to not under or overestimate the proportion of EC.

7.5.3.4 Immunohistochemistry

EC is positive for cytokeratin cocktails, OCT4, CD30 (Fig. 7.13, inset), SALL4, SOX2, and PLAP. The latter is usually weaker than that observed in seminoma. AFP may be focally positive. CD30 may be negative in metastatic EC [99]. In this setting coexpression of OCT4, SALL4, and strong and diffuse expression of cytokeratins may help support the diagnosis [131]. hCG expression is limited to syncytiotrophoblasts [97].

7.5.3.5 Differential Diagnosis

The solid pattern of EC may be confused with seminoma, lymphoma, ST, or solid YST (Table 7.5). YST is also a differential diagnosis in papillary and glandular patterns of EC (Table 7.6). The cytological features of EC may be sufficient to discriminate against other germ cell tumors in most cases; however, difficult cases can be easily resolved with the aid of immunohistochemical studies. Diffuse reactivity for OCT4, CD30, and cytokeratin is usually diagnostic (Table 7.4). Metastatic carcinomas to the testis may pose difficulty in distinguishing them from EC, but the absence of GCNIS, the usually older age of the patients, and commonly an intertubular growth pattern can raise the suspicion of a metastatic process [88].

Table 7.6 Differential diagnosis of testicular tumors with non-somatic glandular and/or microcystic growth pattern

	Key features	Pearls for differential diagnosis
Yolk sac tumor	Most morphologically versatile tumor YST patterns with glandular morphology include the microcystic/reticular, endodermal sinus, papillary and tubulopapillary, polyvesicular vitelline, glandular-alveolar, enteric/endometrioid, macrocystic patterns	Relatively uniform cells with clear or vacuolated to eosinophilic cytoplasm Bland cuboidal, columnar, flattened, or spindle cells Hyaline globules Basement membrane deposition GCNIS present
Embryonal carcinoma	Large infiltrative glands Areas of papillary and glandular pattern Large polygonal, highly pleomorphic cells with amphophilic or clear cytoplasm Necrosis and hemorrhage common Frequent mitoses and apoptosis GCNIS present	Pleomorphism is much more severe in embryonal carcinoma than in YST YST tend to have more variation in the morphologic patterns within one tumor Hyaline globules and basement membrane material suggest YST Frequently both tumors intermingled and closely associated
Rete testis/ epididymal adenocarcinoma	Invasive growth with prominent desmoplasia Centered in the rete testis or epididymis, where transition to malignancy may be seen Solid, papillary, tubulopapillary growth Marked nuclear atypia, mitoses, apoptosis, necrosis Intracytoplasmic or extracellular mucin	Higher grade and desmoplasia should differentiate it from YST Absence of GCNIS separates it from all GCT Location and transition to their benign counterparts helps in its separation from mesothelioma Metastatic process needs to be excluded with clinicopathologic correlation and IHC
Malignant mesothelioma	Epicenter in tunica vaginalis Histologic transition from mesothelial lining Majority pure epithelial or biphasic Papillary, tubulopapillary, glandular, or solid growth patterns Invasion beyond tunica	Tumor epicenter is different from testicular neoplasms GCT occur in younger patients May require IHC to separate from other entities in the differential diagnosis
Metastatic adenocarcinoma	Morphologic features variable depending on primary origin of tumor Tend to show a prominent intertubular and intravascular growth pattern Most common sites: prostate, lung, kidney, GI tract	Usually known history of malignancy Metastases tend to occur in older patients, compared to germ cell tumors High level of suspicion in morphologic features that do not fit working diagnosis May require IHC to establish primary
Leydig cell tumor	Oxyphilic cells, with abundant eosinophilic cytoplasm, round nuclei, prominent nucleoli Cytoplasmic lipofuscin or Reinke crystals may be seen Fibrous, hyalinized edematous, or myxoid stroma Fatty metaplasia, spindle, clear cell, or microcystic changes	Dense eosinophilic cytoplasm separates them from the other entities; however, they may have cytoplasmic clearing Absence of GCNIS separates them from GCT
Sertoli cell tumor	Uniform cuboidal or columnar cells Light eosinophilic to pale cytoplasm with vacuoles Round-ovoid nuclei, inconspicuous nucleoli, rare mitosis Tubular growth is the norm, whether hollow or solid Microcystic pattern not uncommon	Combination of tubular growth and relatively low nuclear grade should differentiate from germ cell tumors Relative monotony of seminomas may be more difficult to differentiate Absence of GCNIS is strong clue

(continued)

Table 7.6 (continued)

	Key features	Pearls for differential diagnosis
Seminoma with tubular pattern	Tubular or microcystic patterns may be seen Fibrous septae dividing sheets of tumor cells Tumor cells evenly spread, no overlap Open chromatin with prominent nucleoli Prominent lymphoid infiltrate and occasional granulomas	Seminomas more monotonous than YST Hyaline globules, papillary formation, basement membrane material suggest YST YST have flattened cells surrounding the cysts; in seminomas cells surrounding the cysts are identical to the rest
Adenomatoid tumor	Well-circumscribed mass, formed by tubules, gland-like irregular spaces, retiform architecture Cells may be cuboidal, flat, ovoid Round nuclei; dense cytoplasm with large vacuoles May show infarction	YST more intraparenchymal, while AT more peripheral YST has obvious malignant cytologic features GCNIS would suggest a germ cell tumor over AT Differentiated from mesothelioma on degree of infiltration and pleomorphism

YST yolk sac tumor, *GCNIS* germ cell neoplasia in situ, *GCT* germ cell tumors, *IHC* immunohistochemistry, *AT* adenomatoid tumor

7.5.3.6 Prognosis

EC is considered an aggressive form of GCT. In fact, the percentage of EC in mixed germ cell tumors has been regarded as a risk factor for metastatic spread [132, 133] (see "Mixed GCT" below).

7.5.4 Yolk Sac Tumor

7.5.4.1 General Aspects

YST is a malignant germ cell tumor that recapitulates the primitive endodermal structures, including embryonic yolk sac, allantois, and extraembryonic mesenchyme. A detailed review of the nomenclature, histogenesis, and histologic types is included in Chap. 6. In the testis, pure YST has a bimodal distribution with most cases occurring in the prepubertal period, while it is uncommon in the postpubertal period. Most reported cases of adult YST have occurred in the third and fourth decade. Elements of YST are present in up to half of mixed germ cell tumors. Clinically, YST presents as a painless palpable testicular mass. Elevation of serum AFP is noted in 95–100 % of cases [134].

7.5.4.2 Macroscopy

YST appears poorly circumscribed and nonencapsulated and has a homogenous gray to white to tan gelatinous cut surface. Hemorrhage, necrosis, and cystic degeneration may be seen.

7.5.4.3 Microscopy

YST shows the most diverse histology among GCT with a multiplicity of architectural patterns, including microcystic (reticular), macrocystic, endodermal sinus, papillary, solid, glandular/alveolar (including so-called endometrioid and enteric), polyvesicular vitelline, myxomatous, sarcomatoid, hepatoid, and parietal. A more detailed description of the histologic types is presented in Chap. 6.

The microcystic or reticular pattern is the most common and consists of anastomosing cords of vacuolated to spindled cells resulting in a meshwork or spider web like appearance (Fig. 7.16). The cystic spaces often show basophilic secretions. Coalescence of the small cysts gives rise to the macrocystic pattern. The endodermal sinus pattern is characterized by the presence of Schiller-Duval bodies, which consist of papillary structures lined by a cuboidal to columnar malignant-appearing epithelium with a distinct central vessel (Fig. 7.17). The papillae are recessed in a cystic space lined by flattened epithelium. Papillary formations with or without fibrovascular cores that project into cysts are the hallmark of the papillary pattern. The lining cells often exhibit a hobnail appearance with high

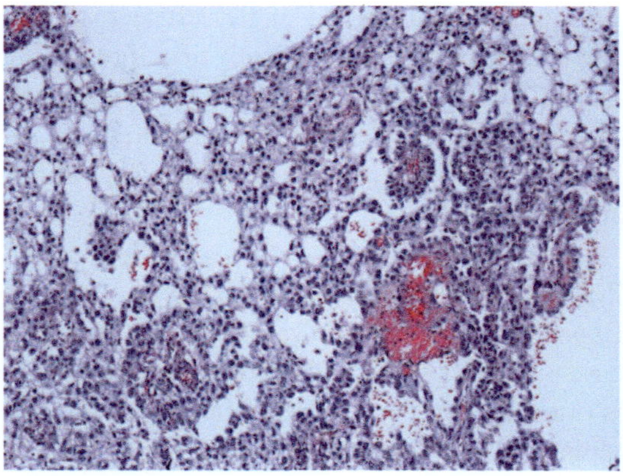

Fig. 7.16 Yolk sac tumor. Yolk sac tumor exhibiting the common microcystic or reticular architectural pattern

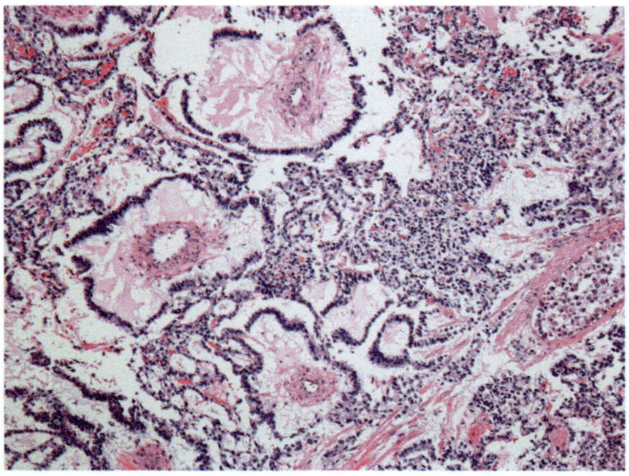

Fig. 7.17 Yolk sac tumor. Yolk sac tumor with an endodermal sinus pattern, characterized by papillary structures lined by a cuboidal to columnar epithelium with a distinct central vessel (Schiller-Duval bodies)

nuclear to cytoplasmic ratios. Exfoliated cells forming cell clusters can be seen in proximity of these papillae. The solid pattern consists of sheets of polygonal cells with variable nuclear yet with well-defined borders and moderate amounts of clear to amphophilic cytoplasm (Fig. 7.18). A variant form of solid YST with scant cytoplasm may have a blastema-like appearance. Glands lined by columnar epithelium with intestinal and secretory-phase endometrium-like features comprise the glandular pattern. These glands may be simple or branching and often present in a background of other YST patterns including micro-

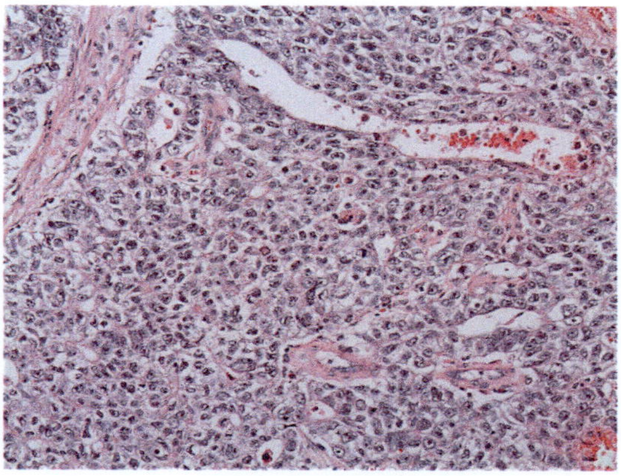

Fig. 7.18 Yolk sac tumor. Yolk sac tumor showing a solid pattern of growth. While exhibiting some degree of pleomorphism, it is considerably less atypical than embryonal carcinoma

cystic, macrocystic, solid, or polyvesicular vitelline. The lining epithelium usually lacks cytological atypia and therefore it could be easily mistaken as teratomatous epithelium. Glandular branching and absence of an encircling smooth muscle layer are features that help differentiate YST from teratoma. The polyvesicular vitelline pattern is exceptional in the testis. It consists of vesicles/cysts lined by a bland epithelium present in an edematous to fibrous stroma. The vesicles may show an area of constriction which gives rise to an eight-shaped structure. The epithelium lining the vesicles often transitions from flat to cuboidal to columnar at the site of constriction. The myxomatous pattern consists of scattered innocuous spindle or stellate cells in a loose, myxoid stroma, which often features a rich capillary network (also called angioblastic pattern). Focal differentiation into skeletal muscle and cartilaginous elements is allowed in YST and should not prompt a diagnosis of teratoma [106]. The term sarcomatoid YST has been used to describe tumors with spindled cellular stroma that retains cytokeratin reactivity. A recent study [135] has proposed that some somatic sarcomatoid malignancies may actually represent sarcomatoid YST. Hepatoid YST consists of cells with abundant eosinophilic cytoplasm, large central nucleus, and prominent nucleolus arranged in nests, cords, or trabeculae. Bile secretion and canaliculi have been observed in these tumors. Parietal YST is characterized by bland neoplastic cells embedded in dense eosinophilic basement membrane material that recapitulates the parietal layer of the murine yolk sac (Reichert's membrane). As noted above, cytological features of YST may vary across histological patterns. Significant nuclear atypia may be seen in solid, glandular, and sarcomatoid patterns of YST. Mitotic activity and single-cell necrosis can also be noted but are typically less prominent than in EC. A characteristic yet not pathognomonic feature of YST cells is the presence of intracytoplasmic eosinophilic globules, which are PAS-positive (Fig. 7.19).

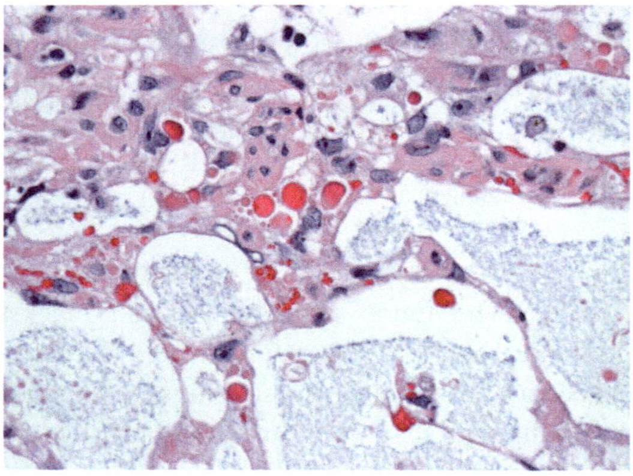

Fig. 7.19 Yolk sac tumor. Characteristic eosinophilic globules

7.5.4.4 Immunohistochemistry

Expression of AFP is variable and in many cases entirely absent. Intense staining for AFP is often seen in hepatoid pattern and in intracytoplasmic eosinophilic globules. Villin, glypican-3, SALL4, and low-molecular-weight keratins are often positive (Fig. 7.20) [82, 136, 137]. Rarely, c-KIT, SOX2, PLAP, and podoplanin positivity may be seen in YST [138]. This could represent a pitfall in the differentiation of the solid variant from seminoma. OCT4 and CD30 are negative. Hep-Par-1 reactivity has been documented in areas with or without hepatoid differentiation [136].

7.5.4.5 Differential Diagnosis

YST is the prototypic testicular neoplasm with glandular differentiation, a differential that is

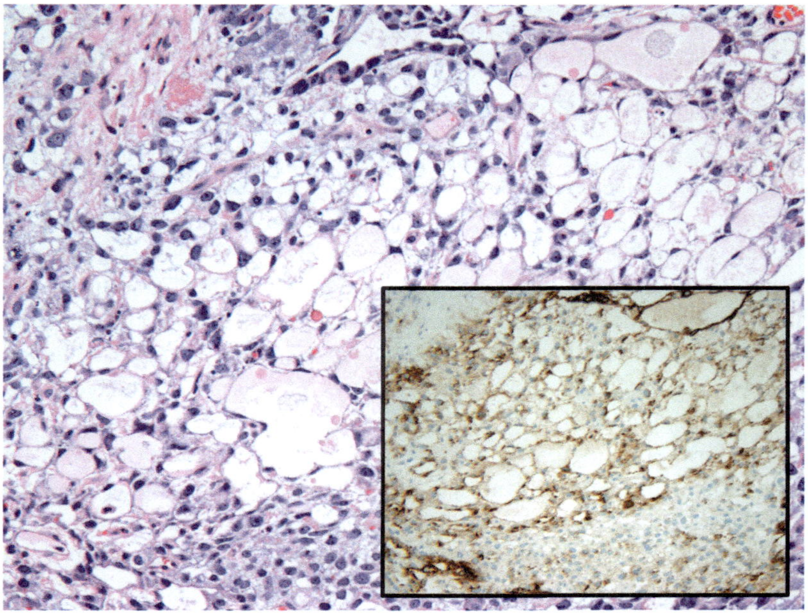

Fig. 7.20 Yolk sac tumor. Microcystic architectural growth pattern of yolk sac tumor, with the *inset* showing characteristic expression of glypican-3 by immunohistochemistry

summarized in Table 7.6. Glandular YST may be confused with immature teratoma, as previously discussed. Solid YST may resemble seminoma. The lack of fibrous septa with lymphocytic infiltration may help in their differentiation. Immunohistochemical reactivity for OCT4 excludes YST and therefore is useful in their distinction. OCT4 may prove useful when glands or papillary structures show cytological atypia that overlaps with EC.

7.5.4.6 Prognosis

Prognostic data for adult pure YST is limited due to the rarity of this entity. A series of 12 patients with stage I and II adult pure YST revealed a similar clinical behavior with respect to other non-seminomatous testicular GCT [139].

7.5.5 Postpubertal-Type Teratoma

7.5.5.1 General Aspects

Teratomas are GCT that display differentiation into somatic elements, including components of variable proportions from the ectoderm, mesoderm, and endoderm layers of the developing embryo. In the testis, they occur in two settings, based on the developmental stage of the surrounding testis: prepubertal or postpubertal. Contrary to their ovarian counterpart, the vast majority of postpubertal testicular teratomas are malignant, independent of the degree of maturity of its constituent elements. This is explained by the different histogenesis in both neoplasms. While in the ovary, teratomatous transformation of the germ cell occurs via parthenogenesis, in the testis it is most often an event that takes place after malignant transformation of a germ cell. As most other GCT in the male gonad, teratoma arises in association with GCNIS and, as explained before, likely corresponds to a terminally differentiated invasive component derived from more primitive forms, such as EC. Thus, postpubertal teratoma in the testis represents a form of differentiation of an already malignant neoplasm with a type II histogenesis. This explains its metastatic potential, independent of its degree of immaturity, and the fact that it is most frequently seen in association with other forms of GCT. Pediatric teratomas are reviewed in Chap. 10. However, occasionally, some teratomas in the postpubertal testis have a histogenesis and morphology comparable to those of the prepubertal gonad (type I) and thus behave in a benign fashion [140]. They are currently classified as prepubertal-type teratomas by the WHO [76] and will be considered separately (see below).

7.5.5.2 Macroscopy

Gross features of teratomas reflect their more prevalent somatic components (Fig. 7.21). Cysts filled with mucinous material or keratinaceous laminated debris are common. However, testicular teratomas tend to be more solid than the ovarian counterpart. More fleshy solid areas usually represent less differentiated components and may be associated with hemorrhage or necrosis. Bone, fat, cartilage, or even teeth material may be seen. Hair is rarely seen, and its presence should suggest a prepubertal-type teratoma.

7.5.5.3 Microscopy

In contrast to mature ovarian teratomas and prepubertal teratomas, postpubertal teratomas lack a well-organized organoid distribution of the different elements and display significant cytologic atypia and mitotic activity. Because of the lack of organoid arrangement, the exact somatic structure being replicated is sometimes difficult to determine. Representation from all three embryological layers is usually, but not always, present (Fig. 7.22). Glandular elements often include enteric-type epithelium, with variable amount of goblet cells, or mucinous glands (Fig. 7.23). Ciliated respiratory-type epithelium is also common. Frequently, glands with cylindrical cells with no particular differentiation are seen. Specialized glandular elements such as the liver, thyroid, or pancreas are rare [141]. Squamous nests are frequently present, displaying varying degrees of keratinization (Fig. 7.24). Commonly

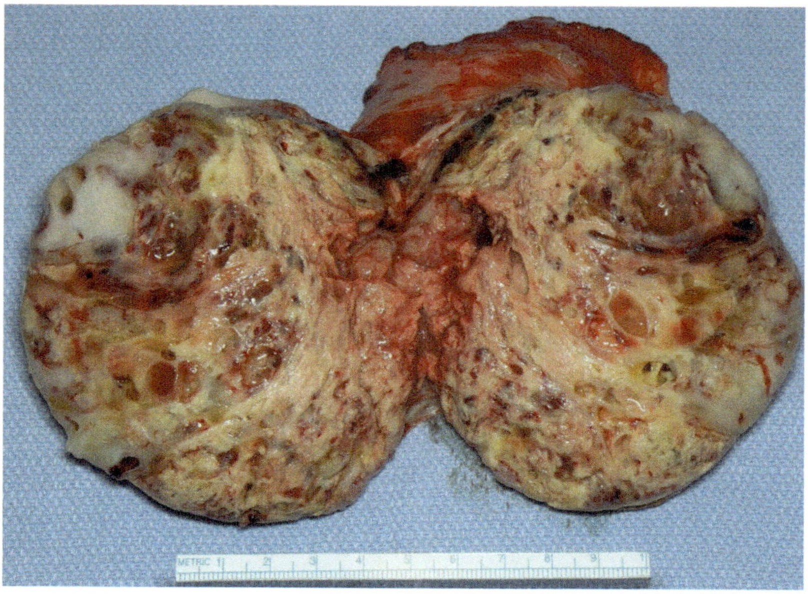

Fig. 7.21 Postpubertal teratoma. Gross image of a mixed germ cell tumor with prominent teratomatous elements characterized by cartilaginous (*upper left*) and numerous cystic areas

encountered mesenchymal elements include the cartilage, skeletal and smooth muscle, and rarely fat. Some of these elements may not be fully mature and may resemble mesenchymal tissue of the fetus and embryo.

More primitive, embryonic-type tissues, such as neuroepithelium and nephroblastic-type tissue, are commonly seen (Fig. 7.25). The presence of these elements does not impact prognosis or diagnostic terminology, unless frank overgrowth is present, in which case one must consider the possibility of a secondary malignancy [135] (see Chap. 12). Neuroectodermal tissue usually includes formation of neural-type tubules, rosettes, or sheets of undifferentiated primitive neuroectodermal small cells. Transition to better-differentiated glial type tissue may be seen. Other neural related tissues include meninges and retina-type pigmented epithelium. Nephroblastoma-type elements include primitive tubules, primitive spindle cells, and blastema elements. As in other types of postpubertal germ cell neoplasia in the testis, GCNIS is seen in adjacent tubules.

Another feature in testicular teratomas that differs from the ovarian counterpart is that all elements, including those with complete maturity, show some degree of cytologic dysplasia. This is easily seen in the epithelial elements, which frequently show nuclear hyperchromasia, irregular chromatin, and mitotic features, and can also be seen in the chondrocytes of the cartilage islands and other mesenchymal elements (Figs. 7.22 and 7.23).

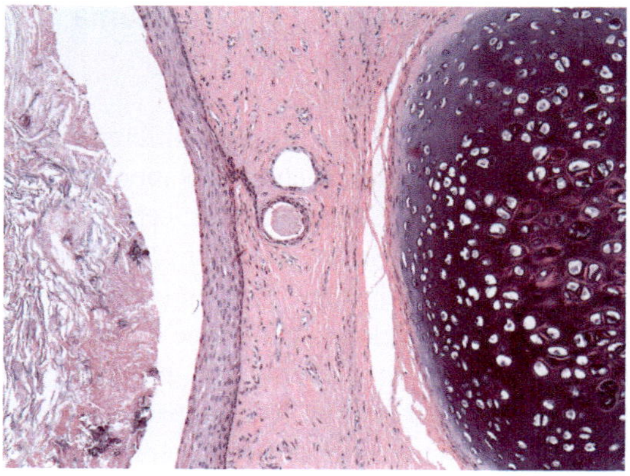

Fig. 7.22 Postpubertal teratoma. Squamous epithelium-lined cystic space is seen adjacent to cartilage

7.5.5.4 Immunohistochemistry

The immunophenotype of teratoma components recapitulates that of the somatic elements that are being reproduced. Pluripotentiality markers are expressed less consistently than in other types of GCT. SALL4 marks up to 80 % of elements within the teratoma. Glypican-3 is seen usually in more immature elements. PLAP and

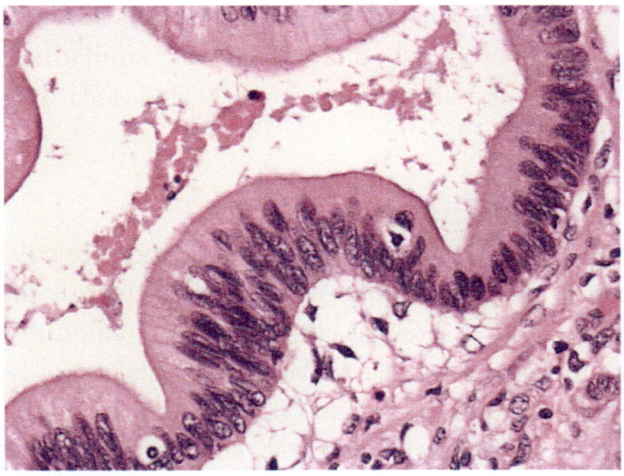

Fig. 7.23 Postpubertal teratoma. Glandular, enteric-type epithelium exhibiting prominent cytologic atypia

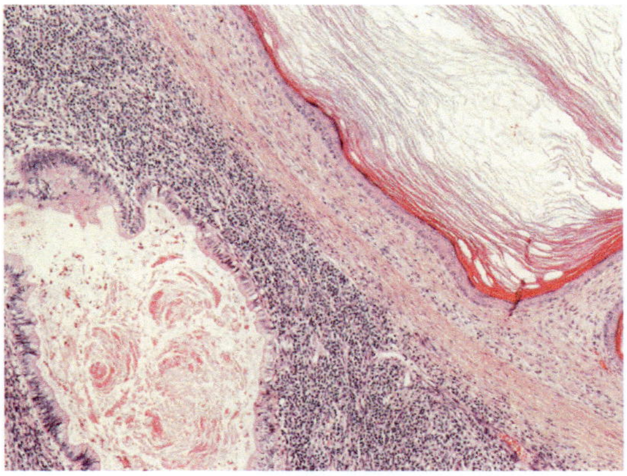

Fig. 7.24 Postpubertal teratoma. Keratinizing squamous epithelium is seen adjacent to glandular elements with nondescript histologic differentiation

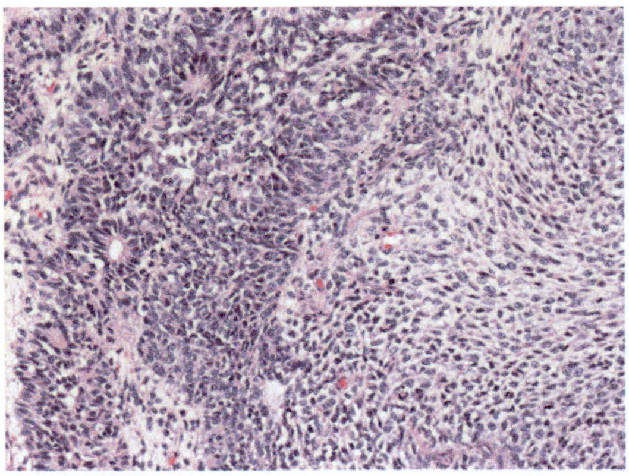

Fig. 7.25 Postpubertal teratoma. Primitive, embryonic-type tissues, such as neuroepithelium, in this case with prominent rosetting, is frequently seen

AFP are usually seen in only a subset of epithelial elements, while OCT4 is usually negative [82]. IMP3, an oncofetal protein that plays an important role in embryogenesis and carcinogenesis, is selectively expressed in postpubertal male teratomas and is negative in the female counterpart [142].

7.5.5.5 Differential Diagnosis

Teratoma needs to be differentiated from other GCT, particularly in the setting of a mixed GCT. While a large component of teratoma is hardly underrecognized, small amounts of somatic glands within extensive areas of EC or YST may be overlooked. Attention should be paid to the organization of the cells within the gland, which contrasts with the disorganized and primitive look of YST and EC elements. Some patterns of YST may be misinterpreted as teratoma, particularly the glandular, alveolar patterns, or the hepatoid patterns. Teratoma admixed with YST usually have abrupt transitions between both elements, while these YST patterns usually merge imperceptibly with more classical patterns. Also, teratomas will usually display other more easily recognizable components, such as cartilage or squamous epithelium. Regardless, misinterpretation of small amounts of teratoma as other GCT or vice versa, in the setting of a mixed GCT, would rarely have a significant clinical impact. Distinction from prepubertal-type teratomas, including dermoid cysts, which carry a markedly different prognosis, is explained below.

7.5.5.6 Prognosis

Postpubertal teratomas are rarely pure, and thus its behavior is usually compounded by that of the other elements associated with it. Cases of pure teratomas have been shown to present with metastases, and metastases from mixed GCT frequently contain teratoma components. Whether this represents and inherent metastatic potential of the teratoma elements or the teratoma represents a maturation process occurring in a metastasis from another GCT (particularly EC), is difficult to determine. Nevertheless, the association of teratoma with metastases and potential death due to disease in patients with GCT is well established. Up to 37 % of pure teratomas present with metastases [143]. Pure teratoma metastases are frequently found in patients that have undergone chemotherapy for metastatic mixed GCT, presumably because other GCT components have a better response to chemotherapy, or, alternatively, because chemotherapy induces other components to differentiate into teratoma. Regardless, outcome is generally favorable. Exceptions include the occurrence of secondary malignancy [135] (see Chap. 12) or cases where complete surgical removal is not possible. In this setting, progressive growth of the teratomatous metastases may result in lethal compression of vital structures with ultimate demise of the patient [144–146].

7.5.6 Choriocarcinoma

7.5.6.1 General Aspects

Choriocarcinoma (CC) is a GCT that shows trophoblastic differentiation and is composed of a variable mixture of mononucleated cytotrophoblastic and multinucleated syncytiotrophoblastic cells. A pure form of the former may be seen, more commonly in the metastatic setting, and occasionally in primary tumors [147]. In the testis, CC is usually a component of a mixed GCT, with pure forms corresponding to less than 1 % of all GCT [148]. Other variants of less common trophoblastic tumors, such as epithelioid trophoblastic tumor and placental site trophoblastic tumor, will be discussed separately.

7.5.6.2 Clinical Presentation

As part of a mixed GCT, they usually present as a painless testicular mass. However, occasionally in this context or more frequently when extensive, predominant, or pure, CC presents as metastatic disease, with symptoms related to hemorrhagic metastasis (hemoptysis, melena, intracranial hemorrhage, etc.) [149–152]. In these cases, a clinically evident primary may not be readily apparent. Serum β-hCG levels are typically high (usually >100,000 mIU/mL). Gynecomastia and hyperthyroidism have been described, and this is secondary to structural and functional similarities between the alpha chain of hCG and TSH and FSH [153–157].

7.5.6.3 Macroscopy

CC are frequently diagnosed while they are still small primary lesions in the testis, given their propensity to present clinically with symptoms of metastatic disease before a testicular mass is discovered. Even at this small size, tumors tend to be extensively necrotic and hemorrhagic, with usually solid residual tumor in the periphery. Larger tumors tend to be extensively cystic.

7.5.6.4 Microscopy

Histologically, classic CC shows a mixture of cyto- and syncytiotrophoblasts (Figs. 7.26 and 7.27). The latter is characterized by multinucleated giant cells with abundant eosinophilic cytoplasm. They may show vacuolated cytoplasm and other degenerative changes. Nuclei within them tend to have dense chromatin and occasional nucleoli. Cytotrophoblasts are polygonal or round cells with well-demarcated cell borders. They are usually small to medium sized, with irregular nuclei, vesicular chromatin, and visible nucleoli. Mitotic activity is easily seen. They usually cluster, forming tight aggregates. Some tumors contain cells of intermediate size, with more abundant and eosinophilic cytoplasm. Because these cells appear to morphologically and immunophenotypically imitate intermediate trophoblasts [158], some authors prefer to refer to all non-syncytiotrophoblast cells as "mononucleated trophoblast cells" to encompass both lines of differentiation (i.e., cytotrophoblast and intermediate trophoblast) [86]. Both mononuclear and multinucleated components are intimately associated, with syncytiotrophoblast cells frequently wrapping or capping aggregates of cytotrophoblast cells. Some cases have a relative paucity of syncytiotrophoblast cells and may appear monophasic, particularly when composed predominantly of the larger mononucleated cells. Extensive areas of necrosis and hemorrhage are frequently seen. Tumor cells tend to project into hemorrhagic areas with columns recapitulating the extravillous growth of trophoblast in the placenta. Vascular invasion is almost invariably seen in tumors with large components of CC. As with most of the other germ cell tumor types, GCNIS is seen in the adjacent seminiferous tubules.

7.5.6.5 Immunohistochemistry

By immunohistochemistry, CC is negative for OCT4 [82]. Syncytiotrophoblastic giant cells are usually positive for β-hCG and glypican-3, while both components tend to show expression of EMA, MUC1, CEA, SALL4, GATA3, and α-inhibin [159]. Contrary to other epithelial GCT, CK7 is expressed in a subset of trophoblasts [160]. p63 is preferentially expressed in cytotrophoblastic cells [161]. hPL is expressed strongly in syncytiotrophoblastic giant cells, but may stain some of the larger mononucleated trophoblasts [158].

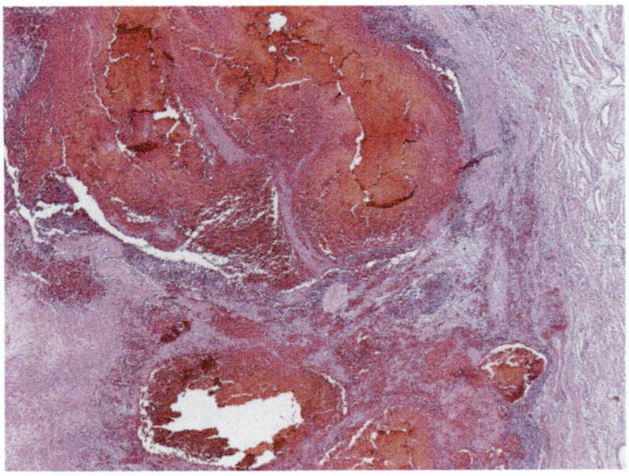

Fig. 7.26 Choriocarcinoma. Choriocarcinoma associated with extensive hemorrhage

7.5.6.6 Differential Diagnosis

The main differential diagnosis is another GCT type associated with syncytiotrophoblastic giant cells. As explained before, it is not uncommon for other types of GCT to be associated with occasional syncytiotrophoblastic giant cells. Since these are not associated with cytotrophoblast, they do not represent a component of CC. This is particularly significant in seminoma, where a misinterpretation of isolated syncytiotrophoblastic cells as CC may exclude a patient from the treatment arm of seminoma. In other GCT types, an overdiagnosis of CC may incorrectly assign a patient an adverse prognosis. Attention should be paid to the morphologic features of the other components and the absence of cytotrophoblasts. The pleomorphism of this component is usually higher than what is seen in YST but less severe than what is seen in EC. Seminoma would present with typical uniformity throughout the tumor mass, with intervening fibrous septa and lymphocytic aggregates. Neither would show the classic wrapping or capping of syncytiotrophoblasts over the cytotrophoblast component. In difficult cases, immu-

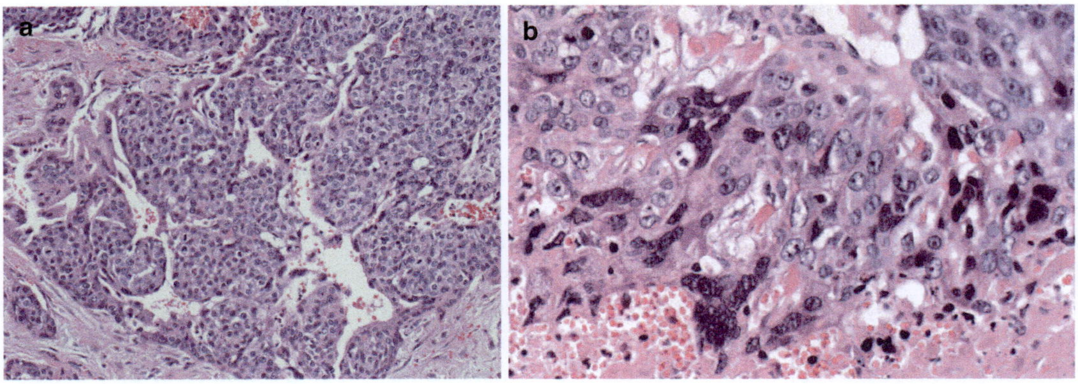

Fig. 7.27 (**a**, **b**) Choriocarcinoma. (**a**) At intermediate magnification, syncytiotrophoblasts characterized by multinucleated giant cells, with abundant eosinophilic cytoplasm, are seen interspersed among polygonal to round cytotrophoblasts with well-demarcated cell borders. (**b**) Higher magnification

nohistochemical stains may be used to correctly identify each component.

7.5.6.7 Prognosis

A CC component within a mixed GCT is usually associated with more aggressive behavior. However, when predominant or exclusive, CC is associated with a particularly ominous prognosis. In a recent study, 15 cases of pure or predominant CC component presented with metastases to distant sites, including the lungs, liver, and brain. About 79 % of these patients died of disease-related complications, with a median survival of 13 months, despite current regimes of chemotherapy [148]. Pulmonary metastases have a better prognosis than other visceral metastases. In the above-referenced study, two patients with exclusive pulmonary metastases were disease-free after 60 and 72 months of follow-up, respectively [148].

7.5.7 Other Trophoblastic Neoplasms

7.5.7.1 Placental Site Trophoblastic Tumor

Contrary to what occurs in the female genital tract, only a handful of cases of testicular placental site trophoblastic tumor have been described, three of them in a metastatic setting [147, 162–166]. The tumors are characterized by the presence of intermediate trophoblast cells exhibiting moderate amounts of dense eosinophilic cytoplasm with occasional vacuolization. The nuclei are smudged, and the majority of cells are mononuclear, with only occasional multinucleation of up to four nuclei per cell. Vessel wall invasion and small areas of hemorrhage are seen. The stroma is myxoid, and Alcian Blue positive. The tumor cells express HPL, and only focal β-hCG. Of the three primary tumors described, one was pure (in a 16-month old boy) [147], while the others were associated with a teratomatous component [163, 166]. Of the three tumors in the metastatic setting, one was chemotherapy naive and occurred in a pulmonary metastasis of a primary mixed GCT [162]. The other presented as a post-chemotherapy retroperitoneal recurrence 4 years after resection of a mixed GCT that included chorciocarcinomatous elements [164]. The third was also a retroperitoneal, post-chemotherapy recurrence of a mixed GCT in a 39-year-old man [165]. In this setting, the main differential diagnosis is a partially regressed CC, as treated CC may show numerous mononuclear intermediate trophoblastic cells. A hemorrhagic background with necrosis would favor the latter (Table 7.7).

7.5.7.2 Epithelioid Trophoblastic Tumor

Originally described in the uterus, epithelioid trophoblastic tumor (ETT) is a rare neoplasm of mononuclear cells that shows differentiation toward the chorionic-type intermediate trophoblast found in the *chorion laeve*. In the testis it was initially described as a component of a metastatic GCT [167], but five recently reported cases included this component in primary tumors [165]. They are characterized by nests of cells with squamoid appearance, with well-defined cytoplasmic membranes and intracytoplasmic vacuoles containing fibrinoid debris. Nuclei are mostly single, with occasional multinucleation. The tumor cells are positive for inhibin, GATA3, p63, PLAP, and variably for β-hCG while negative for SALL4, glypican-3, and OCT4. Serum β-hCG levels were normal or mildly elevated. Most tumors have been part of a mixed GCT with other components, although two cases corresponded to 95 % and 100 % of a recurrence, respectively. The ETT component did not appear to confer a different prognosis to the GCT [165].

7.5.8 Mixed Germ Cell Tumors

With the exception of seminoma, all invasive histologic types described above present more frequently as part of a mixed GCT than as pure forms. Mixed GCT are the most common non-seminomatous GCT. By definition, they contain various combinations of GCNIS-associated tumors and exclude non-GCNIS-associated components such as ST, pediatric YST, and prepubertal-type

Table 7.7 Differential diagnosis of trophoblastic tumors

	Choriocarcinoma	Placental site trophoblastic tumor	Epithelioid trophoblastic tumor	Cystic trophoblastic tumor
Clinicopathologic setting	Primary or metastatic Pre- or post-chemotherapy	Primary or metastatic Pre- or post-chemotherapy	Primary or metastatic Pre- or post-chemotherapy	Metastatic Post-chemotherapy
Neoplastic cell	Cytotrophoblast (CT) and syncytiotrophoblast (SCT)	Implantation site intermediate trophoblast (IT)	Chorionic laeve IT	Likely treated CT and SCT cells, or other GCT cells with induced trophoblastic differentiation
Main morphologic features	Biphasic pattern, with SCT cells wrapping aggregates of CT cells	Large discohesive cells with moderate amount of dense eosinophilic cytoplasm, large nuclei, prominent nucleoli. Infiltrative	Squamoid mononuclear cells with abundant cytoplasm and prominent cell membranes. Fibrinoid material	Mono- or multinucleated cells with abundant eosinophilic cytoplasm, lining cystic cavities, usually associated with teratoma (Fig. 7.46)
Necrosis	+++	++	+/−	−
Hemorrhage	+++	++	+/−	−
Vascular invasion	+++	+++	+/−	−
Immunohistochemistry				
hCG	+	+	+/−	+
hPL	+/−	+	−	+/−
p63	+	−	+	−
Inhibin	+/−	+	+	+
GATA3	+	+	+	+
Proliferative index (MIB1)	>10 %	>10 %	>10 %	<5 %

teratomas. Pure histologic types, including seminoma, associated only with syncytiotrophoblastic giant cells are also not considered mixed GCT.

The most common histologic type present in mixed GCT is EC, which is present in up to 84 % of cases, followed by teratoma (69 %), YST (60 %), and seminoma (39 %). CC is the least frequently present element, with only 17 % of cases containing this element [168]. The most frequent combinations of histologic type are summarized in Table 7.8. EC with teratoma is often cited as the most combination in mixed GCT, followed by EC with seminoma, and EC + YST + teratoma [168] [95, 169, 170]. However, Mosharafa et al. determined that the highest concordance and strongest correlation between histologic types were between teratoma and YST [168].

Grossly, mixed GCT reflect the features of their components. Tumors containing seminoma elements may show solid, tan, lobulated areas, while those containing EC are frequently necrotic and hemorrhagic (Fig. 7.28). Similarly teratoma elements usually confer a multicystic appearance, with mucinous contents. The distribution of elements histologically tends to be rather unpredictable (Fig. 7.29). Seminoma elements tend to concentrate in a specific area, while the other elements are frequently interspersed among each other. EC and YST tend to be spatially close to one another, sometimes intimately admixed. When

Table 7.8 Most common association patterns of elements in mixed germ cell tumors

Incidence of mixed germ cell tumor combinations (adults)	Jacobsen et al. [95] (%)	Mosharafa et al. [168] (%)
T + EC	60	56
T + CC	11	12
T + YST	21	43
T + S	19	22
EC + CC	22	15
EC + YST	29	50
EC + S	33	31
CC + YST	8	9
CC + S	6	4
YST + S	8	19

T teratoma, *EC* embryonal carcinoma, *CC* choriocarcinoma, *YST* yolk sac tumor, *S* seminoma

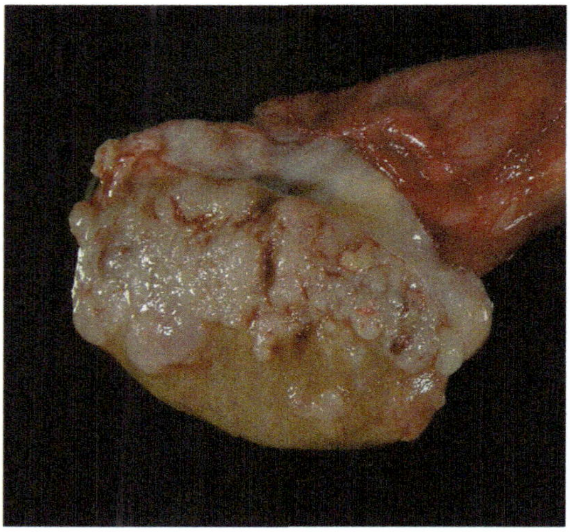

Fig. 7.28 Mixed germ cell tumor. Gross image shows a relatively well-circumscribed mass, with varied features. The smooth, pink-tan cut surface corresponds to an underlying seminoma. Focal hemorrhagic areas represent an embryonal carcinoma component

distributed in a fashion reminiscent of an embryoid body, the terms polyembryoma and diffuse embryoma may be used (see below).

While mixed GCT are usually grouped together under the category of "non-seminomatous GCT," prognostic differences between the different components which are clinically relevant have been identified. Thus, reporting of mixed GCT should include a detail of the different components present and their relative extent, usually expressed as a subjective percentage of the tumor comprised of each component. Of particular importance is the relative amount of EC. Tumors with high percentages of EC are associated with worse prognosis and thus may be ineligible for surveillance [171–173]. This is also true for cases with high percentages of CC [174, 175]. On the other hand, large proportions of YST in the primary tumor are associated with a lower probability of relapse [171]. The presence of teratomatous elements in the primary tumor predicts the presence of residual teratoma metastases in the post-chemotherapy setting [168]. Thus, accurate identification of the different elements in mixed GCT is of utmost importance, as it conveys significant prognostic information. Judicious use of immunohistochemistry panels may help in achieving this objective [82, 176]; however, in our opinion, their use should be reserved for particularly difficult cases and where the results would make clinically significant differences, as most cases can be accurately classified based on their morphology.

7.5.9 Polyembryoma and Diffuse Embryoma

Polyembryomas and diffuse embryomas are GCT that are composed of an EC and a YST component, organized in distinct architectural patterns.

Polyembryomas, specifically, recapitulate day 12 to day 18 embryonic structures by forming "embryoid bodies." Similar to a developing embryo, embryoid bodies have a central plate, which is comprised of an EC component. At the dorsal pole, this structure is surrounded by an empty space limited by a lining of flattened cells, recapitulating the amniotic sac and amnion. The ventral pole has a YST component in a reticular to microcystic architectural pattern, recapitulating the primary embryonic yolk sac. These embryoid bodies are embedded in a loose mesenchymal matrix, such that on low magnification the matrix is usually the predominant component, and often these neoplasms demonstrate a cystic pattern of growth [177]. Non-embryoid elements, if present, constitute less than 10 % of these tumors and range from mature teratomatous elements to trophoblastic cells [178].

Conceptually, polyembryomas may be regarded as a distinct architectural subtype of mixed GCT comprised of an EC and YST com-

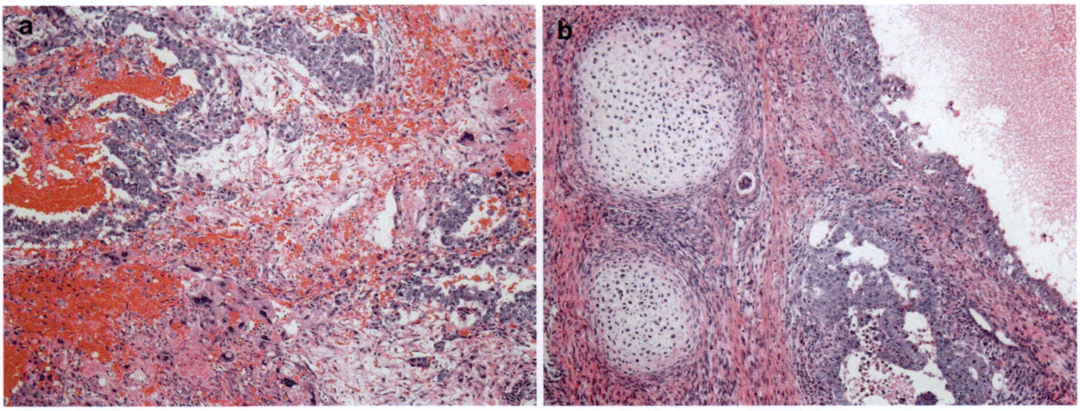

Fig. 7.29 (**a**, **b**) Mixed germ cell tumor. (**a**) Areas with high-grade cytology, corresponding to embryonal carcinoma (*upper left*), are seen juxtaposed adjacent to choriocarcinoma with prominent hemorrhage (*bottom left*). (**b**) Areas of cartilage corresponding to a teratoma are seen adjacent to embryonal carcinoma

ponent (Fig. 7.30). Conversely, they may also be regarded as a form of immature teratomas as the embryoid bodies represent a primitive stage of development [177, 178].

An area of debate is whether the presence of focal microscopic aggregates of an EC or YST element adjacent to an embryoid body is part of the spectrum of polyembryoma or needs to be classified as a mixed GCT. A size cutoff of 3 mm has been arbitrarily proposed, and, as the literature is limited, further studies are required to resolve this question [177, 178].

Diffuse embryomas were first reported by Cardozo de Almeida and Sculley in 1983 [179]. They differ from polyembryomas in that these tumors are comprised of roughly equivalent proportions of YST and EC components and do not form embryoid bodies. Herein, typically, the YST component encircles and often invaginates EC elements, with the latter being arranged in solid, gland-like, and tubulopapillary patterns (Fig. 7.31) [179, 180]. Recent studies have demonstrated the presence of GCNIS component in these tumors suggesting that the latter is the precursor lesion. These studies have therefore made a strong case for classifying diffuse embryomas as postpubertal mixed GCT [181, 182].

Pure forms of both polyembryomas and diffuse embryomas are rare, and these elements are more commonly found as components of mixed GCT [178].

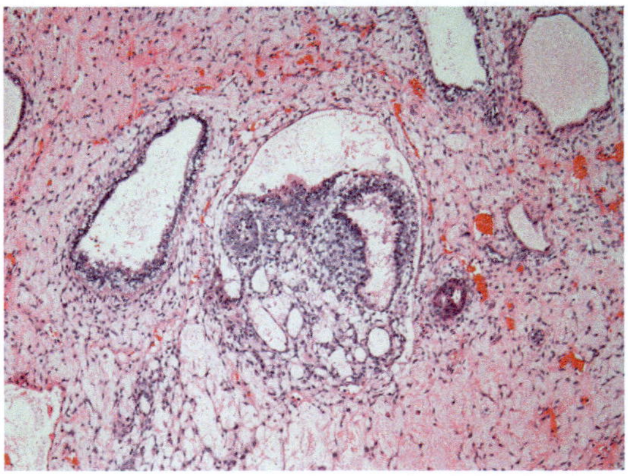

Fig. 7.30 Polyembryoma. A characteristic embryoid body is seen in the center, constituted by a central embryonal carcinoma component, limited by a lining of flattened cells at one pole and a yolk sac tumor component arranged in a microcystic pattern at the opposite pole. The embryoid body is embedded in a loose mesenchymal matrix

7.6 Tumors Not Associated with Germ Cell Neoplasia In Situ

7.6.1 Spermatocytic Tumor (Spermatocytic Seminoma)

7.6.1.1 General Aspects

Spermatocytic tumor (ST) is a rare germ cell tumor that occurs exclusively in the testis. It accounts for 1–2 % of testicular neoplasms [183, 184]. Previously referred to as spermatocytic

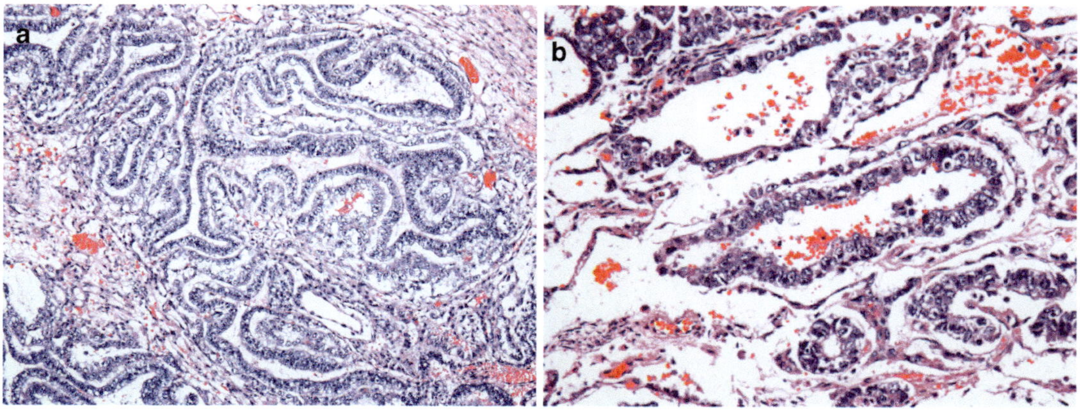

Fig. 7.31 (**a**, **b**) Diffuse embryoma. (**a**) Higher-grade embryonal carcinomatous components are shown organized in gland-like and tubular patterns and are encircled by a yolk sac tumor component. (**b**) Higher power

seminoma, the current terminology of ST has been recently adopted to avoid confusion with the much more common (classic) seminoma, which would carry prognostic and therapeutic implications. Age at presentation ranges from 25 to 87 years; however its peak incidence is in the fifth–sixth decade [185, 186]. ST is not associated with GCNIS, cryptorchidism, or gonadal dysgenesis and is not observed in association with other GCT. Bilaterality is rare with less than ten cases reported in the literature [187] and may be synchronous or sequential in occurrence. Serum markers are not characteristically elevated in ST.

7.6.1.2 Macroscopy

ST is well circumscribed and multilobulated. The cut surface is pale gray or pink tan and shows friable, mucoid, or gelatinous consistency (Fig. 7.32). Cystic change may be noted. Hemorrhage and necrosis may occur in larger tumors. The tumors are limited to the testis with rare exceptions of extratesticular extension reported in tumors with sarcomatous transformation [188, 189].

7.6.1.3 Microscopy

The tumor is arranged in sheets or solid nests (Fig. 7.33a). The stroma is usually scant and may be fibrous or edematous. The tumor cells may show discohesion and intercellular edema which results in a cystic or pseudoglandular appearance. In contrast with seminoma, a granulomatous reaction is extremely rare; nonetheless a discrete lymphocytic infiltrate may be present.

ST is characterized by a polymorphous population of cells composed of three distinct types: small cells with dense chromatin and scant cytoplasm that resemble lymphocytes; intermediate cells with round nucleus, finely granular chromatin and moderate amounts of eosinophilic cytoplasm; and a smaller population of large cells which may be multinucleated (Fig. 7.33b). Medium and large cells may exhibit spireme chromatin distribution, characterized by visible filamentous or cord-like strands of chromatin. Mitotic activity is brisk and atypical forms may be encountered. Rare cases of sarcomatous transformation have been documented in the form of undifferentiated, rhabdomyoblastic, or fibrosarcomatous lesions (see also Chap. 12) [186, 188–192].

7.6.1.4 Immunohistochemistry

ST shows reactivity for SALL4 and c-KIT. OCT4, CD30, keratin, PLAP, and podoplanin are negative [82, 193].

7.6.1.5 Differential Diagnosis

The differential diagnosis of ST includes seminoma, EC, and lymphoma. The presence of GCNIS

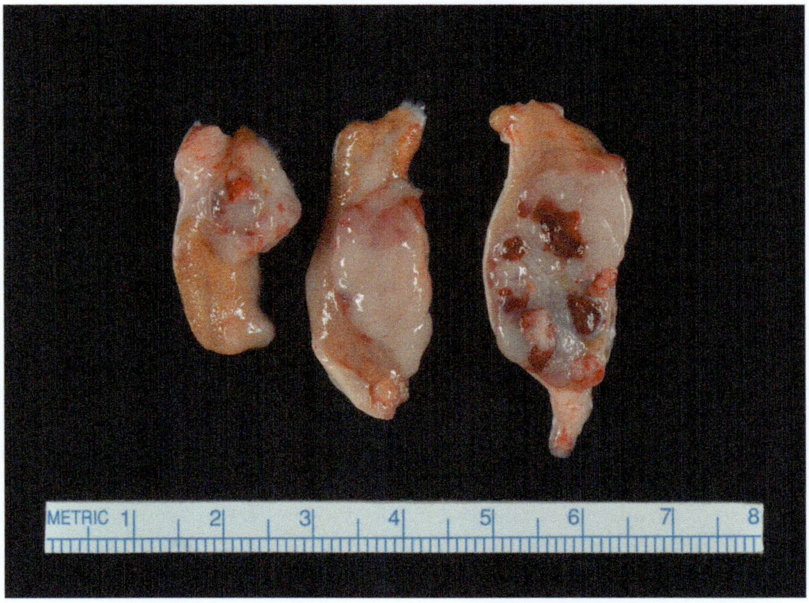

Fig. 7.32 Spermatocytic tumor. Gross image of a spermatocytic tumor showing a tan-pink mass with a mucoid to gelatinous consistency and focal hemorrhagic areas

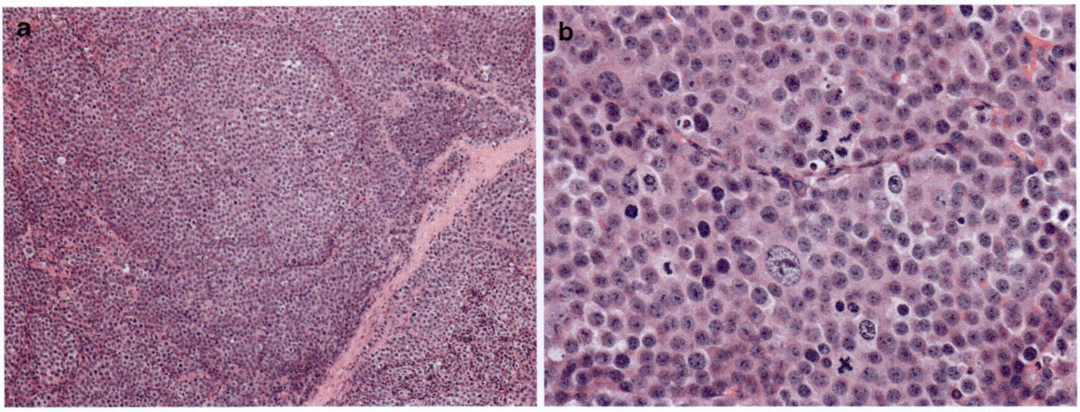

Fig. 7.33 (**a**, **b**) Spermatocytic tumor. Spermatocytic tumor showing a sheetlike growth pattern, in a background of fibrous stroma with minimal lymphocytic infiltration (**a**). Higher magnification reveals three distinct cell populations: lymphocyte-like small cells, intermediate cells, and a population of large cells (**b**)

can help exclude seminoma and EC. The latter should be distinguished from intratubular growth of spermatocytic seminoma which may be noted in few cases of ST. In general a lymphocytic infiltrate is not prominent in ST (see Table 7.5). Difficult cases may require immunohistochemical studies with OCT4 and lymphoid lineage markers.

7.6.1.6 Prognosis

Most ST follows an indolent behavior. Metastases are rare and almost exclusively seen when sarcomatous transformation is present (see Chap. 12) [185]. The metastatic foci are typically composed of sarcomatous elements. Exceedingly rare cases of metastatic spermatocytic seminoma without sarcomatous transformation have been reported [194, 195].

7.6.2 Prepubertal-Type Teratoma in the Postpubertal Testis

Pathologists have for a long time recognized that a subset of testicular teratomas appear different from the classic postpubertal teratomas [196, 197].

These are usually referred to as dermoid cysts [198–200], as their most prevalent component include squamous epithelium and cutaneous adnexae. These tumors are not associated with any other type of GCT, including GNCIS, with the possible exception of carcinoid tumor. Their teratomatous elements are much better organized and reproduce the relationships between elements seen in normal somatic tissues. They also do not display cytologic atypia in the cells of these elements.

Dermoid cysts are grossly usually filled with sebum-type material, admixed with hair and fatty fluid. Microscopically they consist of one or more cavities lined by squamous epithelium with granular layer and abundant keratinization (Fig. 7.34). Every so often, adnexal structures, such as pilosebaceous units or sudoriparous glands, are seen, draining in the cavity through the squamous epithelium. A series of reactive changes can be seen, such as histiocytic aggregates interrupting the squamous lining, or foreign body granulomatous reaction in cases of rupture. Lipogranulomatous reactions are also common. The adjacent seminiferous tubules show normal spermatogenesis, and GCNIS is remarkably absent.

Most published cases of dermoid cyst of the testis include examples that do not have exclusively cutaneous elements [199, 200]. Ciliated epithelium, mucinous epithelium, smooth muscle, fat, glia, bone, meninges, and cartilage have been reported in varying proportions. Recently, Zhang et al. [140] published a series of 25 cases of mature teratomas of the testis; ten of them corresponded to dermoid cysts. However, 15 contained predominantly non-cutaneous elements, including glandular cysts lined by ciliated/respiratory epithelium with goblet cells, surrounded by the smooth muscle, cartilage, and seromucinous glands, resembling a somatic bronchus. Others displayed intestinal-type epithelium surrounded by lamina propria and smooth muscle, recapitulating gut (Fig. 7.35). Fat and meningeal tissue were also present. Interestingly, none of the cases with available follow-up had recurrence or progression of the disease. Finally, in 18 cases tested, no cytogenetic abnormalities of 12p were identified. These data suggest that: (1) These teratomas show a spectrum of conforming elements that goes beyond what's conveyed by the narrowly descriptive term of "dermoid cyst" and may have a wider variety of tissues present. (2) They are benign neoplasms. (3) They are pathogenetically and biologically distinct from classic postpubertal (type II) testicular teratomas, being more likely analogous to prepubertal (type I) teratomas. Recently, the 2016 WHO classification of testicular tumors has included them under the category of prepubertal-type teratomas [76].

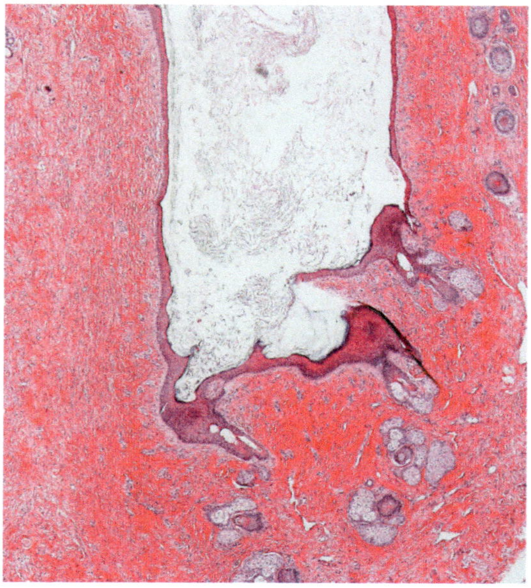

Fig. 7.34 Prepubertal teratoma. A dermoid cyst with prominent adnexal structures which distinguish it from epidermoid cysts. Note the organization of the elements recapitulating the normal skin

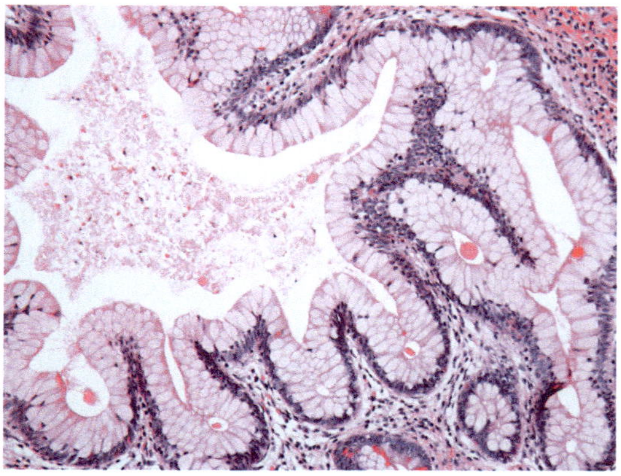

Fig. 7.35 Prepubertal teratoma. Glandular, enteric-type epithelium lacking discernible cytologic atypia. Compare with the cytologic atypia of postpubertal-type teratomas in Fig. 7.23

These mature benign teratomas need to be accurately distinguished from classic (type II) teratomas of the testis, as the former would not need additional therapy and the patient can be reassured of their benign nature. Also, if diagnosed preoperatively, they may be amenable to conservative surgical excision. Attention should be paid to: (1) The more cystic gross pathology of benign mature teratomas, with evidence of hair and sebum, similar to ovarian mature cystic teratoma. (2) The presence of a "polarity of growth," i.e., a structured architecture that recapitulates the microscopic anatomy of skin and sometimes bronchial or enteric elements. (3) The absence of cytologic atypia in the conforming elements, easily seen in classic type II teratomas, in both epithelial and mesenchymal elements (Figs. 7.23 and 7.35). (4) The absence of GCNIS or evidence of testicular dysgenesis in the background seminiferous tubules, with the presence of ongoing spermatogenesis. (5) The lack of association with any other form of GCT

As one can quickly surmise, however, these distinctions can still be challenging on a case-by-case basis and requires a thorough microscopic evaluation of the testicular tumor (Table 7.9). We agree with Zheng et al. [140] when they recommend a conservative approach to diagnosing this entity, since an overdiagnosis of classic (type II) teratoma is probably a more acceptable error than a missed diagnosis. In cases with less than definitive morphology, FISH studies for 12p abnormalities may be valuable in this differential diagnosis.

Another possible variant of mature benign testicular teratomas is *epidermoid cyst* (EpC). As its name implies, EpC is characterized by a keratinizing squamous epithelium-lined cavity, filled with keratinaceous debris (Fig. 7.36a). It is distinguished from dermoid cysts by the absence of adnexal structures and other teratomatous ele-

Table 7.9 Differential diagnosis of prepubertal- and postpubertal-type teratomas

Feature	Prepubertal-type teratoma	Postpubertal-type teratoma
Gross features	Cystic, hair, and keratinaceous debris common	Multicystic alternating with solid. Cysts contain a variety of mucinous, serous, or keratinaceous material. Hair unusual
Background parenchyma	Normal postpubertal, with active spermatogenesis. No significant atrophy beyond area adjacent to tumor	Postpubertal. Scarring, microlithiasis, extensive atrophy
Distribution of elements	Organoid distribution (elements recapitulate architecture of organ that it is trying to emulate – bronchus, skin, or gut)	Random
Teratomatous elements	Predominantly squamous epithelium and cutaneous adnexae, or respiratory-type epithelium, seromucinous glands, cartilage, and muscle recapitulating a bronchus. Intestinal epithelium uncommon	Random, heterogeneous elements Epithelium frequently nondescript. Intestinal epithelium common
Stroma	Normal or reactive. Lipogranulomatous reaction may be present	Primitive, nondescript. May be neoplastic in nature
Other types of GCT present	Usually none. Carcinoid tumors may be present. Rarely associated with prepubertal-type YST	Usually part of a mixed GCT, with EC, YST, seminoma, or choriocarcinoma. Less frequently pure
Cytologic atypia	Absent	Present in epithelial and mesenchymal elements
Mitotic activity	Absent to low	High
GCNIS	Absent	Present
12p abnormalities	Absent	Present

ments (Fig. 7.36b) [201–204]. A teratomatous origin of these cysts cannot be completely excluded, as they could represent a simplified version of a dermoid cyst. Favoring this possibility is the age of presentation, which overlaps with other germ cell tumors, the occasional occurrence of epidermoid cyst with carcinoid tumor [205], and the presence of morphologically indistinguishable cysts within mature teratomas. Their lack of association with GCNIS [206] and 12p abnormalities [207] and their uniformly benign behavior would suggest, similarly to dermoid cysts, a teratomatous process more akin to type I GCT. However, their occurrence in other organs and their occasional location immediately underneath the tunica albuginea and even in the paratesticular region [208] suggest a nonneoplastic origin, probably related to squamous metaplasia of the mesothelium, rete testis, or epididymal epithelium, for at least a subset of these tumors. Indeed, a metaplastic origin for their ovarian equivalent has been suggested [209].

Grossly, EpC are unilocular, well-circumscribed cystic cavity filled and distended by white, doughy laminated contents. Mean size has been reported as 2.0 cm [203]. Microscopically they consist of a fibrous wall internally lined by benign keratinizing squamous epithelium with a granular layer and, as stated, absent skin appendages. Granulomatous reaction to rupture and keratin debris may be seen. On occasion EpC can exhibit stromal osseo-calcific metaplasia, which should not be confused with other teratomatous elements. As stated, some testicular carcinoid tumors may be associated with an adjacent EpC [205].

DC and EpC can present as paratesticular masses without connection to the testis proper [208]. This is especially true with EpC that was derived from the scrotal skin. The main issue is whether the DC/EpC resides in a location well beyond the paratesticular soft tissues and spermatic cord, such that a metastasis from a mature teratoma in a postpubertal patient has to be considered in the differential diagnosis.

7.6.3 Carcinoid Tumor

Carcinoid tumor in the testis can occur in one of the following settings:

1. As a pure primary carcinoid
2. As part of a prepubertal-type teratoma (including dermoid and epidermoid cysts)
3. As part postpubertal-type teratoma
4. As a metastatic lesion

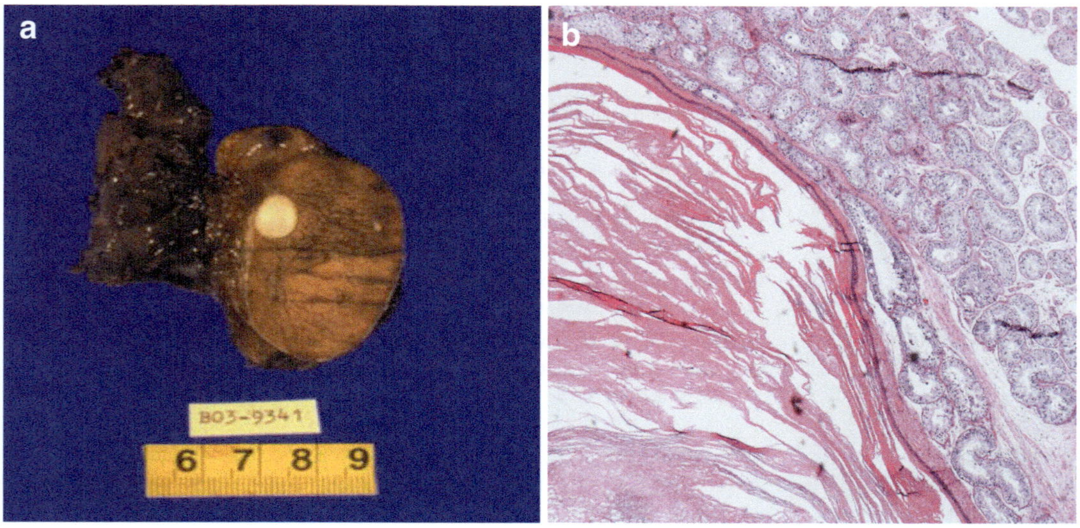

Fig. 7.36 (**a**, **b**) Epidermoid cyst. (**a**) An epidermoid cyst showing a well-circumscribed cystic cavity, filled with keratinaceous debris. (**b**) An epidermoid cyst showing a keratinizing squamous epithelium-lined cavity, filled with keratinaceous debris, adjacent to seminiferous tubules

In published series, pure primary carcinoid corresponds to approximately two-thirds of cases [205, 210–212]. In these cases, evidence to support a germ cell origin is scant, as there is lack of other teratomatous elements or GCNIS. A common origin with Leydig cell tumors has been suggested [213]. When associated teratomatous elements are identified, they are exclusively mature and frequently consist exclusively of an epidermoid or dermoid cyst. These suggest that when teratomatous in origin, carcinoid tumors are most often part of a prepubertal-type teratoma (type I GCT). However, it must be borne in mind that association with GCNIS and 12p abnormalities have been infrequently reported [214, 215]. Furthermore, in Wang et al.'s series, one case associated with a dermoid cyst had subsequent metastasis of YST and EC [205]. These data suggest that a subset of testicular carcinoids may arise in the setting of type II GCT.

Tumors tend to present as unilateral painless masses. Carcinoid syndrome is relatively rare and usually associated with large tumors or metastatic disease [205, 212, 214, 216, 217].

Pathologically, carcinoid tumors of the testis share similar features with those of other sites. Grossly, they are generally solid neoplasms when pure, and solid and cystic when associated with teratoma. They are usually well circumscribed. Mean size in one review of 57 published cases was 3.5 cm; however, pure tumors were larger than those associated with teratoma (4.2 vs. 1.5 cm, respectively) [212]. Histologically, the majority grow with a mixture of an insular or nested pattern, with trabecular pattern (Fig. 7.37). The nests or trabeculae are separated by fibrous septae. Follicular patterns can also be seen, as well as spindle cell differentiation. Cytologically the tumors show characteristic features for neuroendocrine neoplasms, with generally regular round nuclei with salt and pepper chromatin. Abundant eosinophilic cytoplasm is a characteristic. Mitotic activity is usually low. In a recent series, 4 of 29 cases were classified as "atypical carcinoids," based on mitotic activity in 3 cases, and the presence of necrosis in the fourth [205]. Cytologic atypia tends to be focal.

Tumors are consistently positive for one or more neuroendocrine markers, including chromogranin, synaptophysin, neuron-specific enolase, serotonin, gastrin, neurofilament proteins, substance P, and vasoactive intestinal polypeptide. Argentaffin pos-

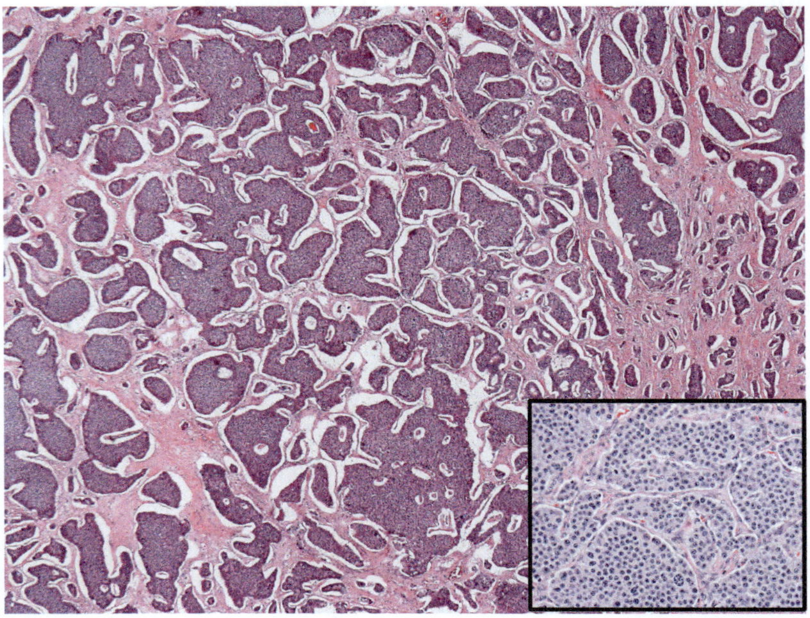

Fig. 7.37 Pure primary carcinoid. Carcinoid tumor with an insular architectural pattern is seen at low magnification. *Inset* shows neuroendocrine features characterized by round nuclei with salt and pepper chromatin, abundant eosinophilic cytoplasm, and occasional mitotic activity

itivity and argyrophillia are present in most cases [210–212, 218, 219].

The main differential diagnosis is metastatic carcinoid tumor, which is usually of gastrointestinal origin. While bilaterality favors metastases, bilateral primary tumors have been described [212]. Sertoli and Leydig cell tumors, adult-type granulosa cell tumors, and paraganglioma may be confused with carcinoid tumors. In cases of morphological ambiguity, a battery of immunostains including cytokeratins, α-inhibin, and neuroendocrine markers is useful in separating these entities. Sertoli and Leydig cell tumors may express neuroendocrine markers, but would be positive for α-inhibin in the vast majority of cases.

Carcinoid tumors tend to have a favorable outcome. Metastases have been reported, sometimes after a prolonged follow-up, and usually to regional lymph nodes, but also to the lung, bone, soft tissue, and liver [212]. Histologic findings cannot predict metastatic behavior, although it is likely that this is more frequent, if not restricted, to "atypical carcinoid" histology [205, 211].

7.6.4 Prepubertal-Type Yolk Sac Tumor

Contrary to prepubertal-type teratoma, prepubertal-type YST occurs almost exclusively in the pediatric testis and thus is reviewed in Chap. 10. However, a recent paper suggests its occurrence in the postpubertal testis, whether pure or in combination with prepubertal-type teratoma [220]. In its most usual setting, it comprises 75–80 % of testicular neoplasms in childhood with half to a third of cases taking place within the first and second year of life. Recent studies have challenged the reported prevalence of pediatric YST, suggesting a reporting bias may have underrepresented cases of teratoma and epidermoid cysts [221]. In contrast to postpubertal cases, prepubertal YST are not associated with GCNIS or isochromosome 12p gains, supporting a type I histogenesis. An association with cryptorchidism and white race is also lacking in prepubertal YST. Histologic features are similar to postpubertal YST, and these are extensively reviewed in Chaps. 6 and 10.

7.7 Tumors with Germ Cell and Sex Cord Stromal Elements

Gonadoblastoma (GB) is the only histologic type in this category in the current WHO classification. Previous editions contained a category of unclassified mixed germ cell-sex cord-stromal tumors, which, unlike gonadoblastomas, were reported in genotypically and phenotypically normal males [222–224]. However, recent literature has demonstrated the lack of staining for PLAP and CD117 in the germ cells, contrary to what would be expected in GCNIS and seminoma, suggesting that these tumors are comprised primarily of neoplastic sex cord-stromal elements, with entrapped nonneoplastic germ cells [177, 225]. Thus, the current WHO classification no longer includes this category. Rare reports of tumors with an architectural pattern of two distinct neoplasms, as would be seen in a collision tumor, have been reported, and it is unclear if these truly represent mixed germ cell-sex cord-stromal tumors [225, 226].

7.7.1 Gonadoblastoma

7.7.1.1 General Concepts

Gonadoblastoma (GB) is a tumor of young patients with gonadal dysgenesis and other DSD, characterized by the presence of neoplastic germ cell and sex cord-stromal elements [227]. To better understand GB, it is important to appreciate some basic concepts of gonadal dysgenesis (GD).

GD is the improper development of the gonad, which, importantly, includes both germ cell and sex cord components [228]. Full understanding of GD is impaired by a complex pathogenesis involving mutations in multiple genes that can manifest in a variety of clinical pictures [229–231] [232–240]. Following the developmental migration of the gonad, a dysgenetic testis can present within the abdomen, inguinal canal, or scrotum. As might be anticipated, a dysgenetic testis will be smaller than normal and is often less white to the naked eye due to the lack of a thick, fibrous tunic. Some represent "streak" gonads in which identification as testis or ovary may only be inferred [241].

Within the dysgenetic testis, there are a plethora of microscopic findings. These include tubules, typically closely positioned to one another and uniformly rounded in appearance, comprised mainly of immature Sertoli cells with varying numbers of germ cells, which often resemble embryonic germ cells (spermatogonia) [39, 241, 242]. Between the tubules reside Leydig cells with scant cytoplasm.

GB might be expected to present in the very young. While it is true that the majority present prior to the age of 20 years, there is as rather wide range of ages from 1 to almost 40 years with a case report described in a 46XY fetus [227, 243]. The most common occurrence of GB is in phenotypical females, who, however, almost invariably have a Y chromosome. GB occurrence in 45X0 Turner syndrome likely represents undetected mosaicisms [244]. Twenty percent of GB occurs in phenotypic males. An infantile uterus is typically present as are fallopian tubes; if bilateral tubes are identified, the patient is likely a phenotypic female, while a male phenotype is more likely if unilateral [227]. Phenotypic male patients commonly have other abnormalities of the genitourinary system, such as cryptorchidism, hypospadias, and gynecomastia.

For GB to develop, the presence of a Y chromosome is necessary, in particular the GBY region, which harbors the *TSPY* gene [245, 246]. Mutations of the sex-determining gene Y (SRY) are also associated with high incidence of GB [244].

When present exclusively in the gonad and treated appropriately, GB has an excellent prognosis. However, at the time of diagnosis, about 50 % have progressed to an invasive germinoma (seminoma or dysgerminoma) and 8 % to other types of GCT [227]. Due to this, GB is considered a precursor for GCT in the setting of GD, specifically in the prepubertal patient.

7.7.1.2 Macroscopy

In the vast majority of reported GB, the tumor develops in an "undeclared" gonad with about one in five developing in either a testis or a streak gonad [227]. The reason for this is due to effacement of the gonad (streak or otherwise) by tumor. If the tumor is pure GB, it displays a tan-yellow or gray and firm appearance that can have a gritty cut surface due to the presence of calcifications. If it is admixed with other GCT, it will exhibit other findings that reflect the type of GCT present (usually seminoma/germinoma) [228].

7.7.1.3 Microscopy

The typical histologic appearance of GB makes the pathologist quickly recognize that the neoplasm is quite unusual, but if the clinical presentation is well understood, the diagnosis can quickly distil into either pure GB or a mix of GB with a more common type of GCT. GB is characterized by round nests of both germ cells and sex cord cells. The germ cells of GB resemble seminoma or GCNIS, but may include a population of less mature looking cells, resembling spermatogonia. They are intermixed with sex cord cells, which resemble Sertoli cells, and are arranged in three patterns, which are usually admixed. In the coronal pattern, the sex cord cells are mostly at the periphery of the nests; in the follicular pattern, they surround individual or small groups of germ cell; and in the Call-Exner-like pattern, they surround globoid basement membrane deposits (Fig. 7.38). The nests are arranged in relatively large lobules surrounded by a fibroinflammatory stroma, peppered with Leydig cells and calcifications, some of which have replaced "regressed" or "burned-out" lobules of GB [228]. The morphology of the invasive GCT, if present, is the usual for each type of invasive GCT. An invasive seminoma/dysgerminoma may be quite subtle and easily overlooked.

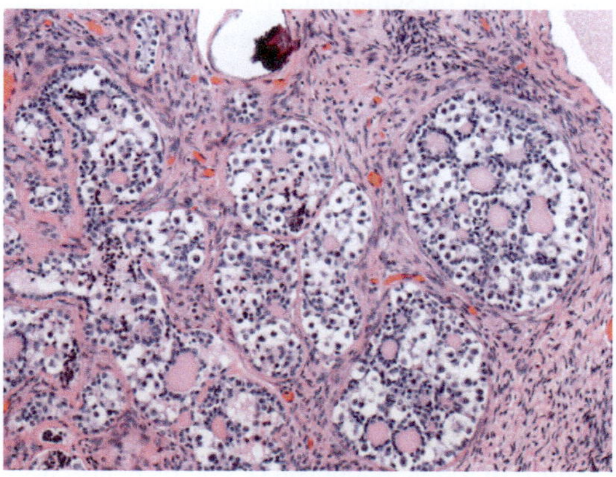

Fig. 7.38 Gonadoblastoma. Nests of germ cells resembling seminoma or GCNIS are surrounded by a fibrous stroma. Focal calcifications likely represent "regressed" lobules of gonadoblastoma. The germ cells surround globoid basement membrane deposits, reminiscent of Call-Exner bodies (Image courtesy of Gary Keeney, MD, Mayo Clinic, Rochester, MN)

7.7.1.4 Immunohistochemistry

Germ cells within GB that resemble germinoma or GCNIS will react with PLAP, OCT4, podoplanin, and CD117 [247]. The smaller ones resembling spermatogonia are usually positive for TSPY and negative for OCT4, while another subpopulation may have an inverse immunophenotype [246]. The sex cord cells are positive for inhibin, vimentin, and WT1 [248].

7.7.1.5 Differential Diagnosis

Sertoli cell nodules resemble GB, particularly if colonized by GCNIS. While they frequently occur in cryptorchidic testes, these are usually nondysgenetic and occur in phenotypically normal males. Sex cord-stromal tumors with entrapped germ cells are also in the differential diagnosis, particularly granulosa cell tumor, whose Call-Exner bodies may be mistaken for GB.

7.7.1.6 Prognosis

Untreated GB carries a high risk of progression to an invasive GCT. In the seminal work by Scully [227], the majority (almost 60 %) of GB were also associated with germinoma with a smattering of EC, YST, and teratoma. Thus, the postulation is that all GB would likely progress to invasive tumors if not removed. As such, if identified on biopsy, gonadectomy may play an important role in the management of patients with possible gonadal dysgenesis. Due to the relatively young age at presentation, and even when mixed with more typical GCT elements, outcomes for gonadoblastomas are excellent [249, 250].

7.8 Germ Cell Tumor Regression

The phenomenon of GCT regression, was initially reported in 1961 by Azzopardi et al., in a series of 17 patients that presented with metastatic germ cell tumors, where sampling of the testicular tissue revealed either the presence of minute foci (<5 mm) of viable neoplasm or complete absence thereof [251]. Subsequent studies by the same group laid the groundwork for defining the histopathologic features of this phenomenon, and its incidence was independently estimated from a series of 61 autopsy cases to comprise about 10 % of cases with metastatic GCT [252, 253].

At present, identification of GCT regression has important clinical implications. Following orchiectomy for metastatic disease, if residual disease or diagnostic evidence of GCT regression is not identified, orchiectomy for presumed contralateral disease must be considered, to prevent the development of recurrences. It may also influence the selection of therapeutic modalities. For instance, in patients with a seminomatous tumor component with a regressed non-seminomatous GCT, the use of radiotherapy alone may be suboptimal [254].

Commonly identified features include the presence of well- to poorly defined scars in a background of testicular atrophy, where the seminiferous tubules are often hyalinized or show a Sertoli cell-only pattern (Fig. 7.39a–e). Scarring by itself is relatively nonspecific as it can be seen secondary to vascular lesions, trauma, or infection. The most specific feature, in this context, is the presence of a residual GCNIS component, which is only seen in approximately half the cases [178, 254–256]. Other features that have been reported include microlithiasis, persistent neovascularization, Leydig cell hyperplasia, and the presence of lymphocytic infiltrates. Microlithiasis must be distinguished from the presence of coarse intratubular calcifications, which is thought to correlate with regressed EC. It is not entirely clear if Leydig cell hyperplasia is a true hyperplasia or a misleading visual impression as a consequence of the atrophy of the surrounding testicular parenchyma. Finally, the presence of lymphocytic infiltrates have been hypothesized to contribute to immune-mediated tumor regression, based on studies that correlate the presence of lymphocytic infiltrates with outcomes in seminomas which were managed with surveillance [254, 257].

Though it appears that seminomas have a greater tendency to undergo spontaneous regression than other GCT, this may be more of a reflection of their higher overall incidence among GCT. Recent studies indicate that CC is not prone to regression [254]. This, on the other hand, may reflect a selection bias as most of these patients likely present with advanced, disseminated disease and receive chemotherapy prior to orchiectomy making it difficult

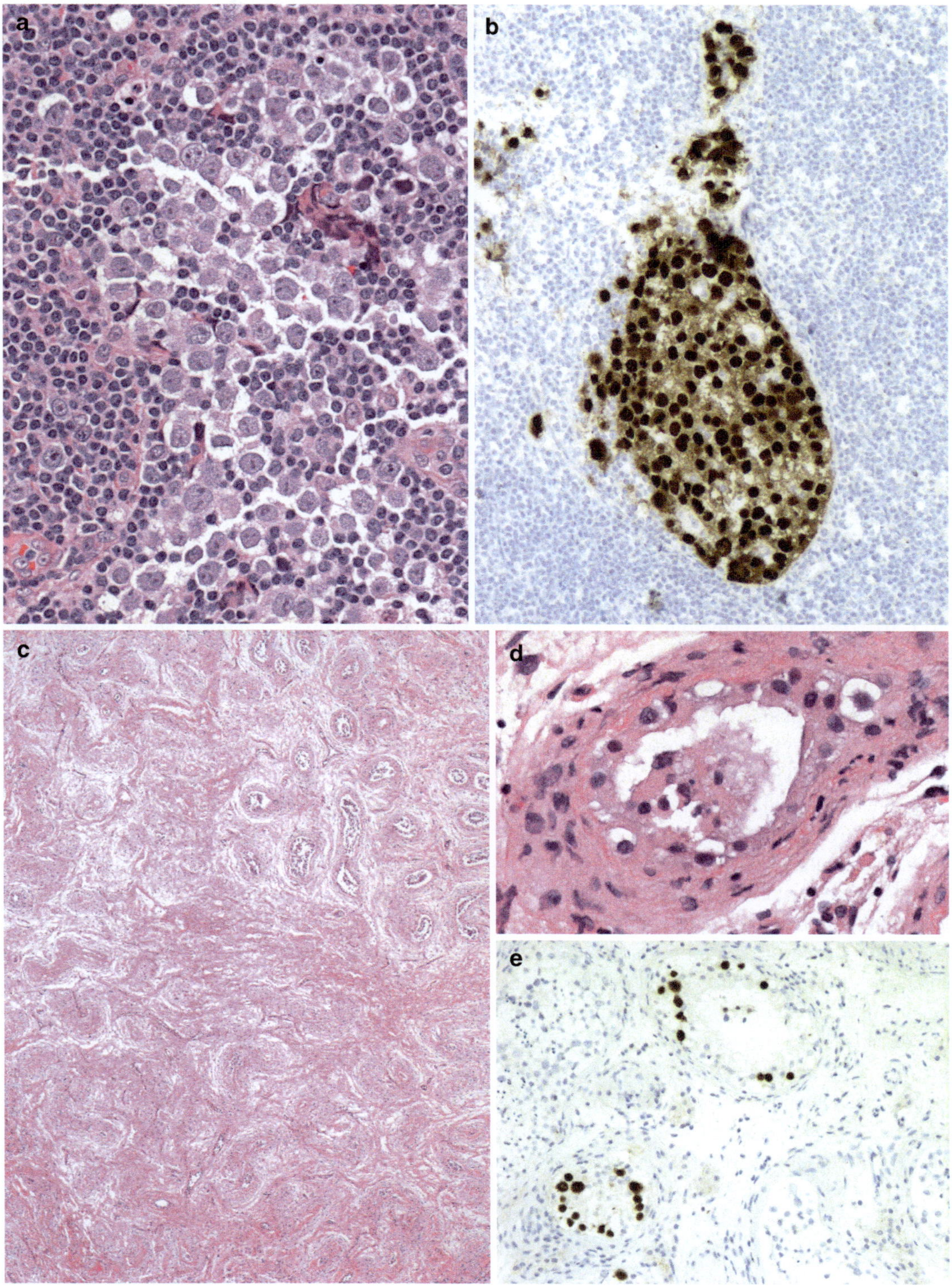

Fig. 7.39 (**a**–**e**) Germ cell tumor regression. Cervical lymph node with a metastatic seminoma (**a**), which showed strong nuclear expression of OCT4 (**b**). Orchiectomy revealed scarring and a background of testicular atrophy (**c**), with hyalinized seminiferous tubules. Focal GCNIS was identified (**d**), which was confirmed by positive OCT4 expression (**e**)

to differentiate therapy-induced regression from spontaneous regression. Finally, no GCT subtype, including teratoma, shows resistance to regression based on a comparison of reported overall incidences and incidences of regression.

7.9 Clinical Presentation

The vast majority of testicular neoplasms present as painless, self-detected masses, while a minority presents with scrotal pain [258]. Aggressive variants of GCT, such as CC, may present with symptoms related to metastatic disease [157], such as hemoptysis or neurological manifestations, before a testicular mass is detected. Some tumors may be detected incidentally, during workup for other symptoms, such as infertility [258].

Additionally, as stated before, a subset of tumors may present with endocrine manifestations, the most common being gynecomastia [100, 101, 259]. This symptom is usually associated with high serum hCG levels, and thus it is more common in tumors containing trophoblastic components, particularly CC, but also other forms containing syncytiotrophoblastic giant cells. Rarely, high β-hCG levels may produce thyrotoxicosis due to TSH-like activity of the hCG [153, 154, 260, 261]. Carcinoid syndrome is unusual in cases of testicular carcinoid [205, 212].

A variety of neurological paraneoplastic syndromes have been described in patients with testicular GCT, the most common one being Ma2 antibody-mediated limbic encephalitis [262–266]. Patients develop a constellation of symptoms that include short-term memory disturbance, epileptic seizures, acute confusional syndrome, personality change, hallucinations, depression, and cognition disturbances. Others present with symptoms more oriented to brainstem, cerebellar, or peripheral nerve dysfunction. The process is believed to be immune-mediated secondary to cytotoxic T cells attacking the neurons. Patients present with elevated levels of Ma antibodies directed against PNMA-2 proteins (also referred to as anti-Ta or anti-Ma2), which play an important role in apoptosis [262]. Neurological symptoms precede the detection of the tumor in the majority of patients, and about a third of them respond to appropriate treatment of the tumor [264]. Frequently, tumors associated with anti-Ma2 antibodies are small, or have undergone regression, as it is believe that the antibodies are a reflection of an effective immune response to the tumor [265–267]. It is still debated whether blind orchiectomy should be performed in symptomatic patients with no clinically documented tumor, as frequently an occult neoplasm may be found [264–266]. Before orchiectomy, clinical exclusion of an extragonadal tumor needs to be performed [268].

Other reported paraneoplastic manifestations of testicular GCT include dermatomyositis [269], polycythemia [270], hypercalcemia [271], membranous glomerulonephritis [272], and Raynaud's phenomenon [273].

7.10 Staging

The American Joint Committee on Cancer and the International Union Against Cancer staging for testicular neoplasms, widely known as the TNM staging system, is the most accepted staging system [274]. T categories are defined by the presence of the following parameters: the presence of invasive vs. in situ tumor, extent of invasive disease (whether tumor invades the spermatic cord, tunica vaginalis, or scrotum), and presence of lymphovascular invasion. Lymphovascular invasion has been extensively demonstrated as a significant prognostic factor, particularly in nonseminomatous GCT (Fig. 7.40) [127, 275–277].

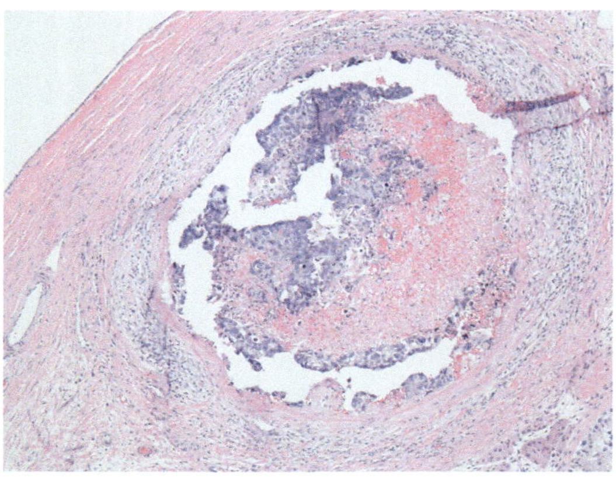

Fig. 7.40 Staging. pT2 disease, characterized by lymphovascular invasion by embryonal carcinoma

In pure seminomas, the association of lymphovascular invasion with relapse is less clear [122, 123, 126, 278] N categories are based on the number of nodes involved and size of metastatic deposits, and M categories depend on whether there is metastatic deposits and, if so, if these involve nonregional nodes, lung, or any other site. The TNM staging for testicular tumors is unique among other organs in that it incorporates serologic marker levels, establishing an additional category labeled S. The S category may thus place two tumors with identical T, N, or M categories into different stage groupings (Table 7.10).

Several shortcomings of the current TNM have been identified and are likely to be

Table 7.10 AJCC/ICCC TNM staging system (7th edition) for testicular GCT

Primary tumor (T):	
pTX	Primary tumor cannot be assessed
pT0	No evidence of primary tumor (regressed GCT)
pTis	Germ cell neoplasia in situ
pT1	Tumor limited to the testis and epididymis without vascular/lymphatic invasion. Tumor may invade into the tunica albuginea, but not tunica vaginalis
pT2	Tumor limited to the testis and epididymis with vascular/lymphatic invasion, or tumor extending through the tunica albuginea into the tunica vaginalis
pT3	Tumor invades the spermatic cord
pT4	Tumor invades the scrotum
Regional lymph nodes (N)	
Clinical	
NX	Regional lymph nodes cannot be assessed
N0	No regional lymph node metastasis
N1	Metastasis with a lymph node mass 2 cm or less in greatest dimension; or multiple lymph nodes, none more than 2 cm in greatest dimension
N2	Metastasis with a lymph node mass more than 2 cm, but not more than 5 cm in greatest dimension; or multiple lymph nodes, any one mass greater than 2 cm, but not more than 5 cm in greatest dimension
N3	Metastasis with a lymph node mass more than 5 cm in greatest dimension
Pathologic	
pNX	Regional lymph nodes cannot be assessed
pN0	No regional lymph node metastasis
pN1	Metastasis with a lymph node mass 2 cm or less in greatest dimension and less than or equal to five nodes positive, none more than 2 cm in greatest dimension
pN2	Metastasis with a lymph node mass more than 2 cm, but not more than 5 cm in greatest dimension; or more than five nodes positive, none more than 5 cm; or evidence of extranodal extension of tumor
pN3	Metastasis with a lymph node mass more than 5 cm in greatest dimension
Serum tumor markers (S)	
SX	Marker studies not available or not performed
S0	Marker study levels within normal limits
S1	LDH <1.5x normal *and* hCG <5000 mlu/ml *and* AFP <1000 ng/ml
S2	LDH 1.5 -10x normal *or* hCG 5000–50,000 mlu/ml *or* AFP 1000–10,000 ng/ml
S3	LDH >10x normal *or* hCG >50,000 ml/ml *or* AFP >10,000 ng/ml
Distant metastasis (M)	
M0	No distant metastasis
M1	Distant metastasis
M1a	Nonregional nodal or pulmonary metastasis
M1b	Distant metastasis other than to nonregional lymph nodes and lung

Table 7.10 (continued)

AJCC/ICCC anatomic stage/prognostic groups

Group	T	N	M	S
Stage 0	pTis	N0	M0	S0
Stage I	pT1–4	N0	M0	SX
Stage IA	pT1	N0	M0	S0
Stage IB	pT2–4	N0	M0	S0
Stage IS	Any pT	N0	M0	S1–3[a]
Stage II	Any pT	N1–3	M0	SX
Stage IIA	Any pT	N1	M0	S0–1
Stage IIB	Any pT	N2	M0	S0–1
Stage IIC	Any pT	N3	M0	S0–1
Stage III	Any pT	Any N	M1	SX
Stage IIIA	Any pT	Any N	M1a	S0–1
Stage IIIB	Any pT	N1–3	M0	S2
	Any pT	Any N	M1a	S2
Stage IIIC	Any pT	N1–3	M0	S3
	Any pT	Any N	M1a	S3
	Any pT	Any N	M1b	Any S

[a]Measured post-orchiectomy

Used with permission of the American Joint Committee on Cancer (AJCC), Chicago, Illinois. The original and primary source for this information is the AJCC Cancer Staging Manual, Seventh Edition (2010) published by Springer Science+Business Media

addressed in future updates of the system[1]. The size of the tumor is not currently included as a staging parameter. However, a review determined that seminomas larger than 4 cm were twice more likely to recur than smaller tumors [123]. Similarly, von der Maase et al. found a significantly higher relapse 4-year relapse rate in patients with tumors larger than 6 cm [279]. The rete testis invasion has also been associated with higher odds of recurrence (Fig. 7.41) [123, 277, 278]. In the same study by Warde et al., patients with tumor invading into the rete testis had a 1.7 higher risk of relapse. Other studies have not found significant differences in survival between tumors with and without the rete testis involvement [126, 280]. Attention should be paid to not equate pagetoid spread of GCNIS to the rete testis epithelium with direct invasion of the rete testis by invasive tumor (Figs. 7.2 and 7.41) [281]. Similarly, there is controversy regarding the adequate stage for a tumor that involves the hilar fat of the testis, but not the spermatic cord. Currently, this tumor should be staged as pT1 or pT2 (depending on the absence or presence of lymphovascular invasion). However, it clearly represents extratesticular extension of the tumor and likely a higher risk of recurrence [281, 282]. Other criticisms to the system include the disproportionate weight conferred to the invasion of the tunica vaginalis and the scrotum, which are relatively rare phenomena [282], and the lack of differentiation of the pattern of involvement of the spermatic cord, which may be involved by direct extension (Fig. 7.42) or lymphovascular invasion [281]. Given the different impact that some features have on seminoma and non-seminomatous germ cell tumors, it may be reasonable to question if the same staging system should be used for both types of tumors.

[1] Since the submission of this manuscript the 8th edition of the AJCC Cancer Staging Manual has been published, addressing some of the issues mentioned in this section. For more information please refer to AJCC Cancer Staging Manual, 8th edition. Amin M et al (ed). Springer 2017.

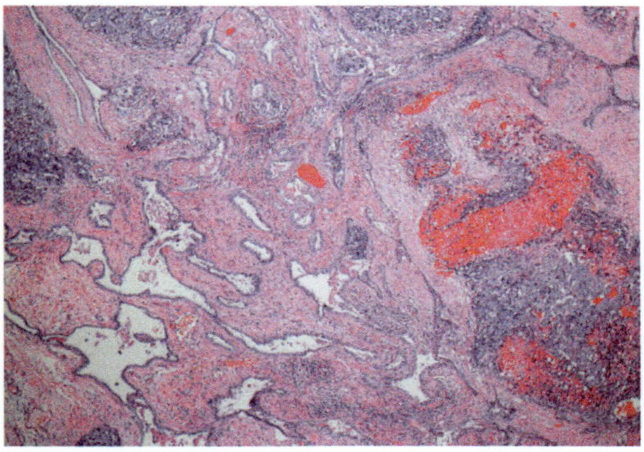

Fig. 7.41 Staging. Embryonal carcinoma demonstrating infiltration into the rete testis

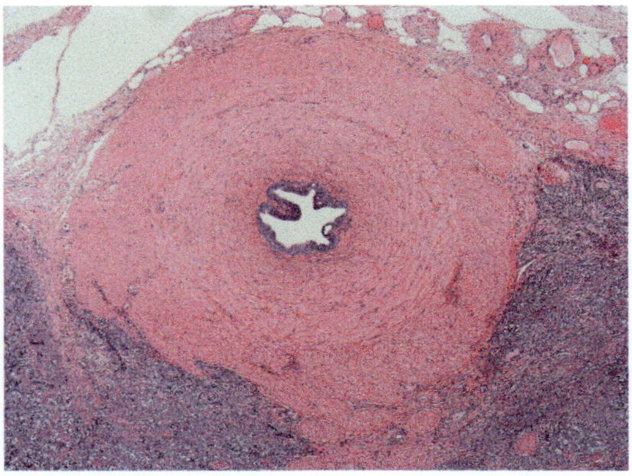

Fig. 7.42 Staging. Seminoma involving the spermatic cord represents pT3 disease

7.11 Handling of Orchiectomy Specimens

Radical orchiectomy specimens usually include the distal segment of the spermatic cord, with attached soft tissue, and the testis and epididymis covered by the parietal layer of the tunica vaginalis. Preferably, orchiectomy specimens should be inspected and at least initially handled in their fresh state. Measurements of the spermatic cord should be obtained. Attention should be paid to the external surface of the tunica vaginalis, to check for tumor invasion. Before opening the tunica vaginalis, the proximal end of the spermatic cord should be transversely sectioned and placed in a separate cassette, to avoid contamination with tumor cells once the tumor is cut [283]. The remaining spermatic cord should be serially sectioned, and any area suspicious for tumor should be submitted. If no gross abnormality is detected, representative sections of the mid and distal portion of the cord should be taken and separately submitted. The parietal layer should then be incised, and the inner lining of the visceral and parietal layers of the tunica vaginalis should be inspected. Any fluid should be noted and described, and measured and collected when appropriate. Attention should be paid to any nodules in the tunica, or evidence of tumor invasion. After recording the three measurements of the testis, this one should be bivalved by longitudinally sectioning from the outer surface toward the hilum. Any tumor should be measured and multifocality or extension to the hilum or rete testis, epididymis, tunica albuginea, or extratesticular soft tissues documented. Larger tumors may require parallel sections to assess for these changes. The cut surface of the tumor should be described, including features such as circumscription and lobulation and the presence of hemorrhage, necrosis, fleshy solid surfaces, cysts and their contents, or cartilaginous material. Sections for histologic examination should include all grossly different areas, including all hemorrhagic and necrotic ones. Because of the importance of finding even a small non-seminomatous component, tumors with the gross appearance of seminoma should be sampled extensively. While small tumors can be easily submitted entirely, larger tumors should be sampled at a minimum of one section per centimeter, usually erring on the side of generous sampling [281]. Additional sections should include those of non-neoplastic parenchyma, including close to and away from the tumor, and the relationship of the tumor with the rete testis, tunica albuginea, and extratesticular structures should be demonstrated in representative sections.

7.12 Pathology of Retroperitoneal Lymph Node Dissection

Retroperitoneal lymph node dissection (RPLND) is the most common surgical procedure for staging of testicular GCT. The rationale behind this is the predictable pattern of spread of testicular tumors, following the lymphatic drainage of the testis. This pattern is identical to all forms of GCT, with the possible exception of CC, which has a propensity to disseminate hematogenously. The primary "landing site" for right testicular tumors is the interaortocaval lymph nodes, followed by the precaval and paracaval nodes, while left testicular tumors tend to drain first into the preaortic and para-aortic lymph nodes, followed by the interaortocaval nodes. Contralateral spread is common in right testicular tumors, while rare in left-sided tumors, except in the setting of bulky disease [284, 285].

In general, RPLND is performed in one of the following settings [286]:

1. Primary RPLND in patients with clinical stage 1 or clinical stage 2 disease non-seminomatous GCT
2. Post-chemotherapy RPLND (PC-RPLND) in patients with a clinical stage 2 or higher non-seminomatous GCT
3. Post-chemotherapy RPLND in patients with seminoma with residual masses
4. Salvage RPLND in the setting of recurrences after any modality of management of non-seminomatous GCT

The above settings are derived from the different management approaches to GCT at different stages, which are extensively discussed in Chap. 5.

7.12.1 Pathologic Findings in Primary RPLND Specimens

The surgical pathologist plays a significant role in defining the staging and management of GCT. Approximately 30 % of clinical stage I testicular GCT will have metastatic disease detected after histopathologic examination of an RPLND specimen [287, 288], or upon recurrence at this site during surveillance [276, 289, 290]. Because of this, management approach of clinical stage I non-seminomatous GCT is still controversial, with primary RPLND being one of the three options available, the other two being active surveillance and one adjuvant cycle of BEP chemotherapy. Advantages of the adjuvant RPLND approach include an accurate staging of the retroperitoneum, marked reduction in the number of cycles and amount of chemotherapy needed long term, the use of chemotherapy exclusively on patients with documented metastatic disease, decrease need of imaging studies for follow-up, and removal of teratoma elements from the retroperitoneum. Disadvantages include exposure to surgery for a large proportion of patients that will not have metastatic disease, risk of retrograde ejaculation and other less common morbidities, need of expert surgeons, and slightly higher levels of relapse compared to BEP (8 % vs. 3 %) [291].

In a recent series [292], the most common histology found in the setting of positive primary RPLND was EC. Out of 183 patients with pathologically positive RPLND, 160 contained this element, and in 99 of those EC was the only element found. YST was present in 50 cases, while seminoma was present in 11 (8 of which had a pure seminoma component). Teratoma was present in 44 cases, but was pure only in 12.

Tumor histologic type does not appear to predict which patients are going to relapse [292, 293]. Other histologic findings, such as extra nodal extension, have been suggested as predictive of higher rates of relapse by some series [294, 295], but not by others [292, 293]. While number of lymph nodes involved and metastatic tumor size define stage and thus impact prognosis and therapy, within the category of pN1 a number or ratio of positive lymph nodes do not appear to confer significant prognostic information [296].

7.12.2 Pathologic Findings in Post-chemotherapy RPLND Specimens

RPLND may be performed in patients that have had previous chemotherapy, usually patients with high-stage clinical disease (N2 or N3). Most RPLND are performed after two cycles of chemotherapy, although it is not uncommon to have it done after three or four cycles [297]. In the post-chemotherapy setting, the histologic findings are not only useful in the staging of the disease but also provide an idea of the response of the tumor to therapy (Table 7.11). Therapy response and volume of residual disease may impact incidence of relapse in these patients.

Three basic histologic patterns may be seen in PC-RPLND specimens:

1. Evidence of tumor response: Usually in the form of necrosis, but also as granulomatous inflammation or fibrosis (Fig. 7.43). However, recent data revealing similar molecular changes between fibrosis and adjacent mature teratoma suggests that fibrosis may actually correspond to residual teratoma and thus should not be considered evidence of tumor response [298].
2. Residual GCT, non-teratomatous types: These include the presence of histologically viable tumor, with EC, seminoma, YST, or CC morphology (Fig. 7.44).
3. Residual teratomatous elements, histologically viable (Fig. 7.45). The rationale for separating teratomatous and non-teratomatous elements resides in the fact that teratoma response to chemotherapy is not expected, and thus, the presence of teratoma does not imply ineffective or insufficient chemotherapy. Surgical removal of teratoma elements is required, as teratomas may grow and impinge on vital structures ("growing teratoma syndrome") [145, 146] or may undergo development of somatic malignancy [135]. The presence of teratoma and YST in the primary tumor is a predictor of teratoma in PC-RPLND specimen [168].

Table 7.11 Pathologic findings in post-chemotherapy RPLND specimens

Evidence of tumor regression:
Fibrosis
Necrosis
Granulomatous inflammation
Residual viable non-teratomatous tumor:
Embryonal carcinoma
Yolk sac tumor
Seminoma
Choriocarcinoma
Mixed GCT
Residual teratoma
Residual special histologies
Somatic-type malignancy
Rhabdomyomatous differentiated tumor
Cystic trophoblastic tumor

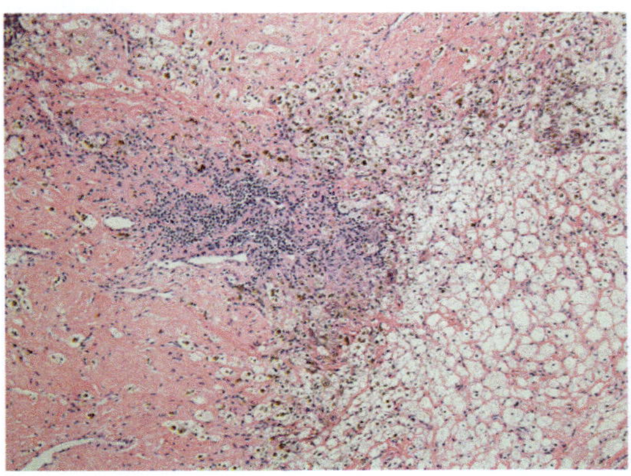

Fig. 7.43 Retroperitoneal lymph node dissection (RPLND). RPLND shows a lymph node with extensive treatment effect characterized by the presence of fibrosis, diffuse histiocytic infiltrates, and focal hemosiderin deposition

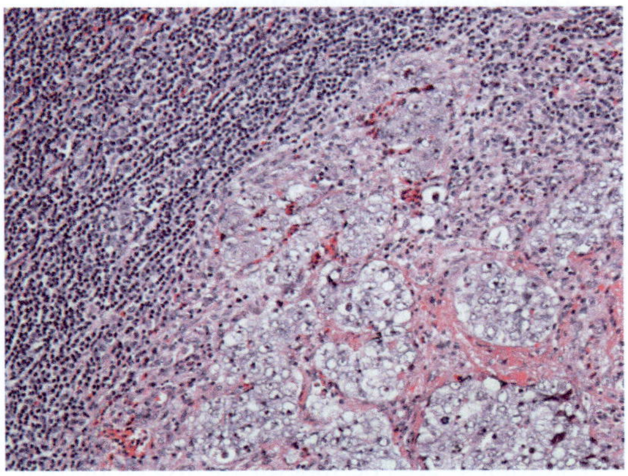

Fig. 7.44 Retroperitoneal lymph node dissection (RPLND). RPLND shows a lymph node with residual embryonal carcinoma

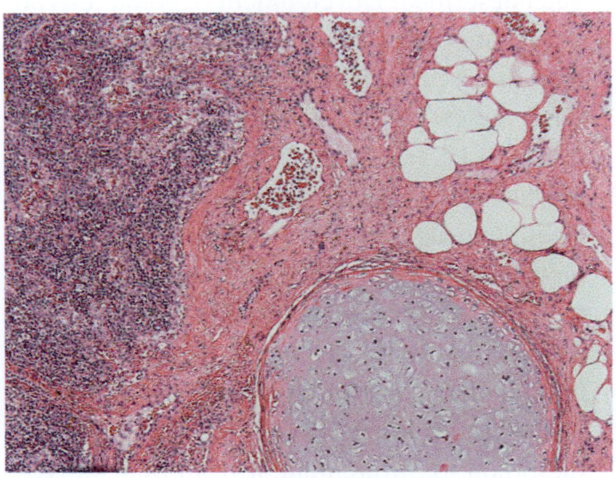

Fig. 7.45 Retroperitoneal lymph node dissection (RPLND). RPLND shows a lymph node with only histologically viable teratoma within a lymph node

In a series from Indiana University of 71 patients that underwent PC-RPLND, fibrosis was found in 51 % of cases, teratoma-only was found in 21 % of cases, and residual viable non-teratoma GCT in 28 % of cases. A 5-year survival for both the fibrosis and teratoma categories was 87 %, compared to 47 % 5-year survival for those with residual viable non-teratoma GCT [297]. In another series of PC-RPLND, this time in patients with pure seminoma associated with elevated levels of AFP, Peterson et al. found viable non-teratoma GCT in 37.5 %, teratoma only in 12 (30 %), and necrosis/fibrosis only in 13 (32.5 %). The histologies of the cases with residual viable non-teratoma GCT included seminoma, EC, YST, sarcoma, and mixed GCT [63].

Post-chemotherapy specimens also may harbor unusual GCT histologies, which the surgical pathologist needs to recognize for appropriate diagnosis and management. These include histologic entities whose development is likely induced or at least potentiated by chemotherapy:

1. *Somatic-type malignancy*: While the development of somatic-type malignancy is not a phenomenon exclusively found in post-chemotherapy specimens, it is more commonly seen in this setting [135]. The most common histologies are peripheral neuroectodermal tumors, rhabdomyosarcoma and other sarcomas, and adenocarcinomas; however, a wide spectrum of somatic histologic types has been described. Development of somatic-type malignancy confers a poor prognosis, particularly in the post-chemotherapy setting. A thorough review of this phenomenon is presented in Chap. 12.
2. *Rhabdomyomatous differentiated tumor*: These unusual tumors, which are explained in larger detailed in Chap. 12, correspond to aggregates of benign terminally differentiated rhabdomyoblasts, in the setting of teratomatous elements, and found in post-chemotherapy RPLND specimens [299]. Its recognition is important to differentiate it from rhabdomyosarcoma arising in teratoma as a somatic-type malignancy.
3. *Cystic trophoblastic tumor (CTT)*: These tumors are characterized by cysts of variable sizes, lined by cells with abundant cytoplasm, smudged nuclei, occasional multinucleation, and cytoplasmic lacunae (Fig. 7.46) [300] (see also Table 7.7). None to minimal mitotic activity is usually seen. The cells are arranged in one or several layers of flattened epithelium, with occasional tufting or micropapillae. Fibrinous material is present in the lumen of the cysts. Immunohistochemical studies show focal β-hCG reactivity. CTT is usually associated only with teratoma elements. Follow-up reveals a relapse rate comparable to that of RPLND specimens with only teratoma elements, in contrast with residual CC, which is associated with high relapse rate and requires additional chemotherapy. Serum β-hCG is only mildly elevated in a few cases. It is thought that CTT represents another form of cytodifferentiation induced by chemotherapy, in this case from original trophoblastic elements. Accurate recognition of this type of tumor is important to avoid overtreatment given to PC-RPLND specimens that show residual CC. Attention should be paid to the lack of biphasic pattern characteristic of CC and the lack of mitotic activity, necrosis, and hemorrhagic background.

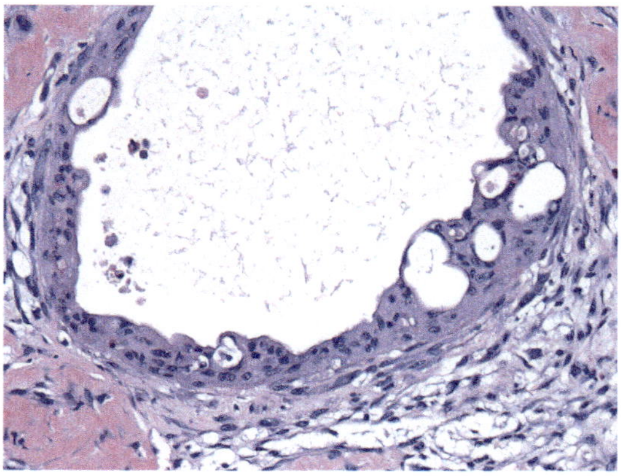

Fig. 7.46 Cystic trophoblastic tumor (CTT). Cystic lesion lined by several layers of flattened epithelium with scant fibrinous material in the lumen. Individual cells have abundant cytoplasm, lack mitotic activity, and show occasional cytoplasmic lacunae (Image courtesy of Gladell Paner, MD, University of Chicago, Chicago, IL)

References

1. McGlynn KA, Devesa SS, Sigurdson AJ, Brown LM, Tsao L, Tarone RE. Trends in the incidence of testicular germ cell tumors in the United States. Cancer. 2003;97(1):63–70. doi:10.1002/cncr.11054.
2. Siegel R, Ma J, Zou Z, Jemal A. Cancer statistics, 2014. CA Cancer J Clin. 2014;64(1):9–29. doi:10.3322/caac.21208.
3. Townsend JS, Richardson LC, German RR. Incidence of testicular cancer in the United States, 1999-2004. Am J Mens Health. 2010;4(4):353–60. doi:10.1177/1557988309356101.
4. Pike MC, Chilvers CE, Bobrow LG. Classification of testicular cancer in incidence and mortality statistics. Br J Cancer. 1987;56(1):83–5.
5. Purdue MP, Devesa SS, Sigurdson AJ, McGlynn KA. International patterns and trends in testis cancer incidence. Int J Cancer. 2005;115(5):822–7. doi:10.1002/ijc.20931.
6. Davies JM. Testicular cancer in England and Wales: some epidemiological aspects. Lancet. 1981;1(8226): 928–32.
7. Moller H, Skakkebaek NE. Risks of testicular cancer and cryptorchidism in relation to socio-economic status and related factors: case-control studies in Denmark. Int J Cancer. 1996;66(3):287–93. doi:10.1002/(SICI)1097-0215(19960503)66:3<287::AID-IJC2>3.0.CO;2-V.
8. Pearce N, Sheppard RA, Howard JK, Fraser J, Lilley BM. Time trends and occupational differences in cancer of the testis in New Zealand. Cancer. 1987;59(9):1677–82.
9. Swerdlow AJ, Skeet RG. Occupational associations of testicular cancer in south east England. Br J Ind Med. 1988;45(4):225–30.
10. Leibovitch I, Baniel J, Rowland RG, Smith Jr ER, Ludlow JK, Donohue JP. Malignant testicular neoplasms in immunosuppressed patients. J Urol. 1996;155(6):1938–42.
11. Social, behavioural and medical factors in the aetiology of testicular cancer: results from the UK study. UK Testicular Cancer Study Group. Br J Cancer. 1994;70(3):513–20.
12. Swerdlow AJ, Huttly SR, Smith PG. Testicular cancer and antecedent diseases. Br J Cancer. 1987;55(1): 97–103.
13. Depue RH, Pike MC, Henderson BE. Estrogen exposure during gestation and risk of testicular cancer. J Natl Cancer Inst. 1983;71(6):1151–5.
14. Gurney J, Shaw C, Stanley J, Signal V, Sarfati D. Cannabis exposure and risk of testicular cancer: a systematic review and meta-analysis. BMC Cancer. 2015;15(1):897. doi:10.1186/s12885-015-1905-6.
15. Trabert B, Sigurdson AJ, Sweeney AM, Strom SS, McGlynn KA. Marijuana use and testicular germ cell tumors. Cancer. 2011;117(4):848–53. doi:10.1002/cncr.25499.
16. Dieckmann KP, Rube C, Henke RP. Association of Down's syndrome and testicular cancer. J Urol. 1997;157(5):1701–4.
17. Dexeus FH, Logothetis CJ, Chong C, Sella A, Ogden S. Genetic abnormalities in men with germ cell tumors. J Urol. 1988;140(1):80–4.
18. Sogge MR, McDonald SD, Cofold PB. The malignant potential of the dysgenetic germ cell in Klinefelter's syndrome. Am J Med. 1979;66(3):515–8.
19. Trabert B, Sigurdson AJ, Sweeney AM, Amato RJ, Strom SS, McGlynn KA. Baldness, acne and testicular germ cell tumours. Int J Androl. 2011;34(4 Pt 2):e59–67. doi:10.1111/j.1365-2605.2010.01125.x.
20. Taran I, Elder JS. Results of orchiopexy for the undescended testis. World J Urol. 2006;24(3):231–9. doi:10.1007/s00345-006-0056-4.
21. Thorup J, McLachlan R, Cortes D, Nation TR, Balic A, Southwell BR, et al. What is new in cryptorchidism and hypospadias – a critical review on the testicular dysgenesis hypothesis. J Pediatr Surg. 2010;45(10):2074–86. doi:10.1016/j.jpedsurg.2010.07.030.
22. Pettersson A, Richiardi L, Nordenskjold A, Kaijser M, Akre O. Age at surgery for undescended testis and risk of testicular cancer. N Engl J Med. 2007;356(18):1835–41. doi:10.1056/NEJMoa067588.
23. Myrup C, Schnack TH, Wohlfahrt J. Correction of cryptorchidism and testicular cancer. N Engl J Med. 2007;357(8):825–7. author reply -7 doi:10.1056/NEJMc071510.
24. Miller A, Seljelik R. Histopathologic classification and natural history of malignant testis tumors in Norway, 1959–1963. Cancer. 1971;28(4):1054–62.
25. Colls BM, Harvey VJ, Skelton L, Thompson PI, Frampton CM. Bilateral germ cell testicular tumors in New Zealand: experience in Auckland and Christchurch 1978–1994. J Clin Oncol. 1996;14(7):2061–5.
26. Osterlind A, Berthelsen JG, Abildgaard N, Hansen SO, Hjalgrim H, Johansen B, et al. Risk of bilateral testicular germ cell cancer in Denmark: 1960–1984. J Natl Cancer Inst. 1991;83(19):1391–5.
27. Fossa SD, Chen J, Schonfeld SJ, McGlynn KA, McMaster ML, Gail MH, et al. Risk of contralateral testicular cancer: a population-based study of 29,515 U.S. men. J Natl Cancer Inst. 2005;97(14):1056–66. doi:10.1093/jnci/dji185.
28. Hemminki K, Chen B. Familial risks in testicular cancer as aetiological clues. Int J Androl. 2006;29(1):205–10. doi:10.1111/j.1365-2605.2005.00599.x.
29. Sonneveld DJ, Sleijfer DT, Schrafford Koops H, Sijmons RH, van der Graaf WT, Sluiter WJ, et al. Familial testicular cancer in a single-centre population. Eur J Cancer. 1999;35(9):1368–73.
30. Westergaard T, Olsen JH, Frisch M, Kroman N, Nielsen JW, Melbye M. Cancer risk in fathers and brothers of testicular cancer patients in Denmark. A population-based study. Int J Cancer. 1996;66(5):627–31. doi:10.1002/(SICI)1097-0215(19960529)66:5<627::AID-IJC8>3.0.CO;2-V.

31. Rapley EA, Crockford GP, Teare D, Biggs P, Seal S, Barfoot R, et al. Localization to Xq27 of a susceptibility gene for testicular germ-cell tumours. Nat Genet. 2000;24(2):197–200. doi:10.1038/72877.
32. Kanetsky PA, Mitra N, Vardhanabhuti S, Li M, Vaughn DJ, Letrero R, et al. Common variation in KITLG and at 5q31.3 predisposes to testicular germ cell cancer. Nat Genet. 2009;41(7):811–5. doi:10.1038/ng.393.
33. Ferlin A, Pengo M, Pizzol D, Carraro U, Frigo AC, Foresta C. Variants in KITLG predispose to testicular germ cell cancer independently from spermatogenic function. Endocr Relat Cancer. 2012;19(1):101–8. doi:10.1530/ERC-11-0340.
34. Karlsson R, Andreassen KE, Kristiansen W, Aschim EL, Bremnes RM, Dahl O, et al. Investigation of six testicular germ cell tumor susceptibility genes suggests a parent-of-origin effect in SPRY4. Hum Mol Genet. 2013;22(16):3373–80. doi:10.1093/hmg/ddt188.
35. Kratz CP, Han SS, Rosenberg PS, Berndt SI, Burdett L, Yeager M, et al. Variants in or near KITLG, BAK1, DMRT1, and TERT-CLPTM1L predispose to familial testicular germ cell tumour. J Med Genet. 2011; 48(7):473–6. doi:10.1136/jmedgenet-2011-100001.
36. Kanetsky PA, Mitra N, Vardhanabhuti S, Vaughn DJ, Li M, Ciosek SL, et al. A second independent locus within DMRT1 is associated with testicular germ cell tumor susceptibility. Hum Mol Genet. 2011;20(15):3109–17. doi:10.1093/hmg/ddr207.
37. Turnbull C, Rapley EA, Seal S, Pernet D, Renwick A, Hughes D, et al. Variants near DMRT1, TERT and ATF7IP are associated with testicular germ cell cancer. Nat Genet. 2010;42(7):604–7. doi:10.1038/ng.607.
38. Ferguson L, Agoulnik AI. Testicular cancer and cryptorchidism. Front Endocrinol (Lausanne). 2013;4:32. doi:10.3389/fendo.2013.00032.
39. Rutgers JL, Scully RE. Pathology of the testis in intersex syndromes. Semin Diagn Pathol. 1987;4(4): 275–91.
40. Cools M, Drop SL, Wolffenbuttel KP, Oosterhuis JW, Looijenga LH. Germ cell tumors in the intersex gonad: old paths, new directions, moving frontiers. Endocr Rev. 2006;27(5):468–84. doi:10.1210/er.2006-0005.
41. Carlsen E, Giwercman A, Keiding N, Skakkebaek NE. Evidence for decreasing quality of semen during past 50 years. BMJ. 1992;305(6854):609–13.
42. Mancini M, Carmignani L, Gazzano G, Sagone P, Gadda F, Bosari S, et al. High prevalence of testicular cancer in azoospermic men without spermatogenesis. Hum Reprod. 2007;22(4):1042–6. doi:10.1093/humrep/del500.
43. Raman JD, Nobert CF, Goldstein M. Increased incidence of testicular cancer in men presenting with infertility and abnormal semen analysis. J Urol. 2005;174(5):1819–22. discussion 22 doi:10.1097/01.ju.0000177491.98461.aa.
44. Skakkebaek NE, Jorgensen N. Testicular dysgenesis and fertility. Andrologia. 2005;37(6):217–8. doi:10.1111/j.1439-0272.2005.00700.x.
45. Eisenberg ML, Li S, Brooks JD, Cullen MR, Baker LC. Increased risk of cancer in infertile men: analysis of U.S. claims data. J Urol. 2015;193(5):1596–601. doi:10.1016/j.juro.2014.11.080.
46. Giwercman A, Muller J, Skakkebaek NE. Carcinoma in situ of the undescended testis. Semin Urol. 1988;6(2):110–9.
47. Skakkebaek NE, Berthelsen JG, Visfeldt J. Clinical aspects of testicular carcinoma in situ. Int J Andrology. 1981;4(Suppl):153–60.
48. Nistal M, Codesal J, Paniagua R. Carcinoma in situ of the testis in infertile men. A histological, immunocytochemical, and cytophotometric study of DNA content. J Pathol. 1989;159(3):205–10. doi:10.1002/path.1711590306.
49. Pryor JP, Cameron KM, Chilton CP, Ford TF, Parkinson MC, Sinokrot J, et al. Carcinoma in situ in testicular biopsies from men presenting with infertility. Br J Urol. 1983;55(6):780–4.
50. Berthelsen JG, Skakkebaek NE. Value of testicular biopsy in diagnosing carcinoma in situ testis. Scand J Urol Nephrol. 1981;15(3):165–8. doi:10.3109/00365598109179596.
51. von der Maase H, Rorth M, Walbom-Jorgensen S, Sorensen BL, Christophersen IS, Hald T, et al. Carcinoma in situ of contralateral testis in patients with testicular germ cell cancer: study of 27 cases in 500 patients. Br Med J (Clin Res Ed). 1986;293(6559):1398–401.
52. von der Maase H. Is a contralateral testicular biopsy in patients with unilateral germ cell testicular cancer indicated as a routine procedure? Acta Oncol. 2005;44(6):523–5. doi:10.1080/02841860500251562.
53. Classen J, Dieckmann KP, Loy V, Bamberg M. Testicular intraepithelial neoplasms (TIN). An indication for radiotherapy? Strahlenther Onkol. 1998;174(4):173–7.
54. Heidenreich A. Contralateral testicular biopsy in testis cancer: current concepts and controversies. BJU Int. 2009;104(9 Pt B):1346–50. doi:10.1111/j.1464-410X.2009.08857.x.
55. Oosterhuis JW, Looijenga LH. Testicular germ-cell tumours in a broader perspective. Nat Rev Cancer. 2005;5(3):210–22. doi:10.1038/nrc1568.
56. Reuter VE. Origins and molecular biology of testicular germ cell tumors. Mod Pathol. 2005;18(Suppl 2):S51–60. doi:10.1038/modpathol.3800309.
57. Miyai K, Yamamoto S, Iwaya K, Asano T, Tamai S, Tsuda H, et al. Allelotyping analysis suggesting a consecutive progression from intratubular germ cell neoplasia to seminoma and then to embryonal carcinoma of the adult testis. Hum Pathol. 2013;44(10):2312–22. doi:10.1016/j.humpath.2013.05.013.
58. Al-Hussain T, Bakshi N, Akhtar M. Intratubular germ cell neoplasia of the testis: a brief review. Adv Anat Pathol. 2015;22(3):202–12. doi:10.1097/pap.0000000000000066.
59. Emerson RE, Ulbright TM. Intratubular germ cell neoplasia of the testis and its associated cancers: the

use of novel biomarkers. Pathology. 2010;42(4):344–55. doi:10.3109/00313021003767355.
60. Berney DM, Lee A, Randle SJ, Jordan S, Shamash J, Oliver RT. The frequency of intratubular embryonal carcinoma: implications for the pathogenesis of germ cell tumours. Histopathology. 2004;45(2):155–61. doi:10.1111/j.1365-2559.2004.01915.x.
61. Berney DM, Lee A, Shamash J, Oliver RT. The association between intratubular seminoma and invasive germ cell tumors. Hum Pathol. 2006;37(4):458–61. doi:10.1016/j.humpath.2005.12.007.
62. Lau SK, Weiss LM, Chu PG. Association of intratubular seminoma and intratubular embryonal carcinoma with invasive testicular germ cell tumors. Am J Surg Pathol. 2007;31(7):1045–9. doi:10.1097/PAS.0b013e31802b8712.
63. Peterson M, Beck S, Bihrle R, Einhorn L, Foster R. Results of retroperitoneal lymph node dissection after chemotherapy in patients with pure seminoma in the orchidectomy specimen but elevated serum alpha-fetoprotein. BJU Int. 2009;104(2):176–8. doi:10.1111/j.1464-410X.2009.08697.x.
64. Srigley JR, Mackay B, Toth P, Ayala A. The ultrastructure and histogenesis of male germ neoplasia with emphasis on seminoma with early carcinomatous features. Ultrastruct Pathol. 1988;12(1):67–86.
65. Czaja JT, Ulbright TM. Evidence for the transformation of seminoma to yolk sac tumor, with histogenetic considerations. Am J Clin Pathol. 1992;97(4):468–77.
66. Hittmair A, Rogatsch H, Hobisch A, Mikuz G, Feichtinger H. CD30 expression in seminoma. Hum Pathol. 1996;27(11):1166–71.
67. Cheville JC, Rao S, Iczkowski KA, Lohse CM, Pankratz VS. Cytokeratin expression in seminoma of the human testis. Am J Clin Pathol. 2000;113(4):583–8. doi:10.1309/3QLC-5MF1-JYXU-A5XX.
68. Willis GW, Hajdu SI. Histologically benign teratoid metastasis of testicular embryonal carcinoma: report of five cases. Am J Clin Pathol. 1973;59(3):338–43.
69. Moran CA, Travis WD, Carter D, Koss MN. Metastatic mature teratoma in lung following testicular embryonal carcinoma and teratocarcinoma. Arch Pathol Lab Med. 1993;117(6):641–4.
70. Kleinsmith LJ, Pierce Jr GB. Multipotentiality of Single Embryonal Carcinoma Cells. Cancer Res. 1964;24:1544–51.
71. Pierce GB, Dixon Jr FJ. Testicular teratomas. I. Demonstration of teratogenesis by metamorphosis of multipotential cells. Cancer. 1959;12(3):573–83.
72. Looijenga LH, Zafarana G, Grygalewicz B, Summersgill B, Debiec-Rychter M, Veltman J, et al. Role of gain of 12p in germ cell tumour development. APMIS. 2003;111:161–71. discussion 72-3
73. Chaganti RS, Houldsworth J. Genetics and biology of adult human male germ cell tumors. Cancer Res. 2000;60(6):1475–82.
74. Rosenberg C, Van Gurp RJ, Geelen E, Oosterhuis JW, Looijenga LH. Overrepresentation of the short arm of chromosome 12 is related to invasive growth of human testicular seminomas and nonseminomas. Oncogene. 2000;19(51):5858–62.
75. Ottesen AM, Skakkebaek NE, Lundsteen C, Leffers H, Larsen J, Rajpert-De ME. High-resolution comparative genomic hybridization detects extra chromosome arm 12p material in most cases of carcinoma in situ adjacent to overt germ cell tumors, but not before the invasive tumor development. Genes Chromosomes Cancer. 2003;38(2):117–25. doi:10.1002/gcc.10244.
76. Moch H, Humphrey P, Ubright TM, Reuter VE. WHO classification of tumours of the urinary system and male genital organs. 4th ed. World Health Organization Classification of Tumours. Lyon: IARC; 2016.
77. Skakkebaek NE. Possible carcinoma-in-situ of the testis. Lancet. 1972;2(7776):516–7.
78. Jacobsen GK, Henriksen OB, von der Maase H. Carcinoma in situ of testicular tissue adjacent to malignant germ-cell tumors: a study of 105 cases. Cancer. 1981;47(11):2660–2.
79. Skakkebaek NE. Atypical germ cells in the adjacent "normal" tissue of testicular tumours. Acta Pathol Microbiol Scand A. 1975;83(1):127–30.
80. Coffin CM, Ewing S, Dehner LP. Frequency of intratubular germ cell neoplasia with invasive testicular germ cell tumors. Histologic and immunocytochemical features. Arch Pathol Lab Med. 1985;109(6):555–9.
81. Soosay GN, Bobrow L, Happerfield L, Parkinson MC. Morphology and immunohistochemistry of carcinoma in situ adjacent to testicular germ cell tumours in adults and children: implications for histogenesis. Histopathology. 1991;19(6):537–44.
82. Ulbright TM, Tickoo SK, Berney DM, Srigley JR. Best practices recommendations in the application of immunohistochemistry in testicular tumors: report from the International Society of Urological Pathology consensus conference. Am J Surg Pathol. 2014;38(8):e50–9. doi:10.1097/PAS.0000000000000233.
83. Rajpert-De Meyts E, Kvist M, Skakkebaek NE. Heterogeneity of expression of immunohistochemical tumour markers in testicular carcinoma in situ: pathogenetic relevance. Virchows Arch. 1996;428(3):133–9.
84. Perry A, Wiley EL, Albores-Saavedra J. Pagetoid spread of intratubular germ cell neoplasia into rete testis: a morphologic and histochemical study of 100 orchiectomy specimens with invasive germ cell tumors. Hum Pathol. 1994;25(3):235–9.
85. Khamu TT, Strutton GM, Kiosoglous A. Pagetoid involvement of vasitis nodosa by intratubular germ cell neoplasia unclassified. Pathology. 2010;42(7):690–2. doi:10.3109/00313025.2010.523688.
86. Ulbright TM, Young RH. Tumors of the testis and adjacent structures. In: AFIP atlas of tumor pathology. 4th ed. Silver Spring: AFP Press; 2013.
87. Mostofi FK, Sesterhenn IA. Pathology of germ cell tumors of testes. Prog Clin Biol Res. 1985;203:1–34.

88. Ulbright TM, Young RH. Metastatic carcinoma to the testis: a clinicopathologic analysis of 26 nonincidental cases with emphasis on deceptive features. Am J Surg Pathol. 2008;32(11):1683–93. doi:10.1097/PAS.0b013e3181788516.
89. Cools M, van Aerde K, Kersemaekers AM, Boter M, Drop SL, Wolffenbuttel KP, et al. Morphological and immunohistochemical differences between gonadal maturation delay and early germ cell neoplasia in patients with undervirilization syndromes. J Clin Endocrinol Metab. 2005;90(9):5295–303. doi:10.1210/jc.2005-0139.
90. Nistal M, Gonzalez-Peramato P, Regadera J, Serrano A, Tarin V, De Miguel MP. Primary testicular lesions are associated with testicular germ cell tumors of adult men. Am J Surg Pathol. 2006;30(10):1260–8. doi:10.1097/01.pas.0000213361.10756.08.
91. Pauls K, Schorle H, Jeske W, Brehm R, Steger K, Wernert N, et al. Spatial expression of germ cell markers during maturation of human fetal male gonads: an immunohistochemical study. Hum Reprod. 2006;21(2):397–404. doi:10.1093/humrep/dei325.
92. Sonne SB, Almstrup K, Dalgaard M, Juncker AS, Edsgard D, Ruban L, et al. Analysis of gene expression profiles of microdissected cell populations indicates that testicular carcinoma in situ is an arrested gonocyte. Cancer Res. 2009;69(12):5241–50. doi:10.1158/0008-5472.CAN-08-4554.
93. Sheikine Y, Genega E, Melamed J, Lee P, Reuter VE, Ye H. Molecular genetics of testicular germ cell tumors. Am J Cancer Res. 2012;2(2):153–67.
94. Donner J, Kliesch S, Brehm R, Bergmann M. From carcinoma in situ to testicular germ cell tumour. APMIS. 2004;112(2):79–88. doi:10.1111/j.1600-0463.2004.apm1120201.x.
95. Krag Jacobsen G, Barlebo H, Olsen J, Schultz HP, Starklint H, Sogaard H, et al. Testicular germ cell tumours in Denmark 1976–1980. Pathology of 1058 consecutive cases. Acta Radiol Oncol. 1984;23(4):239–47.
96. Berney DM, Warren AY, Verma M, Kudahetti S, Robson JM, Williams MW, et al. Malignant germ cell tumours in the elderly: a histopathological review of 50 cases in men aged 60 years or over. Mod Pathol Off J U S Can Acad Pathol, Incl. 2008;21(1):54–9. doi:10.1038/modpathol.3800978.
97. Holzbeierlein JM, Sogani PC, Sheinfeld J. Histology and clinical outcomes in patients with bilateral testicular germ cell tumors: the Memorial Sloan Kettering Cancer Center experience 1950 to 2001. J Urol. 2003;169(6):2122–5. doi:10.1097/01.ju.0000067462.24562.8b.
98. Che M, Tamboli P, Ro JY, Park DS, Ro JS, Amato RJ, et al. Bilateral testicular germ cell tumors: twenty-year experience at M. D. Anderson Cancer Center. Cancer. 2002;95(6):1228–33. doi:10.1002/cncr.10804.
99. Nazeer T, Ro JY, Amato RJ, Park YW, Ordonez NG, Ayala AG. Histologically pure seminoma with elevated alpha-fetoprotein: a clinicopathologic study of ten cases. Oncol Rep. 1998;5(6):1425–9.
100. Yamashiro T, Iraha Y, Kamiya H, Nakayama T, Unten S, Murayama S. Testicular seminoma presenting with mediastinal lymphadenopathy and gynecomastia. Radiat Med. 2007;25(6):303–5. doi:10.1007/s11604-007-0137-3.
101. Kaptanis S, Parvanta L, Beltran L. Testicular seminoma presenting as unilateral gynecomastia. Breast J. 2014;20(4):424–6. doi:10.1111/tbj.12287.
102. Ulbright TME, RM. Neoplasms of the testis. In: Bostwick DCL, editor. Urologic surgical pathology. 3rd ed. Philadelphia: Elsevier; 2014. p. 762–772.
103. Percarpio B, Clements JC, McLeod DG, Sorgen SD, Cardinale FS. Anaplastic seminoma: an analysis of 77 patients. Cancer. 1979;43(6):2510–3.
104. Zuckman MH, Williams G, Levin HS. Mitosis counting in seminoma: an exercise of questionable significance. Hum Pathol. 1988;19(3):329–35.
105. Tickoo SK, Hutchinson B, Bacik J, Mazumdar M, Motzer RJ, Bajorin DF, et al. Testicular seminoma: a clinicopathologic and immunohistochemical study of 105 cases with special reference to seminomas with atypical features. Int J Surg Pathol. 2002;10(1):23–32.
106. Ulbright TMA, Mahul B, Young RH. Tumors of the testis, adnexa, spermatic cord and scrotum. Atlas of tumor pathology. Washington, DC: Armed Forces Institute of Pathology; 1999.
107. Zhou M, Netto G, Epstein J. Uropathology: high-yield pathology. 1st ed. Philadelphia: Elsevier Saunders; 2012.
108. von Hochstetter AR, Sigg C, Saremaslani P, Hedinger C. The significance of giant cells in human testicular seminomas. A clinico-pathological study. Virchows Arch A Pathol Anat Histopathol. 1985;407(3):309–22.
109. Hedinger C, von Hochstetter AR, Egloff B. Seminoma with syncytiotrophoblastic giant cells. A special form of seminoma. Virchows Arch A Pathol Anat Histol. 1979;383(1):59–67.
110. Takeshima Y, Sanda N, Yoneda K, Inai K. Tubular seminoma of the testis. Pathol Int. 1999;49(7):676–9.
111. Talerman A. Tubular seminoma. Arch Pathol Lab Med. 1989;113(11):1204.
112. Ulbright TM, Young RH. Seminoma with tubular, microcystic, and related patterns: a study of 28 cases of unusual morphologic variants that often cause confusion with yolk sac tumor. Am J Surg Pathol. 2005;29(4):500–5.
113. Young RH, Finlayson N, Scully RE. Tubular seminoma. Report of a case. Arch Pathol Lab Med. 1989;113(4):414–6.
114. Zavala-Pompa A, Ro JY, AK e-N, MB A, NG O, Sella A, et al. Tubular seminoma. An immunohistochemical and DNA flow-cytometric study of four cases. Am J Clin Pathol. 1994;102(4):397–401.
115. Ulbright TM, Young RH. Seminoma with conspicuous signet ring cells: a rare, previously uncharacterized morphologic variant. Am J Surg Pathol. 2008;32(8):1175–81. doi:10.1097/PAS.0b013e31816597b0.

116. Henley JD, Young RH, Wade CL, Ulbright TM. Seminomas with exclusive intertubular growth: a report of 12 clinically and grossly inconspicuous tumors. Am J Surg Pathol. 2004;28(9):1163–8.
117. Bernstein ML, Miller RL, Steiner M. The differentiation of malignant lymphoma from anaplastic seminoma in a patient presenting with testicular and jaw swelling. Oral Surg Oral Med Oral Pathol. 1983;56(4):378–87.
118. Raspollini MR. Histologic variants of seminoma mimicking lymphatic malignancies of the testis: a literature review with a report of case series focusing on problems in differential diagnosis. Appl Immunohistochem Mol Morphol. 2014;22(5):348–57. doi:10.1097/PAI.0b013e3182927317.
119. Henley JD, Young RH, Ulbright TM. Malignant sertoli cell tumors of the testis: a study of 13 examples of a neoplasm frequently misinterpreted as seminoma. Am J Surg Pathol. 2002;26(5): 541–50.
120. Stief C, Fizazi K, Evans C, editors. Medical treatment of urological malignancies. International consultation on medical treatment of urological malignancies. Lisbon International Consultation on Urological Disease- European Association of Urology; 2014.
121. Tandstad T, Smaaland R, Solberg A, Bremnes RM, Langberg CW, Laurell A, et al. Management of seminomatous testicular cancer: a binational prospective population-based study from the Swedish norwegian testicular cancer study group. J Clin Oncol Off J Am Soc Clin Oncol. 2011;29(6):719–25. doi:10.1200/JCO.2010.30.1044.
122. Warde P, Gospodarowicz MK, Banerjee D, Panzarella T, Sugar L, Catton CN, et al. Prognostic factors for relapse in stage I testicular seminoma treated with surveillance. J Urol. 1997;157(5):1705–9. discussion 9-10
123. Warde P, Specht L, Horwich A, Oliver T, Panzarella T, Gospodarowicz M, et al. Prognostic factors for relapse in stage I seminoma managed by surveillance: a pooled analysis. J Clin Oncol Off J Am Soc Clin Oncol. 2002;20(22):4448–52.
124. Aparicio J, Maroto P, del Muro X G, Sanchez-Munoz A, Guma J, Margeli M, et al. Prognostic factors for relapse in stage I seminoma: a new nomogram derived from three consecutive, risk-adapted studies from the Spanish Germ Cell Cancer Group (SGCCG). Ann Oncol. 2014;25(11):2173–8. doi:10.1093/annonc/mdu437.
125. Mortensen MS, Lauritsen J, Gundgaard MG, Agerbaek M, Holm NV, Christensen IJ, et al. A nationwide cohort study of stage I seminoma patients followed on a surveillance program. Eur Urol. 2014;66(6):1172–8. doi:10.1016/j.eururo.2014.07.001.
126. Chung P, Daugaard G, Tyldesley S, Atenafu EG, Panzarella T, Kollmannsberger C, et al. Evaluation of a prognostic model for risk of relapse in stage I seminoma surveillance. Cancer Med. 2015;4(1):155–60. doi:10.1002/cam4.324.
127. Rodriguez PN, Hafez GR, Messing EM. Nonseminomatous germ cell tumor of the testicle: does extensive staging of the primary tumor predict the likelihood of metastatic disease? 1986/09/01 ed. J Urol. 1986;136(3):604–8.
128. Reuter VE, Tickoo SK. Tumors of the testis. In: Amin M, et al. editor. Urological patholgy. Philadelphia: Lippincott Williams & Wilkins; 2014. p. 749–803.
129. Kao CS, Ulbright TM, Young RH, Idrees MT. Testicular embryonal carcinoma: a morphologic study of 180 cases highlighting unusual and unemphasized aspects. Am J Surg Pathol. 2014;38(5):689–97. doi:10.1097/PAS.0000000000000171.
130. Nogales FF, Dulcey I, Preda O. Germ cell tumors of the ovary: an update. Arch Pathol Lab Med. 2014;138(3):351–62. doi:10.5858/arpa.2012-0547-RA.
131. Berney DM, Shamash J, Pieroni K, Oliver RT. Loss of CD30 expression in metastatic embryonal carcinoma: the effects of chemotherapy? Histopathology. 2001;39(4):382–5.
132. Gilbert DC, Al-Saadi R, Thway K, Chandler I, Berney DM, Gabe R, et al. Defining a new prognostic index for stage I non-seminomatous germ cell tumors using CXCL12 expression and proportion of embyronal carcinoma. Clin Cancer Res. 2015; doi:10.1158/1078-0432.CCR-15-1186.
133. Porcaro AB, Zecchini Antoniolli S, Novella G, Martignoni G, Bassetto MA, Poli A, et al. Histopathologic risk factors in patients with non-seminomatous germ tumors of the testis in clinical stage 1. Retrospective study of 75 patients. Arch Ital Urol Androl. 2001;73(4):177–80.
134. Talerman A, Haije WG, Baggerman L. Serum alphafetoprotein (AFP) in patients with germ cell tumors of the gonads and extragonadal sites: correlation between endodermal sinus (yolk sac) tumor and raised serum AFP. Cancer. 1980;46(2): 380–5.
135. Magers MJ, Kao CS, Cole CD, Rice KR, Foster RS, Einhorn LH, et al. "Somatic-type" malignancies arising from testicular germ cell tumors: a clinicopathologic study of 124 cases with emphasis on glandular tumors supporting frequent yolk sac tumor origin. Am J Surg Pathol. 2014;38(10):1396–409. doi:10.1097/PAS.0000000000000262.
136. Nogales FF, Quinonez E, Lopez-Marin L, Dulcey I, Preda O. A diagnostic immunohistochemical panel for yolk sac (primitive endodermal) tumours based on an immunohistochemical comparison with the human yolk sac. Histopathology. 2014;65(1):51–9. doi:10.1111/his.12373.
137. Preda O, Nicolae A, Aneiros-Fernandez J, Borda A, Nogales FF. Glypican 3 is a sensitive, but not a specific, marker for the diagnosis of yolk sac tumours. Histopathology. 2011;58(2):312–4. author reply 4-5 doi:10.1111/j.1365-2559.2010.03735.x.
138. Kao CS, Idrees MT, Young RH, Ulbright TM. Solid pattern yolk sac tumor: a morphologic and immunohistochemical study of 52 cases. Am

J Surg Pathol. 2012;36(3):360–7. doi:10.1097/PAS.0b013e31823c510b.
139. Foster RS, Hermans B, Bihrle R, Donohue JP. Clinical stage I pure yolk sac tumor of the testis in adults has different clinical behavior than juvenile yolk sac tumor. J Urol. 2000;164(6):1943–4.
140. Zhang C, Berney DM, Hirsch MS, Cheng L, Ulbright TM. Evidence supporting the existence of benign teratomas of the postpubertal testis: a clinical, histopathologic, and molecular genetic analysis of 25 cases. Am J Surg Pathol. 2013;37(6):827–35. doi:10.1097/PAS.0b013e31827dcc4c.
141. Woods CG, O'Brien BA, Cohen RJ. Benign prostatic glands as a tissue component of testicular teratoma: an uncommon histological finding. Pathology. 2011;43(2):168–9. doi:10.1097/PAT.0b013e3283425bb9.
142. Goodman S, Zhang L, Cheng L, Jiang Z. Differential expression of IMP3 between male and female mature teratomas – immunohistochemical evidence of malignant nature. Histopathology. 2014;65(4):483–9. doi:10.1111/his.12409.
143. Leibovitch I, Foster RS, Ulbright TM, Donohue JP. Adult primary pure teratoma of the testis. The Indiana experience. Cancer. 1995;75(9):2244–50.
144. Spiess PE, Kassouf W, Brown GA, Kamat AM, Liu P, Gomez JA, et al. Surgical management of growing teratoma syndrome: the M. D. Anderson cancer center experience. J Urol. 2007;177(4):1330–4. discussion 4 doi:10.1016/j.juro.2006.11.086.
145. Lee DJ, Djaladat H, Tadros NN, Movassaghi M, Tejura T, Duddalwar V, et al. Growing teratoma syndrome: clinical and radiographic characteristics. Int J Urol. 2014;21(9):905–8. doi:10.1111/iju.12466.
146. Scavuzzo A, Santana Rios ZA, Noveron NR, Jimenez Rios MA. Growing teratoma syndrome. Case Rep Urol. 2014;2014:139425. doi:10.1155/2014/139425.
147. Ulbright TM, Young RH, Scully RE. Trophoblastic tumors of the testis other than classic choriocarcinoma: "monophasic" choriocarcinoma and placental site trophoblastic tumor: a report of two cases. Am J Surg Pathol. 1997;21(3):282–8.
148. Alvarado-Cabrero I, Hernandez-Toriz N, Paner GP. Clinicopathologic analysis of choriocarcinoma as a pure or predominant component of germ cell tumor of the testis. Am J Surg Pathol. 2014;38(1):111–8. doi:10.1097/PAS.0b013e3182a2926e.
149. Ahsaini M, Tazi F, Mellas S, Elammari J, Khalouk A, Stuurman-Wieringa R, et al. Pure choriocarcinoma of the testis presenting with jaundice: a case report and review of the literature. J Med Case Reports. 2012;6:269. doi:10.1186/1752-1947-6-269.
150. Chhieng DC, Jennings TA, Slominski A, Mihm Jr MC. Choriocarcinoma presenting as a cutaneous metastasis. J Cutan Pathol. 1995;22(4):374–7.
151. Guber I, Zografos L, Schalenbourg A. Choroidal metastases in testicular choriocarcinoma, successful treatment with chemo- and radiotherapy: a case report. BMC Urol. 2011;11:24. doi:10.1186/1471-2490-11-24.
152. Joret MO, Starke RM, Scotter J, Heppner P. Metastatic choriocarcinoma induced separate simultaneous intracerebral haemorrhages: a very rare occurrence and its novel association with Klinefelter syndrome. BMJ Case Rep. 2015; published online 12 November 2015. doi:10.1136/bcr-2015-212777.
153. Heda P, Cushing G. Testicular choriocarcinoma presenting as hyperthyroidism. Am J Med. 2013;126(11):e1–2. doi:10.1016/j.amjmed.2013.07.017.
154. Voigt W, Maher G, Wolf HH, Schmoll HJ. Human chorionic gonadotropin-induced hyperthyroidism in germ cell cancer--a case presentation and review of the literature. Onkologie. 2007;30(6):330–4.
155. Caron P, Salandini AM, Plantavid M, Despax R, Bayard F. Choriocarcinoma and endocrine paraneoplastic syndromes. Eur J Med. 1993;2(8):499–500.
156. Moul JW, Moellman JR. Unnecessary mastectomy for gynecomastia in testicular cancer patient. Mil Med. 1992;157(8):433–4.
157. Voskuil JH, Slee PH, Wagenaar SS. The choriocarcinoma syndrome: an emergency. Neth J Med. 1995;46(5):232–5.
158. Manivel JC, Niehans G, Wick MR, Dehner LP. Intermediate trophoblast in germ cell neoplasms. Am J Surg Pathol. 1987;11(9):693–701.
159. Pelkey TJ, Frierson Jr HF, Mills SE, Stoler MH. Detection of the alpha-subunit of inhibin in trophoblastic neoplasia. Hum Pathol. 1999;30(1):26–31.
160. Damjanov I, Osborn M, Miettinen M. Keratin 7 is a marker for a subset of trophoblastic cells in human germ cell tumors. Arch Pathol Lab Med. 1990;114(1):81–3.
161. Emanuel PO, Unger PD, Burstein DE. Immunohistochemical detection of p63 in testicular germ cell neoplasia. Ann Diagn Pathol. 2006;10(5):269–73. doi:10.1016/j.anndiagpath.2006.02.005.
162. Bille A, Girelli L, Colecchia M, Pastorino U. Single synchronous pulmonary metastasis from placental site trophoblastic tumor and teratoma. Tumori. 2015;101(3):e110–1. doi:10.5301/tj.5000290.
163. Petersson F, Grossmann P, Vanecek T, Coric M, Cacic M, Hes O, et al. Testicular germ cell tumor composed of placental site trophoblastic tumor and teratoma. Hum Pathol. 2010;41(7):1046–50. doi:10.1016/j.humpath.2009.10.026.
164. Suurmeijer AJ, Gietema JA, Hoekstra HJ. Placental site trophoblastic tumor in a late recurrence of a nonseminomatous germ cell tumor of the testis. Am J Surg Pathol. 2004;28(6):830–3.
165. Idrees MT, Kao CS, Epstein JI, Ulbright TM. Nonchoriocarcinomatous trophoblastic tumors of the testis: the widening spectrum of trophoblastic Neoplasia. Am J Surg Pathol. 2015;39(11):1468–78. doi:10.1097/PAS.0000000000000509.
166. Leow WQ, Loh HL, Lee LS, Goh CH. A rare case of combined placental site trophoblastic tumour with mature cystic teratoma and mixed germ cell tumour in the testis. Malays J Pathol. 2015;37(2):145–7.

167. Allan RW, Algood CB, Shih IM. Metastatic epithelioid trophoblastic tumor in a male patient with mixed germ-cell tumor of the testis. Am J Surg Pathol. 2009;33(12):1902–5. doi:10.1097/PAS.0b013e3181c03792.
168. Mosharafa AA, Foster RS, Leibovich BC, Ulbright TM, Bihrle R, Einhorn LH, et al. Histology in mixed germ cell tumors. Is there a favorite pairing? J Urol. 2004;171(4):1471–3. doi:10.1097/01.ju.0000116841.30826.85.
169. Mostofi FK. Pathology of germ cell tumors of testis: a progress report. Cancer. 1980;45(7 Suppl):1735–54.
170. Mostofi FK, Sesterhenn IA. Histological typing of testis tumours. In: WHO. World Health Organization, editor. International histological classification of tumours. 2nd ed. Berlin: Springer-Verlag Berlin Heidelberg; 1998.
171. Freedman LS, Parkinson MC, Jones WG, Oliver RT, Peckham MJ, Read G, et al. Histopathology in the prediction of relapse of patients with stage I testicular teratoma treated by orchidectomy alone. Lancet. 1987;2(8554):294–8.
172. Jacobsen GK, Rorth M, Osterlind K, von der Maase H, Jacobsen A, Madsen EL, et al. Histopathological features in stage I non-seminomatous testicular germ cell tumours correlated to relapse. Danish Testicular Cancer Study Group. APMIS. 1990;98(4):377–82.
173. Moul JW, McCarthy WF, Fernandez EB, Sesterhenn IA. Percentage of embryonal carcinoma and of vascular invasion predicts pathological stage in clinical stage I nonseminomatous testicular cancer. Cancer Res. 1994;54(2):362–4.
174. Logothetis CJ, Samuels ML, Selig DE, Ogden S, Dexeus F, Swanson D, et al. Cyclic chemotherapy with cyclophosphamide, doxorubicin, and cisplatin plus vinblastine and bleomycin in advanced germinal tumors. Results with 100 patients. Am J Med. 1986;81(2):219–28.
175. Vaeth M, Schultz HP, Von Der Maase H, Engelholm SA, Krag Jacobsen G, Norgaard-Pedersen B. Prognostic factors in testicular germ cell tumours. Experiences from 1058 consecutive cases. Acta Radiol Oncol. 1984;23(4):271–85.
176. Gopalan A, Dhall D, Olgac S, Fine SW, Korkola JE, Houldsworth J, et al. Testicular mixed germ cell tumors: a morphological and immunohistochemical study using stem cell markers, OCT3/4, SOX2 and GDF3, with emphasis on morphologically difficult-to-classify areas. Mod Pathol. 2009;22(8):1066–74. doi:10.1038/modpathol.2009.66.
177. YoungRH. Testicular tumors--some new and a few perennial problems. Arch Pathol Lab Med. 2008;132(4):548–64. doi:10.1043/1543-2165(2008)132[548:TTNAAF]2.0.CO;2.
178. Ulbright TM. Germ cell tumors of the gonads: a selective review emphasizing problems in differential diagnosis, newly appreciated, and controversial issues. Mod Pathol. 2005;18(Suppl 2):S61–79. doi:10.1038/modpathol.3800310.
179. Cardoso de Almeida PC, Scully RE. Diffuse embryoma of the testis. A distinctive form of mixed germ cell tumor. Am J Surg Pathol. 1983;7(7):633–642.
180. de Peralta-Venturina MN, Ro JY, Ordonez NG, Ayala AG. Diffuse embryoma of the testis. An immunohistochemical study of two cases. Am J Clin Pathol. 1994;102(4):402–5.
181. Lee WA. Diffuse embryoma of the testis: a case report. J Pathol Transl Med. 2008;42(2):103–7.
182. Zamecnik M, Sultani K. Diffuse embryoma with intratubular germ-cell neoplasia. Histopathology. 2004;45(4):427–9. doi:10.1111/j.1365-2559.2004.01943.x.
183. Ghazarian AA, Trabert B, Devesa SS, McGlynn KA. Recent trends in the incidence of testicular germ cell tumors in the United States. Andrology. 2015;3(1):13–8. doi:10.1111/andr.288.
184. Talerman A. Spermatocytic seminoma: clinicopathological study of 22 cases. Cancer. 1980;45(8):2169–76.
185. Aggarwal N, Parwani AV. Spermatocytic seminoma. Arch Pathol Lab Med. 2009;133(12):1985–8. doi:10.1043/1543-2165-133.12.1985.
186. Wetherell D, Lawrentschuk N, Gyomber D. Spermatocytic seminoma with sarcoma: an indication for adjuvant chemotherapy in localized disease. Korean J Urol. 2013;54(12):884–7. doi:10.4111/kju.2013.54.12.884.
187. Kalakoti P, Sahu S, Syed A. Reaction to Koppad SN et al. Bilateral spermatocytic seminoma: an update on "synchronous" and "sequential" presentation of this rare variant of testicular tumour. Pan Afr Med J. 2014;17:248. doi:10.11604/pamj.2014.17.248.4118.
188. True LD, Otis CN, Delprado W, Scully RE, Rosai J. Spermatocytic seminoma of testis with sarcomatous transformation. A report of five cases. Am J Surg Pathol. 1988;12(2):75–82.
189. Trivedi P, Pasricha S, Gupta A. Spermatocytic seminoma associated with undifferentiated sarcoma: a rare case report. Indian J Pathol Microbiol. 2011;54(1):138–40. doi:10.4103/0377-4929.77369.
190. Menon S, Karpate A, Desai S. Spermatocytic seminoma with rhabdomyosarcomatous differentiation: a case report with a review of the literature. J Cancer Res Ther. 2009;5(3):213–5. doi:10.4103/0973-1482.57131.
191. Narang V, Gupta K, Gupta A, Kumar S. Rhabdomyosarcomatous differentiation in a spermatocytic seminoma with review of literature. Indian J Urol. 2012;28(4):430–3. doi:10.4103/0970-1591.105760.
192. Robinson A, Bainbridge T, Kollmannsberger C. A spermatocytic seminoma with rhabdomyosarcoma transformation and extensive metastases. Am J Clin Oncol. 2007;30(4):440–1. doi:10.1097/01.coc.0000180384.40865.f7.
193. Kraggerud SM, Berner A, Bryne M, Pettersen EO, Fossa SD. Spermatocytic seminoma as compared to classical seminoma: an immunohistochemical and DNA flow cytometric study. APMIS. 1999;107(3):297–302.

194. Matoska J, Ondrus D, Hornak M. Metastatic spermatocytic seminoma. A case report with light microscopic, ultrastructural, and immunohistochemical findings. Cancer. 1988;62(6):1197–201.
195. Steiner H, Gozzi C, Verdorfer I, Mikuz G, Bartsch G, Hobisch A. Metastatic spermatocytic seminoma--an extremely rare disease: Part 2. Eur Urol. 2006;49(2):408–9.
196. Dockerty MB, Priestle JT. Dermoid cysts of the testis. J Urol. 1942;48:392–400.
197. Muir I. Benign cystic teratoma (dermoid cyst) of the testicle. Br J Surg. 1952;40(160):144–7.
198. Atkinson RL. Dermoid cyst (benign teratoma) of the testicle. Ill Med J. 1954;105(1):23–6.
199. Ramsay AK, Gurun M, Berney DM, Nairn R. Dermoid cyst of the testis with neural tissue in an adult. Urology. 2012;79(2):e25–6.
200. Ulbright TM, Srigley JR. Dermoid cyst of the testis: a study of five postpubertal cases, including a pilomatrixoma-like variant, with evidence supporting its separate classification from mature testicular teratoma. Am J Surg Pathol. 2001;25(6):788–93.
201. Buckspan MB, Skeldon SC, Klotz PG, Pritzker KP. Epidermoid cysts of the testicle. J Urol. 1985;134(5):960–1.
202. Dieckmann KP, Loy V. Epidermoid cyst of the testis: a review of clinical and histogenetic considerations. Br J Urol. 1994;73(4):436–41.
203. Price Jr EB. Epidermoid cysts of the testis: a clinical and pathologic analysis of 69 cases from the testicular tumor registry. J Urol. 1969;102(6):708–13.
204. Reinberg Y, Manivel JC, Llerena J, Niehans G, Fraley EE. Epidermoid cyst (monodermal teratoma) of the testis. Br J Urol. 1990;66(6):648–51.
205. Wang WP, Guo C, Berney DM, Ulbright TM, Ubright TM, Hansel DE, et al. Primary carcinoid tumors of the testis: a clinicopathologic study of 29 cases.[Erratum appears in Am J Surg Pathol. 2010 Jul;34(7):1075 Note: Ubright, Thomas M [corrected to Ulbright, Thomas M]]. Am J Surg Pathol. 2010;34(4):519–24.
206. Manivel JC, Reinberg Y, Niehans GA, Fraley EE. Intratubular germ cell neoplasia in testicular teratomas and epidermoid cysts. Correlation with prognosis and possible biologic significance. Cancer. 1989;64(3):715–20.
207. Cheng L, Zhang S, MacLennan GT, Poulos CK, Sung MT, Beck SD, et al. Interphase fluorescence in situ hybridization analysis of chromosome 12p abnormalities is useful for distinguishing epidermoid cysts of the testis from pure mature teratoma. Clin Cancer Res. 2006;12(19):5668–72. doi:10.1158/1078-0432.CCR-06-0976.
208. Wegner HE, Herbst H, Dieckmann KP. Paratesticular epidermoid cyst and ipsilateral spermatic cord dermoid cyst: case report and discussion of pathogenesis, diagnosis and treatment. J Urol. 1994;152(6 Pt 1):2101–3.
209. Nogales Jr FF, Silverberg SG. Epidermoid cysts of the ovary: a report of five cases with histogenetic considerations and ultrastructural findings. Am J Obstet Gynecol. 1976;124(5):523–8.
210. Lu C, Zhang Z, Jiang Y, Yang Z, Yang Q, Liao D, et al. Primary pure carcinoid tumors of the testis: clinicopathological and immunophenotypical characteristics of 11 cases. Oncol Lett. 2015;9(5):2017–22. doi:10.3892/ol.2015.3046.
211. Reyes A, Moran CA, Suster S, Michal M, Dominguez H. Neuroendocrine carcinomas (carcinoid tumor) of the testis. A clinicopathologic and immunohistochemical study of ten cases. Am J Clin Pathol. 2003;120(2):182–7.
212. Zavala-Pompa A, Ro JY. el-Naggar A, Ordonez NG, Amin MB, Pierce PD et al. Primary carcinoid tumor of testis. Immunohistochemical, ultrastructural, and DNA flow cytometric study of three cases with a review of the literature. Cancer. 1993;72(5):1726–32.
213. Mai KT, Park PC, Yazdi HM, Carlier M. Leydig cell origin of testicular carcinoid tumour: immunohistochemical and electron microscopic evidence. Histopathology. 2006;49(5):548–9. doi:10.1111/j.1365-2559.2006.02531.x.
214. Merino J, Zuluaga A, Gutierrez-Tejero F, Del Mar SM, Ciani S, Nogales FF. Pure testicular carcinoid associated with intratubular germ cell neoplasia. J Clin Pathol. 2005;58(12):1331–3. doi:10.1136/jcp.2004.022541.
215. Abbosh PH, Zhang S, Maclennan GT, Montironi R, Lopez-Beltran A, Rank JP, et al. Germ cell origin of testicular carcinoid tumors. Clin Cancer Res. 2008;14(5):1393–6. doi:10.1158/1078-0432.CCR-07-4146.
216. Hayashi T, Iida S, Taguchi J, Miyajima J, Matsuo M, Tomiyasu K, et al. Primary carcinoid of the testis associated with carcinoid syndrome. Int J Urol. 2001;8(9):522–4.
217. Park SB, Kim JK, Cho KS. Imaging findings of a primary bilateral testicular carcinoid tumor associated with carcinoid syndrome. J Ultrasound Med. 2006;25(3):413–6.
218. Kato N, Motoyama T, Kameda N, Hiruta N, Emura I, Hasegawa G, et al. Primary carcinoid tumor of the testis: Immunohistochemical, ultrastructural and FISH analysis with review of the literature. Pathol Int. 2003;53(10):680–5.
219. Ordonez NG, Ayala AG, Sneige N, Mackay B. Immunohistochemical demonstration of multiple neurohormonal polypeptides in a case of pure testicular carcinoid. Am J Clin Pathol. 1982;78(6):860–4.
220. Oosterhuis JW, Stoop JA, Rijlaarsdam MA, Biermann K, Smit VT, Hersmus R, et al. Pediatric germ cell tumors presenting beyond childhood? Andrology. 2015;3(1):70–7. doi:10.1111/andr.305.
221. Pohl HG, Shukla AR, Metcalf PD, Cilento BG, Retik AB, Bagli DJ, et al. Prepubertal testis tumors: actual prevalence rate of histological types. J Urol. 2004;172(6 Pt 1):2370–2.
222. Bolen JW. Mixed germ cell-sex cord stromal tumor. A gonadal tumor distinct from gonadoblastoma. Am J Clin Pathol. 1981;75(4):565–73.

223. Matoska J, Talerman A. Mixed germ cell-sex cord stroma tumor of the testis. A report with ultrastructural findings. Cancer. 1989;64(10):2146–53.
224. Talerman A. The pathology of gonadal neoplasms composed of germ cells and sex cord stroma derivatives. Pathol Res Pract. 1980;170(1–3):24–38. doi:10.1016/S0344-0338(80)80153-1.
225. Ulbright TM, Srigley JR, Reuter VE, Wojno K, Roth LM, Young RH. Sex cord-stromal tumors of the testis with entrapped germ cells: a lesion mimicking unclassified mixed germ cell sex cord-stromal tumors. Am J Surg Pathol. 2000;24(4): 535–542.
226. Cheville JC. Classification and pathology of testicular germ cell and sex cord-stromal tumors. Urol Clin North Am. 1999;26(3):595–609.
227. Scully RE. Gonadoblastoma. A review of 74 cases. Cancer. 1970;25(6):1340–56.
228. Ulbright TM, Young RH. Gonadoblastoma and selected other aspects of gonadal pathology in young patients with disorders of sex development. Semin Diagn Pathol. 2014;31(5):427–40.
229. Raymond CS, Parker ED, Kettlewell JR, Brown LG, Page DC, Kusz K, et al. A region of human chromosome 9p required for testis development contains two genes related to known sexual regulators. Hum Mol Genet. 1999;8(6):989–96.
230. Barbaro M, Cools M, Looijenga LHJ, Drop SLS, Wedell A. Partial deletion of the NR5A1 (SF1) gene detected by synthetic probe MLPA in a patient with XY gonadal disorder of sex development. Sex Dev. 2011;5(4):181–7.
231. Paliwal P, Sharma A, Birla S, Kriplani A, Khadgawat R, Sharma A. Identification of novel SRY mutations and SF1 (NR5A1) changes in patients with pure gonadal dysgenesis and 46,XY karyotype. Mol Hum Reprod. 2011;17(6):372–8.
232. Zangen D, Kaufman Y, Zeligson S, Perlberg S, Fridman H, Kanaan M, et al. XX ovarian dysgenesis is caused by a PSMC3IP/HOP2 mutation that abolishes coactivation of estrogen-driven transcription. Am J Hum Genet. 2011;89(4):572–9.
233. Brandt T, Blanchard L, Desai K, Nimkarn S, Cohen N, Edelmann L, et al. 46,XY disorder of sex development and developmental delay associated with a novel 9q33.3 microdeletion encompassing NR5A1. Eur J Med Genet. 2013;56(11):619–23.
234. dos Santos AP, Andrade JGR, Piveta CSC, de Paulo J, Guerra Jr G, de Mello MP, et al. Screening of Y chromosome microdeletions in 46,XY partial gonadal dysgenesis and in patients with a 45,X/46,XY karyotype or its variants. BMC Med Genet. 2013;14:115.
235. Grossi A, Palma A, Zanni G, Novelli A, Loddo S, Cappa M, et al. Multiorgan autoimmunity in a turner syndrome patient with partial monosomy 2q and trisomy 10p. Gene. 2013;515(2):439–43.
236. Norling A, Linden Hirschberg A, Iwarsson E, Persson B, Wedell A, Barbaro M. Novel candidate genes for 46,XY gonadal dysgenesis identified by a customized 1 M array-CGH platform. Eur J Med Genet. 2013;56(12):661–8.
237. Bashamboo A, Brauner R, Bignon-Topalovic J, Lortat-Jacob S, Karageorgou V, Lourenco D, et al. Mutations in the FOG2/ZFPM2 gene are associated with anomalies of human testis determination. Hum Mol Genet. 2014;23(14):3657–65.
238. Boublikova L, Buchler T, Stary J, Abrahamova J, Trka J. Molecular biology of testicular germ cell tumors: unique features awaiting clinical application. Crit Rev Oncol Hematol. 2014;89(3):366–85.
239. Ostrer H. Disorders of sex development (DSDs): an update. J Clin Endocrinol Metab. 2014;99(5):1503–9.
240. Werner R, Merz H, Birnbaum W, Marshall L, Schroder T, Reiz B, et al. 46,XY gonadal dysgenesis due to a homozygous mutation in Desert Hedgehog (DHH) identified by exome sequencing. J Clin Endocrinol Metab. 2015;100(7):E1022–9.
241. Robboy SJ, Miller T, Donahoe PK, Jahre C, Welch WR, Haseltine FP, et al. Dysgenesis of testicular and streak gonads in the syndrome of mixed gonadal dysgenesis: perspective derived from a clinicopathologic analysis of twenty-one cases. Hum Pathol. 1982;13(8):700–16.
242. Rutgers JL. Advances in the pathology of intersex conditions. Hum Pathol. 1991;22(9):884–91.
243. Spear GS, Martin CG. Fetal gonadoblastoid testicular dysplasia. Hum Pathol. 1986;17(5):531–3.
244. Bianco B, Lipay M, Guedes A, Oliveira K, Verreschi IT. SRY gene increases the risk of developing gonadoblastoma and/or nontumoral gonadal lesions in Turner syndrome. Int J Gynecol Pathol. 2009;28(2):197–202. doi:10.1097/PGP.0b013e318186a825.
245. Li Y, Vilain E, Conte F, Rajpert-De Meyts E, Lau YF. Testis-specific protein Y-encoded gene is expressed in early and late stages of gonadoblastoma and testicular carcinoma in situ. Urol Oncol. 2007;25(2):141–6. doi:10.1016/j.urolonc.2006.08.002.
246. Kersemaekers A-MF, Honecker F, Stoop H, Cools M, Molier M, Wolffenbuttel K, et al. Identification of germ cells at risk for neoplastic transformation in gonadoblastoma: an immunohistochemical study for OCT3/4 and TSPY. Hum Pathol. 2005;36(5):512–21.
247. Cools M, Stoop H, Kersemaekers AM, Drop SL, Wolffenbuttel KP, Bourguignon JP, et al. Gonadoblastoma arising in undifferentiated gonadal tissue within dysgenetic gonads. J Clin Endocrinol Metab. 2006;91(6):2404–13. doi:10.1210/jc.2005-2554.
248. Hussong J, Crussi FG, Chou PM. Gonadoblastoma: immunohistochemical localization of Mullerian-inhibiting substance, inhibin, WT-1, and p53. Mod Pathol. 1997;10(11):1101–5.
249. Obata NH, Nakashima N, Kawai M, Kikkawa F, Mamba S, Tomoda Y. Gonadoblastoma with dysgerminoma in one ovary and gonadoblastoma with dysgerminoma and yolk sac tumor in the contralateral ovary in a girl with 46XX karyotype. Gynecol Oncol. 1995;58(1):124–8.
250. Esin S, Baser E, Kucukozkan T, Magden HA. Ovarian gonadoblastoma with dysgerminoma in a 15-year-

old girl with 46, XX karyotype: case report and review of the literature. Arch Gynecol Obstet. 2012;285(2):447–51.
251. Azzopardi JG, Mostofi FK, Theiss EA. Lesions of testes observed in certain patients with widespread choriocarcinoma and related tumors. The significance and genesis of hematoxylin-staining bodies in the human testis. Am J Pathol. 1961;38:207–25.
252. Azzopardi JG, Hoffbrand AV. Retrogression in testicular seminoma with viable metastases. J Clin Pathol. 1965;18:135–41.
253. Bar W, Hedinger C. Comparison of histologic types of primary testicular germ cell tumors with their metastases: consequences for the WHO and the British Nomenclatures? Virchows Arch A Pathol Anat Histol. 1976;370(1):41–54.
254. Balzer BL, Ulbright TM. Spontaneous regression of testicular germ cell tumors: an analysis of 42 cases. Am J Surg Pathol. 2006;30(7):858–65. doi:10.1097/01.pas.0000209831.24230.56.
255. Kontos S, Doumanis G, Karagianni M, Politis V, Simaioforidis V, Kachrilas S, et al. Burned-out testicular tumor with retroperitoneal lymph node metastasis: a case report. J Med Case Reports. 2009;3:8705. doi:10.4076/1752-1947-3-8705.
256. Andhavarapu S, Low B, Raj J, Skinner S, Armenta-Corona J. Regressed testicular seminoma with extensive metastases. Intern J Oncol. 2007;5(1):1–5.
257. Parker C, Milosevic M, Panzarella T, Banerjee D, Jewett M, Catton C, et al. The prognostic significance of the tumour infiltrating lymphocyte count in stage I testicular seminoma managed by surveillance. Eur J Cancer. 2002;38(15):2014–9.
258. Stevenson SM, Lowrance WT. Epidemiology and diagnosis of testis cancer. Urol Clin North Am. 2015;42(3):269–75. doi:10.1016/j.ucl.2015.04.001.
259. Jouin-Caurier D, Chimenes H, Klotz HP. Gynecomastia indicative of an unsuspected testicular tumor (1 case of seminoma--1 case of Leydig cell tumor). Ann Endocrinol (Paris). 1968;29(3):385–94.
260. Arrabal-Polo MA, Jimenez-Pacheco A, Arrabal-Martin M, Moreno-Jimenez J, Gutierrez-Tejero F, Galisteo-Moya R, et al. Hyperthyroidism as a clinical manifestation of a embryonal carcinoma of the testis. Acta Clin Belg. 2012;67(3):214–6. doi:10.1179/acb.67.3.2062658.
261. McCracken EJ, Johnston PC, Lindsay JR, Mulholland C, McAleer JJ, Black RN. Testicular choriocarcinoma: an unusual case of paraneoplastic thyrotoxicosis. QJM. 2012;105(7):675–7. doi:10.1093/qjmed/hcr084.
262. Voltz R. Paraneoplastic neurological syndromes: an update on diagnosis, pathogenesis, and therapy. Lancet Neurol. 2002;1(5):294–305.
263. Dalmau J, Graus F, Villarejo A, Posner JB, Blumenthal D, Thiessen B, et al. Clinical analysis of anti-Ma2-associated encephalitis. Brain. 2004;127(Pt 8):1831–44. doi:10.1093/brain/awh203.
264. Hoffmann LA, Jarius S, Pellkofer HL, Schueller M, Krumbholz M, Koenig F, et al. Anti-Ma and anti-Ta associated paraneoplastic neurological syndromes: 22 newly diagnosed patients and review of previous cases. J Neurol Neurosurg Psychiatry. 2008;79(7):767–73. doi:10.1136/jnnp.2007.118588.
265. Mathew RM, Vandenberghe R, Garcia-Merino A, Yamamoto T, Landolfi JC, Rosenfeld MR, et al. Orchiectomy for suspected microscopic tumor in patients with anti-Ma2-associated encephalitis. Neurology. 2007;68(12):900–5. doi:10.1212/01.wnl.0000252379.81933.80.
266. Tenner L, Einhorn L. Ma-2 paraneoplastic encephalitis in the presence of bilateral testicular cancer: diagnostic and therapeutic approach. J Clin Oncol. 2009;27(23):e57–8. doi:10.1200/JCO.2009.23.3635.
267. Ishikawa H, Kawada N, Taniguchi A, Odachi K, Mizutani A, Asahi M, et al. Paraneoplastic neurological syndrome due to burned-out testicular tumor showing hot cross-bun sign. Acta Neurol Scand. 2015; doi:10.1111/ane.12469.
268. Bosemani T, Huisman TA, Poretti A. Anti-Ma2-associated paraneoplastic encephalitis in a male adolescent with mediastinal seminoma. Pediatr Neurol. 2014;50(4):433–4. doi:10.1016/j.pediatrneurol.2013.12.020.
269. Di Stasi SM, Poggi A, Giannantoni A, Zampa G. Dermatomyositis associated with testicular germ cell cancer. J Urol. 2000;163(1):240.
270. Kaito K, Otsubo H, Usui N, Kobayashi M. Secondary polycythemia as a paraneoplastic syndrome of testicular seminoma. Ann Hematol. 2004;83(1):55–7. doi:10.1007/s00277-003-0745-7.
271. MacDiarmid SA, Norman RW. Extragonadal nonseminomatous germ cell tumour with hypercalcemia, masquerading as renal cell carcinoma: a case report. Can J Surg J Can de chirurgie. 1995;38(1):80–2.
272. Schneider BF, Glass 2nd WF, Brooks CH, Koenig KG. Membranous glomerulonephritis associated with testicular seminoma. J Intern Med. 1995;237(6):599–602.
273. Taillan B, Castanet J, Garnier G, Pesce A, Sanderson F, Fuzibet JG, et al. Paraneoplastic Raynaud's phenomenon. Clin Rheumatol. 1993;12(2):281–2.
274. AJCC cancer staging manual. 7th ed. New York: Springer; 2010.
275. Kollmannsberger C, Tandstad T, Bedard PL, Cohn-Cedermark G, Chung PW, Jewett MA, et al. Patterns of relapse in patients with clinical stage I testicular cancer managed with active surveillance. J Clin Oncol. 2015;33(1):51–7. doi:10.1200/JCO.2014.56.2116.
276. Pizzocaro G, Zanoni F, Salvioni R, Milani A, Piva L, Pilotti S. Difficulties of a surveillance study omitting retroperitoneal lymphadenectomy in clinical stage I nonseminomatous germ cell tumors of the testis. J Urol. 1987;138(6):1393–6.
277. Daugaard G, Gundgaard MG, Mortensen MS, Agerbaek M, Holm NV, Rorth M, et al. Surveillance for stage I nonseminoma testicular cancer: outcomes and long-term follow-up in a population-

based cohort. J Clin Oncol. 2014;32(34):3817–23. doi:10.1200/JCO.2013.53.5831.

278. Kamba T, Kamoto T, Okubo K, Teramukai S, Kakehi Y, Matsuda T, et al. Outcome of different post-orchiectomy management for stage I seminoma: Japanese multi-institutional study including 425 patients. Int J Urol. 2010;17(12):980–7. doi:10.1111/j.1442-2042.2010.02645.x.
279. von der Maase H, Specht L, Jacobsen GK, Jakobsen A, Madsen EL, Pedersen M, et al. Surveillance following orchidectomy for stage I seminoma of the testis. Eur J Cancer. 1993;29A(14):1931–4.
280. Vogt AP, Chen Z, Osunkoya AO. Rete testis invasion by malignant germ cell tumor and/or intratubular germ cell neoplasia: what is the significance of this finding? Hum Pathol. 2010;41(9):1339–44. doi:10.1016/j.humpath.2010.03.005.
281. Berney DM, Algaba F, Amin M, Delahunt B, Comperat E, Epstein JI, et al. Handling and reporting of orchidectomy specimens with testicular cancer: areas of consensus and variation among 25 experts and 225 European pathologists. Histopathology. 2015;67(3):313–24. doi:10.1111/his.12657.
282. Yilmaz A, Cheng T, Zhang J, Trpkov K. Testicular hilum and vascular invasion predict advanced clinical stage in nonseminomatous germ cell tumors. Mod Pathol. 2013;26(4):579–86. doi:10.1038/modpathol.2012.189.
283. Nazeer T, Ro JY, Kee KH, Ayala AG. Spermatic cord contamination in testicular cancer. Mod Pathol. 1996;9(7):762–6.
284. Ray B, Hajdu SI, Whitmore Jr WF. Proceedings: Distribution of retroperitoneal lymph node metastases in testicular germinal tumors. Cancer. 1974;33(2): 340–8.
285. Donohue JP, Zachary JM, Maynard BR. Distribution of nodal metastases in nonseminomatous testis cancer. J Urol. 1982;128(2):315–20.
286. Stephenson AJ, Sheinfeld J. The role of retroperitoneal lymph node dissection in the management of testicular cancer. Urol Oncol. 2004;22(3):225–33. discussion 34-5 doi:10.1016/j.urolonc.2004.04.029.
287. Stephenson AJ, Bosl GJ, Motzer RJ, Kattan MW, Stasi J, Bajorin DF, et al. Retroperitoneal lymph node dissection for nonseminomatous germ cell testicular cancer: impact of patient selection factors on outcome. J Clin Oncol. 2005;23(12):2781–8. doi:10.1200/jco.2005.07.132.
288. Donohue JP, Thornhill JA, Foster RS, Rowland RG, Bihrle R. Retroperitoneal lymphadenectomy for clinical stage A testis cancer (1965 to 1989): modifications of technique and impact on ejaculation. J Urol. 1993;149(2):237–43.
289. Gels ME, Hoekstra HJ, Sleijfer DT, Marrink J, de Bruijn HW, Molenaar WM, et al. Detection of recurrence in patients with clinical stage I nonseminomatous testicular germ cell tumors and consequences for further follow-up: a single-center 10-year experience. J Clin Oncol. 1995;13(5):1188–94.
290. Sogani PC, Perrotti M, Herr HW, Fair WR, Thaler HT, Bosl G. Clinical stage I testis cancer: long-term outcome of patients on surveillance. J Urol. 1998;159(3):855–8.
291. Feldman DR. Treatment options for stage I nonseminoma. J Clin Oncol. 2014;32(34):3797–800. doi:10.1200/JCO.2014.56.8006.
292. Al-Ahmadie HA, Carver BS, Cronin AM, Olgac S, Tickoo SK, Fine SW, et al. Primary retroperitoneal lymph node dissection in low-stage testicular germ cell tumors: a detailed pathologic study with clinical outcome analysis with special emphasis on patients who did not receive adjuvant therapy. Urology. 2013;82(6):1341–6. doi:10.1016/j.urology.2013.04.082.
293. Beck SD, Foster RS, Bihrle R, Cheng L, Donohue JP. Does the histology of nodal metastasis predict systemic relapse after retroperitoneal lymph node dissection in pathological stage B1 germ cell tumors? J Urol. 2005;174(4 Pt 1):1287–90. discussion 90
294. Pizzocaro G, Monfardini S. No adjuvant chemotherapy in selected patients with pathologic stage II nonseminomatous germ cell tumors of the testis. J Urol. 1984;131(4):677–80.
295. Vugrin D, Whitmore WF, Cvitkovic E, Grabstald H, Sogani P, Barzell W, et al. Adjuvant chemotherapy combination of vinblastine, actinomycin D, bleomycin, and chlorambucil following retroperitoneal lymph node dissection for stage II testis tumor. Cancer. 1981;47(5):840–4.
296. Beck SD, Foster RS, Bihrle R, Cheng L, Ulbright TM, Donohue JP. Impact of the number of positive lymph nodes on disease-free survival in patients with pathological stage B1 nonseminomatous germ cell tumor. J Urol. 2005;174(1):143–5. doi:10.1097/01.ju.0000161603.65138.90.
297. Eggener SE, Carver BS, Loeb S, Kondagunta GV, Bosl GJ, Sheinfeld J. Pathologic findings and clinical outcome of patients undergoing retroperitoneal lymph node dissection after multiple chemotherapy regimens for metastatic testicular germ cell tumors. Cancer. 2007;109(3):528–35. doi:10.1002/cncr.22440.
298. Brandli DW, Ulbright TM, Foster RS, Cummings OW, Zhang S, Sweeney CJ, et al. Stroma adjacent to metastatic mature teratoma after chemotherapy for testicular germ cell tumors is derived from the same progenitor cells as the teratoma. Cancer Res. 2003;63(18):6063–8.
299. Clevenger JA, Foster RS, Ulbright TM. Differentiated rhabdomyomatous tumors after chemotherapy for metastatic testicular germ-cell tumors: a clinicopathological study of seven cases mandating separation from rhabdomyosarcoma. Mod Pathol. 2009;22(10):1361–6. doi:10.1038/modpathol.2009.108.
300. Ulbright TM, Henley JD, Cummings OW, Foster RS, Cheng L. Cystic trophoblastic tumor: a nonaggressive lesion in postchemotherapy resections of patients with testicular germ cell tumors. Am J Surg Pathol. 2004;28(9):1212–6.

8 Mediastinal Germ Cell Tumors

Anja C. Roden

8.1 Introduction

Mediastinal germ cell tumors (GCTs) have been described at least since the beginning of the twentieth century [1]. However, for many years there was some controversy whether these tumors are truly of mediastinal origin or whether they represent metastases from an occult gonadal tumor or a subtype of thymoma. Therefore, tumors with morphological features of seminoma were termed, for instance, "seminoma-like tumor" or "seminomatous thymoma" [2, 3]. In the 1970s and 1980s, the concept of primary mediastinal germ cell tumors (PMGCTs) became well established [4–7].

Although PMGCTs are histological and ultrastructural similar to their gonadal counterpart [4], the behavior of at least a subgroup of these tumors is different. For instance, in contrast to gonadal GCT, in the adult mediastinum, nonteratomatous components are regarded as malignant. In addition, PMGCTs have distinct differential diagnoses due to their location. It is important to differentiate PMGCT from these other malignancies in the mediastinum because of differences in treatment and outcome.

8.2 Demographics

PMGCTs are rare and comprise approximately 1–15 % of all mediastinal neoplasms in adults and 11 % in children [8–10]. Even though the mediastinum is the most common primary site of extragonadal GCTs in male patients, PMGCTs only account for approximately 2–5 % of all GCTs [11–13]. According to data from the Surveillance, Epidemiology, and End Results (SEER) 9 registries (1973–2007) [13], 2.1 % of all GCTs (1.8 % of seminomas, 2.4 % of non-seminomatous tumors) in white males and 4.7 % in black males are found in the mediastinum. Similarly, 2.3 % of all GCTs (2.8 % of dysgerminomas, 2.1 % of non-dysgerminomatous GCTs) in white females and 0.8 % in black females are identified in the mediastinum [13]. The incidence rates are 1.3/ 1 million for white males and 0.1/1 million for white females [13]. An epidemiologic study from Germany reported slightly lower incidence rates of mediastinal seminomas, non-seminomatous GCT in males, and non-dysgerminomas in females of 0.11, 0.2, and 0.03 per 1 million people, respectively [14].

In adults, PMGCTs occur predominantly in men; only approximately 9–14 % arise in women [13, 15–17]. The incidence of PMGCT abruptly

A.C. Roden, MD
Department of Laboratory Medicine and Pathology, Mayo Clinic, 200 First Street SW, Rochester, MN 55905, USA
e-mail: Roden.anja@mayo.edu

F.F. Nogales, R.E. Jimenez (eds.), *Pathology and Biology of Human Germ Cell Tumors*, DOI 10.1007/978-3-662-53775-6_8

increases at puberty. Overall, these tumors are more common in older adolescence and postpubertal children with a mean age of 29 years (range, 2–67 years) and peaks of incidence between ages 20 and 25 and around age 35 [13]. However, the age of the patient at the time of tumor diagnosis largely depends on the tumor type. For instance, congenital teratomas and yolk sac tumors occur predominately in very young patients, while seminomas are usually diagnosed in patients 10 years and older [18]. In contrast, sarcomatous elements are rare in PMGCT of children [19]. Similarly, mixed malignant PMGCTs are more common with increasing age. In adults, seminoma is the most common nonteratomatous component; yolk sac tumor, embryonal carcinoma, and choriocarcinoma may also occur. A rare case of placental site trophoblastic tumor presenting as a recurrence 2 years after the resection of a mediastinal teratoma in a 14-year-old male patient has also been described [20].

8.3 Clinical Features

Most mediastinal GCTs (82 % in a radiologic study) occur in the anterior mediastinum [21] and are commonly associated with the thymus [22]. In 14 %, multiple mediastinal compartments are affected [21]. Because of the relative large space in the anterior mediastinum, mediastinal GCTs, especially slow growing mature teratomas or seminomas, are often incidental findings [15, 23, 24]. The clinical presentation of patients with mediastinal GCT depends largely on tumor size. Symptoms such as cough, chest pain, hemoptysis, dyspnea, postobstructive pneumonia, and/or superior vena cava (SVC) syndrome occur in general due to compression of adjacent organs including large airways, great vessels, the heart, and phrenic nerve [25–28]. SVC syndrome, for instance, was described in approximately 10–25 % of anterior mediastinal seminomas [7, 15, 29–31]. Rarely, the tumor can erode into an adjacent bronchus which, in case of a mature teratoma, can lead to expectoration of hair (trichoptysis) or sebaceous debris [32]. Other rare complications include erosion into the pericardium, adjacent vascular structures, or through the skin to form a draining fistula [33, 34]. Painful gynecomastia was described in a patient with seminoma [35]. This patient had high circulating estradiol and beta-HCG levels that normalized after resection of the tumor.

Klinefelter syndrome is the only risk factor that has been identified for PMGCTs. This syndrome was identified in 8–22 % of male patients with PMGCT [27, 28, 36]. A study of 696 men with Klinefelter syndrome of the Danish Cytogenetic Register revealed a 67 times higher risk to develop PMGCT for patients with the syndrome than male patients without it [37]. Patients with PMGCT associated with Klinefelter syndrome are in general younger 4.5–31 years old than patients without the syndrome [27, 28, 37, 38]. In a study by Nichols, patients with Klinefelter syndrome and PMGCT had a median age of 15 years (range, 14–28 years) in contrast to 28 years (range, 18–35 years) in patients with PMGCT without the syndrome [27]. In some patients the diagnosis of Klinefelter syndrome is only established after a PMGCT has been identified. Therefore, a cytogenetic analysis has been recommended for young male patients with PMGCT [39]. The association of Klinefelter syndrome with PMGCTs is thought to be related to a persistent elevation of gonadotropin levels in these patients [37]. Gonadotropins might contribute to the malignant transformation of incompletely migrated primordial cells/germ cells [37]. However, genetic factors on the X chromosome have also been hypothesized [40]. Interestingly, the association with Klinefelter syndrome appears to be specific to PMGCT and pineal GCT but not gonadal or retroperitoneal GCT [38]. Furthermore, Klinefelter syndrome appears to be only associated with non-seminomatous PMGCT including teratoma with seminomatous or embryonal carcinoma and yolk sac tumor, pure teratoma, and yolk sac tumor. The reason for this restricted association of Klinefelter syndrome with non-seminomatous GCT of the mediastinum and pineal gland is not entirely clear. In addition a small subgroup of prepubertal children with PMGCT and Klinefelter syndrome present with precocious puberty due to HCG-producing PMGCT [40].

PMGCTs most commonly metastasize to the lung and bone, but metastases are also found in the liver, spleen, brain, tonsils, and subcutaneous tissue [6]. Only a minority of patients with mediastinal GCT has metachronous testicular tumors. Bokemeyer et al. [41] noted metachronous testicular tumors in only 1.1 % of men with mediastinal GCT. Fossa et al. [42] reported that three (of 15, 20 %) men with mediastinal GCT presented with testicular germ cell neoplasia in situ.

8.4 Imaging Findings

Teratomas usually present as an anterior mediastinal mass on chest X-ray. Calcifications are seen in 25 % of cases [21, 33]; well-formed teeth or bone are very suggestive of the diagnosis. Occasionally, areas of radiolucency suggest fat. Diffuse mediastinal widening or a mediastinal mass partially obscured by pulmonary parenchymal consolidation or cardiomegaly has also been described [21]. CT and MRI better characterize densities within the lesion suggestive of fat, sebaceous material, or cystic elements [21]. A multilocular cystic anterior mediastinal mass with fat content on CT scan is virtually diagnostic of mature teratoma. However, mature teratomas can also present as more heterogeneous masses, containing soft tissue, fluid, fat, and calcium attenuation. On CT scans, the majority of mature teratomas have well-defined margins against the adjacent lung parenchyma with a lobulated contour present in about half of the cases. The most common MR imaging finding is also a heterogeneous mass with signal intensities isointense with muscle, fluid, and fat [21]. Effusions are rather uncommon [21].

Immature teratoma can present as a large unilateral mass with heterogeneous densities and displacing mediastinal structures [43]. Malignant elements may exhibit an irregular thick wall with indistinct margins, obliteration of tissue planes, invasion of mediastinal structures, and/or extensive necrosis [44–46].

Seminomas typically appear as large and bulky, well-marginated, lobulated masses on chest X-ray [29, 47]. While its margins are usually well defined, invasion of the adjacent lung may result in irregular borders. On CT scan, seminomas appear large and coarsely lobulated with a homogeneous attenuation equal to that of soft tissue that may obliterate tissue planes or directly invade adjacent structures [47]. Seminomas show slight contrast enhancement [47]. Areas of low attenuation may also be detected [48]. Ringlike and stippled calcifications within a mediastinal seminoma are uncommon [49]. Rarely, when central necrosis occurs with little residual solid tumor (8 %), the lesion may exhibit extensive unilocular or multilocular low attenuation areas and may mimic cystic anterior mediastinal lesions [50–52]. Metastatic intrathoracic lymphadenopathy may also be observed [53].

Non-seminomatous GCTs manifest as large bulky anterior mediastinal masses that frequently exert mass effect on adjacent thoracic structures. Tumor margins may be well circumscribed or poorly defined [48]. CT typically demonstrates large heterogeneous masses with extensive central areas of low attenuation due to necrosis and hemorrhage. Residual viable tumor in general presents as lobular papillary soft tissue components in the periphery of the lesion that usually enhance [47]. Adjacent mediastinal tissue planes are frequently obliterated, and there may be radiologic findings of mediastinal, lung, or chest wall invasion. Associated pleural and pericardial effusions are common [54]. Metastases to regional lymph nodes may present as mediastinal lymphadenopathy [55, 56].

8.5 Histogenesis

Several hypotheses for the histogenesis of PMGCTs have been proposed:

(i) Primordial germ cells fail to complete the normal migration along the urogenital ridge to the gonadal ridges during embryonal development. This may be due to an abnormality in the primordial germ cell itself or in its microenvironment [57]. Different stages of development of the primordial cells and microenvironmental conditions may determine the final histology of the tumors at these sites [58].

(ii) Germ cells transformed in the testes undergo reverse migration [59, 60]. This hypothesis was supported by the lack of significant differences of chromosome aberrations between gonadal and mediastinal GCTs. For instance, nonrandom chromosomal changes were found to be essentially the same in gonadal and mediastinal GCT, and the incidence of isochromosome 12p was similar between gonadal and mediastinal GCT [59]. However, there are some biological differences between mediastinal non-seminomatous GCTs compared to its gonadal or retroperitoneal counterparts including worse prognosis and higher incidence of yolk sac tumor elements and leukemia in patients with mediastinal GCTs. Although these differences might result from differences in the cell of origin which would dispute this hypothesis, they also could result from the tumor's microenvironment.

(iii) Since mediastinal GCTs are usually associated with the thymus, mediastinal GCT might arise from thymic cells with germ cell potential.

Although morphologically similar, evidence suggests histogenetic differences between primary mediastinal and testicular GCTs [61–63]. For instance, the ploidy of mediastinal tumors is closer to those of testicular GCT of children that are usually diploid [61]; in contrast, testicular GCTs of adults are consistently aneuploidy [58]. A study comparing k-ras-2 gene sequences between mediastinal and testicular seminomas showed that 8 % (1 of 13) of mediastinal seminomas had a k-ras mutation in codon 13, while 15 % (2 of 13) of testicular seminomas had k-ras mutations in codon 12 [62]. Other studies confirmed that k-ras mutations in testicular GCT, if identified, are in codon 12 [64, 65]. Furthermore, weak p53 immunostaining was identified in 31 % of mediastinal seminomas in contrast to 77–90 % of testicular seminomas and 94 % of testicular non-seminomatous GCT [66]. A unique kit gene mutation on exon 17 was identified in about 50 % of primary mediastinal seminomas [63].

8.6 Histologic Classification of Mediastinal Germ Cell Tumors

PMGCTs are classified according to the WHO [67], identical to the classification of gonadal GCTs. PMGCTs are divided into seminomatous tumors (pure), non-seminomatous tumors (yolk sac tumor, embryonal carcinoma, choriocarcinoma, and mixed GCTs), and teratomas. Slightly more (52–60 %) PMGCTs are of non-seminomatous than seminomatous type [13, 68].

While the majority of PMGCT only has one tumor component, a study showed that 34 % of tumors had multiple components, and therefore adequate sampling of the tumor is essential [15]. The most common component of these malignant PMGCTs was a seminoma (88 % of cases), but embryonal carcinomas, malignant teratomas, choriocarcinomas, and yolk sac tumors were also identified.

8.6.1 Teratoma

Teratomas are the most common MGCT accounting for 43–75 % of mediastinal GCTs (Table 8.1) [68, 69]. They usually occur in the anterior medi-

Table 8.1 Frequency of mediastinal germ cell tumors [68, 69]

Mediastinal germ cell tumor	Frequency (% cases)
Teratoma	43–75
Seminoma	10–37
Yolk sac tumor	2–12
Embryonal carcinoma	2–8
Choriocarcinoma	2

astinum but occasionally can be seen in the posterior mediastinum [33, 68, 70]. Mediastinal teratomatous tumors include mature (63 %) and immature teratomas (4 %) and teratomas with other malignant components (i.e., sarcoma, other malignant germ cell elements, or carcinoma) (33 %) [68].

Mature mediastinal teratomas have been described in patients between 1 month and 73-year-olds with a peak incidence in early adulthood (mean 28 years) [33]. On gross examination [33] benign teratomas are usually encapsulated and well circumscribed (Fig. 8.1). The average tumor size is 10.5 cm (range, 2.5–27 cm). Similar to mature teratomas in other locations, cysts might occur within the tumor (Fig. 8.1) [70]. However, in contrast to their gonadal counterpart, monodermal teratomas such as struma ovarii have not been described in the mediastinum.

In contrast to adult gonadal teratomas or congenital/pediatric teratomas, in the adult mediastinum, the distinction between mature and immature teratoma is critical to patient management because immature teratomas have in general a worse prognosis.

At present, there is no grading scheme for extragonadal immature teratomas; however, it has been suggested to report the percentage of immature elements. Teratomas with other malignant germ cell elements (e.g., seminoma, embryonal carcinoma, yolk sac tumor) are regarded as malignant non-seminomatous GCT. Other adverse histologic features include sarcomatous or carcinomatous transformation, or an associated hematologic malignancy (see Chapter 12).

8.6.2 Seminoma

Pure seminomas represent the second most common mediastinal GCT accounting for approximately 10–37 % of all mediastinal GCTs (Table 8.1) [68, 69]. They usually occur in the anterior mediastinum [23, 51]. The vast majority of mediastinal seminomas are identified in men; only rare cases have been reported in women [6, 26, 29, 69, 71, 72]. The reported age ranges between 11 and 79 years with a mean age of 46.5 years in one study [51]; other studies report that the tumor most commonly occurs in the third decade followed by the fourth and second decade [6].

Seminomas vary in size from a few centimeters to over 16 cm in greatest diameter [23]. On gross examination, mediastinal seminomas are usually soft and have a smooth and glistening outer surface or may show a lobulated appearance; the cut surface is coarsely lobular or exhibits a discrete nodular pattern, and the color varies from white to light tan. In some cases, solid areas may alternate with large cystic spaces containing necrotic material, whereas in other areas the tumor may have the appearance of an entirely cystic mass.

Some microscopic features appear to be unique to mediastinal seminomas (Table 8.2) (Fig. 8.2). They frequently involve the thymus, showing cyst formation and thymic epithelial cell hyperplasia [50, 73]. Cystic seminomas may histologically mimic multilocular thymic cysts [50]. These tumors characteristically show areas of lymphoid hyperplasia (Fig. 8.3), cysts lined by squamous epithelium, and cholesterol cleft granulomas. The seminomatous component might be growing along the cystic walls of the tumor making its distinction from thymic follicular hyperplasia or thymoma (Fig. 8.4) difficult. Therefore, a high level of suspicion is necessary in the case of cystic lesions of the thymus, especially if associated with a granulomatous response; extensive sampling of these cystic tumors is critical. Seminomas might be especially difficult to identify in small biopsies from the mediastinum because of cystic changes, and inflammatory and granulomatous response may obscure the diagnostic tumor cells.

In mediastinal seminomas the neoplastic cells can sometimes be arranged in cell nests that are separated by thin fibrovascular septa. The fibrovascular septa will characteristically be infiltrated by a large number of lymphocytes. Foci with giant cells of the syncytiotrophoblast type have been observed in <5 % of mediastinal GCTs [23]. A few cases of anaplastic seminoma have also been reported in the anterior mediastinum [74]. Rarely, in the mediastinum, seminomas can be

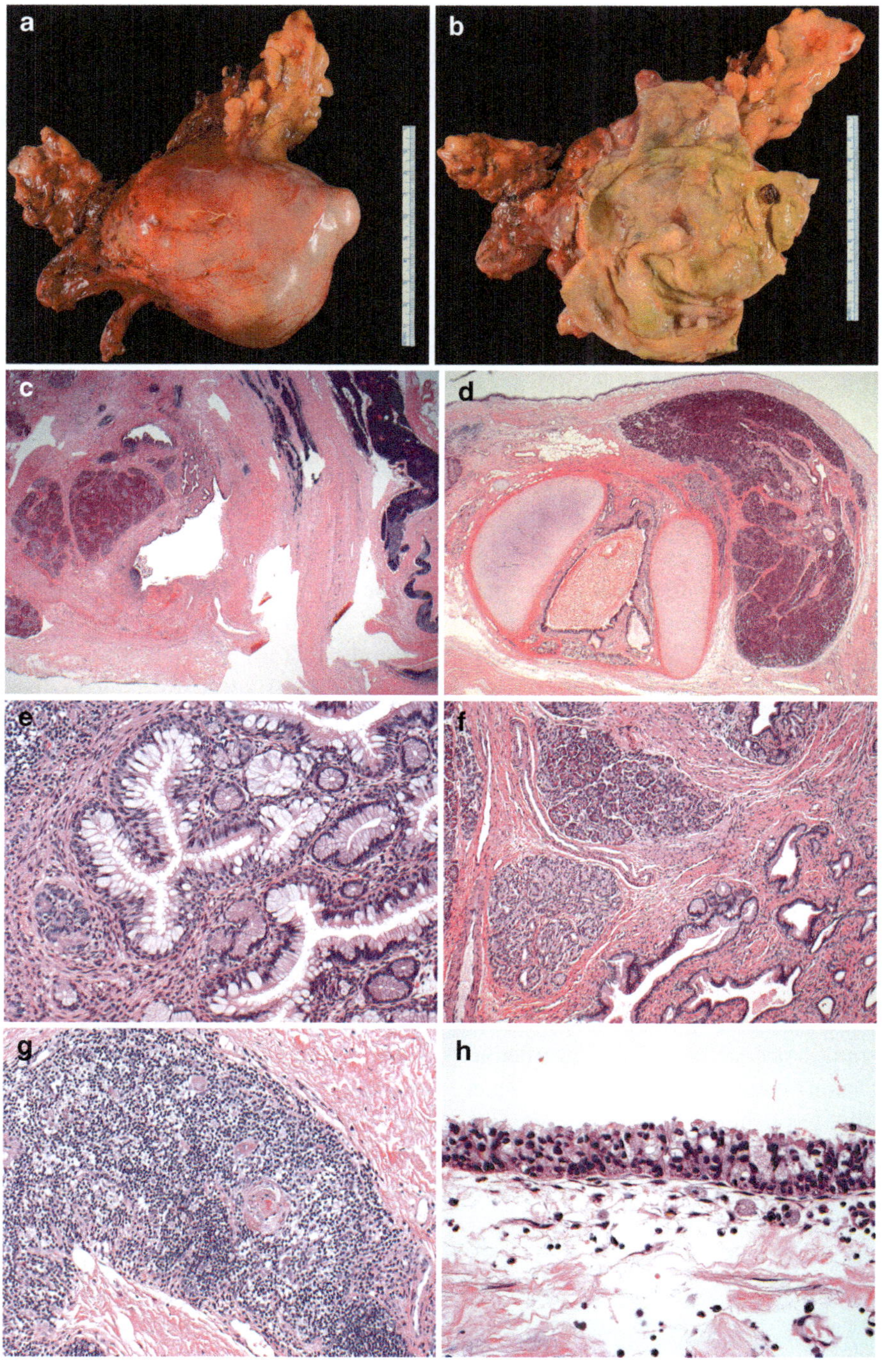

Fig. 8.1 Mature cystic teratoma. (**a**) A well-circumscribed mass with smooth borders has focal yellow tissue at the rim. (**b**) Sectioning reveals a cyst with largely smooth lining and focal small nodular areas. (**c**) On microscopy, the cyst wall is comprised of fibrous tissue with glandular and cystic areas and thymic parenchyma (*upper right*). (**d**) Multiple germ layers are present including the mesoderm (cartilage, adipose tissue) and endoderm (respiratory epithelium (**e**, **f**) and pancreatic tissue (**f**)). This teratoma is in a background of thymic parenchyma (**g**). The cyst is lined by benign ciliated respiratory epithelium (**g**). Magnification × 12.5 (**c**), 20 (**d**), 200 (**e**, **g**), 100 (**f**), 400 (**h**)

Table 8.2 Histopathologic features of primary mediastinal seminomas (Figs. 8.2 and 8.3) [51, 73]

Histopathologic features	Frequency (% cases)
Lymphocytic infiltration	100
Fibrous septa/stroma	91
Prominent tumor cell nucleoli	91
Clear tumor cell cytoplasm	87
Distinct tumor cell borders	87
Non-necrotizing granulomatous inflammation	46–74
Cellular pleomorphism	43
Necrosis	35
Thymic remnants	27
Prominent cystic changes	8
Intercellular edema	4
Syncytiotrophoblasts	4
Mean mitotic count/ten high-power fields	4.4 (range, 0–16)

associated with another non-GCT such as primary leiomyosarcoma [75].

Most seminomas (96 %) of the mediastinum harbor chromosome 12p abnormalities, including 12p amplification (87 %) or isochromosome 12p (65 %) [73].

8.6.3 Other Non-seminomatous Mediastinal Germ Cell Tumors

Other non-seminomatous mediastinal GCTs are rare (Table 8.1).

The histopathologic characteristics of non-seminomatous mediastinal GCTs such as embryonal carcinoma, mixed germ cell tumors, yolk sac tumors (Fig. 8.5), and choriocarcinomas are similar to their gonadal counterparts and are described elsewhere in this book.

8.6.4 Sarcomatoid Component of Mediastinal Germ Cell Tumors

Sarcomatous differentiation of GCT is most frequently seen in the mediastinum and may occur in association with mature teratomas or, less commonly, with other malignant GCTs including immature teratoma, choriocarcinoma, yolk sac tumor, and seminoma [76]. The most common type of heterologous differentiation is rhabdomyosarcoma; other sarcomatous components include angiosarcoma, leiomyosarcoma, glioblastoma multiforme, malignant peripheral nerve sheath tumor, epithelioid hemangioendothelioma, and undifferentiated sarcoma (Fig. 8.6) [76]. Cases with components of chondrosarcoma, osteosarcoma, liposarcoma, malignant fibrous histiocytoma, primitive neuroectodermal tumor, and neuroblastoma have also been reported. Any somatic-type malignancy should be reported because PMGCTs with sarcomatous differentiation are unresponsive to conventional GCT therapy and their prognosis is dismal [76]. An estimate of the involved area may also be helpful (see also Chapter 12).

Because of the poor prognosis of patients with PMGCT with sarcomatous component, its distinction from immature mesenchyme in an immature teratoma is critical. In general, the spindled component of immature teratoma is cytologically bland, is relatively monomorphic, and is typically condensed around teratomatous glands with a concentric, swirling growth pattern. In contrast, sarcomatous differentiation is characterized by an expansile and architectural complex (i.e., intersecting fascicles or storiform pattern) growth with infiltration of the surrounding tissues; usually has a greater degree of nuclear pleomorphism and hyperchromasia, obvious mitotic activity, and obvious malignant heterologous differentiation; and lacks intimately admixed glands. Rhabdomyoblasts can be distinguished from hepatoid yolk sac tumor or other eosinophilic mimickers by the demonstration of expression of smooth muscle markers including desmin and myogenin. In contrast to stromal overgrowth, sarcomatous differentiation shows independent growth by virtue of replacing teratomatous glands or other germ cell elements, thus forming an area of pure sarcoma [77].

Patients with PMGCT with sarcomatous component appear to have a worse prognosis than their gonadal counterparts. Malagon et al. [76] showed that only 18 % of patients with PMGCT

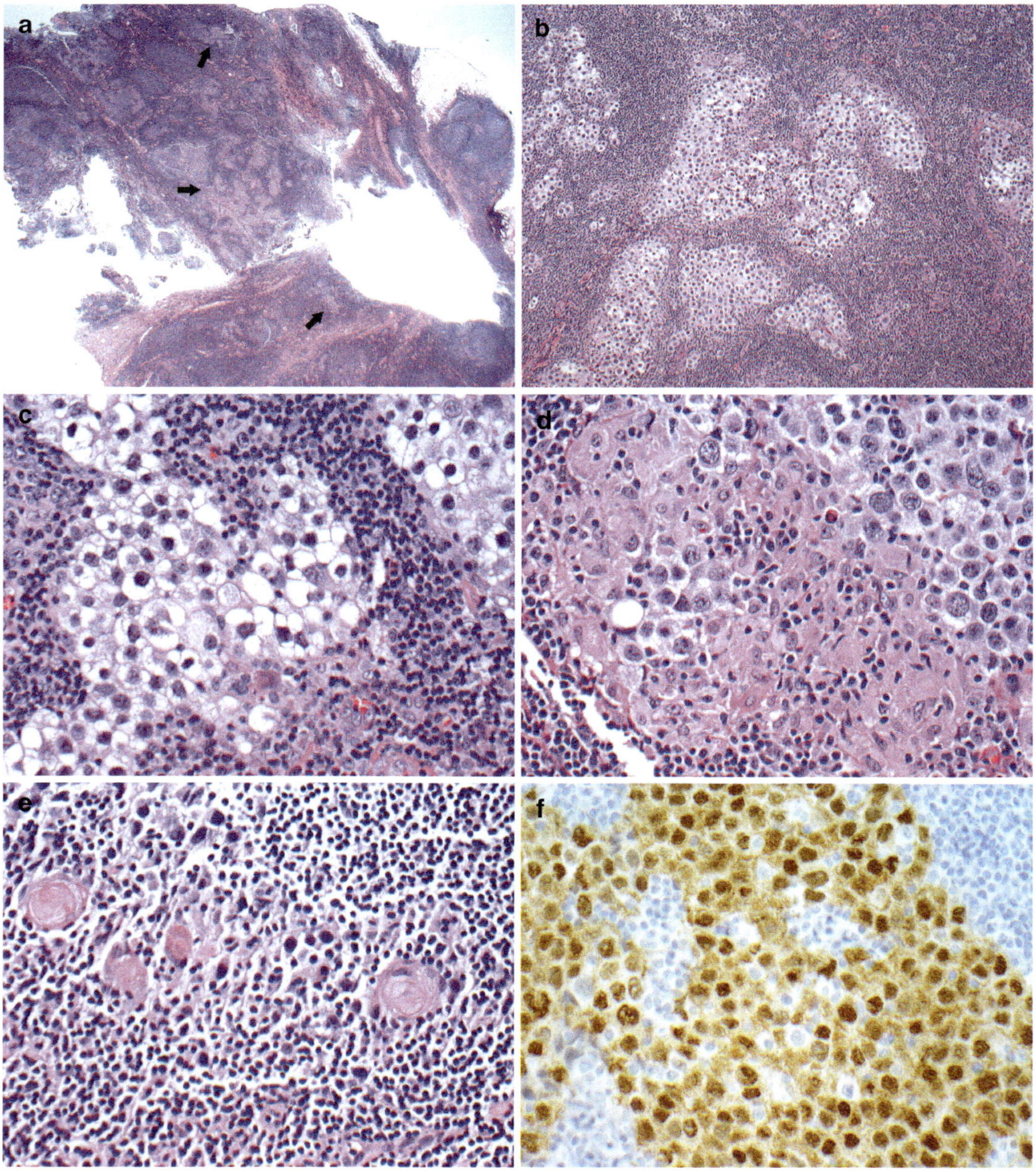

Fig. 8.2 Seminoma arising in the thymus. (**a**) Lymphoid tissue with germinal centers contains scattered nests of atypical cells (*arrows*). (**b**) These irregular nests are comprised of cohesive atypical epithelioid cells, many of which have clear cytoplasm (**c**) and others have more eosinophilic cytoplasm (**d**). Many tumor cells have prominent nucleoli. Occasional Hassall corpuscles within the lymphoid tissue suggest a background of thymic parenchyma (**e**). The neoplastic cells are positive for OCT4 (**f**) and CD117 (**g**) and are negative for CD30 (**h**) and keratin AE1/AE3 (**i**). Magnification × 12.5 (**a**), 100 (**b**), 400 (**c–i**)

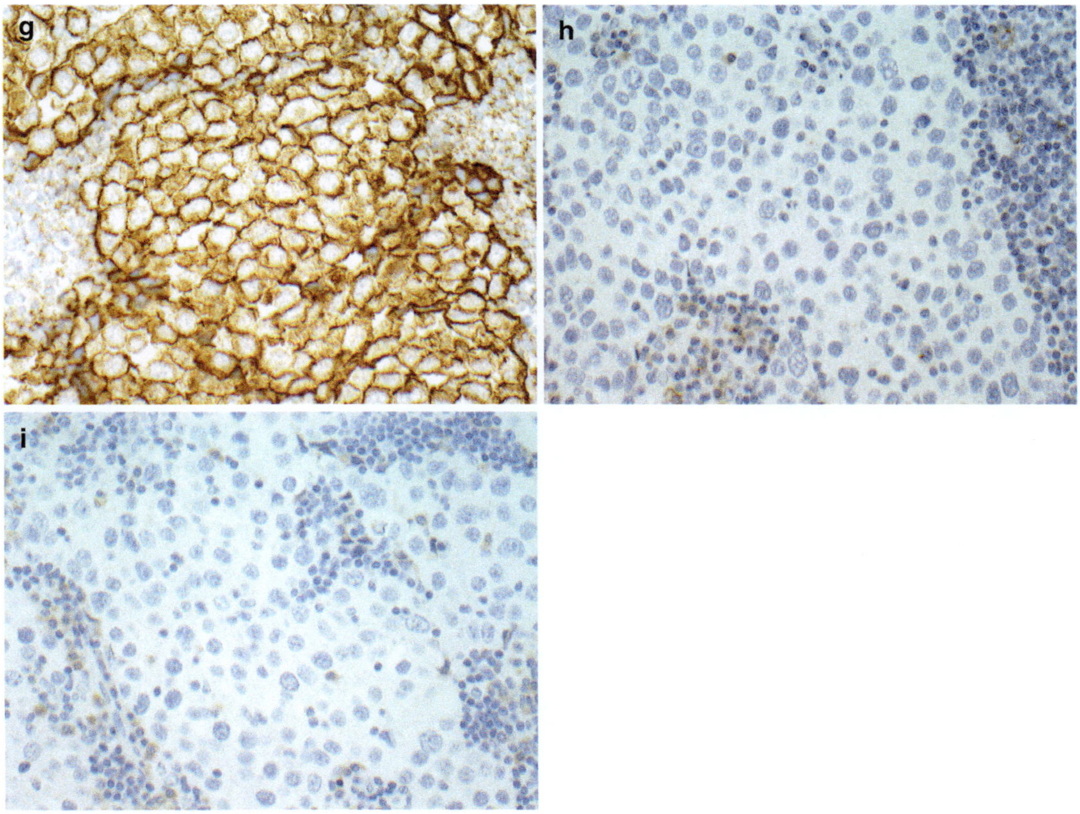

Fig. 8.2 (continued)

with sarcomatous component were alive at 12–42 months, while 82 % had died from disease between 1 and 37 months. Patients with mediastinal tumors usually die due to local compromise of vital structures. In contrast 56 % of patients with testicular GCT with sarcomatous component were alive at 1–72 months, and only 44 % had died from disease between 5 and 96 months. Despite their better prognosis, testicular GCTs with sarcomatous component have a greater tendency for metastases (75 %) compared to mediastinal tumors (18 %). These differences in tumor behavior and prognosis may be due to location and size of the lesion and their resectability. In the study by Malagon et al. [76], the majority of extragonadal tumors were larger and bulkier than the gonadal lesions and were more commonly excised with positive margins. Moreover, the mediastinal location allows tumors to grow much larger before becoming symptomatic, so increasing the potential for malignant transformation. Because of the large size, the majority of PMGCTs were extensively infiltrative at the time of surgery making complete resection very difficult.

8.7 Ancillary Studies

Ancillary studies might be utilized to distinguish PMGCT from other primary or secondary mediastinal tumors. Immunohistochemical and cytogenetic studies are most helpful.

8.7.1 Immunohistochemical Studies (Table 8.3)

In general, the immunophenotype of PMGCT is similar to its gonadal counter. However, there are

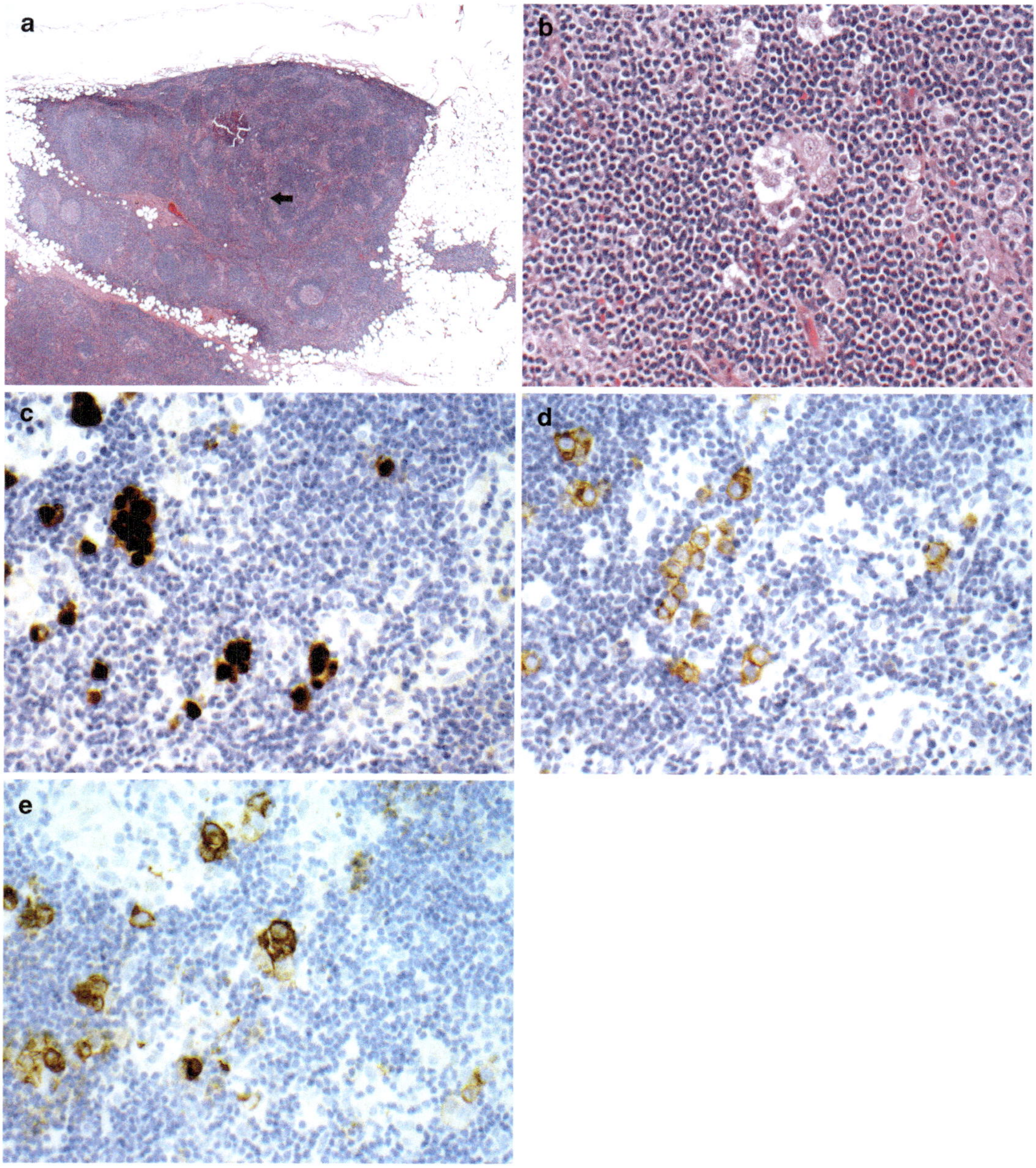

Fig. 8.3 Seminoma. (**a**) In this case there are only occasional single tumor cells or small clusters of tumor cells in a lymphoid background with prominent lymphoid hyperplasia. On low magnification these tumor cells are difficult to identify and might be easily missed (*arrow*). (**b**) On higher magnification small nests of tumor cells are identified. These tumor cells have clear cytoplasm and some have prominent nucleoli. The neoplastic cells are highlighted by OCT3/4 (**c**), CD117 (**d**), and PLAP (**e**). Magnification × 20 (**a**), × 400 (**b**–**e**)

some significant differences in the expression of some antigens, especially on tumor cells of seminomas [78]. For instance, in a study by Suster et al. [78], the low molecular weight keratin CAM 5.2, which shows a strong dot-like paranuclear staining pattern in seminomas, is expressed in 80 % of mediastinal seminomas but only in 20 % of testicular seminomas. Similarly, PLAP is expressed in 92.5 % of mediastinal seminomas (Fig. 8.3) but only in 50 % of testicular seminomas.

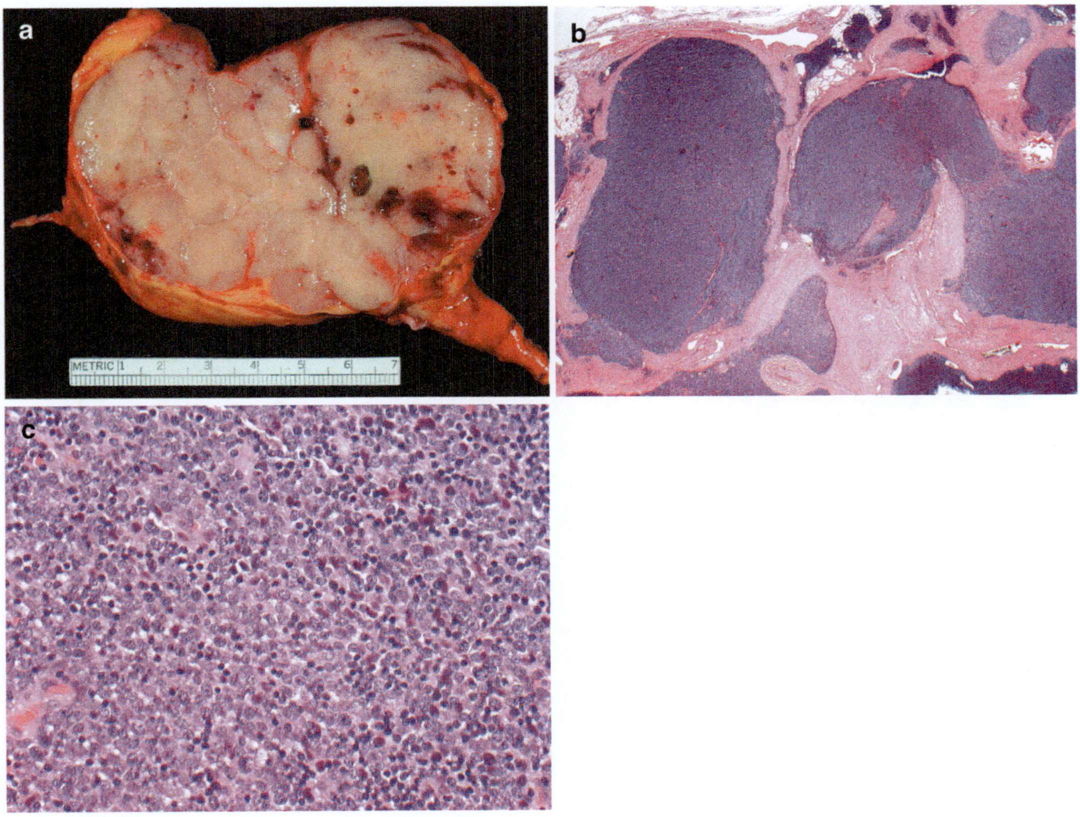

Fig. 8.4 Thymoma, WHO type B2. (**a**) On gross examination, there is a well-circumscribed, tan, lobulated tumor with intervening fibrous septa. (**b**) The lobulated architecture is also apparent on low-power microscopy that reveals hypercellular lobules that are separated by fibrous septa. (**c**) On high magnification there is a mixture of larger epithelial tumor cells and small lymphocytes. Magnification × 12.5 (**b**), × 400 (**c**)

OCT4 has a high sensitivity and specificity for seminoma and embryonal carcinoma [79–81]. Nearly 100 % of seminomas and embryonal carcinomas show nuclear reactivity with OCT4, and the specificity seems superior to other available markers (Fig. 8.2f) [79–81].

Keratins can be expressed in 39–80 % of mediastinal seminomas and therefore might present a pitfall mimicking carcinoma [51, 77, 78]. However, in most cases, the epithelial markers highlight only a small proportion of tumor cells with variable intensities [73].

Strong membranous CD117 (kit) immunoreactivity has been reported in 75–100 % of seminomas (Figs. 8.2g and 8.3d) [82]. However, in the mediastinum, CD117 is not specific to seminomas because it is also expressed in other non-germ cell tumors including small cell carcinoma and adenocarcinomas of the lung [83] and thymic carcinomas [84, 85]. In addition, embryonal carcinoma, yolk sac tumor, and choriocarcinoma may show some degree of weak CD117 reactivity (Fig. 8.5e).

CD30 is expressed in over 80 % of embryonal carcinomas. However, CD30 is also expressed in various hematopoietic malignancies which commonly occur in the mediastinum such as mediastinal large B-cell lymphoma or Hodgkin lymphoma (Fig. 8.7) [86] but also occasionally in yolk sac tumors and rarely seminoma [78].

AFP is not a reliable marker for yolk sac tumors because of its low sensitivity [87]. Furthermore, AFP can be expressed in 60 % of hepatoid adenocarcinomas of the lung [88]. Serum evaluation of AFP and [beta]-human chorionic gonadotropin (beta-HCG) is frequently

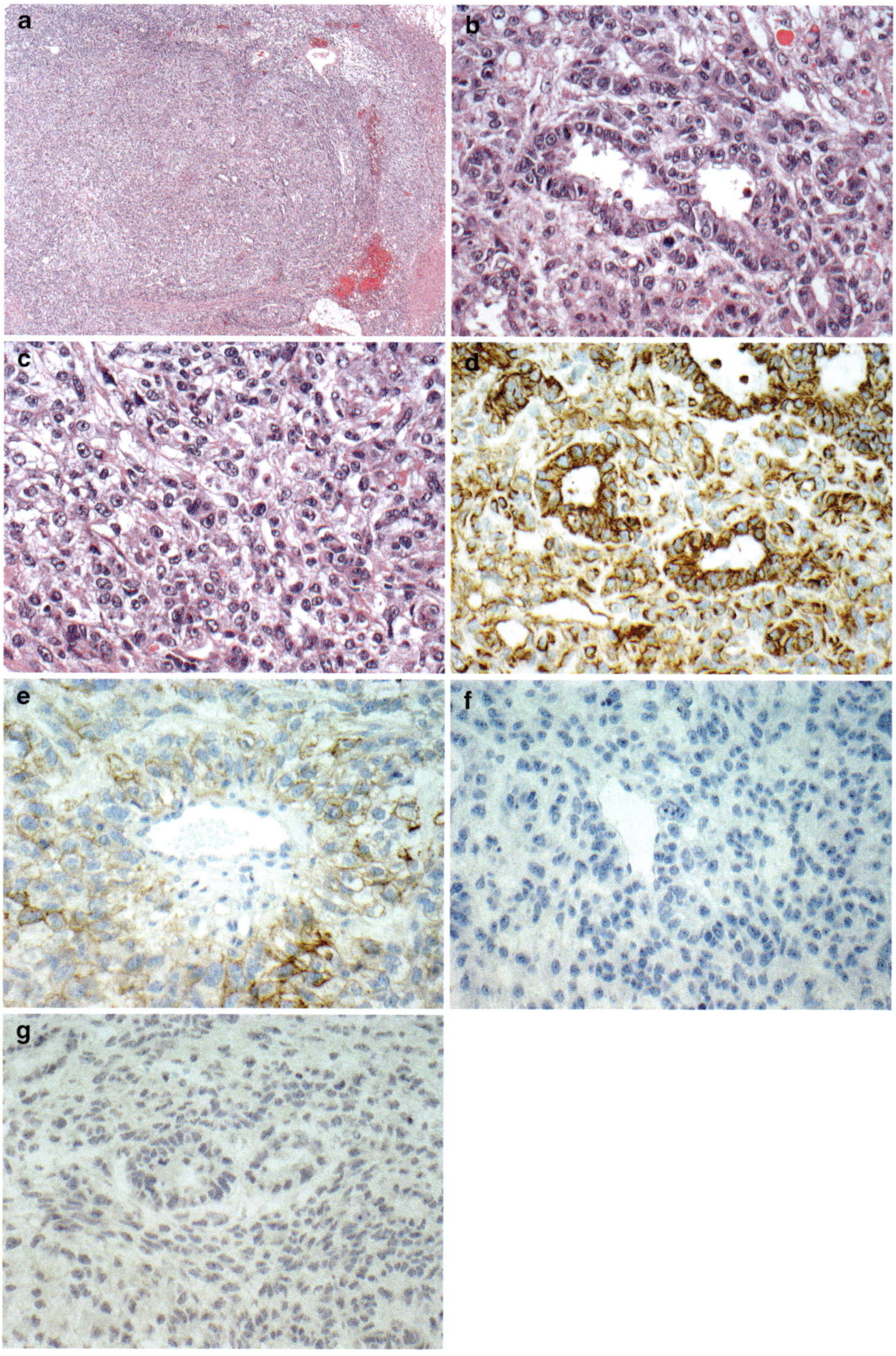

Fig. 8.5 Yolk sac tumor. (**a**) This tumor shows a predominant solid growth pattern with a few glandular structures. (**b**) Schiller-Duval bodies are identified. (**c**) The neoplastic cells are round to oval with clear to eosinophilic cytoplasm, open chromatin, and prominent nucleoli. The tumor cells express keratin AE1/AE3 (**d**), some express weakly CD117 (**e**), and they are negative for PLAP (**f**) and OCT4 (**g**). Magnification × 40 (**a**), × 400 (**b–g**)

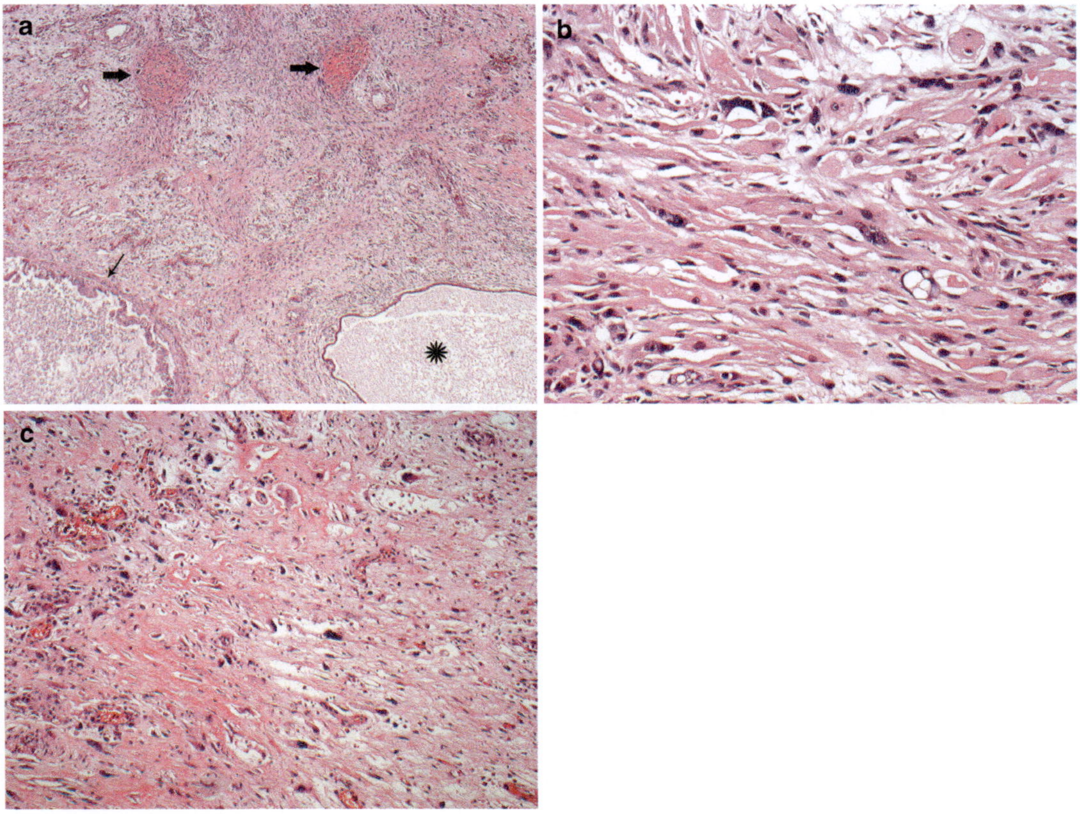

Fig. 8.6 Spindle cell sarcoma arising in teratoma. (**a**) Low magnification view reveals a neoplastic spindle cell proliferation with focal osteoid (*thick arrow*). A cyst lined by glandular epithelium (*thin arrow*) and another cyst lined by bland cuboidal epithelium (*star*) are suggestive of a teratoma. (**b**, **c**) High-power view shows pleomorphic spindle cells with dark nuclear chromatin consistent with a spindle cell sarcoma. Magnification × 40, (**a**), × 400 (**b**, **c**)

more sensitive than immunohistochemistry. However, beta-HCG is also not entirely specific to GCT and can be seen in 10–60 % of adenocarcinomas of the lung [89, 90].

Although placental-like alkaline phosphatase (PLAP) has traditionally been the marker of choice for GCTs (mostly seminoma), in the setting of an "undifferentiated" neoplasm in the mediastinum, its lack of sensitivity, generally high background staining, and the development of newer antibodies have rendered this stain less useful in current diagnostic practice.

8.7.2 Cytogenetic Studies

Bosl et al. [11] found that the isochromosome of chromosome 12 [i(12p)] as evaluated by FISH studies is a useful marker for GCT in males. This was further confirmed in a study by Sung et al. [73] that showed abnormalities of chromosome 12p including 12p amplification and i(12p) in 22 out of 23 cases of mediastinal seminoma. Chaganti et al. [59] performed karyotypic analysis of 13 PMGCT and observed characteristic i(12p) in 69 % cases (9/13).

Table 8.3 Immunophenotype of mediastinal germ cell tumors and their mimickers [Frequency (% positive cases)]

Immunostain	Seminoma	Embryonal carcinoma	Yolk sac tumor	Teratoma	Choriocarcinoma	Lung adenocarcinoma	Small cell carcinoma	Thymic carcinoma	Malignant mesothelioma	NUT midline carcinoma
OCT 4 [73, 115, 150–152]	100	100	0	0	0	N/A[a]	N/A	0	0	0
OCT 3/4 [153–157]	100	82–100	38	0	0	0	N/A	N/A	N/A	0
CD117 [73, 82–85, 150, 157–161]	75–100	77–100	30–59	43	0	17	82	80–86	5	0–25
SALL4 [150, 157]	100	97	100	29	0	N/A	N/A	0	N/A	0
AFP [78, 88, 89, 141, 150, 157, 158]	0	0	56–100	N/A	0	60[b]	N/A	N/A	N/A	0
Beta-HCG [78, 89, 90, 141, 157, 162]	3	33	0	10–60	100	7–60	0	0	0	0
PLAP [73, 78, 141, 150, 157, 163–165]	43–100	59	40	0	0	25–67	N/A	0	15	0
D2–40 [153, 158, 166]	96–100	35	3	N/A	N/A	7	N/A	N/A	96	N/A
CAM 5.2 [73, 78, 157, 161, 165, 167–169]	48–80	100	100	100	100	100	100	100[c]	100	73
Keratin AE1/AE3 [73, 78, 157, 158, 167, 168, 170]	0–43	100	100	100	100	100	50–100	100	100	31
High molecular weight keratin [73, 168, 171–173]	0–39	0	N/A	N/A	N/A	25–82	0	100[d]	89[e]	N/A
CK7 [73, 103, 157, 161, 164, 171, 174, 175]	39–41	100	N/A	97	N/A	79–100	43	80	65–81	40
CK20 [73, 103, 157, 174]	0	0	0	0–100	N/A	10	0	N/A	0	0
EMA [73, 153, 157, 165, 167, 171, 176–178]	2–9	33	29	100[f]	54	95–100	N/A	100	96[e]–98	71

CD5 [84, 157, 165, 175, 179, 180]	0	N/A	N/A	N/A	N/A	10	0	20–70	0–12	33
CD30 [73, 78, 82, 150, 153, 157, 165, 181]	0–2	73–100	0–11	40[g]	0	0	0	N/A	0	12
TTF-1 [103, 157, 161–163, 165, 172, 177, 179, 182–184]	0	0	0	50	N/A	57–89	85–95	0	0	33
Calretinin [167, 173, 177, 185, 186]	9	0	N/A	0	0	4–10	41	N/A	87–100	N/A
WT-1 [157, 179, 182]	N/A	N/A	0	N/A	N/A	0	N/A	5	58–93	0
NUT [157, 187, 188]	19–71%[h]	0	5	N/A	N/A	N/A	N/A	6[i]	N/A	100

[a]N/A, not available
[b]60% of hepatoid lung adenocarcinoma
[c]Thymic carcinoma with clear cell features
[d]Spindle cell thymic carcinoma
[e]Epithelioid malignant mesothelioma
[f]EMA is expressed in all non-neural components of mature teratoma
[g]Presence of CD30 confined to respiratory component, squamous component, GI epithelium, nerve, or cartilage [153], negative in immature teratoma
[h]71% in spermatocytic seminoma, 19 % in conventional seminoma
[i]Tumors in thymic region

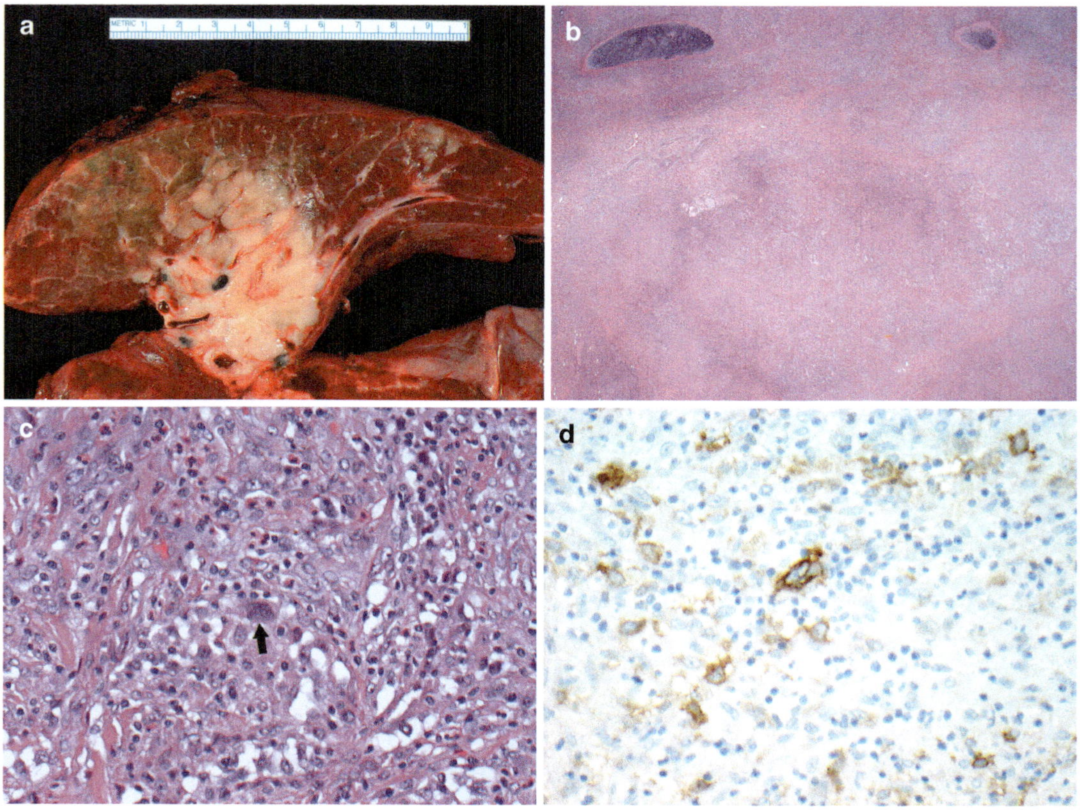

Fig. 8.7 Classical Hodgkin lymphoma, nodular sclerosing type. (**a**) A pneumonectomy specimen contains a 6.0 cm left centrally located, ill-defined, lobulated, yellow-white lung mass. (**b**) On low-power microscopy, there is a fibrotic, vaguely lobulated mass in a central location (note residual hyaline cartilage of a large airway in the upper left and right corners). (**c**). On high power, scattered large, multinucleated Reed-Sternberg cells (*arrow*) are apparent in a background of a mixed inflammatory background predominantly comprised of lymphocytes, eosinophils, macrophages, and plasma cells. The large atypical cells mark with CD30 (**d**). Magnification, × 12.5 (**b**), × 400 (**c**, **d**)

8.8 Differential Diagnosis of Mediastinal Germ Cell Tumors (Tables 8.4 and 8.5)

8.8.1 Germ Cell Tumors Metastatic to the Mediastinum

Although rare, testicular GCT can metastasize to the mediastinum [91]. However, in most mediastinal GCTs, examination of the testes fails to reveal a primary tumor. For instance, in 16 patients with extragonadal GCTs, the testicles did not have any palpable lesion [92]. However, occult testicular tumors were later identified in 10 of 12 patients with retroperitoneal GCT but in none of the mediastinal GCT. Therefore, not all testicular GCTs can be identified by palpation alone. In a study of 20 autopsy cases of mediastinal GCT, Luna and Valenzuela-Tamaris [93] identified only two cases in which the testes contained either an occult tumor or a well-defined testicular scar. Moreover, of 78 autopsies of patients with testicular GCTs, no autopsy showed solely metastases in the anterior mediastinum without involvement of other mediastinal lymph nodes (middle/posterior) [94]. In another such study of 220 cases of metastasizing testicular tumors, no metastases were documented in the mediastinum [95]. These studies emphasize that testicular GCT can metastasize to the mediastinum, but that appears to be rather exceptional, and GCTs in the mediastinum are usually

Table 8.4 Features that might facilitate the distinction between mediastinal germ cell tumors and their mimickers

Diagnosis	Distinguishing feature(s)
Seminoma	Lymphocytic infiltrate
	Cyst formation possible
	Intense staining for OCT4
	Isochromosome 12p abnormalities by FISH
Thymoma	Lobulated architecture
	OCT4 negative
Thymic carcinoma	OCT4 negative
NUT carcinoma	NUT immunostain shows a speckeled nuclear staining pattern
	t(15;19) by FISH, RT-PCR
Metastatic carcinoma	Morphologic features
	Immunophenotype
Synovial sarcoma	t(X;18), FISH, RT-PCR
Lymphoma	Immunophenotype
	Flow cytometry
	B-cell or T-cell receptor rearrangement studies

of primary origin. However, the possibility of mediastinal GCT of testicular origin exists and must be excluded. A careful review of the past medical history, physical examination, and imaging studies are necessary to distinguish between primary and metastatic disease. Moreover, retroperitoneal tumor metastases are virtually always present in cases of testicular seminoma that have metastasized to the mediastinum.

8.8.2 Thymic Cysts

Thymic cysts may be unilocular or multilocular and can potentially mimic teratoma or seminoma. The unilocular cysts, remnants of the third branchial pouch-derived thymopharyngeal duct, may be lined by cuboidal, columnar, or sometimes squamous epithelium and contain thymic tissue within the cyst wall (Fig. 8.8)

Table 8.5 Differential diagnosis of tumors that might occur in the mediastinum by morphologic pattern

Morphologic pattern				
Epithelial-lined cyst	Mixed epithelial and spindled	Poorly differentiated	Papillary	Clear cells
Mature teratoma	Immature teratoma	Embryonal carcinoma	Embryonal carcinoma	Seminoma
Seminoma	Yolk sac tumor	Seminoma	Yolk sac tumor	Yolk sac tumor
Thymic cyst	Malignant mesothelioma	Malignant mesothelioma	Malignant mesothelioma	Thymic clear cell carcinoma
Enteric cyst	Sarcomatoid carcinoma	Thymic carcinoma	Myxopapillary ependymoma	Metastatic Muellerian clear cell carcinoma
Bronchogenic cyst	Synovial sarcoma	Lymphoma		Metastatic renal cell carcinoma
Epidermal inclusion cyst	Pleuropulmonary blastoma	NUT carcinoma, t(15;19)		
Dermoid cyst	Congenital peribronchial myofibroblastic tumor	Epithelioid angiosarcoma		
Branchial cleft cyst	Pulmonary hamartoma	Lung adenocarcinoma		
Thyroglossal duct cyst	Thymoma (WHO type AB)	Melanoma		
Epidermoid cyst	Spindle cell epithelial tumor with thymus-like differentiation (SETTLE)	Carcinoma showing thymus-like differentiation (CASTLE)		
	Metastatic Wilms' tumor			

Modified from [77]

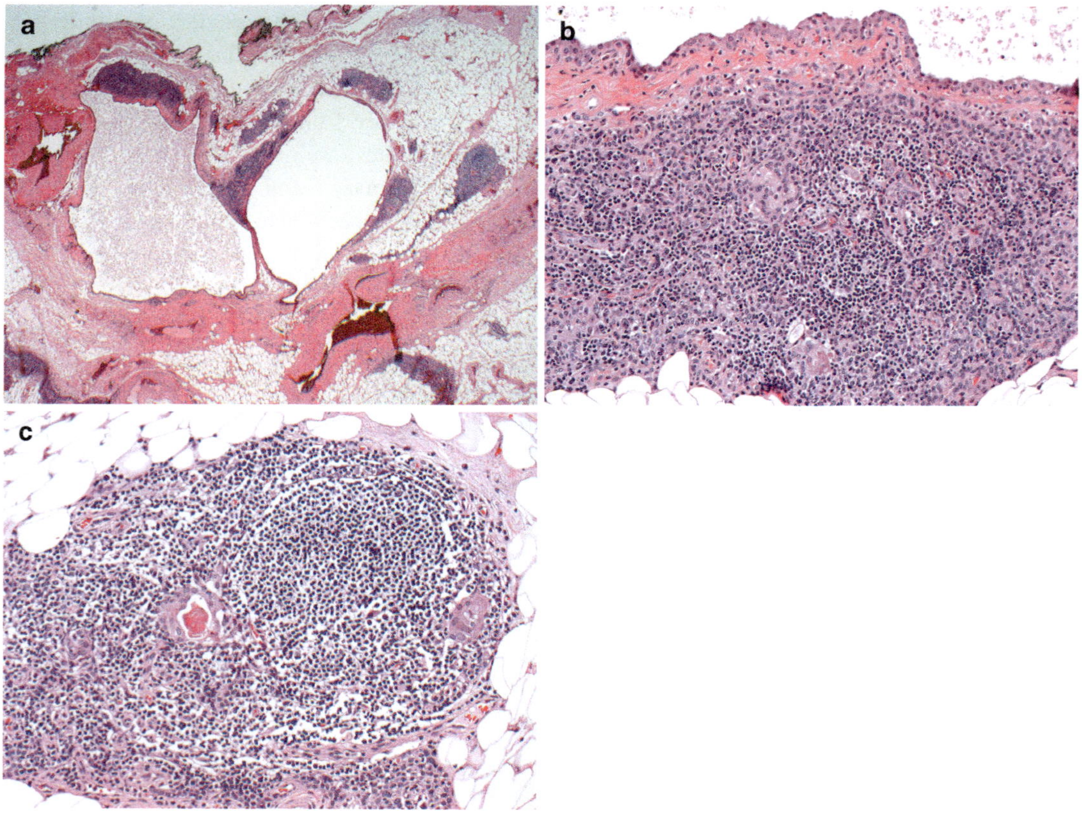

Fig. 8.8 Thymic cyst. (**a**) Cystic structures are surrounded by a lymphoid and adipose tissue. (**b**) The cysts are lined by bland cuboidal cells. (**c**) Benign thymic parenchyma with a Hassall corpuscle is also present in the wall of the cyst. Magnification × 20 (**a**), × 200 (**b**, **c**)

[96]. Multilocular cysts are acquired cystic ductular dilatations due to inflammatory reaction of the thymic parenchyma and are frequently lined by squamous epithelium, but cuboidal or ciliated columnar linings are also described [97, 98]. Cholesterol granulomas and chronic inflammatory infiltrate are commonly found in the cyst wall of multiloculated thymic cysts. The lack of heterologous elements and the presence of thymic tissue underlying the epithelium should aid in the distinction from teratoma. However, seminoma and yolk sac tumors should be carefully excluded because these tumors can show similar cystic changes in the thymus as seen in a multiloculated thymic cyst [50]. Awareness of this secondary thymic change and thorough sampling are critical in excluding an associated GCT.

8.8.3 Enteric Cysts

Enteric cysts are usually found in children and adolescents and are almost exclusively located in the posterior mediastinum (paraesophageal and gastroesophageal) [99–101]. The cyst wall contains a double layer of smooth muscle and can be lined by simple or pseudostratified columnar, squamous, or gastric mucosa.

8.8.4 Bronchogenic Cysts

Bronchogenic cysts, congenital anomalies of foregut origin, can be found in any mediastinal compartment and in any age group [101, 102]. They may closely mimic mature teratomas because they are comprised of respiratory epithelium, smooth muscle, and mature cartilage or mucous glands. Cysts with well or moderate architectural differentiation toward normal tracheobronchial structures, presence of respiratory-type epithelium, lack of enteric-type epithelium, immature elements, atypia, and tumor necrosis favor bronchogenic cyst [103]. The presence of CK7 expression in the absence of staining with CDX2 further supports bronchogenic cyst. In contrast, teratomas have a mixed enteric and respiratory epithelium, and the majority of the glands express CK7, CK20, CDX2, and TTF-1. Moreover, coexpression of CDX2 and TTF-1 was only found in teratoma [103].

8.8.5 Meningocele

Meningoceles are posterior mediastinal cysts which communicate with meninges. In general they occur in infants and children [104]. The clinical/radiographic features are usually characteristic. Microscopically, they might show various amounts of neural tissue and calcification and should not be confused with teratoma.

8.8.6 Pleuropulmonary Blastoma

In children, pleuropulmonary blastoma (PPB) can potentially mimic a GCT because of its biphasic appearance and heterologous mesenchymal differentiation [105, 106]. It can be cystic and/or solid and has sarcomatoid characteristics. It is usually found in the lung but occasionally can be identified in the pleura. The cysts are typically lined by respiratory or cuboidal epithelium and lack the squamous, gastrointestinal, or neuroglial lining often seen in teratoma (Fig. 8.9). The primitive spindled cells of PPB often resemble embryonal rhabdomyosarcoma or fibrosarcoma, patterns which are unusual in teratoma but might be present as sarcomatoid component in other GCTs. Heterologous elements such as cartilage may rarely be seen in PPB. The cyst lining together with the absence of other organized elements helps to distinguish this rare malignant tumor from a primary intrapulmonary teratoma.

8.8.7 Sarcomatoid Carcinoma

Sarcomatoid carcinomas of the lung are a broad spectrum of poorly differentiated non-small cell carcinomas that contain a sarcoma or sarcoma-like component (Fig. 8.10) [107]. The closest mimics of immature teratoma include pleomorphic carcinoma, carcinosarcoma, and pulmonary blastoma. Pleomorphic carcinoma contains a component of morphologically typical non-small cell carcinoma (i.e., squamous cell carcinoma, adenocarcinoma, or large cell carcinoma) admixed with a malignant spindle cell component lacking a specific line of heterologous differentiation. The obvious non-small cell lung carcinoma component should allow distinction from immature teratoma or PMGCT with sarcomatoid component in most cases. Carcinosarcoma, like pleomorphic sarcoma, is comprised of non-small cell carcinoma, but, in contrast to pleomorphic sarcoma, also contains a differentiated sarcomatous component (e.g., chondrosarcoma, rhabdomyosarcoma, osteosarcoma) (Fig. 8.11) [77]. Again, the presence of typical non-small cell lung carcinoma should allow the distinction in most cases. The closest histologic mimic of teratoma is the pulmonary blastoma [108–110]. Despite the name, this is a tumor predominantly of adults characterized by an admixture of fetal-type adenocarcinoma (tubules lined by pseudostratified columnar, nonciliated epithelium with clear to lightly eosinophilic cytoplasm) and embryonic mesenchyme, both of which resemble fetal lung between 10 and 16 weeks of gestation [111]. The glands often have supranuclear or subnuclear vacuoles and sometimes show squamous morular metaplasia creating an endometrioid appearance. The condensation of the spindle cell component around the glands may closely mimic immature teratoma, especially if heterologous differentiation is present. Recognition of the typical fetal-type gland morphology is key to this distinction.

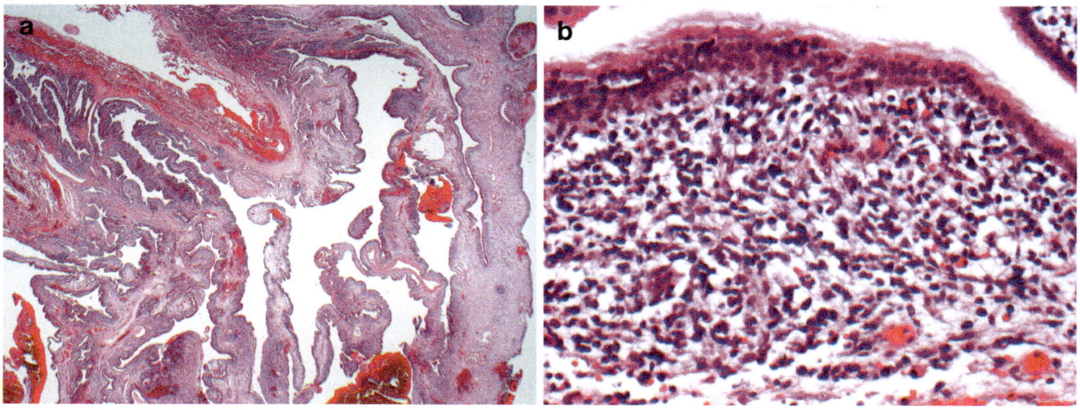

Fig. 8.9 Pleuropulmonary blastoma. (**a**) Low magnification reveals a multicystic tumor. (**b**) The cystic spaces are lined by respiratory epithelium. A population of primitive malignant small round blue cells is beneath the cyst lining. Magnification × 12.5 (**a**), 400 (**b**)

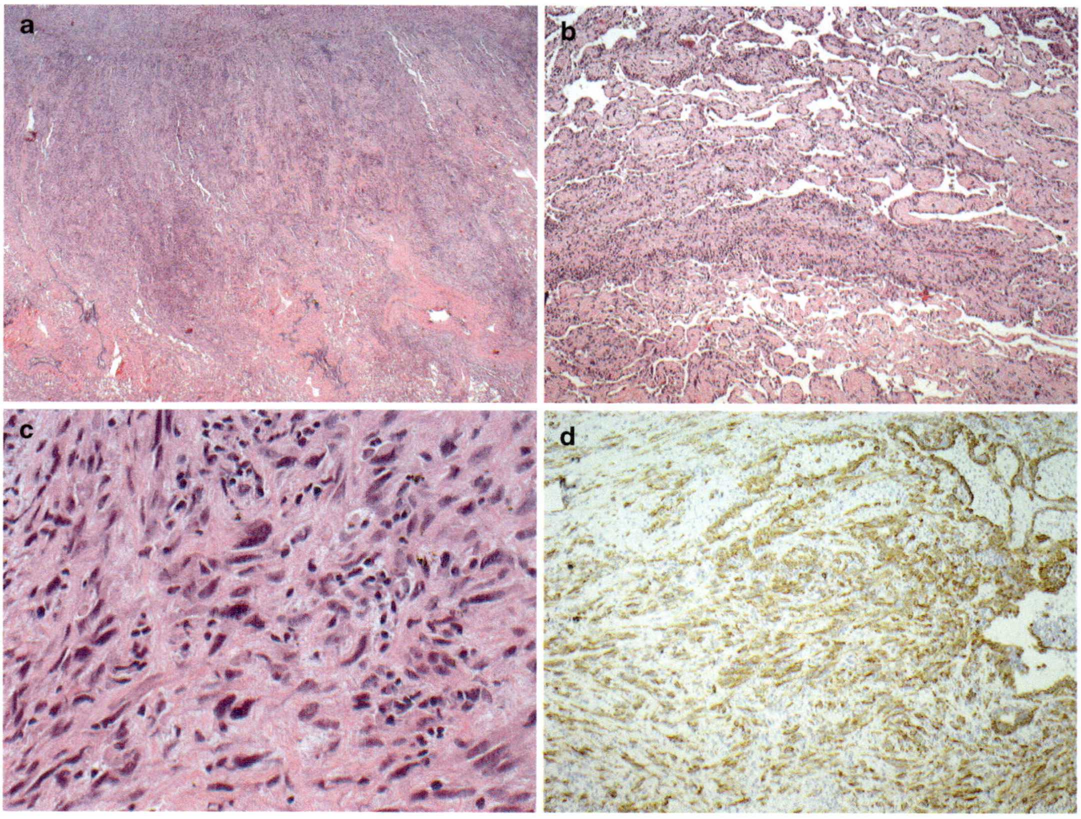

Fig. 8.10 Sarcomatoid carcinoma. (**a**) Low-power view shows a hypercellular malignancy in a fibrotic background. Malignant spindle cells are growing in sheets or along preformed structures, in this case leading to thickening of the pulmonary interalveolar septum (**b**) Neoplastic spindle cells have large nuclei with irregular borders and dark chromatin (**c**) and mark with keratin AE1/AE3 (**d**) Magnification × 20 (**a**), 100 (**b**, **d**), 400 (**c**)

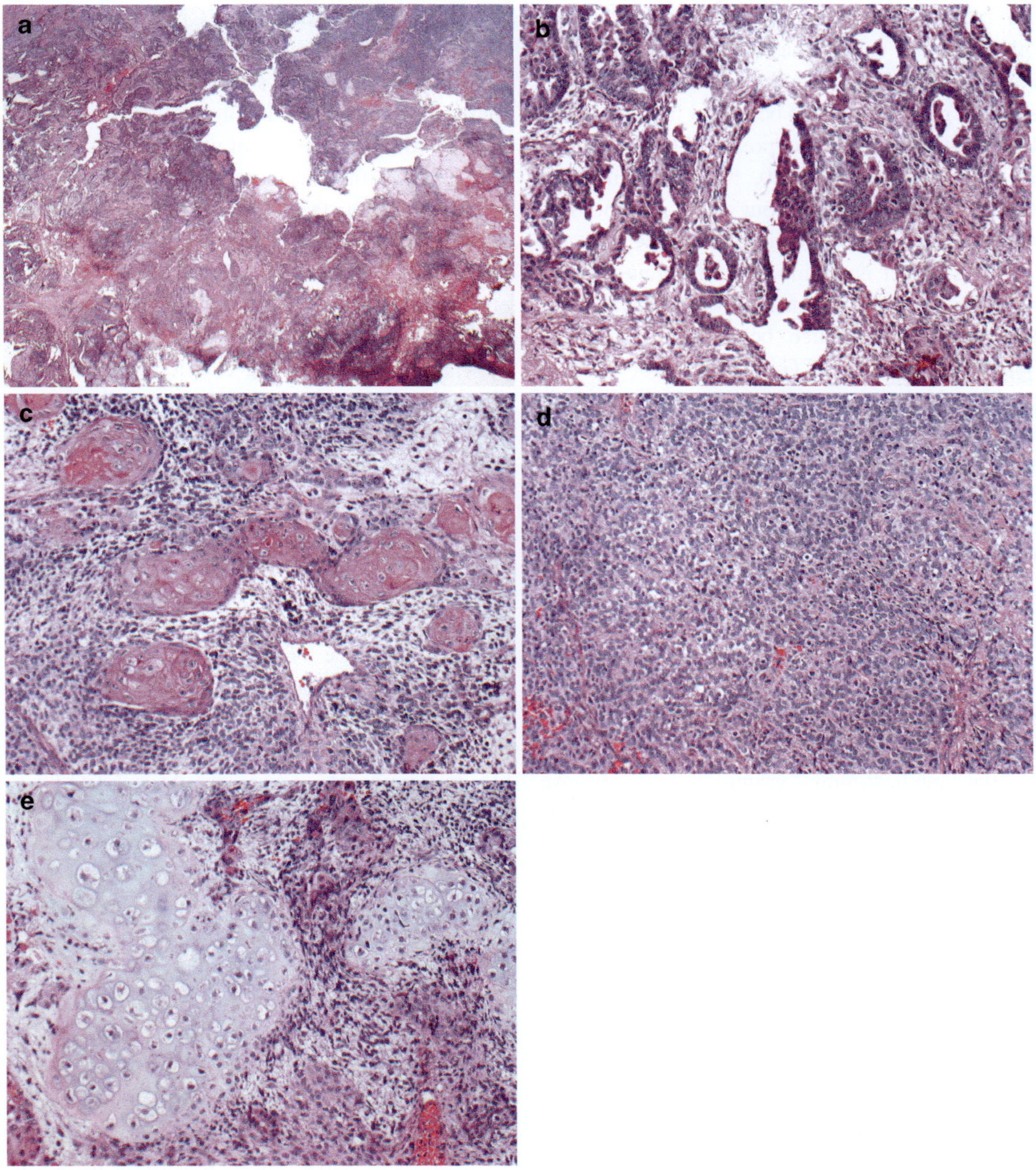

Fig. 8.11 Carcinosarcoma. (**a**) Low-power view shows a biphasic neoplasm that is comprised of an epithelial (glandular, squamous, and undifferentiated) and a mesenchymal (malignant cartilage) component. High-power microscopy reveals neoplastic glands (**b**), focal squamous differentiation (**c**), sheets of neoplastic epithelioid cells to suggest undifferentiated carcinoma (**d**), and neoplastic cartilage (**e**). Magnification × 12.5 (**a**), × 200 (**b**–**e**)

8.8.8 Non-small Cell Carcinoma

Embryonal carcinoma and, at times, seminoma can have significant morphologic overlap with a poorly differentiated carcinoma (e.g., pulmonary, thymic, or metastatic). Metastatic carcinomas with clear cytoplasm, such as clear cell renal cell carcinoma, may further mimic either yolk sac tumor or seminoma. The coexpression of cytokeratin and CD30 is characteristic of embryonal carcinoma, but not entirely specific. OCT4 appears relatively specific for seminoma and embryonal carcinoma [79–81, 112]. Markers of other primary carcinomas such as TTF-1 (lung) and gross cystic disease fluid protein-15 (breast/salivary gland) may also be useful in this setting. Lung carcinomas can express AFP, beta-HCG, and placental lactogen [90, 113], a finding that should not be interpreted out of the morphologic and clinical context as evidence of a GCT.

Lung cancer is an important differential diagnosis of choriocarcinomas in the mediastinum as they also can produce beta-HCG. Therefore, expression of beta-HCG by the tumor cells or positive serology is not diagnostic of choriocarcinoma in the mediastinum. FISH for isochromosome 12p might be helpful since the presence of isochromosome 12p in the tumor cells is diagnostic of GCT and argues against lung cancer. However, the lack of isochromosome 12p does not exclude GCT since not all GCTs harbor isochromosome 12p.

8.8.9 Granulomatous Disease

In the mediastinum, granulomatous disease opens a rather broad differential diagnosis including infection and sarcoidosis. Non-small cell carcinoma or lymphoma can also present with a granulomatous reaction. However, the possibility of a seminoma should at least be considered [77]. In seminoma, close examination will generally reveal scattered neoplastic cells typical of seminoma. Immunostains for OCT4 can be very helpful in highlighting and confirming the diagnosis in cases with minimal disease. Rarely, GCTs other than seminoma may be associated with abundant granulomatous inflammation.

8.8.10 Lymphoma

Lymphoma is relatively common in the mediastinum [77]. Although usually seen in the younger patient, the mediastinum can be involved by lymphoma in any age group. Mediastinal (thymic) large B-cell lymphoma, diffuse large B-cell lymphoma, lymphoblastic lymphoma, anaplastic large cell lymphoma (ALCL), and classic Hodgkin lymphoma may potentially mimic embryonal carcinoma or seminoma. Immunostains for cytokeratin should highlight embryonal carcinoma, and OCT4 may be of utility as a marker of either embryonal carcinoma or seminoma in this setting. Immunostains for CD30 should not be used in isolation as a marker of embryonal carcinoma because of the shared expression in a variety of lymphomas (e.g., Hodgkin lymphoma, ALCL, mediastinal large B-cell lymphoma) [114]. Nodular sclerosing classic Hodgkin lymphoma may mimic subtle patterns of seminoma, particularly on small biopsies or when associated granulomas are present (Fig. 8.7). The syncytial variant of Hodgkin lymphoma may closely mimic an undifferentiated carcinoma such as embryonal carcinoma because of the sheetlike growth pattern. Some hematopoietic neoplasms in this differential may also have either weak or complete absence of CD45 reactivity making their consideration on morphology critical. Other markers of hematopoietic differentiation that may be useful include CD34 and TdT (lymphoblastic lymphoma), EMA and ALK-1 (ALCL), and CD15 and PAX-5 (classic Hodgkin lymphoma).

Detection of Reed-Sternberg or Hodgkin cells that express both CD15 and CD30 favors a diagnosis of Hodgkin lymphoma (Fig. 8.7). Non-Hodgkin lymphoma comprises sheets of neoplastic lymphoid cells and associated reactive lymphocytes and may mimic typical seminoma. However, neoplastic lymphoid cells usually have less distinct cell borders, and their cytoplasm is not usually abundant or optically clear as it is in seminoma. Negative immunoreactivity for germ cell markers with positive immunostaining for lymphoid markers in lymphoma will aid its distinction from seminoma.

8.8.11 Thymoma and Thymic Carcinoma

On biopsy, thymoma, most commonly WHO type B1 or B2, may mimic GCTs with a brisk lymphocytic infiltrate, especially seminoma (Fig. 8.4) [77]. Type B3 thymomas are characterized by sheets of epithelioid cells (Fig. 8.12). Sometimes these cells can have more cytologic atypia and may mimic GCT, especially on a biopsy. In general, seminoma has a rim of clear cytoplasm and one or more prominent nucleoli, features that might help to distinguish it from type B3 thymoma. On resection, thymoma should be recognized by its lobulated growth pattern which is usually not present in GCT (Figs. 8.4 and 8.12). Furthermore, type A and AB thymomas are comprised at least focally of bland-appearing spindle cells. Immunohistochemistry might aid in the diagnosis. A study of 46 PMGCTs including teratoma, seminoma, yolk sac tumor, embryonal carcinoma, and mixed GCTs (teratoma and yolk sac tumor; teratoma, yolk sac tumor, and seminoma) and 22 thymomas (WHO types A, AB, B1, B2, and B3) showed that OCT4 expression was restricted to seminoma, embryonal carcinomas, and mixed GCT with seminomatous component. No OCT4 staining was identified in thymomas, yolk sac tumors, teratomas, and the mixed GCT without a component of seminoma [115].

Thymic carcinoma can be distinguished from seminoma by its often marked cytologic atypia (Figs. 8.13 and 8.14). Thymic carcinomas have a wide morphologic spectrum and may show a specific line of epithelial differentiation (most commonly squamous differentiation, Figs. 8.13 and 8.14). Only rare thymic

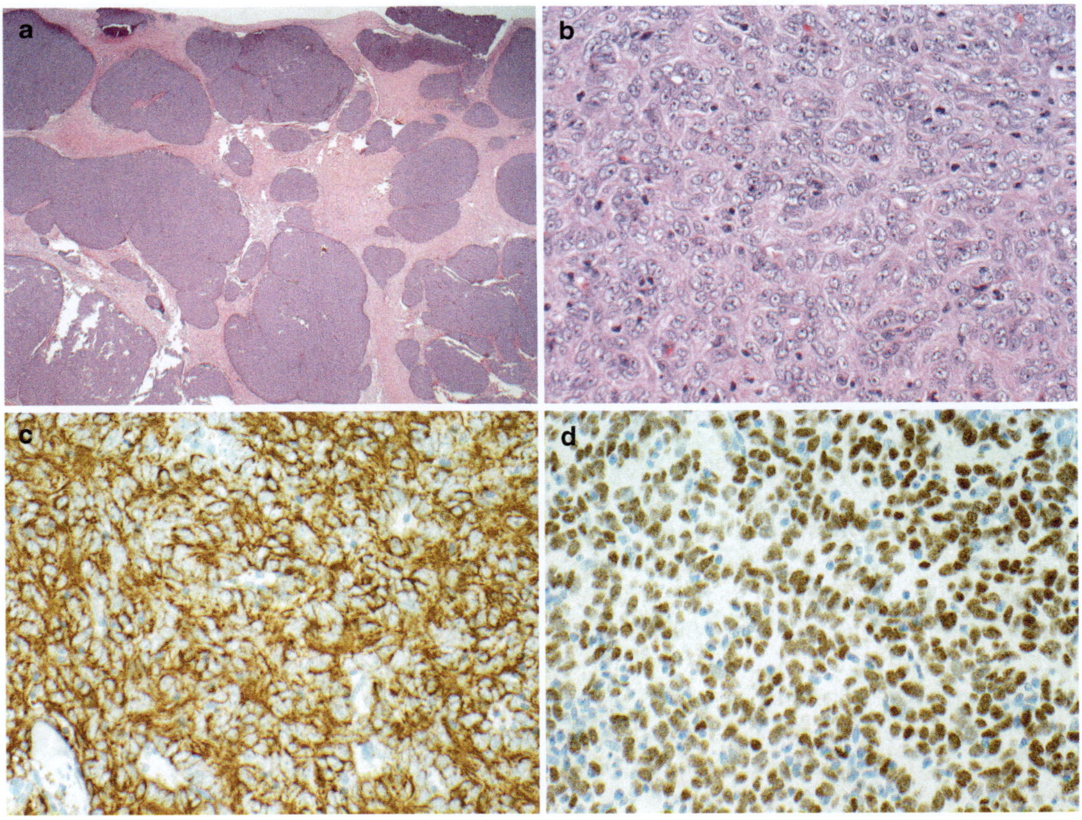

Fig. 8.12 Thymoma, WHO type B3. (**a**) On low power, a lobulated tumor contains cellular lobules that are separated by fibrous bands. (**b**) The lobules are comprised of large polygonal tumor cells that are characterized by open nuclear chromatin and prominent nucleoli. Only occasional small lymphocytes are scattered throughout the tumor. The neoplastic cells are positive for CK5/6 (**c**) and p40 (**d**). Magnification × 12.5 (**a**), × 400 (**b**–**d**)

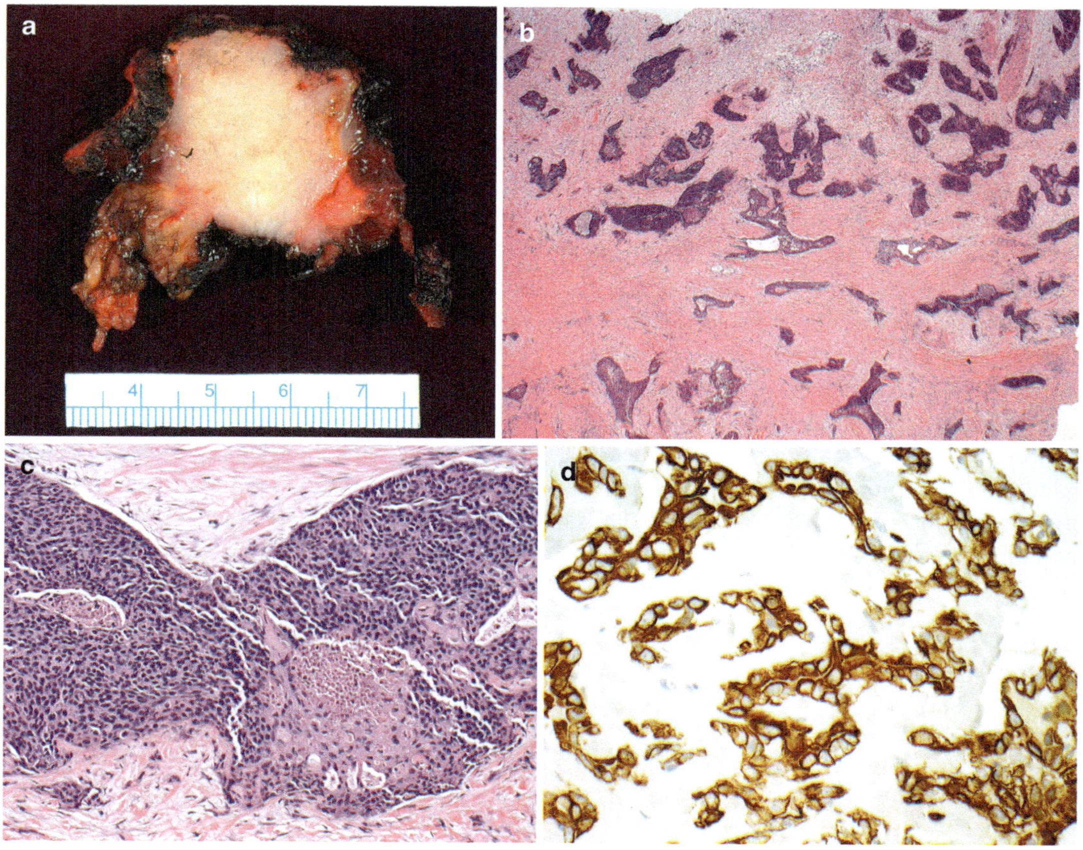

Fig. 8.13 Thymic carcinoma, squamous cell carcinoma. (**a**) The cut surface reveals a spiculated white-tan mass. (**b**) On low-power microscopy, irregular tumor cell nests are growing in a fibrotic background. (**c**) Although most tumor cells appear rather small with a high nuclear-to-cytoplasmic ratio consistent with a high-grade tumor morphology, some tumor cells are characterized by ample eosinophilic cytoplasm suggestive of squamoid differentiation. Necrosis is present. A desmoplastic stromal reaction is apparent around the tumor cell nests. The neoplastic cells are positive for CK5/6 further supporting squamous differentiation. Magnification × 20 (**b**), × 200 (**c**), × 400 (**d**)

carcinomas have prominent clear cytoplasm which may further mimic seminoma or yolk sac tumors [116]. Thymic sarcomatoid carcinomas may closely mimic immature teratoma, particularly those with heterologous differentiation. The malignant epithelial component of thymic sarcomatoid carcinoma should allow distinction from teratoma in most cases. Again, OCT4 immunostain can be helpful to distinguish thymic carcinoma from seminoma and embryonal carcinoma. CD117, however, also stains many thymic carcinomas and therefore is not useful in its distinction from seminomas (Fig. 8.14e).

8.8.12 NUT Carcinoma

Nuclear protein in testis (NUT) carcinomas are rare, aggressive carcinomas that are most commonly located in the mediastinum but also other midline organs and regions. These tumors have a characteristic t(15;19). Patients are usually young with a median age of 16 years; however, more recently this tumor was also identified in older patients. The reported age ranges between 0.1 and 78 years [117–122]. Morphologically, NUT carcinomas may mimic embryonal carcinoma with a lymphoepithelioma-like appear-

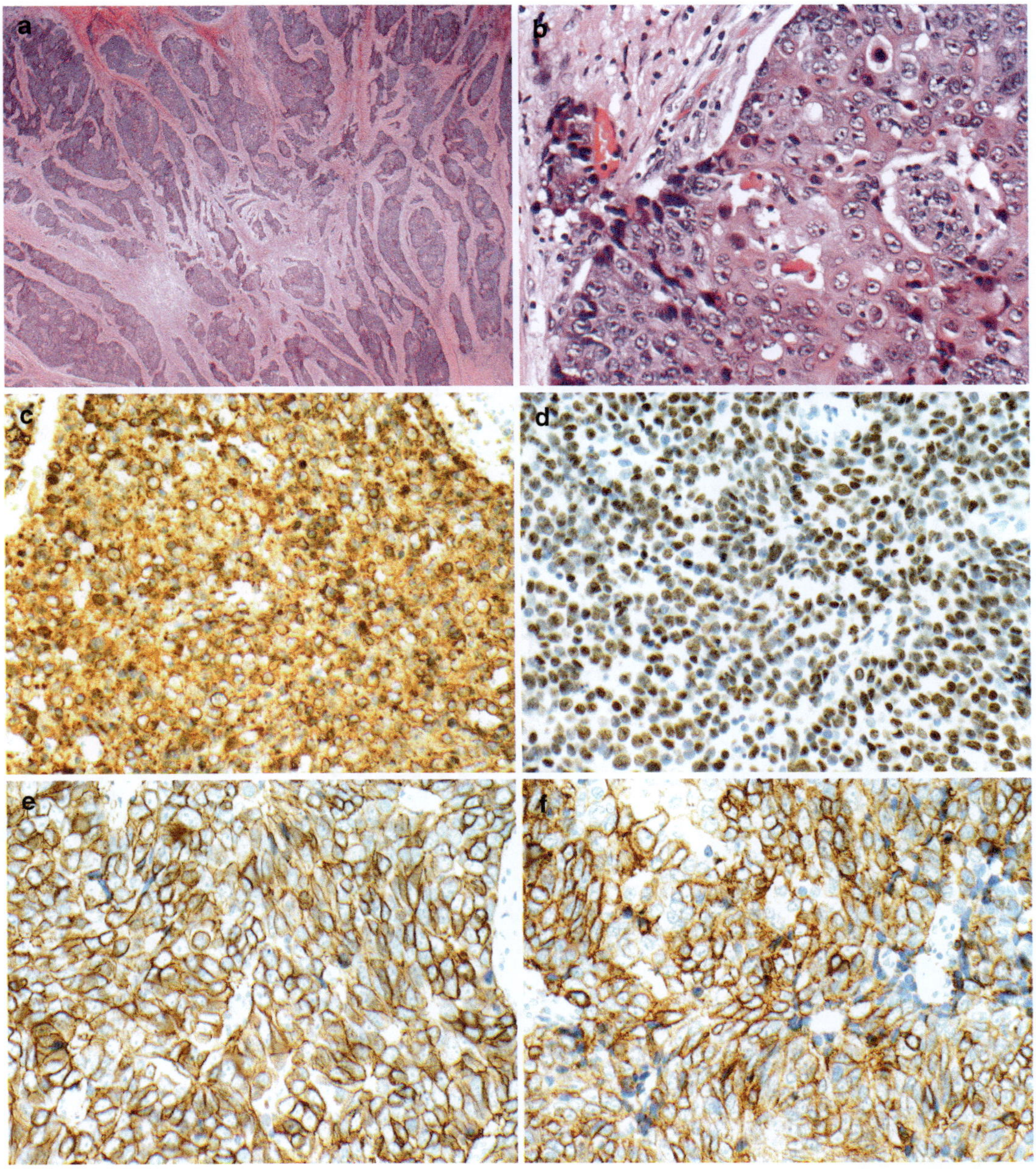

Fig. 8.14 Thymic carcinoma, squamous cell carcinoma. (**a**) This thymic carcinoma is more cellular but also characterized by cell nests growing in a fibrotic background. (**b**) On high power, the neoplastic cells are very atypical but have eosinophilic cytoplasm, and dyskeratotic features are apparent. Squamous differentiation is supported by expression of CK5/6 (**c**) and p40 (**d**) by the neoplastic cells. This thymic carcinoma also expresses CD117 (**e**) and CD5 (**f**). Magnification × 12.5 (**a**), × 400 (**b–f**)

ance. Although may be difficult in biopsies, the presence of squamous differentiation in the NUT carcinoma should help in the distinction from embryonal carcinoma, and diagnoses may be confirmed by NUT immunostain, FISH, or RT-PCR (Fig. 8.15) [123].

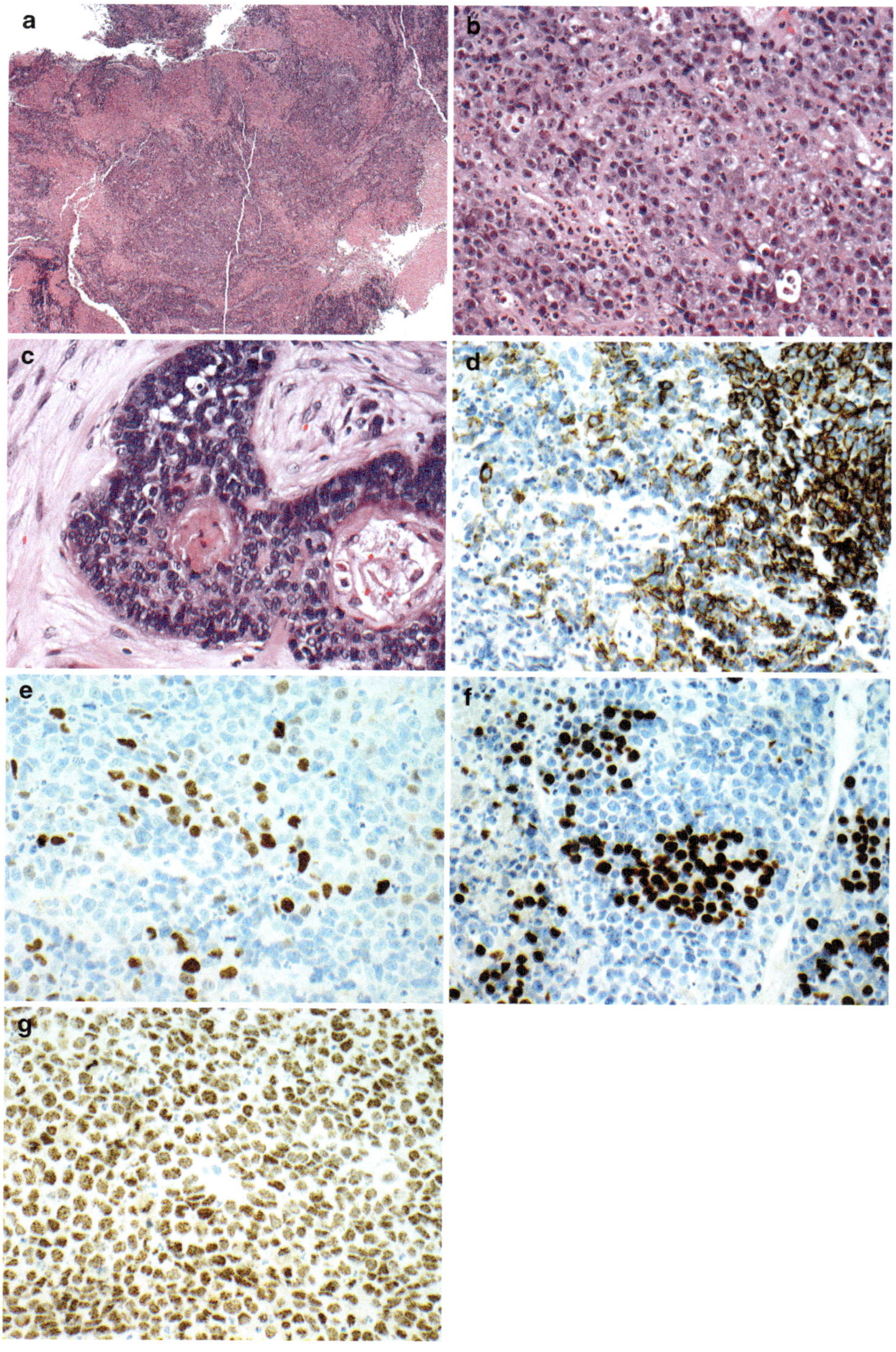

Fig. 8.15 NUT carcinoma. (**a**) On low power, sheets of neoplastic cells are intimately associated with large areas of necrosis. (**b**) The neoplastic cells are of epithelioid, relatively monomorphous cytology with round to oval nuclei, open nuclear chromatin and prominent nucleoli. (**c**) Abrupt keratinization is apparent. The majority of the neoplastic cells is positive for OSCAR keratin (**d**), and subsets of neoplastic cells express p40 (**e**) and/or TTF-1 (**f**). The neoplastic cells are diffusely positive for NUT exhibiting a characteristic speckled staining pattern (**g**). Magnification × 40 (**a**), × 400 (**b**–**g**)

8.8.13 Epithelioid Angiosarcoma

Angiosarcomas, when epithelioid in phenotype, may mimic embryonal carcinoma [124]. The identification of vascular lumen formation, the expression of vascular markers (CD31, CD34, Fli-1), and the absence of OCT4 reactivity should allow the distinction in most cases. Care should be taken in the interpretation of CD31 as intratumoral macrophages, which are typically present after adjuvant therapy, can show strong cytoplasmic reactivity [125].

8.8.14 Synovial Sarcoma

This biphasic sarcoma may occur as a primary mediastinal tumor [126] and could mimic immature teratoma. The spindle cell component of synovial sarcomas has a very distinct appearance with a monomorphic population of spindled cells arranged in tight intersecting fascicles. The glandular component may be focal, but it typically does not demonstrate any obvious mucinous or squamous differentiation as is often seen in teratoma. TLE-1 immunostain and/or molecular confirmation of the synovial sarcoma specific t(X;18) may be helpful.

8.8.15 Malignant Mesothelioma

Malignant mesothelioma has a broad morphologic spectrum that may overlap with several different GCTs, particularly given the shared expression of cytokeratin. Epithelioid mesothelioma can mimic a variety of carcinomas depending on the pattern present (Fig. 8.16). The tubulopapillary and microcystic (adenomatoid) epithelioid patterns may resemble yolk sac tumor, whereas poorly differentiated epithelioid mesothelioma can mimic embryonal carcinoma. Rare mesotheliomas have a myxoid morphology with clusters and cords of epithelioid cells set in pools of myxoid material, a pattern that may appear similar to the myxoid pattern of yolk sac tumor [127]. Biphasic mesothelioma could potentially mimic an immature teratoma if the epithelial component has somewhat bland cytologic features. Heterologous bone and cartilage differentiation has been documented in sarcomatoid mesotheliomas, a feature that could be mixed up with teratoma [128]. Sarcomatoid mesothelioma might mimic sarcomatous heterologous differentiation of immature teratoma. However, immunostains including OCT4 and conventional mesothelial markers together with the clinical setting such as a diffuse growth of the tumor along serosal surfaces should allow distinction from a GCT in most cases.

8.8.16 Congenital Peribronchial Myofibroblastic Tumor

Congenital peribronchial myofibroblastic tumor is a proliferation of cytologically bland myofibroblasts in the lung of newborns that can potentially mimic teratoma when bronchial epithelium and cartilage become entrapped [129]. This proliferation can efface the lung parenchyma, but often involves the interstitium and peribronchovascular areas. Recognition of the interstitial distribution and the entrapped nature of the epithelial and cartilaginous components should allow distinction from teratoma.

8.8.17 Pulmonary Hamartoma

Pulmonary hamartoma is a benign lung neoplasm that is composed of different mesenchymal tissues with entrapped respiratory-type epithelium (Fig. 8.17). The mesenchymal component is typically mature hyaline cartilage, but bone, fat, and smooth muscle may also be present. The tumor is recognized by invagination of respiratory epithelium into the tumor.

8.8.18 Metastatic Melanoma

Metastatic melanoma can mimic any poorly differentiated malignant neoplasm, including embryonal carcinoma. Recognition of the subtle nesting of the neoplastic cells and verification of

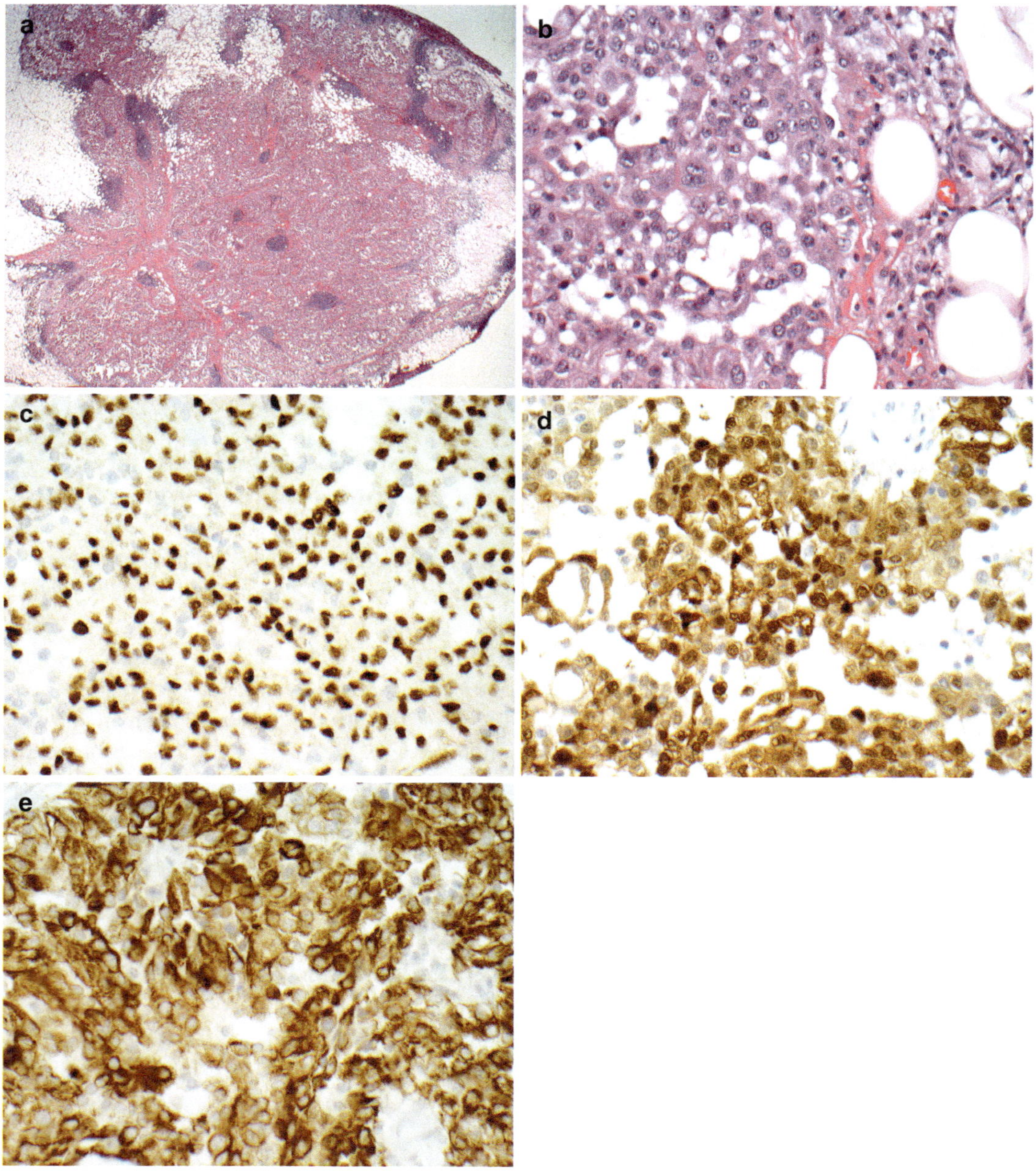

Fig. 8.16 Malignant mesothelioma, epithelioid type. (**a**) Sheets of neoplastic cells are growing in an infiltrative pattern invading into adipose tissue. (**b**) The neoplastic cells are round to oval and are characterized by a fair amount of cytoplasm and round nuclei with open chromatin and conspicuous nucleoli. The neoplastic cells are positive for mesothelial markers including WT-1 (**c**), calretinin (**d**), and CK5/6 (**e**) and lack staining with carcinoma markers such as pCEA, MOC-31, and TTF-1 (not shown). Magnification × 12.5 (**a**), 400 (**b**–**e**)

melanocytic markers in the absence of OCT4 expression should resolve most cases. The absence of a history of a cutaneous melanoma should not preclude this diagnosis because in some cases a primary lesion cannot be identified, possibly secondary to regression.

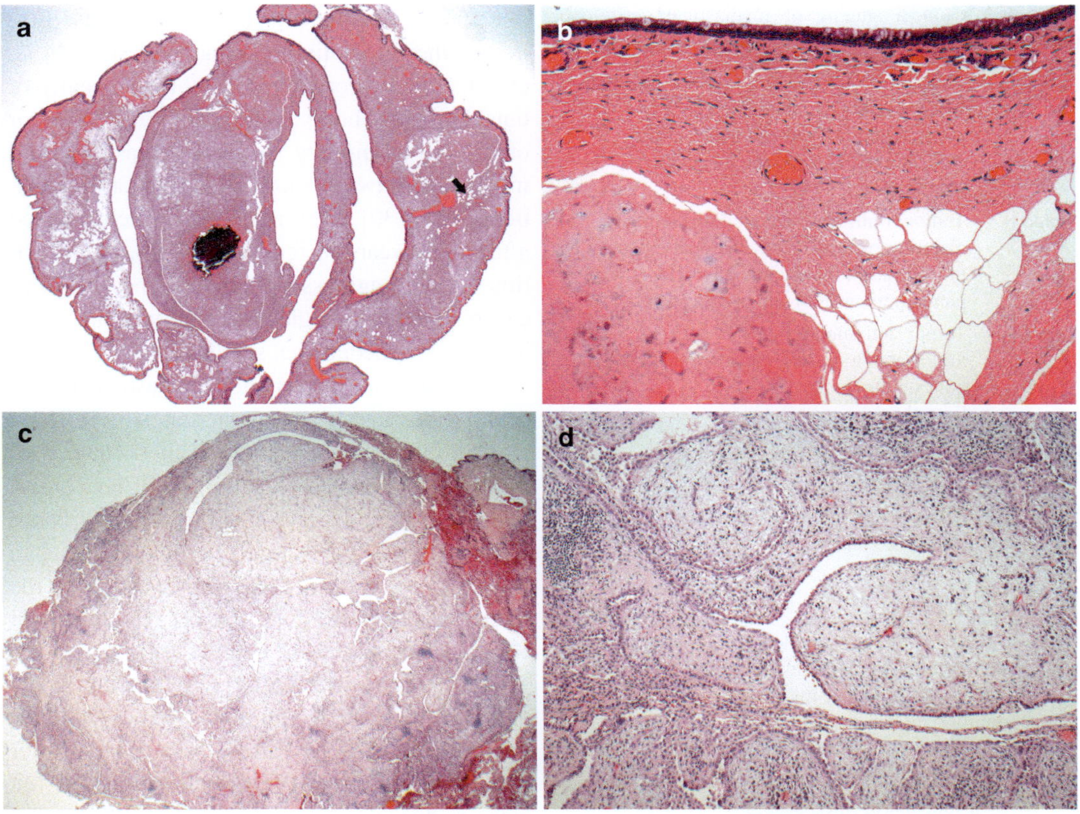

Fig. 8.17 Pulmonary hamartoma: (**a**) A well-circumscribed nodule is predominantly comprised of hyaline cartilage with focal calcification. Adipose tissue is also present (*arrow*). (**b**) Respiratory epithelium invaginates into the nodule. (**c**) This hamartoma is predominantly comprised of bland fibrovascular stroma with focal adipose tissue. (**d**) Bland cuboidal epithelial cells invaginate into the nodule. Magnifications × 12.5 (**a**), 200 (**b**), 40 (**c**), 100 (**d**)

8.9 Staging of Mediastinal Germ Cell Tumors

Clinical and pathologic staging is very important in the prognosis of PMGCTs. Unfortunately, there is not an officially recognized UICC-TNM staging protocol for PMGCTs. The WHO recommends using a modification of the AJCC TNM staging of soft tissue tumors [67].

In 1997, Moran and Suster [68] examined 322 cases of PMGCT and proposed a staging system specifically for these tumors based on the clinical outcome of their cases (see Table 8.6). The author's recommendation was to treat tumors that are confined to the mediastinum without infiltration of adjacent structures (stage I) conservatively, with surgery alone or with surgery, and an added modality based on the histology of the tumor. Lesions of advanced stage (II or III) would require more aggressive treatment with curative intent, whereas palliative treatment was the choice in tumors with extrathoracic metastasis. This staging approach correlated well with the clinical outcome of the patients in that study.

Staging of pediatric extragonadal GCT is summarized in Table 8.7.

8.10 Outcome of Mediastinal Germ Cell Tumors

The 5-year relative survival of patients with mediastinal GCTs is 58 % according to SEER registries [13]. The prognosis depends on the histologic

Table 8.6 Clinical staging of mediastinal germ cell tumors as proposed by Moran and Suster

Stage I	Well-circumscribed tumor with or without focal adhesions to the pleura or pericardium but without microscopic evidence of invasion into adjacent structures
Stage II	Tumor confined to the mediastinum with macroscopic and/or microscopic evidence of infiltration into adjacent structures (i.e., pleura pericardium and great vessels)
Stage III	Tumor with metastases
A	Metastases to intrathoracic organs (lymph nodes, lung, etc.)
B	Extrathoracic metastases

From Moran et al. [68], with permission

Table 8.7 Staging of pediatric extragonadal extracranial GCT as defined by the Children's Oncology Group (COG) [189]

Stage	Characteristic
I	Localized disease; complete resection with no microscopic disease at margins or in regional lymph nodes. Tumor markers must normalize in appropriate half-life after resection. Complete coccygectomy for sacrococcygeal site
II	Microscopic residual disease, capsular invasion, and/or microscopic lymph node involvement. Tumor markers fail to normalize or increase
III	Gross residual disease and gross lymph node involvement (>2 cm)
IV	Distant metastases, including the liver, brain, bone, or lung

subtype with pure mediastinal seminomas having a much better outcome than tumors with a non-seminomatous component.

Mediastinal seminomas respond favorably to radiation therapy and/or cisplatin-based chemotherapy [130]. For instance, patients with mediastinal seminomas treated with cisplatin-based combination chemotherapy have a 5-year survival of 90–100 % and an overall survival of 88–90 % [41, 130–132]. In mediastinal seminomas, factors that have been suggested to be associated with greater rate of progression include patient's age over 35 years, presentation with fever, SVC syndrome, supraclavicular or cervical adenopathy, and radiologic evidence of hilar disease [30]. The most important feature to predict outcome is whether the patient has disease limited to the mediastinum or has widespread disease to adjacent organs of the thoracic cavity or outside of the thorax. Metastasis to the liver or other non-pulmonary visceral metastases and metastases to two or more sites are poor prognostic factors [130]. Recurrences have been reported after many years of remission [133]. A correlation between aggressive behavior and any clinical or histopathologic features has not been identified in mediastinal seminomas [51].

The outcome for primary mediastinal immature teratoma and nonteratomatous GCTs is worse than for their gonadal counterpart. Mediastinal non-seminomatous GCT have only a 40–50 % overall survival after platinum-based chemotherapy and surgery [134–136] and a 5-year survival rate of 48 %, according to the International Germ Cell Cancer Collaborative Group consensus classification [134].

If there is metastatic disease to the lung, liver, or supraclavicular lymph nodes, the overall survival drops to 25 %. Patients who relapse after initial cisplatin-based chemotherapy have an extreme dismal outcome with an overall survival of only 10 %. Surgical resection of residual disease after chemotherapy shows residual viable tumor in 30–47 % of patients [137, 138]. Factors that contribute to inferior survival include an overall poorer response to chemotherapy and higher incidence of degenerative non-germ cell cancer pathology in the residual mass [139]. Age ≥12 years is also suggested as an adverse prognostic factor [77].

Outcomes of these patients are improving with preoperative cisplatin-based combination chemotherapy strategies [130–132, 140]. With neoadjuvant chemotherapy, good prognostic factors include completeness of resection, less than 10 % viable tumor cells, and low-risk group as defined by the International Germ Cell Consensus Classification Group [134].

Primary mediastinal choriocarcinomas have a much worse prognosis than other histologic subtypes because of hematogenous dissemination at the time of diagnosis [141]. However, under current chemotherapeutic regimens, the prognosis of choriocarcinomas has improved [69].

Embryonal carcinomas and yolk sac tumors, whether pure or in association with any other components (seminoma or teratoma), have a similar outcome and are generally regarded as poor prognostic findings.

As in congenital teratomas, the prognosis of PMGCTs in children is significantly affected by tumor stage and completeness of surgical excision [142].

Epidemiologic studies have shown that patients with PMGCTs have an increased risk for death related to hematopoietic malignancies and cardiovascular disorders, but no significant difference in risk of dying from solid cancers compared to patients with gonadal GCTs was identified [12]. In fact 6 % of primary mediastinal non-seminomatous GCT develop hematologic malignancies, the most common being acute megakaryoblastic leukemia (AML-M7) and myeloblastic syndromes [143]. These hematopoietic tumors may involve the PMGCT or be completely extramediastinal. A pathogenetic hypothesis is that hematopoietic stem cells arise in the yolk sac [144]. Another hypothesis is that a teratoma containing all three germ layers with varying degrees of differentiation undergoes malignant transformation to leukemia in the bone marrow. Interestingly, cytogenetic analysis of bone marrow aspirates reveals i(12p) in 38 % of patients [145], a cytogenetic abnormality that is commonly seen in PMGCT, indicating a possible common biologic pathway. GCT-associated acute leukemias are an ominous finding as they are typically refractory to current treatment modalities with a reported survival of less than 2 years in all reported patients. The main differential diagnostic consideration in this setting is a therapy-related myelodysplastic syndrome or acute leukemia following etoposide administration [143]. Therapy-related diseases can be distinguished by their occurrence later in the course (25–60 months), the absence of i(12p), and the possible presence of an etoposide-related translocation such as 11q23 [146–148].

The "growing teratoma syndrome" [139] defines a mediastinal mass that is growing subsequently to neoadjuvant therapy and is associated with secondary cardiopulmonary deterioration precluding safe completion of planned chemotherapy in the presence of declining serum tumor markers. The term was first coined by Logothetis and colleagues [149] who reported patients with non-seminomatous testicular cancer and growing retroperitoneal or lung masses during observation after chemotherapy. In the mediastinum, five (of 188) patients who underwent postchemotherapy surgery for primary mediastinal non-seminomatous GCT were identified [139]. These five men had an average age of 25.8 years (range, 20–33 years). All patients presented at the time of diagnosis with a large symptomatic anterior mediastinal mass and elevated AFP at an average of 9137 ng/mL (range, 791–36,000 ng/mL). HCG was elevated in three patients with a mean of 206 mIU/mL (range, 8–350 mIU/mL). CT revealed evidence of metastatic disease (lung) in one patient at the time of diagnosis, with the remaining patients presenting with disease isolated to the mediastinum. Prechemotherapy biopsies demonstrated mature teratoma in only two patients despite elevated AFP, mature teratoma and foci of non-seminomatous germ cell tumor in two patients and one patient with immature teratoma and foci of non-seminomatous GCT. Three patients had normalized serum tumor markers, and two patients demonstrated rapid serum tumor marker decline before surgery. All five patients underwent complete surgical resection of the mediastinal mass with tumor-free margins. The two patients who had not normalized but demonstrated rapid serum tumor marker decline at the time of surgery had normalized tumor markers by the time of hospital discharge. Surgical pathology of the mediastinal mass demonstrated mature teratoma only and mature teratoma with focal immaturity in one patient each. Pathology was mixed in the other three patients with predominately mature teratoma and focal yolk sac tumor ($n = 1$), non-GCT (angiosarcoma, $n = 1$), and a combination of yolk sac and non-GCT (angiosarcoma, $n = 1$). One patient died of respiratory failure postoperatively. Three of the 4 surviving patients received cisplatin-based chemotherapy after recovery to complete four cycles. Two patients are alive and well 13 and 15 years after initial surgery; however, both have required further resection of metastases during follow-up. The other two patients died of metastatic non-GCT after 20 and 26 months.

References

1. Woolner LB, Jamplis RW, Kirklin JW. Seminoma (germinoma) apparently primary in the anterior mediastinum. N Engl J Med. 1955;252:653–7.
2. Nazari A, Gagnon ED. Seminoma-like tumor of the mediastinum. A case report with 7 year follow-up. J Thorac Cardiovasc Surg. 1966;51:751–4.
3. Kountz SL, Connolly JE, Cohn R. Seminoma-like (orseminomatous) tumors of the anterior mediastinum. Report of four new cases and a review of the literature. J Thorac Cardiovasc Surg. 1963;45:289–301.
4. Levine GD. Primary thymic seminoma--a neoplasm ultrastructurally similar to testicular seminoma and distinct from epithelial thymoma. Cancer. 1973;31:729–41.
5. Cefaro GA, Luzi S, Turriziani A, Salvi G, Marmiroli L. Primary mediastinal seminoma. Br J Urol. 1988;62:461–4.
6. Aygun C, Slawson RG, Bajaj K, Salazar OM. Primary mediastinal seminoma. Urology. 1984;23:109–17.
7. Bush SE, Martinez A, Bagshaw MA. Primary mediastinal seminoma. Cancer. 1981;48:1877–82.
8. Grosfeld JL, Skinner MA, Rescorla FJ, West KW, Scherer 3rd LR. Mediastinal tumors in children: experience with 196 cases. Ann Surg Oncol. 1994;1:121–7.
9. Davis Jr RD, Oldham Jr HN, Sabiston Jr DC. Primary cysts and neoplasms of the mediastinum: recent changes in clinical presentation, methods of diagnosis, management, and results. Ann Thorac Surg. 1987;44:229–37.
10. Jain KK, Bosl GJ, Bains MS, Whitmore WF, Golbey RB. The treatment of extragonadal seminoma. J Clin Oncol. 1984;2:820–7.
11. Bosl GJ, Motzer RJ. Testicular germ-cell cancer. N Engl J Med. 1997;337:242–53.
12. Alanee SR, Feldman DR, Russo P, Konety B. Long-term mortality in patients with germ cell tumors: effect of primary cancer site on cause of death. Urol Oncol. 2014;32:26. e29–15.
13. Stang A, Trabert B, Wentzensen N, Cook MB, Rusner C, Oosterhuis JW, McGlynn KA. Gonadal and extragonadal germ cell tumours in the United States, 1973-2007. Int J Androl. 2012;35:616–25.
14. Rusner C, Trabert B, Katalinic A, Kieschke J, Emrich K, Stang A. Incidence patterns and trends of malignant gonadal and extragonadal germ cell tumors in Germany, 1998-2008. Cancer Epidemiol. 2013;37:370–3.
15. Knapp RH, Hurt RD, Payne WS, Farrow GM, Lewis BD, Hahn RG, Muhm JR, Earle JD. Malignant germ cell tumors of the mediastinum. J Thorac Cardiovasc Surg. 1985;89:82–9.
16. Coskun U, Gunel N, Yildirim Y, Memis L, Boyacioglu ZM. Primary mediastinal yolk sac tumor in a 66-year-old woman. Med Princ Pract. 2002;11:218–20.
17. Shimizu J, Yazaki U, Kinoshita T, Tatsuzawa Y, Kawaura Y, Nonomura A. Primary mediastinal germ cell tumor in a middle-aged woman: case report and literature review. Tumori. 2001;87:269–71.
18. Schneider DT, Calaminus G, Koch S, Teske C, Schmidt P, Haas RJ, Harms D, Gobel U. Epidemiologic analysis of 1,442 children and adolescents registered in the German germ cell tumor protocols. Pediatr Blood Cancer. 2004;42:169–75.
19. Cushing B, Bhanot PK, Watts Jr FB, Hertzler JH, Brough AJ. Rhabdomyosarcoma and benign teratoma. Pediatr Pathol. 1983;1:345–8.
20. Went PT, Dirnhofer S, Stallmach T, Taverna C, Singer G. Placental site trophoblastic tumor of the mediastinum. Hum Pathol. 2005;36:581–4.
21. Moeller KH, Rosado-de-Christenson ML, Templeton PA. Mediastinal mature teratoma: imaging features. AJR Am J Roentgenol. 1997;169:985–90.
22. Weidner N. Germ-cell tumors of the mediastinum. Semin Diagn Pathol. 1999;16:42–50.
23. Weissferdt A, Suster S, Moran CA. Primary mediastinal "thymic" seminomas. Adv Anat Pathol. 2012;19:75–80.
24. Cameron RB, Loehrer PJ, Thomas CR. Neoplasms of the mediastinum. In: DeVita VT, Hellman S, Rosenberg SA, editors. Cancer: principles and practice of oncology. New York: Lippincott Williams & Wilkins; 2001.
25. Recondo J, Libshitz HI. Mediastinal extragonadal germ cell tumors. Urology. 1978;11:369–75.
26. Kiffer JD, Sandeman TF. Primary malignant mediastinal germ cell tumors: a study of eleven cases and a review of the literature. Int J Radiat Oncol Biol Phys. 1989;17:835–41.
27. Nichols CR, Heerema NA, Palmer C, Loehrer Sr PJ, Williams SD, Einhorn LH. Klinefelter's syndrome associated with mediastinal germ cell neoplasms. J Clin Oncol. 1987;5:1290–4.
28. Dexeus FH, Logothetis CJ, Chong C, Sella A, Ogden S. Genetic abnormalities in men with germ cell tumors. J Urol. 1988;140:80–4.
29. Polansky SM, Barwick KW, Ravin CE. Primary mediastinal seminoma. AJR Am J Roentgenol. 1979;132:17–21.
30. Hurt RD, Bruckman JE, Farrow GM, Bernatz PE, Hahn RG, Earle JD. Primary anterior mediastinal seminoma. Cancer. 1982;49:1658–63.
31. Lee YM, Jackson SM. Primary seminoma of the mediastinum. Cancer Control Agency of British Columbia experience. Cancer. 1985;55:450–2.

32. Guibert N, Attias D, Pontier S, Berjaud J, Lavialle-Guillautreau V, Didier A. Mediastinal teratoma and trichoptysis. Ann Thorac Surg. 2011;92:351–3.
33. Lewis BD, Hurt RD, Payne WS, Farrow GM, Knapp RH, Muhm JR. Benign teratomas of the mediastinum. J Thorac Cardiovasc Surg. 1983;86:727–31.
34. Gunes S, Varon J, Walsh G. Mediastinal teratoma presenting as massive hemoptysis in an adult. J Emerg Med. 1997;15:313–6.
35. Nagi DK, Jones WG, Belchetz PE. Gynaecomastia caused by a primary mediastinal seminoma. Clin Endocrinol. 1994;40:545–8. discussion 548-549
36. Hasle H, Jacobsen BB, Asschenfeldt P, Andersen K. Mediastinal germ cell tumour associated with Klinefelter syndrome. A report of case and review of the literature. Eur J Pediatr. 1992;151:735–9.
37. Hasle H, Mellemgaard A, Nielsen J, Hansen J. Cancer incidence in men with Klinefelter syndrome. Br J Cancer. 1995;71:416–20.
38. Beresford L, Fernandez CV, Cummings E, Sanderson S, Ming-Yu W, Giacomantonio M. Mediastinal polyembryoma associated with Klinefelter syndrome. J Pediatr Hematol Oncol. 2003;25:321–3.
39. Aguirre D, Nieto K, Lazos M, Pena YR, Palma I, Kofman-Alfaro S, Queipo G. Extragonadal germ cell tumors are often associated with Klinefelter syndrome. Hum Pathol. 2006;37:477–80.
40. Volkl TM, Langer T, Aigner T, Greess H, Beck JD, Rauch AM, Dorr HG. Klinefelter syndrome and mediastinal germ cell tumors. Am J Med Genet A. 2006;140:471–81.
41. Bokemeyer C, Nichols CR, Droz JP, Schmoll HJ, Horwich A, Gerl A, Fossa SD, Beyer J, Pont J, Kanz L, Einhorn L, Hartmann JT. Extragonadal germ cell tumors of the mediastinum and retroperitoneum: results from an international analysis. J Clin Oncol. 2002;20:1864–73.
42. Fossa SD, Aass N, Heilo A, Daugaard G, Skakkebaek EN, Stenwig AE, Nesland JM, Looijenga LH, Oosterhuis JW. Testicular carcinoma in situ in patients with extragonadal germ-cell tumours: the clinical role of pretreatment biopsy. Ann Oncol. 2003;14:1412–8.
43. Ohtsuka M, Satoh H, Inoue M, Yazawa T, Yamashita YT, Sekizawa K, Hasegawa S. Disseminated metastasis of neuroblastomatous component in immature mediastinal teratoma: a case report. Anticancer Res. 2000;20:527–30.
44. Jung JI, Park SH, Park JG, Lee SH, Lee KY, Hahn ST. Teratoma with malignant transformation in the anterior mediastinum: a case report. Korean J Radiol. 2000;1:162–4.
45. Blomlie V, Lien HH, Fossa SD, Jacobsen AB, Stenwig AE. Computed tomography in primary nonseminomatous germ cell tumors of the mediastinum. Acta Radiol. 1988;29:289–92.
46. Saito A, Watanabe K, Kusakabe T, Abe M, Suzuki T. Mediastinal mature teratoma with coexistence of angiosarcoma, granulocytic sarcoma and a hematopoietic region in the tumor: a rare case of association between hematological malignancy and mediastinal germ cell tumor. Pathol Int. 1998;48:749–53.
47. Rosado-de-Christenson ML, Templeton PA, Moran CA. From the archives of the AFIP. Mediastinal germ cell tumors: radiologic and pathologic correlation. Radiographics. 1992;12:1013–30.
48. Lee KS, Im JG, Han CH, Han MC, Kim CW, Kim WS. Malignant primary germ cell tumors of the mediastinum: CT features. AJR Am J Roentgenol. 1989;153:947–51.
49. Shin MS, Odrezin GT, Van Dyke JA, Ho KJ. Unusual initial calcification of primary and metastatic seminomas. Detection by computed tomography. Chest. 1991;99:1543–5.
50. Moran CA, Suster S. Mediastinal seminomas with prominent cystic changes. A clinicopathologic study of 10 cases. Am J Surg Pathol. 1995;19:1047–53.
51. Moran CA, Suster S, Przygodzki RM, Koss MN. Primary germ cell tumors of the mediastinum: II. Mediastinal seminomas--a clinicopathologic and immunohistochemical study of 120 cases. Cancer. 1997;80:691–8.
52. Levine GD, Rosai J. Thymic hyperplasia and neoplasia: a review of current concepts. Hum Pathol. 1978;9:495–515.
53. Strollo DC, Rosado-de-Christenson ML. Primary mediastinal malignant germ cell neoplasms: imaging features. Chest Surg Clin N Am. 2002;12:645–58.
54. Nichols CR. Mediastinal germ cell tumors. Clinical features and biologic correlates. Chest. 1991;99:472–9.
55. Moran CA, Suster S, Koss MN. Primary germ cell tumors of the mediastinum: III. Yolk sac tumor, embryonal carcinoma, choriocarcinoma, and combined nonteratomatous germ cell tumors of the mediastinum--a clinicopathologic and immunohistochemical study of 64 cases. Cancer. 1997;80:699–707.
56. Levitt RG, Husband JE, Glazer HS. CT of primary germ-cell tumors of the mediastinum. AJR Am J Roentgenol. 1984;142:73–8.
57. Glenn OA, Barkovich AJ. Intracranial germ cell tumors: a comprehensive review of proposed embryologic derivation. Pediatr Neurosurg. 1996;24:242–51.
58. Oosterhuis JW, Looijenga LH. Testicular germ-cell tumours in a broader perspective. Nat Rev Cancer. 2005;5:210–22.
59. Chaganti RS, Rodriguez E, Mathew S. Origin of adult male mediastinal germ-cell tumours. Lancet. 1994;343:1130–2.
60. Chaganti RS, Houldsworth J. Genetics and biology of adult human male germ cell tumors. Cancer Res. 2000;60:1475–82.
61. Oosterhuis JW, Rammeloo RH, Cornelisse CJ, De Jong B, Dam A, Sleijfer DT. Ploidy of malignant

mediastinal germ-cell tumors. Hum Pathol. 1990;21:729–32.
62. Przygodzki RM, Moran CA, Suster S, Khan MA, Swalsky PA, Bakker A, Koss MN, Finkelstein SD. Primary mediastinal and testicular seminomas: a comparison of K-ras-2 gene sequence and p53 immunoperoxidase analysis of 26 cases. Hum Pathol. 1996;27:975–9.
63. Przygodzki RM, Hubbs AE, Zhao FQ, O'Leary TJ. Primary mediastinal seminomas: evidence of single and multiple KIT mutations. Lab Investig. 2002;82:1369–75.
64. Moul JW, Theune SM, Chang EH. Detection of RAS mutations in archival testicular germ cell tumors by polymerase chain reaction and oligonucleotide hybridization. Genes Chromosom Cancer. 1992;5:109–18.
65. Mulder MP, Keijzer W, Splinter TA, Bos JL. Frequent occurrence of activated ras oncogenes in seminomas but not in nonseminomatous germ cell tumors. Recent Results Cancer Res. 1991;123:125–31.
66. Lewis DJ, Sesterhenn IA, McCarthy WF, Moul JW. Immunohistochemical expression of P53 tumor suppressor gene protein in adult germ cell testis tumors: clinical correlation in stage I disease. J Urol. 1994;152:418–23.
67. Moreira AL, Chan JKC, Looijenga LHJ, Marx A, Stroebel P, Ulbright TM, Wick R. Germ cell tumors of the mediastinum. In: Travis WD, Brambilla E, Burke AP, et al., editors. World Health Organization classification of tumours pathology and genetics tumours of the lung, pleura, thymus, and heart. Lyon: IARC press; 2015. p. 244–66.
68. Moran CA, Suster S. Primary germ cell tumors of the mediastinum: I. Analysis of 322 cases with special emphasis on teratomatous lesions and a proposal for histopathologic classification and clinical staging. Cancer. 1997;80:681–90.
69. Takeda S, Miyoshi S, Ohta M, Minami M, Masaoka A, Matsuda H. Primary germ cell tumors in the mediastinum: a 50-year experience at a single Japanese institution. Cancer. 2003;97:367–76.
70. Peterson CM, Buckley C, Holley S, Menias CO. Teratomas: a multimodality review. Curr Probl Diagn Radiol. 2012;41:210–9.
71. el-Domeiri AA, Hutter RV, Pool JL, FW Jr F. Primary seminoma of the anterior mediastinum. Ann Thorac Surg. 1968;6:513–21.
72. Brown K, Collins JD, Batra P, Steckel RJ, Kagan AR. Mediastinal germ cell tumor in a young woman. Med Pediatr Oncol. 1989;17:164–7.
73. Sung MT, Maclennan GT, Lopez-Beltran A, Zhang S, Montironi R, Cheng L. Primary mediastinal seminoma: a comprehensive assessment integrated with histology, immunohistochemistry, and fluorescence in situ hybridization for chromosome 12p abnormalities in 23 cases. Am J Surg Pathol. 2008;32:146–55.
74. Friedman SJ, Belbel RL, Martin NJ. Extragonadal mediastinal seminoma. Mil Med. 1988;153:32–4.
75. Moran CA, Suster S, Perino G, Kaneko M, Koss MN. Malignant smooth muscle tumors presenting as mediastinal soft tissue masses. A clinicopathologic study of 10 cases. Cancer. 1994;74:2251–60.
76. Malagon HD, Valdez AM, Moran CA, Suster S. Germ cell tumors with sarcomatous components: a clinicopathologic and immunohistochemical study of 46 cases. Am J Surg Pathol. 2007;31:1356–62.
77. McKenney JK, Heerema-McKenney A, Rouse RV. Extragonadal germ cell tumors: a review with emphasis on pathologic features, clinical prognostic variables, and differential diagnostic considerations. Adv Anat Pathol. 2007;14:69–92.
78. Suster S, Moran CA, Dominguez-Malagon H, Quevedo-Blanco P. Germ cell tumors of the mediastinum and testis: a comparative immunohistochemical study of 120 cases. Hum Pathol. 1998;29:737–42.
79. Emerson RE, Ulbright TM. The use of immunohistochemistry in the differential diagnosis of tumors of the testis and paratestis. Semin Diagn Pathol. 2005;22:33–50.
80. Hattab EM, Tu PH, Wilson JD, Cheng L. OCT4 immunohistochemistry is superior to placental alkaline phosphatase (PLAP) in the diagnosis of central nervous system germinoma. Am J Surg Pathol. 2005;29:368–71.
81. Jones TD, Ulbright TM, Eble JN, Baldridge LA, Cheng L. OCT4 staining in testicular tumors: a sensitive and specific marker for seminoma and embryonal carcinoma. Am J Surg Pathol. 2004;28:935–40.
82. Leroy X, Augusto D, Leteurtre E, Gosselin B. CD30 and CD117 (c-kit) used in combination are useful for distinguishing embryonal carcinoma from seminoma. J Histochem Cytochem. 2002;50:283–5.
83. Butnor KJ, Burchette JL, Sporn TA, Hammar SP, Roggli VL. The spectrum of Kit (CD117) immunoreactivity in lung and pleural tumors: a study of 96 cases using a single-source antibody with a review of the literature. Arch Pathol Lab Med. 2004;128:538–43.
84. Nakagawa K, Matsuno Y, Kunitoh H, Maeshima A, Asamura H, Tsuchiya R. Immunohistochemical KIT (CD117) expression in thymic epithelial tumors. Chest. 2005;128:140–4.
85. Pan CC, Chen PC, Chiang H. KIT (CD117) is frequently overexpressed in thymic carcinomas but is absent in thymomas. J Pathol. 2004;202:375–81.
86. Ferreiro JA. Ber-H2 expression in testicular germ cell tumors. Hum Pathol. 1994;25:522–4.
87. Ramalingam P, Malpica A, Silva EG, Gershenson DM, Liu JL, Deavers MT. The use of cytokeratin 7 and EMA in differentiating ovarian yolk sac tumors from endometrioid and clear cell carcinomas. Am J Surg Pathol. 2004;28:1499–505.

88. Haninger DM, Kloecker GH, Bousamra Ii M, Nowacki MR, Slone SP. Hepatoid adenocarcinoma of the lung: report of five cases and review of the literature. Mod Pathol. 2014;27:535–42.
89. Dirnhofer S, Freund M, Rogatsch H, Krabichler S, Berger P. Selective expression of trophoblastic hormones by lung carcinoma: neurendocrine tumors exclusively produce human chorionic gonadotropin alpha-subunit (hCGalpha). Hum Pathol. 2000;31:966–72.
90. Boucher LD, Yoneda K. The expression of trophoblastic cell markers by lung carcinomas. Hum Pathol. 1995;26:1201–6.
91. White JE, Fincher RM, D'Cruz IA. Pericardial metastasis from testicular seminoma: appearance and disappearance by echocardiography. Am J Med Sci. 1991;301:182–5.
92. Bohle A, Studer UE, Sonntag RW, Scheidegger JR. Primary or secondary extragonadal germ cell tumors? J Urol. 1986;135:939–43.
93. Luna MA, Valenzuela-Tamariz J. Germ-cell tumors of the mediastinum, postmortem findings. Am J Clin Pathol. 1976;65:450–4.
94. Johnson DE, Appelt G, Samuels ML, Luna M. Metastases from testicular carcinoma. Study of 78 autopsied cases. Urology. 1976;8:234–9.
95. Hornak M, Ondrus D, Matoska J, Carsky S. Postchemotherapy surgery in nonseminomatous testicular tumors. Eur Urol. 1996;29:325–30.
96. Suster S, Rosai J. Multilocular thymic cyst: an acquired reactive process. Study of 18 cases. Am J Surg Pathol. 1991;15:388–98.
97. Michal M, Havlicek F. Pseudo-epitheliomatous hyperplasia in thymic cysts. Histopathology. 1991;19:281–2.
98. Suster S, Barbuto D, Carlson G, Rosai J. Multilocular thymic cysts with pseudoepitheliomatous hyperplasia. Hum Pathol. 1991;22:455–60.
99. Chitale AR. Gastric cyst of the mediastinum. A distinct clinicopathological entity. J Pediatr. 1969;75:104–10.
100. Sabiston Jr DC, Scott Jr HW. Primary neoplasms and cysts of the mediastinum. Ann Surg. 1952;136:777–97.
101. Salyer DC, Salyer WR, Eggleston JC. Benign developmental cysts of the mediastinum. Arch Pathol Lab Med. 1977;101:136–9.
102. Ramenofsky ML, Leape LL, McCauley RG. Bronchogenic cyst. J Pediatr Surg. 1979;14:219–24.
103. Roma A, Varsegi M, Magi-Galluzzi C, Ulbright T, Zhou M. The distinction of bronchogenic cyst from metastatic testicular teratoma: a light microscopic and immunohistochemical study. Am J Clin Pathol. 2008;130:265–73.
104. Abell MR. Mediastinal cysts. AMA Arch Pathol. 1956;61:360–79.
105. Dehner LP. Pleuropulmonary blastoma is THE pulmonary blastoma of childhood. Semin Diagn Pathol. 1994;11:144–51.
106. Priest JR, McDermott MB, Bhatia S, Watterson J, Manivel JC, Dehner LP. Pleuropulmonary blastoma: a clinicopathologic study of 50 cases. Cancer. 1997;80:147–61.
107. Kerr KM, Pelosi G, Austin JHM, et al. Sarcomatoid carcinoma. In: Travis WD, Brambilla E, Burke AP, et al., editors. World Health Organization classification of tumours pathology and genetics tumours of the lung, pleura, thymus, and heart. Lyon: IARC press; 2015. p. 88–94.
108. Koss MN, Hochholzer L, O'Leary T. Pulmonary blastomas. Cancer. 1991;67:2368–81.
109. Nakatani Y, Dickersin GR, Mark EJ. Pulmonary endodermal tumor resembling fetal lung: a clinicopathologic study of five cases with immunohistochemical and ultrastructural characterization. Hum Pathol. 1990;21:1097–107.
110. Nakatani Y, Kitamura H, Inayama Y, Kamijo S, Nagashima Y, Shimoyama K, Nakamura N, Sano J, Ogawa N, Shibagaki T, Resl M, Mark EJ. Pulmonary adenocarcinomas of the fetal lung type: a clinicopathologic study indicating differences in histology, epidemiology, and natural history of low-grade and high-grade forms. Am J Surg Pathol. 1998;22:399–411.
111. Yousem SA, Wick MR, Randhawa P, Manivel JC. Pulmonary blastoma. An immunohistochemical analysis with comparison with fetal lung in its pseudoglandular stage. Am J Clin Pathol. 1990;93:167–75.
112. Cheng L. Establishing a germ cell origin for metastatic tumors using OCT4 immunohistochemistry. Cancer. 2004;101:2006–10.
113. Arnould L, Drouot F, Fargeot P, Bernard A, Foucher P, Collin F, Petrella T. Hepatoid adenocarcinoma of the lung: report of a case of an unusual alpha-fetoprotein-producing lung tumor. Am J Surg Pathol. 1997;21:1113–8.
114. Higgins JP, Warnke RA. CD30 expression is common in mediastinal large B-cell lymphoma. Am J Clin Pathol. 1999;112:241–7.
115. Jung SM, Chu PH, Shiu TF, Wu HH, Kuo TT, Chu JJ, Lin PJ. Expression of OCT4 in the primary germ cell tumors and thymoma in the mediastinum. Appl Immunohistochem Mol Morphol. 2006;14:273–5.
116. Hasserjian RP, Klimstra DS, Rosai J. Carcinoma of the thymus with clear-cell features. Report of eight cases and review of the literature. Am J Surg Pathol. 1995;19:835–41.
117. Bauer DE, Mitchell CM, Strait KM, Lathan CS, Stelow EB, Luer SC, Muhammed S, Evans AG, Sholl LM, Rosai J, Giraldi E, Oakley RP, Rodriguez-Galindo C, London WB, Sallan SE, Bradner JE, French CA. Clinicopathologic features and long-term outcomes of NUT midline carcinoma. Clin Cancer Res. 2012;18:5773–9.
118. French CA, Kutok JL, Faquin WC, Toretsky JA, Antonescu CR, Griffin CA, Nose V, Vargas SO, Moschovi M, Tzortzatou-Stathopoulou F, Miyoshi I,

Perez-Atayde AR, Aster JC, Fletcher JA. Midline carcinoma of children and young adults with NUT rearrangement. J Clin Oncol. 2004;22:4135–9.
119. Kees UR, Mulcahy MT, Willoughby ML. Intrathoracic carcinoma in an 11-year-old girl showing a translocation t(15;19). Am J Pediatr Hematol Oncol. 1991;13:459–64.
120. Lee AC, Kwong YI, Fu KH, Chan GC, Ma L, Lau YL. Disseminated mediastinal carcinoma with chromosomal translocation (15;19). A distinctive clinicopathologic syndrome. Cancer. 1993;72:2273–6.
121. Toretsky JA, Jenson J, Sun CC, Eskenazi AE, Campbell A, Hunger SP, Caires A, Frantz C, Hill JL, Stamberg J. Translocation (11;15;19): a highly specific chromosome rearrangement associated with poorly differentiated thymic carcinoma in young patients. Am J Clin Oncol. 2003;26:300–6.
122. Vargas SO, French CA, Faul PN, Fletcher JA, Davis IJ, Dal Cin P, Perez-Atayde AR. Upper respiratory tract carcinoma with chromosomal translocation 15;19: evidence for a distinct disease entity of young patients with a rapidly fatal course. Cancer. 2001;92:1195–203.
123. Evans AG, French CA, Cameron MJ, Fletcher CD, Jackman DM, Lathan CS, Sholl LM. Pathologic characteristics of NUT midline carcinoma arising in the mediastinum. Am J Surg Pathol. 2012;36:1222–7.
124. Fletcher CD, Beham A, Bekir S, Clarke AM, Marley NJ. Epithelioid angiosarcoma of deep soft tissue: a distinctive tumor readily mistaken for an epithelial neoplasm. Am J Surg Pathol. 1991;15:915–24.
125. McKenney JK, Weiss SW, Folpe AL. CD31 expression in intratumoral macrophages: a potential diagnostic pitfall. Am J Surg Pathol. 2001;25:1167–73.
126. Suster S, Moran CA. Primary synovial sarcomas of the mediastinum: a clinicopathologic, immunohistochemical, and ultrastructural study of 15 cases. Am J Surg Pathol. 2005;29:569–78.
127. Shia J, Qin J, Erlandson RA, King R, Illei P, Nobrega J, Yao D, Klimstra DS. Malignant mesothelioma with a pronounced myxoid stroma: a clinical and pathological evaluation of 19 cases. Virchows Arch. 2005;447:828–34.
128. Yousem SA. Malignant mesothelioma. Arch Pathol Lab Med. 1987;111:503.
129. Travis WD. Congenital peribronchial myofibroblastic tumour. In: Travis WD, Brambilla E, Burke AP, et al., editors. World Health Organization classification of tumours pathology and genetics tumours of the lung, pleura, thymus, and heart. Lyon: IARC press; 2015. p. 119–21.
130. Bokemeyer C, Droz JP, Horwich A, Gerl A, Fossa SD, Beyer J, Pont J, Schmoll HJ, Kanz L, Einhorn L, Nichols CR, Hartmann JT. Extragonadal seminoma: an international multicenter analysis of prognostic factors and long term treatment outcome. Cancer. 2001;91:1394–401.
131. Bokemeyer C, Hartmann JT, Fossa SD, Droz JP, Schmol HJ, Horwich A, Gerl A, Beyer J, Pont J, Kanz L, Nichols CR, Einhorn L. Extragonadal germ cell tumors: relation to testicular neoplasia and management options. APMIS. 2003;111:49–59. discussion 59-63
132. Hartmann JT, Nichols CR, Droz JP, Horwich A, Gerl A, Fossa SD, Beyer J, Pont J, Kanz L, Einhorn L, Bokemeyer C. Prognostic variables for response and outcome in patients with extragonadal germ-cell tumors. Ann Oncol. 2002;13:1017–28.
133. Biester RJ, Lippert MC, Mills SE. Late recurrence of a seminoma. J Urol. 1987;137:749–50.
134. International Germ Cell Consensus Classification: a prognostic factor-based staging system for metastatic germ cell cancers. International Germ Cell Cancer Collaborative Group. J Clin Oncol. 1997;15:594–603.
135. Kesler KA, Rieger KM, Hammoud ZT, Kruter LE, Perkins SM, Turrentine MW, Schneider BP, Einhorn LH, Brown JW. A 25-year single institution experience with surgery for primary mediastinal nonseminomatous germ cell tumors. Ann Thorac Surg. 2008;85:371–8.
136. Rivera C, Arame A, Jougon J, Velly JF, Begueret H, Dahan M, Riquet M. Prognostic factors in patients with primary mediastinal germ cell tumors, a surgical multicenter retrospective study. Interact Cardiovasc Thorac Surg. 2010;11:585–9.
137. Vuky J, Bains M, Bacik J, Higgins G, Bajorin DF, Mazumdar M, Bosl GJ, Motzer RJ. Role of postchemotherapy adjunctive surgery in the management of patients with nonseminoma arising from the mediastinum. J Clin Oncol. 2001;19:682–8.
138. Kesler KA, Rieger KM, Ganjoo KN, Sharma M, Fineberg NS, Einhorn LH, Brown JW. Primary mediastinal nonseminomatous germ cell tumors: the influence of postchemotherapy pathology on long-term survival after surgery. J Thorac Cardiovasc Surg. 1999;118:692–700.
139. Kesler KA, Patel JB, Kruter LE, Birdas TJ, Rieger KM, Okereke IC, Einhorn LH. The "growing teratoma syndrome" in primary mediastinal nonseminomatous germ cell tumors: criteria based on current practice. J Thorac Cardiovasc Surg. 2012;144:438–43.
140. Fizazi K, Tjulandin S, Salvioni R, Germa-Lluch JR, Bouzy J, Ragan D, Bokemeyer C, Gerl A, Flechon A, de Bono JS, Stenning S, Horwich A, Pont J, Albers P, De Giorgi U, Bower M, Bulanov A, Pizzocaro G, Aparicio J, Nichols CR, Theodore C, Hartmann JT, Schmoll HJ, Kaye SB, Culine S, Droz JP, Mahe C. Viable malignant cells after primary chemotherapy for disseminated nonseminomatous germ cell tumors: prognostic factors and role of postsurgery chemotherapy--results from an international study group. J Clin Oncol. 2001;19:2647–57.
141. Moran CA, Suster S. Primary mediastinal choriocarcinomas: a clinicopathologic and immunohistochemical

study of eight cases. Am J Surg Pathol. 1997;21:1007–12.
142. Schneider DT, Calaminus G, Reinhard H, Gutjahr P, Kremens B, Harms D, Gobel U. Primary mediastinal germ cell tumors in children and adolescents: results of the German cooperative protocols MAKEI 83/86, 89, and 96. J Clin Oncol. 2000;18:832–9.
143. Hartmann JT, Nichols CR, Droz JP, Horwich A, Gerl A, Fossa SD, Beyer J, Pont J, Fizazi K, Einhorn L, Kanz L, Bokemeyer C. Hematologic disorders associated with primary mediastinal nonseminomatous germ cell tumors. J Natl Cancer Inst. 2000;92:54–61.
144. Albany C, Einhorn LH. Extragonadal germ cell tumors: clinical presentation and management. Curr Opin Oncol. 2013;25:261–5.
145. Motzer RJ, Amsterdam A, Prieto V, Sheinfeld J, Murty VV, Mazumdar M, Bosl GJ, Chaganti RS, Reuter VE. Teratoma with malignant transformation: diverse malignant histologies arising in men with germ cell tumors. J Urol. 1998;159:133–8.
146. Felix CA. Leukemias related to treatment with DNA topoisomerase II inhibitors. Med Pediatr Oncol. 2001;36:525–35.
147. Ohshima A, Miura I, Chubachi A, Hashimoto K, Nimura T, Utsumi S, Takahashi N, Hayashi Y, Seto M, Ueda R, Miura AB. 11q23 aberration is an additional chromosomal change in de novo acute leukemia after treatment with etoposide and mitoxantrone. Am J Hematol. 1996;53:264–6.
148. Stanulla M, Wang J, Chervinsky DS, Aplan PD. Topoisomerase II inhibitors induce DNA double-strand breaks at a specific site within the AML1 locus. Leukemia. 1997;11:490–6.
149. Logothetis CJ, von Eschenbach AC, Samuels ML, Trindade A, Johnson DE. Doxorubicin, mitomycin, and 5-FU (DMF) in the treatment of hormone-resistant stage D prostate cancer: a preliminary report. Cancer Treat Rep. 1982;66:57–63.
150. Liu A, Cheng L, Du J, Peng Y, Allan RW, Wei L, Li J, Cao D. Diagnostic utility of novel stem cell markers SALL4, OCT4, NANOG, SOX2, UTF1, and TCL1 in primary mediastinal germ cell tumors. Am J Surg Pathol. 2010;34:697–706.
151. Lau SK, Chang KL. OCT4: a sensitive and specific immunohistochemical marker for metastatic germ cell tumors. Adv Anat Pathol. 2006;13:76–9.
152. Mills AF, Lanfranchi M, Wein RO, Mukand-Cerro I, Pilichowska M, Cowan J, Bedi H. NUT midline carcinoma: a case report with a novel translocation and review of the literature. Head Neck Pathol. 2014;8:182–6.
153. Iczkowski KA, Butler SL, Shanks JH, Hossain D, Schall A, Meiers I, Zhou M, Torkko KC, Kim SJ, MacLennan GT. Trials of new germ cell immunohistochemical stains in 93 extragonadal and metastatic germ cell tumors. Hum Pathol. 2008;39:275–81.
154. Looijenga LH, Stoop H, de Leeuw HP, de Gouveia Brazao CA, Gillis AJ, van Roozendaal KE, van Zoelen EJ, Weber RF, Wolffenbuttel KP, van Dekken H, Honecker F, Bokemeyer C, Perlman EJ, Schneider DT, Kononen J, Sauter G, Oosterhuis JW. POU5F1 (OCT3/4) identifies cells with pluripotent potential in human germ cell tumors. Cancer Res. 2003;63:2244–50.
155. Howitt BE, Brooks JD, Jones S, Higgins JP. Identification and characterization of 2 testicular germ cell markers, Glut3 and CyclinA2. Appl Immunohistochem Mol Morphol. 2013;21:401–7.
156. Santagata S, Ligon KL, Hornick JL. Embryonic stem cell transcription factor signatures in the diagnosis of primary and metastatic germ cell tumors. Am J Surg Pathol. 2007;31:836–45.
157. Nakamura H, Tsuta K, Tsuda H, Katsuya Y, Naka G, Iizuka T, Igari T. NUT midline carcinoma of the mediastinum showing two types of poorly differentiated tumor cells: A case report and a literature review. Pathol Res Pract. 2015;211:92–8.
158. Kao CS, Idrees MT, Young RH, Ulbright TM. Solid pattern yolk sac tumor: a morphologic and immunohistochemical study of 52 cases. Am J Surg Pathol. 2012;36:360–7.
159. Bode PK, Barghorn A, Fritzsche FR, Riener MO, Kristiansen G, Knuth A, Moch H. MAGEC2 is a sensitive and novel marker for seminoma: a tissue microarray analysis of 325 testicular germ cell tumors. Mod Pathol. 2011;24:829–35.
160. Lau SK, Weiss LM, Chu PG. D2-40 immunohistochemistry in the differential diagnosis of seminoma and embryonal carcinoma: a comparative immunohistochemical study with KIT (CD117) and CD30. Mod Pathol. 2007;20:320–5.
161. Parikh SA, French CA, Costello BA, Marks RS, Dronca RS, Nerby CL, Roden AC, Peddareddigari VG, Hilton J, Shapiro GI, Molina JR. NUT midline carcinoma: an aggressive intrathoracic neoplasm. J Thorac Oncol. 2013;8:1335–8.
162. Weissferdt A, Moran CA. Anaplastic thymic carcinoma: a clinicopathologic and immunohistochemical study of 6 cases. Hum Pathol. 2012;43:874–7.
163. Saad RS, Landreneau RJ, Liu Y, Silverman JF. Utility of immunohistochemistry in separating thymic neoplasms from germ cell tumors and metastatic lung cancer involving the anterior mediastinum. Appl Immunohistochem Mol Morphol. 2003;11:107–12.
164. Moch H, Oberholzer M, Dalquen P, Wegmann W, Gudat F. Diagnostic tools for differentiating between pleural mesothelioma and lung adenocarcinoma in paraffin embedded tissue. Part I: immunohistochemical findings. Virchows Arch. 1993;423:19–27.
165. Suzuki S, Kurabe N, Ohnishi I, Yasuda K, Aoshima Y, Naito M, Tanioka F, Sugimura H. NSD3-NUT-expressing midline carcinoma of the lung: first characterization of primary cancer tissue. Pathol Res Pract. 2014;211(5):404-8
166. Chu AY, Litzky LA, Pasha TL, Acs G, Zhang PJ. Utility of D2-40, a novel mesothelial marker, in

the diagnosis of malignant mesothelioma. Mod Pathol. 2005;18:105–10.
167. Comin CE, Novelli L, Boddi V, Paglierani M, Dini S. Calretinin, thrombomodulin, CEA, and CD15: a useful combination of immunohistochemical markers for differentiating pleural epithelial mesothelioma from peripheral pulmonary adenocarcinoma. Hum Pathol. 2001;32:529–36.
168. Morice WG, Ferreiro JA. Distinction of basaloid squamous cell carcinoma from adenoid cystic and small cell undifferentiated carcinoma by immunohistochemistry. Hum Pathol. 1998;29:609–12.
169. Suster S, Moran CA. Spindle cell thymic carcinoma: clinicopathologic and immunohistochemical study of a distinctive variant of primary thymic epithelial neoplasm. Am J Surg Pathol. 1999;23:691–700.
170. Wick MR, Swanson PE, Manivel JC. Placental-like alkaline phosphatase reactivity in human tumors: an immunohistochemical study of 520 cases. Hum Pathol. 1987;18:946–54.
171. Cheville JC, Rao S, Iczkowski KA, Lohse CM, Pankratz VS. Cytokeratin expression in seminoma of the human testis. Am J Clin Pathol. 2000; 113:583–8.
172. Rekhtman N, Ang DC, Sima CS, Travis WD, Moreira AL. Immunohistochemical algorithm for differentiation of lung adenocarcinoma and squamous cell carcinoma based on large series of whole-tissue sections with validation in small specimens. Mod Pathol. 2011;24:1348–59.
173. Carella R, Deleonardi G, D'Errico A, Salerno A, Egarter-Vigl E, Seebacher C, Donazzan G, Grigioni WF. Immunohistochemical panels for differentiating epithelial malignant mesothelioma from lung adenocarcinoma: a study with logistic regression analysis. Am J Surg Pathol. 2001; 25:43–50.
174. Chu P, Wu E, Weiss LM. Cytokeratin 7 and cytokeratin 20 expression in epithelial neoplasms: a survey of 435 cases. Mod Pathol. 2000;13:962–72.
175. Kuo TT, Chan JK. Thymic carcinoma arising in thymoma is associated with alterations in immunohistochemical profile. Am J Surg Pathol. 1998; 22:1474–81.
176. Lau SK, Weiss LM, Chu PG. Differential expression of MUC1, MUC2, and MUC5AC in carcinomas of various sites: an immunohistochemical study. Am J Clin Pathol. 2004;122:61–9.
177. Comin CE, Dini S, Novelli L, Santi R, Asirelli G, Messerini L. h-Caldesmon, a useful positive marker in the diagnosis of pleural malignant mesothelioma, epithelioid type. Am J Surg Pathol. 2006;30:463–9.
178. Truong LD, Mody DR, Cagle PT, Jackson-York GL, Schwartz MR, Wheeler TM. Thymic carcinoma. A clinicopathologic study of 13 cases. Am J Surg Pathol. 1990;14:151–66.
179. Pan CC, Chen PC, Chou TY, Chiang H. Expression of calretinin and other mesothelioma-related markers in thymic carcinoma and thymoma. Hum Pathol. 2003;34:1155–62.
180. Chu PJ, Best PM. Molecular cloning of calcium channel alpha(2)delta-subunits from rat atria and the differential regulation of their expression by IGF-1. J Mol Cell Cardiol. 2003;35:207–15.
181. Lau SK, Thomas P, Weiss LM. Immunohistochemical evaluation of CON6D/B5: a new CD30 monoclonal antibody. Appl Immunohistochem Mol Morphol. 2010;18:273–7.
182. Ordonez NG. The diagnostic utility of immunohistochemistry in distinguishing between epithelioid mesotheliomas and squamous carcinomas of the lung: a comparative study. Mod Pathol. 2006; 19:417–28.
183. Masai K, Tsuta K, Kawago M, Tatsumori T, Kinno T, Taniyama T, Yoshida A, Asamura H, Tsuda H. Expression of squamous cell carcinoma markers and adenocarcinoma markers in primary pulmonary neuroendocrine carcinomas. Appl Immunohistochem Mol Morphol. 2013;21:292–7.
184. Folpe AL, Gown AM, Lamps LW, Garcia R, Dail DH, Zarbo RJ, Schmidt RA. Thyroid transcription factor-1: immunohistochemical evaluation in pulmonary neuroendocrine tumors. Mod Pathol. 1999;12:5–8.
185. Lugli A, Forster Y, Haas P, Nocito A, Bucher C, Bissig H, Mirlacher M, Storz M, Mihatsch MJ, Sauter G. Calretinin expression in human normal and neoplastic tissues: a tissue microarray analysis on 5233 tissue samples. Hum Pathol. 2003;34:994–1000.
186. Movahedi-Lankarani S, Kurman RJ. Calretinin, a more sensitive but less specific marker than alpha-inhibin for ovarian sex cord-stromal neoplasms: an immunohistochemical study of 215 cases. Am J Surg Pathol. 2002;26:1477–83.
187. Kao CS, Badve SS, Ulbright TM. The utility of immunostaining for NUT, GAGE7 and NY-ESO-1 in the diagnosis of spermatocytic seminoma. Histopathology. 2014;65:35–44.
188. Gokmen-Polar Y, Cano OD, Kesler KA, Loehrer PJ, Badve S. NUT midline carcinomas in the thymic region. Mod Pathol. 2014;27:1649–56.
189. National Cancer Institute. Extragonadal extracranial GCT staging from the COG. 2014. Childhood Extracranial Germ Cell Tumors Treatment (PDQ®). http://www.cancer.gov/cancertopics/pdq/treatment/extra-cranial-germ-cell/HealthProfessional/Page2#_385.

Central Nervous System Germ Cell Tumors

9

M. Adelita Vizcaino and Fausto J. Rodriguez

9.1 Introduction

Germ cell tumors (GCT) of the central nervous system (CNS) have a predilection for midline locations (particularly pineal and pituitary glands), affect mostly children and young adults, and show clinical and histological similarities to their more common counterparts arising in the gonads [1–3]. They are second to the mediastinum in frequency of extragonadal GCT and may in fact be among the most curable primary brain tumors [4, 5]. Optimal management of these neoplasms involves a multidisciplinary approach, in view of the complexities of surgery, availability of markers for diagnosis, response to both chemotherapy and radiotherapy, and possible complications secondary to treatment [2, 6]. In this chapter we discuss the clinical and pathologic features of GCT with a focus on properties specific to the CNS.

M.A. Vizcaino, MD
Department of Cellular and Tissue Biology, Faculty of Medicine, UNAM, Mexico City, Mexico

Department of Pathology, Johns Hopkins University School of Medicine, Baltimore, MD, USA

F.J. Rodriguez, MD (✉)
Department of Pathology, Division of Neuropathology and Sydney Kimmel Comprehensive Cancer Center, Johns Hopkins University School of Medicine, Sheikh Zayed Tower, Room M2101, 1800 Orleans Street, Baltimore, MD 21231, USA
e-mail: frodrig4@jhmi.edu

9.2 Epidemiology

GCT arising in the CNS are relatively rare, representing approximately 1 % of all malignant CNS tumors [7]. In East Asia, including Japan and South Korea, the incidence traditionally has been thought to be higher, reported to comprise 2.1–14 % of all primary intracranial neoplasms, with Asian races associated with a higher rate of CNS GCT than other races [8, 9]. However, a recent analysis of four tumor registries from Japan and the USA showed no statistical differences between both populations; incidences by anatomic site and gender found in the USA were similar to those observed in the Japanese population [4].

CNS GCT mainly affect the young, with up to 90 % of the cases developing before the age of 20, and a peak incidence in teenagers [3, 5]. Males are more likely to develop intracranial germ cell tumors, particularly for those located in the pineal region [4, 10]. Conversely, the incidence rate of pituitary GCT is slightly higher for females than males [4, 6].

A link has been described between GCT and specific genetic disorders associated with chromosomal abnormalities, such as Down and Klinefelter syndromes. In fact, GCT are the most common brain neoplasms among these patients [2, 11]. It is known that Down syndrome is associated with lower frequencies of solid cancers, but interestingly, half of the CNS neoplasms aris-

F.F. Nogales, R.E. Jimenez (eds.), *Pathology and Biology of Human Germ Cell Tumors*,
DOI 10.1007/978-3-662-53775-6_9

ing in this syndrome are GCT [12, 13]. Recent reports indicate that patients with Klinefelter syndrome have an increased relative risk for the development of malignant extragonadal GCT, supporting that sex chromosome aneuploidies participate in germ cell tumorigenesis [14].

9.3 Classification

Under the current WHO classification of tumors of the CNS, five histopathological subtypes of intracranial germ cell tumors are recognized: germinoma, embryonal carcinoma, teratoma, yolk sac tumor, and choriocarcinoma (Table 9.1) [15]. Nonetheless, due to the excellent response to therapy and prognosis of the germinoma group, germ cell tumors are broadly grouped into two main categories: pure germinomas and non-germinomatous germ cell tumors (NGGCT) [9, 16, 17].

Other proposed classification systems separate histologic variants into different therapeutic groups based on their prognosis. Indeed, the Japanese Pediatric Brain Tumor Study Group divided GCT into three prognostic groups according to the histological subtypes: good (pure germinoma, mature teratoma), intermediate (germinoma with elevated levels of β-HCG, extensive or multifocal germinoma, immature teratoma, mixed tumors composed mainly of germinoma or teratoma), and poor (choriocarcinoma, yolk sac tumor, embryonal carcinoma, and mixed tumors composed mainly of non-germinomatous components) (Table 9.2) [1, 18–20].

Table 9.1 WHO classification of primary CNS germ cell tumors [43]

Germinoma
Embryonal carcinoma
Yolk sac tumor
Choriocarcinoma
Teratoma
Immature
Mature
With malignant transformation
Mixed germ cell

Table 9.2 Therapeutic classification of GCT by the Japanese Pediatric Brain Tumor Study Group [44]

Best prognosis
1. Germinoma
2. Mature teratoma
Intermediate prognosis
1. Germinoma with syncytiotrophoblast
2. Teratoma
Immature
With malignant transformation
3. Mixed GCT with predominant germinoma and/or teratoma
Poor prognosis
1. Choriocarcinoma
2. Yolk sac tumor
3. Embryonal carcinoma
4. Mixed GCT with predominant choriocarcinoma, yolk sac tumor, and/or embryonal carcinoma

9.4 Location

Primary CNS GCT are usually midline-located tumors. Most affect the pineal, neurohypophysis, or suprasellar region and to a lesser extent, the basal ganglia [21–23]. Metastatic cases account for a minority. All these structures are situated around the third ventricle, overlying the hypothalamus. Although rare, dissemination along ventricular walls can be seen at initial presentation [2, 3, 24].

Between 5 and 10 % of intracranial GCT present as bifocal lesions or with simultaneous involvement of pineal gland and neurohypophysis, most of which are germinomas [1, 5, 10, 18]. This presentation is still a curious but unexplained phenomenon [25]. Some authors assume that metastatic spread is more likely than simultaneous development based on the demographic data and the high ventricular seeding rate [24, 26]. However, bifocal tumors without evidence of dissemination on spinal magnetic resonance imaging (MRI) and negative cerebrospinal fluid (CSF) cytology, as well as tumor markers, are often considered locoregional disease rather than metastatic [27].

Other anatomic areas that may be involved less frequently include cerebral hemispheres, lateral ventricles [28], thalamus [18], corpus callosum, cerebellopontine angle, cerebellum, optic pathway [29], and medulla oblongata [1].

9.4.1 Clinical Features

Intracranial GCT may present in an insidious manner, and many patients show a long symptomatic period before definite diagnosis [23, 24]. Symptoms in patients with intracranial GCT include headache, nausea and vomiting, polyuria and/or polydipsia, visual disturbances, fatigue, weight loss or poor growth, and precocious puberty. The initial clinical presentation varies according to patient's age, tumor location, and size. Pineal GCT usually manifest with signs of increased intracranial pressure/CSF obstruction (headache, nausea, vomiting, and papilledema) and often require shunting. Somnolence is seen in up to half of patients and ataxia, seizures, and behavioral changes in a subset. Endocrinopathies and disturbances in sexual development such as precocious puberty in patients with isolated pineal region tumors are less common [17, 18]. The cause of precocious puberty in pineal GCT is only partially understood. Interestingly, choriocarcinomas are more likely to present with precocious puberty than any other type of GCT [10, 24].

Conversely, suprasellar tumors usually present with an endocrinopathy, such as central diabetes insipidus, hypopituitarism, hypothyroidism, adrenal insufficiency, precocious puberty, delayed sexual development, or retarded growth. Patients rarely present with signs of increased intracranial pressure. Ophthalmic symptoms are quite common in all patients. Pineal region tumors usually cause photosensitivity or diplopia, with up to a third of patients having a component of Parinaud syndrome (impaired upward gaze, convergence nystagmus, and impaired papillary response). A subset of patients with suprasellar tumors develop visual changes due to tumor entrapment of the optic nerve or chiasm [1, 17]. In patients with GCT of the basal ganglia, progressive hemiparesis is also a common complaint [23, 24]. A sudden onset of intracranial bleeding is often an initial, dramatic presentation of choriocarcinoma.

9.4.2 Neuroimaging

Germinomas usually appear as masses with relatively uniform density (or signal intensity) except for variable intratumoral cysts. They show increase density on computed tomography (CT) scans, with a somewhat blurred border. Pineal calcification, uncommon in children less than 10 years of age, is a useful clue to the diagnosis of a GCT [10]. On MRI, germinoma appears as a well-defined mass with slightly low signal intensity on T1-weighted images and high intensity on T2-weighted images. Pineal teratomas appear as heterogeneous well-demarcated masses with occasional calcifications, irregular cysts, or fatty tissue and thus are identifiable on both CT and MRI. Differentiation from pineal parenchymal tumors is not always easy [1]. Pituitary GCT may affect the neurohypophysis exclusively or extend into the intrasellar region. Signal characteristics are quite similar to those of pineal tumors. Loss of the posterior pituitary "bright spot" is useful in the assessment and monitoring of suprasellar germinoma.

For pineal lesions, the main differential diagnoses are pineal parenchymal tumors and low-grade gliomas. For sellar GCT, differential diagnoses are craniopharyngioma, Langerhans cell histiocytosis, hypophysitis, sarcoidosis, and low-grade gliomas [1, 2].

9.4.3 Tumor Biomarkers

Currently, stereotactic or endoscopic biopsy is the gold standard for diagnosing CNS GCT. However, some NGGCT produce tumor markers that are helpful to identify the histological subtype, such as alpha-fetoprotein (AFP) and beta-subunit human chorionic gonadotropin (β-HCG). These secreted markers can be measured both in serum and CSF [1, 3, 5, 18]. Even when the role of ventricular CSF oncoprotein assays is evolving; lumbar CSF remains the most useful and reliable source of fluid for clinical guidance in the diagnosis, management, and tumor response of CNS GCT, since it is more sensitive than radiologic changes [30].

Any detectable elevation of AFP (serum >5–10 ng/dL; CSF >2–5 ng/dL) or marked CSF elevation of β-HCG (typically more than 100–200 IU) may be considered diagnostic of NGGCT without histologic confirmation [5]. Germinomas with syncytiotrophoblastic giant cells also produce hCG, but the CSF titer is lower than that of choriocarcinoma in most cases. Mixed GCT, according to the histological elements they contain, produce variable amounts of tumor markers [1, 5]. sKIT, a soluble form of the c-KIT has been proposed as a useful tumor marker for CNS germinomas, since the levels in this tumor are significantly higher, and its CSF concentration is particularly increased in patients with subarachnoid dissemination.

9.5 Diagnostic Pathology

Except in cases where tumor markers are unequivocally elevated in CSF and/or serum, histologic confirmation is necessary for diagnosis of CNS GCT [5]. It must be emphasized that GCT often have mixed histologic subtypes, and consequently, the description of different components and relative amount is recommended in pathology reports.

9.5.1 Germinoma

Pure germinomas account for most CNS GCT and are histologically identical to their gonadal counterparts of seminoma (testes) and dysgerminoma (ovaries). They are composed of large, undifferentiated cells arranged in monomorphous sheets, nests, or lobules outlined by thin fibrovascular septa. Germinoma cells have an abundant clear cytoplasm due to glycogen accumulation, as well as round, vesicular, and centrally positioned nuclei with prominent, squared, or rectangular-shaped nucleoli (Figs. 9.1 and 9.2). These cytological features are retained in lumbar or ventricular CSF samples. Mitoses are easily identified and may be conspicuous, but necrosis is infrequent. Strands of connective tissue variably infiltrated by non neoplastic lymphocytes are common. However, some germinomas show an extensive lymphoplasmacytic reaction or granulomatous response. A subset of germinomas may contain variable amounts of syncytiotrophoblastic giant cells which express β-HCG, human placental lactogen (hPL), and cytokeratins. Germinomas with syncytiotrophoblastic elements are not as aggressive as choriocarcinomas and therefore should not be confused (Fig. 9.3). Nonetheless, they seem to have a higher recurrence rate following radiation therapy than germinomas without these cells [5].

The most consistent immunohistochemical pattern in germinomas includes a strong nuclear labeling for SALL4 [31] and OCT4 [32], cell membrane staining for CD117 (KIT), and membrane/cytoplasmic positivity for PLAP (See Chap. 4). Some markers await validation, for example, HESRG [33].

9.5.2 Non-germinomatous Germ Cell Tumors

NGGCT comprise a unique group of neoplasms displaying several forms of differentiation. They include pure or mixed populations of teratoma, embryonal carcinoma, yolk sac tumor, and choriocarcinoma [5, 34].

9.5.3 Embryonal Carcinoma

Embryonal carcinomas are composed of large, mitotically active cells that proliferate in cohesive nests and sheets, sometimes forming papillae or irregular gland-like spaces, and frequent areas of coagulative necrosis (Fig. 9.4). Tumor

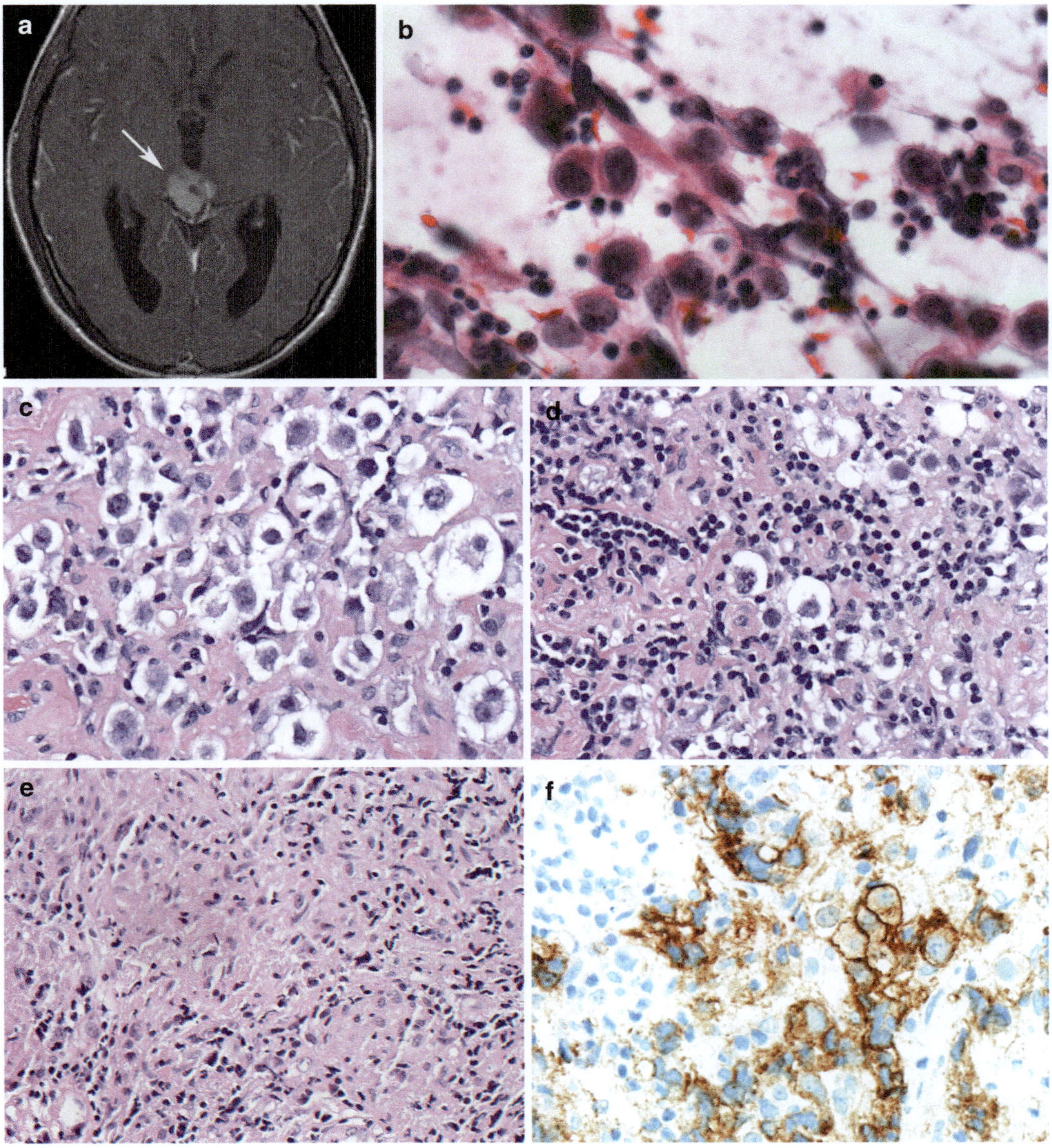

Fig. 9.1 Germinoma of the pineal region. The pineal region is a frequent anatomic site for germinoma which forms contrast enhancing masses on MRI (*arrow*) (**a**). The cytologic features of germinoma include the presence of discohesive cells with large nuclei and squared nucleoli (**b**). Abundant pale cytoplasm is usually evident on histologic sections (**c**). A variable amount of associated chronic inflammation is usual (**d**). On occasion associated granulomatous inflammation may be conspicuous as in other anatomic sites (**e**). Strong immunolabeling with PLAP is typical (**f**)

cells show enlarged, irregular nuclei with prominent nucleoli, as well as abundant, clear cytoplasm, the latter being strongly and diffusely positive for cytokeratins and for CD30. They also share immunoreactivity for PLAP and OCT4 with germinomas, but KIT is negative [18].

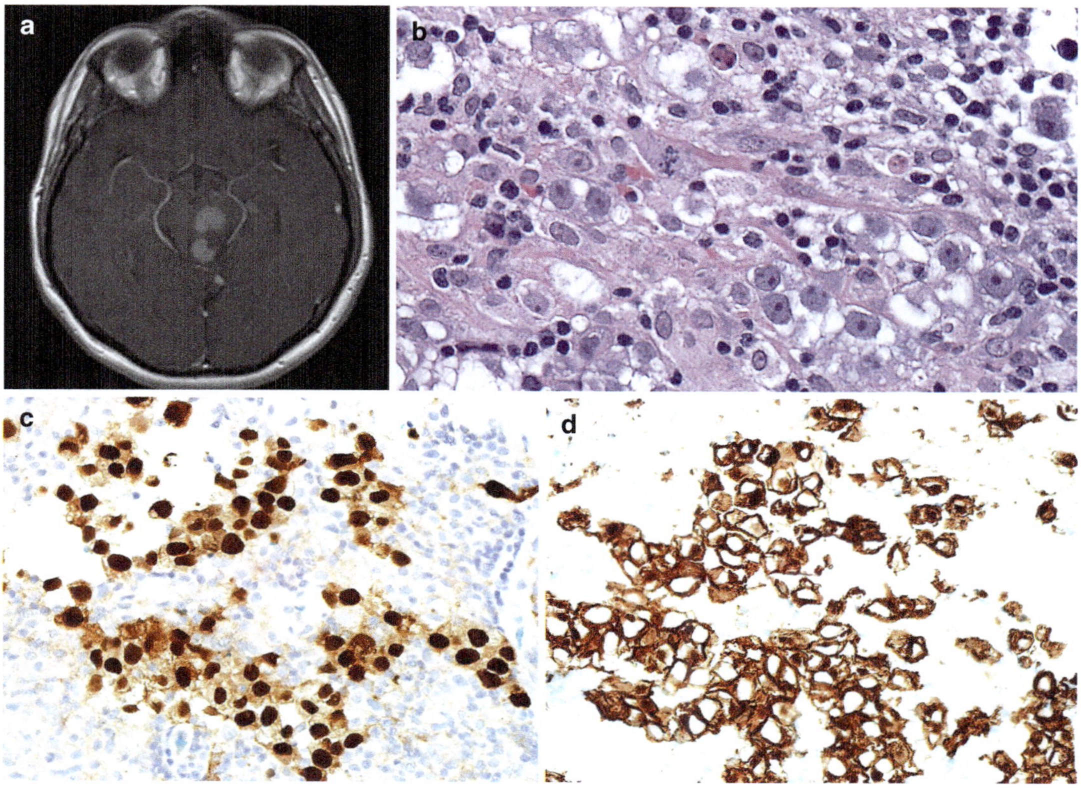

Fig. 9.2 Germinoma involving the midbrain. On occasion germinomas may involve the midbrain in a more diffuse form or a in a multinodular fashion, which may cause diagnostic confusion (**a**). Mitotic activity and apoptosis are frequent in germinomas (**b**). Additional immunohistochemical features of germinoma include OCT4 (**c**) and KIT expression (**d**)

9.5.4 Yolk Sac Tumor

Yolk sac/endodermal sinus tumors are composed of primitive epithelial cells in a myxoid matrix resembling extraembryonic mesoblast. Epithelial elements may demonstrate several growth patterns but are more commonly arranged about an intervening meshwork of irregular tissue spaces (reticular pattern) (Fig. 9.5). Infrequently, cuboidal yolk sac tumor cells line delicate fibrovascular projections to form distinctive papillae known as Schiller-Duval bodies [18]. A diagnostic feature that is not always seen is the presence of brightly eosinophilic, PAS-positive/diastase-resistant hyaline globules within the cytoplasm of tumor cells or in the stroma. Mitotic activity varies considerably, but necrosis is uncommon. AFP, SALL4, and Glypican-3 immunoreactivity is characteristic. Conversely, yolk sac tumors are negative for KIT and OCT 4.

9.5.5 Teratoma

Teratomas differentiate along ectodermal, mesodermal, and/or endodermal lines, usually with tissues representing all three germ cell layers. They are classically divided into mature, immature, and teratoma with malignant transformation. Distinction may be therapeutically relevant [18].

Mature teratomas are exclusively composed of differentiated "adult-like" tissue elements representing ectoderm, mesoderm, and endoderm, with scant or absent mitotic activity. The most common ectodermal components identified include

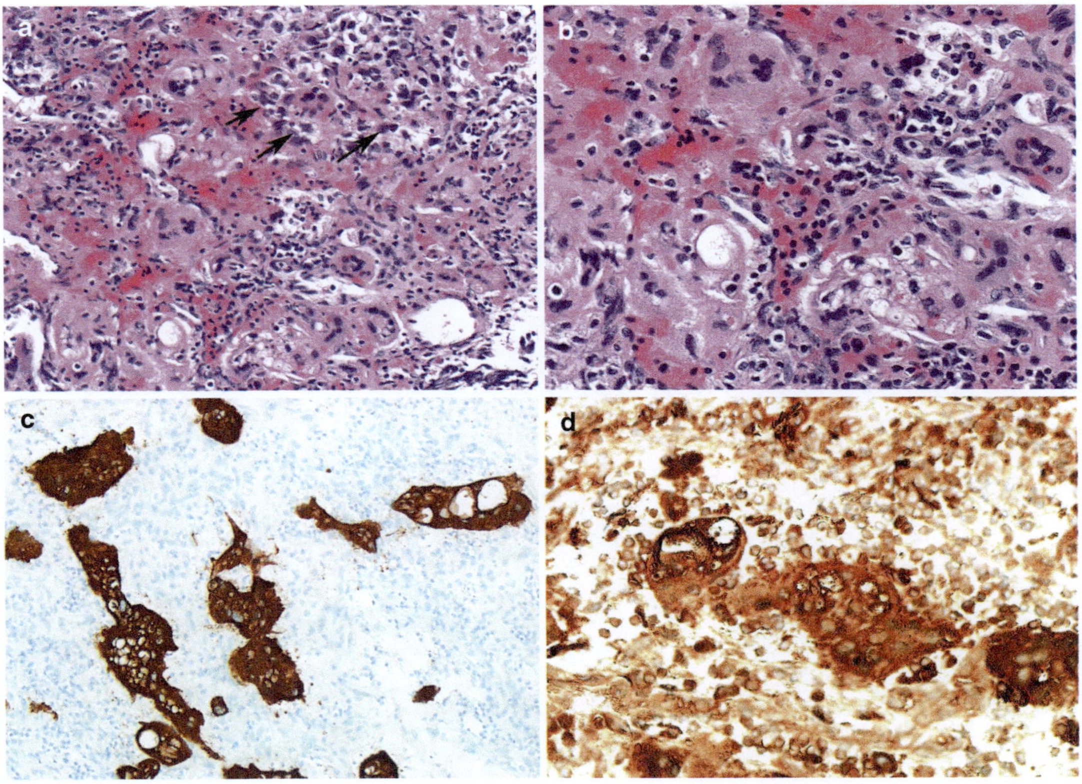

Fig. 9.3 Syncytiotrophoblast giant cells in germinoma. A subset of germinomas contain variable numbers of syncytiotrophoblast-like giant cells (**a**, **b**) which may be associated with B-HCG elevations. Cytokeratin (**c**) and B-HCG (**d**) expression may be detected by immunohistochemistry

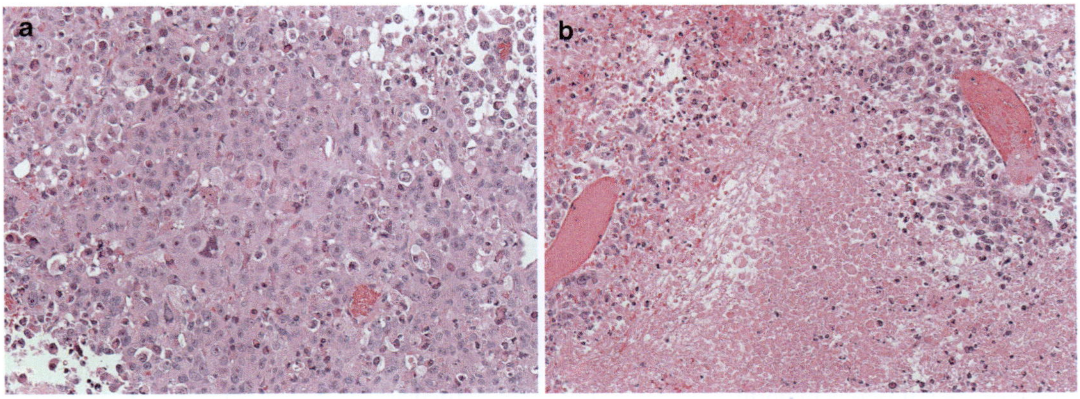

Fig. 9.4 Embryonal carcinoma. Embryonal carcinomas are characterized by the presence of large cells with significant cytologic atypia, pseudopapillae (**a**), as well as necrosis (**b**)

skin with appendages and brain, while mesodermal constituents include the cartilage, bone, fat, and muscle (Fig. 9.6). Cysts lined by respiratory or enteric epithelia are usually the endodermal elements, with some lesions also containing pancreatic or hepatic tissue [18]. *Immature teratomas* contain incompletely differentiated elements that resemble fetal tissue. Teratomas are classified as

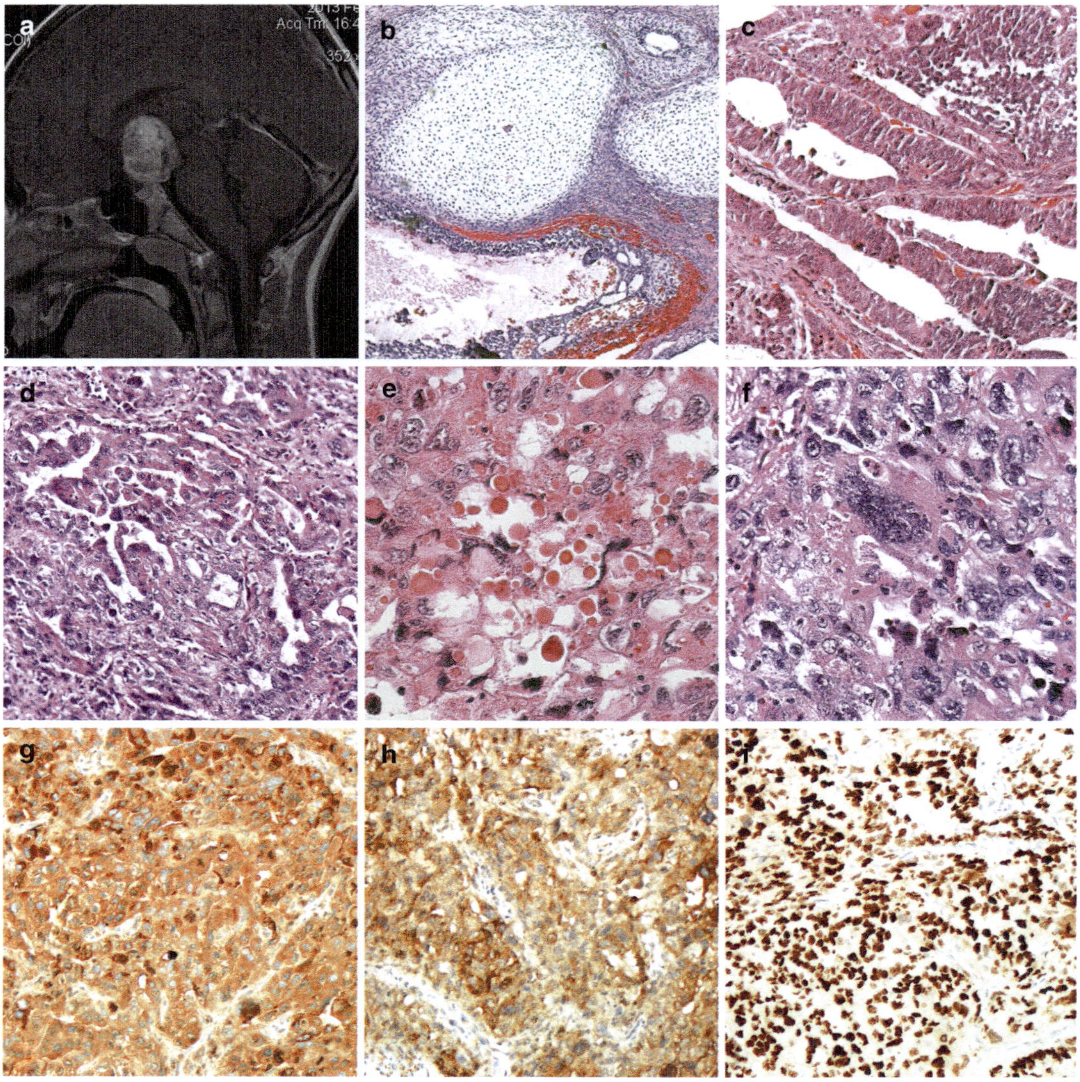

Fig. 9.5 Non-germinomatous germ cell tumor (NGGCT) with immature teratoma and yolk sac components**.** NGGCT can present as large suprasellar masses on imaging (**a**). An immature teratoma component was characterized by the presence of cartilage (**b**) and neuroepithelium (**c**). A dominant yolk sac component was characterized by tubules (**d**), hyaline droplets (**e**), and pleomorphic giant cells (**f**), a finding recognized in gonadal examples. Immunohistochemical stains demonstrated the yolk sac component to express AFP (**g**), glypican-3 (**h**), and SALL4 (**i**)

immature even if only a minority is composed of these less differentiated tissues, and mitotic activity is evident. Hypercellular and mitotically active stroma, reminiscent of embryonic mesenchyme, and primitive neuroectoderm mimicking neuroepithelial structures and developing neural tube are particularly common (Fig. 9.7). Clefts lined by melanotic neuroepithelium, representing abortive retinal differentiation, may also be encountered. The differential diagnosis of the latter involves the enigmatic pineal anlage tumor (Fig. 9.8). *Teratoma with malignant transformation*

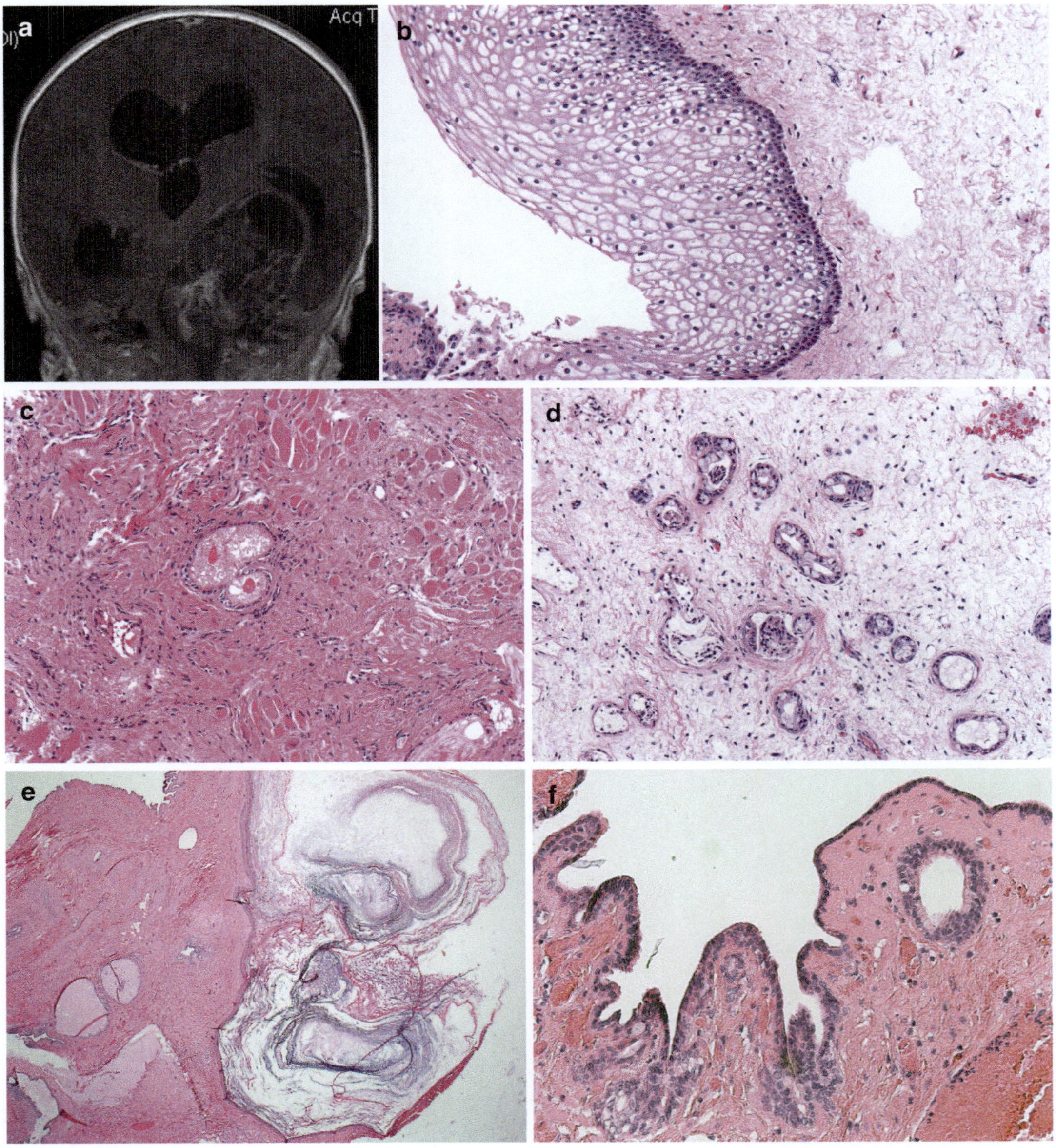

Fig. 9.6 Intracranial mature teratoma. This mature teratoma formed a large heterogeneous mass in the skull base (T1-weighted MRI post-contrast) (**a**). Well-differentiated squamous epithelium is a feature of most teratomas (**b**). Mesenchymal tissues, including skeletal muscle (**c**) as well as adnexal glands (**d**), are variably present in these tumors. Epidermoid cyst-like structure (e) and pigmented epithelium (f) in a mature teratoma of the pineal gland

refers to teratomas containing a conventional somatic cancer. Rhabdomyosarcoma, undifferentiated sarcoma, squamous cell carcinoma, or adenocarcinoma are the most frequent (see also chapter 12).

9.5.6 Choriocarcinoma

Choriocarcinomas are characterized by trophoblastic differentiation. Identification of both cytotrophoblastic elements and

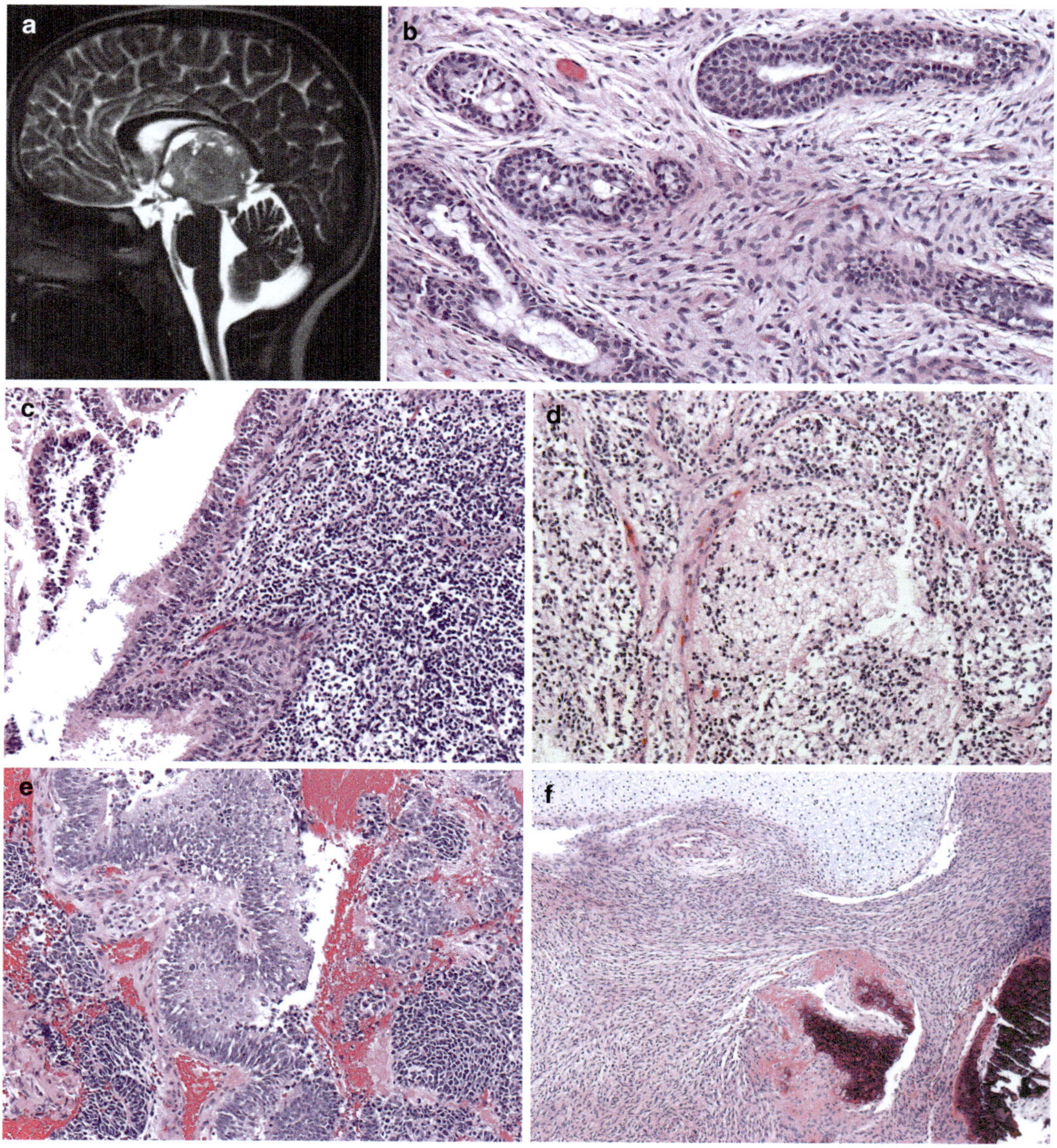

Fig. 9.7 Immature teratoma of the pineal gland. This immature teratoma presented as a large pineal region mass (T2-weighted MRI) (**a**). This tumor contained glandular epithelium in a cellular stroma (**b**), neuroepithelium (**c**) and nodules of neuronal differentiation (**d**), neural tube-like structures (**e**), as well as mesenchymal elements such as cartilage and bone (**f**) which are frequent in these tumors

syncytiotrophoblastic giant cells is required for diagnosis (Fig. 9.9). The latter typically have abundant, basophilic cytoplasm and contain multiple, hyperchromatic nuclei. The neoplastic syncytiotrophoblast surrounds or partially drapes cohesive masses of large mononucleated cells with vesicular nuclei and clear or acidophilic cytoplasm, which represent the cytotrophoblastic component. Ectatic stromal vascular channels, blood lakes, and extensive hemorrhagic necrosis are typical. Immunoreactivity for β–HGC and HPL in the cytoplasm of syncytiotrophoblastic giant cells is characteristic.

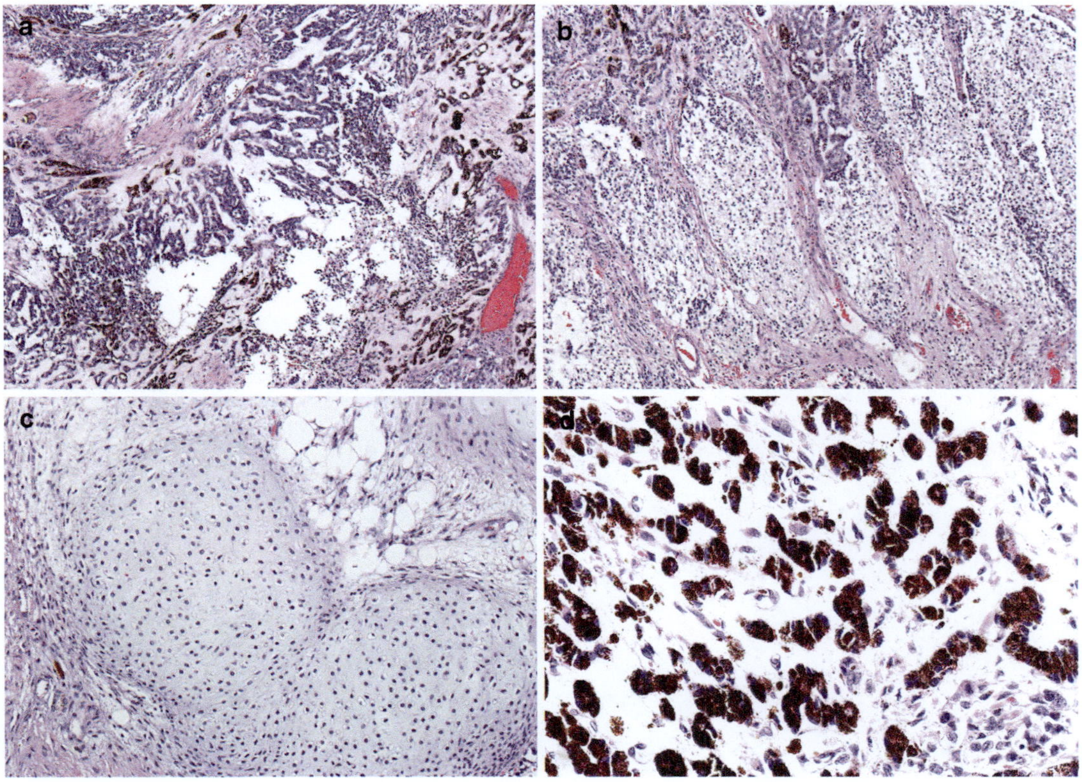

Fig. 9.8 Pineal anlage tumor**.** Anastomosing epithelial cords are frequent in the pineal anlage tumor. A clue to the diagnosis is the variable pigmentation (**a**). Neuronal differentiation usually takes the form of nodules reminiscent of those found in medulloblastoma (**b**). Cartilage as a mesenchymal component is another frequent feature (**c**). Pigmented epithelium is the hallmark of the pineal anlage tumor (**d**) and usually separates it from immature teratoma, the main entity in the differential diagnosis

9.6 Staging

GCT, especially germinomas, tend to disseminate throughout the neuroaxis [10]; thus, complete CNS staging is necessary in these neoplasms. All patients with a suspected CNS GCT require an extensive metastatic evaluation including brain and spine MRI with gadolinium, measurement of AFP and HCG levels in both serum and CSF, CSF cytology, and evaluation of pituitary and hypothalamic function. Visual field examinations for suprasellar or hypothalamic tumors and baseline neuropsychologic examinations are also recommended [18].

Although there is no specific staging system which has been uniformly accepted for primary CNS GCT, the TM system for medulloblastomas has been applied [5]. M0 are those tumors without evidence of metastatic disease (MRI of brain and spine + CSF cytological examination). M1 are those with free-floating tumor cells. M2 and M3 patients are those with lump disease in the spine or subarachnoid space intracranially [10].

9.7 Molecular Pathology

Activating mutations in KIT and a few chromosomal abnormalities have been reported in pure germinomas. However, the biology and molecular mechanisms of intracranial GCT, as well as the impact on patients' survival are still unestablished [3, 16, 35, 36]. The KIT protein is activated by dimerization upon binding of its known ligand, stem cell factor (SCF), which plays a role

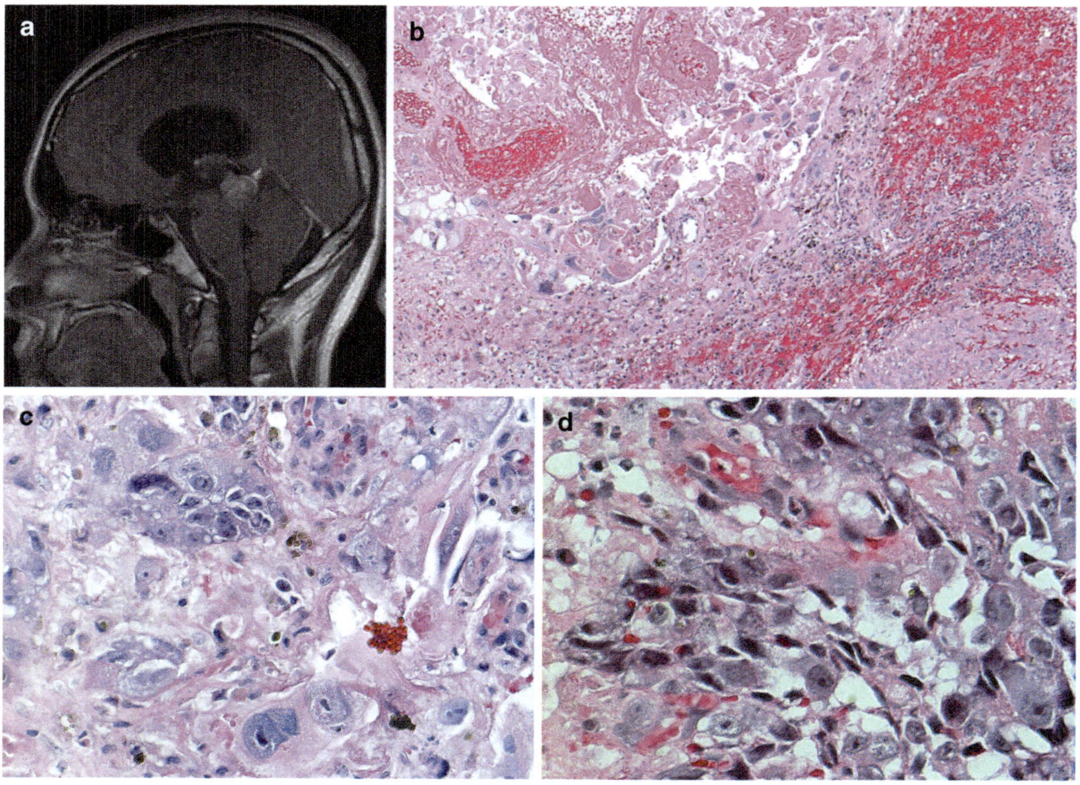

Fig. 9.9 Choriocarcinoma of the pineal gland. Intracranial choriocarcinomas are rare but usually involve the typical sites such as the pineal gland (**a**). Extensive hemorrhage may be a feature (**b**). An admixture of cytotrophoblast and syncytiotrophoblast is diagnostic (**c**, **d**)

in cell proliferation and differentiation through several signaling pathways. *KIT* mutations found in tumors are gain of function, resulting in constitutive KIT activation [37]. Fukushima et al. reported a high frequency of mutually exclusive *KIT* and *RAS* mutations in pure CNS germinomas as they are in the testes [16] and are also associated with increased *KIT* expression and chromosomal instability. The most common cytogenetic aberrations in this study were gains of 21q and X, followed by gains of 1q and 12p. Chromosomal loses were less frequent. Unlike testicular GCT, isochromosome 12p does not appear to be a frequent event in CNS GCT [16, 38–41].

More recently, Wang et al. studied 62 intracranial GCT by next-generation sequencing, as well as SNP and expression arrays. The main finding was that 53 % of CNS GCT contained somatic mutations in at least one of the genes involved in *KIT/RAS* or *AKT/mTOR* pathways [36]. The *KIT/RAS* signaling pathway was mutated in more than 50 % intracranial GCT, including somatic mutations in *KIT*, *KRAS*, and *NRAS* (downstream KIT mediators) and *CBL* (known negative regulator). When looking at histologic subtypes, *KIT* was mutated in 16 CNS GCT, but not in NGGCT. Additionally, *KIT* was overexpressed in the majority of pure germinomas but rarely in NGGCT. As in the study by Fukushima et al., *KRAS/NRAS* and *KIT* mutations were mutually exclusive genetic events in these tumors. *CBL*, encoding a RING finger ubiquitin E3 ligase, was the third most frequently mutated gene in CNS GCT. Somatic alterations in the *AKT/mTOR* pathway included copy number gains of the *AKT-1* locus at 14q32.33, with associated overexpression. Finally, this study identified somatic mutations in *BCORL1, MTOR, TP53, SPTA1, KDM2A*, and *LAMA4,* as well as germline mutations in *CDK5RAP2, JMJD1C,*

USP35, and *PCDH15*. Of interest, *JMJD1C* functions as a chromatin modifier gene and appears to have a role in germinal tissue development [36].

In another study using SNP arrays, Therashima et al. identified 8q13.2 gains containing the *PRDM14* gene. PRDM14 is a transcriptional regulator of primordial germ cells (PGCs) and has been found to be overexpressed in other cancers, suggesting that it may be important in the biology of intracranial GCT [3].

9.7.1 Prognosis

The prognosis of intracranial GCT depends strongly on the histologic subtype. In general, germinomas have an excellent prognosis, with most series suggesting 5-year progression-free survival (PFS) rates and even cure in over 90 % of patients. Germinomas with syncytiotrophoblastic giant cells may carry a less favorable prognosis than pure germinomas, although this has not been consistent in all series [25]. In contrast, NGGCT, including mixed GCT, have a poorer prognosis, with reported survival rates ranging between 40 and 70% [10]. Among NGGCT, mature teratomas have the best prognosis. Elevated markers in serum and/or CSF also have an effect in prognosis at the time of diagnosis. In particular, the presence of residual disease at the end of treatment and high AFP levels (>1000 ng/mL) at diagnosis are significant predictors of recurrence [2, 5].

9.8 Treatment

All CNS GCT but mature teratomas are considered malignant neoplasms and are therefore not surgically curable tumors. Historically, morbidity associated with surgical resection was common, and shunt placement was usually performed followed by radiation therapy [1]. At the present time, neurosurgery is required for tissue sampling, tumor cytoreduction, and/or treatment of symptoms and signs, particularly hydrocephalus. Even when anatomic location of most CNS GCT represent surgical challenges, surgery at the moment is relatively safe with approaches involving microsurgical dissection and endoscopy. The role of surgical resections, both total and subtotal, is unclear in pure germinomas but is thought to be of benefit for disease control in NGGCT and may be curative for mature teratomas [3, 5].

Although standard management of CNS GCT remains unsettled, approaches typically involve a combination of surgery, chemotherapy, and radiotherapy to prevent tumor dissemination [42]. The specific components of these therapies vary according to the clinical diagnostic category, particularly between pure germinoma and NGGCT [5, 30].

References

1. Fujimaki T. Central nervous system germ cell tumors: classification, clinical features, and treatment with a historical overview. J Child Neurol. 2009;24(11):1439–45. doi:10.1177/0883073809342127.
2. Murray MJ, Horan G, Lowis S, Nicholson JC. Highlights from the Third International Central Nervous System Germ Cell Tumour symposium: laying the foundations for future consensus. Ecancermedicalscience. 2013;7:333. doi:10.3332/ecancer.2013.333.
3. Terashima K, Yu A, Chow WY, Hsu WC, Chen P, Wong S, Hung YS, Suzuki T, Nishikawa R, Matsutani M, Nakamura H, Ng HK, Allen JC, Aldape KD, Su JM, Adesina AM, Leung HC, Man TK, Lau CC. Genome-wide analysis of DNA copy number alterations and loss of heterozygosity in intracranial germ cell tumors. Pediatr Blood Cancer. 2014;61(4):593–600. doi:10.1002/pbc.24833.
4. McCarthy BJ, Shibui S, Kayama T, Miyaoka E, Narita Y, Murakami M, Matsuda A, Matsuda T, Sobue T, Palis BE, Dolecek TA, Kruchko C, Engelhard HH, Villano JL. Primary CNS germ cell tumors in Japan and the United States: an analysis of 4 tumor registries. Neuro-Oncology. 2012;14(9):1194–200. doi:10.1093/neuonc/nos155.
5. Thakkar JP, Chew L, Villano JL. Primary CNS germ cell tumors: current epidemiology and update on treatment. Med Oncol. 2013;30(2):496. doi:10.1007/s12032-013-0496-9.
6. Hoffman HJ, Otsubo H, Hendrick EB, Humphreys RP, Drake JM, Becker LE, Greenberg M, Jenkin D. Intracranial germ-cell tumors in children. J Neurosurg. 1991;74(4):545–51. doi:10.3171/jns.1991.74.4.0545.
7. Central brain tumor registry of the United States statistical report supplement. Neuro-Oncology. 2014;16.
8. Goodwin TL, Sainani K, Fisher PG. Incidence patterns of central nervous system germ cell tumors: a SEER Study. J Pediatr Hematol Oncol. 2009;31(8):541–4. doi:10.1097/MPH.0b013e3181983af5.

9. Jennings MT, Gelman R, Hochberg F. Intracranial germ-cell tumors: natural history and pathogenesis. J Neurosurg. 1985;63(2):155–67. doi:10.3171/jns.1985.63.2.0155.
10. Packer RJ, Cohen BH, Cooney K. Intracranial germ cell tumors. Oncologist. 2000;5(4):312–20.
11. Arens R, Marcus D, Engelberg S, Findler G, Goodman RM, Passwell JH. Cerebral germinomas and Klinefelter syndrome. A review. Cancer. 1988;61(6):1228–31.
12. Maeda Y, Yoshikawa K, Kajiwara K, Ideguchi M, Amano T, Saka M, Nomura S, Fujii M, Suzuki M. Intracranial yolk sac tumor in a patient with Down syndrome. J Neurosurg Pediatr. 2011;7(6):604–8. doi:10.3171/2011.3.PEDS10500.
13. Sugimoto K, Ideguchi M, Sadahiro H, Yoshikawa K, Goto H, Nomura S, Fujii M, Suzuki M. Yolk sac tumor of the bilateral basal ganglia in a patient with Down syndrome. Brain Tumor Pathol. 2013;30(4):247–52. doi:10.1007/s10014-012-0134-9.
14. Queipo G, Aguirre D, Nieto K, Pena YR, Palma I, Olvera J, Chavez L, Najera N, Kofman-Alfaro S. Intracranial germ cell tumors: association with Klinefelter syndrome and sex chromosome aneuploidies. Cytogenet Genome Res. 2008;121(3–4):211–4. doi:10.1159/000138887.
15. Louis D, Ohgaki H, Wiestler OD, Cavenee WK. WHO classification of tumours of the central nervous system. 4th ed. Lyon: IARC Press; 2007.
16. Fukushima S, Otsuka A, Suzuki T, Yanagisawa T, Mishima K, Mukasa A, Saito N, Kumabe T, Kanamori M, Tominaga T, Narita Y, Shibui S, Kato M, Shibata T, Matsutani M, Nishikawa R, Ichimura K, Intracranial Germ Cell Tumor Genome Analysis Consortium (iGCT Consortium). Mutually exclusive mutations of KIT and RAS are associated with KIT mRNA expression and chromosomal instability in primary intracranial pure germinomas. Acta Neuropathol. 2014;127:(6):911–25. doi:10.1007/s00401-014-1247-5.
17. Sethi RV, Marino R, Niemierko A, Tarbell NJ, Yock TI, MacDonald SM. Delayed diagnosis in children with intracranial germ cell tumors. J Pediatr. 2013;163(5):1448–53. doi:10.1016/j.jpeds.2013.06.024.
18. Echevarria ME, Fangusaro J, Goldman S. Pediatric central nervous system germ cell tumors: a review. Oncologist. 2008;13(6):690–9. doi:10.1634/theoncologist.2008-0037.
19. Lee D, Suh YL. Histologically confirmed intracranial germ cell tumors; an analysis of 62 patients in a single institute. Virchows Archiv Int J Pathol. 2010;457(3):347–57. doi:10.1007/s00428-010-0951-3.
20. Matsutani M, Sano K, Takakura K, Fujimaki T, Nakamura O, Funata N, Seto T. Primary intracranial germ cell tumors: a clinical analysis of 153 histologically verified cases. J Neurosurg. 1997;86(3):446–55. doi:10.3171/jns.1997.86.3.0446.
21. Ozelame RV, Shroff M, Wood B, Bouffet E, Bartels U, Drake JM, Hawkins C, Blaser S. Basal ganglia germinoma in children with associated ipsilateral cerebral and brain stem hemiatrophy. Pediatr Radiol. 2006;36(4):325–30. doi:10.1007/s00247-005-0063-4.
22. Phi JH, Cho BK, Kim SK, Paeng JC, Kim IO, Kim IH, Kim DG, Jung HW, Kim JE, Wang KC. Germinomas in the basal ganglia: magnetic resonance imaging classification and the prognosis. J Neuro-Oncol. 2010;99(2):227–36. doi:10.1007/s11060-010-0119-7.
23. Sonoda Y, Kumabe T, Sugiyama S, Kanamori M, Yamashita Y, Saito R, Ariga H, Takai Y, Tominaga T. Germ cell tumors in the basal ganglia: problems of early diagnosis and treatment. J Neurosurg Pediatr. 2008;2(2):118–24. doi:10.3171/PED/2008/2/8/118.
24. Phi JH, Kim SK, Lee YA, Shin CH, Cheon JE, Kim IO, Yang SW, Wang KC. Latency of intracranial germ cell tumors and diagnosis delay. Childs Nerv Syst ChNS Off J Int Soc Pediatr Neurosurg. 2013;29(10):1871–81. doi:10.1007/s00381-013-2164-y.
25. Weksberg DC, Shibamoto Y, Paulino AC. Bifocal intracranial germinoma: a retrospective analysis of treatment outcomes in 20 patients and review of the literature. Int J Radiat Oncol Biol Phys. 2012;82(4):1341–51. doi:10.1016/j.ijrobp.2011.04.033.
26. Cuccia V, Alderete D. Suprasellar/pineal bifocal germ cell tumors. Childs Nerv Syst ChNS Off J Int Soc Pediatr Neurosurg. 2010;26(8):1043–9. doi:10.1007/s00381-010-1120-3.
27. Phi JH, Kim SK, Lee J, Park CK, Kim IH, Ahn HS, Shin HY, Kim IO, Jung HW, Kim DG, Paek SH, Wang KC. The enigma of bifocal germ cell tumors in the suprasellar and pineal regions: synchronous lesions or metastasis? J Neurosurg Pediatr. 2013;11(2):107–14. doi:10.3171/2012.10.PEDS11487.
28. Tsugu H, Oshiro S, Ueno Y, Abe H, Komatsu F, Sakamoto S, Matsumoto S, Nabeshima K, Fukushima T, Inoue T. Primary yolk sac tumor within the lateral ventricle. Neurol Med Chir. 2009;49(11):528–31.
29. DiLuna ML, Two AM, Levy GH, Patel T, Huttner AJ, Duncan CC, Piepmeier JM. Primary, non-exophytic, optic nerve germ cell tumors. J Neuro-Oncol. 2009;95(3):437–43. doi:10.1007/s11060-009-9941-1.
30. Legault G, Allen JC. Potential role of ventricular tumor markers in CNS germ cell tumors. Pediatr Blood Cancer. 2013;60(10):1647–50. doi:10.1002/pbc.24620.
31. Mei K, Liu A, Allan RW, Wang P, Lane Z, Abel TW, Wei L, Cheng H, Guo S, Peng Y, Rakheja D, Wang M, Ma J, Rodriguez MM, Li J, Cao D. Diagnostic utility of SALL4 in primary germ cell tumors of the central nervous system: a study of 77 cases. Mod Pathol Off J United States Can Acad Pathol Inc. 2009;22(12):1628–36. doi:10.1038/modpathol.2009.148.
32. Hattab EM, Tu PH, Wilson JD, Cheng L. OCT4 immunohistochemistry is superior to placental alkaline phosphatase (PLAP) in the diagnosis of central nervous system germinoma. Am J Surg Pathol. 2005;29(3):368–71.
33. Wanggou S, Jiang X, Li Q, Zhang L, Liu D, Li G, Feng X, Liu W, Zhu B, Huang W, Shi J, Yuan X, Ren C. HESRG: a novel biomarker for intracranial germinoma and embryonal carcinoma. J Neuro-Oncol. 2012;106(2):251–9. doi:10.1007/s11060-011-0673-7.
34. Villano JL, Virk IY, Ramirez V, Propp JM, Engelhard HH, McCarthy BJ. Descriptive epidemiology of

central nervous system germ cell tumors: nonpineal analysis. Neuro-Oncology. 2010;12(3):257–64. doi:10.1093/neuonc/nop029.
35. Sakuma Y, Sakurai S, Oguni S, Hironaka M, Saito K. Alterations of the c-kit gene in testicular germ cell tumors. Cancer Sci. 2003;94(6):486–91.
36. Wang L, Yamaguchi S, Burstein MD, Terashima K, Chang K, Ng HK, Nakamura H, He Z, Doddapaneni H, Lewis L, Wang M, Suzuki T, Nishikawa R, Natsume A, Terasaka S, Dauser R, Whitehead W, Adekunle A, Sun J, Qiao Y, Marth G, Muzny DM, Gibbs RA, Leal SM, Wheeler DA, Lau CC. Novel somatic and germline mutations in intracranial germ cell tumours. Nature. 2014;511(7508):241–5. doi:10.1038/nature13296.
37. Kamakura Y, Hasegawa M, Minamoto T, Yamashita J, Fujisawa H. C-kit gene mutation: common and widely distributed in intracranial germinomas. J Neurosurg. 2006;104(3 Suppl):173–80. doi:10.3171/ped.2006.104.3.173.
38. Schneider DT, Zahn S, Sievers S, Alemazkour K, Reifenberger G, Wiestler OD, Calaminus G, Gobel U, Perlman EJ. Molecular genetic analysis of central nervous system germ cell tumors with comparative genomic hybridization. Mod Pathol Off J United States Can Acad Pathol Inc. 2006;19(6):864–73. doi:10.1038/modpathol.3800607.
39. Okada Y, Nishikawa R, Matsutani M, Louis DN. Hypomethylated X chromosome gain and rare isochromosome 12p in diverse intracranial germ cell tumors. J Neuropathol Exp Neurol. 2002;61(6):531–8.
40. Lemos JA, Barbieri-Neto J, Casartelli C. Primary intracranial germ cell tumors without an isochromosome 12p. Cancer Genet Cytogenet. 1998;100(2):124–8.
41. Sukov WR, Cheville JC, Giannini C, Carlson AW, Shearer BM, Sinnwell JP, Ketterling RP. Isochromosome 12p and polysomy 12 in primary central nervous system germ cell tumors: frequency and association with clinicopathologic features. Hum Pathol. 2010;41(2):232–8. doi:10.1016/j.humpath.2009.07.017.
42. Kim JW, Kim WC, Cho JH, Kim DS, Shim KW, Lyu CJ, Won SC, Suh CO. A multimodal approach including craniospinal irradiation improves the treatment outcome of high-risk intracranial nongerminomatous germ cell tumors. Int J Radiat Oncol Biol Phys. 2012;84(3):625–31. doi:10.1016/j.ijrobp.2011.12.077.
43. Louis DN, Ohgaki H, Wiestler OD, Caveneee WK, Ellison DW, Figarella-Branger D, Perry A, Reifenberger G, von Deimling A. WHO Classification of Tumours of the Central Nervous System. Revised 4th edition. Lyon, IARC, 2016.
44. Fujjimaki T. Central nervous system germ cell tumors: classification, clinical features, and treatment with a historical overview. J Child Neurol. 2009;24:1439.

10 Pediatric Germ Cell Tumors

Eduardo Zambrano, Danielle Vargas De Stefano, and Miguel Reyes-Múgica

Germ cell tumors (GCT) are relatively rare in the pediatric age group, representing only 1–3 % of childhood tumors [1]. Pediatric GCT comprise a remarkably diverse group, with significant variability in age and site of presentation, clinical behavior, and histology [2–4]. Although they share a common origin from progenitor germ cells, markedly different types of GCT may develop due to variations from normal differentiation (i.e., gonadal GCT) and/or aberrant migration (i.e., extragonadal GCT), most commonly occurring in midline locations (mediastinal, retroperitoneal, sacrococcygeal, genital, or cranial) [1, 5–7]. Most correspond to types 0 and I of Oosterhuis and Looijenga's classification (see Chap. 3).

10.1 Epidemiology

GCT may occur at any age; however, a bimodal age distribution is more commonly observed, with a first peak between birth and 4 years of age (pediatric GCT proper), and a second one beginning with the onset of puberty and continuing through the third and fourth decades [2, 3, 6, 8]. In children, extragonadal sites predominate, accounting for 50 % of cases, compared with adults in whom only 10 % are extragonadal [1].

The majority of pediatric GCT (type I GCT) are benign, with mature teratomas being the most common [9–11]. Approximately 20 % of pediatric GCT are malignant, representing approximately 3 % of all pediatric cancers [3, 12], although the rate of malignancy varies by age of presentation and anatomical site [1, 12, 13]. The majority of malignant GCT in children are yolk sac tumors (YST) [4, 6].

GCT are the most common neoplasm in the newborn accounting for 35–40 % of all tumors in the first month of life [4, 6]. Most GCT in the fetal and neonatal periods are teratomas. Only approximately 5 % of neonatal GCT contain a malignant component, usually YST [6]. Overall, sacrococcygeal tumors are the most common perinatal GCT,

E. Zambrano, MD, MSc
Lucile Packard Children's Hospital Stanford, Stanford University Medical Center, 300 Pasteur Drive, Rm H1501K, Stanford, CA 94305, USA
e-mail: edzambra@stanford.edu

D.V. De Stefano, MD
Department of Pathology, Stanford University Medical Center, 300 Pasteur Drive, Stanford, CA 94305, USA
e-mail: dstefano@stanford.edu

M. Reyes-Múgica, MD (✉)
Chief of Pathology and Head of Laboratories, Children's Hospital of Pittsburgh of UPMC. University of Pittsburgh School of Medicine. Department of Pathology B256, One Children's Hospital Drive, Pittsburgh, PA 15224, USA
e-mail: miguel.reyes@chp.edu

F.F. Nogales, R.E. Jimenez (eds.), *Pathology and Biology of Human Germ Cell Tumors*,
DOI 10.1007/978-3-662-53775-6_10

Table 10.1 Most frequent GCT in the pediatric population by age group, anatomical site, and histologic type

Age group	Anatomical site	Usual histologic type	Comments
Newborn/infancy	Sacrococcygeal (40 %) Ovary mediastinal > abdominal (15–20 % vs. 5 %) Cervicofacial (rare <5 %)	Teratoma	GCT are the most common neoplasm in newborns (35–40 % of tumors during first of month of life); only 5 % are malignant (most commonly YST)
Childhood	Testes	Mostly YST with low (5 %) metastatic rate	Usually benign/indolent behaviors when presenting in gonadal sites Gonadal seminoma and dysgerminoma are rare and frequently associated with gonadal dysgenesis
	Ovary	Teratoma (40 %) Mature cystic teratoma	
	Mediastinal	Approximately 15 % are malignant. Most common malignant histology is YST in girls and younger boys and mixed histology in older boys	
Puberty/adolescence	Testes	Postpubertal teratomas have higher metastatic potential than prepubertal	GCT are the most common solid tumor in adolescent males
		Higher incidence of embryonal carcinoma and mixed non-seminomatous tumors	
	Ovary	Mature cystic teratoma	

GCT germ cell tumors, *YST* yolk sac tumor

accounting for 40 % of the total [4]. GCT rank as fourth or fifth most frequent malignant neoplasm in patients below 14 years of age, after neuroblastoma, rhabdomyosarcoma, Wilms tumor, and retinoblastoma [6, 14]. Table 10.1 presents a summary of the most common types of GCT affecting the pediatric population by age group and anatomical sites of involvement

10.1.1 Risk Factors for Developing GCT

GCT lack familial distribution and are thought to arise from sporadic genetic mutations [3]. Several common chromosomal mutations have been identified that may represent random occurrences although some common environmental risk factors have been reported. Maternal exposure to various chemicals and solvents may be associated with an elevated risk of GCT in the offspring; however, this has not been proven conclusively [3, 15].

Cryptorchidism is associated with 3- to 9-fold increased risk of GCT (most commonly seminoma) compared to the general male population [3, 16–18]. Early orchiopexy is associated with a lower incidence of developing a testicular germ cell tumor [3, 19].

There is an increased risk for development of gonadal GCT in certain disorders of sex development (DSD), with an incidence reported as high as 30 % in patients with gonadal dysgenesis, and 5–10 % for undervirilization syndromes [3, 20–24]. These risks are thought to increase significantly with age. Therefore, prophylactic gonadectomy during childhood is recommended [3, 25, 26].

10.1.1.1 Neoplastic Risk in Disorders of Sex Development

The risk in each group of DSD is difficult to evaluate, because the reported prevalence per diagnostic group varies considerably and also because statistical data from literature reviews are based on gonadectomies performed during the previous decades, mainly prophylactically in early childhood; therefore, the real incidence of GCT may be higher. In addition, an accurate risk for malignant transformation in DSD patients is hard to predict because of two major problems: first, the confusing terminology and classification systems referred to the different forms of DSD, in which several synonyms and eponyms are used in literature, and definitions for the terminology are often lacking from bibliographical sources; second, there are no well-established criteria for the identification of malignant germ cells, especially in young children. This is specifically due to the phenomenon of delay of germ cell maturation, which might result in overdiagnosis of germ cell neoplasia *in situ* (GCNIS) [20].

10.1.1.2 Gonadal Tumors in Patients with DSD

Seminoma (if the gonad is considered a testis)/dysgerminoma (if the gonad is considered an ovary) and the non-seminomatous germ cell tumors are by far the most frequently occurring tumors in patients with DSD [24, 25, 27–33]. However, other gonadal (benign and malignant) neoplasms have sporadically been reported in patients with DSD, often in combination with the above-mentioned GCT. The development of invasive GCT is always preceded by the presence of an *in situ* neoplastic lesion. In gonads of DSD patients, the precursor of cancer may be GCNIS in testicular tissue or gonadoblastoma (GB), in those without obvious testicular differentiation or in patients with testicular dysgenesis [26, 34, 35]. Because gonadectomies in patients with DSD have often been performed prophylactically, most of the encountered changes in germ cells are benign or *in situ* malignant conditions. Recently, detailed histologic investigation of gonads of DSD patients led to the identification of the putative precursor of GB, which was referred to as undifferentiated gonadal tissue [25].

10.1.1.3 The Prevalence of GCT in Patients with DSD

The prevalence of GCT is increased in patients with DSD containing Y chromosome material in their genome and especially the chromosome Y "GBY" region, which is related to the presence of the TSPY gene, the most likely candidate [36, 37]. The presence of SRY or other sex determining genes is irrelevant in this context. Specifically DSD patients with gonadal dysgenesis (DSD with maldeveloped gonads, including streak gonads and testicular dysgenesis) or undervirilization (DSD with abnormal androgen function) are at risk. Traditionally, the prevalence of GCT in patients with gonadal dysgenesis is estimated at around 30 % [25] and in patients with undervirilization syndromes at 5–10 % [22, 23]. However, reported prevalence numbers per diagnostic group may vary considerably.

A clear insight into the prevalence of GCT in patients with gonadal dysgenesis (either complete or partial) is hampered by confusing nomenclature, which is most pronounced for this patient category, as well as an overestimated incidence of GCNIS in some series, reported as high as 91 % [38] to 100 % [39] that probably corresponds to a state of arrested or delayed maturation of the germ cells. GCT in patients with gonadal dysgenesis are frequently found at a very young age, during the first year of life [25, 29, 33, 40] or may even at birth [41]. Nearly all the *in situ* neoplastic lesions in patients with gonadal dysgenesis are GB, leading to seminoma/dysgerminoma in 92 % of the cases. A high risk of GB exists when sex determination is disrupted at an early stage of Sertoli cell differentiation (due to abnormalities in SRY, SOX9, or WT1). It must be remembered that development of GB requires the GBY region of the Y chromosome, which is in and of itself sufficient to lead to this neoplasm. Early Sertoli cell development is also disturbed in patients with mixed gonadal dysgenesis, who also carry a high risk of developing GB as precursor lesion. The same is true for patients with 9p deletions, likely related to the loss of DMRT1 [42]. Careful histological analysis of gonadal tissue of DSD patients revealed that undifferentiated gonadal tissue is the most likely precursor stage of GB [26]. The GCNIS lesion accounts for only 8 % of precursor lesions in patients with gonadal dysgenesis and is probably only encountered in the presence of testicular tissue [25].

In the group of undervirilized patients (DSD with abnormal androgen function), the overall prevalence of GCT approximates 2.3 % [25]. In the androgen insensitivity syndrome (AIS), the reported prevalence of GCT has varied from 5–10 % [22, 23] to 22 % [43], although in more recent series of prophylactic gonadectomies, the estimated prevalence is 5.5 % [28, 34, 44–51]. Tumor prevalence in AIS markedly increases after puberty and reaches 33 % by 50 years of age [28]. Although data are limited, the risk seems to be markedly higher in the partial form (PAIS) (15 %) [34, 44, 49] than in the complete variant (CAIS) (0.8 %) [28, 34, 44, 47–49]. This difference may be explained by the fact that there is a rapid and total loss of germ cells by apoptosis in CAIS patients, starting from the age of 1 year, whereas PAIS patients maintain their germ cell population at about two-thirds of the normal number at puberty [34]. The risk of cancer in the PAIS patients is influenced by the anatomical localization of the gonad, being the highest in abdominal sites and the lowest in scrotal localization [42, 52]. In contrast to patients with gonadal dysgenesis, nearly all the reported tumors in the group of patients with undervirilization syndromes are GCNIS lesions (81 %) or seminomas (19 %), probably because these defects have occurred in late gonadal development. For other causes of undervirilization, the development of GCT is exceptional: one tumor is reported in a series of six patients with 17β-hydroxysteroid dehydrogenase deficiency (17 %) [34]; no tumors were found in a series of three patients with 5α-reductase deficiency [47] and of two patients with Leydig cell hypoplasia [34].

The risk for GCT development in patients with DSD can be grouped into four categories: high, intermediate, low, and unknown. High-risk patients include those with gonadal dysgenesis (DSD with maldeveloped gonads) [20], intra-abdominal gonads and the GBY region in their genome, including Frasier [53] and Denys–Drash syndromes and also patients with PAIS and non-scrotal gonads. The percentages found in the literature vary from 15 to 60 %. At intermediate risk are patients with the Y+ (GBY+) Turner syndrome and those with 17β-hydroxysteroid dehydrogenase deficiency, gonadal dysgenesis (harboring the Y chromosome), or PAIS, the two latter categories with scrotal gonads. The low-risk group includes patients with CAIS as well as patients with ovotesticular DSD (in which the gonads mostly consist of well-differentiated ovarian and testicular tissue) [54] and those with Turner syndrome lacking an apparent Y chromosome. The unknown category includes 5α -reductase deficiency, Leydig cell hypoplasia and specific gene mutations for which there are insufficient or no data for proper analysis [52, 55].

An interesting question is why GB forms in some patients and GCNIS in others. Hersmus et al. [42] hypothesized that this is due to the specific microenvironment, especially the absence of functional Sertoli cells, leading to female development. In other words, GCNIS can only be formed at a certain level of testicular development. Thus, GB and GCNIS are simply two variants of the same defect, being in fact a continuum, of which the phenotypic presentation is determined by the microenvironment, the level of virilization [42].

10.2 Classification of Pediatric GCT by Histologic Type

The totipotential nature of germ cells allows a wide variety of tumors. Different histologic patterns are common in a single tumor, and 25 % of childhood tumors contain more than one histology, which is often a malignant histology with coexisting teratoma [12]. Since many of the entities described below are illustrated in other chapters such as those dealing with the ovary and the testis, we have included only images relevant to the pediatric age group or with specific clinical relevance.

10.2.1 Teratomas

The most common GCT in the pediatric population, teratomas are often composed of multiple embryologic layers, arising from multipotent cells, containing one or more embryonic germ layers (ectoderm, mesoderm, endoderm) of tissue foreign to the anatomic site of origin [56]. In its usual definition, tissue elements of all three blastoderm layers (endoderm, mesoderm, and ectoderm) are present; however, teratomas also occur in which only one or two of the germ layers are present (referred to as monodermal or bidermal teratomas, respectively) [6]. They are divided into both mature and immature forms. Mature teratomas are benign; they are commonly cystic and may contain several well-differentiated cell types (i.e., skin, hair, teeth, thyroid, gastric mucosa, brain tissue). Immature teratomas are less differentiated and more reminiscent of embryonal tissue, frequently including neuroectoderm typically in the form of neuroepithelial rosettes and tubules and often mixed with mature tissue [3, 6, 57].

A grading system has been developed for immature teratomas, based on the amount of neuroepithelial tissue present. Mature teratomas containing only fully differentiated tissue are considered grade 0 and immature teratomas range from grade 1 (composed of <10 % immature neuroepithelium) to grade 3 (>50 % immature neuroepithelium present). In adult ovarian teratomas, the finding of grade 2 or 3 immature teratoma predicts the likelihood of metastasis and confers a worse prognosis [57]. In children, however, it is not clear that the same relationship of grade to outcome of immature teratomas applies. A Pediatric Oncology Group study concluded that surgery alone was curative in children and adolescents with stage I immature teratomas of any grade and that chemotherapy should be reserved for cases of relapse [58–60]. However, an association between grade 3 immature teratoma and microscopic foci of YST (83 %) has been demonstrated, emphasizing the need for thorough histologic evaluation of tumors [59].

10.2.2 Dysgerminomas and Seminomas

These GCT result from abnormal premeiotic differentiation [3]. Seminoma or dysgerminoma has a reduced ability for further differentiation and is rare before puberty but occurs at gonadal sites in adolescence [1]. Testicular seminoma and ovarian dysgerminomas are analogous malignant tumors that occur more commonly in young adults. When they occur in younger children, it is typically in association with gonadal dysgenesis [3].

10.2.3 Choriocarcinoma and Yolk Sac Tumors

Choriocarcinomas and YST result from abnormal postmeiotic extraembryonic differentiation [3].

YST is the most frequent malignant histological type in pediatric GCT and is common at the sacrococcygeal, retroperitoneal, mediastinal, and prepubertal testicular locations [1, 4, 6]. In neonatal GCT, YST most commonly occurs within a teratoma; about 5 % of neonatal teratomas have a yolk sac component [4]. In older infants and young children, YST more commonly occurs as a pure YST, not as a component of a teratoma [6]. AFP elevation is seen in the presence of YST and serves as a marker of persistent or recurrent disease, especially in infants between 6 months and 2 years of age, when the normally elevated newborn levels ranging from 41,000 to 160,000 ng/ml decrease. AFP levels are reported up to 87 ng/ml within 95.5 interval assuming a logarithmic normal distribution, decreasing to a mean of 8 ng/ml at 2 years of age before finally reaching the normal adult serum level of 0–6 ng/ml (vide infra) [61].

Choriocarcinoma is rare in childhood. It is even rarer in the infant and is usually thought to originate from a focus of choriocarcinoma that has arisen in the placenta as a variant of gestational trophoblastic disease [1, 6]. Choriocarcinoma is composed of cytotrophoblast, syncytiotrophoblast, and extravillous trophoblast, often mixed in random fashion surrounding areas of hemorrhage and necrosis. Vascular invasion is a common feature of choriocarcinomas [6]. The disease is rapidly fatal if left untreated [62]. Given that placental choriocarcinoma can also metastasize to the mother (up to 60 % of cases), if an infantile choriocarcinoma is diagnosed, the mother should also be screened for the disease [6].

10.2.4 Embryonal Carcinoma

Embryonal carcinoma possesses the ability to differentiate into embryonic and extraembryonic tumors [1, 7]. It most frequently presents in postpubertal GCT [9].

10.2.5 Gonadoblastoma

GB are benign tumors that contain both germ cells and stromal cells. As mentioned above, neonates can also present with GB in patients with dysgenetic gonads.

10.3 Classification of Pediatric GCT by Anatomical Site

10.3.1 Extragonadal GCT

10.3.1.1 Sacrococcygeal Teratomas

Sacrococcygeal teratomas [63] (SCT) are the most common extragonadal GCT in the pediatric population. They are the most common GCT in newborns and infants, occurring in approximately 1:35,000 live births. Anatomically, they have been classified into four types according to the degree of their external vs. intrapelvic/intra-abdominal extension [3, 64]. Clinically, however, SCT usually fit one of two distinct scenarios: those presenting with large predominantly external and benign (>90 %) lesions in the neonatal period or those presenting between birth and 4 years that typically have more intrapelvic/intra-abdominal tumor involvement. The latter group of tumors is much more likely to be malignant (60–90 %). It has been theorized that the absence of visible external tumor leads to a delay in diagnosis and therefore a higher incidence of cancer due to malignant degeneration. Symptoms in these patients are often the result of pelvic enlargement with compression of the bladder or rectum [3].

Large SCT can cause symptoms in utero secondary to mass effect, fetal hydrops, maternal polyhydramnios, fetal dystocia, tumor rupture (Fig. 10.1) or shunting with high-output cardiac failure, and maternal polyhydramnios. Fetal resection or other fetal intervention (cyst drainage, laser ablation, alcohol sclerosis) may be necessary. Cesarean section delivery is recommended [1, 65, 66].

The association of presacral teratoma with anal stenosis/anorectal malformation and sacral defects is known as the Currarino triad, an autosomal dominant disorder [67], secondary to mutations in the HLXB9 homeobox gene [68, 69].

Factors that Predict Recurrence of SCT

Recurrence of disease post resection ranges from 2 to 35 %, averaging 12.5 % [70]. About 50 % of patients with immature or mature teratoma that recur have a malignant component at recurrence. Recurrence is more likely among patients with immature teratoma (33 % recurrence) than for patients with mature teratoma (10 % recurrence) [6]. In a retrospective review of all teratomas diagnosed during childhood, grade of immaturity correlated with risk of recurrence; higher grade of immature teratoma

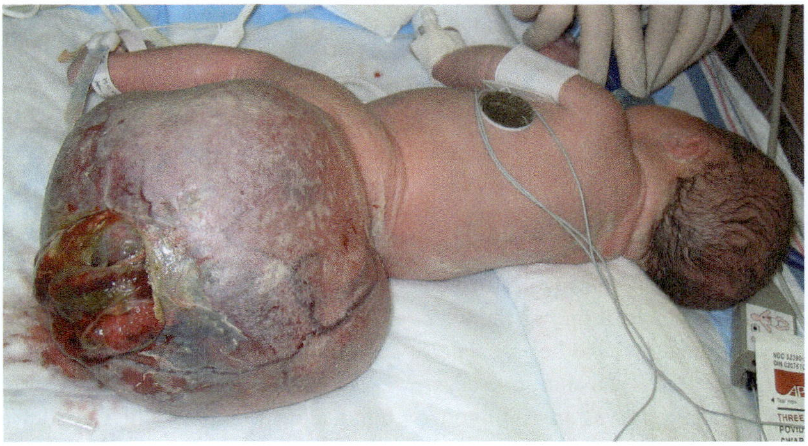

Fig. 10.1 Neonate with a giant sacrococcygeal teratoma that ruptured during delivery. Note the thin skin coverage with prominent vascular markings and the area of rupture showing solid and cystic components forming the tumor mass

was more likely to relapse than lower grade (grade 0, mature), 10 %; grade 1, 14 %; grade 2, 21 %; grade 3, 31 %) [6].

10.3.1.2 Thoracic and Mediastinal Germ Cell Tumors

Mediastinal pediatric GCT are extremely rare, representing 5 % of all GCT and 6–18 % of all pediatric mediastinal tumors [2, 3]. Thoracic GCT are more common than abdominal GCT in the newborn period (15–20 % vs. 5 %) [6, 71]. The majority of these tumors are located in the anterior mediastinum and originate in the thymus, though they can be found to rise from the posterior mediastinum, heart, or epicardial structures [6].

Older children typically present with chest pain, precocious puberty or facial fullness, and vascular congestion as a result of caval obstruction and superior vena cava syndrome. In younger children, respiratory distress is more common and often accompanied by fever [2, 3].

Approximately 15 % of pediatric mediastinal GCT are malignant and carry the worst prognosis of all germ cell tumors [6]. YST is the predominant histology in girls as well as younger boys, whereas older boys have mixed histology in over 50 % [1, 72]. An association with Klinefelter syndrome and certain hematologic disorders has been described [3, 73].

10.3.1.3 Abdominal and Retroperitoneal Germ Cell Tumors

Abdominal GCT (retroperitoneal, gastric, other abdominal viscera) account for only 5 % of all GCT [71]. They can be both intraperitoneal and retroperitoneal, and complete extirpation is the main treatment [6]. The vast majority present within the first 5 years of life, and especially the first year of life (one half of them occur during the first year of life and 73 % occur before 5 years of age) [6]. Most present during infancy with a mass and pain as the common symptoms; weight loss, constipation, and acute abdomen may also occur [1]. There is a 2:1 female predominance [3].

The majority of these tumors are benign: mature and immature teratomas predominate, with malignancy rates up to 15–24 % [70, 74, 75]. The most frequent malignant histology is YST (63 %), but choriocarcinoma and mixed tumors also occur [71].

Fetus In Fetu

The term *fetus in fetu* (FIF) is attributed to Meckel in 1800 and describes the inclusion of one fetus inside of another. Since then, around 100 cases have been reported. The embryologic origins of FIF are unclear, and it is not decided whether FIF is a monochorionic, monozygotic twin of the host or rather a well- differentiated teratoma (fetiform teratoma) [6]. Differentiating between FIF and teratoma can be challenging; Spencer suggested that a FIF must have one or more of the following conditions: (1) be enclosed within a distinct sac, (2) be partially or completely covered by normal skin, (3) have grossly recognizable anatomic parts, (4) be attached to the host by only a few relatively large blood vessels, and (5) either be located immediately adjacent to one of the sites of attachment of conjoined twins or be associated with the neural tube or the gastrointestinal system [6, 76]. Most cases present as an asymptomatic abdominal mass during infancy. Although 80 % of FIF occurs in the upper retroperitoneum, FIF has also been reported in the liver, sacrum, pelvis, scrotum, external genitalia, mediastinum, and oropharynx. We have seen examples attaining a highly advanced degree of fetal development (Figs. 10.2 and 10.3) and even spontaneous limb movement (Fig. 10.4) [77]. The treatment is surgical resection [6, 78].

10.3.1.4 Cervicofacial Teratomas

GCT of the cervical and facial regions represent 5 % of all GCT during childhood, but GCT in this site most commonly present during the prenatal or perinatal period [79]. Nearly all are mature or immature teratomas. Approximately one-third presents with airway obstruction [80]. Giant fetal tumors have been noted with hydrops fetalis, which may lead to fetal demise [81].

10.3.2 Gonadal GCT

10.3.2.1 Testes

Pediatric testicular GCT are rare and occur in a bimodal distribution, with a small peak in the first 3 years of life and a larger peak in adolescence (Figs. 10.5 and 10.6) [1]. They are more frequently diagnosed in adolescents and young adults and are the most common solid tumor in adolescent males [9, 82]. They have distinct histologic and biologic differences associated with each age group (pre- and postpubertal) [1, 56, 83, 84].

In the prepubertal male, testicular GCT frequently present as a mass or testicular enlargement with approximately 10 % associated with a hydrocele [3, 85]. Postpubertal testicular GCT typically behave more aggressively than prepubertal GCT [9].

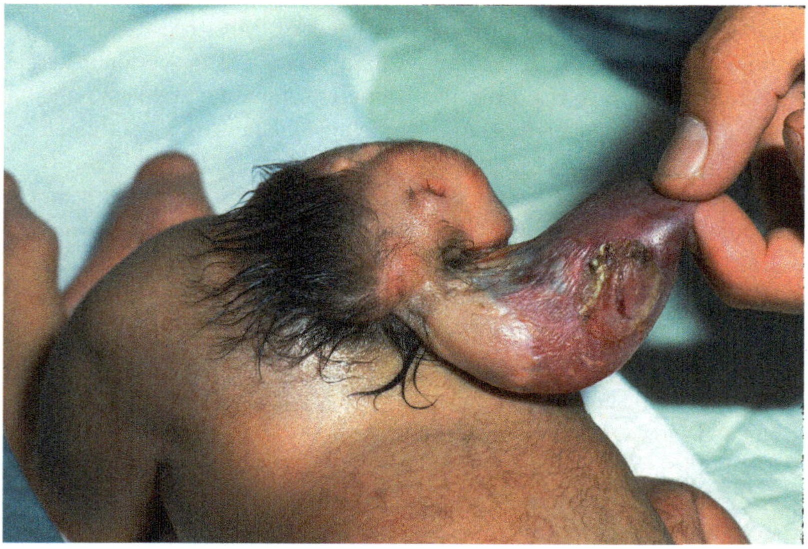

Fig. 10.2 Neonate with an example of *fetus in fetu* located in the lumbosacral area. The abnormal growth shows advanced fetiform development, featuring half of a face, with hair in a distribution similar to the scalp, forehead, the left eye, and partial development of the nose and ear. The right side of the face is continuous with a membranous sac containing neuroepithelial and other teratomatous elements

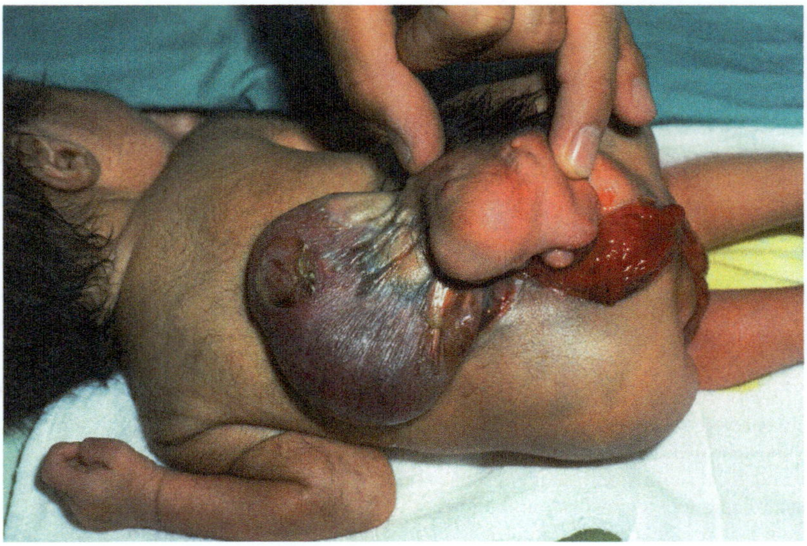

Fig. 10.3 Another view from the lesion shown in Fig. 10.2. The incomplete half of a face is flanked by a membranous sac on the left and a fleshy mass of teratomatous elements on the lower sacral area on the right

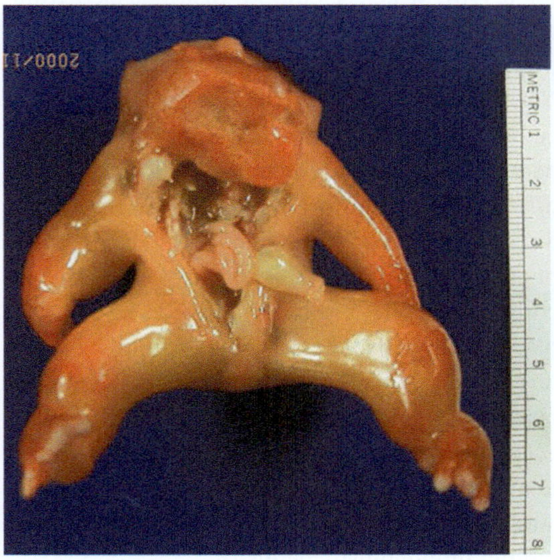

Fig. 10.4 *Fetus in fetu* with remarkably advanced fetiform anatomy showing a head with anencephalic features, a thorax and abdomen, four limbs, and digit formation. This case showed intrauterine movement of the limbs in ultrasound analysis [77]

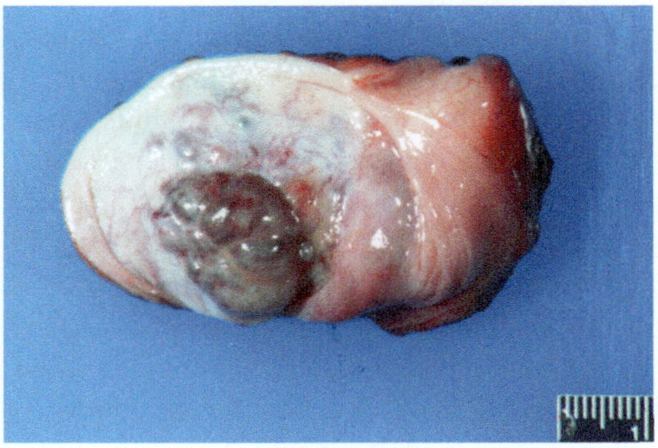

Fig. 10.5 Intratesticular mixed germ cell tumor in an adolescent. The tumor is located in the center of the testis showing a lobulated surface with hemorrhage and prominent vascularity

A review of recent pathology-based studies demonstrates that 70–75 % of prepubertal testicular tumors are benign [9–11]. Testicular GCT in prepubertal males are typically pure YST with a low incidence of metastasis (approximately 5 %) [9]. Pure teratoma is also common in children, accounting for about 40 % of such testicular tumors, but in the prepubertal patient uniformly exhibits a benign behavior [9, 86, 87]. In contrast, adolescents have a higher incidence of pure embryonal carcinoma and mixed GCT non-seminomatous, which behave more aggressively than their prepubertal counterparts, with 20–30 % presenting with metastatic disease at diagnosis [9, 11, 83]. Additionally, postpubertal teratomas are typically part of a mixed non-seminomatous GCT and have a higher potential for metastasis

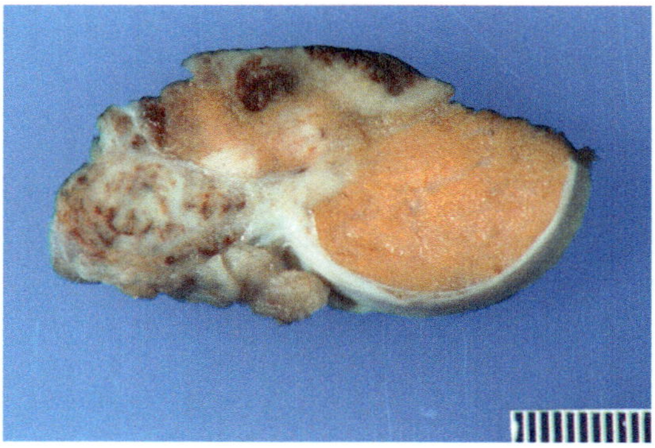

Fig. 10.6 Cross section of the tumor shown in Fig. 10.5. The mass is relatively well circumscribed, featuring several nodules of white neoplastic tissue with areas of prominent vascularity and focal infiltration into the testicular parenchyma. Hemorrhagic foci are evident on the upper portion of the lesion

and malignant degeneration than prepubertal teratomas [9, 10].

In a recent study comparing pediatric, adolescent, and adult testicular GCT, higher rates of pure seminomas were observed in adults, compared with children and adolescents; adolescents had higher proportion of mixed tumors, while younger pediatric patients were more likely to harbor pure YST or teratoma. In this study, adolescents had significantly more advanced AJCC Stage Group at presentation, and they also had lower event-free and overall survivals compared with pediatric or adult patients. As mentioned by the authors, the difference in histology may partially account for the higher stage at presentation observed in adolescents. However, adolescent patients were observed to suffer statistically significantly worse event-free survival than children or adults even when attempting to statistically account for stage and histology. A possible explanation for this difference may include the surge in hormonal stimulation occurring in adolescence, which leads to a deregulation of the complex mitosis–meiosis switch in pre-existing testicular carcinoma *in situ* [9].

In young patients other tumors, such as Sertoli cell tumors and paratesticular rhabdomyosarcomas, are more common than germ cell tumors. The exact incidence of malignancy in prepubertal testicular masses is not known, but in one series, 74 % were benign with 48 % teratomas and only 5 % YST [1, 84, 85].

10.3.2.2 Ovary

The ovary is the most common site for GCT after infancy. More than 80 % of all ovarian tumors are benign, with many of these having predominantly cystic components [1, 88]. Most ovarian GCT in children are teratomas, followed by dysgerminomas, YST, embryonal carcinomas, and mixed tumors [3, 89]. Presenting symptoms are pain, lower abdominal fullness, and less commonly an acute abdomen from torsion or tumor rupture [1].

10.3.2.3 Placenta

Primary GCT, although rarely, may also occur in the placenta [90] and even in the umbilical cord [91]. The most frequently reported type is teratoma [90], which should be distinguished from an "amorphous fetus." Other forms of non-trophoblastic primary placental neoplasia include YST [92, 93], chorangioma, and foci of hepatocellular (hepatocellular adenoma) [94] or adrenal parenchyma, which may represent a challenging differential diagnosis. Placental teratomas can be distinguished from the nonneoplastic fetus amorphus by the presence in the latter of an umbilical cord and skeletal organization [93]. YST in the placenta show the characteristic histological features seen in these tumors in other locations. Hepatocellular adenomas usually express a characteristic morphology and immunohistochemical markers, such as Hep Par1. The same is true for adrenal rests, which can be highlighted by alpha inhibin immunohistochemistry.

10.4 Interpretation of Serum Tumor Markers

Proteins secreted by certain subtypes of GCT may be used as tumor markers, and their pattern and degree of elevation can provide an indication of the likely tumor histology. AFP is generally elevated in patients with YST, although low levels of AFP (<100 mg/l) can be observed in immature teratomas (perhaps due to occult microscopic foci of YST within the tumor) [6]. However, because AFP is also normally synthesized by fetal liver, yolk sac, and gastrointestinal tract, interpretation of AFP levels during the neonatal period must incorporate knowledge of age-related norms [6, 7, 61, 95]. AFP is elevated in all infants at birth but drops to normal levels over the course of the first 2 years of life, as its synthesis in the liver ceases. The half-life of AFP varies with age during the first months of life and stabilizes at 5–7 days by 9 months of life [95]. In adult men with metastatic testicular cancer, the degree of elevation of the tumor markers is of prognostic value, and the failure of tumor markers to decline appropriately when undergoing treatment indicates likely resistance to chemotherapy; however, neither of these factors has consistently been shown to be prognostic in children [6, 96].

When an elevated AFP level is detected, alternative possibilities should be considered including synthesis by liver tumors such as hepatoma, hepatoblastoma, and even mesenchymal hamartoma of the liver [97], or diseases such as hypothyroidism, folate deficiencies, autoimmune disorders, acquired immunodeficiency disorder, congenital heart defects, cystic fibrosis, and platelet aggregation disorders [6].

hCG is a peptide hormone produced in pregnancy, which is made by the embryo soon after conception and later by the placental syncytiotrophoblast [6]. The beta subunit of hCG serves as a marker of syncytiotrophoblasts when they are present in the tumor, typically a choriocarcinoma, where it can be significantly elevated. Its half-life is 16 h [7].

10.5 Molecular Genetics of Pediatric GCT

The genomic alterations seen in malignant GCT of infants and children are generally distinct from those occurring in tumors from postpubertal patients [4, 6, 76, 98]. In the prepubertal period, pure teratomas of the testis or of extragonadal sites nearly always exhibit normal genomic profiles, which contrasts sharply with the universally abnormal cytogenetic profile of postpubertal teratomas arising as a component of a mixed malignant GCT [6, 64, 86].

Adolescent and adult GCT tend to be aneuploid. The most consistent chromosomal aberration in adolescent/adult malignant GCT is the overrepresentation of chromosome 12p (80 %) [3, 99–101], which is found in all histologic subtypes and primary sites (ovarian, testicular, and extragonadal) [6]. Abnormalities of chromosomes 7 and 8 have also been found in up to 70 % of adolescent and adult testicular GCT [101].

Unlike teratomas, cytogenetic and other genomic aberrations are consistently reported in YST on infants and children. Array CGH profiles on patients aged <5 years have shown that most teratomas have normal profiles, with occasional loss of chromosome 20p as the only recurrent change. In contrast, YST show abnormal profiles, including gains of chromosomes 1q, 3p, and 20q, and loss of chromosomes 1p, 6q, and 18q [102].

References

1. Rescorla FJ. Pediatric germ cell tumors. Semin Pediatr Surg. 2012;21(1):51–60. doi:10.1053/j.sempedsurg.2011.10.005.
2. Billmire DF. Malignant germ cell tumors in childhood. Semin Pediatr Surg. 2006;15(1):30–6. doi:10.1053/j.sempedsurg.2005.11.006.
3. Horton Z, Schlatter M, Schultz S. Pediatric germ cell tumors. Surg Oncol. 2007;16(3):205–13. doi:10.1016/j.suronc.2007.07.005.
4. Isaacs Jr H. Perinatal (fetal and neonatal) germ cell tumors. J Pediatr Surg. 2004;39(7):1003–13.
5. Brown NJ. Teratomas and yolk-sac tumours. J Clin Pathol. 1976;29(11):1021–5.
6. Frazier AL, Weldon C, Amatruda J. Fetal and neonatal germ cell tumors. Semin Fetal Neonatal Med. 2012;17(4):222–30. doi:10.1016/j.siny.2012.05.004.
7. Rescorla FJ. Pediatric germ cell tumors. Semin Surg Oncol. 1999;16(2):144–58.
8. Ries LAG, Smith MA, Gurney JG, Linet M, Tamra T, Young JL, et al. Cancer incidence and survival among children and adolescents: United States SEER Program 1975-1995. NIH Pub No 99-4649:179pp - 179pp. 1999. http://www.cabdirect.org/abstracts/20083240999.html;jsessionid=A80086F82EDB83B2CE8ECC87D4BD11D3.
9. Cost NG, Lubahn JD, Adibi M, Romman A, Wickiser JE, Raj GV, et al. A comparison of pediatric, adolescent, and adult testicular germ cell malignancy. Pediatr

Blood Cancer. 2014;61(3):446–51. doi:10.1002/pbc.24773.
10. Pohl HG, Shukla AR, Metcalf PD, Cilento BG, Retik AB, Bagli DJ, et al. Prepubertal testis tumors: actual prevalence rate of histological types. J Urol. 2004;172(6 Pt 1):2370–2.
11. Taskinen S, Fagerholm R, Aronniemi J, Rintala R, Taskinen M. Testicular tumors in children and adolescents. J Pediatr Urol. 2008;4(2):134–7. doi:10.1016/j.jpurol.2007.10.002.
12. Gobel U, Schneider DT, Calaminus G, Haas RJ, Schmidt P, Harms D. Germ-cell tumors in childhood and adolescence. GPOH MAKEI and the MAHO study groups. Ann Oncol Off J Eur Soc Med Oncol/ESMO. 2000;11(3):263–71.
13. Young Jr JL, Ries LG, Silverberg E, Horm JW, Miller RW. Cancer incidence, survival, and mortality for children younger than age 15 years. Cancer. 1986;58(2 Suppl):598–602.
14. Isaacs Jr H. Congenital and neonatal malignant tumors. A 28-year experience at Children's Hospital of Los Angeles. Am J Pediatr Hematol/Oncol. 1987;9(2):121–9.
15. Shu XO, Nesbit ME, Buckley JD, Krailo MD, Robinson LL. An exploratory analysis of risk factors for childhood malignant germ-cell tumors: report from the Childrens Cancer Group (Canada, United States). Cancer Causes Control CCC. 1995;6(3):187–98.
16. Depue RH, Pike MC, Henderson BE. Estrogen exposure during gestation and risk of testicular cancer. J Natl Cancer Inst. 1983;71(6):1151–5.
17. Morrison AS. Cryptorchidism, hernia, and cancer of the testis. J Natl Cancer Inst. 1976;56(4):731–3.
18. Swerdlow AJ, Huttly SR, Smith PG. Testicular cancer and antecedent diseases. Br J Cancer. 1987;55(1):97–103.
19. Pettersson A, Richiardi L, Nordenskjold A, Kaijser M, Akre O. Age at surgery for undescended testis and risk of testicular cancer. N Engl J Med. 2007;356(18):1835–41. doi:10.1056/NEJMoa067588.
20. Nistal M, Paniagua R, Gonzalez-Peramato P, Reyes-Mugica M. Perspectives in pediatric pathology, Chapter 5. Gonadal dysgenesis. Pediatr Dev Pathol. 2015;18(4):259–78. doi:10.2350/14-04-1471-PB.1.
21. Nistal M, Paniagua R, Gonzalez-Peramato P, Reyes-Mugica M. Perspectives in pediatric pathology, Chapter 6. Male undermasculinization. Pediatr Dev Pathol. 2015;18(4):279–96. doi:10.2350/14-04-1465-PB.1.
22. Rutgers JL, Scully RE. Pathology of the testis in intersex syndromes. Semin Diagn Pathol. 1987;4(4):275–91.
23. Rutgers JL, Scully RE. The androgen insensitivity syndrome (testicular feminization): a clinicopathologic study of 43 cases. Int J Gynecol Pathol Off J Int Soc Gynecol Pathol. 1991;10(2):126–44.
24. Verp MS, Simpson JL. Abnormal sexual differentiation and neoplasia. Cancer Genet Cytogenet. 1987;25(2):191–218.
25. Cools M, Drop SL, Wolffenbuttel KP, Oosterhuis JW, Looijenga LH. Germ cell tumors in the intersex gonad: old paths, new directions, moving frontiers. Endocr Rev. 2006;27(5):468–84. doi:10.1210/er.2006-0005.
26. Cools M, Stoop H, Kersemaekers AM, Drop SL, Wolffenbuttel KP, Bourguignon JP, et al. Gonadoblastoma arising in undifferentiated gonadal tissue within dysgenetic gonads. J Clin Endocrinol Metab. 2006;91(6):2404–13. doi:10.1210/jc.2005-2554.
27. Hughes IA, Houk C, Ahmed SF, Lee PA, International Consensus Conference on Intersex organized by the Lawson Wilkins Pediatric Endocrine Society/European Society for Paediatric Endocrinology Consensus. Consensus statement on management of intersex disorders. J Pediatr Urol. 2006;2(3):148–62. doi:10.1016/j.jpurol.2006.03.004.
28. Manuel M, Katayama PK, Jones Jr HW. The age of occurrence of gonadal tumors in intersex patients with a Y chromosome. Am J Obstet Gynecol. 1976;124(3):293–300.
29. Robboy SJ, Miller T, Donahoe PK, Jahre C, Welch WR, Haseltine FP, et al. Dysgenesis of testicular and streak gonads in the syndrome of mixed gonadal dysgenesis: perspective derived from a clinicopathologic analysis of twenty-one cases. Hum Pathol. 1982;13(8):700–16.
30. Savage MO, Lowe DG. Gonadal neoplasia and abnormal sexual differentiation. Clin Endocrinol. 1990;32(4):519–33.
31. Schellhas HF. Malignant potential of the dysgenetic gonad. Part 1. Obstet Gynecol. 1974;44(2):298–309.
32. Scully RE. Gonadoblastoma. A review of 74 cases. Cancer. 1970;25(6):1340–56.
33. Troche V, Hernandez E. Neoplasia arising in dysgenetic gonads. Obstet Gynecol Surv. 1986;41(2):74–9.
34. Cools M, van Aerde K, Kersemaekers AM, Boter M, Drop SL, Wolffenbuttel KP, et al. Morphological and immunohistochemical differences between gonadal maturation delay and early germ cell neoplasia in patients with undervirilization syndromes. J Clin Endocrinol Metab. 2005;90(9):5295–303. doi:10.1210/jc.2005-0139.
35. Kersemaekers AM, Honecker F, Stoop H, Cools M, Molier M, Wolffenbuttel K, et al. Identification of germ cells at risk for neoplastic transformation in gonadoblastoma: an immunohistochemical study for OCT3/4 and TSPY. Hum Pathol. 2005;36(5):512–21. doi:10.1016/j.humpath.2005.02.016.
36. Lau Y, Chou P, Iezzoni J, Alonzo J, Komuves L. Expression of a candidate gene for the gonadoblastoma locus in gonadoblastoma and testicular

seminoma. Cytogenetics and Cell Genetics. 2000;91(1–4):160–4.
37. Lau YF. Gonadoblastoma, testicular and prostate cancers, and the TSPY gene. Am J Hum Genet. 1999;64(4):921–7.
38. Slowikowska-Hilczer J, Romer TE, Kula K. Neoplastic potential of germ cells in relation to disturbances of gonadal organogenesis and changes in karyotype. J Androl. 2003;24(2):270–8.
39. Slowikowska-Hilczer J, Szarras-Czapnik M, Kula K. Testicular pathology in 46,XY dysgenetic male pseudohermaphroditism: an approach to pathogenesis of testis cancer. J Androl. 2001;22(5):781–92.
40. Pena-Alonso R, Nieto K, Alvarez R, Palma I, Najera N, Erana L, et al. Distribution of Y-chromosome-bearing cells in gonadoblastoma and dysgenetic testis in 45,X/46,XY infants. Mod Pathol Off J United States Can Acad Pathol Inc. 2005;18(3):439–45. doi:10.1038/modpathol.3800293.
41. Donahoe PK, Crawford JD, Hendren WH. Mixed gonadal dysgenesis, pathogensis, and management. J Pediatr Surg. 1979;14(3):287–300.
42. Hersmus R, de Leeuw BH, Wolffenbuttel KP, Drop SL, Oosterhuis JW, Cools M, et al. New insights into type II germ cell tumor pathogenesis based on studies of patients with various forms of disorders of sex development (DSD). Mol Cell Endocrinol. 2008;291(1–2):1–10. doi:10.1016/j.mce.2008.02.028.
43. Morris JM, Mahesh VB. Further observations on the syndrome, "testicular feminization". Am J Obstet Gynecol. 1963;87:731–48.
44. Ahmed SF, Cheng A, Dovey L, Hawkins JR, Martin H, Rowland J, et al. Phenotypic features, androgen receptor binding, and mutational analysis in 278 clinical cases reported as androgen insensitivity syndrome. J Clin Endocrinol Metab. 2000;85(2):658–65. doi:10.1210/jcem.85.2.6337.
45. Armstrong GR, Buckley CH, Kelsey AM. Germ cell expression of placental alkaline phosphatase in male pseudohermaphroditism. Histopathology. 1991;18(6):541–7.
46. Bangsboll S, Qvist I, Lebech PE, Lewinsky M. Testicular feminization syndrome and associated gonadal tumors in Denmark. Acta Obstet Gynecol Scand. 1992;71(1):63–6.
47. Cassio A, Cacciari E, D'Errico A, Balsamo A, Grigioni FW, Pascucci MG, et al. Incidence of intratubular germ cell neoplasia in androgen insensitivity syndrome. Acta Endocrinol. 1990;123(4):416–22.
48. Gourlay WA, Johnson HW, Pantzar JT, McGillivray B, Crawford R, Nielsen WR. Gonadal tumors in disorders of sexual differentiation. Urology. 1994;43(4):537–40.
49. Muller J. Morphometry and histology of gonads from twelve children and adolescents with the androgen insensitivity (testicular feminization) syndrome. J Clin Endocrinol Metab. 1984;59(4):785–9. doi:10.1210/jcem-59-4-785.
50. Ramani P, Yeung CK, Habeebu SS. Testicular intratubular germ cell neoplasia in children and adolescents with intersex. Am J Surg Pathol. 1993;17(11):1124–33.
51. Regadera J, Martinez-Garcia F, Paniagua R, Nistal M. Androgen insensitivity syndrome: an immunohistochemical, ultrastructural, and morphometric study. Arch Pathol Lab Med. 1999;123(3):225–34. doi:10.1043/0003-9985(1999)123<0225:AIS>2.0.CO;2.
52. Looijenga LH, Hersmus R, Oosterhuis JW, Cools M, Drop SL, Wolffenbuttel KP. Tumor risk in disorders of sex development (DSD). Best Pract Res Clin Endocrinol Metab. 2007;21(3):480–95. doi:10.1016/j.beem.2007.05.001.
53. Gwin K, Cajaiba MM, Caminoa-Lizarralde A, Picazo ML, Nistal M, Reyes-Mugica M. Expanding the clinical spectrum of Frasier syndrome. Pediatr Dev Pathol. 2008;11(2):122–7. doi:10.2350/07-01-0209.1.
54. Nistal M, Paniagua R, Gonzalez-Peramato P, Reyes-Mugica M. 7- Ovotesticular DSD (True Hermaphroditism). Pediatr Dev Pathol. 2015;18(5):345–52. doi:10.2350/14-04-1466-PB.1.
55. Lee PA, Houk CP, Ahmed SF, Hughes IA, International Consensus Conference on Intersex organized by the Lawson Wilkins Pediatric Endocrine Society and the European Society for Paediatric Endocrinology. Consensus statement on management of intersex disorders. International Consensus Conference on Intersex. Pediatrics. 2006;118(2):e488–500. doi:10.1542/peds.2006-0738.
56. Dehner LP. Gonadal and extragonadal germ cell neoplasia of childhood. Hum Pathol. 1983;14(6):493–511.
57. Norris HJ, Zirkin HJ, Benson WL. Immature (malignant) teratoma of the ovary: a clinical and pathologic study of 58 cases. Cancer. 1976;37(5):2359–72.
58. Cushing B, Giller R, Ablin A, Cohen L, Cullen J, Hawkins E, et al. Surgical resection alone is effective treatment for ovarian immature teratoma in children and adolescents: a report of the pediatric oncology group and the children's cancer group. Am J Obstet Gynecol. 1999;181(2):353–8.
59. Heifetz SA, Cushing B, Giller R, Shuster JJ, Stolar CJ, Vinocur CD, et al. Immature teratomas in children: pathologic considerations: a report from the combined Pediatric Oncology Group/Children's Cancer Group. Am J Surg Pathol. 1998;22(9):1115–24.
60. Marina NM, Cushing B, Giller R, Cohen L, Lauer SJ, Ablin A, et al. Complete surgical excision is effective treatment for children with immature teratomas with or without malignant elements: a Pediatric Oncology Group/Children's Cancer Group Intergroup Study. J Clin Oncol Off J Am Soc Clin Oncol. 1999;17(7):2137–43.

61. Blohm ME, Vesterling-Horner D, Calaminus G, Gobel U. Alpha 1-fetoprotein (AFP) reference values in infants up to 2 years of age. Pediatr Hematol Oncol. 1998;15(2):135–42.
62. Blohm ME, Gobel U. Unexplained anaemia and failure to thrive as initial symptoms of infantile choriocarcinoma: a review. Eur J Pediatr. 2004;163(1):1–6. doi:10.1007/s00431-003-1361-1.
63. González-Crussí F. Extragonadal teratomas. Atlas of tumor pathology. Second Series. Armed Forces Institute of Pathology: Washington, D.C; 1982.
64. Altman RP, Randolph JG, Lilly JR. Sacrococcygeal teratoma: American Academy of Pediatrics Surgical Section Survey-1973. J Pediatr Surg. 1974;9(3):389–98.
65. Benachi A, Durin L, Vasseur Maurer S, Aubry MC, Parat S, Herlicoviez M, et al. Prenatally diagnosed sacrococcygeal teratoma: a prognostic classification. J Pediatr Surg. 2006;41(9):1517–21. doi:10.1016/j.jpedsurg.2006.05.009.
66. Makin EC, Hyett J, Ade-Ajayi N, Patel S, Nicolaides K, Davenport M. Outcome of antenatally diagnosed sacrococcygeal teratomas: single-center experience (1993-2004). J Pediatr Surg. 2006;41(2):388–93. doi:10.1016/j.jpedsurg.2005.11.017.
67. Currarino G, Coln D, Votteler T. Triad of anorectal, sacral, and presacral anomalies. AJR Am J Roentgenol. 1981;137(2):395–8. doi:10.2214/ajr.137.2.395.
68. Hagan DM, Ross AJ, Strachan T, Lynch SA, Ruiz-Perez V, Wang YM, et al. Mutation analysis and embryonic expression of the HLXB9 Currarino syndrome gene. Am J Hum Genet. 2000;66(5):1504–15.
69. Ross AJ, Ruiz-Perez V, Wang Y, Hagan DM, Scherer S, Lynch SA, et al. A homeobox gene, HLXB9, is the major locus for dominantly inherited sacral agenesis. Nat Genet. 1998;20(4):358–61. doi:10.1038/3828.
70. De Backer A, Madern GC, Hakvoort-Cammel FG, Haentjens P, Oosterhuis JW, Hazebroek FW. Study of the factors associated with recurrence in children with sacrococcygeal teratoma. J Pediatr Surg. 2006;41(1):173–81; discussion 181. doi:10.1016/j.jpedsurg.2005.10.022.
71. Billmire D, Vinocur C, Rescorla F, Colombani P, Cushing B, Hawkins E, et al. Malignant retroperitoneal and abdominal germ cell tumors: an intergroup study. J Pediatr Surg;2003;38(3):315–8; discussion 318. doi:10.1053/jpsu.2003.50100.
72. Billmire D, Vinocur C, Rescorla F, Colombani P, Cushing B, Hawkins E, et al. Malignant mediastinal germ cell tumors: an intergroup study. J Pediatr Surg. 2001;36(1):18–24. doi:10.1053/jpsu.2001.19995.
73. Hartmann JT, Nichols CR, Droz JP, Horwich A, Gerl A, Fossa SD, et al. Hematologic disorders associated with primary mediastinal nonseminomatous germ cell tumors. J Natl Cancer Inst. 2000;92(1):54–61.
74. Jones NM, Kiely EM. Retroperitoneal teratomas – potential for surgical misadventure. J Pediatr Surg 2008;43(1):184–6; discussion 187. doi:10.1016/j.jpedsurg.2007.09.041.
75. Luo CC, Huang CS, Chu SM, Chao HC, Yang CP, Hsueh C. Retroperitoneal teratomas in infancy and childhood. Pediatr Surg Int. 2005;21(7):536–40. doi:10.1007/s00383-005-1424-7.
76. Spencer R. Parasitic conjoined twins: external, internal (fetuses in fetu and teratomas), and detached (acardiacs). Clin Anat. 2001;14(6):428–44. doi:10.1002/ca.1079.
77. Jones DC, Reyes-Mugica M, Gallagher PG, Fricks P, Touloukian RJ, Copel JA. Three-dimensional sonographic imaging of a highly developed fetus in fetu with spontaneous movement of the extremities. J Ultrasound Med. 2001;20(12):1357–63.
78. Escobar MA, Rossman JE, Caty MG. Fetus-in-fetu: report of a case and a review of the literature. J Pediatr Surg. 2008;43(5):943–6. doi:10.1016/j.jpedsurg.2008.01.061.
79. Kerner B, Flaum E, Mathews H, Carlson DE, Pepkowitz SH, Hixon H, et al. Cervical teratoma: prenatal diagnosis and long-term follow-up. Prenat Diagn. 1998;18(1):51–9.
80. Azizkhan RG, Haase GM, Applebaum H, Dillon PW, Coran AG, King PA, et al. Diagnosis, management, and outcome of cervicofacial teratomas in neonates: a Childrens Cancer Group study. J Pediatr Surg. 1995;30(2):312–6.
81. Hirose S, Sydorak RM, Tsao K, Cauldwell CB, Newman KD, Mychaliska GB et al. Spectrum of intrapartum management strategies for giant fetal cervical teratoma. J Pediatr Surg. 2003;38(3):446–50; discussion 450. doi:10.1053/jpsu.2003.50077.
82. Howlader N, Noone AM, Krapcho M, Garschell J, Miller D, Altekruse SF et al. SEER cancer statistics review, 1975-2012. National Cancer Institute, Bethesda, MD. 2015. http://seer.cancer.gov/csr/1975_2012/.
83. Bussey KJ, Lawce HJ, Olson SB, Arthur DC, Kalousek DK, Krailo M, et al. Chromosome abnormalities of eighty-one pediatric germ cell tumors: sex-, age-, site-, and histopathology-related differences – a Children's Cancer Group study. Genes Chromosom Cancer. 1999;25(2):134–46.
84. Schneider DT, Calaminus G, Koch S, Teske C, Schmidt P, Haas RJ, et al. Epidemiologic analysis of 1,442 children and adolescents registered in the German germ cell tumor protocols. Pediatr Blood Cancer. 2004;42(2):169–75. doi:10.1002/pbc.10321.
85. Schlatter M, Rescorla F, Giller R, Cushing B, Vinocur C, Colombani P et al. Excellent outcome in patients with stage I germ cell tumors of the testes: a study of the Children's Cancer Group/Pediatric Oncology Group. J Pediatr Sur. 2003;38(3):319–24; discussion 324. doi:10.1053/jpsu.2003.50101.
86. Gobel U, Calaminus G, Engert J, Kaatsch P, Gadner H, Bokkerink JP, et al. Teratomas in infancy and childhood. Med Pediatr Oncol. 1998;31(1):8–15.
87. Shukla AR, Woodard C, Carr MC, Huff DS, Canning DA, Zderic SA, et al. Experience with testis sparing surgery for testicular teratoma. J Urol. 2004;171(1):161–3. doi:10.1097/01.ju.0000101185.90327.b4.

88. Oltmann SC, Fischer A, Barber R, Huang R, Hicks B, Garcia N. Pediatric ovarian malignancy presenting as ovarian torsion: incidence and relevance. J Pediatr Surg. 2010;45(1):135–9. doi:10.1016/j.jpedsurg.2009.10.021.
89. Billmire D, Vinocur C, Rescorla F, Cushing B, London W, Schlatter M et al. Outcome and staging evaluation in malignant germ cell tumors of the ovary in children and adolescents: an intergroup study. J Pediatr Surg. 2004;39(3):424–9; discussion 429.
90. Wang L, Du X, Li M. Placental teratoma. A case report and review of the literature. Pathol Res Pract. 1995;191(12):1267–70. discussion 1270
91. Del Sordo R, Fratini D, Cavaliere A. Teratoma of umbilical cord: a case report and literature review. Pathologica. 2006;98(4):224–8.
92. Drut R, Mortera M, Drut RM. Yolk sac tumor of the placenta in Wiedemann-Beckwith syndrome. Pediatr Dev Pathol. 1998;1(6):534–7.
93. Matsuyama TA, Kushima M, Yamochi-Onizuka T, Ota H. Placental yolk sac tumor with divergent endodermal differentiation. Int J Gynecol Pathol Off J Int Soc Gynecol Pathol. 2004;23(4):398–402.
94. DeNapoli TS. Coexistent chorangioma and hepatic adenoma in one twin placenta: a case report and review of the literature. Pediatr Dev Pathol. 2015; doi:10.2350/14-12-1592-CR.1.
95. Wu JT, Book L, Sudar K. Serum alpha fetoprotein (AFP) levels in normal infants. Pediatr Res. 1981;15(1):50–2.
96. Mazumdar M, Bajorin DF, Bacik J, Higgins G, Motzer RJ, Bosl GJ. Predicting outcome to chemotherapy in patients with germ cell tumors: the value of the rate of decline of human chorionic gonadotrophin and alpha-fetoprotein during therapy. J Clin Oncol Off J Am Soc Clin Oncol. 2001;19(9):2534–41.
97. Cajaiba MM, Sarita-Reyes C, Zambrano E, Reyes-Mugica M. Mesenchymal hamartoma of the liver associated with features of Beckwith-Wiedemann syndrome and high serum alpha-fetoprotein levels. Pediatr Dev Pathol. 2007;10(3):233–8. doi:10.2350/06-07-0128.1.
98. Looijenga LH, Stoop H, de Leeuw HP, de Gouveia Brazao CA, Gillis AJ, van Roozendaal KE, et al. POU5F1 (OCT3/4) identifies cells with pluripotent potential in human germ cell tumors. Cancer Res. 2003;63(9):2244–50.
99. Atkin NB, Baker MC. i(12p): specific chromosomal marker in seminoma and malignant teratoma of the testis? Cancer Genet Cytogenet. 1983;10(2):199–204.
100. Sandberg AA, Meloni AM, Suijkerbuijk RF. Reviews of chromosome studies in urological tumors. III. Cytogenetics and genes in testicular tumors. J Urol. 1996;155(5):1531–56.
101. Summersgill B, Goker H, Weber-Hall S, Huddart R, Horwich A, Shipley J. Molecular cytogenetic analysis of adult testicular germ cell tumours and identification of regions of consensus copy number change. Br J Cancer. 1998;77(2):305–13.
102. Veltman I, Veltman J, Janssen I, Hulsbergen-van de Kaa C, Oosterhuis W, Schneider D, et al. Identification of recurrent chromosomal aberrations in germ cell tumors of neonates and infants using genomewide array-based comparative genomic hybridization. Genes Chromosom Cancer 2005;43(4):367–76. doi:10.1002/gcc.20208.

Germ Cell Tumors of Miscellaneous Extragonadal Sites

11

Alejandro Rubio-Fernández, Juan Antonio Retamero, and Francisco F. Nogales

The occurrence of pluripotential tumors in diverse midline extragonadal locations can be explained by the migration of germ cells, which originate in the dorsal wall of the yolk sac and migrate during the fourth to sixth week of development through the dorsal midline toward their final destination in the developing gonads [1]. Ectopic primitive germ cells may give rise to extragonadal germ cell tumors (EGGCTs), and consequently they occur usually in children or young adults (type I GCT), in medial or paramedial locations. In this pathogenetic situation, their precursor lesions are unknown, and tumors are often difficult to situate in the various types of pluripotential tumors, as they belong, according to specific site, sex, and age of presentation to any pathogenetic type (see Chap. 3). Primitive germ cells undergo meiosis at the same time in both female and male embryos in either gonadal or extragonadal sites, [2, 3] thus explaining the blurred boundaries between different types in extragonadal tumors of both sexes. Primitive germ cells are the origin of both seminomatous and non-seminomatous tumors, belonging to types I and II, in the Oosterhuis and Looijenga classification (see Chap. 3).

However, the germ cell origin model does not provide an adequate explanation for either those GCTs found in locations distant from the midline or those presenting as secondary patterns of various somatic tumors, often with an endodermal origin. These would correspond to type VI pluripotential tumors originating from cells similar to induced pluripotential somatic stem cells (iPSC) (see Chap. 3). These tumors, where somatic malignances develop non-seminomatous GCT, are particularly prevalent in organs of endodermal derivation, such as the lung, stomach, liver, and urinary tract, where the secondary differentiated GCT patterns frequently are YST of glandular or hepatoid pattern, thus indicating a continuum of endodermal differentiations included under the concept of primitive endodermal tumors [4]. This group possibly includes some teratoid carcinosarcomas, fetal adenocarcinoma of the lung, gastric adenocarcinoma with hepatoid and clear cell differentiation, hepatoblastoma, combined hepatocellular cholangiocarcinoma with stem cell features, some instances of urothelial cell carcinomas, and hepatoid adenocarcinomas from several locations, including examples of liver metastases.

A. Rubio-Fernández, MD
Department of Pathology, Hospital San Pedro de Alcántara Cáceres, 10003 Cáceres, Spain
e-mail: alejandrorubiofernandez@gmail.com

J.A. Retamero, MD, MRCPsych
F.F. Nogales, MD, PhD (✉)
Department of Pathology, University of Granada Medical School, 18012 Granada, Spain
e-mail: jaretamero@gmail.com; fnogales@ugr.es

F.F. Nogales, R.E. Jimenez (eds.), *Pathology and Biology of Human Germ Cell Tumors*,
DOI 10.1007/978-3-662-53775-6_11

In this chapter we will analyze the clinicopathologic features of EGGCT from the most relevant extragonadal locations with the exception of those already reviewed in Chaps. 8, 9, and 10. We will not include instances of type 0 GCT, such as epignathi and parasitic twins addressed in Chap. 10.

Incidence of EGGCT has been reviewed in Chap. 2. However, the true incidence of EGGCT is difficult to calculate as usually only malignant tumors are registered and the presence of benign lesions omitted. Histopathological reports of EGGCT are few [5] with most reviews and case reports being clinically focused and, more often than not, with incomplete histopathologic data. EGGCTs represent 3.3 % of all GCTs; of these, 1.5 % are found in the central nervous system (CNS) (see Chap. 9), 0.9 % occur in mediastinum and thorax (see Chap. 8), and 0.8 % occur in the abdomen and pelvis. Unspecified locations represent only 0.1 % [6]. Most EGGCTs outside the mediastinum, CNS, and prostate are non-seminomas.

11.1 Head and Neck

The head and neck region accounts for 5 % of all EGGCT [7].

11.1.1 Orbit

GCTs in the orbit are rare. A recent review analyzing a large number of patients over a 35-year span did not find a single case [8]. However, some 50 cases of orbital teratomas have been reported [9]. They are more common in neonates and infants but have been reported in adults; they seem to affect females more than males and the right orbit more than the left [10]. The typical clinical course is of rapid growth after birth, although rarely they may grow slowly [11]. The eye is normally functional although mass-related ocular abnormalities, such as proptosis, exposure keratosis, optic nerve atrophy, or cataracts, may occur [10]. The association of orbital teratoma and anophthalmia is rare [12]. Histologically, teratomas are typically mature and may have similar tissues to those encountered in gonadal teratomas [13]. Malignant teratomas seldom occur and should be differentiated from teratoid carcinosarcoma, which has sarcomatous change [14]. Teratoma with a neuroblastic component has been reported [15].

Yolk sac tumors (YST) present very infrequently in the orbit, with only 16 cases reported to date [16–23]. They typically affect infants under the age of 3, but have also been reported in the fetus [19]. AFP levels are typically raised and imaging studies reveal a soft tissue density mass. Early surgery and chemotherapy should be performed in order to preserve the eye and vision, to encourage orbitofacial development and for cosmetic effect [24]. Frequently, the eye develops correctly, but vision can be impaired, and exenteration may be necessary.

11.1.2 Oropharynx, Nose, and Paranasal Sinuses

GCTs arising in the oral and nasopharyngeal cavity are uncommon, the most frequent being benign teratoma in infants and children. Epignathus is a graphic variant, and etymologically refers to teratomas arising from the jaw, but also is frequently used to describe teratomas arising from the oropharynx. These tumors would correspond to type 0 in the Oosterhuis and Looijenga classification (see Chap. 3) and should be considered a parasitic twin (fetus in fetu) [25]. They are also dealt with in Chap. 10. Although rare, teratomas presenting with immature elements have been described in the neonatal period [26].

Sinonasal teratoid carcinosarcomas (SNTCS) are lesions of debatable germ cell origin; histologically they are a heterogeneous combination of epithelial, mesenchymal, and neural elements. They are rare, aggressive neoplasms with high mortality and recurrence rates and with a predilection for males in their fifth decade [27]. Although typically present in the sinonasal region, they may occur in the nasopharynx and oral cavity [28]. Clinically, their presentation includes nasal obstruction and bleeding, headaches, and visual disturbances. Histologically, they show an admixture of benign and malignant epithelial, mesenchymal, and primitive neuroectodermal components. The benign elements are well-differentiated acinar or glandular structures and clear cell "fetal type" squamous epithelium [29], and the malignant components are atypical epithelia, mesenchyme with fibroblastic, rhabdomyoblastic, and chondroid elements, as well as rosette-forming primitive neural components. The main argument against their germ cell origin is the absence of concomitant GCT elements [28–30]. It is possible that they correspond to type VI GCT. The only case reported in association with YST areas [30] had an extra copy of chromosome 12p13 [30]. A further report [31] lacked GCT elements but showed trisomy 12 and 1p deletion. Their differences with teratoma are difficult to establish, since they may rarely have a trilaminar differentiation and show coexistence of benign and malignant elements. For this reason, it is possible that cases in the literature described as immature teratomas in the sinonasal region may correspond to SNTCS [15].

SNTCS have a poor prognosis; there are recurrences in about a quarter of patients, a tenth develop metastases, and only half of the patients survive beyond 3–4 years. Due to their poor outcome, vigorous treatment with surgery and radiotherapy is recommended [27].

11.1.3 Thyroid

The presence of thyroid tissue closely related to other teratoid elements and clearly surrounded by a pseudocapsule appears to be the most significant criterium for establishing a thyroidal origin for teratomas [32, 33].

Thyroid teratomas have been reported in all age groups, but are more frequent in neonates and children. The majority of pediatric thyroid teratomas have a mature histology, but complications are frequent, due to compression of the airway or other vital neck structures, particularly in neonates. Benign thyroid teratomas have a similar gender distribution, and at presentation, there is no gender-related age difference. Their size is variable, ranging 2–13 cm, with a mean of 6 cm [34]. Histologically, benign teratomas display a wide variety of tissue types, including different types of epithelia and pilosebaceous and adnexal structures, as well as lung, hepatic, and pancreatic tissue intermixed with mesenchymal structures such as skeletal and striated muscle and adipose and embryonic connective tissue. The presence of a prominent neural tissue component such as mature glia and choroid or immature neural elements is common.

Malignant teratomas tend to occur in older age groups [35]. There are approximately 30 reported cases occurring between the ages of 8–68 with a mean age of 28. They are more frequent in females; indeed about three quarters of all reported cases of malignant teratomas with immature neural components were found in females, occasionally associated with pregnancy [32, 35–37].

The differential diagnosis of thyroid teratomas includes lesions such as branchial cleft cyst, thyroglossal cyst, congenital goiter, and cystic hygroma. Immature and malignant teratomas, owing to their abundant immature neural components, need to be distinguished from other malignant small cell tumors [32, 38].

The presence of other primary GCT in the thyroid gland is exceedingly rare. There is a single case report of YST in the thyroid in a 10-year-old girl [39] and an additional case of a *mixed* GCT in the thyroid of a 35-year-old man who presented with a thyroid mass containing areas of embryonal carcinoma and choriocarcinoma, but with no primary testicular tumor [40].

11.1.4 Parotid Gland

The presence of GCT in the parotid gland is anecdotal. The few cases reported are mainly mature teratomas [41–46] and occur in the first and second decades of life. Fine needle aspiration may be useful in diagnosis [46].

There are two reports of YST of the parotid gland, both in girls [47, 48] and presenting as a rapidly growing mass with raised serum AFP levels.

11.2 Lung

Although teratomas are frequently found in the anterior mediastinum (see Chap. 8), pulmonary teratomas are less frequent, with approximately 80 cases reported to date [49–56]. They present as slowly growing masses [54], are slightly more common in females [51], and have a median age at diagnosis of 28.5 years, ranging from 10 months to 68 years [57], and a variable tumor size; some cases occupied the entire lung [52].

Clinical symptoms are diverse and include productive cough, sweating, loss of weight, lassitude, recurrent hemoptysis, bronchiectasis, empyema, fever, headache, and pleuritic chest pain [50, 58–60]. Trichoptysis (expectoration of hairs) is pathognomonic of pulmonary teratoma, but only occurs in 13–15 % of cases [50, 55, 61]. They are often misdiagnosed as pneumonia, lung abscess, tuberculosis, hydatid cyst, or aspergilloma [50, 59, 62]. Recurrent hemoptysis may be related to the enzymatic digestion from teratomatous pancreatic tissue [50, 58–60, 63]. YST-related hematological neoplasms, similar to those occurring in mediastinal GCT, are also found in the lung [64].

Pulmonary teratomas have a protracted course; the time from onset of symptoms to diagnosis is usually lengthy [58]. Although prognosis is favorable after surgical resection in most cases, some malignant immature teratomas (30 % of pulmonary teratomas) with occasional lymph node metastases or lymphovascular invasion have been reported, often in female patients [50, 53, 56, 65].

Interestingly, a link between mediastinal and pulmonary teratomas seems to exist, which could be explained by the putative origin of the latter, thought to arise from the aberrant migration of third pharyngeal pouch tissues, a thymic anlage [50, 51, 54, 57, 58, 60, 62]. Moreover, thymus has been found in several pulmonary teratomas [58, 63, 66]. Furthermore, teratomas of the lung show a predilection for the anterior aspect of the upper lobes (65 %), particularly the left, mostly in the S3 segment, related to the location of aberrant thymic tissue [50, 58, 62]. The finding of intrapulmonary thymomas lends support to this hypothesis [57]. Also, adhesions or connections between the pulmonary tumor and the mediastinum have been found, further complicating the clarification of the primary origin [51, 57, 59, 63]. The finding of a direct bronchial communication with the teratoma (50 %) has been proposed as a reliable sign for a pulmonary origin [49, 50, 65, 66]. The relationship between thymus and type II GCT is reviewed in Chap. 3.

The differential diagnosis should include pulmonary hamartoma, bronchogenic cyst, cystic adenomatoid malformation, and cystic lymphangiomas [51, 61]. Although hamartoma is composed predominantly of mesenchymal tissue, especially cartilage, smooth muscle, and adipose tissue, it can contain respiratory epithelium. The peripheral situation and the characteristic coin-like appearance in radiologic studies of hamartoma help in the differential diagnosis. In contrast, pulmonary teratomas show a more heterogeneous cystic and solid appearance with foci of calcification and peripheral radiolucent areas [51, 62].

Primary choriocarcinoma of the lung is a rare condition with only some 35 cases reported [67–69]. They occur more frequently in the right lung [70] and present with symptoms such as diffuse alveolar hemorrhage, namely, "choriocarcinoma syndrome" (see Sect. 11.7) [71], recurrent hemoptysis, persistent cough, massive loss of body weight, chest pain, dyspnea, hemothorax, gynecomastia, testicular atrophy, and loss of libido [72, 73]. Indeed, the presence of gynecomastia in a male patient with bilateral pulmonary nodules is highly suggestive of choriocarcinoma [73]. Pulmonary choriocarcinoma must be strongly suspected in women older than 35 with hemoptysis and previous molar pregnancies [69]. As expected, β-hCG serum levels are usually elevated [67, 68, 70–74].

A bimodal age distribution at diagnosis suggests a different pathogenesis for each group of choriocarcinomas [67, 74].

The first group corresponds to women of reproductive age, frequently with a history of previous pregnancy, miscarriage, or gestational trophoblastic disease. Indeed, the presence of trophoblastic cells in pulmonary arteries in autopsies of women with the aforementioned conditions has been found [71, 75], supporting the idea of pulmonary embolization of trophoblastic cells as a probable origin for primary pulmonary choriocarcinoma in these patients [70].

The second group is predominantly composed of men in their fifth or sixth decades with a history of smoking [74]. A concurrent conventional pulmonary carcinoma component (adeno-, squamous-, large cell carcinoma) is often observed, intermixed with the choriocarcinomatous areas [70, 72]. The relationship between carcinomas and choriocarcinoma is supported by the thyroid transcription factor (TTF-1) immunoexpression present in syncytiotrophoblastic-like cells of malignant lung neoplasms [76]. Furthermore, a variable β-hCG immunoexpression has been described in both components [72, 74]; 6 % of lung carcinomas can produce β-hCG, with most cases corresponding to giant and large cell carcinomas [72]. Thus, giant and large cell carcinomas with prominent multinucleated giant cells and β-hCG production are the major differential diagnoses of pulmonary choriocarcinoma. Choriocarcinoma of the lung does not represent a GCT and is of gestational origin in women of fertile age and of metaplastic origin in older men with a history of smoking and concurrent pulmonary epithelial malignancy. This group has a poor response to chemotherapy and a bad prognosis [70].

YST and fetal adenocarcinomas. While teratoma of the lung may represent a classic example of a GCT related to embryological development, pulmonary fetal adenocarcinoma (PFA) highlights our current understanding of pluripotential states in endodermally derived somatic malignancies. PFA was described in 1982 [77] as "pulmonary blastoma with argyrophil cells and lacking sarcomatous features" and is also known as pulmonary endodermal tumor resembling fetal lung [78], among other terminologies. The tumor shows a complex branching pattern of tubules resembling fetal lung of the glandular period lined by glycogen-rich, vacuolated epithelial cells, similar to glandular YST elsewhere, especially in its well-differentiated form. This fact may account for a large number of the reported

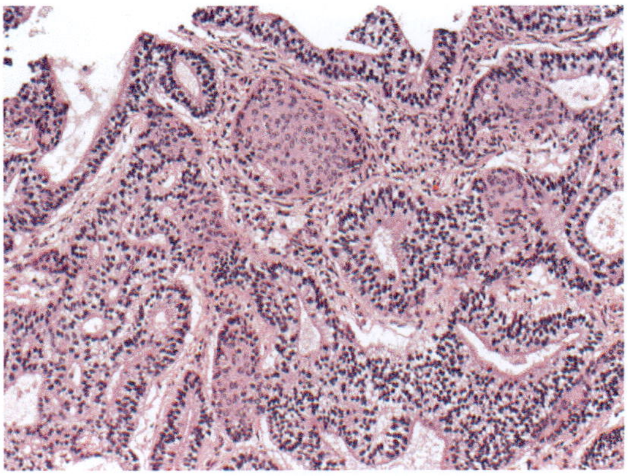

Fig. 11.1 Fetal type adenocarcinoma of the lung. Branching tubules lined by cylindrical epithelium with prominent vacuolation displaying morular aggregates with cells exhibiting optically clear nuclei

primary pulmonary YST [79–81]. Indeed, PFA shows expression of endodermal and pluripotentiality markers such as CDX2, AFP, glypican-3, and SALL4, analogous to YST [4]. However, the differentiated form of this tumor shows formation of morules (Fig. 11.1) with biotin-rich, optically clear nuclei identical to those found in low-grade tumors of the endometrium, thyroid, etc. but never occur in YST [82]. In contrast, pulmonary blastoma has an atypical mesenchymal component and does not show such a marked morular formation [83]. PFA represents an endodermal primitive tumor of pulmonary differentiation, possibly not derived from germ cells but from pulmonary stem cells that can also differentiate into other cell lines. This would explain the various associated patterns such as conventional-type adenocarcinoma, large cell neuroendocrine carcinoma, small cell carcinoma, enteric adenocarcinoma, and choriocarcinoma, as well as the TTF-1 positivity described in the mixed forms of PFA [84, 85].

11.3 Stomach

Gastric teratomas account for less than 1 % of teratomas in children [86, 87] and 1.6 % of all abdominal teratomas [88, 89]. Approximately 110 cases have been reported to date [88]. Gastric teratomas nearly always occur in infants, especially in neonates [90] (type I GCT), although there are cases reported in older children and adults [86]. The predominance of these tumors in male children is striking [87].

Clinically, gastric teratomas usually present as palpable masses with associated symptoms such as abdominal distention, vomiting, hematemesis, melena, respiratory distress, anemia, fever, weakness, abdominal pain, eating problems, constipation, failure to thrive, and spontaneous gastric perforation mimicking meconium peritonitis [86, 88, 91].

Gastric teratomas are essentially benign neoplasms, and although several so-called immature teratomas have been described in the stomach [92], only one case in an 83-year-old man has shown malignant transformation [93]. Recurrence has been associated with incomplete surgical resection or to the presence of a YST component [94, 95]. An example with a "focus of Wilms tumor with predominant epithelial component" [96] and another one with "small foci of neuroblastoma" [97] has been reported. A further association with Beckwith-Wiedemann syndrome and gliomatosis peritonei has been observed [90, 98], presenting as a scrotal sac mass [99]. Furthermore, a tumor composed of independent components of adenocarcinoma and immature "neuroepitheliomatous" monodermal teratoma has been reported [100]. Some of these random polydifferentiated neoplasms show a strong similarity with the multiple GCT patterns originated from somatic neoplasms in the female genital tract addressed in Chap. 6 that possibly arise from iPSC found in somatic tumors, representing type VI GCT.

Differential diagnosis includes diverse lesions such as mesenteric lymphangioma, pancreatic cyst, mesoblastic nephroma, neuroblastoma, Wilms tumor, hepatoblastoma, pancreatoblastoma, rhabdomyosarcoma, pheochromocytoma, angiomyolipoma, liposarcoma, and retroperitoneal teratoma [101, 102].

Gastric teratomas present as large exogastric (65 %), endogastric (9 %), or mixed endo/exogastric (26 %) [103] masses reaching up to 22–23 cm in size [90, 92]. Although they are found in any portion of the stomach, the greater curvature and posterior wall are the most common sites [89]. This could be explained by the more rapid growth of the posterior margin of the stomach in its expansion from the central part of the primitive anterior intestine [104].

As previously noted, immature tissues are not infrequent in gastric teratomas and show a good prognosis after complete surgical resection without neoadjuvant therapy [87, 91] [86], unless a concurrent yolk sac component exists [94, 95].

Gastric choriocarcinomas are rare tumors, accounting for approximately 0.08 % of all gastric cancers [105]. They are usually associated with a gastric adenocarcinoma (70 %) [105], sometimes with a transition between both components. Choriocarcinoma arises from adenocarcinoma as a phenomenon of tumor heterogeneity, as shown by the presence of identical *TP53* mutations and similar genomic imbalances in both gastric adenocarcinomas and coexisting choriocarcinoma [106, 107]. The existence of pure forms could be explained by an overgrowth of the choriocarcinomatous area [108]. Heterogeneity is reflected by cases of pure gastric adenocarcinoma with choriocarcinomatous metastases. Moreover, in examples of gastric adenocarcinoma with choriocarcinomatous areas, a divergent pattern of metastases of both components has been observed [108]. Other differentiated patterns of GCT such as YST [106], hepatoid [109], or neuroendocrine components [110] can occur in association with gastric choriocarcinoma.

Gastric choriocarcinomas occur in the same age groups as adenocarcinoma and present as grossly bulky masses [108] with a necrotic and hemorrhagic appearance, an average size of 7 cm (range 2–18 cm), and a location similar to gastric adenocarcinoma [105]. Clinical symptoms include gynecomastia, precocious puberty, and, especially, gastrointestinal bleeding [108]. Endoscopic biopsies are rarely diagnostic due to the necrotic and hemorrhagic quality of these tumors [105, 111]. Although gastric choriocarcinomas show elevated serum β-hCG, the diagnosis of choriocarcinoma must be histopathological, since raised β-hCG has also been observed in otherwise conventional gastric adenocarcinomas without choriocarcinomatous features [112].

Gastric choriocarcinomas usually infiltrate beyond the muscularis propria; thus, they are frequently diagnosed at an advance stage with the presence of metastases to the lymph nodes (68–87 %), liver (45–66 %), peritoneum (15–23 %), or lung (8–28.3 %) [105, 111]. Hepatic failure due to tumor metastasis (29 %) is considered the most frequent cause of death in these patients [105]; gastric choriocarcinomas have a poor prognosis with most deaths occurring within 6 months of diagnosis [105, 108].

YST and AFP-producing gastric carcinomas have been found in association with fetal gastrointestinal, fetal hepatic, and yolk sac tumor differentiation [113, 114]. Gastric adenocarcinoma histology overlaps with both endodermal primitive and somatic YST growth patterns such as hepatoid and glandular gastrointestinal [4]. Additionally, some AFP-producing gastric carcinomas coexpress glypican-3 as well as pluripotency/stem cell factors such as SALL4 [115], thus revealing a YST immunophenotype [116]. This suggests that endodermal differentiation plays a major role in these tumors through reprogramming and later transdifferentiation [117]. AFP-producing gastric carcinomas account for 2.7–5.4 % of all carcinomas of the stomach [115], occurring most often in men, with an average age of 61. They have the same topography as conventional gastric adenocarcinoma [118], but they have been also reported in association with Barrett's esophagus [119].

As previously mentioned, AFP-producing gastric carcinomas present a variable combination of histological patterns of endodermal differentiation. Clear cell areas are considered to represent a differentiation into fetal gut (Fig. 11.2) [115]. Hepatoid differentiation is the most characteristic pattern [120] and can represent more than 50 % of the tumor [114]. Hepatoid areas express AFP (80 %), glypican-3 (56 %), HepPar-1 (69 %), SALL4 (47 %), as well as palate, lung, and nasal epithelium protein (16 %), neuroendocrine markers, polyclonal CEA and have a reduced CDX2 expression [114, 120–122]. Thus, hepatoid AFP-producing clear cell gastric carcinomas mimic hepatocellular carcinoma and combined hepatocellular-cholangiocarcinoma metastases, especially when accompanied by a conventional gastric adenocarcinoma component. Although SALL4 was initially considered a helpful marker in differentiation, due to its expression in the former and its absence in the latter [114, 115], recent studies refute this idea, relating the presence of SALL4 simply to stem cell pluripotentiality features and poor prognosis [123] (see Sect. 11.7).

Hepatoid AFP-producing gastric carcinomas are aggressive tumors with a short median overall survival time (6 months) and low 3-year survival rate (22.6 %) [122]. Indeed, a morphological analysis of AFP-producing gastric car-

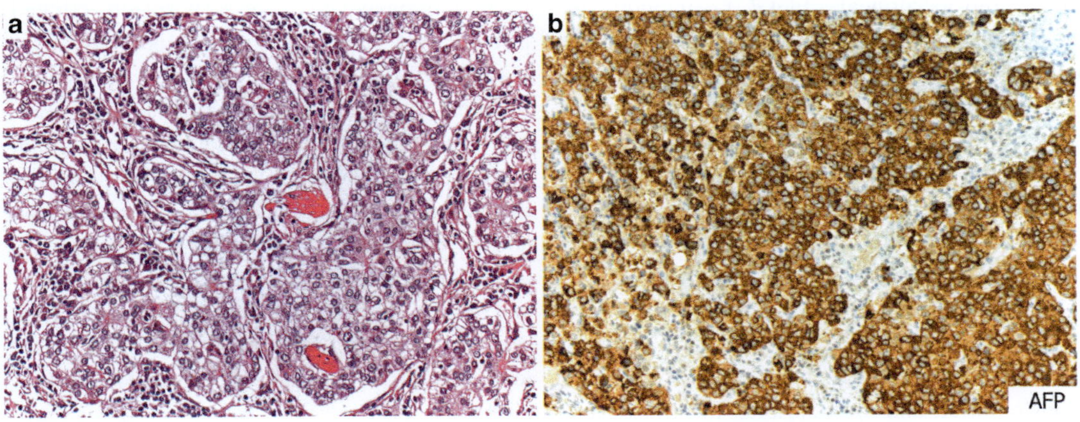

Fig. 11.2 (**a**) Clear cell adenocarcinoma of the stomach showing characteristic polyhedral vacuolated cells intensely positive for AFP (**b**)

cinomas revealed that although these tumors usually contain a variable proportion of conventional adenocarcinomas, they also have glandular YST and hepatoid areas that seem related to a capacity for submucosal invasion and a more advanced stage [118]. Pure gastric YST is extremely rare; invariably it has been described in combination with adenocarcinoma [106, 124–127]. Clinically, AFP-producing gastric carcinomas are aggressive neoplasms with high rates of lymphatic (72.2 %) and venous invasion (88.9 %), lymph node metastases (72.2 %), and synchronous or metachronous liver metastases (63.9–75.6 %) [118, 121]. In our experience, metastases to the ovary, especially when they are unilateral, can prompt a misdiagnosis of primary ovarian YST.

Histogenetically, these tumors belong to type VI in the Oosterhuis and Looijenga classification of pluripotential tumors.

11.4 Retroperitoneum

Although the retroperitoneum has been classically considered a location for EGGCTs (1–6.4 %) [128, 129], including teratomas [128], choriocarcinomas [130] YST [129], and seminomas [131], probably most retroperitoneal GCTs in postpubertal patients are metastases from primary gonadal lesions, rather than primaries [131]. Diagnostic criteria for rare primary retroperitoneal EGGCT include an encapsulated neoplasm without lymph node involvement or a high retroperitoneal neoplasm with adjacent lymph node involvement but no tumor in the lower aortic, iliac, or pelvic lymph nodes [132]. Testicular GCTs metastatic to retroperitoneum are covered in Chaps. 7 and 10.

11.5 Spleen

Only rare examples of primary splenic GCT have been reported and are restricted to clinical and radiological case reports with incomplete histology [133, 134]. Splenogonadal fusion may be associated with concomitant GCT such as embryonal carcinoma and seminoma [135].

11.6 Pancreas

The retroperitoneal location of the head, neck, and body of the pancreas can be a source of confusion in the diagnosis of primary pancreatic EGGCTs, and a possible infiltration by a retroperitoneal EGGCT into the pancreas must always be ruled out. However, rare pancreatic EGGCTs have been reported to date; almost all are teratomas [136–140].

Pancreatic teratomas have been described predominantly as cystic masses in the head, neck, body, tail, or uncinate process in patients with an age range from 4 months to 74 years [137, 138]. Male to female ratio is 1.5:1 and most tumors are larger than 2 cm [138]. Pancreatic teratomas are usually symptomatic, causing nausea, vomiting, anorexia, weight loss, fatigue, fever, and abdominal and back pain [137, 141]. Histologically, most teratomas reported are cystic lesions with a squamous lining epithelium, sebaceous structures, and lymphoid and connective tissue, although other combinations of ectodermal, endodermal, and mesodermal elements have been described, including an immature teratoma [142].

Teratomas should be distinguished from a broad spectrum of pancreatic cystic lesions such as lymphoepithelial and epidermoid cysts arising in an intrapancreatic accessory spleen, ciliated foregut cyst [143], and mucinous cystic neoplasms [138].

Three cases of choriocarcinoma with seemingly incomplete clinicopathological evaluation have been reported [144–146]. Additionally, three cases of pancreatic YST have been reported and were radiological studied with minimal histological detail. Two were aggressive with hepatic and lymph node metastases and elevated AFP serum levels. The third was a mixed GCT with a teratomatous component in a pediatric patient [136, 139, 147].

Only metastases of seminoma to the pancreas have been reported [148].

11.7 Liver

Primary teratomas of the liver account for less than 1 % of hepatic tumors, with approximately 30 reported cases [149, 150]. They occur in children, the majority females below the age of 3 [149]. Hepatic teratomas are usually incidental findings [149] and often located in the falciform ligament [151], where they may become as large as 20 cm [150] and cause florid symptomatic cholangitis [152]. Teratomatous tissues may be associated with aggressive neuroendocrine tumors [153] and with a small proportion of hepatoblastomas [154], and some are chemotherapy related [155]. They are variably intermixed or in abrupt transition with hepatoblastoma tissue [150, 156, 157]. However, the presence of mesenchymal components such as osteoid, mesoblastic-like myxoid areas, smooth muscle cells, and squamous epithelium alone does not justify a diagnosis of teratoid hepatoblastoma [155]. The immunohistochemical expression of AFP, glypican-3, HepPar-1, and SALL4 [158] in hepatoblastomas is shared with YST, pointing toward an overlapping histology and phenotype among embryonal endodermal tumors.

Hepatic choriocarcinomas have been reported in both adults and children [159, 160]. Clinically, choriocarcinoma syndrome is defined by a massive spontaneous hemorrhage of choriocarcinomatous nodules [161] in visceral extragonadal locations such as the liver and lung; rupture may occur which could cause a potentially lethal hemoperitoneum [162] related to the invasion of small blood vessels [71]. In such cases, diagnosis of hepatic choriocarcinoma is difficult; there is a substantial risk of hemorrhage if a needle biopsy is performed; furthermore, this may yield only blood and necrotic tissue.

YST, hepatocellular carcinomas, and hepatoid adenocarcinomas**.** Only 20 cases of primary hepatic YST have been reported [163, 164]; this is not surprising as it is a challenging diagnosis due to the remarkable similarities of YST and pure or predominant endodermal somatic patterns (i.e., hepatoid, glandular) [4] and hepatocellular carcinoma, sharing many histological features, such as a trabecular pattern, hematopoiesis, hyaline globules, and bile secretion. Moreover, pure or mixed hepatoid gastric carcinomas, which frequently metastasize to the liver, sometimes have a hepatoid component (see Sect. 11.3) which is indistinguishable from hepatoid YST. Immunohistochemically, these lesions coexpress markers such as AFP, glypican-3 [116], and HepPar-1 [165]. Although SALL4 was initially described as a marker of fetal gut differentiation, thus supporting the diagnosis of YST and hepatoid gastric adenocarcinoma versus hepatocellular carcinoma [115], later studies have shown conspicuous SALL4 expression in the latter, associated with a poor prognosis [123, 165].

In conclusion, some forms of hepatoblastoma, hepatoid gastric adenocarcinoma, hepatocellular carcinoma, and metastatic hepatoid gastric adenocarcinoma constitute similar entities sharing a common endodermal differentiated phenotype identical to differentiated, somatic YST.

11.8 Adrenal Gland

Approximately 20 primary teratomas have been reported in the adrenal gland. They account for 0.7–3 % of primary adrenal tumors [166–168] and are more frequent in infants or children, occurring only rarely in adults. The age of presentation ranges from 2 to 62 years, with a median age of 31 [168, 169]. They are more often located on the right side [170], and the male to female ratio is approximately 1:2 [169]. They are usually asymptomatic, although they may cause abdominal distension, intestinal obstruction, abdominal or low back pain [171], or torsion. If rupture occurs, a leakage to the abdominal cavity is possible, causing a granulomatous reaction with adhesions, hemorrhage, and finally shock [169]. Primary adrenal teratomas have been described mostly as cystic voluminous masses with a maximum size of 38 cm [169], with no reports of any immature variants. Since some teratomas may have an extensive adipose tissue component, the differential diagnosis should include other adrenal lipomatous tumors such as myelolipoma, lipoma, and angiomyolipoma [166–168]. As in the pancreas, it is necessary to discard a primary retroperitoneal origin for these lesions.

No cases of YST, seminoma, or other malignant GCT have been reported in the adrenal gland.

11.9 Kidney

Around 30 cases of renal teratomas have been reported [172–182]. In a recent review, they comprised 0.19 % of all kidney tumors [179]. The age of presentation spans from 6 weeks to 71 years, with no gender predilection, and tumor sizes range from 3.5 to 12 cm. As in other locations, renal teratomas are occasionally related to developmental disorders such as renal dysplasia [175, 180] and ectopic [174] or horseshoe kidney [172, 173, 178, 181].

Similarly to the testis and ovary [183], at least ten examples of carcinoid tumor associated to teratoma have been reported in the kidney [172, 173, 176, 181], with a concurrent horseshoe kidney in some [172, 173, 181] and one synchronous clear cell renal cell carcinoma [181]. Clinically, retroperitoneal teratomas with secondary kidney invasion may be mistaken for renal primaries.

Primary immature teratomas should be differentiated from teratoid Wilms tumor. This nephroblastoma variant shows a predominance of mature heterologous elements composed of adipose, epithelial tissue (Fig. 11.3), skeletal muscle, and cartilage, rarely with neural differentiation [184]. They should be differentiated from ovarian immature teratomas with metanephric areas.

There are some reports of parenchymal renal choriocarcinomas [185, 186] that may have a gestational origin, as demonstrated by the presence of paternal DNA [290]. Clear cell and urothelial carcinomas with trophoblastic components may occur [187, 188].

Only four instances of primary renal YST have been reported [177, 187, 189, 190], one showing venous invasion [190]. Urothelial tumors of the pelvis may present a prominent glandular YST component that may adopt a papillary pattern reminiscent of papillary renal carcinoma (Fig. 11.4) [187], representing type VI tumors. No seminomas have been reported in the kidney.

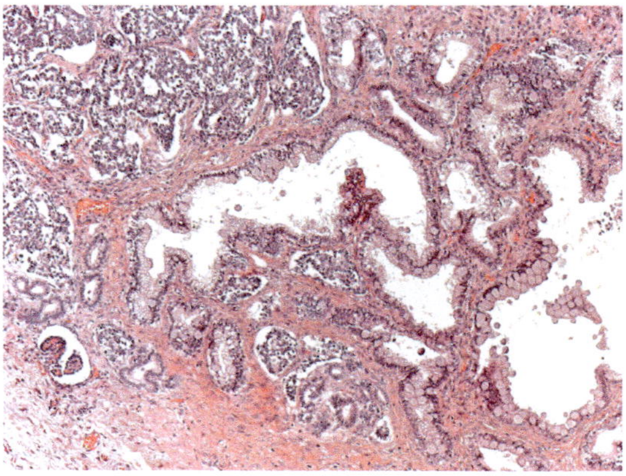

Fig. 11.3 Wilms tumor elements coexisting in this field with intestinal type mucinous epithelium. Elsewhere, numerous neural rosettes and cartilage were present

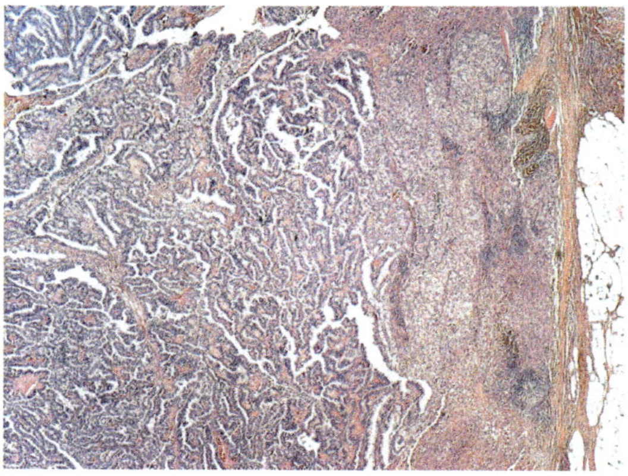

Fig. 11.4 Papillary glandular YST in the renal pelvis simulate papillary renal cell adenocarcinoma

11.10 Prostate, Seminal Vesicles and Serosal Cavities

The prostate is a rare location for EGGCT where characteristic gonadal type II GCT, are found, including seminoma, mixed GCT, YST,

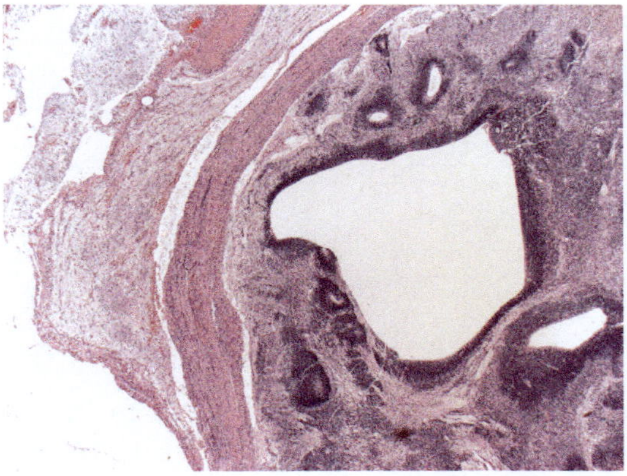

Fig. 11.5 Pericardial immature teratoma in an infant with marked neural differentiation

and teratoma. They probably originate from primordial germ cells trapped on their migratory path in the developing prostate. This possibility is also true for serosal GCTs, including pericardium (Fig. 11.5) [191, 192], peritoneum, omentum, etc. [193–196]. In teratomas of peritoneal locations, the possibility of an ovarian parasitic mature cystic teratoma must be excluded [197].

Fifteen prostatic GCTs have been reported to date [198, 199] presenting with a median age of 40.8 years. They are usually large masses (5–10 cm) that occupy the totality of the prostate gland and frequently invade adjacent structures (bladder neck, seminal vesicle). Clinical symptoms include hematuria, dysuria, obstructive symptoms, hematospermia, and pelvic pain. A curious association with Klinefelter syndrome (also see Chap. 3), which is more frequently related to mediastinal GCTs (see Chap. 8), has been reported [199]. Indeed, according to some authors, a diagnosis of EGGCT in a young male patient should be followed by a clinical/genetic work-up for Klinefelter syndrome [154]. Histologically, the reported cases are pure seminomas, YST, teratomas, and mixed GCT, one of which with embryonal carcinoma. A further case was associated with angiosarcoma as a somatic malignant change [198]. Two seminomas were initially misdiagnosed as prostatic adenocarcinoma and rhabdomyosarcoma [200]. Prostatic invasion from testicular GCT is rare event [201–203].

11.11 Urinary Bladder

Teratomas also occur in the urinary bladder with at least ten cases reported [204]. They are sometimes associated with a rectovesical fistula with hypospadias and bladder diverticulum. Clinical symptoms may include the alarming occurrence of pilimiction (trichiura), that is, passing hairs in the urine [205], more frequently due to an ovarian benign cystic teratoma fistula into the bladder.

As previously noted, the reported cases of primary choriocarcinoma of the urinary bladder [206–208] are examples of the much more common diffuse or focal trophoblastic differentiation in an otherwise usual urothelial carcinoma, which may present elevated serum and urine levels of hCG. A choriocarcinomatous tumor of the bladder showed an excess of copies in 12p on molecular analysis, being interpreted as a sign of i(12p) mutation and thus justifying a diagnosis of primary vesical choriocarcinoma [208]. However, a recent genomic study highlighted similar 12p gains in urothelial carcinomas [209].

Two vesical YST, excluding those in the urachus, have been reported, one in a 1-year-old patient [210] and the other, which had an unusual presentation, in a 31-year-old female [211]. We have seen one instance of urothelial carcinoma associated with glandular YST (Figs 11.6a–c). In these cases, the glandular component should be differentiated from adenocarcinoma. When associated with urothelial carcinoma, they represent type VI tumors.

Hepatoid carcinomas have been reported in the bladder, and they should be differentiated from hepatoid YST [212].

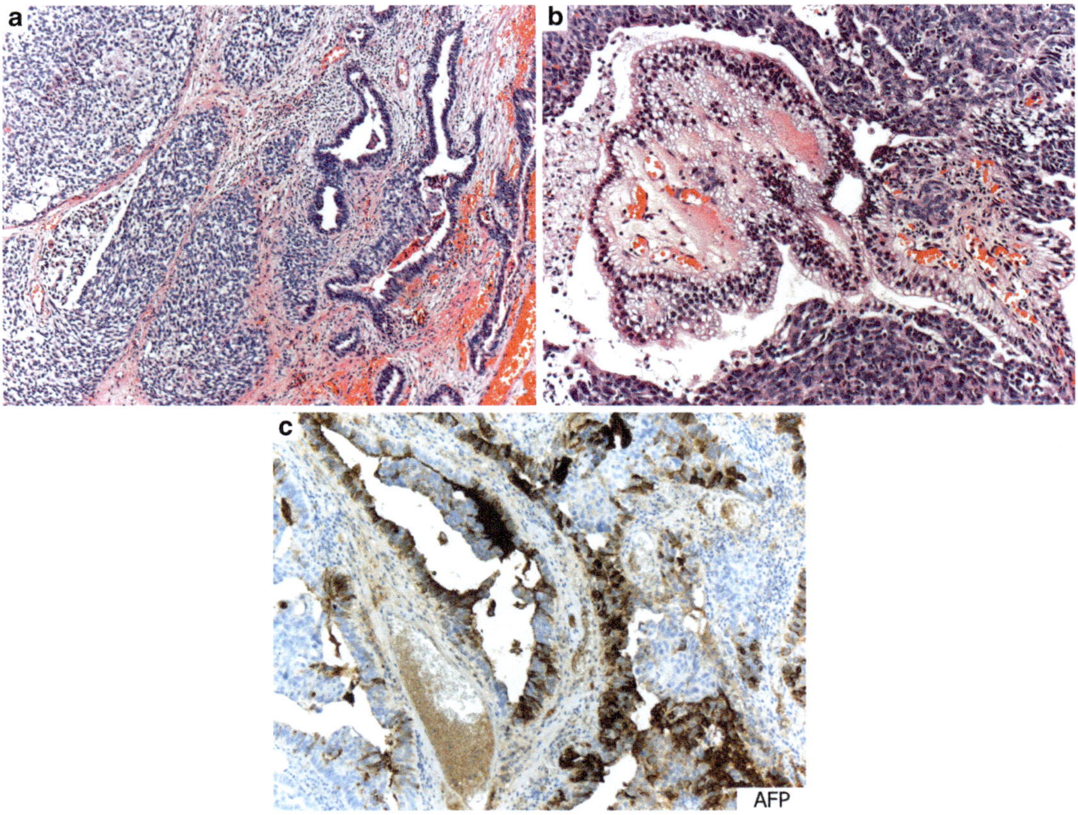

Fig. 11.6 (**a**) High-grade urothelial carcinoma with branching glandular formations of YST. (**b**) Interface between high-grade urothelial carcinoma and glandular YST with characteristic vacuolation. Glandular areas overexpress AFP (**c**)

11.12 Urachus

The urachus is a 5–6 cm vestigial remnant of the human allantois. The latter is an endodermal evagination of the developing hindgut which becomes surrounded by the mesodermal connecting stalk; thus, the urachus may represent a structure retaining some degree of pluripotency. Nine GCTs have been reported in the urachus: three YST and six teratomas [213–216].

The six reported cases of teratoma affected females between 8 and 53 (median age, 26 years), presenting as urachal cysts or sinuses, with a benign histology [216]. Interestingly, one of the patients presented an associated anti-N-methyl-D-aspartate receptor autoimmune encephalitis, a clinical complication often reported in association with teratomas [217, 218].

YST of the urachus presented as large infraumbilical masses (12–19 cm) in two children (7 months and 2 years) [213, 214] and in a 44-year-old woman [215], with a connection to the urinary bladder dome by a pedicle.

11.13 Other Miscellaneous Sites

Teratomas have been reported in the anorectal region [219], vermiform appendix [220], biliary tree [221], gallbladder [222, 223], bone [224], small bowel [225], and large bowel [226, 227], including a fetiform teratoma associated with the intestinal duplication [228], penis [229], and skin [230, 231].

Several cases of carcinoma with choriocarcinomatous features have been described in other locations such as the breast [232], skin [233], gallbladder [234], jejunum [235], colon [236], or rectum [237].

Apparent primary YST have been reported in the biliary tree [238] and penis [239].

Acknowledgment Dr. A. Roden, Mayo Clinic, Rochester, USA, Provided Fig. 11.1

References

1. Mamsen LS, Brochner CB, Byskov AG, Mollgard K. The migration and loss of human primordial germ stem cells from the hind gut epithelium towards the gonadal ridge. Int J Dev Biol. 2012;56(10–12):771–8. doi:10.1387/ijdb.120202lm.
2. McLaren A. Germ cells and germ cell sex. Philos Trans R Soc Lond B Biol Sci. 1995;350(1333):229–33. doi:10.1098/rstb.1995.0156.
3. Upadhyay S, Zamboni L. Ectopic germ cells: natural model for the study of germ cell sexual differentiation. Proc Natl Acad Sci U S A. 1982;79(21):6584–8.
4. Nogales FF, Preda O, Nicolae A. Yolk sac tumours revisited. A review of their many faces and names. Histopathology. 2012;60(7):1023–33. doi:10.1111/j.1365-2559.2011.03889.x.
5. Gonzalez-Crussi F. Extragonadal teratomas. Atlas of tumor pathology. 2nd Series. Washington, DC: Armed Forces Institute of Pathology; 1982.
6. Arora RS, Alston RD, Eden TO, Geraci M, Birch JM. Comparative incidence patterns and trends of gonadal and extragonadal germ cell tumors in England, 1979 to 2003. Cancer. 2012;118(17):4290–7. doi:10.1002/cncr.27403.
7. Lack EE. Extragonadal germ cell tumors of the head and neck region: review of 16 cases. Hum Pathol. 1985;16(1):56–64.
8. Bonavolonta G, Strianese D, Grassi P, Comune C, Tranfa F, Uccello G, et al. An analysis of 2,480 space-occupying lesions of the orbit from 1976 to 2011. Ophthal Plast Reconstr Surg. 2013;29(2):79–86. doi:10.1097/IOP.0b013e31827a7622.
9. Holmes T. Congenital tumor removed from the orbit. Trans Pathol Soc London. 1863;14:248.
10. Singh M, Singh U, Gupta A, Zadeng Z. Primary orbital teratoma with tooth in an adult: a rare association with cataract and corectopia. Orbit. 2013;32(5):327–9. doi:10.3109/01676830.2013.814682.
11. Levin ML, Leone Jr CR, Kincaid MC. Congenital orbital teratomas. Am J Ophthalmol. 1986;102(4):476–81.
12. Chawla B, Chauhan K, Kashyap S. Mature orbital teratoma with an ectopic tooth and primary anophthalmos. Orbit. 2013;32(1):67–9. doi:10.3109/01676830.2012.747212.
13. Sesenna E, Ferri A, Thai E, Magri AS. Huge orbital teratoma with intracranial extension: a case report. J Pediatr Surg. 2010;45(5):e27–31. doi:10.1016/j.jpedsurg.2010.01.031.
14. Mahesh L, Krishnakumar S, Subramanian N, Babu K, Biswas J. Malignant teratoma of the orbit: a clinicopathological study of a case. Orbit. 2003;22(4):305–9.
15. Wilson DJ, Dailey RA, Wobig JL, Dimmig JW. Neuroblastoma within a congenital orbital teratoma. Arch Ophthalmol. 2002;120(2):213–5.
16. Katz N, Ruymann F, Margo C, Popejoy L, LaPiana F. Endodermal sinus tumor (yolk-sac carcinoma) of the orbit. J Pediat Ophthalmol Strabismus. 1981;19(5):270–4.
17. Margo CE, Folberg R, Zimmerman LE, Sesterhenn IA. Endodermal sinus tumor (yolk sac tumor) of the orbit. Ophthalmology. 1983;90(12):1426–32.
18. Bresters D, Zwaan CM, Veerman AJ, Leemans CR, Westerveld GJ, van der Linden JC. A three-year-old girl with a yolk sac tumor in the orbit/maxillary sinus. Med Pediatr Oncol. 2003;40(1):70–1. doi:10.1002/mpo.1375.
19. Lee J-C, Jung S-M, Chao A-S, Hsueh C. Congenital mixed malignant germ cell tumor involving cerebrum and orbit. J Perinat Med. 2003;31(3):261–5.
20. Dragan LR, Aghaian E, Vora R, Kim GE, Seiff SR. Orbital, middle cranial fossa, and pterygopalatine fossa yolk sac tumor in an infant. Ophthal Plast Reconstr Surg. 2004;20(6):469–71.
21. Mogaddam AS, Memar B, Aledavood A, Eslampoor A. Isolated orbital endodermal sinus tumor. Ophthal Plast Reconstr Surg. 2007;23(6):477–9.
22. Alkatan HM, Al-Kofide A, Al-Hussain H. Yolk sac tumor: histopathologic report of 2 cases. Can J Ophthalmol/J Can d'Ophtalmol. 2008;43(1):125–6.
23. Kiratli H, Balcı KE, Güler G. Primary orbital endodermal sinus tumor (yolk sac tumor). J AAPOS. 2008;12(6):623–5.
24. Gnanaraj L, Skibell BC, Coret-Simon J, Halliday W, Forrest C, DeAngelis DD. Massive congenital orbital teratoma. Ophthal Plast Reconstr Surg. 2005;21(6):445–7.
25. Nogales Jr FF, Gasca L, Llamas R, Blanco A, Sanz JF. Case Report: epignathus-clinical, radiologic, and pathologic considerations. Int J Gynaecol Obstet. 1977;15(1):41–3.
26. Firat C, Aytekin AH, Akatli AN, Karadag A, Samdanci E. Surgical management of immature teratoma involving the oral cavity and orbit in a neonate. J Craniofac Surg. 2014;25(6):e578–80. doi:10.1097/SCS.0000000000001171.
27. Misra P, Husain Q, Svider PF, Sanghvi S, Liu JK, Eloy JA. Management of sinonasal teratocarcinosarcoma: a systematic review. Am J Otolaryngol. 2014;35(1):5–11. doi:10.1016/j.amjoto.2013.04.010.
28. Carrizo F, Pineda-Daboin K, Neto AG, Luna MA. Pharyngeal teratocarcinosarcoma: review of the literature and report of two cases. Ann Diagn Pathol. 2006;10(6):339–42. doi:10.1016/j.anndiagpath.2006.03.006.
29. Heffner DK, Hyams VJ. Teratocarcinosarcoma (malignant teratoma?) of the nasal cavity and paranasal sinuses a clinicopathologic study of 20 cases. Cancer. 1984;53(10):2140–54.
30. Thomas J, Adegboyega P, Iloabachie K, Mooring JW, Lian T. Sinonasal teratocarcinosarcoma with

yolk sac elements: a neoplasm of somatic or germ cell origin? Ann Diagn Pathol. 2011;15(2):135–9. doi:10.1016/j.anndiagpath.2010.01.004.
31. Vranic S, Caughron SK, Djuricic S, Bilalovic N, Zaman S, Suljevic I, et al. Hamartomas, teratomas and teratocarcinosarcomas of the head and neck: report of 3 new cases with clinico-pathologic correlation, cytogenetic analysis, and review of the literature. BMC Ear Nose Throat Disord. 2008;8:8. doi:10.1186/1472-6815-8-8.
32. Kimler SC, Muth WF. Primary malignant teratoma of the thyroid. Case report and literature review of cervical teratomas in adults. Cancer. 1978;42(1):311–7.
33. Riedlinger WFJ, Lack EE, Robson CD, Rahbar R, Nosé V. Primary thyroid teratomas in children: a report of 11 cases with a proposal of criteria for their diagnosis. Am J Surg Pathol. 2005;29(5):700–6.
34. Thompson LDR, Rosai J, Heffess CS. Primary thyroid teratomas. Cancer. 2000;88(5):1149–58.
35. Pérez-Mies B, Zapata RMR, García-Fernández E, Serrano MN. Malignant teratoma of the thyroid in a pregnant woman. Ann Diagn Pathol. 2010;14(4):264–7.
36. Ueno NT, Amato RJ, Ro JJ, Weber RS. Primary malignant teratoma of the thyroid gland: report and discussion of two cases. Head Neck. 1998;20(7):649–53.
37. Tsang RW, Brierley JD, Asa SL, Sturgeon JF. Malignant teratoma of the thyroid: Aggressive chemoradiation therapy is required after surgery. Thyroid. 2003;13(4):401–4.
38. Akoojee SB. Teratoma of the thyroid gland. S Afr Med J. 1980;57(3):93–4.
39. Furtado LV, Leventaki V, Layfield LJ, Lowichik A, Muntz HR, Pysher TJ. Yolk sac tumor of the thyroid gland: a case report. Pediatr Dev Pathol. 2011;14(6):475–9.
40. Wierzbicka-Chmiel J, Chrószcz M, Słomian G, Kajdaniuk D, Zajęcki W, Borgiel-Marek H, et al. Mixed germ cells tumour primarily located in the thyroid-a case report. Endokrynol Pol. 2012;63(5):388–90.
41. Shadid EA, O'neil E, RT G. Benign teratoid tumor of the parotid: Case Report. Plast Reconstr Surg. 1975;55(3):363–5.
42. Rose P, Howard E. Congenital teratoma of the submandibular gland. J Pediatr Surg. 1982;17(4):414–6.
43. Choi E-C, Jin J-B, Kim J-Y, Hong W-P, Kim M-J, Park Y-K. Dermoid cyst of the parotid gland. Yonsei Med J. 1988;29(2):199–203.
44. McKiernan D, Koay B, Vinayak B, Gould S, Freeland A. A case of submandibular teratoma. J Laryngol Otol. 1995;109(10):992–4.
45. Pirodda A, Ferri GG, Truzzi M, Cavicchi O. Benign cystic teratoma of the parotid gland. Otolaryngol Head Neck Surg. 2001;125(4):429–30.
46. Ohta M, Imamura Y, Mori M, Maegawa H, Kojima A, Fujieda S. Benign cystic teratoma of the parotid gland. Acta Cytol. 2009;53(4):427–30.
47. Viva E, Zorzi F, Annibale G, Stefini S, Baronchelli C, Bonetti MF. Endodermal sinus (yolk sac) tumor of the parotid gland: a case report. Int J Pediatr Otorhinolaryngol. 1992;24(3):269–74.
48. Sredni ST, da Cunha IW, de Carvalho Filho NP, Magrin J, Pinto CAL, Lopes LF. Endodermal sinus tumor of the parotid gland in a child. Pediatr Dev Pathol. 2004;7(1):77–80.
49. Rana SS, Swami N, Mehta S, Singh J, Biswal S. Intrapulmonary teratoma: an exceptional disease. Ann Thorac Surg. 2007;83(3):1194–6. doi:10.1016/j.athoracsur.2006.07.072.
50. Cai C, Zeng Y, Chen H, Gu Y, Zeng Q, Zhong N. Fibrobronchoscopic evidence of endobronchial hairs in intrapulmonary teratoma with hemoptysis but without trichoptysis. Am J Med Sci. 2008;336(5):441–4. doi:10.1097/MAJ.0b013e318163a313.
51. Siddiqui FA, Jain A, Maheshwari V, Beg MH. FNA diagnosis of teratoma lung: A case report. Diagn Cytopathol. 2010;38(10):758–60. doi:10.1002/dc.21318.
52. Dar RA, Mushtaque M, Wani SH, Malik RA. Giant intrapulmonary teratoma: a rare case. Case Rep Pulmonol. 2011;2011:298653. doi:10.1155/2011/298653.
53. Giunchi F, Segura JJ. Primary malignant teratoma of lung: report of a case and review of the literature. Int J Surg Pathol. 2012;20(5):523–7. doi:10.1177/1066896911431454.
54. Sawant AC, Kandra A, Narra SR. Intrapulmonary cystic teratoma mimicking malignant pulmonary neoplasm. BMJ Case Rep. 2012; doi:10.1136/bcr.02.2012.5770.
55. Kang HS, Lee HY, Kang HH, Park CK, Lee SH, Moon HS. Intrapulmonary teratoma presenting with trichoptysis. J Thorac Oncol. 2013;8(1):126–7. doi:10.1097/JTO.0b013e3182762d07.
56. Barman S, Mandal KC, Kumar R, Biswas SK, Mukhopadhyay M, Mukhopadhyay B. Congenital cystic lesions of lung in the paediatric population: a 5-year single institutional study with review of literature. Afr J Paediatr Surg. 2015;12(1):66–70. doi:10.4103/0189-6725.150987.
57. Iwasaki T, Iuchi K, Matsumura A, Sueki H, Yamamoto S, Mori T. Intrapulmonary mature teratoma. Jpn J Thorac Cardiovasc Surg. 2000;48(7):468–72.
58. Jamieson MP, McGowan AR. Endobronchial teratoma. Thorax. 1982;37(2):157–9.
59. Smith CJ. A teratoma of the lung containing teeth. Ann R Coll Surg Engl. 1967;41(5):413–22.
60. Steier KJ. Benign cystic teratoma of the lung. Postgrad Med. 1988;83(4):85–6. 91
61. Agarwal R, Srinivas R, Saxena AK. Trichoptysis due to an intrapulmonary teratoma. Respir Care. 2007;52(12):1779–81.
62. Khan JA, Aslam F, Fatimi SH, Ahmed R. Cough, fever and a cavitary lung lesion--an intrapulmonary teratoma. J Postgrad Med. 2005;51(4):330–1.
63. Joo M, Kang YK, Lee HK, Lee HS, Yum HK, Bang SW, et al. Intrapulmonary and gastric

teratoma : report of two cases. J Korean Med Sci. 1999;14(3):330–4.
64. Kakkar N, Vasishta RK, Banerjee AK, Garewal G, Deodhar SD, Bambery P. Primary pulmonary malignant teratoma with yolk sac element associated with hematologic neoplasia. Respiration. 1996;63(1):52–4.
65. Zenker D, Aleksic I. Intrapulmonary cystic benign teratoma: a case report and review of the literature. Ann Thorac Cardiovasc Surg. 2004;10(5):290–2.
66. Asano S, Hoshikawa Y, Yamane Y, Ikeda M, Wakasa H. An intrapulmonary teratoma associated with bronchiectasia containing various kinds of primordium: a case report and review of the literature. Virchows Arch. 2000;436(4):384–8.
67. Serno J, Zeppernick F, Jakel J, Schrading S, Maass N, Meinhold-Heerlein I, et al. Primary pulmonary choriocarcinoma: case report and review of the literature. Gynecol Obstet Invest. 2012;74(2):171–6. doi:10.1159/000336784.
68. Di Crescenzo V, Laperuta P, Napolitano F, Carlomagno C, Garzi A, Vitale M. An unusual case of primary choriocarcinoma of the lung. BMC Surg. 2013;13(Suppl 2):S33. doi:10.1186/1471-2482-13-S2-S33.
69. Gvinianidze L, Panagiotopoulos N, Woo WL, Borg E, Lawrence D. The challenging management of lung choriocarcinoma. J Thorac Dis. 2014;6(10):E220–2. doi:10.3978/j.issn.2072-1439.2014.09.18.
70. Corpa Rodriguez ME, Fernandez Lahera J, Guadalajara Labajo H, Vazquez Pelillo JC, de Serrano M NM, Garcia Sanchez-Giron J. Choriocarcinoma of the lung. Arch Bronconeumol. 2009;45(3):153–5. doi:10.1016/j.arbres.2008.02.007.
71. Shintaku M, Hwang MH, Amitani R. Primary choriocarcinoma of the lung manifesting as diffuse alveolar hemorrhage. Arch Pathol Lab Med. 2006;130(4):540–3. doi:10.1043/1543-2165(2006)130[540:PCOTLM]2.0.CO;2.
72. Chen F, Tatsumi A, Numoto S. Combined choriocarcinoma and adenocarcinoma of the lung occurring in a man: case report and review of the literature. Cancer. 2001;91(1):123–9.
73. Hadgu A, Tindni A, Panda M. Primary pulmonary choriocarcinoma in a male. BMJ Case Rep. 2010; doi:10.1136/bcr.02.2010.2712.
74. Ikura Y, Inoue T, Tsukuda H, Yamamoto T, Ueda M, Kobayashi Y. Primary choriocarcinoma and human chorionic gonadotrophin-producing giant cell carcinoma of the lung: are they independent entities? Histopathology. 2000;36(1):17–25.
75. Vegh GL, Szigetvari I, Soltesz I, Major K, Batorfi J, Dancso J, et al. Primary pulmonary choriocarcinoma: a case report. J Reprod Med. 2008;53(5):369–72.
76. Berthod G, Bouzourene H, Pachinger C, Peters S. Solitary choriocarcinoma in the lung. J Thorac Oncol. 2010;5(4):574–5. doi:10.1097/JTO.0b013e3181cbf372.
77. Kradin RL, Young RH, Dickersin GR, Kirkham SE, Mark EJ. Pulmonary blastoma with argyrophil cells and lacking sarcomatous features (pulmonary endodermal tumor resembling fetal lung). Am J Surg Pathol. 1982;6(2):165–72.
78. Mardini G, Pai U, Chavez AM, Tomashefski Jr JF. Endobronchial adenocarcinoma with endometrioid features and prominent neuroendocrine differentiation. A variant of fetal adenocarcinoma. Cancer. 1994;73(5):1383–9.
79. Arora S, Patel H, Mir P, Mishra S, Kalra A, Sawhney H, et al. A bilateral primary yolk sac tumor of the lung associated with chromosome 3 polysomy: understanding the Oct 3/4 and Sox 2 interaction. Cancer Biol Ther. 2013;14(1):1–4. doi:10.4161/cbt.22624.
80. Pelosi G, Petrella F, Sandri MT, Spaggiari L, Galetta D, Viale G. A primary pure yolk sac tumor of the lung exhibiting CDX-2 immunoreactivity and increased serum levels of alkaline phosphatase intestinal isoenzyme. Int J Surg Pathol. 2006;14(3):247–51. doi:10.1177/1066896906290657.
81. Ramirez Gil ME, Martinez Vallina P, Menal Munoz P, Hernandez Ferrandez J, Munoz Gonzalez G, De Andres JJ R. Pulmonary pure yolk-sac tumor. A rare anatomopathological entity. Arch Bronconeumol. 2011;47(3):157–8. doi:10.1016/j.arbres.2010.03.007.
82. Cameselle-Teijeiro J, Alberte-Lista L, Chiarelli S, Buritica C, Goncalves L, Gonzalez-Campora R, et al. CD10 is a characteristic marker of tumours forming morules with biotin-rich, optically clear nuclei that occur in different organs. Histopathology. 2008;52(3):389–92. doi:10.1111/j.1365-2559.2007.02911.x.
83. Sekine S, Shibata T, Matsuno Y, Maeshima A, Ishii G, Sakamoto M, et al. Beta-catenin mutations in pulmonary blastomas: association with morule formation. J Pathol. 2003;200(2):214–21. doi:10.1002/path.1352.
84. Morita S, Yoshida A, Goto A, Ota S, Tsuta K, Yokozawa K, et al. High-grade lung adenocarcinoma with fetal lung-like morphology: clinicopathologic, immunohistochemical, and molecular analyses of 17 cases. Am J Surg Pathol. 2013;37(6):924–32. doi:10.1097/PAS.0b013e31827e1e83.
85. Suzuki M, Yazawa T, Ota S, Morimoto J, Yoshino I, Yamanaka S, et al. High-grade fetal adenocarcinoma of the lung is a tumour with a fetal phenotype that demonstrates diverse differentiation including high-grade neuroendocrine carcinoma: a clinicopathological, immunohistochemical and mutational study of 20 cases. Histopathology. 2015; doi:10.1111/his.12711.
86. Oviedo A, Chou P, Gonzalez-Crussi F. Pathologic quiz case. Abdominal mass in a neonate. Arch Pathol Lab Med. 2001;125(3):445–6. doi:10.1043/0003-9985(2001)125<0445:PQCAMI>2.0.CO;2.
87. Gupta DK, Srinivas M, Dave S, Agarwala S, Bajpai M, Mitra DK. Gastric teratoma in children. Pediatr Surg Int. 2000;16(5–6):329–32.

88. Sugandhi N, Gupta AK, Bhatnagar V. Gastric teratoma: unusual location and difficulties in diagnosis. Trop Gastroenterol. 2012;33(1):75–7.
89. Cairo MS, Grosfeld JL, Weetman RM. Gastric teratoma: unusual cause for bleeding of the upper gastrointestinal tract in the newborn. Pediatrics. 1981;67(5):721–4.
90. Singh S, Rawat J, Ahmed I. Immature extragastric teratoma of infancy: a rare tumour with review of the literature. BMJ Case Rep. 2011; doi:10.1136/bcr.12.2010.3674.
91. Park WH, Choi SO, Kim JI. Congenital gastric teratoma with gastric perforation mimicking meconium peritonitis. J Pediatr Surg. 2002;37(5):E11.
92. Tripathy BB, Saha K, Shukla RM, Mukhopadhyay M, Mukhopadhyay B. Gastric teratoma in a female infant. J Indian Med Assoc. 2010;108(10):694–6.
93. Matsukuma S, Wada R, Daibou M, Watanabe N, Kuwabara N, Abe H, et al. Adenocarcinoma arising from gastric immature teratoma. Report of a case in an adult and a review of the literature. Cancer. 1995;75(11):2663–8.
94. Ukiyama E, Endo M, Yoshida F, Tezuka T, Kudo K, Sato S, et al. Recurrent yolk sac tumor following resection of a neonatal immature gastric teratoma. Pediatr Surg Int. 2005;21(7):585–8. doi:10.1007/s00383-005-1404-y.
95. Bhattacharya NK, Bandyopadhyay R, Chatterjee U, Basu AK. Recurrence of immature gastric teratoma with peritoneal gliomatosis. Eur J Pediatr Surg. 2010;20(6):419–20. doi:10.1055/s-0030-1254108.
96. Mohta A, Sengar M, Neogi S, Khurana N. Gastric teratoma with predominant nephroblastic elements. Pediatr Surg Int. 2010;26(9):923–5. doi:10.1007/s00383-010-2670-x.
97. Hook S, Spicer R, Williams J, Grier D, Lowis S, Foot A, et al. Severe anemia in a 25-day-old infant due to gastric teratoma with focal neuroblastoma. Am J Perinatol. 2003;20(5):233–7. doi:10.1055/s-2003-42343.
98. Coulson WF. Peritoneal gliomatosis from a gastric teratoma. Am J Clin Pathol. 1990;94(1):87–9.
99. Yeo DM, Lim GY, Lee YS, Sohn DW, Chung JH. Gliomatosis peritonei of the scrotal sac associated with an immature gastric teratoma. Pediatr Radiol. 2010;40(7):1288–92. doi:10.1007/s00247-009-1506-0.
100. Govender D, Serra S, Chetty R. Collision tumor of the stomach: an unusual combination of adenocarcinoma and immature "neuroepitheliomatous" teratoma. Int J Surg Pathol. 2012;20(5):507–10. doi:10.1177/1066896911429941.
101. Sharma A, Arora R, Gupta R, Dinda AK. Immature gastric teratoma in an infant: report of a case and review of the literature. Indian J Pathol Microbiol. 2010;53(4):868–70. doi:10.4103/0377-4929.72018.
102. Valenzuela-Ramos MC, Mendizabal-Mendez AL, Rios-Contreras CA, Rodriguez-Montes CE. Pediatric gastric teratoma. J Radiol Case Rep. 2010;4(10):6–13. doi:10.3941/jrcr.v4i10.491.
103. Yoon SE, Goo HW, Jun S, Lee IC, Yoon CH. Immature gastric teratoma in an infant: a case report. Korean J Radiol. 2000;1(4):226–8.
104. Greco P, Cutrona D, Russo A, Grasso G. A case of gastric teratoma. J Pathol. 1979;129(3):121–3. doi:10.1002/path.1711290303.
105. Kobayashi A, Hasebe T, Endo Y, Sasaki S, Konishi M, Sugito M, et al. Primary gastric choriocarcinoma: two case reports and a pooled analysis of 53 cases. Gastric Cancer. 2005;8(3):178–85. doi:10.1007/s10120-005-0332-9.
106. Satake N, Chikakiyo M, Yagi T, Suzuki Y, Hirose T. Gastric cancer with choriocarcinoma and yolk sac tumor components: case report. Pathol Int. 2011;61(3):156–60. doi:10.1111/j.1440-1827.2010.02635.x.
107. Tate G, Suzuki T, Nemoto H, Kishimoto K, Hibi K, Mitsuya T. Allelic loss of the PTEN gene and mutation of the TP53 gene in choriocarcinoma arising from gastric adenocarcinoma: analysis of loss of heterozygosity in two male patients with extragonadal choriocarcinoma. Cancer Genet Cytogenet. 2009;193(2):104–8. doi:10.1016/j.cancergencyto.2009.04.005.
108. Liu Z, Mira JL, Cruz-Caudillo JC. Primary gastric choriocarcinoma: a case report and review of the literature. Arch Pathol Lab Med. 2001;125(12):1601–4. doi:10.1043/0003-9985(2001)125<1601:PGC>2.0.CO;2.
109. Motoyama T, Aizawa K, Fujiwara T, Endoh Y, Watanabe H. Coexistence of choriocarcinoma and hepatoid adenocarcinoma in the stomach. Pathol Int. 1994;44(9):716–21.
110. Hirano Y, Hara T, Nozawa H, Oyama K, Ohta N, Omura K, et al. Combined choriocarcinoma, neuroendocrine cell carcinoma and tubular adenocarcinoma in the stomach. World J Gastroenterol. 2008;14(20):3269–72.
111. Imai Y, Kawabe T, Takahashi M, Matsumura M, Komatsu Y, Hamada E, et al. A case of primary gastric choriocarcinoma and a review of the Japanese literature. J Gastroenterol. 1994;29(5):642–6.
112. Mori H, Soeda O, Kamano T, Tsunekawa K, Ueda N, Yoshida A, et al. Choriocarcinomatous change with immunocytochemically HCG-positive cells in the gastric carcinoma of the males. Virchows Arch A Pathol Anat Histol. 1982;396(2):141–53.
113. Bourreille J, Metayer P, Sauger F, Matray F, Fondimare A. Existence d'alpha foeto proteine au cours d'un cancer secondaire du foie d'origine gastrique. Presse Med. 1970;78(28):1277–8.
114. Osada M, Aishima S, Hirahashi M, Takizawa N, Takahashi S, Nakamura K, et al. Combination of hepatocellular markers is useful for prognostication in gastric hepatoid adenocarcinoma. Hum Pathol. 2014;45(6):1243–50. doi:10.1016/j.humpath.2014.02.003.
115. Ushiku T, Shinozaki A, Shibahara J, Iwasaki Y, Tateishi Y, Funata N, et al. SALL4 represents fetal gut differentiation of gastric cancer, and is diagnostically useful in distinguishing hepatoid gastric

carcinoma from hepatocellular carcinoma. Am J Surg Pathol. 2010;34(4):533–40. doi:10.1097/PAS.0b013e3181d1dcdd.
116. Nogales FF, Quinonez E, Lopez-Marin L, Dulcey I, Preda O. A diagnostic immunohistochemical panel for yolk sac (primitive endodermal) tumours based on an immunohistochemical comparison with the human yolk sac. Histopathology. 2014;65(1):51–9. doi:10.1111/his.12373.
117. Fujii Y, Yoshihashi K, Suzuki H, Tsutsumi S, Mutoh H, Maeda S, et al. CDX1 confers intestinal phenotype on gastric epithelial cells via induction of stemness-associated reprogramming factors SALL4 and KLF5. Proc Natl Acad Sci U S A. 2012;109(50):20584–9. doi:10.1073/pnas.1208651109.
118. Kinjo T, Taniguchi H, Kushima R, Sekine S, Oda I, Saka M, et al. Histologic and immunohistochemical analyses of alpha-fetoprotein--producing cancer of the stomach. Am J Surg Pathol. 2012;36(1):56–65. doi:10.1097/PAS.0b013e31823aafec.
119. Tanigawa H, Kida Y, Kuwao S, Uesugi H, Ojima T, Kobayashi N, et al. Hepatoid adenocarcinoma in Barrett's esophagus associated with achalasia: first case report. Pathol Int. 2002;52(2):141–6.
120. Ye MF, Tao F, Liu F, Sun AJ. Hepatoid adenocarcinoma of the stomach: a report of three cases. World J Gastroenterol. 2013;19(27):4437–42. doi:10.3748/wjg.v19.i27.4437.
121. Liu X, Cheng Y, Sheng W, Lu H, Xu X, Xu Y, et al. Analysis of clinicopathologic features and prognostic factors in hepatoid adenocarcinoma of the stomach. Am J Surg Pathol. 2010;34(10):1465–71. doi:10.1097/PAS.0b013e3181f0a873.
122. Yang J, Wang R, Zhang W, Zhuang W, Wang M, Tang C. Clinicopathological and prognostic characteristics of hepatoid adenocarcinoma of the stomach. Gastroenterol Res Pract. 2014;2014:140587. doi:10.1155/2014/140587.
123. Liu TC, Vachharajani N, Chapman WC, Brunt EM. SALL4 immunoreactivity predicts prognosis in Western hepatocellular carcinoma patients but is a rare event: a study of 236 cases. Am J Surg Pathol. 2014;38(7):966–72. doi:10.1097/PAS.0000000000000218.
124. Bihari C, Rastogi A, Chandan KN, Yadav V, Panda D. Gastric adenocarcinoma with yolk sac tumor differentiation and liver metastasis of yolk sac tumor component. Case Rep Oncol Med. 2013;2013:923596. doi:10.1155/2013/923596.
125. Napaki S. Combined yolk sac tumour and adenocarcinoma of the oesophago-gastric junction. Pathology. 2004;36(6):589–92.
126. Puglisi F, Damante G, Pizzolitto S, Mariuzzi L, Guerra S, Pellizzari L, et al. Combined yolk sac tumor and adenocarcinoma in a gastric stump: molecular evidence of clonality. Cancer. 1999;85(9):1910–6.
127. Suzuki T, Kimura N, Shizawa S, Yabuki N, Yamaki T, Sasano H, et al. Yolk sac tumor of the stomach with an adenocarcinomatous component: a case report with immunohistochemical analysis. Pathol Int. 1999;49(6):557–62.
128. Lack EE, Travis WD, Welch KJ. Retroperitoneal germ cell tumors in childhood. A clinical and pathologic study of 11 cases. Cancer. 1985;56(3):602–8.
129. Murat E, Dagdemir A, Bilgici MC, Sullu Y. Primary yolk sac tumor of the retroperitoneal region. Contemp Oncol (Pozn). 2013;17(6):530–2. doi:10.5114/wo.2013.38916.
130. Moss JF, Slayton RE, Economou SG. Primary retroperitoneal pure choriocarcinoma. Two long-term complete responders from a rare fatal disease. Cancer. 1988;62(6):1053–4.
131. Sung MT, MacLennan GT, Cheng L. Retroperitoneal seminoma in limited biopsies: morphologic criteria and immunohistochemical findings in 30 cases. Am J Surg Pathol. 2006;30(6):766–73.
132. Wegenke JD, Chuprevich TW, Nilsson TE. Retroperitoneal seminoma. J Urol. 1977;117(2):262–3.
133. Daftary M, Barnett RN. Malignant teratoma of the spleen. Yale J Biol Med. 1971;43(4–5):283–7.
134. Ghinescu C, Sallami Z, Jackson D. Choriocarcinoma of the spleen--a rare cause of atraumatic rupture. Ann R Coll Surg Engl. 2008;90(3):W12–4. doi:10.1308/147870808X257229.
135. Imperial SL, Sidhu JS. Nonseminomatous germ cell tumor arising in splenogonadal fusion. Arch Pathol Lab Med. 2002;126(10):1222–5. doi:10.1043/0003-9985(2002)126<1222:NGCTAI>2.0.CO;2.
136. Kersting S, Janot MS, Munding J, Suelberg D, Tannapfel A, Chromik AM, et al. Rare solid tumors of the pancreas as differential diagnosis of pancreatic adenocarcinoma. JOP. 2012;13(3):268–77.
137. Lane J, Vance A, Finelli D, Williams G, Ravichandran P. Dermoid cyst of the pancreas: a case report with literature review. J Radiol Case Rep. 2012;6(12):17–25. doi:10.3941/jrcr.v6i12.1152.
138. Scheele J, Barth TF, Wittau M, Juchems M, Henne-Bruns D, Kornmann M. Cystic teratoma of the pancreas. Anticancer Res. 2012;32(3):1075–80.
139. Wang J, Zheng Z, Qiu Y, Tou J, Liu W, Xiong Q, et al. Primary mixed germ cell tumor arising in the pancreatic head. J Pediatr Surg. 2013;48(1):e21–4. doi:10.1016/j.jpedsurg.2012.10.054.
140. Yoon WJ, Lee JK, Lee KH, Ryu JK, Kim YT, Yoon YB. Cystic neoplasms of the exocrine pancreas: an update of a nationwide survey in Korea. Pancreas. 2008;37(3):254–8. doi:10.1097/MPA.0b013e3181676ba4.
141. Tucci G, Muzi MG, Nigro C, Cadeddu F, Amabile D, Servadei F, et al. Dermoid cyst of the pancreas: presentation and management. World J Surg Oncol. 2007;5:85. doi:10.1186/1477-7819-5-85.
142. Kela M, Singh S, Lahoti B. Immature cystic teratoma of head of pancreas. J Indian Assoc Pediatr Surg. 2008;13(2):85–7. doi:10.4103/0971-9261.43037.
143. Kohzaki S, Fukuda T, Fujimoto T, Hirao K, Matsunaga N, Hayashi K, et al. Case report: ciliated

foregut cyst of the pancreas mimicking teratomatous tumour. Br J Radiol. 1994;67(798):601–4. doi:10.1259/0007-1285-67-798-601.
144. Childs CC, Korsten MA, Choi HS, Schwarz R, Fisse RD. Pancreatic choriocarcinoma presenting as inflammatory pseudocyst. Gastroenterology. 1985;89(2):426–31.
145. Ramachandran BS, Murugesan M, Ali M, Padmanabhan P. Primary pancreatic choriocarcinoma presenting as pancreatitis. JOP. 2012;13(2):217–8.
146. Rao M, Chen Y, Zhu Y, Huang Z, Zhang L. Primary pancreatic choriocarcinoma revealed on FDG PET/CT. Clin Nucl Med. 2015;40(1):76–8. doi:10.1097/RLU.0000000000000584.
147. Zhang B, Gao S, Chen Y, Wu Y. Primary yolk sac tumor arising in the pancreas with hepatic metastasis: a case report. Korean J Radiol. 2010;11(4):472–5. doi:10.3348/kjr.2010.11.4.472.
148. Stoos-Veic T, Tadic M. Endoscopic ultrasound and fine needle aspiration for the diagnosis of extragonadal seminoma metastatic to the pancreas. Endoscopy. 2014;46(Suppl 1):UCTN:E522. doi:10.1055/s-0034-1377596.
149. Certo M, Franca M, Gomes M, Machado R. Liver teratoma. Acta Gastroenterol Belg. 2008;71(2):275–9.
150. Fried I, Rom-Gross E, Finegold M, Simanovsky N, Revel-Vilk S, Ben-Neriah Z, et al. Concomitant massive hepatic teratoma, hypofibrinogenemia, and adrenal neuroblastoma in an infant: case report. Pediatr Dev Pathol. 2015; doi:10.2350/13-08-1361-CR.1.
151. Ayyappan AP, Singh SE, Shah A. Mature cystic teratoma in the falciform ligament of the liver. J Postgrad Med. 2007;53(1):48–9.
152. Rahmat K, Vijayananthan A, Abdullah B, Amin S. Benign teratoma of the liver: a rare cause of cholangitis. Biomed Imaging Interv J. 2006;2(3):e20. doi:10.2349/biij.2.3.e20.
153. Nirmala V, Chopra P, Machado NO. An unusual adult hepatic teratoma. Histopathology. 2003;43(3):306–8.
154. McKenney JK, Heerema-McKenney A, Rouse RV. Extragonadal germ cell tumors: a review with emphasis on pathologic features, clinical prognostic variables, and differential diagnostic considerations. Adv Anat Pathol. 2007;14(2):69–92. doi:10.1097/PAP.0b013e31803240e6.
155. Saxena R. Practical hepatic pathology. A Diagnostic Approach: A Volume in the Pattern Recognition Series. Elsevier Saunders, Philadelphia; 2011.
156. Conrad RJ, Gribbin D, Walker NI, Ong TH. Combined cystic teratoma and hepatoblastoma of the liver. Probable divergent differentiation of an uncommitted hepatic precursor cell. Cancer. 1993;72(10):2910–3.
157. Moll A, Krenauer A, Bierbach U, Till H, Hirsch W, Leuschner I, et al. Mixed hepatoblastoma and teratoma of the liver in a 3-year-old child: a unique combination and clinical challenge. Diagn Pathol. 2009;4:37. doi:10.1186/1746-1596-4-37.
158. Gnemmi V, Leteurtre E, Sudour-Bonnange H, Devisme L, Guettier C, Buob D, et al. SALL4 is a marker of the embryonal subtype of hepatoblastoma. Histopathology. 2013;63(3):425–8. doi:10.1111/his.12187.
159. Kim SN, Chi JG, Kim YW, Dong ES, Shin HY, Ahn HS, et al. Neonatal choriocarcinoma of liver. Pediatr Pathol. 1993;13(6):723–30.
160. Sekine R, Hyodo M, Kojima M, Meguro Y, Suzuki A, Yokoyama T, et al. Primary hepatic choriocarcinoma in a 49-year-old man: report of a case. World J Gastroenterol. 2013;19(48):9485–9. doi:10.3748/wjg.v19.i48.9485.
161. Baagar K, Khan FY, Alkuwari E. Choriocarcinoma syndrome: a case report and a literature review. Case Rep Oncol Med. 2013;2013:697251. doi:10.1155/2013/697251.
162. Bakhshi GD, Borisa AD, Bhandarwar AH, Tayade MB, Yadav RB, Jadhav YR. Primary hepatic choriocarcinoma: a rare cause of spontaneous haemoperitoneum in an adult. Clin Pract. 2012;2(3):e73. doi:10.4081/cp.2012.e73.
163. Xu AM, Gong SJ, Song WH, Li XW, Pan CH, Zhu JJ, et al. Primary mixed germ cell tumor of the liver with sarcomatous components. World J Gastroenterol. 2010;16(5):652–6.
164. Warren M, Thompson KS. Two cases of primary yolk sac tumor of the liver in childhood: case reports and literature review. Pediatr Dev Pathol. 2009;12(5):410–6. doi:10.2350/08-09-0521.1.
165. Gonzalez-Roibon N, Katz B, Chaux A, Sharma R, Munari E, Faraj SF, et al. Immunohistochemical expression of SALL4 in hepatocellular carcinoma, a potential pitfall in the differential diagnosis of yolk sac tumors. Hum Pathol. 2013;44(7):1293–9. doi:10.1016/j.humpath.2012.10.017.
166. Lam KY, Lo CY. Adrenal lipomatous tumours: a 30 year clinicopathological experience at a single institution. J Clin Pathol. 2001;54(9):707–12.
167. Lam KY, Lo CY. Teratoma in the region of adrenal gland: a unique entity masquerading as lipomatous adrenal tumor. Surgery. 1999;126(1):90–4. doi:10.1067/msy.1999.98924.
168. Zhao J, Sun F, Jing X, Zhou W, Huang X, Wang H, et al. The diagnosis and treatment of primary adrenal lipomatous tumours in Chinese patients: a 31-year follow-up study. Can Urol Assoc J. 2014;8(3–4):E132–6. doi:10.5489/cuaj.977.
169. Li Y, Zhong Z, Zhao X. Primary mature teratoma presenting as an adrenal tumor in a child. Urology. 2011;78(3):689–91. doi:10.1016/j.urology.2010.12.022.
170. Yao W, Li K, Xiao X, Zheng S, Chen L. Neonatal suprarenal mass: differential diagnosis and treatment. J Cancer Res Clin Oncol. 2013;139(2):281–6. doi:10.1007/s00432-012-1316-x.
171. Ciftci I, Cihan T, Koksal Y, Ugras S, Erol C. Giant mature adrenal cystic teratoma in an infant. Acta Inform Med. 2013;21(2):140–1. doi:10.5455/aim.2013.21.140-141.
172. Armah HB, Parwani AV. Primary carcinoid tumor arising within mature teratoma of the kidney: report

of a rare entity and review of the literature. Diagn Pathol. 2007;2:15. doi:10.1186/1746-1596-2-15.
173. Armah HB, Parwani AV, Perepletchikov AM. Synchronous primary carcinoid tumor and primary adenocarcinoma arising within mature cystic teratoma of horseshoe kidney: a unique case report and review of the literature. Diagn Pathol. 2009;4:17. doi:10.1186/1746-1596-4-17.
174. Idrissi-Serhrouchni K, El-Fatemi H, El madi A, Benhayoun K, Chbani L, Harmouch T, et al. Primary renal teratoma: a rare entity. Diagn Pathol. 2013;8:107. doi:10.1186/1746-1596-8-107.
175. Jacob S, Ghosh D, Rawat P, Bhatti W, Singh VP. Primary intra-renal teratoma associated with renal dysplasia: an unusual entity. Indian J Surg. 2013;75(Suppl 1):77–9. doi:10.1007/s12262-011-0356-9.
176. Kuroda N, Katto K, Tamura M, Shiotsu T, Hes O, Michal M, et al. Carcinoid tumor of the renal pelvis: consideration on the histogenesis. Pathol Int. 2008;58(1):51–4. doi:10.1111/j.1440-1827.2007.02188.x.
177. Liu YC, Wang JS, Chen CJ, Sung PK, Tseng HH. Intrarenal mixed germ cell tumor. J Urol. 2000;164(6):2020–1.
178. Mochizuki K, Ohno Y, Tokai Y, Kanematsu T, Okada M, Kamitamari A, et al. Congenital intrarenal teratoma arising from a horseshoe kidney. J Pediatr Surg. 2006;41(7):1313–5. doi:10.1016/j.jpedsurg.2006.03.019.
179. Okani CO, Aramide KO, Edegbe FO, Ogun GO, Okolo CA, Salako BL, et al. Intrarenal teratoma: report of two cases seen in a 31-year retrospective study at the University College Hospital. Ibadan Niger J Med. 2014;23(3):271–5.
180. Otani M, Tsujimoto S, Miura M, Nagashima Y. Intrarenal mature cystic teratoma associated with renal dysplasia: case report and literature review. Pathol Int. 2001;51(7):560–4.
181. Sun K, You Q, Zhao M, Yao H, Xiang H, Wang L. Concurrent primary carcinoid tumor arising within mature teratoma and clear cell renal cell carcinoma in the horseshoe kidney: report of a rare case and review of the literature. Int J Clin Exp Pathol. 2013;6(11):2578–84.
182. Yavuz A, Ceken K, Alimoglu E, Akkaya B. Mature cystic renal teratoma. Iran J Radiol. 2014;11(1):e11260. doi:10.5812/iranjradiol.11260.
183. Rouzbahman M, Clarke B. Neuroendocrine tumors of the gynecologic tract: select topics. Semin Diagn Pathol. 2013;30(3):224–33. doi:10.1053/j.semdp.2013.06.007.
184. Sultan I, Ajlouni F, Al-Jumaily U, Al-Ashhab M, Hashem H, Ghandour K, et al. Distinct features of teratoid Wilms tumor. J Pediatr Surg. 2010;45(10):e13–9. doi:10.1016/j.jpedsurg.2010.06.035.
185. Kheradmand A. Choriocarcinoma presenting as bilateral renal tumor: a case report. Urol J. 2005;2(1):52–3.
186. Usta TA, Karacan T, Ozyurek E, Naki MM, Omeroglu SN, Demirkiran F. Primary renal artery choriocarcinoma causing secondary renovascular hypertension. Int J Surg Case Rep. 2014;5(12):1197–9. doi:10.1016/j.ijscr.2014.11.056.
187. Preda O, Dema A, Iacob M, Goyenaga P, Dulcey I, Aneiros Fernandez J, et al. Urothelial carcinoma of the renal pelvis with simultaneous trophoblastic and malignant clear cell endodermal-type differentiation. Virchows Arch. 2012;460(3):353–6. doi:10.1007/s00428-012-1211-5.
188. Zettl A, Konrad MA, Polzin S, Ehsan A, Riedmiller H, Muller-Hermelink HK, et al. Urothelial carcinoma of the renal pelvis with choriocarcinomatous features: genetic evidence of clonal evolution. Hum Pathol. 2002;33(12):1234–7. doi:10.1053/hupa.2002.129208.
189. Kumar Y, Bhatia A, Kumar V, Vaiphei K. Intrarenal pure yolk sac tumor: an extremely rare entity. Int J Surg Pathol. 2007;15(2):204–6. doi:10.1177/1066896906295776.
190. Lin S, Li X, Sun C, Feng S, Peng Z, Huang S, et al. CT findings of intrarenal yolk sac tumor with tumor thrombus extending into the inferior vena cava: a case report. Korean J Radiol. 2014;15(5):641–5. doi:10.3348/kjr.2014.15.5.641.
191. Cohen R, Mirrer B, Loarte P, Navarro V. Intrapericardial mature cystic teratoma in an adult: case presentation. Clin Cardiol. 2013;36(1):6–9. doi:10.1002/clc.22051.
192. Milovanovic V, Lukac M, Krstic Z. Intrapericardial immature teratoma in a newborn: a case report. Cardiol Young. 2014;24(1):164–6. doi:10.1017/S1047951113000097.
193. Chiba T, Iwami D, Kikuchi Y. Mesenteric teratoma in an 8-month-old girl. J Pediatr Surg. 1995;30(1):120.
194. Jones MA, Clement PB, Young RH. Primary yolk sac tumors of the mesentery. A report of two cases. Am J Clin Pathol. 1994;101(1):42–7.
195. Tangour-Bouaicha M, Bel Haj Salah M, Ben Brahim E, Ben Othmene M, Douggaz A, Sassi S, et al. Primary peritoneal yolk sac tumour. A case report. Ann Pathol. 2010;30(5):378–81. doi:10.1016/j.annpat.2010.07.004.
196. Hegde P. Extragonadal omental teratoma: a case report. J Obstet Gynaecol Res. 2014;40(2):618–21. doi:10.1111/jog.12198.
197. Khoo CK, Chua I, Siow A, Chern B. Parasitic dermoid cyst of the pouch of Douglas: a case report. J Minim Invasive Gynecol. 2008;15(6):761–3. doi:10.1016/j.jmig.2008.07.014.
198. Lee HM, Song SY, Park JO, Kim BH. Primary immature teratoma of the prostate with angiosarcoma component: its unusual response to chemotherapy. Int J Urol. 2006;13(3):305–7. doi:10.1111/j.1442-2042.2006.01286.x.
199. Paner GP, Aron M, Hansel DE, Amin MB. Non-epithelial neoplasms of the prostate. Histopathology. 2012;60(1):166–86. doi:10.1111/j.1365-2559.2011.04020.x.
200. Hashimoto T, Ohori M, Sakamoto N, Matsubayashi J, Izumi M, Tachibana M. Primary seminoma

of the prostate. Int J Urol. 2009;16(12):967–70. doi:10.1111/j.1442-2042.2009.02403.x.
201. Adachi Y, Rokujyo M, Kojima H, Nagashima K. Primary seminoma of the seminal vesicle: report of a case. J Urol. 1991;146(3):857–9.
202. Fairey AE, Mead GM, Murphy D, Theaker J. Primary seminal vesicle choriocarcinoma. Br J Urol. 1993;71(6):756–7.
203. Yao XD, Hong YP, Ye DW, Wang CF. Primary yolk sac tumor of seminal vesicle: a case report and literature review. World J Surg Oncol. 2012;10:189. doi:10.1186/1477-7819-10-189.
204. Jain SK, Kaza RC, Vindal A, Bains L. Vesical dermoid: an unusual presentation. Int J Urol. 2010;17(9):824–5. doi:10.1111/j.1442-2042.2010.02596.x.
205. Sabnis RB, Bradoo AM, Desai RM, Bhatt RM, Randive NU. Primary benign vesical teratoma. A case report. Arch Esp Urol. 1993;46(5):444–5.
206. Dennis PM, Turner AG. Primary choriocarcinoma of the bladder evolving from a transitional cell carcinoma. J Clin Pathol. 1984;37(5):503–5.
207. Sievert K, Weber EA, Herwig R, Schmid H, Roos S, Eickenberg HU. Pure primary choriocarcinoma of the urinary bladder with long-term survival. Urology. 2000;56(5):856.
208. Hanna NH, Ulbright TM, Einhorn LH. Primary choriocarcinoma of the bladder with the detection of isochromosome 12p. J Urol. 2002;167(4):1781.
209. Hurst CD, Platt FM, Taylor CF, Knowles MA. Novel tumor subgroups of urothelial carcinoma of the bladder defined by integrated genomic analysis. Clin Cancer Res. 2012;18(21):5865–77. doi:10.1158/1078-0432.CCR-12-1807.
210. Taylor G, Jordan M, Churchill B, Mancer K. Yolk sac tumor of the bladder. J Urol. 1983;129(3):591–4.
211. Mui WH, Lee KC, Chiu SC, Pang CY, Chu SK, Man CW, et al. Primary yolk sac tumour of the urinary bladder: a case report and review of the literature. Oncol Lett. 2014;7(1):199–202. doi:10.3892/ol.2013.1670.
212. El-Bahrawy M. Alpha-fetoprotein-producing non-germ cell tumors of the urological system. Rev Urol. 2011;13(1):14–9.
213. D'Alessio A, Verdelli G, Bernardi M, DePascale S, Chiarenza SF, Giardina C, et al. Endodermal sinus (yolk sac) tumor of the urachus. Eur J Pediatr Surg. 1994;4(3):180–1. doi:10.1055/s-2008-1066097.
214. Huang HY, Ko SF, Chuang JH, Jeng YM, Sung MT, Chen WJ. Primary yolk sac tumor of the urachus. Arch Pathol Lab Med. 2002;126(9):1106–9. doi:10.1043/0003-9985(2002)126<1106:PYSTOT>2.0.CO;2.
215. Romero-Rojas AE, Messa-Botero OA, Melo-Uribe MA, Diaz-Perez JA, Chinchilla-Olaya SI. Primary yolk sac tumor of the urachus. Int J Surg Pathol. 2011;19(5):658–61.doi:10.1177/1066896909356923.
216. Kingo PS, Hoyer S, Marinovskij E, Rawashdeh YF. Paediatric urachal benign teratoma: a case report and review of the literature. APMIS. 2014;122(9):761–5. doi:10.1111/apm.12216.
217. Dulcey I, Cespedes MU, Ballesteros JL, Preda O, Aneiros-Fernandez J, Clavero PA, et al. Necrotic mature ovarian teratoma associated with anti-N-methyl-D-aspartate receptor encephalitis. Pathol Res Pract. 2012;208(8):497–500. doi:10.1016/j.prp.2012.05.018.
218. Thomas A, Rauschkolb P, Gresa-Arribas N, Schned A, Dalmau JO, Fadul CE. Anti-N-methyl-D-aspartate receptor encephalitis: a patient with refractory illness after 25 months of intensive immunotherapy. JAMA Neurol. 2013;70(12):1566–8. doi:10.1001/jamaneurol.2013.3205.
219. Terada T. Histopathologic study of the rectum in 1,438 consecutive rectal specimens in a single Japanese hospital: I. benign lesions. Int J Clin Exp Pathol. 2013;6(3):349–57.
220. Cotton MH, Blake JR. Dermoid cyst: a rare tumour of the appendix. Gut. 1986;27(3):334–6.
221. Demircan M, Uguralp S, Mutus M, Kutlu R, Mizrak B. Teratoma arising from anomalous common bile ducts: a case report. J Pediatr Surg. 2004;39(4):e1–2.
222. Naim M. Immature embryoid teratoma of the gall bladder: case of a primary primitive neoplasm. BMJ Case Rep. 2009; doi:10.1136/bcr.01.2009.1455.
223. Torikai M, Tahara H, Kaji T, Shimono R, Yano T, Yoshioka T, et al. Immature teratoma of gallbladder associated with gliomatosis peritonei, a case report. J Pediatr Surg. 2007;42(10):E25–7. doi:10.1016/j.jpedsurg.2007.08.001.
224. Koh JS, Park JH, Kang CH. A primary extragonadal teratoma of the proximal humerus. J Korean Med Sci. 2009;24(5):989–91. doi:10.3346/jkms.2009.24.5.989.
225. Joshi P, Parelkar S, Shetty S, Sanghvi B, Mundada D, Kapadnis S, et al. Mature duodenal teratoma in a neonate with exomphalos minor with intestinal obstruction: first case report. Eur J Pediatr Surg. 2014;24(2):187–9. doi:10.1055/s-0032-1330848.
226. Zalatnai A, Dubecz S, Harka I, Banhidy Jr F. Malignant teratoma of the left colon associated with chronic ulcerative colitis. Virchows Arch A Pathol Anat Histopathol. 1987;411(1):61–5.
227. Wilkinson N, Cairns A, Benbow EW, Donnai P, Buckley CH. Dermoid cyst of the caecum. Histopathology. 1996;29(2):186–8.
228. Barker GM, Choudhury SR, Nicholls G, Whittle MR, Raafat F, Mitra SK. Fetiform teratoma arising from a tubular intestinal duplication. A probable cause of ruptured exomphalos. Pediatr Surg Int. 1997;12(2–3):204–5.
229. Hu J, Nagao K, Tai T, Ozaki Y, Kobayashi H, Iuchi A, et al. Congenital primary penile teratoma in a child. Urology. 2014;83(6):1404–6. doi:10.1016/j.urology.2014.01.044.
230. Camacho F. Benign cutaneous cystic teratoma. J Cutan Pathol. 1982;9(5):345–51.
231. Tsai TF, Chuan MT, Hsiao CH. A cystic teratoma of the skin. Histopathology. 1996;29(4):384–6.
232. Zhu Y, Liu M, Li J, Jing F, Linghu R, Guo X, et al. Breast carcinoma with choriocarcinomatous features:

a case report and review of the literature. World J Surg Oncol. 2014;12:239. doi:10.1186/1477-7819-12-239.

233. Northcutt AD, Tschen JA. Choriocarcinoma-like squamous cell carcinoma: a new variant expressing human chorionic gonadotropin. Am J Dermatopathol. 2015;37(2):171–4. doi:10.1097/DAD.0000000000000042.
234. Albores-Saavedra J, Cruz-Ortiz H, Alcantara-Vazques A, Henson DE. Unusual types of gallbladder carcinoma. A report of 16 cases. Arch Pathol Lab Med. 1981;105(6):287–93.
235. Iyomasa S, Senda Y, Mizuno K, Miyake H, Amemiya T, Yamaguchi J, et al. Primary choriocarcinoma of the jejunum: report of a case. Surg Today. 2003;33(12):948–51. doi:10.1007/s00595-003-2626-z.
236. Jiang L, Wu JT, Peng X. Primary choriocarcinoma of the colon: a case report and review of the literature. World J Surg Oncol. 2013;11:23. doi:10.1186/1477-7819-11-23.
237. Verbeek W, Schulten HJ, Sperling M, Tiesmeier J, Stoop H, Dinjens W, et al. Rectal adenocarcinoma with choriocarcinomatous differentiation: clinical and genetic aspects. Hum Pathol. 2004;35(11):1427–30. doi:10.1016/j.humpath.2004.06.005.
238. Munghate GS, Agarwala S, Bhatnagar V. Primary yolk sac tumor of the common bile duct. J Pediatr Surg. 2011;46(6):1271–3. doi:10.1016/j.jpedsurg.2011.02.001.
239. Nakazawa R, Soneda S, Kinoshita A, Kitagawa H, Koike J, Chikaraishi T. A case report: primary extragonadal yolk sac tumor of penile shaft in a 2-year-old child. Int J Urol. 2009;16(4):413–5. doi:10.1111/j.1442-2042.2009.02264.x.

12 Somatic-Type Malignancy in Germ Cell Tumors

Nooshin K. Dashti and Rafael E. Jimenez

12.1 Introduction

The presence of somatic-type malignancy (STM) within a preexisting germ cell tumor (GCT) is defined as a GCT that "develops a distinct secondary component that resembles a somatic-type malignant neoplasm, as seen in other organs and tissues (e.g., sarcomas and carcinomas)" [1]. In the literature it is also referred to as "teratoma with secondary malignant component," [2] "non-germ cell malignancy," or more recently and appropriately "STM arising from GCT" (STM-GCT) [3]. The term "teratoma with malignant transformation" is discouraged, particularly in the setting of testicular GCT, as it would imply a benign nature of the background teratoma. The phenomenon of STM-GCT impacts both the prognosis and the management of patients with germ cell neoplasms, and its recognition is of utmost importance.

STM has been described in a wide spectrum of GCT. Its occurrence in benign mature cystic teratomas of the ovary (type IV GCT of the Oosterhuis and Looijenga classification [4]) has been long recognized. It is also a known complication of mixed GCT (type II GCT) of the testis, where it is most frequently associated with a postpubertal-type teratoma component. The development of a high-grade sarcoma is a well-known albeit infrequent complication of spermatocytic tumor (type III GCT). Finally, its occurrence in type I (prepubertal type teratomas) has been reported, and it may be actually more common than thought, as the recognition of this type of GCT in postpubertal patients is rather recent. STM may occur at the primary site, at metastatic sites, or at both sites, and it has been reported both in gonadal and extragonadal GCT. The frequency, pathogenesis, clinical presentation, and prognostic and therapeutic implications of STM-GCT vary depending on the abovementioned scenarios (Table 12.1).

12.2 Pathogenesis

The pathogenesis of STM is a matter of debate. The most accepted theory is that STM arises from corresponding somatic elements in teratomas, given that in this GCT type, somatic differentiation has already occurred, independently of whether it is originally benign (as in ovarian or prepubertal testicular teratomas) or already

N.K. Dashti, MD
Mayo Clinic, Rochester, USA
e-mail: Dashti.Nooshin@mayo.edu

R.E. Jimenez, MD (✉)
Laboratory Medicine and Pathology, College of Medicine, Mayo Clinic, Rochester, USA
e-mail: Jimenez.Rafael@mayo.edu

F.F. Nogales, R.E. Jimenez (eds.), *Pathology and Biology of Human Germ Cell Tumors*,
DOI 10.1007/978-3-662-53775-6_12

Table 12.1 Comparison of somatic-type malignancies arising in different settings across the spectrum of germ cell neoplasia

	Type I GCT	Type II GCT	Type III GCT	Type IV GCT
Frequency	Rare	Uncommon	Rare	Common
Location	Testis Ovary Extragonadal	Testis Ovary Extragonadal	Testis	Ovary
Preexisting GCT	Prepubertal type teratoma Yolk sac tumor	Postpubertal type teratoma Seminoma/dysgerminoma/germinoma Mixed GCT	Spermatocytic tumor	Mature cystic teratoma Immature teratoma
Most frequent STM histologies	Adenocarcinoma Sarcomas	Rhabdomyosarcoma and other sarcomas PNET Adenocarcinoma	Undifferentiated sarcoma Rhabdomyosarcoma	Squamous cell carcinoma (overwhelming majority) Adenocarcinoma Sarcomas Melanoma
Molecular signature	Unknown	12p abnormalities	Unknown	Isodisomy (homozygosity)

GCT germ cell tumor, *STM* somatic-type malignancy, *PNET* primitive neuroectodermal tumor malignant (as in postpubertal testicular teratomas) [1, 5]. Coexistence of teratoma and STM in the majority of reported cases supports this hypothesis. Anomalies of chromosome 12, notably duplication of the short arm of chromosome 12 (isochromosome 12p), a hallmark of invasive type II GCT, have been documented in somatic malignancies arising from teratoma, either at the primary site or metastatic location (Fig. 12.1) [6–8], and in gonadal and extragonadal tumors [9]. This finding is also well documented in hematologic malignancies derived from mediastinal GCT [10–14]. Similarly, the classical finding of homozygosity (isodisomy) of mature cystic ovarian teratomas has been demonstrated in somatic adenocarcinomas arising within them [15]. This evidence supports the notion of a metachronous origin of STM from an already neoplastic germ cell. The somatic malignancy thus likely arises from the activation of oncogenes that normally play a role in the development of these tumors at their normal sites. For example, rearrangements of chromosome 2, region 2q34–37, present in rhabdomyosarcomas [16], have been also identified in the rhabdomyosarcoma component of STM [17]. Similarly, genetic alterations in 11q24, which have been frequently reported in Ewing's sarcoma and PNET, were found in a case of PNET arising from a GCT [17]. Chromosome 5 abnormalities such as del(5q), classically associated with hematologic malignancies, have been identified in leukemias originating from GCT [11, 17]. Finally, loss of heterozygosity reported at 11p13 locus in Wilms' tumor has been identified in a nephroblastoma arising from testicular GCT [18]. Thus, it appears that the molecular mechanisms associated with neoplastic progression of usual somatic malignancies are also common in those arising from germ cell neoplasia. It is not clear whether the triggers of these mechanisms are the same ones as in regular sites or whether they differ among the different cells of origin and locations of teratomas. Senescence of tissues may explain the development of STM in certain cases, particularly those associated with benign GCT. STM in benign cystic teratomas of the ovary, for example, is more often diagnosed in the fifth to sixth decade of life [19], which contrasts with the usual presentations of these tumors in adolescence and early adulthood, suggesting that a certain elapsing time is necessary for the activation of these oncogenetic mechanisms.

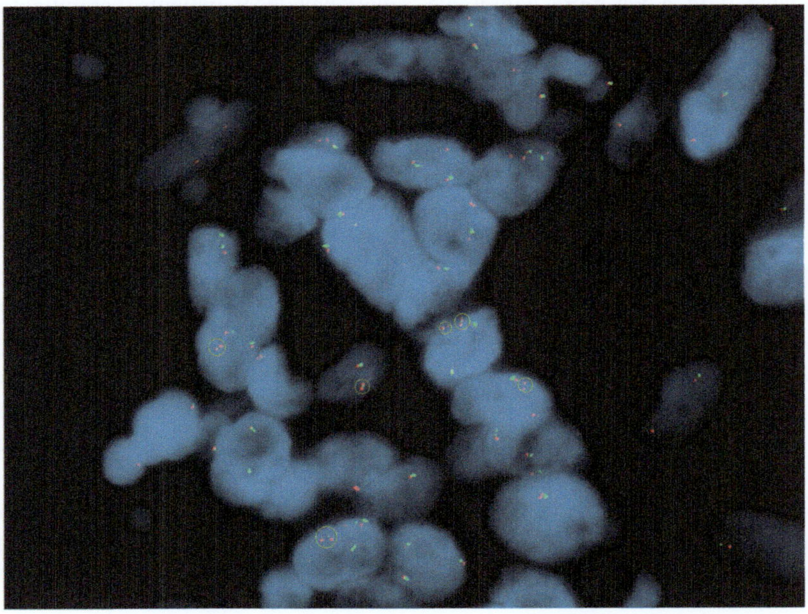

Fig. 12.1 Isochromosome 12p in metastatic adenocarcinoma (same patient as Fig. 12.3). The tumor cells had two copies of 12p (*red dots*) attached to the centromere (*green dot*), as identified with fluorescent in situ hybridization

Another mechanism of development of STM-GCT is the overgrowth of immature elements. While the presence of immature tissue elements is relatively common, particularly in type II GCT, an expansile growth of these primitive tissues is associated with more aggressive behavior and metastases from the overgrown component. Examples of primitive tissues that may show overgrowth include neuroepithelium, rhabdomyoblasts, and nephrogenic blastema, and thus, when present, a diagnosis of primitive neuroectodermal tumor (PNET), rhabdomyosarcoma, or nephroblastoma, respectively, is considered. The definition of "overgrowth" is arbitrary and has been traditionally defined in type II testicular tumors as a low-power field (40x) of pure immature elements [1]. However, while this definition appears necessary in type II tumors where immature elements are frequently part of the spectrum of tissues present in these teratomas, the presence of any amount of developmentally immature tissue, in particular neuroepithelium, is considered an adverse finding in ovarian teratomas [20, 21]. Interestingly, the presence of these immature elements, even if extensive, has not been historically considered STM at these sites but rather evidence of "immaturity." Regardless, the finding is frequently associated with a far more aggressive clinical behavior than the original teratoma.

The occurrence of STM in patients that lacked teratomatous component in their primary or metastatic GCT has prompted other histogenetic theories [17, 22–28]. Some studies have suggested yolk sac tumor (YST) origin as an alternative in these cases [3, 23, 29–31]. Spindle cell sarcomas may arise from sarcomatoid YST by a process of epithelial to mesenchymal metaplasia. Similarly, some intestinal-type adenocarcinomas may arise of progressive differentiation of glandular YST [3]. The development of a high-grade sarcoma component is also a rare complication of spermatocytic tumors, an uncommon scenario where this tumor is associated with malignant clinical behavior (see also Chap. 7) [32]. The mechanism by which these GCT develop a somatic phenotype is unknown but is likely related to the pluripotential nature of germ cells and the activation of differentiation pathways.

The role of chemotherapy in the pathogenesis of STM-GCT, specifically in patients with type II GCT, is not clear. The majority of cases of STM occur in the metastatic and post-chemotherapy setting [3]. However, the extended use of chemotherapy in current management of GCT increases the number of patients with STM who have previously received chemotherapy. Additionally, STM occurs also in patients that did not receive chemotherapy [3, 17, 22, 33]. The presence of STM in metastatic sites without a corresponding counterpart in the primary site has led authors to propose the development of STM from totipotential germ cells at the metastatic site [25, 34]. By destroying the more aggressive tumor components, chemotherapy may select the more indolent slow-growing elements, which after further genetic changes may be responsible for the formation of biologically aggressive STM and late recurrence of GCT [3, 35]. Hematologic malignancies arising from mediastinal GCT were thought to be due to chemotherapy or radiation for a long time. Occurrence of hematologic malignancies in patients that did not receive irradiation or chemotherapy argues against this statement [11, 13, 36]. Further, in contrast to treatment-related leukemia, these GCT-derived somatic-type hematologic malignancies develop earlier [37]. Similarly, STM associated with intracranial GCT was thought to be treatment related. Documented STM in treatment-naïve GCT and the relatively brief interval between initial diagnosis and transformation supports an origin independent of therapy. Despite theories such as partial differentiation of totipotential germ cells with concomitant malignant transformation, tumor arising from differentiated teratomatous elements [23], or dedifferentiation similar to the phenomenon that occurs in liposarcoma and chondrosarcoma [32], the transformation mechanism remains unsettled.

12.3 Histologic Diagnosis

Recognition of a malignant somatic component in GCT depends on the type of malignant component. In general, carcinomatous malignancies are recognized by usual morphologic criteria applied to carcinomas in other locations. Overt cytological atypia, brisk mitotic activity, infiltrative and confluent growth, desmoplastic reaction, and invasive borders are part of such criteria. These criteria are more easily recognizable when the background GCT is benign. However, recognizing these features may prove problematic in the background of a type II GCT, as some elements interpreted as somatic may actually correspond to variants of these GCT, particularly YST. Non-seminomatous GCT usually have a prominent, reactive stroma, which may be confused with desmoplasia. Additionally, the inherent cytologic atypia invariably present in teratomatous elements of type II tumors makes the recognition of a carcinomatous component more difficult [27]. Thus, the addition of a quantitative criterion in carcinomas may be useful to establish a diagnosis of STM in the setting of type II neoplasms [1, 3]. Similarly, because the morphologic features of some sarcomas overlap with normal embryonal or fetal tissue frequently present in teratomas, particularly in type II GCT, it is necessary to establish additional criteria for the diagnosis of a somatic mesenchymal or neuroectodermal malignancy arising in a GCT. As stated above, the most commonly used criterion is the presence of an expansile component exclusively filling at least one low-power microscopic field (40× magnification) [1, 3]. Thus, for example, a nodule composed of embryonal-appearing skeletal muscle would be considered part of a teratoma if it involves less than one low-power field, while it would be considered rhabdomyosarcomatous transformation if the nodule involves a larger area. Even in the setting of high-grade stromal atypia, most authors favor needing a quantitative criterion to diagnose a stromal somatic malignancy [3, 23]. Melanocytic and hematologic neoplasms are usually diagnosed by extrapolating diagnostic criteria applied elsewhere.

12.4 STM in Type I GCT

The occurrence of STM in pediatric teratomas is well documented. Biskup et al. reported on nine cases of STM associated with pure teratomas (two sacrococcygeal and seven ovarian tumors) [38]; eight of the nine were children and adolescents. Another series reported 14 cases of STM in children and adolescents [39]. While based on the age of the patients and described histology, some of their cases may correspond to type II and type IV tumors; at least some of them were likely type I. STM histologies reported in these series included adenocarcinomas, rhabdomyosarcoma, other sarcomas, neuroendocrine carcinoma, astrocytoma, and neuroblastoma. Sites included ovary, retroperitoneum, sacrococcygeal, and mediastinal. One interesting case of an adenocarcinoma arising in a testicular dermoid cyst in a 52-year-old patient was reported [40]. The patient had had the testicular mass since childhood and developed sudden enlargement of the mass and metastatic disease. The depicted pathology is classical of dermoid cyst and convincingly shows the adenocarcinoma arising from mucinous epithelium within the teratoma. The presence of the mass since childhood, aside from being consistent with a type I neoplasm, underscores the importance of senescence in the development of STM in this setting. A metastasizing PNET has been reported arising in an immature teratoma of a 20-month-old [41]. As stated, STM in type I GCT may occur more frequently than thought, as the occurrence of type I tumors in postpubertal patients has only been recently recognized [42], and reported series do not allow to confidently separate type II from type I teratomas. For example, in a series of GCT associated with sarcomatous STM, three patients with a sarcomatous component had testicular tumors that encompassed exclusively teratomatous elements and thus could have represented type I neoplasms [34]. The overlapping morphology between type I and type IV tumors makes this issue even more likely when dealing with ovarian neoplasms [4].

12.5 STM in Type II GCT

12.5.1 Testicular Tumors

By far, the majority of cases of STM occurring in type II tumors correspond to testicular neoplasms and is in this setting where most of the experience with this phenomenon has been developed. The incidence is estimated to range from 3 to 6.6 % [17, 23, 43, 44]. In one of the earliest studies [43], 580 GCT were reviewed, and teratoma with "malignant transformation" was found in 17 cases, while in another study [23] teratoma with STM was identified in 11 cases of a total of 269 GCT reviewed. In a later series [24], of 607 GCT reviewed, 21 patients had teratoma with STM; 11 cases (54 %) of those had STM in the primary tumor. Thus, STM may develop either in the primary GCT or in a metastatic deposit and may develop in treatment-naïve tumors or in the post-chemotherapy setting.

The majority of cases of STM-GCTs have an associated teratoma component; however, up to 30 % may not have a recognizable teratoma neither in the primary nor in the metastatic tumor [3]. In a recent series, both glandular and spindle cell tumors had intermediate morphologic and immunophenotypic features between glandular and sarcomatoid YST and somatic adenocarcinomas and sarcomas, respectively [3]. This suggests that at least a proportion of STM-GCT cases may arise from YST. This would not be surprising, given the morphologic plasticity of YST.

Sarcomas, particularly rhabdomyosarcomas, are the most common STM to be reported in type II neoplasms (Fig. 12.2). In a recent series from five institutions, sarcomas represented 37 % of cases of STM, with rhabdomyosarcomas representing 13.5 % of the total [33]. Other sarcoma histologies reported include leiomyosarcoma, myxoid liposarcoma, chondrosarcoma, and malignant peripheral nerve sheath tumor. However, some of these sarcomas may actually correspond to sarcomatoid YST and thus do not represent true STM. In a recent study, of 68 sarcomas, 24 were reclassified as sarcomatoid YST and five as sarcomatoid carcinomas [3]. Adenocarcinomas (Fig. 12.3) are the most common epithelial neoplasms, representing 16 % of cases in the series mentioned above [33]. Other carcinomas include squamous cell carcinoma, neuroendocrine carcinomas, renal cell carcinoma, and hepatocellular carcinoma. While carcinoid tumors are currently classified as a type of monodermal teratoma [1], an alternative approach would be to consider them as a form of STM. PNET are also common, representing 31 % of cases in the abovementioned series (Fig. 12.4) [33]. Other primitive tumors include neuroblastoma and nephroblastoma (Fig. 12.5). Mixed histologies are also encountered. A detailed list of reported histologies in STM of the testis is presented in Table 12.2.

The type of histology impacts the prognosis. Rhabdomyosarcoma histology is associated with a better prognosis, while PNET is associated with the worst [33]. Elapsed time between diagnosis of the GCT and the diagnosis of STM appears to correlate with histology of STM. The vast majority of PNET and rhabdomyosarcoma are diagnosed concomitantly to or within two years of GCT diagnosis, while most adenocarcinomas are diagnosed after two years of diagnosis of the GCT [22, 33, 55]. This suggests that senescence and perhaps exposure to therapy may be more important risk factors in the development of epithelial STM and less important in the development of sarcomatous or primitive histologies. Correlation between grade of STM and aggressive behavior is controversial. One series did not find a correlation with prognosis according to the grade of glandular tumors but did find it with sarcomas [3], while another one did not find a difference in behavior between low- and high-grade sarcomas [34].

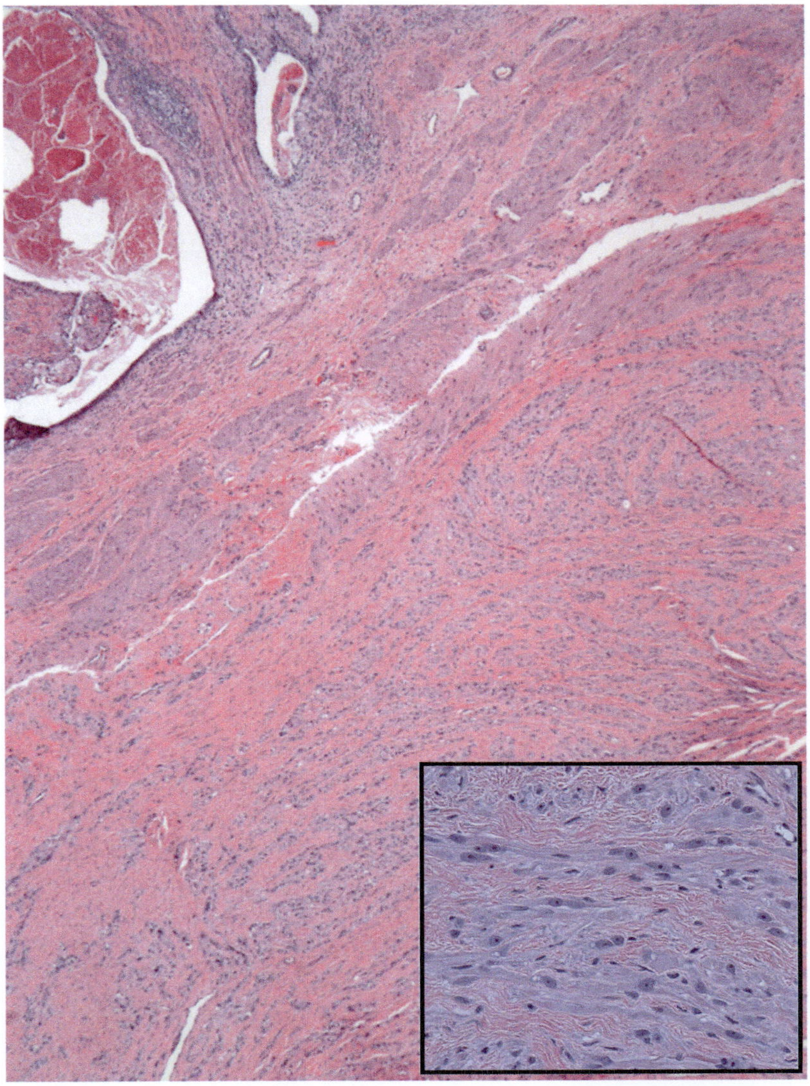

Fig. 12.2 Rhabdomyosarcoma, occupying more than one low-power field of this teratoma. Inset: high-power view, revealing polygonal to *spindle-shaped* cells with hyperchromatic nuclei and cytoplasmic cross striations

Overall, the presence of STM confers patients with a detrimental impact on survival across all stages. Clinical stage I and metastatic good-risk patients with STM-GCT had approximately 10 % and 20 % reduction in overall survival, respectively, compared to patients with pure GCT [33]. However, the site where STM is present is also important. STM present in primary tumors is associated with better prognosis than STM developed in metastatic deposits [2, 56]. Partially reflecting this, elapsed time from diagnosis of GCT to diagnosis of STM also impacts prognosis, with the best prognosis associated with STM diagnosed at the same time as the GCT [33]. The presence of STM diagnosed after therapy for GCT is associated with dismal prognosis, particularly if developed more than two years after GCT diagnosis [3, 33]. This is particularly significant, since STM-GCT represent approximately 23 % of late recurrences (i.e., after two years) in patients with testicular GCT [35, 57]. Patients who do

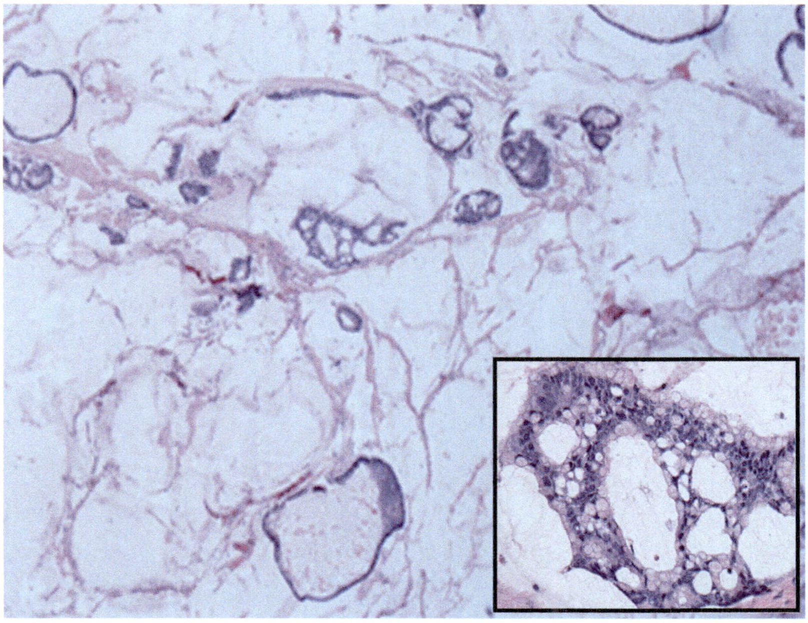

Fig. 12.3 Adenocarcinoma in retroperitoneal lymph nodes, 23 years after a diagnosis of mixed germ cell tumor. Tumor associated with abundant extracellular mucin. Inset: high-power view showing intracellular apical mucin in the neoplastic cells. (Same patient as Fig. 12.1)

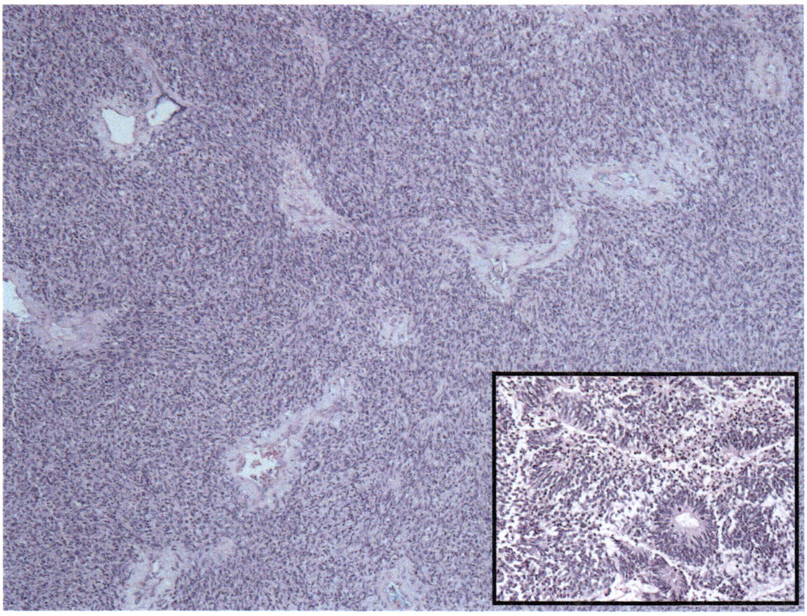

Fig. 12.4 Primitive neuroectodermal tumor, *small round blue cells* in broad sheets, occupying more than one low-power field. Inset: higher-power view shows occasional rosette formation

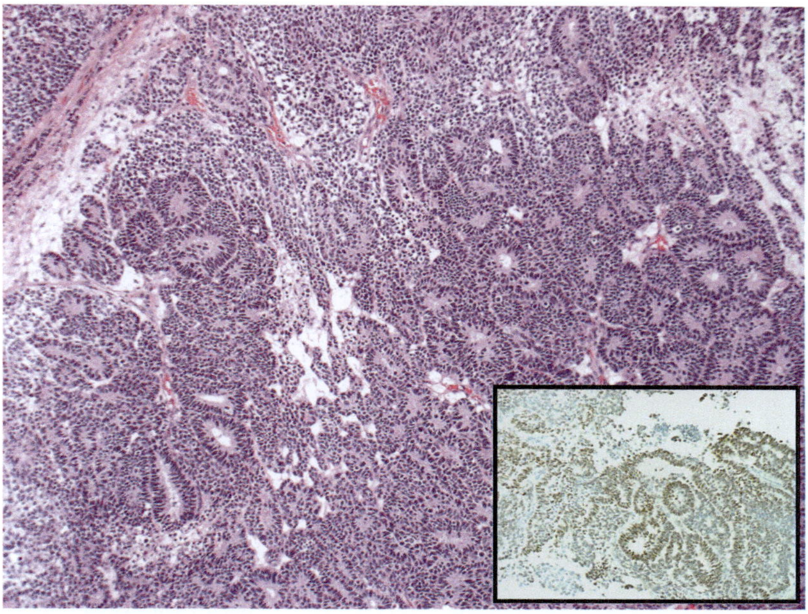

Fig. 12.5 Wilms' tumor with undifferentiated blastema, fibroblast-like stroma, and epithelial elements including abortive tubules. Inset: WT stain showing nuclear positivity in the epithelial component

not respond to initial therapy, experience relapse, or have metastatic or disseminated disease have poor prognosis [2, 22, 58].

12.5.2 Ovarian Tumors

The published experience with STM in ovarian type II GCT is limited to occasional case reports. This, however, may be a reflection of the much more uncommon occurrence of these tumors in the ovary, compared to the testis. Type II GCT can usually be inferred if the STM arises in a background of a mixed GCT or associated with non-teratomatous elements, like dysgerminoma. Similarly to the testicular counterpart, the majority of the reported histologies correspond to sarcomas. These include a case of a 33-year-old with an ovarian dysgerminoma associated with a fibrosarcoma component [59], a case of dysgerminoma with a rhabdomyosarcoma in a 14-year-old girl [60], and another case of rhabdomyosarcoma in a 23-year-old associated with dysgerminoma and teratoma [61]. Additionally, in the series of sarcomatous STM-GCT mentioned above [34], two of the three ovarian GCT with STM contained mixed germ cell elements, including mature teratoma and embryonal carcinoma with leiomyosarcoma, and dysgerminoma and immature teratoma with rhabdomyosarcoma. Collective evidence on this particular setting is quite scarce to draw significant conclusions about prognosis and treatment. Further, published series not always include enough information to allow retrospective identification of a type II teratoma and differentiate it from a type I or type IV teratoma or to exclude the possibility of its occurrence in a phenotypic female with an underdiagnosed Y chromosome mosaicism (see Chap. 6).

12.6 STM in Type III GCT

Spermatocytic tumors (ST) are rare and the presence of an associated sarcomatous component is even rarer. ST is not associated with other GCT and usually has a favorable prognosis (see Chap. 7). In reported cases of ST with a sarcomatous component, undifferentiated spindle cell sarcoma and rhabdomyosarcomas have been mentioned (Fig. 12.6). The presence of a sarcomatous component is associated with poor prognosis and metastatic disease [32, 62, 63].

Table 12.2 Reported STM histologies in primary or metastatic GCT of the testis

STM histology	References
Rhabdomyosarcoma	Colecchia [2], Guo [25], Malagon [34], Motzer [17], Necchi [22], Donadio [45], Comiter [24]
Adenocarcinoma	Colecchia [2], Necchi [22], Motzer [17], Donadio [45], El Mesbahi [46]
Squamous cell carcinoma	Ahmed [43]
Neuroendocrine carcinomas	Colecchia [2], Wang [47], Reyes [48], Necchi [22]
PNET	Ganjoo [49], Necchi [22], Mohanty [50], Comiter [24], Motzer [17], Donadio [45], Colecchia [2]
Nephroblastoma	Necchi [22], Ulbright [23], Colecchia [2], Emerson [18]
Well-differentiated liposarcoma	Colecchia [2], Necchi [22]
Leiomyosarcoma	Colecchia [2], Ahmed [43], Necchi [22], Comiter [24]
Myxoid leiomyosarcoma	Malagon [34]
Chondrosarcoma	Colecchia [2], Necchi [22], Comiter [24]
Angiosarcoma	Ulbright [51], Malagon [34]
Neuroblastoma	Colecchia [2], Ulbright [23]
Malignant fibrous histiocytoma	Ahmed [43]
Glioma	Ahmed [43]
Malignant peripheral nerve sheath tumor	Colecchia [2], Comiter [24], Necchi [52]
Gemistocytic astrocytoma	Colecchia [2]
Choroid plexus tumor	Colecchia [2], Necchi [22]
Microcystic meningioma	Allen [53]
Sarcoma not otherwise specified	Colecchia [2], Malagon [34], Necchi [22]
Dendritic cell tumor	Necchi [22]
Hemangioendothelioma	Necchi [22]
Malignant giant cell tumor	Ulbright [23]
Hepatocellular carcinoma	Jain [54]
PNET and choroid plexus tumor	Colecchia [2]
Gemistocytic astrocytoma and choroid plexus teratoma	Colecchia [2]
Rhabdomyosarcoma and adenocarcinoma	Colecchia [2]
Nephroblastoma and rhabdomyosarcoma	Colecchia [2]
PNET and rhabdomyosarcoma	Colecchia [2]
Rhabdomyosarcoma and undifferentiated sarcoma	Ganjoo [49]
Rhabdomyosarcoma and Ewing's sarcoma/primitive neuroectodermal tumor	Ganjoo [49]
Rhabdomyosarcoma and small round blue cell tumor not otherwise characterized	Ganjoo [49]
Osteogenic sarcoma and rhabdomyosarcoma	Motzer [17]
Rhabdomyosarcoma and primitive neuroectodermal tumor	Motzer [17]
Rhabdomyosarcoma and chondrosarcoma	Motzer [17]
Rhabdomyosarcoma and squamous cell carcinoma	Motzer [17]

GCT germ cell tumor, *STM* somatic-type malignancy, *PNET* primitive neuroectodermal tumor

Published experience with this phenomenon is limited to case reports and small series. The largest one reported five cases of ST with sarcomatous "transformation," four undifferentiated sarcomas and one rhabdomyosarcoma. Two (possibly three) of the patients died of metastatic disease [32]. Another series reported two cases of ST with sarcomatous component. The sarcomatous element in one case was rhabdomyosarcoma, while the other case had primitive mesenchymal spindle cell sarcoma. Both cases were older than 40 years; their sarcomatous component metastasized and had a poor outcome

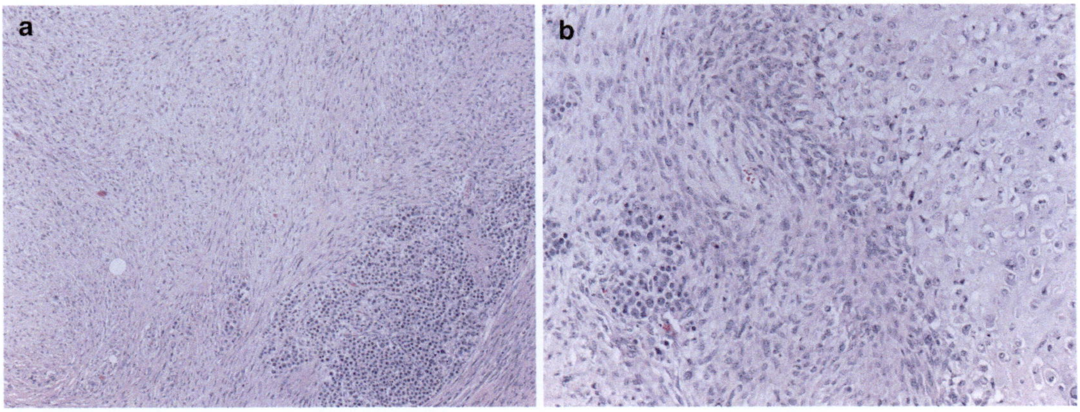

Fig. 12.6 Spermatocytic tumor (type III GCT) with associated malignant spindle cell (**a**) and cartilaginous (**b**) components (Pictures courtesy of Dr. Thomas Ulbright, Indiana University)

despite aggressive treatment. In their description of the ST component, the authors point out slight differences with classic description, including high mitotic rate and atypical mitotic figures [64]. One case report presented an ST with undifferentiated sarcomatous component in a 43-year-old male. The tumor was resected but chemotherapy was not given. The patient developed a recurrent scrotal mass and multiple bilateral lung metastases 9 months later. A chemotherapy regimen of cisplatin, bleomycin, and etoposide was initiated, but the patient died after 1 month [65]. Similarly, another case report presented an ST in a 51-year-old male with rhabdomyosarcoma component, metastasis to the lungs, liver and retroperitoneal lymph nodes, and death 2 months after the diagnosis [62].

The exact mechanism explaining the origin of sarcomatous component is not clear. As expected, teratomatous elements were absent in all reported cases. True et al. suggested that the sarcomatous components are an expression of anaplastic transformation of the ST [32]. One could also theorize that the sarcomatous component is a result of the pluripotential features of the neoplastic germ cell, although it is not clear why only mesenchymal neoplasms arise in this setting. Due to aggressive behavior of ST with sarcomatous component, additional treatment is warranted, although no specific modality is favored based on the limited experience. These tumors are rare and the effectiveness of chemotherapy and/or radiotherapy is not clear [66, 67].

12.7 STM in Type IV GCT

Mature cystic teratomas of the ovary (type IV GCT) can also be complicated by STM, and this phenomenon is extensively reviewed in Chap. 6. A few salient aspects will be discussed here.

STM occurs in 1.5–3 % of mature cystic teratomas [19, 68]. Contrary to what occurs in type II GCT, the majority of STM arising in mature cystic teratomas of the ovary are squamous cell carcinomas (Fig. 12.7) [69]. They seem to affect elderly women in their fifth and sixth decade [19]. Other epithelial neoplasms that may be found within mature cystic teratomas include adenocarcinoma (Fig. 12.8) [70], neuroendocrine carcinoma, and transitional cell carcinoma [71]. Sarcomas are much less frequent and include osteosarcoma [71, 72], rhabdomyosarcoma, angiosarcoma (Fig. 12.9), malignant fibrous histiocytoma, chondrosarcoma, spindle cell sarcoma, and undifferentiated sarcoma [73, 74]. Malignant melanomas are much less common than metastatic melanomas to the ovary [75, 76]. A thorough list of histologies reported in STM in mature cystic ovarian teratomas is presented in Chap. 6 in Table 6.5.

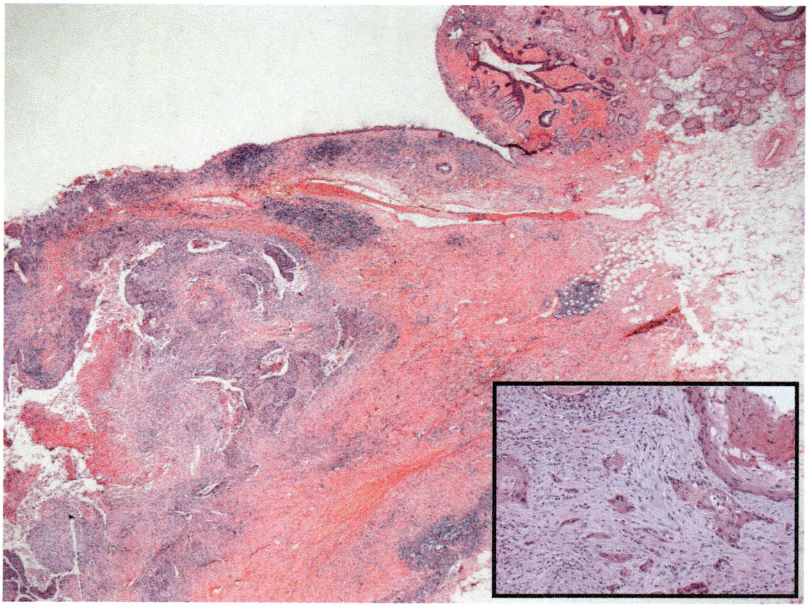

Fig. 12.7 Well-differentiated squamous cell carcinoma arising in a mature cystic teratoma of the ovary. Inset: high-power view of invasive nests of squamous cells and keratin formation

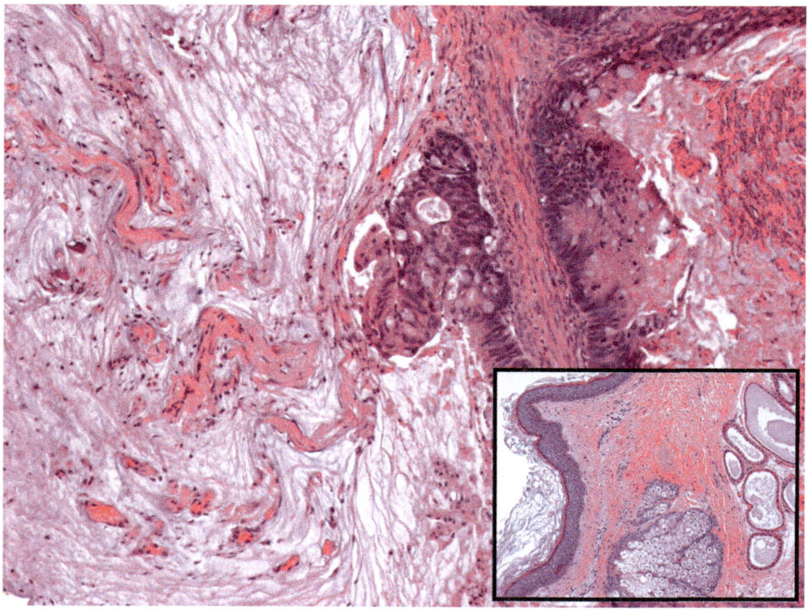

Fig. 12.8 Mucinous adenocarcinoma arising in mature cystic teratoma of the ovary. Malignant mucinous epithelium with extravasated mucin. Inset: the tumor arose within a classic dermoid cyst

Some have proposed that the different histological spectrum between STM in ovarian and testicular teratomas reflects the differences in tissues present in the teratomas from both gonads [77]. While this may be the case, it is also likely that the inherent biological behavior of the background tumor impacts the type of STM that develops in both settings. Thus, tumors associated with tissue senescence, like squamous cell carcinomas and adenocarcinomas, are more likely to occur in benign tumors that can remain occult for prolonged periods of time (i.e., ovarian) and would be less likely to develop in a malignant

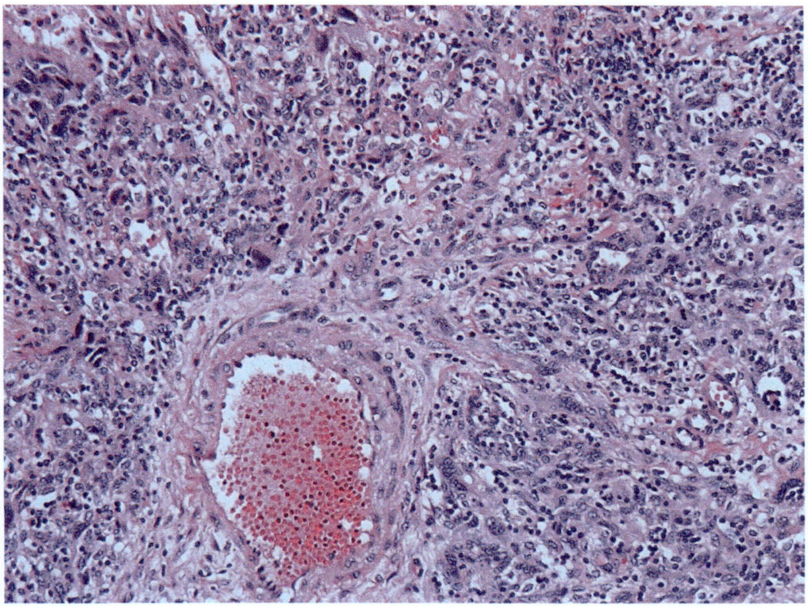

Fig. 12.9 Angiosarcoma arising in a mature cystic teratoma of the ovary. Anastomosing vascular spaces lined by cytologically atypical endothelial cells

tumor that would manifest early or advance rapidly (i.e., testis).

As stated above, a different approach to the one in testicular tumors has been followed in ovaries, regarding the presence of immature elements in mature cystic teratomas. While in testicular tumors a small amount of immature neuroepithelium is irrelevant, a diagnosis of PNET is rendered when this reaches an expansile growth measuring at least one low-power field [1]. In ovarian tumors any amount of immature neuroepithelium is diagnostically relevant but is not considered STM even if expansile and occupying more than one low-power field, but rather is described as "immaturity" and renders a teratoma as an "immature teratoma" [20, 21]. This different approach is more based on historical and definitional reasons, but the difference in the biologic behavior of the background teratomas in both settings and their different pathogenesis may provide also biological fundament. Irrespective of terminology, larger amounts of immature elements are associated with more aggressive behavior, which constitutes the rationale for the "grading" of immature teratomas (see Chap. 6). STM has also been described in ovarian immature teratomas. Due to higher frequency of immature neural tissue in immature teratomas, secondary somatic malignancies are often neural in origin [78]. Sarcomas, particularly rhabdomyosarcomas, can also occur. Similar to testicular tumors, its diagnosis is based on the presence of "overgrowth," although a definitive criterion for its diagnosis has not been established [21].

12.8 STM in Extragonadal GCT

Extragonadal GCT in mediastinum, intracranial (pineal gland), retroperitoneum, and sacral region may undergo transformation to STM. Extragonadal GCTs most likely correspond to type I or type II GCT [4]. However, it is difficult to retrospectively determine in published reports the type of GCT described, although likely both types are represented in most series. Reports on STM in mediastinum are by far more prevalent, with retroperitoneum and intracranial (pineal gland) sites being far less common. Sarcomatous transformation is the most common encountered STM in extragonadal GCT, and rhabdomyosarcoma is its most common subtype [9, 17, 34, 45].

In mediastinum, other reported histologies include PNET [45], adenocarcinoma [45, 79], squamous cell carcinoma [79], osteosarcoma [79], anaplastic small-cell cancer [45], angiosarcoma [34, 79, 80], MPNST [24, 34], leiomyosarcoma [34], epithelioid hemangioendothelioma [34], undifferentiated sarcoma [34], myxoid liposarcoma [34], malignant "triton" tumor [34], sarcoma accompanied by non-Hodgkin's lymphoma [17], sarcoma accompanied by acute nonlymphocytic leukemia [17], squamous cell carcinoma [79], liposarcoma [79, 80], osteosarcoma [79], malignant schwannoma [80], carcinoid tumor [80], glioblastoma multiforme (Fig. 12.10) [34], and hematologic malignancies. Hematologic malignancies seem to be quite specific to mediastinal GCT with no reported cases of transformation to hematologic malignancy in other locations [2, 11–13]. Additional information on mediastinal tumors with STM may be found in Chap. 8.

GCT of the retroperitoneum may show STM such as rhabdomyosarcoma [17, 34, 39], adenocarcinoma [17, 81], liposarcoma [39], chondrosarcoma [39], Wilms' tumor [82], and PNET [83].

Cases of intracranial (pineal gland) GCT with STM are rather sparse. Rhabdomyosarcoma [84–87] and adenocarcinoma [86] have been described. The majority of these cases were associated with teratoma. YST was the second most common intracranial GCT associated with STM [85, 87].

Similar to gonadal sites, poor prognosis has been observed in extragonadal STM-GCT [17, 24, 34]. Mediastinal STM-GCT are detected in more advanced stages with higher rates of progression, metastasis, and relapse. They are less amenable to complete resection with clean borders. They tend to be larger, bulkier, and poorly circumscribed, involving complex vital organs, making radical surgery difficult and inducing comorbidities such as cardiac tamponade and superior vena cava syndrome [17, 34, 79]. Similarly, intracranial GCT with STM tend to have metastases at the time of diagnosis, are less amenable to complete resection, and generally show very poor prognosis [86]. Extragonadal site was an adverse prognostic factor on multivariate analysis in a large series [33]. Extragonadal

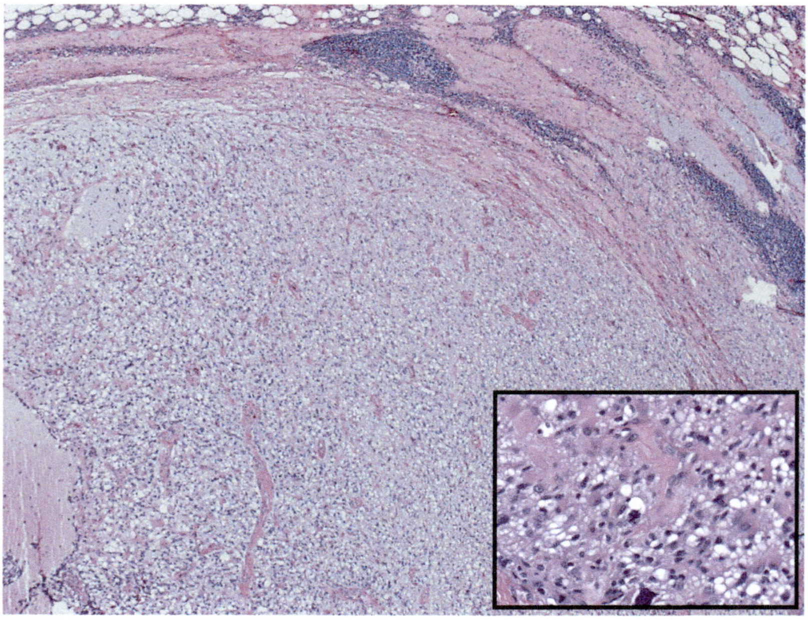

Fig. 12.10 Glioblastoma multiforme arising in a teratoma of the mediastinum. Inset: high-power view showing the neoplastic glial cells with an atypical mitotic figure

STM-GCT should be distinguished from metastasis of a gonadal GCT. This is particularly troublesome when the gonadal primary has undergone regression.

12.9 Therapeutic Implications

Development of STM has major clinical implications. Despite excellent prognosis of GCT with surgery and chemotherapy, the presence of somatic malignant component is associated with dismal prognosis [17, 24, 34, 88]. Incomplete surgical removal is consistently shown as an adverse prognostic factor. In a recent series, 35 % of STM-GCT with clinical stage I that were treated with primary retroperitoneal lymph node dissection (RPLND) had viable STM in the lymph nodes [33]. Based on this finding, the authors advocate primary RPLND in cases with STM in the primary tumor. Overall, prompt surgical resection of the primary tumor, any metastasis, and residual post-chemotherapy tumor are the mainstay of treatment [52]. Surgery should be considered in all patients regardless of stage.

Even though investigators universally advocate timely radical surgery as the most important treatment step, the role and type of chemotherapy regimen is less clear. Conventional cisplatin-based chemotherapy has been shown to be ineffective in patients that develop STM associated with otherwise responsive GCT [2, 17, 22–24, 58, 88]. It is not clear if histology-driven chemotherapy provides benefit. In cases that are not amenable to surgery, higher doses of cisplatin-based chemotherapy will cause severe toxicity, and, considering the resistance of STM, it may not provide much therapeutic benefit [3]. Tailored chemotherapy may be considered in relatively chemosensitive histology such as rhabdomyosarcoma and PNET, but further studies are needed [17, 45, 46, 89]. STM with multiple histologic types poses another therapeutic challenge. Their treatment is more complicated and their prognosis is usually worse than single transformed histology [22]. Some authors recommend a cisplatin-based regimen as both initial and salvage therapy especially if there is still evidence of the presence of cisplatin-sensitive elements by biopsy or tumor marker levels. Rescue chemotherapy oriented to GCT may provide salvage in persisting disease cases [52, 55]. In the above-mentioned series, a definitive recommendation on the type of chemotherapy could not be rendered despite being the largest series published so far, given the multiple subgroups in the study with different histologies and treatment regiments [33].

Refractoriness to cisplatin-based chemotherapy or a mixed response, with regression in one site, and no change or progression in another site, should raise suspicion of STM transformation. Decrease or normalization of tumor biomarker levels, despite objective disease progression as evidenced clinically or by imaging, may be another telltale sign of transformation to STM. Suspicion should be high in such cases. Serum tumor markers are not optimal surrogates of treatment response, and close radiologic follow-up of tumor size is thus warranted.

No standardized treatment plan has been devised for STM arising from either immature or mature ovarian teratoma. Reported treatments for mature ovarian teratoma include surgical treatment [90], surgical treatment in combination with chemotherapy [91], external radiation, radionucleotide therapy, chemotherapy only, and both radiation and chemotherapy [19].

As in gonadal cases, surgery with chemotherapy remains the mainstay treatment in mediastinal and intracranial STM-GCT. Additional radiation therapy has been tried in some cases [34, 86].

12.10 Differential Diagnosis

12.10.1 GCT Elements

Atypia seen in teratoma can mimic STM. This is particularly problematic in type II teratomas, as teratomatous elements invariably display significant degrees of cytologic atypia [27, 77]. Atypical features may also be exacerbated by prior treatment [92]. The degree of atypia, expansile growth, and infiltrative borders are key features in differentiating between microscopic foci of atypia and STM. Overgrowth and replacement of adjacent conventional germ cell elements should be taken into consideration.

Differentiating adenocarcinoma and glandular YST may be diagnostically challenging as well. Magers et al. excluded ten cases from their series initially diagnosed as somatic-type adenocarcinoma, as they represented glandular YST or indeterminate glandular tumors by morphology and immunohistochemistry [3]. Glandular YST tends to be positive for glypican 3 and/or AFP but not always [93, 94]. Unlike adenocarcinomas in STM, glandular YST usually lacks reaction to EMA or CK7. Both entities express CDX2 and both are often positive for SALL4. In cases that lack a clear-cut pattern of staining, the dominant combination of stains may be considered. Similarly, sarcomatoid YST may be misinterpreted as sarcomatoid tumors [3]. In the series mentioned above, of 68 cases originally classified as sarcoma, 24 were reclassified as sarcomatoid YST. A panel of stains would be helpful in ambiguous cases. Sarcomatoid YST shows positivity for both AE1/AE3 and glypican 3.

Hepatocellular carcinoma arising in a teratoma, while rare, may pose a difficult differential diagnosis [54]. First and foremost, it must be differentiated from benign hepatic tissue within a teratoma. Glypican 3 is considered helpful in differentiating benign hepatic cells from hepatocellular carcinoma, as its expression is associated with early events in hepatocarcinogenesis [95]. However, it may not be as useful in the setting of GCT, given its expression in hepatoid YST [96]. Conversely, SALL4 may be positive in hepatocellular carcinomas [97]. Serum tumor markers and the presence of multiple morphologic patterns of YST favor hepatoid YST, while infiltrative pattern in the stroma, nuclear atypia, trabecular, and acinar or pseudoglandular arrangement would favor hepatocellular carcinoma [54].

Embryonal carcinoma may sometimes mimic somatic carcinomas or vice versa. Attention to the distribution of the carcinoma elements (scattered, versus localized), and the classic immunophenotype of CD30-positive, EMA negative in embryonal carcinoma should resolve most of the difficult cases.

12.10.2 Post-therapy Changes

In 2009, Clevenger et al. reported seven cases of a highly differentiated rhabdomyomatous proliferation in the setting of post-chemotherapy RPLND [98]. Contrary to rhabdomyosarcomas arising in teratomas, these tumors were characterized by cells with abundant eosinophilic cytoplasm with occasional cross striations. They displayed mild cytologic atypia and no necrosis, mitotic activity or primitive appearing component. All but one was associated with typical teratoma. Of six patients with follow-up, none had evidence of progressive or recurrent sarcoma, although some had recurrence of teratomatous elements. The authors concluded that the phenomenon is that of cytodifferentiation induced by chemotherapy, a phenomenon that has been seen in treated somatic rhabdomyosarcomas, also associated with a good prognosis [99]. Recognition of this type of tumor is important to avoid confusion with rhabdomyosarcoma arising in teratoma, which would carry a significantly worse prognosis (Fig. 12.11).

12.10.3 Metastatic, Non-GCT-Related Neoplasms

When the teratomatous elements are relatively minor or completely effaced, the distinction between STM and metastatic non-GCT-related tumor is difficult. Thorough sampling to find the teratomatous elements is essential and should be undertaken in tumors with somatic histology at gonadal and extragonadal sites likely to harbor GCT. Late recurrence up to 30 years after the initial treatment has been reported in testicular tumors [100, 101]. The association of a new STM to a testicular GCT that was treated years prior may not be readily evident to the physician at the time of presentation. Differentiating STM derived from GCT and de novo somatic malignancies is challenging. Complete physical exam and imag-

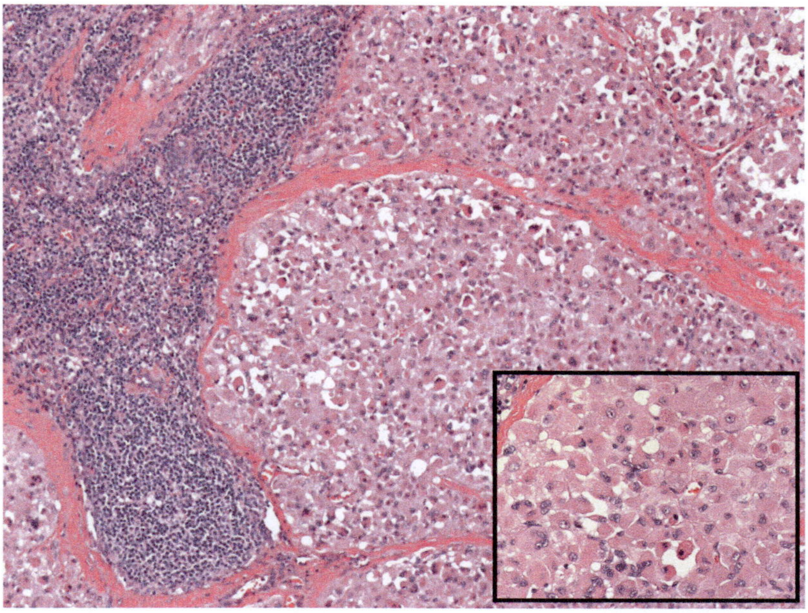

Fig. 12.11 Differentiated rhabdomyomatous tumor. This patient had rhabdomyosarcoma arising in a testicular teratoma and underwent four cycles of chemotherapy before RPLND. Inset: notice the "maturation" of the tumor cells, with globoid appearance, lack of mitosis activity, and degenerative-type atypia

ing studies should be performed in search for possible germ cell primary. The i12p testing may shed light in this workup when a type II tumor is suspected [7]. Conversely, ample sampling of a teratoma is essential in making an accurate diagnosis as undersampling poses a pitfall of missing STM components. The differential diagnosis of ovarian melanoma arising in mature cystic ovarian teratoma and the much more common melanoma metastatic to the ovary usually relies on the unilaterality, presence of junctional change, and clinical exclusion of another primary in the former [75, 76, 102].

Bibliography

1. Moch H, Humphrey P, Ubright TM, Reuter VE. WHO classification of tumours of the urinary system and male genital organs. 4th ed. World Health Organization Classification of Tumours. Lyon: IARC; 2016.
2. Colecchia M, Necchi A, Paolini B, Nicolai N, Salvioni R. Teratoma with somatic-type malignant components in germ cell tumors of the testis: a clinicopathologic analysis of 40 cases with outcome correlation. Int J Surg Pathol. 2011;19(3):321–7.
3. Magers MJ, Kao C-S, Cole CD, Rice KR, Foster RS, Einhorn LH, et al. "Somatic-type" malignancies arising from testicular germ cell tumors: a clinicopathologic study of 124 cases with emphasis on glandular tumors supporting frequent yolk sac tumor origin. Am J Surg Pathol. 2014;38(10):1396–409.
4. Oosterhuis JW, Looijenga LH. Testicular germ-cell tumours in a broader perspective. Nat Rev Cancer. 2005;5(3):210–22. doi:10.1038/nrc1568.
5. Bahrami A, Ro JY, Ayala AG. An overview of testicular germ cell tumors. Arch Pathol Lab Med. 2007;131(8):1267.
6. Kernek KM, Brunelli M, Ulbright TM, Eble JN, Martignoni G, Zhang S, et al. Fluorescence in situ hybridization analysis of chromosome 12p in paraffin-embedded tissue is useful for establishing germ cell origin of metastatic tumors. Mod Pathol. 2004;17(11):1309–13.
7. Kum JB, Ulbright TM, Williamson SR, Wang M, Zhang S, Foster RS, et al. Molecular genetic evidence supporting the origin of somatic-type malignancy and teratoma from the same progenitor cell. Am J Surg Pathol. 2012;36(12):1849–56.
8. Korski K, Breborowicz D, Filas V, Breborowicz J, Grygalewicz B, PIENKOWSKA-GRELA B. A case of primary testicular germ cell tumor with rhabdomyosarcoma metastases as an example of applying the FISH method to diagnostic pathology. Apmis. 2007;115(11):1296–301.
9. Malagón HD, Montiel DP. Mediastinal germ cell tumours. Diagn Histopathol. 2010;16(5):228–36.

10. Ladanyi M, Samaniego F, Reuter VE, Motzer RJ, Jhanwar SC, Bosl GJ, et al. Cytogenetic and immunohistochemical evidence for the germ cell origin of a subset of acute leukemias associated with mediastinal germ cell tumors. J Natl Cancer Inst. 1990;82(3):221–7.
11. Nichols CR, Roth BJ, Heerema N, Griep J, Tricot G. Hematologic neoplasia associated with primary mediastinal germ-cell tumors. N Engl J Med. 1990;322(20):1425–9.
12. deMent SH. Association between mediastinal germ cell tumors and hematologic malignancies: an update. Hum Pathol. 1990;21(7):699–703.
13. Hartmann JT, Nichols CR, Droz J-P, Horwich A, Gerl A, Fossa SD, et al. Hematologic disorders associated with primary mediastinal nonseminomatous germ cell tumors. J Natl Cancer Inst. 2000;92(1):54–61.
14. Orazi A, Neiman RS, Ulbright TM, Heerema NA, John K, Nichols CR. Hematopoietic precursor cells within the yolk sac tumor component are the source of secondary hematopoietic malignancies in patients with mediastinal germ cell tumors. Cancer. 1993;71(12):3873–81.
15. Kerr SE, Flotte AB, McFalls MJ, Vrana JA, Halling KC, Bell DA. Matching maternal isodisomy in mucinous carcinomas and associated ovarian teratomas provides evidence of germ cell derivation for some mucinous ovarian tumors. Am J Surg Pathol. 2013;37(8):1229–35. doi:10.1097/PAS.0b013e31828f9ecb.
16. Rodriguez E, Reuter VE, Mies C, Bosl GJ, Chaganti R. Abnormalities of 2q: a common genetic link between rhabdomyosarcoma and hepatoblastoma? Genes Chromosomes Cancer. 1991;3(2):122–7.
17. Motzer RJ, Amsterdam A, Prieto V, Sheinfeld J, Murty V, Mazumdar M, et al. Teratoma with malignant transformation: diverse malignant histologies arising in men with germ cell tumors. J Urol. 1998;159(1):133–8.
18. Emerson RE, Ulbright TM, Zhang S, Foster RS, Eble JN, Cheng L. Nephroblastoma arising in a germ cell tumor of testicular origin. Am J Surg Pathol. 2004;28(5):687–92.
19. Hirakawa T, Tsuneyoshi M, Enjoji M. Squamous cell carcinoma arising in mature cystic teratoma of the ovary: clinicopathologic and topographic analysis. Am J Surg Pathol. 1989;13(5):397–405.
20. Nogales FF, Dulcey I, Preda O. Germ cell tumors of the ovary: an update. Arch Pathol Lab Med. 2014;138(3):351–62. doi:10.5858/arpa.2012-0547-RA.
21. Oliva E, Young RH. Germ cell tumours of the ovary: selected topics. Diagn Histopathol. 2014;20(9):364–75.
22. Necchi A, Colecchia M, Nicolai N, Piva L, Catanzaro M, Biasoni D, et al. Towards the definition of the best management and prognostic factors of teratoma with malignant transformation: a single-institution case series and new proposal. BJU Int. 2011;107(7):1088–94.
23. Ulbright TM, Loehrer PJ, Roth LM, Einhorn LH, Williams SD, Clark SA. The development of non-germ cell malignancies within germ cell tumors. A clinicopathologic study of 11 cases. Cancer. 1984;54(9):1824–33.
24. Comiter CV, Kibel AS, Richie JP, Nucci MR, Renshaw AA. Prognostic features of teratomas with malignant transformation: a clinicopathological study of 21 cases. J Urol. 1998;159(3):859–63.
25. Guo CC, Punar M, Contreras AL, Tu S-M, Pisters L, Tamboli P, et al. Testicular germ cell tumors with sarcomatous components: an analysis of 33 cases. Am J Surg Pathol. 2009;33(8):1173–8.
26. Ulbright TM. Gonadal teratomas: a review and speculation. Adv Anat Pathol. 2004;11(1):10–23.
27. Ulbright TM. Germ cell tumors of the gonads: a selective review emphasizing problems in differential diagnosis, newly appreciated, and controversial issues. Mod Pathol. 2005;18:S61–79.
28. Ulbright TM. Germ cell neoplasms of the testis. Am J Surg Pathol. 1993;17(11):1075–91.
29. Michael H, Ulbright T, Brodhecker C. The pluripotential nature of the mesenchyme-like component of yolk sac tumor. Arch Pathol Lab Med. 1989;113(10):1115–9.
30. Ulbright T, Roth L, editors. Recent developments in the pathology of germ cell tumors. Semin Diagn Pathol. 1987;4:304–19.
31. Ulbright TM, Michael H, Loehrer PJ, Donohue JP. Spindle cell tumors resected from male patients with germ cell tumors. A clinicopathologic study of 14 cases. Cancer. 1990;65(1):148–56.
32. True LD, Otis CN, Delprado W, Scully RE, Rosai J. Spermatocytic seminoma of testis with sarcomatous transformation. A report of five cases. Am J Surg Pathol. 1988;12(2):75–82.
33. Giannatempo P, Pond GR, Sonpavde G, Albany C, Loriot Y, Sweeney CJ, et al. Treatment and clinical outcomes of patients with teratoma with somatic-type malignant transformation: an International collaboration. J Urol. 2015; doi:10.1016/j.juro.2015.12.082.
34. Malagón HD, Valdez AMC, Moran CA, Suster S. Germ cell tumors with sarcomatous components: a clinicopathologic and immunohistochemical study of 46 cases. Am J Surg Pathol. 2007;31(9):1356–62.
35. Michael H, Lucia J, Foster RS, Ulbright TM. The pathology of late recurrence of testicular germ cell tumors. Am J Surg Pathol. 2000;24(2):257–73.
36. Bokemeyer C, Nichols CR, Droz J-P, Schmoll H-J, Horwich A, Gerl A, et al. Extragonadal germ cell tumors of the mediastinum and retroperitoneum: results from an international analysis. J Clin Oncol. 2002;20(7):1864–73.

37. Nichols C. Mediastinal germ cell tumors. Clinical features and biologic correlates. Chest J. 1991;99(2):472–9.
38. Biskup W, Calaminus G, Schneider DT, Leuschner I, Gobel U. Teratoma with malignant transformation: experiences of the cooperative GPOH protocols MAKEI 83/86/89/96. Klin Padiatr. 2006;218(6):303–8. doi:10.1055/s-2006-942272.
39. Terenziani M, D'Angelo P, Bisogno G, Boldrini R, Cecchetto G, Collini P, et al. Teratoma with a malignant somatic component in pediatric patients: the Associazione Italiana Ematologia Oncologia Pediatrica (AIEOP) experience. Pediatr Blood Cancer. 2010;54(4):532–7.
40. Kasai T, Moriyama K, Tsuji M, Uema K, Sakurai N, Fujii Y. Adenocarcinoma arising from a mature cystic teratoma of the testis. Int J Urol. 2003;10(9):505–9.
41. Kato K, Ijiri R, Tanaka Y, Toyoda Y, Chiba K, Kitami K. Testicular immature teratoma with primitive neuroectodermal tumor in early childhood. J Urol. 2000;164(6):2068–9.
42. Zhang C, Berney DM, Hirsch MS, Cheng L, Ulbright TM. Evidence supporting the existence of benign teratomas of the postpubertal testis: a clinical, histopathologic, and molecular genetic analysis of 25 cases. Am J Surg Pathol. 2013;37(6):827–35. doi:10.1097/PAS.0b013e31827dcc4c.
43. Ahmed T, Bosl GJ, Hajdu SI. Teratoma with malignant transformation in germ cell tumors in men. Cancer. 1985;56(4):860–3.
44. Little Jr J, Foster R, Ulbright T, Donohue J. Unusual neoplasms detected in testis cancer patients undergoing post-chemotherapy retroperitoneal lymphadenectomy. J Urol. 1994;152(4):1144–9.
45. Donadio AC, Motzer RJ, Bajorin DF, Kantoff PW, Sheinfeld J, Houldsworth J, et al. Chemotherapy for teratoma with malignant transformation. J Clin Oncol. 2003;21(23):4285–91.
46. El Mesbahi O, Terrier-Lacombe M-J, Rebischung C, Theodore C, Vanel D, Fizazi K. Chemotherapy in patients with teratoma with malignant transformation. Eur Urol. 2007;51(5):1306–12.
47. Wang WP, Guo C, Berney DM, Ubright TM, Hansel DE, Shen R, et al. Primary carcinoid tumors of the testis: a clinicopathologic study of 29 cases. Am J Surg Pathol. 2010;34(4):519–24.
48. Reyes A, Moran CA, Suster S, Michal M, Dominguez H. Neuroendocrine Carcinomas (Carcinoid Tumor) of the Testis A Clinicopathologic and Immunohistochemical Study of Ten Cases. Am J Clin Pathol. 2003;120(2):182–7.
49. Ganjoo KN, Foster RS, Michael H, Donohue JP, Einhorn LH. Germ cell tumor associated primitive neuroectodermal tumors. J Urol. 2001;165(5):1514–6.
50. Mohanty SK, Balani JP, Parwani AV. Primitive neuroectodermal tumor arising in a testicular teratoma with retroperitoneal metastasis: report of an interesting case with review of literature. Urology. 2007;70(4):812. e7-. e10
51. Ulbright TM, Clark SA, Einhorn LH. Angiosarcoma associated with germ cell tumors. Hum Pathol. 1985;16(3):268–72.
52. Necchi A, Colecchia M, Nicolai N, Mego M, De Giorgi U, Mikuz G, et al. editors. Somatic malignant differentiation in adult male germ-cell tumors (GCTs): preliminary evidences from the INTera database (International Project for Teratoma with Malignant Transformation). J Clin Oncol. 2009;27(15S):e16013.
53. Allen EA, Burger PC, Epstein JI. Microcystic meningioma arising in a mixed germ cell tumor of the testis: a case report. Am J Surg Pathol. 1999;23(9):1131.
54. Jain D, Mittal S, Gupta M. Hepatocellular carcinoma arising in a testicular teratoma: a distinct and unrecognized secondary somatic malignancy. Ann Diagn Pathol. 2013;17(4):367–71.
55. Necchi A, Colecchia M, Nicolai N, Carbone A, Piva L, Biasoni D, et al. 952 teratoma with malignant transformation in adult male germ-cell cancer: a large retrospective review at fondazione irccs istituto nazionale dei tumori (int) of milan. Eur Urol Supp. 2009;8(4):358.
56. Michael H, Hull MT, Ulbright TM, Foster RS, Miller KD. Primitive neuroectodermal tumors arising in testicular germ cell neoplasms. Am J Surg Pathol. 1997;21(8):896–904.
57. George DW, Foster RS, Hromas RA, Robertson KA, Vance GH, Ulbright TM, et al. Update on late relapse of germ cell tumor: a clinical and molecular analysis. J Clin Oncol. 2003;21(1):113–22.
58. Spiess PE, Pisters LL, Liu P, Pettaway CA, Kamat AM, Gomez JA et al., editors. Malignant transformation of testicular teratoma: a chemoresistant phenotype. Urol Oncol Semin Orig Investig. 2008;26:595–9: Elsevier.
59. Alvarado-Cabrero I, Valencia-Cedillo R, Mohs-Alfaro M, De Anda-Gonzalez J. Ovarian dysgerminoma associated with fibrosarcoma: a case report. Int J Gynecol Pathol. 2011;30(5):466–9. doi:10.1097/PGP.0b013e318217139e.
60. Akhtar M, Bakri Y, Rank F. Dysgerminoma of the ovary with rhabdomyosarcoma Report of a case. Cancer. 1989;64(11):2309–12.
61. Kabukcuoglu F, Sungun A, Senturk BA, Evren I, Ilhan R. Mixed germ cell tumor of the ovary with sarcomatous component. Pathol Oncol Res. 2001;7(1):60–2.
62. Matoska J, Talerman A. Spermatocytic seminoma associated with rhabdomyosarcoma. Am J Clin Pathol. 1990;94(1):89–95.
63. Robinson A, Bainbridge T, Kollmannsberger C. A spermatocytic seminoma with rhabdomyosarcoma transformation and extensive metastases. Am J Clin

Oncol. 2007;30(4):440–1. doi:10.1097/01.coc.0000180384.40865.f7.
64. Floyd C, Ayala AG, Logothetis CJ, Silva EG. Spermatocytic seminoma with associated sarcoma of the testis. Cancer. 1988;61(2):409–14.
65. Trivedi P, Pasricha S, Gupta A. Spermatocytic seminoma associated with undifferentiated sarcoma: a rare case report. Indian J Pathol Microbiol. 2011;54(1):138–40. doi:10.4103/0377-4929.77369.
66. Jeong Y, Cheon J, Kim TO, Lim DH, Lee S, Cho YM, et al. Conventional cisplatin-based combination chemotherapy is effective in the treatment of metastatic spermatocytic seminoma with extensive rhabdomyosarcomatous transformation. Cancer Res Treat. 2014; doi:10.4143/crt.2014.049.
67. Wetherell D, Lawrentschuk N, Gyomber D. Spermatocytic seminoma with sarcoma: an indication for adjuvant chemotherapy in localized disease. Korean J Urol. 2013;54(12):884–7. doi:10.4111/kju.2013.54.12.884.
68. Amérigo J, Nogales FF, Fernandez-Sanz J, Oliva H, Velasco A. Squamous cell neoplasms arising from ovarian benign cystic teratoma. Gynecol Oncol. 1979;8(3):277–83.
69. Pins MR, Young RH, Daly WJ, Scully RE. Primary squamous cell carcinoma of the ovary: report of 37 cases. Am J Surg Pathol. 1996;20(7):823–33.
70. Manson C, Cross P, Herd M, Wake C. Mixed adeno- and squamous carcinoma in a mature cystic teratoma in a young woman. Histopathology. 1993;23(5):481–2.
71. Allam-Nandyala P, Bui MM, Caracciolo JT, Hakam A. Squamous cell carcinoma and osteosarcoma arising from a dermoid cyst-a case report and review of literature. Int J Clin Exp Pathol. 2010;3(3):313.
72. Ngwalle KE, Hirakawa T, Tsuneyoshi M, Enjoji M. Osteosarcoma arising in a benign dermoid cyst of the ovary. Gynecol Oncol. 1990;37(1):143–7.
73. Kefeli M, Kandemir B, Akpolat I, Yildirim A, Kokcu A. Rhabdomyosarcoma arising in a mature cystic teratoma with contralateral serous carcinoma: case report and review of the literature. Int J Gynecol Pathol. 2009;28(4):372–5.
74. Morency EG, Lerner D, Garcia R, Kalir T. High-grade sarcoma masquerading as growing teratoma syndrome after resection of ovarian immature teratoma: report of a case. Int J Gynecol Pathol. 2012;31(3):276–9.
75. Davis GL. Malignant melanoma arising in mature ovarian cystic teratoma (dermoid cyst). Report of two cases and literature analysis. Int J Gynecol Pathol. 1996;15(4):356–62.
76. McCluggage WG, Bissonnette JP, Young RH. Primary malignant melanoma of the ovary: a report of 9 definite or probable cases with emphasis on their morphologic diversity and mimicry of other primary and secondary ovarian neoplasms. Int J Gynecol Pathol. 2006;25(4):321–9.
77. Mikuz G, Colecchia M. Teratoma with somatic-type malignant components of the testis. A review and an update. Virchows Arch. 2012;461(1):27–32.
78. Clinkard D, Khalifa M, Osborned R, Bouffet E. Successful management of medulloblastoma arising in an immature ovarian teratoma in pregnancy. Gynecol Oncol. 2011;120(2):311–2.
79. Moran CA, Suster S. Primary germ cell tumors of the mediastinum. Cancer. 1997;80(4):681–90.
80. Kolodziejski L, Duda K, Niezabitowski A, Dyczek S, Staniec B. Occurrence of malignant non-germ cell components in primary mediastinal germ cell tumours. Eur J Surg Oncol (EJSO). 1999;25(1):54–60.
81. Wang L-J, Chu S-H, Ng K-F, Wong Y-C. Adenocarcinomas arising from primary retroperitoneal mature teratomas: CT and MR imaging. Eur Radiol. 2002;12(6):1546–9.
82. Carney JA. Wilms' tumor and renal cell carcinoma in retroperitoneal teratoma. Cancer. 1975;35(4):1179–83.
83. Gatcombe HG, Assikis V, Kooby D, Johnstone PA. Primary retroperitoneal teratomas: a review of the literature. J Surg Oncol. 2004;86(2):107–13.
84. Preissig SH, Smith MT, Huntington HW. Rhabdomyosarcoma arising in a pineal teratoma. Cancer. 1979;44(1):281–4.
85. Shokry A, Janzer RC, Hochstetter ARV, Yasargil MG, Hedinger C. Primary intracranial germ-cell tumors: a clinicopathological study of 14 cases. J Neurosurg. 1985;62(6):826–30.
86. Freilich RJ, Thompson SJ, Walker RW, Rosenblum MK. Adenocarcinomatous transformation of intracranial germ cell tumors. Am J Surg Pathol. 1995;19(5):537–44.
87. Glass R, Culbertson C. Teratoma of pineal gland with choriocarcinoma and rhabdomyosarcoma. Arch Pathol. 1946;41:552–5.
88. Rice KR, Magers MJ, Beck SD, Cary KC, Einhorn LH, Ulbright TM, et al. Management of germ cell tumors with somatic type malignancy: pathological features, prognostic factors and survival outcomes. J Urol. 2014;192(5):1403–9. doi:10.1016/j.juro.2014.05.118.
89. Ehrlich Y, Beck S, Ulbright T, Cheng L, Brames M, Andreoiu M, et al. Outcome analysis of patients with transformed teratoma to primitive neuroectodermal tumor. Ann Oncol. 2010;21(9):1846–50.
90. Iwasa A, Oda Y, Kaneki E, Ohishi Y, Kurihara S, Yamada T, et al. Squamous cell carcinoma arising in mature cystic teratoma of the ovary: an immunohistochemical analysis of its tumorigenesis. Histopathology. 2007;51(1):98–104.
91. Kleinman GM, Young RH, Scully RE. Primary neuroectodermal tumors of the ovary: a report of 25 cases. Am J Surg Pathol. 1993;17(8):764–78.
92. Davey DD, Ulbright TM, Loehrer PJ, Einhorn LH, Donohue JP, Williams SD. The significance of atypia within teratomatous metastases after chemotherapy

for malignant germ cell tumors. Cancer. 1987;59(3):533–9.
93. Nogales FF, Quinonez E, Lopez-Marin L, Dulcey I, Preda O. A diagnostic immunohistochemical panel for yolk sac (primitive endodermal) tumours based on an immunohistochemical comparison with the human yolk sac. Histopathology. 2014;65(1):51–9. doi:10.1111/his.12373.
94. Preda O, Dema A, Iacob M, Goyenaga P, Dulcey I, Aneiros Fernandez J, et al. Urothelial carcinoma of the renal pelvis with simultaneous trophoblastic and malignant clear cell endodermal-type differentiation. Virchows Arch. 2012;460(3):353–6. doi:10.1007/s00428-012-1211-5.
95. Yao M, Yao D-F, Bian Y-Z, Zhang C-G, Qiu L-W, Wu W, et al. Oncofetal antigen glypican-3 as a promising early diagnostic marker for hepatocellular carcinoma. Hepatobiliary Pancreat Dis Int. 2011;10(3):289–94.
96. Preda O, Nicolae A, Aneiros-Fernández J, Borda A, Nogales FF. Glypican 3 is a sensitive, but not a specific, marker for the diagnosis of yolk sac tumours. Histopathology. 2011;58(2):312–4. doi:10.1111/j.1365-2559.2010.03735.x.
97. Gonzalez-Roibon N, Katz B, Chaux A, Sharma R, Munari E, Faraj SF, et al. Immunohistochemical expression of SALL4 in hepatocellular carcinoma, a potential pitfall in the differential diagnosis of yolk sac tumors. Hum Pathol. 2013;44(7):1293–9. doi:http://dx.doi.org/10.1016/j.humpath.2012.10.017.
98. Clevenger JA, Foster RS, Ulbright TM. Differentiated rhabdomyomatous tumors after chemotherapy for metastatic testicular germ-cell tumors: a clinicopathological study of seven cases mandating separation from rhabdomyosarcoma. Mod Pathol. 2009;22(10):1361–6.
99. Arndt CA, Hammond S, Rodeberg D, Qualman S. Significance of persistent mature rhabdomyoblasts in bladder/prostate rhabdomyosarcoma: results from IRS IV. J Pediatr Hematol Oncol. 2006;28(9):563–7. doi:10.1097/01.mph.0000212978.21372.97.
100. Shahidi M, Norman AR, Dearnaley DP, Nicholls J, Horwich A, Huddart RA. Late recurrence in 1263 men with testicular germ cell tumors. Cancer. 2002;95(3):520–30.
101. Baniel J, Foster RS, Gonin R, Messemer JE, Donohue JP, Einhorn LH. Late relapse of testicular cancer. J Clin Oncol. 1995;13(5):1170–6.
102. Gupta D, Deavers MT, Silva EG, Malpica A. Malignant melanoma involving the ovary: a clinicopathologic and immunohistochemical study of 23 cases. Am J Surg Pathol. 2004;28(6):771–80.

Index

F.F. Nogales, R.E. Jimenez (eds.), *Pathology and Biology of Human Germ Cell Tumors*,
DOI 10.1007/978-3-662-53775-6

CPI Antony Rowe
Eastbourne, UK
May 31, 2019